WORKPLACE WRITING

Planning, Packaging, and Perfecting Communication

SHARON J. GERSON
DeVry University (Retired)

STEVEN M. GERSON
Johnson County Community College

Prentice Hall
Upper Saddle River, New Jersey
Columbus, Ohio

Library of Congress Cataloging-in-Publication Data

Gerson, Sharon J.,
Workplace writing : planning, packaging, and perfecting communication / Sharon J. Gerson, Steven M. Gerson.
p. cm.
Includes bibliographical references and index.
ISBN 978-0-13-159969-7
1. Business writing. 2. Business communication. I. Gerson, Steven M., II. Title.

HF5718.3.G47 2010
651.7'4—dc22

2008049524

Editor in Chief: Vernon Anthony
Acquisitions Editor: Gary Bauer
Editorial Assistant: Megan Heintz
Production Coordination: Kelli Jauron, S4Carlisle Publishing Services
Project Manager: Christina Taylor
Senior Operations Supervisor: Pat Tonneman
Art Director: Diane Ernsberger
Interior and Cover Design: Ilze Lemesis
Manager, Image Resource Center, Rights and Permissions: Zina Arabia
Manager, Cover Visual Research and Permissions: Karen Sanatar
Image Permission Coordinator: Frances Toepfer
Director of Marketing: David Gesell
Marketing Manager: Leigh Ann Sims
Marketing Assistant: Les Roberts

Photo credits: DAJ/Getty Images Inc./Stockbyte, Royalty Free, p. 264; Robert Fried/Robert Fried Photography, p. 177; Steve Gorton/Dorling Kindersley Media Library, pp. 177, 203; Getty Images/Digital Vision, pp. 176, 212; Getty Images, Inc./Stockbyte, Royalty Free, pp. 175, 177, 508; Pando Hall/Getty Images/Digital Vision, p. 238; Justin Horrock/istockphoto.com, cover; Jiang Jin/SuperStock, Inc., p. 177; Frank La Bus/Pearson Education/PH College, p. 180; Antony Nagelmann/Getty Images, Inc./Taxi, p. 180; Photolibrary.com, p. 175; Andreas Pollok/Getty Images, Inc./Stone Allstock, p. 181; Richard Ross/Getty Images, Inc./Stone Allstock, p. 175; Russell Sadur/Dorling Kindersley Media Library, p. 175; Chris Schmidt/istockphoto.com, cover; Mark Scott/Getty Images, Inc./Digital Vision, p. 340; Triangle Images/Getty Images, Inc./Digital Vision, p. 182; Matthew Ward/Dorling Kindersley Media Library, p. 172

This book was set in Sabon by S4Carlisle Publishing Services and was printed and bound by Courier/Kendallville. The cover was printed by Courier/Kendallville.

Pearson Education Ltd., London
Pearson Education Singapore Pte. Ltd.
Pearson Education Canada, Inc.
Pearson Education—Japan
Pearson Education Australia Pty. Limited
Pearson Education North Asia Ltd. , Hong Kong
Pearson Educación de Mexico, S.A. de C.V.
Pearson Education Malaysia Pte. Ltd.

Prentice Hall
is an imprint of

10

ISBN-13: 978-0-13-159969-7
ISBN-10: 0-13-159969-0

For our daughters, Stacy and Stefani

BRIEF CONTENTS

CONTENTS

Chapter 13 Persuasive Workplace Communication 368

Chapter 14 Designing Web Sites 396

PREFACE

FROM THE AUTHORS

Welcome to the first edition of *Workplace Writing: Planning, Packaging, and Perfecting Communication.* This reader-friendly textbook combines easy-to-follow instructions for producing all forms of workplace communication with interesting scenarios and examples featuring real people facing communication challenges on the job.

For over 30 years, the writing process—prewriting, writing, and rewriting—has been the standard for teaching students how to write effectively. Many of us were taught in K–12 how to write using this method, or we were introduced to the writing process in college. In this textbook, we build upon this method with a unique approach that applies the writing process to both oral and written communication, and uses terminology that relates to the world of work. We call this the P^3 process.

The P^3 process was thoroughly developed in the classroom to provide students with an understandable structure that helps them think about and create effective workplace communication. This unique approach to the communication process consists of three parts: **planning, packaging,** and **perfecting.** These three terms suggest to the writer that writing can be considered a product, much like the products that people buy and sell in business. In other words, people can plan their documents, package them, and then perfect the product for the reader. We bring the P^3 process to life by using real businesspeople facing real communication challenges at work. Throughout the textbook, we show real-world writing samples as they are written based on the three stages of our process:

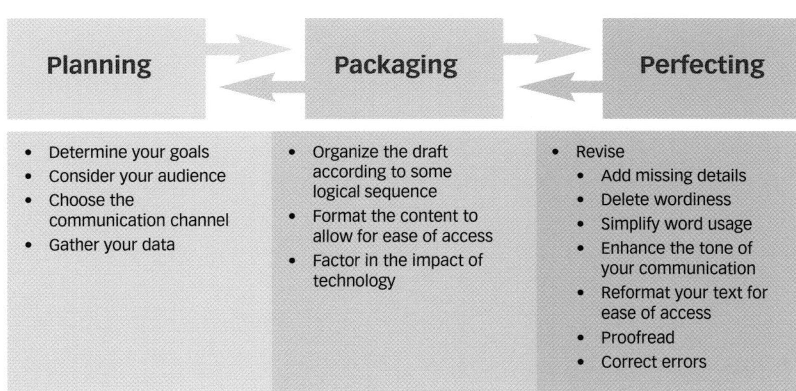

In the **planning** stage of the P^3 process, we show how the writer determines goals, audience, and communication channel. Then we show a planning technique the writer uses to gather data for the correspondence.

In the **packaging** stage of the P^3 process, we show how the writer organizes his or her rough draft and formats the text for ease of access. We also show how the writer receives suggestions for revision.

In the **perfecting** stage of the P^3 process, we show how the rough draft has been revised along with writer commentary. This perfecting stage illustrates the finished product of the P^3 process: a written document that communicates successfully with its intended audience.

THE P^3 PROCESS IN ACTION

The P^3 process seeks to accomplish key goals, including

- Providing criteria for different communication channels
- Giving examples of different types of communication
- Teaching an effective procedure to help students succeed in the classroom and in their careers
- Showing the importance of revision and how it can be achieved
- Revealing the thought processes and the approach to writing and revising followed by businesspeople
- Bringing to life the principles of workplace communication

Nicole's Communication Challenge

Nicole says, "As a new hire at ImageSkill, I need to make my first presentation to a potential client, Greenfield City Management. I have to write a proposal in response to their RFP (request for proposal) and make a PowerPoint presentation to the city council.

Though I am a trained public relations expert, making a professional presentation orally and in writing still is a challenge for me. I am challenged by the scope and importance of the project. My manager says that to accomplish the task, I must do the following:

- **Make an initial client contact**—Through telephone calls, e-mail messages, networking, or a preliminary meeting, I will get a general idea of what the client needs.

Nicole Met Her Challenge

To meet her communication challenge, Nicole used the P^3 process.

Packaging

In Chapter 4, you were introduced to Nicole Stefani who had been hired by ImageSkill as a public relations expert. Her first major job as a new hire is to help the city of Greenfield improve their public relations and promote an effective overall image. See Chapter 4, pages 107–108, to learn how Nicole gathered her data and planned her communication. Now she needs to *package* her written communication. She writes a rough draft, organizing her text according to importance.

Nicole says, "To package the proposal, I prioritized the answers from the interviews and focused on which responses were most prevalent. My goal was to organize both written and oral comments according to a 'least important/most important' presentation. That is, I offered the clients options for improving their public relations: the first options were the most cost-effective but the least promising; the latter options were more costly but more rewarding for the city." Figure 5.7 shows how Nicole packaged her rough draft for review by her boss.

To see Marc's revision suggestions, go to Chapter 6 (Perfecting Workplace Communication), pages 157–158.

In order to highlight and bring to life the P^3 process, we begin each chapter with a profile of a businessperson. Following the profile, we present a workplace communication challenge faced by the individual. For example, Chapter 4 presents "Nicole's Communication Challenge." At the end of the chapter we return to this scenario in a section called "Nicole Met Her Challenge," where we demonstrate how the featured person utilized the P^3 process in order to communicate effectively. Students receive insight into the person's communication challenges by learning how the writer met and overcame difficulties through application of the planning, packaging, and perfecting process.

With our approach to workplace communication presented in the textbook, we provide the teacher and the student with the following exciting tools:

- Real people, who are easy to relate to, encountering on-the-job communication challenges.
- The actual planning, packaging, and perfecting of documents written and created by these businesspeople.
- "Conversations" with these featured people explaining how they met and overcame challenges by relying on the P^3 process.

THE WORKPLACE WRITING LEARNING SYSTEM

Addressing the Needs of the Evolving Workplace

Today's workplace is constantly evolving. This text addresses the important topical issues encountered in workplace communication today including

- Multiculturalism and the global economy
- New trends in business
- Communication in dispersed team settings
- Strategies for ethical workplace communication
- Collaboration and teamwork, especially with wikis and other technological tools
- Dozens of types of electronic, hard-copy, and oral communication channels
- Persuasive communication techniques for meeting the needs of the audience
- Up-to-date job search information
- Oral communication and the workplace
- Instant messaging and blogging in the workplace

Pedagogical Features

Each chapter in the text contains a variety of helpful pedagogical features that enhance the narrative, including

- **Frequently Asked Questions** (FAQs) boxes provide answers to questions people ask about topical issues.

FAQs

Q: Why is the use of PowerPoint so important in the workplace?

A: Widely used by businesspeople, educators, and trainers, PowerPoint is among the most prevalent forms of presentation technology. "According to Microsoft, 30 million PowerPoint presentations take place every day: 1.25 million every hour" (Mahin). Employees in education, business, industry, technology, and government use PowerPoint not only for oral presentations but also as hard-copy text.

Q: Does everyone like PowerPoint? Aren't there any negative attitudes toward this technology?

A: Not everyone likes PowerPoint. Opposition to the use of this technology, however, usually stems from the following problems:

- Dull PowerPoint slides, lacking in variety and interest
- The use of Microsoft's standardized templates
- "Death by Bullet Point," caused by an excessive dependence on bullets

However, these challenges can be overcome easily through techniques discussed in this chapter.

- **Technology Tips** boxes show students how to use Microsoft 2007.

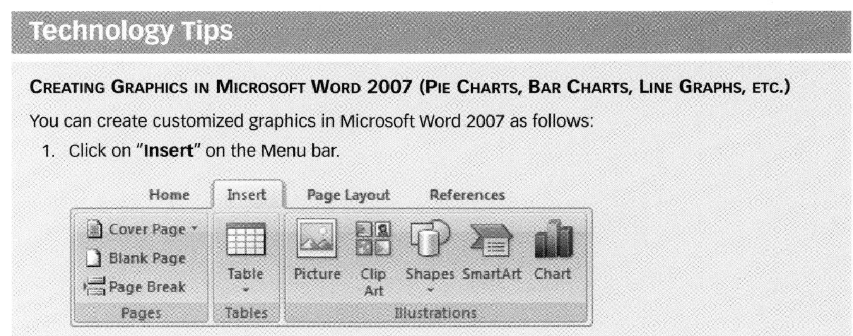

Technology Tips

CREATING GRAPHICS IN MICROSOFT WORD 2007 (PIE CHARTS, BAR CHARTS, LINE GRAPHS, ETC.)

You can create customized graphics in Microsoft Word 2007 as follows:

1. Click on "**Insert**" on the Menu bar.

PERFECTING CHECKLIST

_____ 1. Have you added information for clarity?
- Answer the reporter's questions (Who, What, When, Where, Why, How).
- Provide specific details.
- Avoid connotative words; use denotative words.

_____ 2. Have you achieved conciseness?
- Delete dead words and phrases.
- Limit sentence length.
- Delete "be" verbs.

_____ 3. Did you simplify words for conciseness and easier understanding?

_____ 4. Have you moved information for emphasis?

_____ 5. Did you reformat your text for ease of access?
- Limit paragraph length.
- Use highlighting techniques.
- Insert graphics to clarify complex ideas.

_____ 6. Did you enhance tone to build rapport?

_____ 7. Have you proofread to correct errors?

- **Checklists** guide students through the perfecting stage of their writing.

example

Either the salespeople or the warehouse worker deserv*es* raises.
Not only the warehouse worker but also the salespeople deserv*e* raises.
Neither the salespeople nor the warehouse worker deserv*es* raises.

- **Real-World Examples** show students illustrations of real-world documents.

BEFORE	AFTER
Shun Words	**Concise Version**
came to the conclus*ion*	concluded (or decided, ended, stopped)
with the excep*tion* of	except (or but)
make revis*ions*	revise
investiga*tion* of the	investigate (or look at, review, assess)
consider implementat*ion*	implement (or use)
utiliza*tion* of	use

- **Before and After Examples** demonstrate to students the importance of perfecting.

- **Document examples written for different audiences** provide examples of a document written first for a lay audience and then rewritten and perfected for specialists.
- **Web companion references** direct students to additional samples, interactive activities, and exercises.
- **Topical chapter references** provide links to related topics in other chapters.
- **Dot.Com** updates provide up-to-date information about online sources.

Ethics in Workplace Communication

Because of the importance of ethics in the workplace, we have placed an ethical case study at the end of each chapter. Based on these case studies, students can discuss ethical considerations and make decisions about ethical strategies to follow in the workplace.

BUILDING WORKPLACE SKILLS

End-of-Chapter Individual and Collaborative Activities

The end-of-chapter activities provide students with opportunities to apply chapter principles. In addition, because students learn in a variety of ways, we provide activities that engage students through memorization, critical thinking, collaboration, writing, oral presentations, and research. For the first time in any workplace communication textbook, students will be able to develop communication skills by working through activities directed to their specific fields of study. These role-playing exercises help students prepare for actual career-related communication challenges.

End-of-Chapter Cases, Applications, and Assignments

- Case Studies

> **MEETING WORKPLACE COMMUNICATION CHALLENGES**
>
> **CASE STUDY**
>
> The City of Oak Springs, Iowa, needs to improve a ten-mile stretch of road that runs east and west through the town. The winding, two-lane road was built in 1976. Since then, the city has grown with several businesses and new housing along this road, and

- Ethics in the Workplace Case Studies

> **ETHICS IN THE WORKPLACE CASE STUDY**
>
> Alyssa Adams is a sales representative for a biomedical software company. She frequently travels throughout the country to meet with potential clients. In the past year, her company created a new policy for dealing with clients: sales representatives are limited to a maximum entertainment expenditure reimbursement of $25 a year per client.

- Individual and Team Projects

> **INDIVIDUAL AND TEAM PROJECTS**
>
> Complete the activities by doing the following:
>
> - **Oral presentations**—Give a three- to five-minute briefing to share the results of your findings with your colleagues.
> - **Written**—Write a brief (1/2 page) memo or e-mail to your teacher, highlighting your findings.
>
> 1. Visit employees from your college's or university's business services. These could include payroll, administrative computing, human resources, accounts payable, benefits, budget, catering, the bookstore, printing services, records, admissions, loans, and so forth. After interviewing these employees, respond to the following assignments either by writing a brief memo or e-mail to your instructor, or by giving a three- to five-minute oral presentation to your class.

- Degree-Specific Assignments (tailor specific assignments to specific degree programs) focusing on accounting, finance, marketing, business information systems, and management

> **DEGREE-SPECIFIC ASSIGNMENTS**
>
> Restful Inn Hoteliers is a growing international company with corporate headquarters in Seattle, WA; Boston, MA; Barcelona, Spain; and Singapore. As it builds more hotels around the world, Restful Inn needs to hire employees skilled in hotel management, business information systems, accounting, food and beverage management, international sales and marketing, and finance.

- Problem-Solving Think Pieces

> **PROBLEM-SOLVING THINK PIECES**
>
> 1. In an interview, a company benefits manager said that she spent over 50 percent of her workday on communication issues. These included the following:
>   - Consulting face-to-face with staff, answering their questions about retirement, health insurance, and payroll deductions
>   - Meeting weekly with human resources colleagues

- Web Workshops

> **WEB WORKSHOP**
>
> 1. **Corporate Mission Statements.** Go online to research a corporate Web site's mission/vision page. Learn what the company's mission statement reveals about the worth it places on communication skills between customers and co-workers. Share your findings with the class.
>   - **Oral presentations.** Give a three- to five-minute briefing to share the results of your research with your colleagues.
>   - **Written.** Write a brief ($\frac{1}{2}$ page) memo highlighting your findings.

- Quiz Questions

> **QUIZ QUESTIONS**
>
> 1. Explain why good oral and written communication is essential to your success in business.
> 2. What are four communication channels you can use in business?
> 3. What four channels are almost always used in workplace communication?
> 4. In what directions do you communicate in business?
> 5. What challenges confront you when you write or speak laterally or vertically within an organization?

ONLINE RESOURCES TO ENHANCE WORKPLACE WRITING

MyTechCommLab www.mytechcommlab.com

This comprehensive resource is available at no additional cost with purchase of a new text. MyTechCommLab provides a wide array of multimedia tools, all in one place, and all designed specifically for workplace writers.

- **80 Model Documents,** most with interactive activities and annotations selected from a variety of professions and purposes (letters, memos, career correspondence, proposals, reports, instructions and procedures, descriptions and definitions, Web site, and presentation).
- **50 Interactive Documents** include rollover annotations highlighting purpose, audience, design, and other critical topics.
- **Grammar, Mechanics, and Writing Help:** If students need more practice in basic grammar and usage, MyTechCommLab's grammar diagnostics will generate a study plan linked to the thousands of test items in ExerciseZone, with results tracked by Pearson's exclusive Grade Tracker.
- **Document Design Resources: A Writing Process Tutorial** leads students through each stage of the writing process—from prewriting to final formatting. A new **tutorial on Writing Formal Reports** offers step-by-step guidance for creating one of the most common document types in technical communication and working with sources.
- **Research Help: Research Navigator™** helps students research quickly and efficiently. Our program is complete with extensive help on the research process and includes four exclusive databases of credible and reliable source material—EBSCO's ContentSelect Academic Journal Database, The New York Times Search-by-Subject Archive, the FT.com archives, and a "Best of the Web" Link Library.

To order *Workplace Writing: Planning, Packaging, and Perfecting Communication* with a MyTechCommLab access code, order ISBN: 0-13-815859-2

A stand-alone access code can be purchased online at www.prenhall.com.

MyWritingLab www.mywritinglab.com

MyWritingLab is an online learning system that provides better writing practice through diagnostic assessment and progressive exercises to move students from literal comprehension to critical thinking and writing. With this better practice model, students develop the skills needed to become better writers!

To order *Workplace Writing: Planning, Packaging, and Perfecting Communication* with a MyWritingLab access code, order ISBN: 0-13-246162-5

A stand-alone access code can be purchased online at www.prenhall.com.

COMPANION WEBSITE: A WEALTH OF ONLINE MATERIALS

The Companion Website contains a wealth of cases, exercises, activities, and documents that have been developed for each chapter. Go to www.prenhall.com/gerson.

Online materials for each chapter in the text include the following:

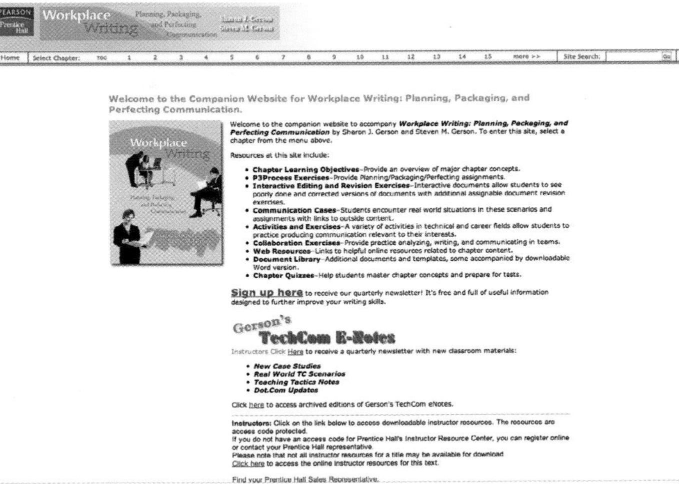

- **Chapter Learning Objectives**—Overview of major chapter concepts.
- *P*³ **Process Exercises**—Planning/Packaging/Perfecting assignments.
- **Interactive Editing and Revision Exercises**—Interactive documents allow students to see poorly done and corrected versions of documents with additional assignable document revision exercises.
- **Communication Cases**—Students encounter real-world situations with links to outside content and a student response box for students to send answers to the professor.
- **Activities and Exercises**—Activities specific to a variety of technical and career fields allow students to practice producing communication relevant to their interests.
- **Collaboration Exercise**—Assignments designed to provide practice writing and communicating in teams.
- **Web Resources**—Links to helpful online resources related to chapter content.
- **Document Library**—Additional documents.
- **Chapter Quizzes**—Self-grading, multiple-choice quizzes help students master chapter concepts and prepare for tests.

INSTRUCTOR RESOURCES

Instructor's Manual

The Instructor's Manual is loaded with helpful teaching notes for your classroom, including answers to the chapter quiz questions, a test bank, and instructor notes for the assignments and activities located on the Companion Website.

Workplace Communication Newsletter

In addition to this textbook and the online instructor and student resources, you can have up-to-date information about the field of workplace communication in our quarterly newsletters. Contact Prentice Hall at www.prenhall.com/gerson to obtain a free subscription to this enlightening newsletter.

Download Instructor Resources from the Instructor Resource Center

The downloadable Instructor's Resources include the following components:

- Test Generator
- PowerPoint Lecture Presentation Package
- Instructor's Manual

The Instructor's Manual, PowerPoint lecture presentations, and TestGenerator are available for download by instructors. Go to www.prenhall.com, click on the "download instructor resources" link, and then click "Register Today" to receive an instructor access code. Within 48 hours of registering you will receive a confirming e-mail including an instructor access code. Once you have received your code, locate your text in the online catalog and click on the "Instructor Resources" button on the left side of the catalog product page. Once you have logged in, you can access instructor material for all Prentice Hall textbooks.

ACKNOWLEDGMENTS

We would like to thank the following reviewers for their helpful comments on the manuscript: Dr. Jennifer L. Bowie, Georgia State University; Shanti Bruce, Nova Southeastern University; Jill Hersh, Farleigh Dickinson University; Mary Lee Stephenson Huffer, PhD, Lake Sumter Community College; Howard Kerner, Nova Southeastern University; Nick Linardopoulos, Drexel University; Earl Holmer, Missouri State University; Dirk Remley, Kent State University; Gaye Winter, The University of Southern Mississippi.

We would also like to thank our Editor, Gary Bauer, for his continued support and encouragement of our creative endeavors. In addition, we want to thank Megan Heintz, Editorial Assistant, and Christina Taylor, Project Manager at Prentice Hall, as well as Kelli Jauron, Project Manager at S4Carlisle, for their efforts, patience, and creativity in helping us bring out this first edition. We especially want to thank our "real people" for the professional insights they offered us about their communication challenges: Dr. Buddy Ramos, Stefani Gerson, Stacy Gerson, Phil Wegman, Linda Freeman, Mary Woltkamp, Tom Woltkamp, Dr. Georgia Nesselrode, and Shannon Conner. Finally, we want to thank Jeanne Conner of JConner Photography for her artistry.

Sharon J. Gerson and Steven M. Gerson

ABOUT THE AUTHORS

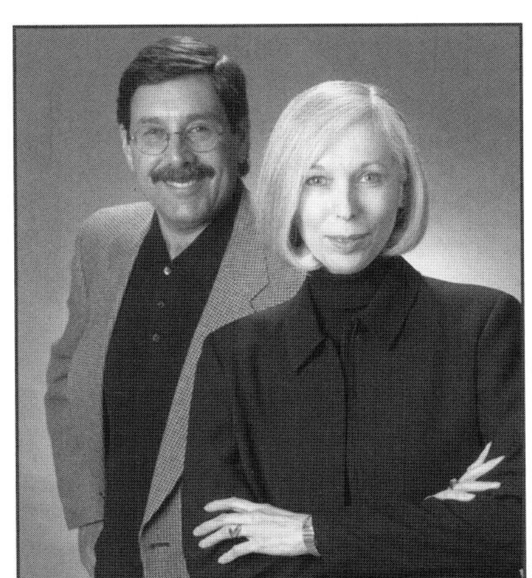

SHARON AND STEVE GERSON are dedicated career professionals who have a combined total of over 70 years teaching experience at the college and university level. They have taught technical writing, business writing, professional writing, and technical communication to thousands of students, attended and presented at dozens of conferences, written numerous articles, and published several textbooks, including *Technical Communication: Process and Product* (sixth edition), *The Red Bridge Reader* (third edition, co-authored by Kin Norman), *Writing That Works: A Teacher's Guide to Technical Writing* (second edition), and *Workplace Communication: Process and Product* (first edition).

In addition to their academic work, Sharon and Steve are involved in business and industry through their business, Steve Gerson Consulting. In this business, they have worked for companies such as Sprint, AlliedSignal–Honeywell, General Electric, JCPenney, Avon, the Missouri Department of Transportation, H&R Block, Mid America Regional Council, and Commerce Bank. Their work for these businesses includes writing, editing, and proofreading many different types of technical documents, such as proposals, marketing collateral, reports, and instructions.

Steve also has presented hundreds of hands-on workshops on technical/business writing, business grammar in the workplace, oral presentations in the workplace, and business etiquette. Over 10,000 business and governmental employees have benefited from these workshops. For the past decade, Steve has worked closely with K–12 teachers. He has presented many well-attended, interactive workshops to give teachers useful tips about technical writing in the classroom.

Both Steve and Sharon have been awarded for teaching excellence and are listed in *Who's Who Among America's Teachers*. Steve is a Society for Technical Communication Fellow. In 2003, Steve was named Kansas Professor of the Year by the Carnegie Foundation for the Advancement of Education.

CHAPTER 1

Communicating in the Workplace

▸REAL PEOPLE *in the workplace*

Dr. Manuel "Buddy" Ramos is a senior client executive for CIBER, Inc., an international system integration consultancy with value-priced software and support services for both private and government clients. CIBER serves businesses from over 60 U.S. offices, 20 European offices, and four offices in Asia. Operating in 18 countries, with 8,000 employees, CIBER and its IT specialists continuously build and upgrade clients' software capabilities.

Buddy's job at CIBER is to develop relationships as a business advisor and to partner with his clients. He serves as the main point of contact for ongoing communications through the sales and overall account service. Buddy must know the client's strategic business and technology goals and then communicate how and when a client can benefit from CIBER solutions.

Buddy doesn't work in an office. He telecommutes, working from his home, airports, and hotels in cities throughout the world.

Buddy spends **95 to 98 percent of his workday communicating externally with clients or internally with account teams.** Basically, his entire job is spent communicating, both in writing and orally.

To complete his job duties, Buddy spends about four to five hours a day speaking to clients. He accomplishes this using many different communication channels. At any given time, Buddy is on his cell phone, engaged in teleconference calls with his client teams and clients. Simultaneously, he and his account teams are using CIBER's intranet e-mail system. This allows Buddy to e-mail his team members, suggesting to them what points they might elaborate on with the client. Frequently, Buddy and his coworkers are also online, using Internet metasearch options to find answers to their clients' questions. That's an example of multitasking to achieve business communication.

When Buddy isn't on the phone, he is in face-to-face meetings with customers—locally, regionally, nationally, and internationally. Not only must he know his subject matter and know how to work collaboratively with people, but also he must be familiar with business practices and cultures from Argentina to Alabama.

Once Buddy has completed his meetings, either in person or on the phone, he has to supervise the 100-page proposals that will be sent to customers and the reports that will be sent to the home office.

CHAPTER GOALS

When you complete this chapter, you will be able to do the following:

1. Understand the purposes of workplace communication.
2. Recognize the importance of workplace communication in your career.
3. Communicate effectively both within and outside an organization.
4. Communicate effectively to management, co-workers, and subordinates in a business.
5. Choose the appropriate formal or informal tone to use when writing or speaking.
6. Recognize your communication style.
7. Distinguish among the communication channels.
8. Analyze trends in the modern workplace.
9. Consider ethics and apply ethical strategies to your words and actions in the workplace.
10. Use the *P*³ process to communicate effectively.

WWW

To learn more about workplace communication, visit our Web Companion at http://www.prenhall.com/gerson.

Buddy's job is to help people implement new software solutions for their business operations. To accomplish this task, he must be an expert communicator and proficient at

- Using multiple technologies and communication channels (Internet research tools, intranet e-mail systems, and cell phone teleconferencing abilities)
- Writing long proposals to customers, follow-up letters, and short reports to co-workers
- Solving end-user problems
- Understanding different cultures
- Working collaboratively with management, customers, vendors, and team members

As Buddy says, "Communication is everything in my job."◀

Buddy's Communication Challenge

Buddy says, "In response to an RFP (request for proposal), my goal is to provide a winning, cost-effective proposal to implement Oracle/PeopleSoft's 'Campus Solutions' software. This software will be used by a multiple-campus university system for a variety of administrative applications, including admissions, student records, financial aid, and academic advising. I define my role as that of air traffic controller of the proposal. As senior client executive, I manage all of the participants in the project and ensure that all of the proposal pieces (work, personnel, timelines, qualifications, references) are accounted for. My staff includes a project director, project manager, proposal specialist, and five consultants.

My challenges include the following:

1. Handling the logistics of communication when working with a dispersed team. Though I'm home-based in Kansas, the project director lives in Los Angeles, and the proposal specialist lives in Boise, Idaho. The team communicates via telephone calls, instant messages, e-mail, and conference calls. In addition, when the team needs to meet face-to-face, we travel as needed.
2. Meeting client needs. For this proposal, involving multiple campuses in a large, university system, my team and I work with over 30 key administrative users, CIOs, CFOs, vice presidents, registrars, and financial aid directors.
3. Multitasking. The team is working on other projects simultaneously.

4. Writing the proposal and reviewing multiple drafts before presentation to the client.
5. Winning the account is the ultimate challenge. Once the proposal has been written, I must make sure that the project director has developed a plan that meets the customer's requirements."

See pages 27–29 to learn how Buddy met his communication challenge.

PURPOSES OF WORKPLACE COMMUNICATION

Whether you are an employer or an employee, a customer or a vendor, you will be involved with workplace communication. You will write business correspondence and speak to colleagues, clients, or a salesperson. Knowing how to communicate successfully in a work environment will help you to express your point of view and influence people.

What are the purposes of workplace communication? When will you be writing or speaking on the job? Consider these possibilities:

- As a Computer Information Systems employee, you work at a 1-800 hotline help desk. A call comes through from a concerned client. Your job not only is to speak politely and professionally to the customer but also to follow-up with an **e-mail** documenting your responses.
- As a trust officer in a bank, one of your jobs is to make proposals to potential clients. In doing so, you will write a **proposal** about your bank's services, and you will use **PowerPoint** to make an **oral presentation** for this client.
- As a customer, you have just celebrated your 6-year-old child's birthday at a local pizza parlor. Unfortunately, the pizzas for the 15 guests were cold, the service was rude, and the promised entertainment arrived late. You now need to write a **complaint letter** to the store's management, recounting your experience.
- As the manager of human resources, one of your major responsibilities is to document your training staff's job accomplishments. To do so, you must write year-end **progress reports** for the employees, which will be used to justify their raises.
- As an entrepreneur, you want to advertise your new catering business. To do so, you plan to write **brochures** (to be distributed locally) and create a **Web site** (to expand your business opportunities).
- As a member of a sorority, fraternity, or college organization, you plan a fund-raising activity for philanthropic purposes. To accomplish this task, you must write **letters of inquiry** to local businesses to acquire pricing bids for food, entertainment, and facilities rental.
- As a mid-level manager, you are in charge of a team of employees who work in different cities, have varied job titles, and work different shifts. To maintain constant communication about a proposal the team is creating for distribution to the marketing department, you use a **wiki** so everyone can interact and make suggestions about the proposal's content.
- As a recently graduated accounting major, it is time to get a job. You need to write an effective **resume** and **letter of application** to show corporations what an outstanding asset you will be to their work environment. Then, you will need to **interview** well and write a **follow-up letter**.
- As a manager, you need to stay in contact with your sales representatives who frequently call on customers or visit companies in the region. You use both **text messages** and **instant messages** to communicate with these employees.
- Your employer wants you to create and maintain a **blog** to reach a new market segment for the company's products.

In the business world, you will communicate to different audiences, for different purposes, using different channels of communication.

THE IMPORTANCE OF WORKPLACE COMMUNICATION

Why is workplace communication important? At work, your primary job is not necessarily writing, or is it? For example, you might be an accountant, computer technician, training facilitator, salesperson, buyer, supervisor, account administrator, or administrative assistant. Your employers will expect expertise from you in those areas of specialization. As a computer technician, for instance, you have been hired because of your knowledge of hardware and software. As an accountant, you have been hired because of your knowledge of accounts payables and receivables. However, a major part of every job is an employee's ability to communicate.

To succeed on the job, you need to write and speak effectively to others—constantly. Gaston Caperton, president of the National Commission on Writing for America's Families, Schools, and Colleges, says that "Writing is a fundamental professional skill. Most of the new jobs in the years ahead will emphasize writing. If students want professional work in service firms, in banking, finance, insurance, and real estate, they must know how to communicate on paper clearly and concisely" ("Writing Skills"). In other words, workplace communication is important for many reasons.

Operating a Business

Workplace communication is not a frill or an occasional occurrence. Rather, it is a major component of your job. The National Commission on Writing for America's Families, Schools, and Colleges is a blue-ribbon group of leaders from public schools, higher education, and the business and writing communities. This commission surveyed human resource directors from 150 leading American corporations with a combined workforce of more than 10 million employees in the United States and combined annual revenues of $4 trillion ("Writing Skills"). Based on their research, the commission concluded the following:

- "People who cannot write and communicate clearly will not be hired, and if already working, are unlikely to last long enough to be considered for promotion. Half of responding companies reported that they take writing into consideration when hiring professional employees and when making promotion decisions. 'In most cases, writing ability could be your ticket in . . . or it could be your ticket out,' said one respondent. Commented another: 'You can't move up without writing skills.'"
- "Two-thirds of salaried employees in large American companies have some writing responsibility. 'All employees must have writing ability. . . . Manufacturing documentation, operating procedures, reporting problems, lab safety, waste-disposal operations—all have to be crystal clear,' said one human resource director."
- "Eighty percent or more of the companies in the services and the finance, insurance, and real estate (FIRE) sectors, the corporations with greatest employment growth potential, assess writing during hiring. 'Applicants who provide poorly written letters wouldn't likely get an interview,' commented one insurance executive" ("Writing Skills").

Workplace communication (written and oral) allows you to manage your products, services, employees, and customers. Through letters, reports, e-mail, and teleconferences, you will communicate about how you manufacture your products, market your services, manage your staff, deliver your goods, meet deadlines, and explain to employees and customers how to follow procedures correctly. Through verbal communication, you will represent your company to civic leaders, clients, and vendors.

Using Time

In addition to serving valuable purposes on the job, getting a job, or meeting your needs as a customer, communication is important because it is time-consuming. Just imagine how much of your time at work will be spent communicating with others.

To clarify the use of different workplace communication channels, look at Figure 1.1. In a 2004 survey of approximately 120 companies employing over 8 million people, the

FIGURE 1.1 **Channels "Almost Always" Used in Workplace Communication**

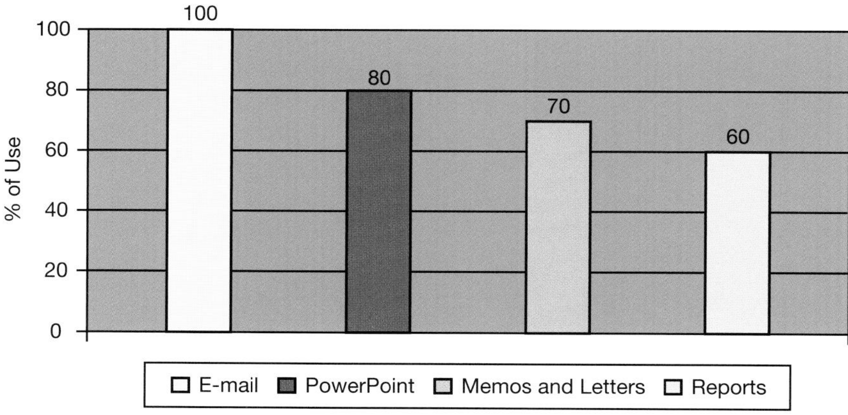

National Commission on Writing found that employees "almost always" use different forms of writing, including e-mail messages, PowerPoint, memos, letters, and reports ("Writing: A Ticket to Work").

Though you will spend a great deal of time writing on the job, you will spend even more time communicating in other manners: speaking, listening, and reading. Calculate the time you will spend verbally communicating in meetings, on the phone with vendors and clients, walking to and from the elevator with your co-workers, and on collaborative work teams while discussing how to complete a project. For example, you will use oral communication skills when speaking to a customer in your office or in the showroom. You will need good communication skills to convince your boss to let you miss two workdays while you coach your child's soccer team at an out-of-town tournament. You will need to communicate face-to-face constantly with co-workers. You also will need to use effective oral communication when you represent your company at the local speaker's bureau. When you are not writing at work, you will spend much of your time speaking, listening, and reading.

Costing Money

You have heard the expression before—time is money. Here are four simple ways of looking at the cost of your workplace communication.

- **Cost of correspondence**—A recent study by Dartnell's Institute of Business Research says that the "average cost of producing and mailing a letter is $19.92" ("Business Identity"). That factors in the time it takes a worker to write the letter as well as the cost of the paper, printing, and stamp. If one letter costs almost $20, imagine how much an entire company's correspondence might cost annually, including every employee's e-mail, letters, memos, and reports.
- **Percentage of salary**—Consider how much of your salary is being paid for your communication skills. Let's say you make $35,000 a year and spend 20 percent of your time writing (as do many employees in the workplace). Your company is paying you approximately $7,000 just to write. That does not include the additional time you spend on oral communication.

  If you are not communicating effectively on the job, then you are asking your bosses to pay you a lot of money for substandard work. Your time spent communicating, both in writing and orally, is part of your salary—and part of your company's expenditures.
- **Cost of training**—Corporations spend money to improve their employees' writing skills. The National Commission on Writing for America's Families, Schools, and Colleges reported that "More than 40 percent of responding firms offer or require training for salaried employees with writing deficiencies. 'We're likely to send out 200–300 people annually for skills upgrade courses like "business writing" or "technical writing," said one respondent.' Based on survey responses,

the Commission estimates that remedying deficiencies in writing costs American corporations as much as $3.1 billion annually" ("Writing Skills").

- **Generating income**—Your communication skills do more than just cost the company money; these talents also can earn money for both you and the company. A well-written sales letter, flier, brochure, proposal, or Web site can generate corporate income. Good written communication is not just part of your salary—it helps pay your wages.

Additional information:

See Chapter 13, "Persuasive Workplace Communication," Chapter 14, "Designing Web Sites," and Chapter 19, "Internal and External Proposals."

Building Interpersonal and Business Relationships

Your workplace does not just focus on money—the bottom line. A major component of a successful company is the environment it develops, the tone it expresses, the atmosphere it creates. Successful companies know that effective communication, both written and oral, creates a better workplace. These "soft skills" make customers want to shop with you and employees work for you.

Your oral and written communication reflects something about you. E-mail messages, letters, memos, or telephone skills are a photograph of you and your company. If you write well, you are telling your audience that you can think logically and communicate your thoughts clearly. When your writing is grammatically correct, or when your telephone tone of voice is calm and knowledgeable, you prove to your audience that you are a professional. If, however, you write poorly, you give your audience a different picture of yourself as a worker or customer. You reveal that you neither think clearly nor communicate your thoughts effectively. Worse, if your telephone or face-to-face conversations are loud, confrontational, or poorly structured, you will reveal a lack of professionalism. Workplace communication is an extension of your interpersonal communication skills, and co-workers or customers will judge your competence based on what you say and how you say it.

Additional information:

See Chapter 7, "Oral Presentations and Nonverbal Communication in the Workplace."

Successful Corporate Communication

A corporation's attitude toward communication paints a picture. For example, the corporate logo of Commerce Bank (Missouri, Kansas, and Illinois) is "Ask, Listen, Solve" (http://www.commercebank.com/). Commerce Bank is conveying corporate values to its employees and customers. The bank realizes that to meet customer needs, it must rely on the simple communication techniques of asking questions and listening to responses. Commerce Bank emphasizes these values by providing training seminars to their employees, teaching them that "responsiveness," "caring," "courtesy," and "relationship and rapport building" are keys to the bank's success. These goals not only relate to how employees must communicate with customers, but also how employees must communicate with their co-workers.

McDonald's seeks to create a similar corporate tone. This international corporate leader promotes its "People Promise and People Vision," stating, "*We're not just a hamburger company serving people; we're a people company serving hamburgers.*" Among other techniques for achieving this goal, McDonald's states that its employees must "communicate openly, listening for understanding and valuing diverse opinions" ("People Principles").

What these companies, and many other successful corporations, have in common is a business strategy that allows customers and employees to communicate with each other more effectively. People need to work well with each other in order for a company to succeed and in order for customers to achieve satisfaction. Through successful written and oral communication, employees can build better relationships with both clients and co-workers.

Developing a Corporate Image and Accountability

There has never been a more important time in the history of international business for corporations to improve their images and to demonstrate their accountability. The early years of the twenty-first century were bad times for many stockholders, employees, and corporate managers. Newspapers and televisions daily reported on the improprieties of former executives from Tyco, Enron, WorldCom, Adelphia, RiteAid, and ImClone.

To combat or avoid negative perceptions, a company must use effective workplace communication. Many companies promote responsible business practices to their employees and stakeholders. Microsoft's Web site advertises its "Integrity and honesty"; "Accountability for commitments, results, and quality to customers, shareholders, partners, and employees"; and "Customer trust [achieved] through the quality of our products and our responsiveness and accountability to customers and partners" ("Microsoft Standards").

Black & Veatch, an engineering, construction, and consulting firm, states in its Web site that it

- values its "shareholders, and we share the results of the company's success with those who produce it"
- "strives to be sincere, fair and forthright, treating others with dignity and respecting their individual differences, feelings and contributions"
- "sets high performance expectations and hold ourselves accountable for the quality of our work and the results we achieve as individuals, as team members, and as a company" ("Vision, Mission & Values").

These published statements in the companies' corporate Web sites are ways they seek to develop successful corporate images as well as to be accountable to their employees and communities. Accountability must be a major goal of your workplace communication.

COMMUNICATING UP, DOWN, AND ACROSS ORGANIZATIONS

In the workplace, communication occurs between you and your bosses, colleagues, subordinates, external vendors, customers, and stakeholders. Whether you are looking for a job or are already employed, you will need to communicate to diverse audiences, as follows:

- Internally (within a company)
- Externally (outside your company)
- Laterally (to colleagues and customers)
- Vertically (up to management and down to subordinates)

Communicating Internally and Externally, Laterally and Vertically

Imagine the following scenario. A customer requests two types of information from a company: quantity pricing for a product and product support (user instructions, guarantees, and delivery options). This letter of inquiry is sent to the company's sales manager, the customer's primary contact. To acquire the information needed, the sales manager sends an e-mail internally and laterally to the customer service department, asking for clarity on guarantees and delivery options. The sales manager also sends an e-mail internally and laterally to the corporate documentation department, asking for a hard-copy user's manual. In addition, the sales manager writes an external e-mail to the product vendor, asking for pricing quotes.

Once the sales manager receives the requested information, he or she responds to the customer with a follow-up letter and writes an internal report up to the supervisor. What has occurred is a typical business scenario involving multiple-channel flows of communication. Correspondence (letters, e-mail, and reports) is sent internally and externally, laterally and vertically, as shown in Figure 1.2.

Additional information:

See Chapter 3, "Meeting the Needs of the Audience."

External Communication

When communicating externally, you might write or speak to customers, vendors, auditors, news agencies, civic groups, lawyers, city/county/state agencies, potential employees, and colleagues from other corporations with whom you are working.

Purposes

Communicating to Customers. In some instances, external communication reaches out to customers, making them aware of your products or services. You advertise your prices, warranties, delivery methods, corporate credentials, and product or service options (sizes, shapes, material of construction, colors, caloric content, timelines and deadlines, or duration of service).

In other situations, communicating externally to customers requires that you solve problems and resolve conflicts. Suppose a customer complains about late deliveries or poor

FIGURE 1.2 **Multiple Channels of Communication**

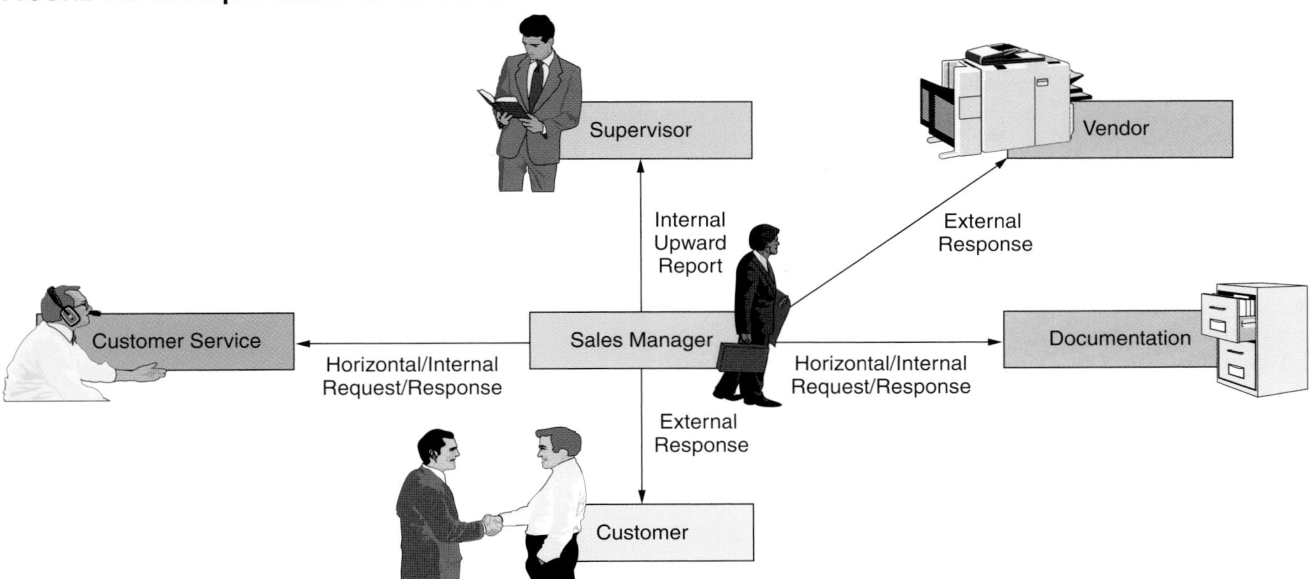

product manufacturing. Your job in writing externally is to keep this customer happy for future purposes or to resolve the issue short of litigation.

Communicating to Prospective Employees. Sometimes your external communication is at job fairs, college recruitment visits, or in one-on-one interviews. In these instances, you want to attract the most talented new employees. You also need to remember that the employee you chose *not* to hire could one day be a colleague of yours at another company.

Communicating to Civic Groups. Maybe you have been asked to speak, as a company representative, to civic groups, school gatherings, or to community members at churches, mosques, or synagogues. Your goal might be to promote your company, industry, product, or service, or answer questions, quell public disapproval, and persuade the audience of your company's good intentions.

Communicating to Lawyers, Auditors, or City/County/State Agencies. Your state's department of natural resources checks your company's air and water emissions quarterly for potentially dangerous pollutants. An external auditor reviews your books for compliance to accounting regulations. Your county tax appraiser reassesses your building. A lawyer writes a letter to your company suggesting that your new product's logo is too similar to another company's logo and that you might be infringing upon his or her client's copyright. In these instances, you must write external letters or reports that clearly inform the audience of your company's position.

Vendors. Communicating externally to vendors, you might write RFPs (requests for proposals) for new services or products. In such cases, you ask for pricing bids, descriptions of services, timetables for delivery, and corporate credentials. You could communicate to vendors requesting increased orders for new business. In contrast, you might need to write to a vendor and cancel that company's service.

Colleagues at Other Companies. Your company has teamed with other companies on a project. You are collaborating with colleagues in public relations, management, human resources, and regulatory compliance. In these instances, you will need to communicate externally to these colleagues through e-mail messages, letters, or intranet sites.

Corporate Stakeholders. Another challenge you might have is communicating externally to stakeholders in your company. These could include stock owners and community residents. Through your company's Web site or through e-mail messages, you will share information to stockholders about mergers, acquisitions, stock price increases or decreases, layoffs, hirings, promotions and demotions, retirements, and changes in company facilities (building expansions, additions, or relocations). Community members also care about any changes in your company (strategies, locations, employee status, and so on). What affects your company affects the community.

See Figure 1.3 for an example of a letter to an external audience.

This letter written to an external audience, a customer, is informative, pleasant, and accessible. The writer enthusiastically communicates the details about the new Web site and is a good follow-up to the completion of the project.

Note how Zach Johnson writes an internal e-mail about the Web site design to his manager in Figure 1.5 on p. 12. In contrast to content in this letter, in the e-mail message, Zach includes technical details and abbreviations familiar to his internal audience.

FIGURE 1.3 Letter to Client from a Business

WeDesign4U

1329 Embarcadero Dr.
San Francisco, CA 75925

October 15, 2009

Robert Taylor
uSkate, Inc.
3289 Palo View
Phoenix, AZ 29030

Subject: Report on uSkate.Com Web Site Completion

Your uSkate Web site is complete. Our Web development team has created what we believe will be the most interactive, colorful, and exciting Web site on the net. Let me highlight a few of the key components of your site's content.

- Easy navigation and full coverage of online shopping through the following headings:
  - Brand name prebuilts
  - Equipment—trucks, decks, bearings, and wheels
  - Protective Gear—helmets, pads, and guards
  - Clothing—jackets, caps, shirts, jeans, and shorts
- Interactive links to skateboarding news and competitions
- Free participatory blogs to engage your audience
- Opportunities for pop-up advertisements
- Videos of recent competitions and trick examples
- Trick tips from the most renowned skateboarders
- Online newsletters
- Contact information with your national and local sites, e-mail addresses, fax number, and phone numbers

We're most excited about the site's visual appeal. Our Web development team used your company's logo colors on the site for brand recognition. The site's home page opens with a short, action-packed music video that refreshes every week. Pop-up advertisements allow your customers to choose from the latest fashions and gear options.

Knowing you wanted your Web site up and running before the holiday season, we beat the deadline by 25 days to give you a head start in the marketplace. The link to your site is uSkate.com. Click and enjoy.

Lead Web Developer
Zach Johnson
zjohnson@wedesign4u.com
(512) 555-1212, ext. 1290

Internal Communication

Communicating internally, you might write or speak laterally to co-workers within your department and colleagues in other departments; vertically down to subordinates whom you manage; and vertically up to your supervisors or the company's CEO (as shown in Figure 1.4).

Purposes When writing or speaking laterally and vertically within an organization, you are confronted with at least two challenges: level of detail and tone (formal or informal). How much does the reader know or want to know? What tone should you take when writing or speaking?

Communicating Up to Management. You might need to communicate upward to your immediate supervisor, his or her boss, other managers within the company, and eventually the CEO of a corporation.

- **Level of detail**—Generally speaking, your audience's distance from a topic can determine the amount of detail you need to provide. For example, if the topic of discussion is not something your audience typically works with on a day-to-day basis, he or she might only want a briefing. Upper-level management is busy with incoming correspondence from many different departments. This audience will profit from a quick, clear overview.

  Immediate supervisors are responsible for budgeting money, facilities, and human resources. This audience not only will be more knowledgeable about your topic but also needs more detail than his or her boss.
- **Tone—Informal to Formal**—When communicating up to management, you must use a pleasant, diplomatic tone. You can be firm and confident in striving to persuade the audience to accept your point of view. Let evidence support your case. However, you cannot use a commanding tone.

FIGURE 1.4 Vertical and Lateral Communication within a Company

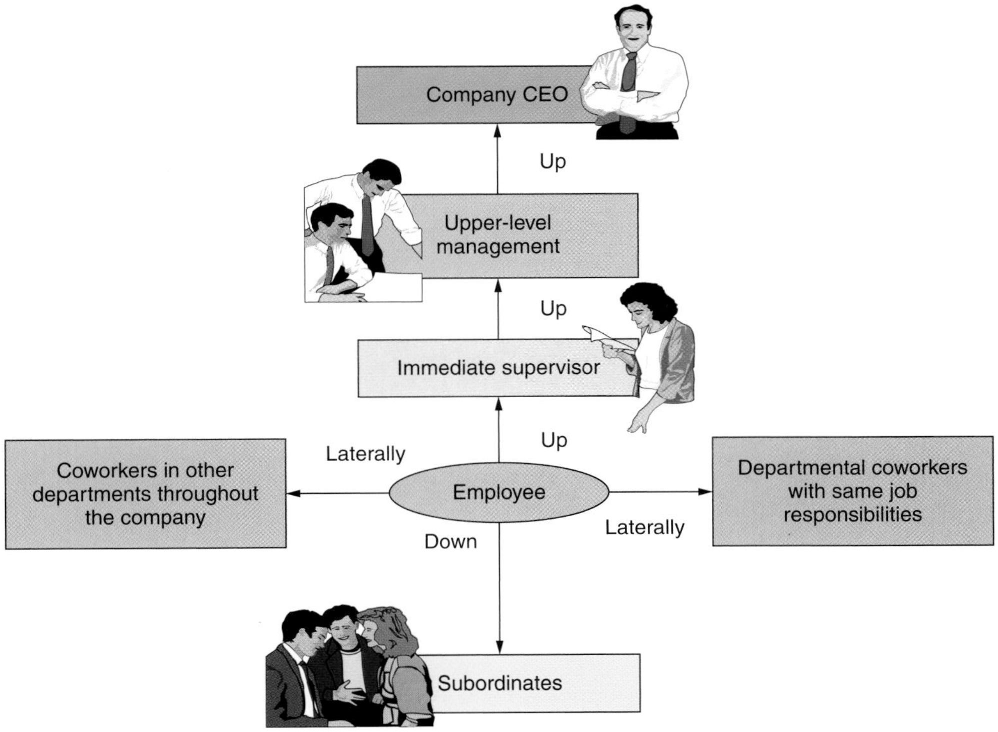

If you are writing solely to your immediate supervisor, and if that person is someone you have worked closely and well with, you could use a relatively informal tone. In contrast, you probably should avoid an informal tone or one that is too personal when communicating with upper-level management and a CEO. Using professional formality is the safe choice. See Table 1.1.

Figure 1.5 is an e-mail message sent from a subordinate to upper-level management. Note the contrast to Figure 1.3, which was written on the same subject by the same writer but to a different audience.

Communicating Laterally to Coworkers. Laterally, you will communicate within your department and to colleagues in other departments. For example, you might work on a team project with marketing specialists from your department, employees who have the same job title and level of experience as you. In addition, your team might consist of employees from other departments—a Web designer from information technology, an accountant, and a human resources employee. These are lateral colleagues because they do not manage you nor do you manage them.

- **Level of detail—**Team members have equal levels of importance in a project but different levels of knowledge about different topics. The IT Webmaster, from the previous scenario, is skilled in technology, the accountant can number crunch, the

Table 1.1 Level of Detail and Tone Appropriate for Management

Audience	Immediate Supervisor	Upper-Level Management	CEO
Knowledge of Subject Matter	High	Low	Lowest
Appropriate Tone	Informal	More formal	Most formal
Level of Detail	High	Low	Lowest

FIGURE 1.5 E-Mail Message to Management

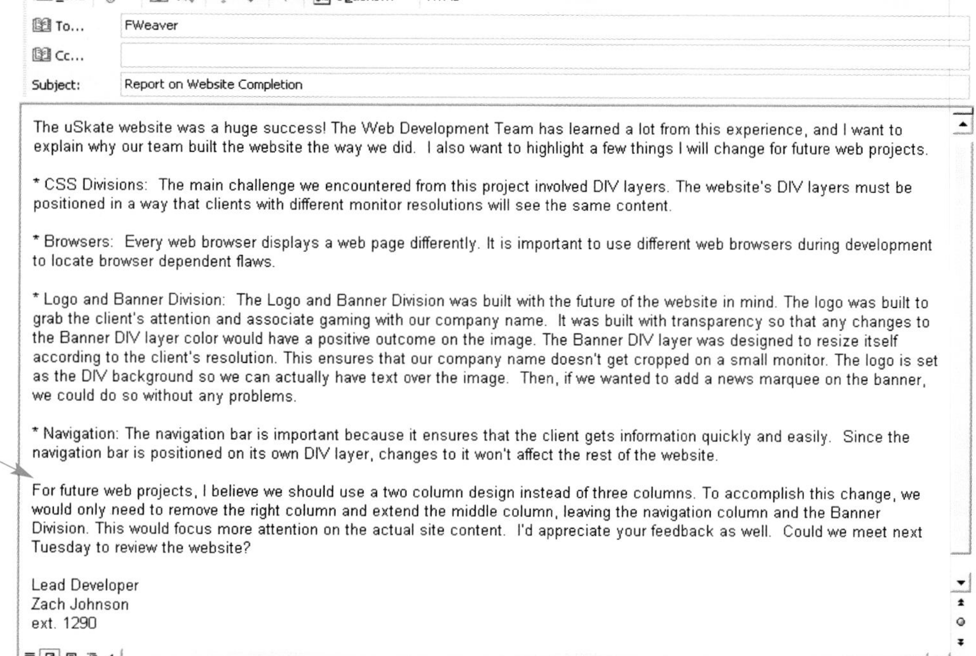

Send | Options... | HTML

To... FWeaver

Cc...

Subject: Report on Website Completion

The uSkate website was a huge success! The Web Development Team has learned a lot from this experience, and I want to explain why our team built the website the way we did. I also want to highlight a few things I will change for future web projects.

* CSS Divisions: The main challenge we encountered from this project involved DIV layers. The website's DIV layers must be positioned in a way that clients with different monitor resolutions will see the same content.

* Browsers: Every web browser displays a web page differently. It is important to use different web browsers during development to locate browser dependent flaws.

* Logo and Banner Division: The Logo and Banner Division was built with the future of the website in mind. The logo was built to grab the client's attention and associate gaming with our company name. It was built with transparency so that any changes to the Banner DIV layer color would have a positive outcome on the image. The Banner DIV layer was designed to resize itself according to the client's resolution. This ensures that our company name doesn't get cropped on a small monitor. The logo is set as the DIV background so we can actually have text over the image. Then, if we wanted to add a news marquee on the banner, we could do so without any problems.

* Navigation: The navigation bar is important because it ensures that the client gets information quickly and easily. Since the navigation bar is positioned on its own DIV layer, changes to it won't affect the rest of the website.

For future web projects, I believe we should use a two column design instead of three columns. To accomplish this change, we would only need to remove the right column and extend the middle column, leaving the navigation column and the Banner Division. This would focus more attention on the actual site content. I'd appreciate your feedback as well. Could we meet next Tuesday to review the website?

Lead Developer
Zach Johnson
ext. 1290

In contrast to Zach's letter to uSkate on page 10, this internal e-mail contains technical information and details omitted from the external letter. Recognizing whether your audience is internal or external helps you to choose the correct style, tone, amount of content, and type of document.

Table 1.2 Level of Detail and Tone Appropriate for Lateral Communication

Audience	Co-workers within department	Co-workers in other departments
Knowledge of Subject Matter	High	Low
Appropriate Tone	Informal	Informal
Level of Detail	Low	High

Table 1.3 Level of Detail and Tone Appropriate for Subordinates

Audience	Long-term subordinates	New-hire subordinates
Knowledge of Subject Matter	High	Low
Appropriate Tone	Informal	Informal
Level of Detail	Low	High

human resources employee knows hiring regulations, and you and your marketing colleagues can sell. To ensure equal knowledge among all members of a team, provide sufficient detail.

- **Tone—Informal**—An informal tone is appropriate for lateral communication to co-workers in a department and team members from other departments. When you are speaking or writing to people with whom you work closely and communicate frequently, use a friendly and informal tone. See Table 1.2.

Communicating Down to Subordinates. The word "down" is not meant to convey a negative connotation. You should not "speak down" to subordinates, berating or belittling them. "Down" merely reflects a subordinate's position in an organization's hierarchy. You will write or speak vertically downward to employees who work under your supervision.

- **Level of detail**—The level of detail required when communicating to subordinates depends on the audience's length of job tenure. If a subordinate has worked in the department for several years, this long-term employee will have knowledge about the department's mission, terminology, and procedures. This long-term subordinate will need fewer details to complete most projects. In contrast, if the department has hired an employee recently, this new hire will need more information about departmental activities.
- **Tone—Informal**—An informal tone is appropriate when communicating to subordinates. When you are speaking or writing to people with whom you work closely and communicate frequently, use a friendly and informal tone. However, your tone will vary depending on the situation. If you need to criticize or reprimand a subordinate for a first infraction, the most successful motivational approach will be friendly and positive "constructive criticism." In contrast, if a subordinate has continuous problems, you can use a more assertive tone. See Table 1.3.

RECOGNIZING YOUR COMMUNICATION STYLE

During a recent conversation with a city employee (Chloe), we learned about a boss who continues to face the computer screen on his desk when Chloe enters the office for a discussion. Chloe does not want to bother or anger the boss by commenting on the rudeness of never maintaining eye contact. However, Chloe realizes that she is being too passive. To be a successful communicator, you should be aware of your personal communication style. When you understand your own communication style, you will be able to create positive

Table 1.4 Characteristics of Communication Styles
(Sherman)

Characteristics	Aggressive	Passive	Assertive
Mottoes and Beliefs	"Everyone should be like me." "I am never wrong."	"Don't make waves." "Don't disagree."	"I have rights, and so do others."
Communication Style	Close minded Poor listener	Always agrees Hesitant	Active listener States ideas directly
Characteristics	Domineering Bullying	Apologetic Does not express self	Confident Nonjudgmental
Behavior	Puts others down Bossy	Asks permission unnecessarily A fence sitter	Firm Action oriented
Nonverbal Cues	Points with jabbing finger Frowns	Looks down Slouches	Direct eye contact Confident and relaxed posture
Verbal Cues	"Don't ask why. Just do it." Verbal abuse	"I'll try to do what you suggest. . . ."	"I choose to. . . ." "What are my options?"
Confrontation and Problem Solving	Must win arguments Threatens	Avoids or postpones Needs supervision	Negotiates Confronts problems when they occur
Feelings Felt	Anger Hostility	Powerlessness	Enthusiasm Well-being
Effects	Wastes time and energy micromanaging Provokes others	Loses self-esteem Builds dependent relationships	Achieves self-esteem and confidence

impressions on others (adapted from Myers-Briggs Type Indicator). Three basic communication styles are aggressive, passive, and assertive (see Table 1.4).

Most people use a combination of communication styles. At times, being aggressive in business is necessary, such as when a decision must be made rapidly or during emergencies. However, being passive is good when an issue is minor or when you are less powerful than the other person. Though the assertive style is the ideal method of communication, at times being passive or aggressive can be beneficial. Being aware of your communication style and altering it for situations and audiences will make you a successful communicator in the workplace. Consider the following workplace scenarios and determine whether you should react aggressively, passively, or assertively:

- You are in a meeting with 12 people. Your co-worker Sandy enters the conference room 20 minutes late, talking loudly on a cell phone, and interrupting the flow of conversation in the meeting. You know that Sandy can behave confrontationally when he is reprimanded. However, you also realize you have to retain control of the meeting. How should you relate to Sandy?
- You work at a help desk. You are on the telephone with a customer who is angry about his recent billing. You explain the charges, but the customer continues to speak loudly and use inappropriate language. How should you handle this situation, being careful not to offend the customer?
- You overhear two of your co-workers relating private and personal information about another co-worker. You know that your supervisor has warned employees about gossiping in the workplace. However, you have to continue to work with these fellow employees. How should you handle this situation politely yet firmly and encourage your co-workers to refrain from gossiping?

USING DIFFERENT COMMUNICATION CHANNELS

Workplace communication takes many different forms. Not only will you communicate in writing and orally, but also you will rely on various types of correspondence and types of technology, each dependent upon the audience, purpose, and situation. To communicate successfully in the workplace, you must adapt to many different communication channels.

For example, sometimes you might need to make an informal request for information. If you need an update from a vendor on delivery status, a brief e-mail message should suffice. Maybe you will need to make a telephone call to ensure that a client has received an order. Chatting with a co-worker face-to-face about a team project is another common channel of informal communication. In other instances, formal reports, letters, or proposals are more appropriate communication channels.

Table 1.5 gives you examples of different communication channels, both oral and written.

Table 1.6 illustrates how different writers and speakers might use various channels to communicate effectively to both internal and external audiences.

Table 1.5 Communication Channels

Written Communication	Oral Communication
• Job information (resumes, letters of application, follow-up letters) • E-mail • Memos • Letters • Reports • Proposals • Business plans • Fliers • Brochures • Newsletters • Web sites • Blogs (Web logs) • Instant messages • Text messages • Intranet and extranet sites	• Leading meetings • Conducting interviews • Making sales calls • Managing others • Participating in teleconferences and videoconferences • Facilitating training sessions • Participating in collaborative team projects • Providing customer service • Making telephone calls • Leaving voice mail messages • Making presentations at conferences or to civic organizations • Participating in interpersonal communication at work • Conducting performance reviews

Table 1.6 Communication Channels—Audience and Purpose

Writers/Speakers	Type of Communication and Communication Channel	Purposes	Internal or External
Human resources/training	Procedures—either hard copy or electronic (intranet, extranet)	Help employees and staff perform tasks	Internal
Administrative assistants	Meeting minutes	Document what has been said and what is planned	Internal
Marketing personnel	Brochures, sales letters, and phone calls	Promote new products or services	External
Customers	Inquiry or complaint letters or phone calls	Ask about/complain about products or services	External
Quality assurance	Investigative, incident, or status reports	Report to regulatory agencies about events	External
Vendors	Phone calls and e-mail	Update clients on new prices, products, or services	External
Management	Oral presentations to department staff	Update employees on mergers, acquisitions, layoffs, raises, or site relocations	Internal
Everyone (management, employees, clients, vendors, governmental agencies)	E-mail, instant messages, text messages, blogs	All conceivable purposes	Internal and external

TRENDS IN THE WORKPLACE

In the twenty-first century, work patterns and careers have changed dramatically in response to advances in communication technology. The modern workplace is being redefined by numerous changes in the business culture.

Technology

As recently as the early 1990s, workplace communication was dominated by letters, memos, and reports—hard-copy correspondence. The emergence of new technologies, however, has changed the way people communicate in the workplace. Web sites, intranet and extranet sites, blogging (Web log journals), e-mail messages, text messaging, instant messaging, wikis, Webinars, and more have impacted the size, speed, and tone of business correspondence.

E-mail messages, text messages, and instant messages not only tend to be shorter than reports and letters, but also these forms of electronic communication are informal channels, affecting the message's tone. People tend to "chat" online, and they can do so synchronously. E-mail allows for and encourages "real time" communication. People around the corner and around the world can write to each other via e-mail and respond within seconds. Through Web sites, intranets and extranets, and e-mail, you also can access large amounts of data quickly and inexpensively. Electronic communication brings the world closer and allows people to communicate more rapidly.

However, electronic communication also has drawbacks. E-mail and instant messaging, though allowing people to communicate quickly, also can be distracting. Spam, pop-up advertisements, and junk messages clutter our work environments. Electronic communication also contributes to what some call "desk rage" ("Work Trends"). The chatty nature of e-mail can lead to e-mail overload, quick responses, irritability, and anger. In addition, companies are monitoring their staff's e-mail usage and using e-mail messages to substantiate poor on-the-job performance (Beem).

Globalization

Where does Dell Computers, home-based in Round Rock, Texas, sell its products? The answer is simple—globally. John Deere, with its world headquarters in Moline, Illinois, has offices in Africa, Australia, China, India, South America, and throughout Europe. Cerner, the "leading U.S. supplier of healthcare information technology solutions," is located in Kansas City, Kansas (Cerner). From the Cerner Web site, you can access corporate information in English, German, Spanish, and French, due to its international business. Wal-Mart, from its home base in Bentonville, Arkansas, directs a worldwide corporation of "2,980 retail units and . . . more than 550,000 associates in Argentina, Brazil, Canada, China, Costa Rica, El Salvador, Guatemala, Honduras, Japan, Mexico, Nicaragua, Puerto Rico, and the United Kingdom" (Wal-Mart). Because Garmin sells its GPS (Global Positioning Systems) products globally, it chose Houston Rockets international basketball star Yao Ming to be the company spokesperson. Starbucks, Coca-Cola, and McDonalds likewise have offices everywhere.

You are likely to work for a company that does business internationally. One of the major effects of technological advances, such as the Internet, is e-commerce and the globalization of modern business. The Internet allows potential clients worldwide to purchase products or services online. This means that you might need to communicate effectively with vendors, clients, and co-workers from many different countries and cultures. Due to the growth of a global economy, you will use e-mail, instant messaging, text messaging, corporate intranets, faxes, teleconferences, and videoconferences to communicate with a diverse workforce and clientele. When you conduct business internationally, you must consider cultural distinctions, language barriers, and behavior constraints. You might even need to provide content in different languages. To learn more about the global impact on workplace communication, you can refer to Chapter 3 in this textbook.

Additional information:

See Chapter 3, "Meeting the Needs of the Audience."

Changing Workplace

The modern workplace is changing in many ways. Employers hire and serve an increasingly culturally diverse population. In addition, the workforce is in enormous flux. The baby boom generation has started to retire. Participants in a "Capturing Knowledge from the Retiring Workforce" program at the 2005 GigaWorld IT Forum contended that "fully 45% of public employees in the U.S. will be at retirement age in the next five years" (Adrian). This will create a "knowledge drain" and force businesses to consider options for using these retired workers either in mentoring programs or as part-time consultants (Adrian).

In contrast to the retiring baby boomers, the workforce will be inhabited by Generation Y, Echo Boomers, and Millenials, the generation of workers born between 1976 and 1995 ("Generation Y" 1). These "tech savvy" employees are familiar "with every gadget imaginable" and can "multitask, talk, walk, listen and type, and text" simultaneously (Safer). This generation, "born into the digital revolution," comfortably uses collaborative communication channels such as Facebook, MySpace, and blogs. In addition, they want constant access to information from "anywhere at any time from any device in any media" ("Generation Y" 1). To accommodate these employees, the workplace must become increasingly technological.

Part-time workers comprise a growing portion of the workforce. According to one study, temporary workers comprise 20 percent of the workforce, an increase of over 400 percent since 1980. At this rate, the study suggests that approximately half of the U.S. workforce could be employed part-time or on a contract basis before 2010 (Boyett and Snyder 1).

Other employees are choosing to telecommute from off-site locations, and they are doing so at an increasing rate of 20 percent per year. Due to new technology, almost 60 percent of today's workforce is now performing jobs "for which physical location is no longer critical" (Boyett and Snyder 4). Fax machines, wireless handheld computers (PDAs), cell phones, teleconferences, and videoconferences allow virtual teams, virtual workplaces, and entrepreneurs working out of their homes to conduct business while physically long distances apart.

Last, employers have become more sensitive to their employees' family commitments. This leads to on-site daycare centers and flex-time schedules, allowing staff to arrive earlier and leave earlier. Many employers also provide on-site concierge services which let employees drop off dry cleaning, buy show tickets, and make personal travel arrangements while at work.

Career Changes

Companies continue to downsize, restructure, send work offshore, and lay off employees. "Rough estimates suggest that the U.S. has lost 400,000 to 500,000 information-technology-processing jobs to outsourcing over the last few years" (Tyson). These actions affect employees' sense of loyalty and their perception of work. Realizing that they might not work for a company until retirement, many workers choose short-term contract jobs, freelance work, and entrepreneurial options. The following statistics reflect the rise in entrepreneurship:

- "In the 1980s, only 1 or 2 percent of graduating MBAs wanted to start out as entrepreneurs; today 10 to 20 percent want to be their own bosses."
- "'Control of destiny' is the most commonly cited career goal of young entrepreneurs, and 87 percent of entrepreneurs say the reason they left corporate life was to gain more control over their lives" ("America's Young Entrepreneurs").

Additional information:

See Chapter 9, "Electronic Communication: E-mail Messages, Instant Messages, Text Messages, and Blogging."

ETHICS IN THE WORKPLACE

Ethical considerations—doing the right thing—are an important part of business. You must consider the appropriateness of your behavior and words. As you communicate in the workplace, you are representing your company whether you are in the building or off-site with a vendor or a customer. You might travel to a conference in a different city to represent your company; alternatively, you might have to make presentations to the city council. At work, you might have to meet with subordinates or other co-workers. In any situation, you should follow the standards for business decorum and comport yourself appropriately. Inappropriate behavior, words, or dress could be considered unethical.

Decisions based on ethical considerations often are in a "grey area" and dependent on how the subsequent actions, words, or behavior are perceived and by whom. If you find yourself in this grey area and need to determine whether or not something is ethical, consider the following:

1. Where the action takes place (in the workplace or off-site)
2. When the action takes place (during working hours, after work, or when you are away from the office but representing your company)
3. Who sees or hears the action or words (a person in authority at the company or a subordinate, co-worker, vendor, or customer)
4. Whether or not the action or words "subjectively" offend (jokes e-mailed to all employees will offend some people but not others)
5. Whether or not the action or words "objectively" offend (racial epithets offend everyone)
6. If the action is an isolated incident (Bob only got drunk at one Christmas party versus Bob coming to work drunk on a weekly basis)
7. If the company has a policy prohibiting the action or words
8. Whether or not the person holds a position of power over other people involved
9. If the action or words are expressed when the person is a "public" face of the company
10. Whether or not the appropriate action was egregious (the degree or level of inappropriateness)

Why Business Ethics Are Important

Business schools and management experts stress the importance of ethics. Many problems can occur when businesses fail to maintain ethical standards. These problems can include dissatisfied customers, large legal judgments, prison terms, anti-trust litigation, loss of goodwill, lost sales, fines, and bankruptcies. In addition to these external results of poor ethical standards, poor business ethics can result in internal failures. Organizational problems include high employee turnover, poor work performance, and a stressful work environment (Bottorff). Employees do not want to work in corporations where bosses ask them to perform illegal, immoral, or unethical actions. Knowing this, however, does not make working or communicating in the workplace easy. Ethical dilemmas exist in corporations. What should you do when confronted with ethical challenges?

One way to solve these dilemmas is by checking your actions against three concerns: legal, practical, and ethical. For example, if you plan to communicate sales literature for a new product, will your text be

1. Legal, focusing on liability, negligence, and consumer protection laws?
2. Practical, since dishonest business communication backfires and can cause the company to lose sales or to suffer legal expenses?
3. Ethical, written to promote customer welfare and avoid deceiving the end user? (Bremer et al. 76–77)

These are not necessarily three separate issues. Each interacts with the other. Laws are based on ethics and practical applications.

Legalities

If you are uncertain regarding ethics, consult a lawyer. When asked to communicate information that profits the company but deceives the customer, you might question where your loyalties lie. The boss pays the bills, but your customers also might be your next-door neighbors. Such conflicts exist and challenge all employees. What do you do? Trust your instincts, and trust the laws. Laws are written to protect the customer, company, and employees. If you believe you are being asked to do something illegal, seek legal counsel.

Practicalities

Though it might appear to be in the best interests of the company to hide potentially damaging information from customers, this is not the case. As a business communicator, your goal is candor. This means that you must be truthful in stating the facts. It also means that you must not keep silent about facts that are potentially dangerous (Girill 178–79).

The ultimate goal of a company is not just making a profit, but making money the right way—"good ethics is good business" (Guy 9). What good is it to earn money from a customer who will never buy from you again, or who will sue for reparation? That's not practical.

Ethicalities

Every industry and professional organization faces the challenges of defining and abiding by ethical standards. You can search the Internet and find ethical codes of conduct for many organizations, including the Direct Marketing Association, the Institute of Electrical and Electronics Engineers, the Association for Computing Machinery, the American Institute of Certified Public Accountants, and the International Society for Performance Improvement, to name a few.

Ten Questions to Ask When Confronting an Ethical Dilemma

You have encountered a situation at work which does not have a clear-cut solution. You have asked yourself whether or not the situation is legal, practical, and ethical, but you still do not have an answer. Like many employees, you might feel isolated when confronted by apparently unethical behavior. What should you do next? Consider these questions:

- Does your company have a policy regarding the situation?
- Have you discussed the situation with your boss?
- Have you met with the director of human resources for guidance?
- Does your company have an ombudsman who provides support or training?
- What do your colleagues say about the situation?
- Do you belong to a professional organization with published policies regarding ethical behavior?
- Do you have a moral authority with whom you can confer?
- How has your organization dealt with similar situations in the past?
- Does your company have a way to communicate anonymously about ethical issues, such as a drop box or secure e-mail address?
- Has your company offered training about ethical behavior in the workplace?

A Sample Challenge to Ethical Behavior in the Workplace

You are a biomedical equipment salesperson. Your company has created a new piece of equipment to be marketed worldwide. Part of the sales literature that your boss tells you to share with potential customers contains the following sentence:

> **NOTE:** Our product has been tested for defects and safety by trained technicians.

When read literally, this sentence is true. The product has been tested, and the technicians are trained. However, you know that the product has been tested for only 24 hours by technicians trained on-site without knowledge of international regulations. As a loyal employee, are you required to do as your boss requests? Even though the statement is not completely true, can you legally include it in your sales literature?

The answer to both questions is no. You have an ethical responsibility to write the truth. Your customers expect it, and it is in the best interests of your company. Of equal importance, including the sentence in your sales literature is illegal. Though the sentence

FIGURE 1.6 The IABC Code of Ethics

International Association of Business Communication
Code of Ethics for Professional Communicators

Preface

Because hundreds of thousands of business communicators worldwide engage in activities that affect the lives of millions of people, and because this power carries with it significant social responsibilities, the International Association of Business Communication developed the Code of Ethics for Professional Communicators.

The Code is based on three different yet interrelated principles of professional communication that apply throughout the world.

These principles assume that just societies are governed by a profound respect for human rights and the rule of law; that ethics, the criteria for determining what is right and wrong, can be agreed upon by members of an organization; and that understanding matters of taste requires sensitivity to cultural norms.

These principles are essential:

- Professional communication is legal.
- Professional communication is ethical.
- Professional communication is in good taste.

Recognizing these principles, members of IABC will:

- engage in communication that is not only legal but also ethical and sensitive to cultural values and beliefs;
- engage in truthful, accurate, and fair communication that facilitates respect and mutual understanding; and,
- adhere to the following articles of the IABC Code of Ethics for Professional Communicators.

Because conditions in the world are constantly changing, members of IABC will work to improve their individual competence and to increase the body of knowledge in the field with research and education. ("Code of Ethics")

is essentially true, it implies something that is false. Readers will assume that the product has been thoroughly tested by technicians who have been correctly trained. Thus, the sentence deceives the readers. Such comments are "actionable under law" if they lead to false impressions (Wilson). If you fail to properly disclose information, including dangers, warnings, cautions, or notes like the sentence in question, then your company is legally liable.

Strategies for Ethical Workplace Communication

If you encounter situations like the sample challenge to ethical behavior in the workplace, the International Association of Business Communication (IABC) (see Figure 1.6) and the Society for Technical Communication (STC) (see Figure 1.7) provide two sources for ethical standards.

Strategies for Communicating Ethically

When writing or speaking, apply the following strategies to your communication.

Use Language and Visuals with Precision You are writing a sales brochure for a Florida resort hotel. You state that the hotel is "in easy walking distance to the Gulf." The hotel actually is located a half mile from the beach. The road is unpaved and uphill.

FIGURE 1.7 **STC Code for Communicators**

STC Code for Communicators

As a technical communicator, I am the bridge between those who create ideas and those who use them. Because I recognize that the quality of my services directly affects how well ideas are understood, I am committed to excellence in performance and the highest standards of ethical behavior.

I value the worth of the ideas I am transmitting and the cost of developing and communicating those ideas. I also value the time and effort spent by those who read, or see, or hear my communication.

I therefore recognize my responsibility to communicate technical information truthfully, clearly, and economically.

My commitment to professional excellence and ethical behavior means that I will

- Use language and visuals with precision.
- Prefer simple, direct expression of ideas.
- Satisfy the audience's need for information, not my own need for self-expression.
- Hold myself responsible for how well my audience understands my message.
- Respect the work of colleagues, knowing that a communication problem may have more than one solution.
- Strive continually to improve my professional competence.
- Promote a climate that encourages the exercise of professional judgment and that attracts talented individuals to careers in technical communication.

("Code for Communicators")

BEFORE

FIGURE 1.8 **Imprecise Depiction of Sales Growth**

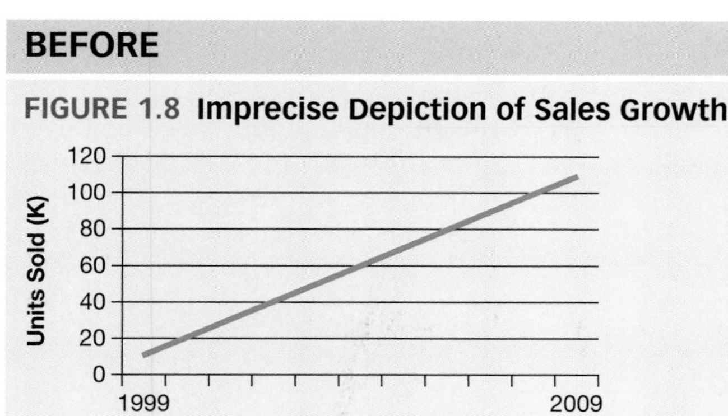

AFTER

FIGURE 1.9 **Precise Depiction of Sales Growth**

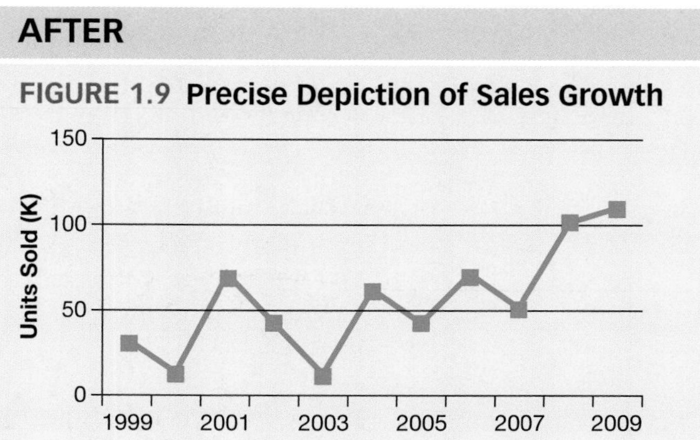

Though the original sales line implies ease of access, for many people this uphill walk could prove to be difficult. In this case, you would not be using words precisely. Information in workplace communication must be presented accurately.

Precision also is required when you use visuals to convey information. Look at Figures 1.8 and 1.9. Both figures show that XYZ's sales have risen. However, the first figure suggests that sales have risen consistently and dramatically. The second figure is more accurate and honest. It shows sales growth, but it also depicts peaks and valleys. As with language, the writer is ethically responsible for presenting visual information precisely.

Prefer Simple, Direct Expression of Ideas While writing a proposal, you might be asked to use legal language to help your company avoid legal problems. Following is a typical warranty that legally protects your company.

Company Warranty Information

> Acme's liability for damages from any cause whatsoever, including fundamental breach, arising out of this Statement of Limited Warranty, or for any other claim related to this product, shall be limited to the greater of $10,000 or the amount paid for this product at the time of the original purchase, and shall not apply to claims for personal injury or damages to personal property caused by Acme's negligence, and in no event shall Acme be liable for any damages caused by your failure to perform your responsibilities under this Statement of Limited Warranty, or for loss of profits, lost savings, or other consequential damages, or for any third-party claims.

Will your audience be able to understand the long sentence structure and complex words? If not, readers will become frustrated and fail to recognize what is covered under the warranty. Although you are following your boss's directions, you are not communicating clearly. Writing can be legally binding and easy to understand. The first requirement does not negate the second. As a successful writer, you should help your audience understand the message by using simple words and direct expression of ideas whenever possible.

Satisfy the Audience's Need for Information, Not Your Own Need for Self-Expression The previously mentioned warranty might be considered sophisticated in its word usage and, thus, "more professional." The warranty, however, does not communicate what the audience needs. Such elaborate and convoluted writing will only satisfy one's need for self-expression; that is not the goal of effective workplace communication.

E-mail presents another opportunity for unethical behavior. Because e-mail is so easy to use, many company employees use it too frequently. They write e-mail for business purposes, but they also use e-mail when writing to family and friends. Employees know not to abuse their company's metered mail by using corporate envelopes and stamps to send in their gas bills or write thank-you notes to Aunt Rose. These same employees, however, will abuse the company's e-mail system (on company time) by writing e-mail messages to relatives coast to coast. They are not satisfying their business colleagues' need for information. In contrast, they are using company-owned e-mail systems for their own self-expression (Hartman and Nantz 60). That can be considered unethical.

Hold Yourself Responsible for How Well the Audience Understands the Message As a workplace communicator, you must place the audience first. When you write a precise proposal or brochure, using simple words and syntax, you can take credit for helping the reader understand your text. Conversely, you must also accept responsibility if the reader fails to understand. Ethical standards for successful communication require the writer to always remember the readers—the people who read corporate annual reports to understand how their stocks are doing, who read brochures before making travel arrangements, or who read proposals to determine whether to purchase a service. An ethical writer remembers that these people will be confused by inaccessible stock reports and misled by inaccurate proposals or brochures. Take the time to check your facts. Present your information precisely, and communicate clearly so that your audience is safe and satisfied. You are responsible for your message (Barker et al.).

Observe Copyright Laws The precise use of language and visuals also involves intellectual property laws as they relate to the Internet. "Every element of a Web page—text, graphics, and HTML code—is protected by U.S. and International copyright laws, whether or not a formal copyright application has been submitted" (LeVie 20–21). If you and your company "borrow" from an existing Internet site, thus infringing upon that site's copyright, you can be assessed actual or statutory damages.

In addition to financial damages, your company, if violating intellectually property laws, could lose customers, damage its reputation, and lose future capital investments. To solve these problems and to protect your own property rights, you should

- Assume that any information on the Internet is covered under copyright protection laws unless proven otherwise.
- Obtain permission for use from the original creator of graphics or text.
- Cite the source of your information.
- Create your own graphics and text.
- Copyright any information you create.
- Place a copyright notice at the bottom of your Web site (Johnson 17).

Additional information:

See Chapter 8, "Visual Aids in Workplace Communication."

Avoid Plagiarism If you borrow ideas from published work, give the other author credit. Taking words, graphics, and ideas without attributing your source through a footnote or parenthetical citation is unethical.

However, avoiding plagiarism in the workplace is not always clear cut. According to Jessica Reyman, "Communicators commonly perform a variety . . . of composing activities that could be considered plagiarism in the context of the classroom" but are not necessarily plagiarism in the workplace (61). For example, workplace communicators often use boilerplate information and templates—text that has been used before within their company. They rely on "existing designs and layouts"; write collaboratively; and cut, paste, and repurpose corporate content (Reyman 61). Do you need to cite content that has already been used? The answer is yes . . . and no.

Additional information:

See Chapter 16, "Research and Documentation."

For example, if you are writing a proposal to a new client, and you draw from your company's library of existing proposals for content, that is acceptable use of boilerplate content and templates. You do not need to cite the source of your information. However, if you surf the Internet, find a proposal from some other company, and use their content, that is plagiarism and unethical behavior. You must cite the source of this original information.

Respect Your Audience's Privacy Since so much of today's communication to co-workers, colleagues, and clients takes place online, you need to consider the ethical dilemmas presented by electronic communications on the Internet. When writing e-mail or distributing information through a Web site, consider confidentiality and courtesy.

- **Confidentiality**—The 1974 Privacy Act allows "individuals to control information about themselves and to prevent its use without consent" (Turner 59). The Electronic Communication Privacy Act of 1986, which applies all federal wiretap laws to electronic communication, states that e-mail messages can be disclosed only "with the consent of the senders or recipients" (Turner 60). However, both of these laws can be abused easily on the Internet. For example, neither law specifically defines "consent." Data such as your credit records can be accessed without your knowledge by anyone with the right hardware and software. Your confidentiality can be breached easily.

  In addition, the 1986 Electronic Communication Privacy Act fails to define "the sender." In the workplace, a company owns the e-mail system, just as it owns other more tangible items such as desks, computers, and file cabinets. Companies have been held liable for electronic messages sent by employees. Thus, many corporations consider the contents of one's e-mail and one's e-mailbox company property, not the property of the employee.

With ownership comes the right to inspect an employee's messages. Although you might write an e-mail message assuming that your thoughts are confidential, this message can be monitored without your knowledge (Hartman and Nantz 61).

These instances might be legal, but are they ethical? Though a company can eavesdrop on your e-mail or access your life's history through databases, that does not mean they should. As an employee or corporate manager, you should respect another's right to confidentiality. Ethically, you should avoid the temptation to read someone else's e-mail or to access data about an individual without his or her consent, unless confronted by a performance or personnel issue. Companies should inform their employees that e-mail may be monitored.

- Courtesy—As noted, e-mail is not as private as you might believe. Whatever you write in your e-mail message can be read by others. This is especially true if you hit the "reply to all" button, which will send an e-mail message to all readers copied in the cc: line. Therefore, be very careful about what you say in e-mail. Avoid offending co-workers by "flaming" (writing discourteous messages, usually typed in all caps). Before you assess a co-worker's ideas or ability, and before you criticize your employer, remember that common courtesy, respect for others, is ethical (Adams et al. 328).

As a business person, you will always be confronted by a multitude of options, such as loyalty to your company, responsible citizenship, need for a salary, accountability to your client and your co-workers, and personal integrity. You must weigh the issues—ethically, legally, and practically—and then write and speak according to your conscience.

Web Sources for Additional Information about Ethics

To learn more about ethics in today's workplace, access the Web sites in Table 1.7. On these sites, you will find ethical case studies, corporate guidelines, researched articles, frequently asked questions, and regulations.

FAQs

Q: Do writers actually follow a process when they compose correspondence?

A: Most good writers follow a process.

It's like plotting your route before a trip. Sure, someone can get in a car without a map, head west (or east or north or south) and find the destination without getting lost, but mapping the route before a trip ensures that you won't get lost and waste time.

There's no one way to plot your destination. Planning might entail only a quick outline—a few brief notes that list the topics you plan to cover and the order in which you'll cover them. This way, you will know where you're going before you get there.

In addition to creating both brief and sometimes much more detailed outlines, an important part of planning is considering your audience. By considering the readers, writers can decide how much detail, definition, or explanation is needed. In fact, thinking about the audience can even help writers determine how many examples or illustrations to include and what details need to be removed from the document.

After planning, good writers always package their text by writing a rough draft, formatting the text and adding graphics as needed.

All writing can be improved. Improving text requires perfecting. Word processing programs make this essential step in the process easier. Word processing programs let you add, delete, and reformat text. Today, it's impossible to type a document on a word processor without considering the terms highlighted in color (spelling and grammar errors). Thus, editing is an integral part of writing.

Table 1.7 Ethics Web Resources

Organization	URL
Online Ethics Center at the National Academy of Engineering	http://www.onlineethics.diamax.com/
European Business Ethics Network	http://www.eben-net.org/
International Business Ethics Institute	http://www.business-ethics.org/
Society for Business Ethics	http://www.societyforbusinessethics.org/

USING THE P^3 PROCESS TO MEET COMMUNICATION CHALLENGES

Communication is a major part of your daily work experience. A well-written memo, letter, report, or e-mail message or a well-constructed oral presentation gets the job done and makes you and your company look good. In contrast, poorly written correspondence and poorly conceived presentations waste time and create negative images.

Recognizing the importance of workplace communication does not ensure that your correspondence will be successful. How do you effectively write the memo, letter, or report? How do you successfully produce the finished oral presentation? For some people, producing effective correspondence or presentations simply involves getting out a piece of paper or turning on a computer and leaping in. This is not the case for most business professionals, however. For most writers, the following might occur when they begin to compose correspondence:

1. **Blank page syndrome**—Writers stare at the blank page or screen until beads of sweat form on their foreheads, and they encounter the dreaded blank page syndrome.

   or

2. **Lack of focus**—They write but, in doing so, they wander about aimlessly for several pages, lacking direction and focus.

In the first instance, writers suffer from writer's block, which prevents them from placing words on the page or screen. In the second situation, they fill up the page or screen with words but never convey a logical thought.

What writers need is a solution to these problems. The writer who suffers from blank page syndrome needs help generating information. The writer who wanders aimlessly needs help organizing, formatting, and revising the content. The P^3 process helps writers overcome workplace communication challenges. For over 30 years, the writing process—prewriting, writing, and rewriting—has been the technique to follow for effective writing.

The P^3 process is a thoroughly developed and unique approach to the writing process and consists of three parts: planning, packaging, and perfecting. These three terms suggest that writing can be considered a product, much like the products that people buy and sell in business. In other words, people can plan their documents, package them, and then perfect the product for the reader.

The P^3 Approach—An Overview

The P^3 approach to effective communication is a recursive, process approach to communication. That is, once you have completed planning, moved on to packaging, and then to perfecting, you might need to start the process again to ensure success. Often you will find yourself planning, packaging, and perfecting simultaneously.

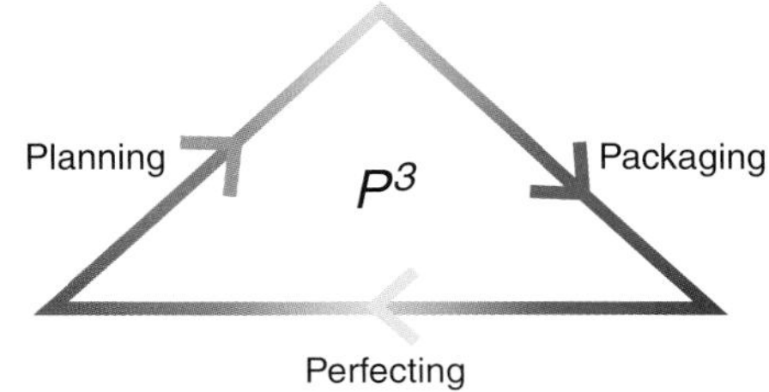

Planning

Before you can write your document or give your oral presentation, you must have something to say. By planning, you spend time generating information *prior* to writing the correspondence or composing the notes for your presentation. In the planning stage, you

- Determine your goals (What do you hope to achieve in the communication?).
- Consider audience (Who will read your correspondence or hear your comments?).
- Choose the appropriate communication channel (Will you write a memo, letter, e-mail, or report; will you give a formal speech using PowerPoint or an informal briefing?).
- Gather data (What content do you need for your communication?).

Packaging

Once you have gathered your data and determined your objectives, the next step is to package the communication into a draft. To draft your document, do the following:

- Organize the draft according to some logical sequence.
- Design your document, formatting the text to allow for ease of access.
- Factor in the impact of technology.

Perfecting

The final step, and one that is essential to successful communication, is to rewrite your draft, making it as perfect as it can be. This step requires that you revise the rough draft. Revision allows you to perfect your memo, letter, e-mail, report, or oral presentation so you can write or speak effectively. Doing so requires that you follow a seven-step procedure:

- Add missing details for clarity.
- Delete wordiness for conciseness.
- Simplify your word usage.
- Enhance the tone of your communication.
- Reformat your text for ease of access.
- Practice the speech or reread and review your text for content.
- Proofread and correct any grammatical or textual errors.

Additional information:

See Chapter 4, "Planning Workplace Communication," Chapter 5, "Packaging Workplace Communication through Effective Document Design," and Chapter 6, "Perfecting Workplace Communication."

The following table illustrates the three parts of the P^3 communication process.

The P^3 Communication Process

Planning	Packaging	Perfecting

• Determine your goals
• Consider your audience
• Choose the communication channel
• Gather your data
	• Organize the draft according to some logical sequence
	• Format the content to allow for ease of access
	• Factor in the impact of technology
		• Revise
		  • Add missing details
		  • Delete wordiness
		  • Simplify word usage
		  • Enhance the tone of your communication
		  • Reformat your text for ease of access
		  • Proofread
		  • Correct errors

Buddy Met His Challenge

To meet his communication challenge, Buddy used the P^3 process.

Planning

To plan his proposal, Buddy first considered the following:

- Goals—respond to RFP and produce a winning proposal
- Audience—30 university decision-makers
- Channels—e-mail, IM, teleconferences, proposal, letter
- Data—drawn from boilerplate, responses to the RFP, and interviews

According to Buddy, "Planning entails establishing team member tasks and a timeline based on the due date and working backwards. For example, if the proposal is due on June 12, my team decides what needs to be done and how much time is required to meet this deadline.

The project director, proposal specialist, and I have a conference call after receiving the RFP to determine our objectives and to gather data for the proposal. Another team member, the attorney, provides a legal review and develops exceptions to the terms and conditions.

The primary team members receive their tasks/assignments.

- I prepare a proposal planning matrix, an executive summary, cover letter, and pricing information.
- The proposal specialist assembles the proposal with boilerplate corporate information.
- The project director is responsible for the guts of the proposal (the plan, the number of resources, and timeline).

All team members have target due dates and each does his or her own job. The team members formally touch base once a week in a conference call to review their progress, along with informal IM and e-mail everyday." Figure 1.10 is Buddy's "Proposal Planning Matrix."

Packaging

To package the proposal, each team member submits parts of the proposal in rough draft form. The proposal specialist compiles all parts of the rough draft and sends the draft back with questions and comments. She also designs and packages the proposal to customize it for the intended audience, including the college's logo. Figure 1.11 shows the original rough draft for one part of Buddy's proposal.

Buddy's co-worker suggested the following to improve this brief rough draft: "Please validate or reword the last sentence to make sure you have stated it correctly. You need to add all of the relevant information to clarify our expertise in the field. You could revise the last sentence by adding details to show our software solutions for other Texas higher education clients. By illustrating your comments with our solutions for other schools, CTU will easily see how valuable we are and that we can meet their software needs."

Perfecting

"It's my ultimate job to read for 'one voice,' to make sure that the proposal flows," Buddy says. The proposal specialist and Buddy read to make sure that this proposal doesn't "pull from other proposals" since they use boilerplate text from other proposals. They never want to include proprietary information from other colleges. The proposal specialist reads for quality assurance to make sure that the table of contents has correct page numbers, that headings are consistent, and that content is grammatically correct. According to Buddy, "I know I must add details to clarify how CIBER will meet the client's needs and emphasize my company's expertise." Figure 1.12 shows the perfected text for one part of Buddy's proposal.

FIGURE 1.10 **Proposal Planning Matrix**

Fri. June 8 1:30 p.m. EST/ 12:30 p.m. CST	Team kickoff call 1:30 p.m. EST/12:30 p.m. CST	888-555-5555	Confirm resources and writing assignments. Review answers to submitted questions.
Section #	**Section Title**	**Responsibilities/Comments/Assignments**	**Due Dates**
Tab 1	Table of Contents	Mary	Done
Tab 2	Introduction, Executive Overview	Buddy and Mike	COB June 28, 2009
Tab 3	Supplier Information	CIBER–Mary and Buddy to complete the majority of the questions in this section. Oracle to respond to the following questions: 4.1.3 Product Growth–Mike 4.3.1 Relationship with University–Mike 4.4.1 Pennsylvania Reference–Mike 4.7 Value Added Services–CRM–Chris 4.8 Relevant Supporting Documentation–Melanie	COB June 27, 2009 June 28 June 28 June 28 June 28 June 28
Tab 4	Specifications/ Qualifications	5.1–Oracle Provided Spreadsheet–Oracle to provide the majority of the spreadsheet responses CIBER–John–to provide the implementation services response Mary to combine the two responses	Done June 27 June 28 June 29
Tab 5	Pricing Information	Oracle–Mike CIBER–Buddy	June 29 June 29
Tab 6	Agreements	CIBER Legal Oracle–Melanie to provide Oracle's exceptions and current contract	Done June 28
Tab 7	Alternative Proposal	Oracle–Mike to provide CRM piece	June 28
Tab 8	Signature Page	CIBER –Mary and Mike	June 28
Tab 9	Agreements	CIBER–Sample Contract–Mary Oracle Sample Contracts–Melanie	Done June 28
Tab 10	Appendix A–CIBER's Org Chart	Mary	Done
Tab 11	Appendix B–Audited Financial Statements	CIBER Oracle–Melanie	Done June 27
		Proposal shipped to customer	**Thur. July 12** for receipt on Fri., July 13

FIGURE 1.11 **Original Rough Draft of One Section of the Proposal**

Proven experience implementing Campus Solutions 8.9 or higher and other requested software solutions

Response:

Alabama, North Wyoming University, Southern Idaho University, and the University of Corpus Christi are a sample of CIBER's Campus Solutions 8.9 experience. Most notably is the University of Corpus Christi System because it is similar to the CTU System RFP where CIBER is providing upgrade services to one campus and implementing CS at another.

FIGURE 1.12 Perfected Text

Proven experience implementing Campus Solutions 8.9 or higher and other requested software solutions

Response:

8.9 Experience

Alabama, North Wyoming University, Southern Idaho University, and the University of Corpus Christi are a sample of CIBER's Campus Solutions 8.9 experience.

Campus Solutions Experience in Texas:

As the primary partner with the University of South Texas, the University of South Texas Health Science Center, and the University of Corpus Christi System, CIBER has acquired a strong reputation for knowing what drives the needs of public universities in Texas. This will be advantageous to Central Texas University (CTU) Systems because CIBER has already delivered on the needs that are common across Texas schools. When it comes to providing for what CTU institutions need in Texas, such as meeting requirements for Coordinating Board reporting or the Texas Guaranteed Student Loan Corporation, we do not want CTU Systems to have to "reinvent the wheel" when a solution already exists. One of the other key elements in CIBER's strategy for the project will be to use as many members of the University of North Texas System and University of Corpus Christi System consulting teams as possible.

The most notable relevant experience is the University of Corpus Christi (UCC) System because it is similar to the CTU System RFP where CIBER is providing upgrade services to one campus and implementing CS at another.

The University of Corpus Christi System is comprised of four separate campuses. These campuses operate as separate institutions academically and have traditionally used separate student administration systems. However, as a single system, UCC needed one integrated software system that would create efficiency in processing and integrated reporting across all four campuses, while still allowing the four campuses to operate independently. Integrated reporting, allowing UCC to match the numbers in their student system and financials systems, and increased service level to constituents, rather than the desire to decrease administration costs, were the driving forces behind this implementation.

CIBER is currently upgrading the University of Corpus Christi from Oracle PS 8.0 to PS 8.9.

Buddy says, "I added details, including information about other Texas clients to verify CIBER's expertise in software solutions. I know it's the details that sell."

"Then I included specific information about how CIBER will meet CTU's system needs, emphasizing lowered costs."

CHAPTER HIGHLIGHTS

1. Knowing how to communicate effectively in the workplace will help you express your point of view and influence people.
2. In the workplace, employees frequently spend time writing e-mail messages, reports, memos, and letters.
3. Your time spent communicating, both orally and in writing, is part of your salary and part of your company's expenditures.
4. Workplace communication is an extension of your interpersonal communication skills, and co-workers or customers will judge your competence based on what you say and how you say it.
5. You will communicate frequently both internally in the business and externally to customers, vendors, and stakeholders.
6. When writing or speaking laterally and vertically, you have to choose level of detail and informal or formal tone.

7. Knowing your communication style—aggressive, passive, or assertive—helps you to be a successful workplace communicator.
8. To communicate successfully in the workplace, you must adapt to many different channels of communication.
9. To be a successful workplace communicator, follow the P^3 process of planning, packaging, and perfecting.
10. When you communicate in business, your text must be legal, practical, and ethical.

WWW *Find additional workplace communication exercises, samples, and interactive activities at http://www.prenhall.com/gerson*

MEETING WORKPLACE COMMUNICATION CHALLENGES

CASE STUDIES

1. Due to problems getting new clients, Deer Creek's management decided to hire a Manager of Corporate Communication. This person's job was to
   - Act as a liaison between Deer Creek's engineers, city/county/state government agencies, and end-user clients.
   - Prepare proposals (gather information through meetings and write the reports).
   - Make oral presentations to potential clients.

   You have been asked to write the job description for this position, which will be advertised in the local newspaper, in Deer Creek's employment office, and online at the company's Web site: DeerCreekConstruction.com.

Assignment

Research the position of Manager of Corporate Communication (either online, in the Occupational Outlook Handout, or at a local company's work site). Find out the salary range and educational requirements for this job. More importantly, based on what you have learned in this chapter about workplace communication, what additional job responsibilities will this position entail? What skills should this Manager of Corporate Communication have? Based on your decisions, write the job description.

2. The letter on page 31 is flawed in many ways (ethics, tone, audience recognition, etc.). Based on what you have learned in this chapter, explain why and how this letter is flawed. Then rewrite the letter to improve it.

3. As director of APT Training Institute, Shelly Stine has been hired as an external consultant by Home and Hearth Bank to help its employees create and implement their Individual Development Plans (IDPs). Bank employees have attended Ms. Stine's 360-Degree Assessment Workshops where they learned how to get feedback on their job performance from their supervisors, co-workers, and subordinates. They also provided self-evaluations.

Once the 360-Degree Assessments were completed, employees submitted them to Ms. Stine who, with the help of her staff, developed the bank employees' IDPs. Shelly sent the IDPs to the employees, prefaced by a cover letter.

Assignment

On pages 32–34 are three cover letters (**Letters A, B, and C**) that Ms. Stine sent to Sharon Baker, the bank's account executive.

TASCO
PO Box 2110
Shawnee Mission, IA 56207

November 15, 2009

Dear Sir,

Tasco is an HVAC manufacturer's representative for HVAC equipment, but we can cut out the middle man and save you lots of bucks. Though we usually provide wholesalers and retailers the products that their customers need for their homes and businesses, if you work with us directly, you won't have to pay those store prices that you know have been increased just for company profits—*at your expense.* **Tasco** covers many lines of manufacturers.

Our manufacturers carry many different styles of products for homeowners, including

- Gravity Wall Furnaces
- Vented Console Heaters
- Vented Free Heaters (with Manual and Thermostat Bulbs)
- Counter Flow Furnaces
- Vented Console Radiant Front Heaters
- Blue Flame Heaters
- Console Direct Vent Heaters
- LP kits, blower kits, etc.

Unlike our competitors, who tend to ship products late and do not stand behind any damages that may occur during shipment, we ship products in five business days or less. Similarly, our freight charges run around $50–$60 per crate, while our competitors will charge you an arm and a leg, plus many of their products are made overseas where standards are low.

Due to the winter months here in the Midwest, it is recommended that most homeowners replace their heating and air conditioning units if their models are older than fifteen years old. In old neighborhoods like yours, you probably need to replace your furnace **NOW** before cold weather hits. If models are old or made by companies which do not have standards as good as ours, older homeowners like you are on borrowed time and do not know when their old model will decide to give up at the most unexpected time (i.e., when the weather is below zero or the temperature is above 100 degrees).

Please call with any questions or comments at 815-555-2121. Our office staff would be available to answer any questions. We are here just to help you.

Sincerely,

Jack Henry

Read the letters and complete the following assignments:

- Decide whether Ms. Stine's tone is formal or informal, and give examples to prove your point.
- Decide whether Ms. Stine's communication style is passive, aggressive, or assertive, and give examples to prove your point.

Letter A

APT Training Institute
2425 Oleander Dr.
New Orleans, LA 72291

February 26, 2009

Sharon Baker, Acct. Exec.
Home and Hearth Bank
1092 Turtle Hill Road
Evening Star, GA 53321

Sharon,

I've attached your IDP. It will help you overcome several problem areas I encountered when reviewing your job responsibilities. I strongly suggest that you address these issues ASAP.

Here's what you must focus on:

- Delegating responsibility more effectively, in contrast to your somewhat lackadaisical approach to management
- Listening more closely to your supervisors when they make suggestions—(you probably remember that you didn't always even listen to me during the workshop)
- Turning in work on time vs. missing deadlines, a problem that John, one of your co-workers, said really ticked him off
- Setting goals that you should be able to achieve

Well, those ideas ought to keep you busy, Sharon. Just remember, I'm here to help. And I'm very proud of the extent to which I am skilled in the areas that you most need to address.

Sincerely,

Shelly

Attachment

- Decide if Ms. Stine's tone and communication style is correct for the audience, and give examples to prove your point.
- Rewrite the letter or letters to improve the tone and communication style.

4. Amir Aksarben works in corporate communication for Prismatic Consulting Engineering. He has had the job for two months. He needs to respond to an RFP (request for proposal), posted by The Oceanview City Council. The RFP is asking for bids to improve Oceanview's flood control. Before Amir can respond to the RFP, he needs to gather more information, as follows:

   - Amir needs to communicate with Oceanview, inquiring about specific problems the city is encountering and their timeframe for making a decision. Amir's contact at Oceanview is Sally Howser, who went to college with Amir and worked with him on several school-related team projects.
   - He needs to communicate with his boss, Randy Towner, the owner and founder of Prismatic, determining what other projects Prismatic is engaged in, which

Letter B

APT Training Institute
2425 Oleander Dr.
New Orleans, LA 72291

February 26, 2009

Sharon Baker, Account Executive
Home and Hearth Bank
1092 Turtle Hill Road
Evening Star, GA 53321

Dear Ms. Baker:

In response to a request from Home and Hearth's Human Resource Management, I have attached your Individual Development Plan (IDP). I truly hope that you find many of APT's assessments beneficial to your continuing quest for job improvement. Let me apologize up front if any of the suggestions upset you in any way. APT only wants what is best for clients. I realize that sometimes criticism, though perhaps justified by the IDP process, might seem harsh. However, APT feels that if clients analyze the IDP findings objectively, the APT suggestions could help to overcome potential problem areas. I greatly hope that you will approach APT's suggestions with the goal of self-improvement in mind.

Please consider focusing on the following:

- Effective delegation of responsibility, trying to use the skills of your very talented subordinates
- Considering your supervisors' suggestions, many of which I believe could benefit the department goals
- Striving to meet deadlines, remembering that whenever a deadline is missed, this could impact another employee's job duties
- Setting goals in accordance to the bank's mission statement

Ms. Baker, with all due respect to you, I hope these ideas meet with your approval. Though I am not an expert in bank management, I think I could help you meet your IDP plans. Please consider setting up an appointment at your convenience with me or my staff. Finally, thank you for allowing me the pleasure of working with you to create an IDP that I believe should suit your needs.

Sincerely,

Shelly Stine

Attachment

engineers and architects Randy would like to assign to the construction crew for Oceanview's project, and which other Prismatic employees Randy would like to work on the proposal writing team.

- Amir must ask his colleagues in other Prismatic departments (engineering, accounting, and drafting) for input regarding the proposal's content (timeframe to complete the job, costs, and schematics). All of these colleagues have worked on similar projects in the past.

Letter C

APT Training Institute
2425 Oleander Dr.
New Orleans, LA 72291

February 26, 2009

Sharon Baker, Account Executive
Home and Hearth Bank
1092 Turtle Hill Road
Evening Star, GA 53321

Dear Ms. Baker:

Thank you for meeting with me last week to work on your attached Individual Development Plan (IDP). I appreciate the open and honest discussion we had about your strengths and areas needing improvement. In our meeting, we addressed criticisms from your subordinates and supervisors, but also we focused on constructive options.

Following are work-related topics that APT can help you address:

- Delegating responsibility effectively. By using the skills of your talented staff, you can empower employees.
- Practicing APT's "Win/Win Listening Skills." Considering your supervisors' suggestions will help you meet department goals.
- Meeting deadlines. A missed deadline negatively impacts business and employee workload. This fact was documented in your last month's personnel assessment, when your missed deadline for the Stanhope proposal cost the company approximately $15,000 in potential business.
- Setting goals. APT can help you negotiate your 1-year and 5-year plans, in accordance with the bank's mission statement.

Ms. Baker, based on your 360-Degree Assessment, which you signed and approved, you recognize areas needing improvement. With your assistance, APT will work with you to achieve your bank's high standards. We have the skills; I know you have the desire. Please make an appointment at your convenience with me or my staff. Thank you for allowing me to work with you on your IDP.

Sincerely,

Shelly Stine

Attachment

- He needs to tell his newly hired summer intern to start making phone calls to the city and county permit regulators to ensure that Prismatic has the appropriate work permits on file.

Assignment

Based on the previously noted information, answer the following questions, explaining your decisions:

a. Which communication channel should Amir use in each of the previous instances and why?

b. Which tone (formal or informal) should Amir use in each of the previous communication situations and why?
c. How much detail will Amir need to give each of the different audiences and why?

ETHICS IN THE WORKPLACE CASE STUDY

According to ethical standards, visuals should be precise. Look at the following bar chart. It shows that during one quarter, the company lost over $50,000. However, the visual could be considered misleading, since the bar showing a loss is not drawn to the same scale as the other bars.

Question: Based on what you have read in this chapter regarding ethics, is the following visual acceptable or not?

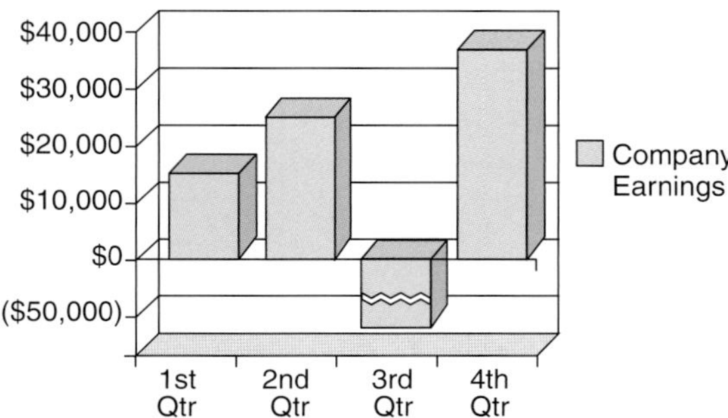

INDIVIDUAL AND TEAM PROJECTS

Complete the activities by doing the following:

- **Oral presentations**—Give a three- to five-minute briefing to share the results of your findings with your colleagues.
- **Written**—Write a brief (1/2 page) memo or e-mail to your teacher, highlighting your findings.

1. Visit employees from your college's or university's business services. These could include payroll, administrative computing, human resources, accounts payable, benefits, budget, catering, the bookstore, printing services, records, admissions, loans, and so forth. After interviewing these employees, respond to the following assignments either by writing a brief memo or e-mail to your instructor, or by giving a three- to five-minute oral presentation to your class.

   a. **Time.** How much time do your college's or university's employees spend writing and communicating orally to students, co-workers, subordinates, and management?

   b. **Types of writing.** Ask employees from your college or university what they have written at work this week.
   - What types of communication did they write (e-mail, letters, memos, reports, Web pages, and so on)?
   - What were their goals in writing (sell, document, request, order, etc.)?
   - Who were their audiences (were they writing laterally or vertically, internally or externally, formal or informal communication)?

   c. **Communication channels.** Ask employees from your college or university what channels they used to communicate at work this week (e-mail, faxes, cell phones, teleconferences, videoconferences, PDAs, Web sites, etc.). Create a table showing the amount of usage for each communication channel.

   d. **Oral communication.** Ask employees from your college or university what types of oral communication they were involved in at work this week (formal or informal), to whom they spoke (internal, external, lateral, or vertical), and what their goals were.

2. ***Visits to professionals.*** Visit professionals from your area of interest or degree program. Interview these employees to learn the following information:
   - What they have written at work
   - Which communication channels they use
   - Whether they communicate internally or externally, laterally or vertically, formally or informally
   - Which ethical concerns they have confronted
   - How their business has changed, in terms of technology's impact or globalization
3. ***Self-evaluation.***
   - Calculate how much time you spend writing and communicating orally with customers, co-workers, subordinates, management, or colleagues, either on the job or in a school-related organization (club, fraternity, sorority, team, etc.). What percentage of your job or organization activity involves communication?
   - Make a list of what you have written at work or for your school-related organization this week. What types of communication did you write, who were your audiences, and what were the goals of your communication? What percentage of your job or organization activity involves written communication?
   - Make a list of what types of oral communication you have used on the job or in your school-related organization this week. Were you in meetings, speaking on the phone, working with people face-to-face, leaving voice mail messages, and so on? What percentage of your job or organization activity involves oral communication?
   - What communication channels have you used for workplace communication or a school-related club this week? Did you use voice mail, cell phones, faxes, e-mail, PDAs, pagers, Web searches, or teleconferencing, along with your computer's word processing packages? What technologies are you using now to communicate that you did not use last year? How is technology changing the way you communicate?
   - What is your communication style? Are you passive, aggressive, or assertive? Give examples to prove your point.
4. ***Your career choice.*** Research your career choice or other potential career opportunities. To do so, visit your college's or university's career placement service, look at newspaper ads, visit with human resources managers, network with employees already working in the field, or go online to search Internet job sources. Then, determine what types of communication skills are required for your potential job.

DEGREE-SPECIFIC ASSIGNMENTS

Restful Inn Hoteliers is a growing international company with corporate headquarters in Seattle, WA; Boston, MA; Barcelona, Spain; and Singapore. As it builds more hotels around the world, Restful Inn needs to hire employees skilled in hotel management, business information systems, accounting, food and beverage management, international sales and marketing, and finance.

In addition to loving travel and working with people, these new hires must have excellent communication skills—written, oral, and nonverbal. Some of these employees will work directly with hotel visitors (hotel management, sales and marketing). Others will work with staff and vendors (accounting, food and beverage management, business information systems, and finance). All will need to write well and be skilled verbal communicators.

Management will be on the floor, greeting customers and helping with the hotel's check-in, customer needs, and hotel operations. In addition, management will also have to write employee evaluation reports, response letters to customers, daily e-mail to corporate headquarters, feasibility reports to corporate headquarters regarding hotel projections, and letters to vendors regarding products and services.

Sales and marketing will promote their hotel through many different methods. These could include sales letters, fliers, or brochures to previous customers; television and radio advertisements; and speeches to civic leaders.

Accounting, food and beverage management, and finance will write reports for government auditors, vendors, and staff. In addition to writing and responding to daily e-mail, these employees also will be on the phone constantly.

Business information technology employees will develop Internet, intranet, and extranet systems to connect clients, vendors, and staff.

As an added benefit, all of these employees will be allowed to relocate to Restful Inn hotels throughout the world. Thus, they will encounter multicultural communication challenges.

Assignment

Write job descriptions for any or all of the positions at Restful Inn. To do so, research job descriptions online, at your college's job placement center, or in a library. Include appropriate salary ranges and educational requirements. Most importantly, specify the communication responsibilities, based on what you have learned in this chapter.

PROBLEM-SOLVING THINK PIECES

1. In an interview, a company benefits manager said that she spent over 50 percent of her workday on communication issues. These included the following:
   - Consulting face-to-face with staff, answering their questions about retirement, health insurance, and payroll deductions
   - Meeting weekly with human resources colleagues
   - Collaborating with project team members
   - Preparing and writing quarterly reports to HR supervisors
   - Teleconferencing with third-party insurance vendors regarding new services and/or costs
   - E-mailing supervisors and staff, in response to questions
   - Calling and responding to telephone calls
   - Faxing information as requested
   - Writing letters to vendors and staff to document services

Though she had to use various methods of both written and oral communication, the communication channels each have benefits and drawbacks. E-mailing, for example, has pluses and minuses (convenience over depth of discussion, perhaps). Think about each of the communication options in the previous list. Using the following matrix, list the benefits of each particular type of communication versus the drawbacks. Then, consider how you would solve the problem.

Communication Channel	Benefits	Drawbacks	Possible Solutions
One-on-one discussions			
Group meetings			
Collaborative projects			
Written reports			
Teleconferences			
E-mail			
Phone calls			
Faxes			
Letters			

2. Restful Inn Hoteliers is an international company with corporate headquarters in Seattle, WA; Boston, MA; Barcelona, Spain; and Singapore. Restful Inn employees are skilled in hotel management, accounting, food and beverage management, business information systems, international sales and marketing, and finance. These employees also must have excellent communication skills—written, oral, and nonverbal.
   - What types of written, oral, and nonverbal communication experiences do you think these employees will face on the job?
   - What do you think they will have to write, to whom, and why?
   - What types of oral presentations might they need to make, to whom, and why?
   - What daily verbal and nonverbal communications might they be involved in, to whom, and why?

   Use the following matrix to prepare your answers.

Job Title	Written Communication	Oral Communication	Nonverbal Communication
Hotel management			
Accounting			
Food and beverage management			
International sales and marketing			
Finance			
Business information systems			

3. Tamara Jones is a receptionist at a bank. Her job is to greet clients, direct them to the appropriate bankers, and to answer questions about bank procedures. When she needs to speak to a customer or coworker, she's always available and pleasant. However, her job has periods of downtime when she waits for the next customer to arrive.

During this downtime, she occasionally plays computer games like solitaire or hearts. However, she's gotten tired of the games that are loaded on her bank computer. To enhance her "gaming" options, Tamara has downloaded numerous computer games that require a lot of computer space and memory.

She had been complaining about the slowness of her computer. As the manager of information technology at the bank, you were doing an inventory of bank computers and found that Tamara's computer memory was overloaded with games.

Is is ethical for Tamara to be using bank time and equipment to play computer games? Also, is it ethical for Tamara to be downloading external software? Justify your answer based on information provided in this chapter.

4. CompToday computer hardware company must abide by the Sarbanes-Oxley Act, passed by Congress in response to accounting scandals. This Act specifically mandates the following related to documentation standards:

Section 103: Auditing, Quality Control, and Independence Standards and Rules.

Companies must "prepare, and maintain for a period of not less than 7 years, audit work papers, and other information related to any audit report, in sufficient detail to support the conclusions reached in such report."

Section 401(a): Disclosures in Periodic Reports; Disclosures Required.

"Each annual and quarterly financial report . . . must be presented so as not to contain an untrue statement or omit to state a material fact necessary in order to make the pro forma financial information not misleading."

Beverly Warden, technical documentation specialist at CompToday, is responsible for managing the Sarbanes-Oxley reports. She is being confronted by the following ethical issues:

a. To help Beverly prepare the first annual report, her chief financial officer (CFO) has given her six months of audits (January through June). These prove that the company is meeting its accounting responsibilities. However, Beverly's report covers the entire year, including July through December. Section 103 states that the report must provide "sufficient detail to support the conclusions reached in [the] report."

   Ethical Questions

   Are the company's first six month's audits sufficient? If Beverly writes a report stating that her company is in compliance, is she abiding by her Communicator's Ethical Principles, which state that a writer's work is "consistent with laws and regulations"?

   What are her ethical, practical, and legal responsibilities?

   - Share your findings in an oral presentation.
   - Write a letter, memo, report, or e-mail to your teacher stating your opinion regarding this issue.

b. Beverly's CFO also has told her that, during the year, the company fired an outside accounting firm and hired a new one to audit the company books. The first firm expressed concerns about several bookkeeping practices. The newly hired firm, providing a second opinion after reviewing the books, concluded that all bookkeeping practices were acceptable. The CFO sees no reason to mention the first firm.

   Ethical Questions

   Section 401(a) states that reports must contain no untrue statements or omit to state a material fact. Ethical principles also say that her writing must be truthful and accurate, to the best of her ability. Beverly can report factually that the new accounting firm finds no bookkeeping errors. Should she also report the first accounting firm's assessment? Is that a material fact? If she omits any mention of the first accounting firm, as her boss suggests, is she meeting both her writer's responsibilities and the needs of Sarbane-Oxley?

   What are her ethical, practical, and legal responsibilities?

   - Share your findings in an oral presentation.
   - Write a letter, memo, report, or e-mail to your teacher stating your opinion regarding this issue.

WEB WORKSHOP

1. **Corporate Mission Statements.** Go online to research a corporate Web site's mission/vision page. Learn what the company's mission statement reveals about the worth it places on communication skills between customers and co-workers. Share your findings with the class.
   - **Oral presentations.** Give a three- to five-minute briefing to share the results of your research with your colleagues.
   - **Written.** Write a brief ($^1/_2$ page) memo highlighting your findings.
2. **Corporate Ethics.** Go online to research corporate or organizational codes of ethics. To do so, look in your area of interest or expertise, or research a company that you might want to work for. What do these companies or organizations say about the importance of ethics, and what are their ethical codes of conduct?
   - Compare and contrast various codes of conduct to what this chapter presents—where are there similarities and differences, inclusions and omissions?
   - Write a code of ethics for your job area, career choice, or field of interest.

QUIZ QUESTIONS

1. Explain why good oral and written communication is essential to your success in business.
2. What are four communication channels you can use in business?
3. What four channels are almost always used in workplace communication?
4. In what directions do you communicate in business?
5. What challenges confront you when you write or speak laterally or vertically within an organization?
6. What are three basic communication styles?
7. When is the aggressive style of communication helpful in business?
8. List some strategies for achieving ethical communication.
9. What are the three parts of the P^3 process?
10. Explain how the workplace is changing.

CHAPTER 2

Collaborating in the Workplace

▸REAL PEOPLE *in the workplace*

Shelly Smith is a product reliability team leader at Century Software, a company that designs, develops, sells, and maintains software systems for payroll, benefits, and insurance.

Why does Century use teams? Ms. Smith says, "Teams are important when a company is trying to improve procedures. Change is always difficult. This is especially true when changes impact individual workers. To avoid surprises, you must have buy-in from all stakeholders. The team needs to understand the current situation, have a chance to give their input, know what's coming, and understand how any changes will affect them personally."

Ms. Smith's goal as team leader is to ensure product reliability. Since product development depends on different departments within Century nationwide, Shelly formed a cross-functional team. Her group consists of 15 employees from software development, design, quality assurance, sales, and customer service.

Ms. Smith spends between 15 to 20 hours a week as team leader. Her job includes acquiring and maintaining a budget, facilitating the team meetings and team tasks, and keeping the team on track. Plus, she must stay in touch with her team members. Century uses four electronic communication channels to ensure a consistent, collaborative workforce: teleconferences, videoconferences, Webinars, and wikis.

- **Teleconferences**—Shelly, who works in Century's Birmingham office, has team members who work in San Francisco, Salt Lake City, Cleveland, Columbus, Rochester, and Baltimore. To communicate with her dispersed team members, "We have a conference call just to touch base. Sometimes we have a formal agenda, and sometimes I just ask, 'What's going on?' A casual, weekly teleconference allows us to stay up to date on issues facing us individually or as a group. We collectively understand that collaborating is good for problem solving."
- **Videoconferences**—You can't communicate effectively with 15 people on the telephone. While teleconferencing works well for Shelly and smaller groups, when she needs to communicate with the entire team about corporate-wide issues that affect policy, budget, personnel, and strategic planning, face-to-face meetings might be the optimum solution. However, transporting numerous people to a central location is neither time-efficient nor cost-effective. A three-hour meeting might require two days of travel plus hotel, food, and air fares. To save time and money, Century uses videoconferences.

CHAPTER GOALS

When you complete this chapter, you will be able to do the following:

1. Work effectively with other employees in business.
2. Develop collaborative skills to interact successfully with other people in the workplace.
3. Understand that teamwork encourages diverse opinions, contributes to understanding, empowers members, and encourages collegiality.
4. Resolve conflicts in teams by setting guidelines, encouraging equal discussion and involvement, and discouraging taking sides.
5. Collaborate successfully by selecting an effective team leader, determining goals, identifying and analyzing problems, determining potential improvements, verifying the solutions, breaching the gaps to achieve human performance improvement, and completing the project.

WWW

To learn more about working effectively in teams, visit our Web Companion at http://www.prenhall.com/gerson.

- **Webinars**—In the course of software development, training is required so team members can learn new applications. Webinars allow Shelly's team to participate in online training seminars without incurring travel expenses. Century's Webinars allow Shelly's dispersed team to participate in live group discussions and training from their individual computers.
- **Wikis**—Wikis allow Shelly's team members to write collaboratively. They can add, remove, and edit content online. Plus, a wiki lets them track the history of a document as it is revised. Wikis help the dispersed team to create effective reports.

Shelly says that she spends approximately 50 percent of her work time communicating via e-mail messages, telephone calls, and teleconferences. For efficiency, cost savings, and consistent communication to a geographically dispersed workforce, Century has found that multiple, electronic channels help team members achieve their communication goals.◀

Shelly's Communication Challenge

Shelly says, "To succeed, my team sets long-term and short-term goals. Its ultimate goal is to achieve a 10 percent reduction in customer complaints about its software. That reduction is the benchmark, the kind of measurable level that all management wants to see. The smaller goals included forming the team, setting team term limits, identifying the specific tasks needed to achieve reliability, drafting instructions, and formally releasing these procedures, with management's approval.

Currently, my team and I are working on a beta version of software for employee benefits and insurance. The team's challenge is to develop the software, write the codes, design the help screens, and write the instructional text for end users. My challenge is to get timely input from all team members and package the end product."

See pages 53–55 to learn how Shelly met her communication challenge.

THE IMPORTANCE OF TEAMWORK

Companies have found that teamwork enhances productivity. Teammates help and learn from each other. They provide checks and balances. Through teamwork, employees can develop open lines of communication to ensure that projects are completed successfully.

Collaboration

In business and industry, many user manuals, reports, proposals, PowerPoint presentations, and Web sites are team written. Teams consist of engineers; graphic artists; marketing specialists; corporate employees in legal, delivery, production, sales, and accounting departments; and management. These collaborative team projects extend beyond the company. A corporate team also will work with subcontractors from other corporations. The collaborative efforts include communicating with companies in other cities and countries through teleconferences, faxes, and e-mail. Modern workplace communication requires the participation of "communities of practice": formal and informal networks of people who collaborate on projects based on common goals, interests, initiatives, and activities (Fisher and Bennion 278).

The National Association of Colleges and Employers lists the "Top Ten Skills Employers Want" (see Table 2.1). Notice where interpersonal, teamwork, oral communication, and written communication skills appear in relation to other skills.

The Problems with "Silo Building"

Working well with others requires collaboration versus "silo building." The *silo* has become a metaphor for departments and employees that behave as if they have no responsibilities outside their areas. They build bunkers around themselves, failing to collaborate with others. In addition, they act as if no other department's concerns or opinions are valuable.

Such "stand-alone" departments or people isolate themselves from the company as a whole and become inaccessible to other departments. They "focus narrowly" (Hughes 9). This creates problems. Poor accessibility and poor communication "can cause duplicate efforts, discourage cooperation, and stifle cross-pollination of ideas" (Hughes 9).

To be effective, companies need "open lines of communication within and between departments" (Hughes 9). The successful employee must be able to work collaboratively with others to share ideas. In the workplace, teamwork is essential.

Why Teamwork Is Important

Teamwork benefits employees, corporations, and consumers. By allowing all constituents a voice in project development, teamwork helps to create effective workplaces and ensures product integrity.

Diversity of Opinion When you look at problems individually, you tend to see issues from a limited perspective—*yours*. In contrast, teams offer many points of view. For instance, if

Table 2.1 Top Ten Qualities/Skills Employers Want

Skill	Rating
1. Communication Skills (Verbal and Written)	4.7
2. Honesty/Integrity	4.7
3. Teamwork Skills (works well with others)	4.6
4. Strong Work Ethic	4.5
5. Analytical Skills	4.4
6. Flexibility/Adaptability	4.4
7. Interpersonal Skills (relates well to others)	4.4
8. Motivation/Initiative	4.4
9. Computer Skills	4.3
10. Detail Oriented	4.1
(5-point scale, where 1 = not at all important and 5 = extremely important)	

("Top Ten Skills Employers Want")

a team has members from accounting, public relations, customer service, engineering, and information technology, then that diverse group can offer diverse opinions. You should always look at a problem from various angles.

Checks and Balances Diversity of opinion also provides the added benefit of checks and balances. Rarely should one individual or one department determine outcomes. When a team consists of members from different disciplines, those members can say, "Wait a minute. Your idea will negatively impact my department. We had better stop and reconsider."

Broad-Based Understanding If decisions are made in a silo, by a small group of like-minded individuals, then these conclusions might surprise others in the company. Surprises are rarely good. You always want buy-in from the majority of your stakeholders. An excellent way to achieve this is through team projects. When multiple points of view are shared, a company benefits from broad-based knowledge. Improved communication allows people to see the bigger picture.

Empowerment Collaboration gives people from varied disciplines an opportunity to provide their input. When groups are involved in the decision-making process, they have a stake in the project. This allows for better morale and productivity.

Team Building Everyone in a company should have the same goals—corporate success, customer satisfaction, and quality production. Team projects encourage shared visions, a better work environment, a greater sense of collegiality, and improved performance. Employees can say, "We are all in this together, working toward a common goal."

DIVERSE TEAMS . . . DISPERSED TEAMS

Collaborative projects will depend on diverse team members and dispersed team members.

Diverse Teams

Teams will be diverse, consisting of people from different areas of expertise. Your teams will be made up of engineers, graphic artists, accountants, technical writers, financial advisors, human resource employees, and others. In addition, the team will consist of people who are different ages, genders, cultures, and races.

Dispersed Teams

In a global economy, members of a team project might not be able to work together, face to face. Team members might be located across time and space. They could work in different cities, states, time zones, countries, or different shifts. For example, you might work for your company in New York, while members of your team work for the company at other sites in Chicago, Denver, and Los Angeles. This challenge to collaboration is compounded when you also must team with employees at your company's sites in India, Mexico, France, and Japan. According to a 2005 report, 41 percent of employees at the top international corporations live outside the borders of their company's home country (Nesbitt and Bagley-Woodward 25).

In addition, employees might work from their homes as telecommuters versus in the office. According to Forrester Research, the U.S. teleworker population grew approximately 7 percent to 44 million from 2003 to 2004. By 2008, 75 percent of businesses invested significantly in mobile communication. As companies become more familiar with electronic technologies, these companies will become mobile enterprises. Forrester Research defines a mobile enterprise as one that connects vendors, businesses, employees, and customers from any location ("Achieving the Promise").

USING GROUPWARE TO COLLABORATE IN VIRTUAL TEAMS

When employees are dispersed geographically, getting all team members together is costly in terms of time and money. Companies solve this problem by forming virtual, remote teams that collaborate using electronic communication tools—groupware. Groupware consists of software and hardware that helps companies cut down on travel costs, allows for telecommuting, and facilitates communication for employees located in different cities and countries.

Groupware includes the following types of hardware and software (Nesbit and Bagley-Woodward 28):

- Electronic conferencing tools such as Webinars, listservs, chat systems, message and discussion boards or forums, videoconferences, and teleconferences (see Figure 2.1 for a sample discussion forum). A new technology allows for "telepresences"—a virtual meeting format usually configured as a conference table; three large video screens big enough for life-sized, head-and-shoulders images of employees from distant locations; high-definition video; and multiple-channel audio. Many companies, such as Heinz, General Electric, PepsiCo, and Wachovia, are using telepresence technology for product briefings, training courses, and strategy sessions (Finney).
- Electronic management tools. For example, Digital Dashboards is project management software that schedules, tracks, and charts the steps in a project. Another example of an electronic management tool is Microsoft Outlook's electronic calendaring. This allows you to send a meeting request to dispersed team members, check the availability of meeting attendees, reschedule meetings electronically, forward meeting requests, and cancel a meeting—without ever visiting with your team members face-to-face.
- Electronic communication tools for writing and sending documents. These include tools like instant messaging, e-mail, blogs, intranets and extranets, and wikis.

Collaborative Writing Tools

Wikis

What's a wiki? It is the latest in technology designed for collaborative writing. A wiki "is a website that allows the visitors . . . to easily add, remove, and otherwise edit and change available content, and typically without the need for registration. This ease of interaction and operation makes a wiki an effective tool for mass collaborative authoring" ("Wiki"). In addition, wikis let collaborative writers track "the history of a document as it is revised." Whenever a team member edits text in the "wiki, that new text becomes the current version, while older versions are stored" (Mader). The largest example of a wiki is Wikipedia, "the free encyclopedia that anyone can edit." "Wikipedia is an encyclopedia collaboratively written by many of its readers" ("What Is Wikipedia?").

Wikis have entered the workplace. Prentice Hall, the world's largest publishing company, is using wiki technology to create a community-written textbook. *We Are Smarter Than Me* is the first networked book on business. Collaborators used wiki technology to write a "book on how the emergence of community and social networks will change the future rules of business" ("We Are Smarter Than Me").

Who's Using Wikis? Many companies use wikis for collaborative writing projects. Yahoo, for example, uses a wiki. Eric Baldeschwieler, Director of Software Development of Yahoo, says, "Our development team includes hundreds of people in various locations all over the world, so web collaboration is VERY important to us." Cmed runs pharmaceutical clinical trials and develops new technology. In this heavily regulated environment, wikis "improved communication and increased the quantity (and through peer review, the quality) of documentation." Cingular Wireless "project

FIGURE 2.1 **Discussion Forum**

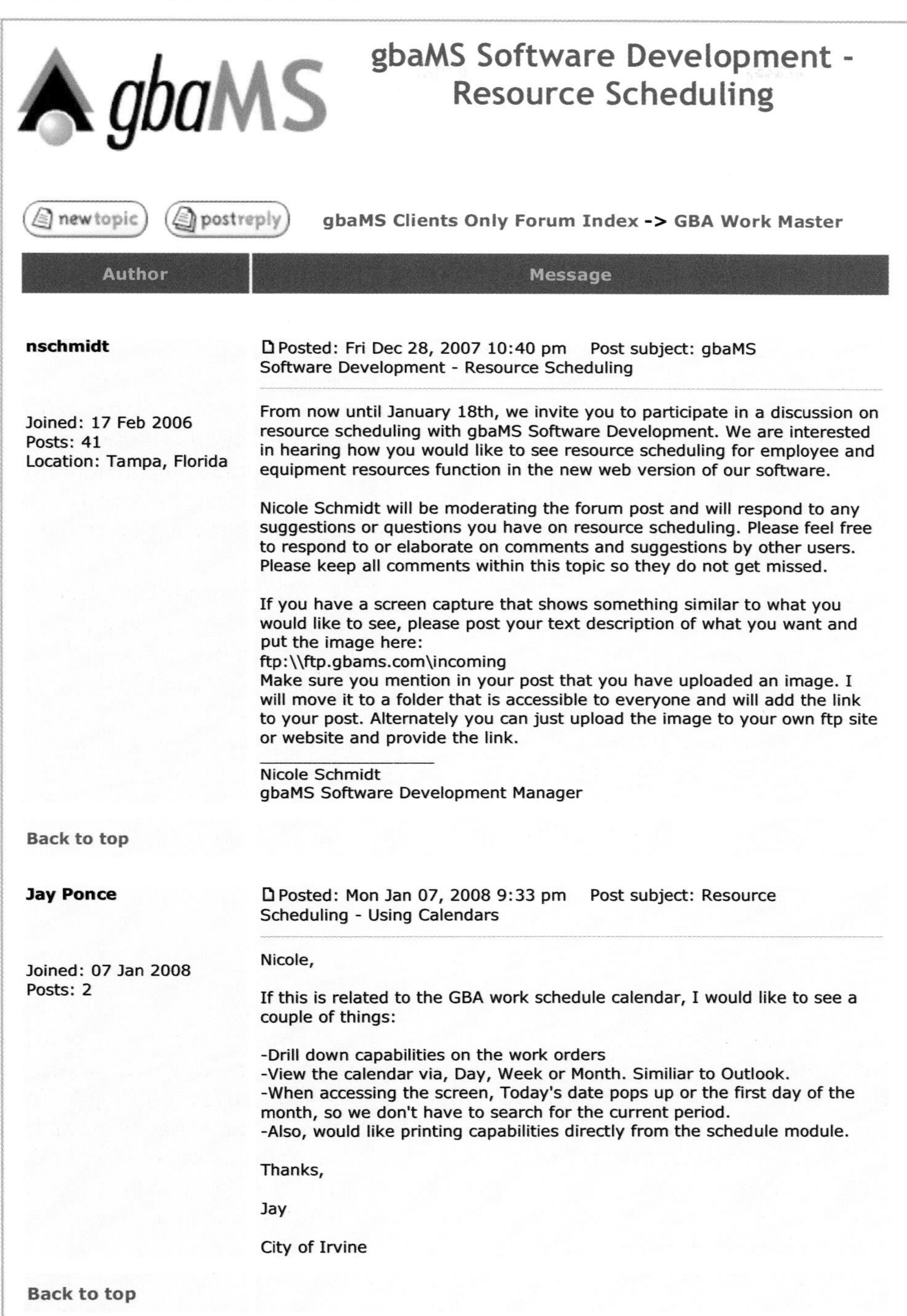

gbaMS

gbaMS Software Development - Resource Scheduling

new topic   post reply   gbaMS Clients Only Forum Index -> GBA Work Master

Author	Message
nschmidt Joined: 17 Feb 2006 Posts: 41 Location: Tampa, Florida	Posted: Fri Dec 28, 2007 10:40 pm Post subject: gbaMS Software Development - Resource Scheduling From now until January 18th, we invite you to participate in a discussion on resource scheduling with gbaMS Software Development. We are interested in hearing how you would like to see resource scheduling for employee and equipment resources function in the new web version of our software. Nicole Schmidt will be moderating the forum post and will respond to any suggestions or questions you have on resource scheduling. Please feel free to respond to or elaborate on comments and suggestions by other users. Please keep all comments within this topic so they do not get missed. If you have a screen capture that shows something similar to what you would like to see, please post your text description of what you want and put the image here: ftp:\\ftp.gbams.com\incoming Make sure you mention in your post that you have uploaded an image. I will move it to a folder that is accessible to everyone and will add the link to your post. Alternately you can just upload the image to your own ftp site or website and provide the link. Nicole Schmidt gbaMS Software Development Manager
Back to top	
Jay Ponce Joined: 07 Jan 2008 Posts: 2	Posted: Mon Jan 07, 2008 9:33 pm Post subject: Resource Scheduling - Using Calendars Nicole, If this is related to the GBA work schedule calendar, I would like to see a couple of things: -Drill down capabilities on the work orders -View the calendar via, Day, Week or Month. Similiar to Outlook. -When accessing the screen, Today's date pops up or the first day of the month, so we don't have to search for the current period. -Also, would like printing capabilities directly from the schedule module. Thanks, Jay City of Irvine
Back to top	

Reprinted with permission of GBA Master Series, Inc.

managers have been encouraged to utilize the site for any issues needing collaborating efforts in lieu of emails." Disney used a wiki "for an engineering team that was rearchitecting the Go.com portal. During this time we found TWiki to be a very effective means of posting and maintaining development specs and notes as well as pointers to resources." Texas Instrument's "India design centre" uses wikis to manage all project-specific information, such as documenting ideas, plans, and status; sharing information with other teams across various work sites; and updating information and content to team members ("Twiki").

How Can You Use a Wiki? In your dispersed teams, whether virtual, remote, or mobile, you might use wikis in the following ways to create collaborative documents:

Additional information:

See Chapter 14, "Designing Web Sites."

- **Create Web sites**—Wikis help team members easily add pages, insert graphics, create hyperlinks, and add simple navigation.
- **Project development with peer review**—A wiki makes it easy for team members to write, revise, and submit projects, since all three activities can take place in the wiki.
- **Group authoring**—Wikis allow group members to build and edit a document. This creates a sense of community within a group, allows group members to build on each other's work, and provides immediate, asynchronous access to all versions of a document.
- **Track group projects**—Each wiki page lets you track how group members are developing their contributions. The wiki also lets you give feedback and suggest editorial changes.

Figures 2.2, 2.3, and 2.4 are illustrations of a wiki, edited text in a wiki, and a record of edited versions.

Google Documents

Another collaborative writing tool you can use easily is Google Documents, which is useful for document sharing, collaborating on group projects, and publishing to the World Wide Web. Google Documents is free to anyone with a Web browser and an Internet connection. This electronic tool allows for seamless integration of its collaborative features and user-friendly interface.

Using Google Documents, you and your group members can edit Word documents, as well as RTF and HTML files. Teams can work on a document at the same time. Changes made by one writer will be seen by all team members instantly.

FIGURE 2.2 Wiki Page

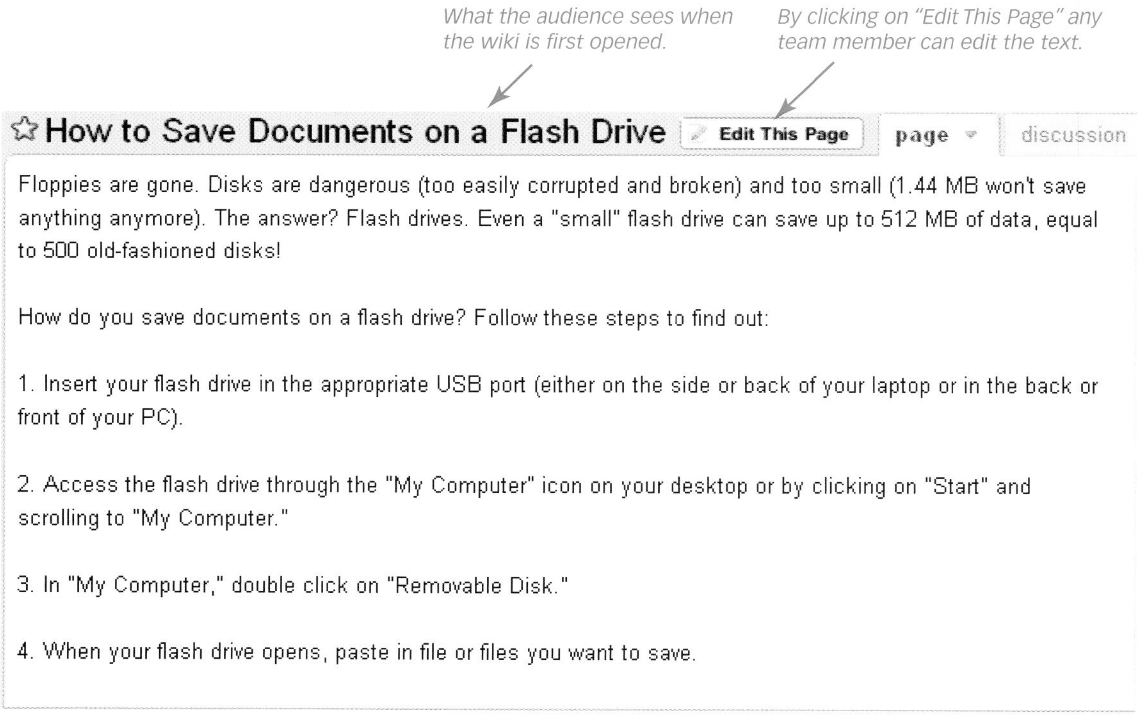

What the audience sees when the wiki is first opened.
By clicking on "Edit This Page" any team member can edit the text.
How to Save Documents on a Flash Drive
Edit This Page
page
discussion
Floppies are gone. Disks are dangerous (too easily corrupted and broken) and too small (1.44 MB won't save anything anymore). The answer? Flash drives. Even a "small" flash drive can save up to 512 MB of data, equal to 500 old-fashioned disks!
How do you save documents on a flash drive? Follow these steps to find out:
1. Insert your flash drive in the appropriate USB port (either on the side or back of your laptop or in the back or front of your PC).
2. Access the flash drive through the "My Computer" icon on your desktop or by clicking on "Start" and scrolling to "My Computer."
3. In "My Computer," double click on "Removable Disk."
4. When your flash drive opens, paste in file or files you want to save.

Reprinted with permission of Wikispaces.

FIGURE 2.3 Wiki in "Edit" Mode

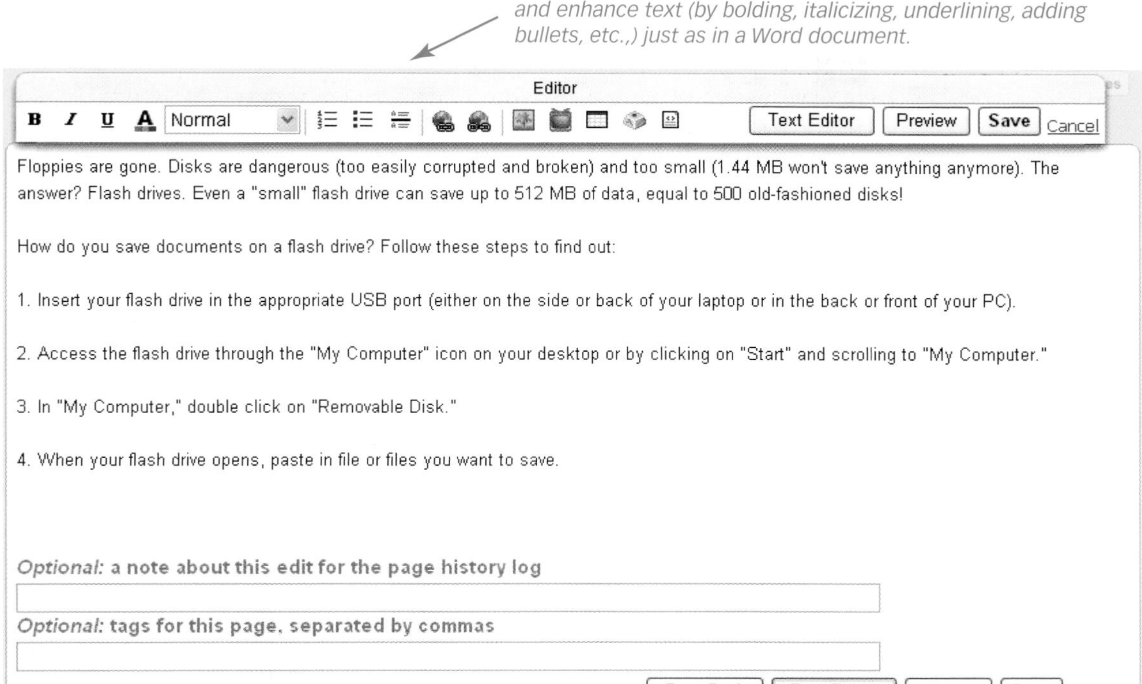
The "editing" screen allows team members to add, delete, and enhance text (by bolding, italicizing, underlining, adding bullets, etc.,) just as in a Word document.
Editor
Normal
Text Editor
Preview
Save
Cancel
Floppies are gone. Disks are dangerous (too easily corrupted and broken) and too small (1.44 MB won't save anything anymore). The answer? Flash drives. Even a "small" flash drive can save up to 512 MB of data, equal to 500 old-fashioned disks!
How do you save documents on a flash drive? Follow these steps to find out:
1. Insert your flash drive in the appropriate USB port (either on the side or back of your laptop or in the back or front of your PC).
2. Access the flash drive through the "My Computer" icon on your desktop or by clicking on "Start" and scrolling to "My Computer."
3. In "My Computer," double click on "Removable Disk."
4. When your flash drive opens, paste in file or files you want to save.
Optional: a note about this edit for the page history log
Optional: tags for this page, separated by commas
Save Draft
Text Editor
Preview
Save
Cancel

Reprinted with permission of Wikispaces.

FIGURE 2.4 Record of Edited Versions

This screen allows team members to compare all versions of the text, thus seeing what has been added, deleted, or enhanced.
How to Save Documents on a Flash Drive
page
discussion
history
Date
Compare
Author
Comment
Aug 23, 2008 11:37 am
select
sgerson
Aug 23, 2008 11:36 am
select
sgerson
Aug 23, 2008 11:09 am
select
sgerson
1-3 of 3

Reprinted with permission of Wikispaces.

Google Documents provide these benefits:

- View a document's revision history.
- Return to earlier versions.
- Edit and view a document.

- Add new team members or delete writers.
- Post documents to a blog or publish a document to a Web page.

CHALLENGES TO EFFECTIVE TEAMWORK

Any collaborative activity is challenging to manage: team members do not show up for class or work; one student or employee monopolizes the activity while another individual snoozes; people exert varying amounts of enthusiasm and ability; personalities clash. Some people fight over everything. Occasionally, when a boss participates on a team, employees fear speaking openly. Some team members will not stay on the subject. One team member will not complete her assignments. Group dynamics are difficult and can lead to performance gaps.

Human Performance Improvement

Human Performance Improvement (HPI) focuses on "root cause analysis" to assess and overcome the barriers inherent in teamwork. To close performance gaps, HPI analyzes the following possible causes for collaborative breakdowns:

1. **Knowledge**—Perhaps employees do not know how to perform a task. They have never acquired the correct skills or do not understand which skills are needed to complete the specific job. Varying skills of team members can impede the group's progress.
2. **Resources**—Think of these possibilities: tools are broken or missing; the department is out of funds; you do not have enough personnel to do the job; the raw material needed for the job is below par; you ordered one piece of machinery but were shipped something different; you needed 100 items but have only 50 in stock. To complete a project, you often have to solve problems with resources.
3. **Processes**—For teams to succeed in collaborative projects, everyone must know his or her responsibilities. Who reports to whom? How will these reports be handled (orally, in writing)? Who does what job? Are responsibilities shared equally? Structure, of some sort, is needed to avoid chaos, lost time, inefficiency, hurt feelings, and many other challenges to teamwork. To achieve successful collaboration, the team should set and maintain effective procedures.
4. **Information**—A team needs up-to-date and accurate information to function well. If required database information is late or incorrect, then the team will falter. If the information is too high tech for some of the team members, a lack of understanding may undermine the team effort.
5. **Support**—To succeed in any project, a team needs support. This could be financial, attitudinal, or managerial. When managers from different departments are fighting "turf wars" over ownership of a project, teams cannot succeed. Teams need enough money for staffing, personnel, or equipment.
6. **Wellness**—A final consideration involves the team's health and well-being. People get sick or miss work for health reasons. People have car accidents. If a teammate must miss work for a day or an extended period, this will negatively impact the team's productivity. Stress and absences can lead to arguments, missed deadlines, erratic work schedules, and poor quality.

HPI Intervention Techniques

After assessing root causes that challenge a team's success, HPI creates intervention options. These might include the following:

- Improved compensation packages
- Employee recognition programs
- Revised performance appraisals
- Improved employee training

- Simulations
- Mentoring or coaching
- Restructured work environments to enhance ergonomics
- Safety implementations
- Strategic planning changes
- Improved communication channels
- Health and wellness options—lectures, on-site fitness consultants, incentives to lose weight, and therapist and social worker interventions

People need help in order to work more effectively with each other. A progressive company recognizes these challenges and steps in to help.

CONFLICT RESOLUTION IN COLLABORATIVE PROJECTS

To ensure that team members work well together and that projects are completed successfully, consider these approaches to conflict resolution:

1. **Choose a team leader**—Sometimes, team leaders are chosen by management; sometimes, team leaders emerge from the group by consensus. However this person gains the position, he or she can serve many valuable purposes. The team leader becomes "point person," the individual whom all can turn to for assistance. He or she can solve problems, seek additional resources, or organize the team effort. For instance, a team leader can give the team direction, interface with management, and/or act as the team's mediator.
2. **Set guidelines**—One reason that conflicts occur is because people do not know what to expect or what is expected of them. However, if expectations are clear, then several major sources of conflict can be resolved.

   For example, one simple conflict might be related to time. A team member could be unaware of when the meeting will end and schedule another meeting. If that team member then has to leave the first meeting early, disrupting the team's progress, this can cause a conflict.

   To solve this problem, set guidelines. Hold an initial meeting (online or teleconferenced for remote, virtual teams) to define goals and establish guidelines, establish project milestones, or create schedules for synchronous dialogues. Communicate to all team members (before the meeting via e-mail or early in a project) how long the project will last. Also, clarify the team's goals, the chain of command (if one exists), and each team member's responsibilities.
3. **Ensure that all team members have compatible hardware and software**—This is especially important for virtual, remote teams. To communicate successfully, all team members need access to the same e-mail platform. Some members should not use Yahoo or MSN or Hotmail while others use Outlook. This would cause communication challenges if software is incompatible. The problem is further heightened when video or telephone equipment is different.
4. **Encourage equal discussion and involvement**—A team's success demands that everyone participate. A team leader should encourage involvement and discussion. All team members should be mutually accountable for team results, including planning, writing, editing, proofreading, and packaging the finished project. Be sure that everyone is allowed a chance to give input.

   Conflicts also arise when one person monopolizes the work. If one person speaks excessively, others will feel left out and disregarded. A team leader should ensure equal participation. He or she should call on others for their opinions and ask for additional input from the team.

   In addition, team leaders must limit an overly aggressive team member's participation by saying, "Thanks, John, for your comments. Now, let's see what others have to say." Or, "Wait one second, John. I'll come back to you after we've heard from a few others."

5. **Discourage taking sides**—Discussion is necessary, but conflict will arise if team members take sides. An "us against them" mentality will harm the team effort. You can avoid this pitfall by seeking consensus, tabling issues, creating subcommittees (each of these points is discussed later), or asking for help from an outside source (boss or teacher, for example).
6. **Seek consensus**—Not every member of the team needs to agree on a course of action. However, a team cannot go forward without majority approval. To achieve consensus, your job as team leader is to listen to everyone's opinion, seek compromise, and value diversity. Conflict can be resolved by allowing everyone a chance to speak. Once everyone has spoken, then take a vote.
7. **Table topics when necessary**—If an issue is so controversial that it cannot be agreed upon, take a time-out. Tell the team, "Let's break for a few minutes. Then we can reconvene with fresh perspectives." Maybe you need to table the topic for the next meeting. Sometimes, conflicts need a cooling-off period.
8. **Create subcommittees**—If a topic cannot be resolved, teammates are at odds, or sides are being taken, create a subcommittee to resolve the conflict. Let a smaller group tackle the issue and report back to the larger team.
9. **Find the good in the bad**—Occasionally, one team member comes to a meeting with an agenda. This person does not agree with the way things have been handled in the past or the way things are being handled presently. You do not agree, nor do other team members. However, you cannot resolve this issue simply by saying, "That's not how we do things." A disgruntled team member will not accept such a limited viewpoint.

   As team leader, seek compromise. Let the challenging team member speak. Discuss each of the points of dissension. Allow for input from the team. Some of the ideas might have more merit than you originally assumed.
10. **Deal with individuals individually**—From time to time, a team member will cause problems for the group. The teammate might speak out of turn or say inappropriate things. These could include off-color or off-topic comments. A team member might cause problems for the group by habitually showing up late, missing meetings, or monopolizing discussions.

    To handle these conflicts, avoid pointing a finger of blame at this person during the meeting. Do not react aggressively or impatiently. Doing so will lead to several problems:
    - Your reaction might call more attention to this person. Sometimes people come to meetings late or speak out in a group *just* to get attention. If you react, you might give the individual exactly what he wants.
    - Your reaction might embarrass this person.
    - Your reaction might make you look unprofessional.
    - Your reaction might deter others from speaking out. You want an open environment, allowing for a free exchange of ideas.

    Speak to any offending team members individually. This could be accomplished at a later date, in your office, or during a coffee break. Speaking to the person later and individually might defuse the conflict.
11. **Stay calm**—Act professionally when dealing with conflict. To resolve conflicts, speak slowly, keep your voice steady and quiet, and stay seated (rising will look too aggressive). You also might want to take notes. This will provide you with a record of the discussion.
12. **Remove, reassign, or replace if necessary**—Finally, if a team member cannot be calmed, cannot agree with the majority, or has too many other conflicts, your best course of action might be to remove, reassign, or replace this individual.

CHECKLIST FOR COLLABORATION

_____ 1. Have you chosen a team leader (or has a team leader been assigned)?

_____ 2. Do all participants understand the team's goal and their individual responsibilities?

_____ 3. Does the team have a schedule, complete with milestones and target due dates?

_____ 4. Does the team have compatible hardware and software?

_____ 5. In planning the team's project, did you seek consensus?

_____ 6. Have all participants been allowed to express themselves?

_____ 7. If conflicts occurred, did you table topics for later discussion or additional research?

_____ 8. Did you encourage diversity of opinion?

_____ 9. Have you avoided confronting people in public, choosing to meet with individuals privately to discuss concerns?

_____ 10. If challenges continue, have you reassigned team members?

Shelly Met Her Challenge

To meet her communication challenge, Shelly used the P^3 process, as follows:

Planning

To plan her proposal, Shelly first considered the following:

- Goals—create online help instructions for insurance end users
- Audience—corporate employees who use Century Software for insurance and benefit plans
- Channels—wikis for collaboratively written instructions by the team members
- Data—drawn from software developers and usability teams

According to Shelly, "Planning for a team project requires that I create a new wiki site for the team's input. Once I create this site and add all team members, making sure that they have compatible software, I begin the project by doing the following:

1. Meet with human resources personnel to determine what information they will need from the employees to establish personalized accounts.
2. Interview the software developers to obtain required data for the online help screens.
3. Interview the software designers to decide how many fields will be required on each screen.

To help the team start drafting their instructional text, I place informational steps in the wiki site. Then, I e-mail the dispersed team to say, 'I've put some material into our wiki space. You can begin inputting suggestions and revisions. I look forward to your participation in this collaborative project.' Writing this e-mail prior to the project ensures that I achieve equal participation from all team members." Figure 2.5 is Shelly's initial wiki site with instructional information.

(Continued)

FIGURE 2.5 Initial Wiki Site with Instructional Information for Team Members to Evaluate

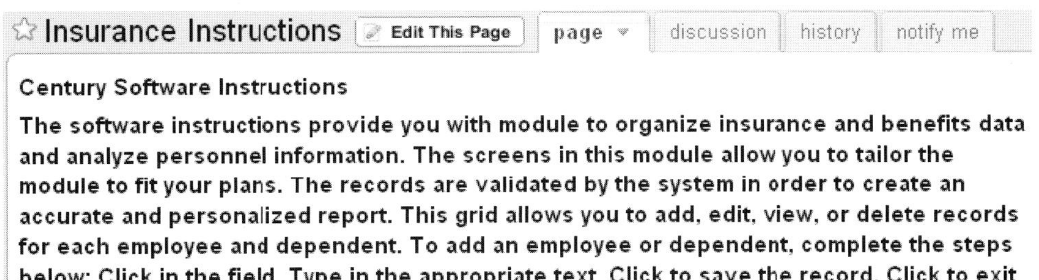

Reprinted with permission of Wikispaces.

Packaging

To package the instructions, each team member edited Shelly's original text online. By using a wiki, Shelly and her team members could see a history of each participant's suggested changes. The wiki site highlighted inserted text and deleted text in different colors to emphasize where changes had been made. In this way, Shelly and her team could decide which version best suited their end users' needs. Figure 2.6 is the wiki with team members' suggested revisions.

Perfecting

Shelly and her team know how important it is to perfect their documents. "Usability testing and reports from customer complaints showed us that our clients needed more than text to understand our software. Customers valued screen shots that 'walked them through the instructions.' After input from all of the team members, we picked out the best version of text. Then, working from the actual online help screens, I created screen captures to illustrate the procedural steps. We ran beta versions of our revised, perfected instructions and found that customer complaints decreased by 10 percent. This meets our performance goals. Good communication is great for business," says Shelly. Figure 2.7 shows Shelly's finished instructions with visual aids.

FIGURE 2.6 Wiki with Team Members' Suggestions

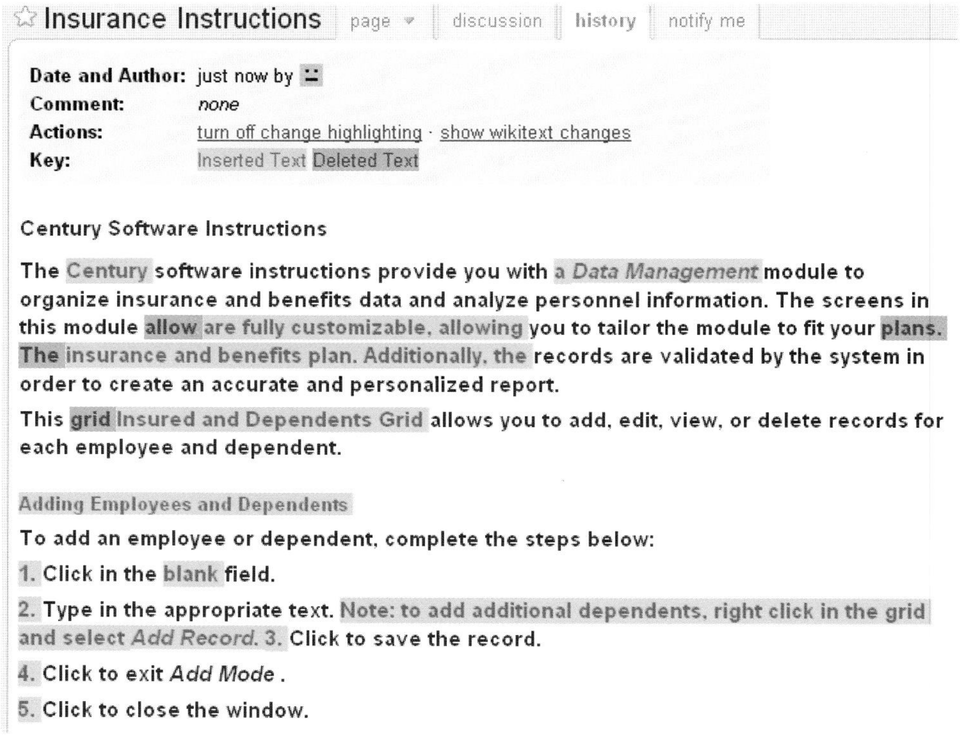

Insurance Instructions | page | discussion | history | notify me

Date and Author: just now by
Comment: *none*
Actions: turn off change highlighting · show wikitext changes
Key: Inserted Text Deleted Text

Century Software Instructions

The Century software instructions provide you with a *Data Management* module to organize insurance and benefits data and analyze personnel information. The screens in this module allow are fully customizable, allowing you to tailor the module to fit your plans. The insurance and benefits plan. Additionally, the records are validated by the system in order to create an accurate and personalized report.

This grid Insured and Dependents Grid allows you to add, edit, view, or delete records for each employee and dependent.

Adding Employees and Dependents

To add an employee or dependent, complete the steps below:

1. Click in the blank field.

2. Type in the appropriate text. Note: to add additional dependents, right click in the grid and select *Add Record*. 3. Click to save the record.

4. Click to exit *Add Mode* .

5. Click to close the window.

Reprinted with permission of Wikispaces.

FIGURE 2.7 **Perfected Instructions with Visual Aids**

Century Software Instructions

The Century software instructions provide you with a *Data Management* module to organize insurance and benefits data and analyze personnel information. The screens in this module are fully customizable, allowing you to tailor the module to fit your insurance and benefits plan. Additionally, the records are validated by the system in order to create an accurate and personalized report.

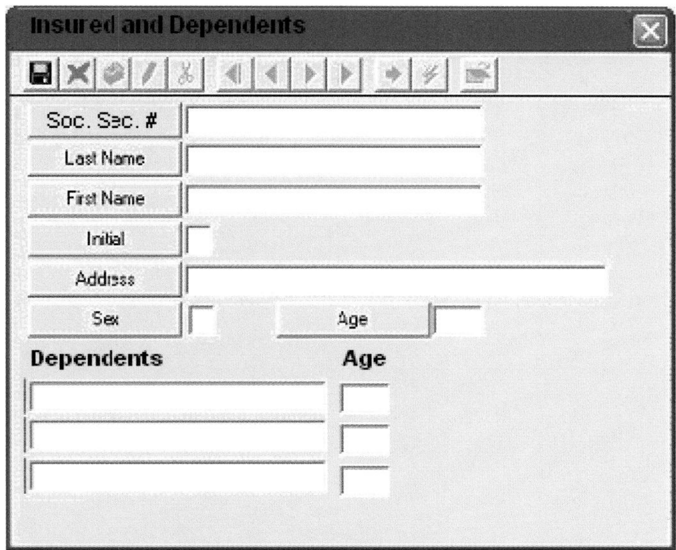

Shelly says, "My goal is to write instructions that all end users can follow. My team's input and the usability testing allowed me to make sure the instructions and visuals 'spoke' to the end user. Wikis are a great way for my team to interact and saved us a lot of time and resources."

This Insured and Dependents Grid allows you to add, edit, view, or delete records for each employee and dependent.

Adding Employees and Dependents

To add an employee or dependent, complete the following steps:

1. Click in the blank field.
2. Type in the appropriate text.
   Note: To add additional dependents, right click in the grid and select *Add Record.*
3. Click to save the record.
4. Click to exit *Add Mode.*
5. Click to close the window.

CHAPTER HIGHLIGHTS

1. Businesses expect you to be able to work with others on team projects.
2. Teamwork is one of the most important skills according to the National Association of Colleges and Employers.
3. Problems occur if people isolate themselves from other departments and colleagues, becoming inaccessible "silos." This leads to poor communication.
4. Teams encourage diversity of opinions, provide checks and balances, encourage broad-based understandings, empower employees, and create collegiality.

5. Companies often use teams to anticipate problems and solve them to produce quality products and services.
6. Human Performance Improvement (HPI) focuses on why and how collaborative team efforts break down.
7. HPI studies the causes of group failures by assessing "gaps" in knowledge, motivation, resources, procedures, information, support, and wellness.
8. Teams encounter conflicts, which must be resolved for a project to succeed.
9. To resolve conflicts, set guidelines, encourage equal discussion and involvement, discourage taking sides, table topics or create subcommittees when necessary, remain positive, deal with individuals constructively, and remove or reassign team members if necessary.
10. Teams should follow precise strategies to succeed: Choose teams carefully, choose an effective team leader, identify problems, analyze these problems, determine potential improvements, verify the solutions, breach the gaps to achieve Human Performance Improvement, and complete the project collaboratively.

WWW *Find additional teamwork exercises, samples, and interactive activities at http://www.prenhall.com/gerson.*

MEETING WORKPLACE COMMUNICATION CHALLENGES

CASE STUDIES

1. Future Promise, a not-for-profit organization working with underprivileged teens, is planning to create an agency Web site. To do so, it has formed a 12-person team, consisting of the agency's accountant, sports and recreation director, public relations manager, counselor, training facilitator, graphic artist, computer and information systems director, two local high school principals, two local high school students, and a representative from the mayor's office. Jeannie Kort, the PR manager, is acting as team leader.

The team needs to determine the Web site's content, design, and levels of interactivity. Jeannie's boss, Brent Searing, has given the team a deadline and a few components that must be included in the site:

- College scholarship opportunities
- After-school intramural sports programs
- Job-training skills (resume building and interviewing)
- Service learning programs to encourage civic responsibility
- Future Promise's 800-Hotline (for suicide prevention, STD information, depression, substance abuse, and peer counseling)
- Additional links (for donors, sponsors, educational options, job opportunities, etc.)

Assignment

Form a team (or become a member of a team as assigned by your instructor). As a team, how can you best accomplish the previously defined task? To complete this job, research the topic. Use the Internet for source content, and brainstorm possible approaches. Once you have formed a plan of action, do any or all of the following:

- Write an e-mail or memo to your instructor (the CEO) explaining the following:
  - Why you are writing
  - What you are writing about

    - What exactly your team's plan of action is—what your recommendations are for constructing this Web site
    - When you plan to complete the project (provide tentative due dates and measurables)
- Give a three- to five-minute oral presentation reporting on your findings. This can be a team or individual project. If it's a team report, divide the speech equally by the number of people involved.
- Create the text and/or the Web site for Future Promise.

2. Quick and Sure Delivery (QSD) has not been either quick or sure lately. Customer complaints are up 23 percent this quarter (QSD only receives an average of three complaints a month). Clients are telling customer service representatives that deliveries are arriving up to 10 hours later than promised. In addition, delivered goods are being left unattended outside homes and businesses. This has led to damages due to rain; in addition, on at least five instances, delivered packages have been stolen.

Since QSD guarantees that packages will be handed directly to a home or business owner and never left unattended, these occurrences are actionable under the law. More importantly, QSD's reputation is being harmed. Already, word is getting around, and customers are taking their business elsewhere. Business is down 12 percent this month, when compared to last year at the same time (QSD made 1,578 deliveries during the month last year, charging an average of $27 per delivery). Something must be done.

Assignment

Form a team to study this problem. Your team's goal will be to improve QSD's quality and performance. First, quantify the damage done (based on the previous numbers). Then do the following:

- Analyze the problem(s). To do so, in the group, brainstorm possible causes. What "gaps" might exist (personnel, checks and balances, management assessments, missing incentives, vehicle problems, weather, etc.)?
- Invent or envision solutions. How would you solve the problems? Consider "Human Performance Improvement" issues, as discussed in this chapter.
- Plan your approach. To do so, establish verifiable measures of success (including timeframes and quantifiable actions).

Once you have made your decisions, report your team's findings as follows:

- Write an e-mail or memo to your instructor explaining
    - Why you are writing
    - What you are writing about
    - What exactly your team's plan of action is—what your recommendations are for solving QSD's problem
    - When you plan to complete the project (provide tentative due dates and measurables)
- Give a three- to five-minute oral presentation reporting on your findings. This can be a team or individual project. If it's a team report, divide the speech equally by the number of people involved.

ETHICS IN THE WORKPLACE CASE STUDY

Mary Madero has been employed at Commercial Savings and Loans for many years. She frequently communicates via e-mail with colleagues in the bank and at other savings and loans organizations. Recently, Mary was elected president of the ASLTS (the Association for Savings and Loans Teller Supervisors). As ASLTS president, she is involved in fundraising for charity, community service outreach projects, organization picnics, and the

ASLTS monthly bowling league. To complete these projects, she works with a team of loan teller supervisors.

Sometimes Mary gets e-mail during the workday that relates to her new position as president. In addition, at home, she must answer ASLTS e-mail communication from her team. In the evenings and on weekends, she uses her company laptop for additional work related to her position as teller supervisor at the savings and loans.

Question: Is it ethical for Mary to use her company's e-mail account, or should she use her personal e-mail account for ASLTS business? Is it ethical for Mary to use the company laptop, or should she be using her own laptop for ASLTS business? Justify your answer based on information provided in Chapter 1.

INDIVIDUAL AND TEAM PROJECTS

1. Attend a meeting. This could be at your church, synagogue, or mosque; a city council meeting; your school's, college's, or university's board of trustees meeting; or a meeting at your place of employment. Was the meeting successful? Did it have room for improvement? To help answer these questions, use the following Conflict Resolution in Team Meetings Matrix. Then report your findings to your professor or classmates, either orally or in writing.

Conflict Resolution in Team Meetings Matrix			
Goals	Yes	No	Comments
1. Were meeting *guidelines* clear?			
2. Did the meeting facilitator *encourage equal discussion and involvement*?			
3. Were the meeting's attendees *discouraged from taking sides*?			
4. Did the meeting facilitator *seek consensus*?			
5. Were *topics tabled* if necessary?			
6. Were *subcommittees created* if necessary?			
7. Did the meeting facilitator *find the good in the bad*?			
8. Did the meeting facilitator *deal with individuals* individually?			
9. Did the meeting's facilitator *stay calm*?			

2. Have you been involved in a team project at work or at school? Perhaps you and your classmates teamed to write a proposal, research Web sites, create a Web site, or perform mock job interviews. Maybe you were involved in a team project for another class. Did the team work well together? If so, analyze how and why the team succeeded. If the team did not function effectively, why not? Analyze the "gaps" between what should have been and what was. To help you with this analysis, use the following Human Performance Index Matrix. Then, report your findings to your instructor or classmates either orally or in writing.

Human Performance Index Matrix			
Potential Gaps	Yes	No	Comments
1. Did teammates have equal and appropriate levels of *knowledge* to complete the task?			
2. Did teammates have equal and appropriate levels of *motivation* to complete the task?			
3. Did teammates have sufficient *resources* to complete the task?			
4. Did teammates understand their roles in the *process* needed to complete the task?			
5. Did the teammates have sufficient and up-to-date *information* to complete the task?			
6. Did the teammates have sufficient *support* to complete the task?			
7. Did *wellness* issues affect the team's success?			

DEGREE-SPECIFIC ASSIGNMENTS

Individually or in small groups, visit employees who work in your major field. Once you and your teammates have visited these individuals, asked your questions (see the following assignments), and completed your research, share your findings as follows:

- **Oral presentations**—As a team, give a three- to five-minute briefing to share the results of your research with your colleagues.
- **Oral presentations**—Invite employee representatives from other work environments to share with your class their responses to your questions.
- **Written**—Write a team memo, letter, or report about your findings.

Assignment

1. Ask employees at the sites you visit about the challenges they face with conflict resolution.
2. Ask employees at the sites you visit what "gaps" they see between what the company or agency experiences and what the company or agency hopes to achieve. What intervention methods do they think could help close these gaps?
3. Ask employees at the sites you visit about their experiences in team meetings. How are team meetings successfully managed? What causes poorly led team meetings?
4. Use the Internet and/or your library to research companies that rely on teamwork. Focus on which industries these companies represent and the goals of their team projects. You could also consider the challenges they encounter, their means of resolving conflicts, the numbers of individuals on each team, and whether the teams are cross-functional.
5. What technologies (such as wikis, teleconferences, videoconferences, Webinars, etc.) do employees in your degree field use for collaboration? How often do employees use technology for collaborative work, what are the benefits of these technologies, and what are the problems technology creates for collaborative work?

PROBLEM-SOLVING THINK PIECES

You are the team leader of a project at work that has been ongoing for a year. During the year, the team has met weekly, every Wednesday at 8:00 a.m. It is now time to assess the team's successes and areas needing improvement. Your goal will be to recommend changes as needed before the team begins its second year on this project. You have encountered these problems:

- One team member, Caroline Jensen, has missed meetings. She almost always missed at least one meeting a month. Occasionally, she would miss two or three in a row. You have met with this employee to ask for an explanation. She says she has had childcare issues. These have forced her to use the company's flex-time option, allowing her to come to work later than usual, at 9:00 a.m.
- Another team member, Carlos Rodriquez, tends to talk a lot during the meetings. He has good things to say, but he speaks his mind very loudly and interrupts others as they are speaking. He also elaborates on his points in great detail, even when the point has been made.
- A third team member, Sharon Mitchell, almost never provides her input during the meetings. She will e-mail comments later or talk to people during breaks. Her comments are valid and on-topic, but not everyone gets to hear what she says.
- A fourth team member, Craig Mabrito, is very impatient during the meetings. This is evident from his verbal and nonverbal communication. He grunts, slouches, drums on the table, and gets up to walk around while others are speaking.
- A fifth employee, Julie Jones, is overly aggressive. She is confrontational, both verbally and physically. Julie points her finger at people when she speaks, raises her voice to drown out others as they speak, and uses sarcasm as a weapon. Julie also crowds people, standing very close to them when speaking.

How will you handle these challenges? Either individually or as a team, decide on a course of action. For example, try this approach:

- Analyze the problem(s). To do so, brainstorm. What "gaps" might exist causing these problems?
- Invent or envision solutions. How would you solve the problems? Consider "Human Performance Improvement" issues, as discussed in this chapter.
- Plan your approach. To do so, establish verifiable measures of success (including timeframes and quantifiable actions).
- Write an e-mail or memo to each team member, explaining how he or she can work more effectively within the team.

WEB WORKSHOP

Using an Internet search engine, go online to research ways in which companies use teamwork. Access the following Web links (type "teamwork" in the Web site search engines to restrict your research).

- Best Practice Database. http://www.bestpracticedatabase.com.
- FastCompany. http://www.fastcompany.com/
- Best Manufacturing Practices. http://www.bmpcoe.org.

After accessing the sites and reading corporate spotlights on teamwork, assess your findings as follows:

1. Compare and contrast the ways in which different companies use teams to improve their productivity.
2. Decide which company has used teams most effectively. Explain your decisions.
3. Compare the team practices you discover online to the approaches discussed in this chapter.

QUIZ QUESTIONS

1. Why is teamwork important?
2. What is a "silo"?
3. What makes collaborating in teams difficult?
4. What is groupware?
5. How do companies benefit from using groupware?
6. What is a wiki?
7. Why is teamwork important in business?
8. How does Human Performance Improvement deal with performance gaps?
9. List ten ways to resolve conflicts in team meetings.
10. What is a dispersed team?

CHAPTER 3

Meeting the Needs of the Audience

▸REAL PEOPLE *in the workplace*

Phil Wegman, Program Director of Skills Enhancement for the Center for Business and Technology, sighed deeply and said, "I receive two to three calls *every day* from companies desperate for Spanish language training. They need to teach their supervisors how to communicate more effectively with customers as well as with employees for whom English is a second language."

Phil's client base is enormous, covering literally hundreds of fields. These include dental and medical, criminal justice, public safety, transit, government, and education, as well as industries such as hospitality management (hotels and restaurants), construction, casinos, manufacturing, warehousing, banking, retail, childcare, and accounting.

Police officers, nurses, physicians, and paramedics, for example, have asked Phil to provide them occupational language training that is "work specific and real life." Then, when an emergency situation occurs, the safety personnel can ask, in Spanish, French, German, or Chinese, "Where does it hurt?"

McDonald's, Hardee's, and Burger King supervisors need to know enough Spanish to be able to say to their staff, "Here's how you ring up this sale," "Here's how you make this meal," or "Here's how you clean this piece of equipment." Of equal importance, companies ask Phil to teach their supervisors how to answer customer questions or to take customer orders. Many fast-food restaurants have on staff a worker who is fluent in a second language. However, the pace is so hectic in fast food that the restaurants can't always pull that one employee off the job to handle the situation. Thus, *all* supervisors need the proper language skills to provide direct commands, make simple statements, or answer common questions in a foreign language.

There is one more key component to language knowledge, Phil states. "When a company teaches its employees a different language, the company honors that culture. The company tells its client base, 'We're trying to learn who you are and what makes you unique.'" The company is learning cross-cultural information to minimize barriers; the company is saying to its staff and customers, "We respect who you are and we want to work with you. That's just good business." ◂

CHAPTER GOALS

When you complete this chapter, you will be able to do the following:

1. Recognize the audience's knowledge of the subject matter, role, and diversity so you can write and speak more successfully.
2. Define terms that might be confusing for different audiences.
3. Consider your audience's role (management, lateral colleague, subordinate, vendor, or customer) when you speak or write.
4. Recognize the diversity of your audience, including gender, race, religion, age, sexual orientation, physical limitations, and culture.
5. Consider the importance of multicultural and cross-cultural audiences in both your written and oral expression.
6. Avoid biased language in relation to your audience's age, physical limitations, and gender.
7. Achieve audience involvement to build rapport and motivate the reader. You can achieve audience involvement if you avoid commands, ask questions, use positive words, employ "you usage," focus on audience benefit, and personalize text with names.

WWW

To learn more about audience, visit our Web Companion at http://www.prenhall.com/gerson.

Phil's Communication Challenge

According to Phil, "Arctic Cooling Technologies is a manufacturer of cooling towers and air-cooled condensers for power generation, industrial, refrigeration, and heating, ventilating, and air conditioning (HVAC) markets. To help manufacture cooling towers, equipment, and parts, Arctic employs several non-native workers for whom English is a second language. These employees are excellent workers, but the company realizes that language barriers can cause problems with productivity, quality, and safety. Because of this, the company called me and asked for my help to provide them customized language training. Arctic's 30 Spanish-speaking manufacturing employees needed to improve their English skills so they could communicate with their supervisors, understand the oral and written instructions related to machine operation, and communicate with the company's human resources department regarding their employee benefits, retirement packages, and 401(k)s. The English-speaking supervisors needed to learn 'workable Spanish,' enough knowledge to say, 'Good morning,' 'Hello,' 'Goodbye,' or to ask, 'How are you?' 'Any questions?' or 'How can I help?'

My challenge was to write a proposal to Arctic responding to their request for language training. Before I could write the proposal, I needed to discover

- What my audience's level of education was
- Whether their knowledge of English was beginning, intermediary, or advanced
- How much time the company wanted to devote to language training
- If the company would provide the employees release time for training
- Whether the training would be mandatory or voluntary
- How much money the company would allocate for training
- How the company planned to assess the success of the training classes"

See pages 87–89 to learn how Phil met his communication challenge.

RECOGNIZING THE AUDIENCE

In the business world, you will never write or speak in a vacuum. When you write your memo, e-mail message, letter, report, brochure, or Web site text, *someone* will read it. When you give an oral briefing, convene a meeting, communicate with customers in a salesroom, or make a speech at a conference, *someone* will be listening. The question is, "Who?"

- Who is your audience?
- What does this reader or listener know?
- What does this reader or listener not know?

- What must you write or say to ensure that your audience understands your point?
- How do you communicate to more than one person (multiple audiences)?
- What is your audience's position in relation to your job title?
- What diversity issues (gender, age, sexual orientation, cultural, multicultural) must you consider?
- What tone should you use when communicating?
- How do you ensure audience involvement?

If you do not know the answers to these questions, you might fail to communicate. Your letter may contain jargon or acronyms the reader will not understand. The tone of the memo may be inappropriate for management, for your subordinates, or for your customers. Your verbal communication might not factor in your audience's unique culture and language. To communicate successfully, you must recognize your audience's level of understanding. You must factor in your audience's unique traits, which could have an impact on your communication success. These could include many variables, as shown in Table 3.1.

Knowledge of the Subject Matter

What does your audience know about the subject matter? Do they work closely with you on the project? That would make them *specialists* in your field. Does the audience have general knowledge of the subject matter, even though their expertise is elsewhere? That would make the audience *semi-specialists* in your field. Is the audience totally unknowledgeable about the subject matter? That would make them a *lay audience*. Finally, could your audience be a combination of these types? If so, you would be confronted with a *multiple audience*.

Specialists

Specialists work in a field in which they display expertise. They might work directly with you in your department, or they might work in a similar capacity for another company. Wherever they work, they are your colleagues because they share your educational background, work experience, or level of understanding. If you are a computer programmer, for example, another computer programmer who is working on the same or similar system is a specialist. If you are an accountant working with tax laws, other accountants are specialists.

Table 3.1 Audience Variables

Knowledge of the Subject Matter • Specialists • Semi-specialists • Lay audience • Multiple readers and listeners
Roles • Management • Coworker • Subordinate • Customer • Vendor • Other business professionals
Issues of Diversity • Gender • Age • Race and/or religion • Sexual orientation • Disabilities • Language and/or culture of origin—*multicultural* or *cross-cultural*

Once you recognize that your audience consists of specialists, what does this tell you? These readers have the following characteristics:

- They are experts in the field you are writing about. If you write an e-mail message to one of your department colleagues about a project you two are working on, your associate is a specialist. If you write a letter to a vendor requesting specifications for one of their company's systems, that reader is a specialist.
- Specialists understand the terminology of their field. Communication with other specialists can include jargon, acronyms, and abbreviations.
- Specialists require minimal detail regarding standard procedures or scientific, mathematical, or technical theories.
- Specialists read to discover new knowledge or for updates regarding the status of a project.
- Specialists need little background information regarding a project's history or objectives unless the specific subject matter of the correspondence is new to them. If, for example, you are writing a status report to your first-line supervisor, who has been involved in a project since its inception, then you will not need to flesh out the project's history.

Figure 3.1 is a letter written from a specialist (a state department of transportation employee) to another specialist (a contractor).

Semi-Specialists

Semi-specialists include coworkers in other departments. Semi-specialists also might include bosses, subordinates, or colleagues who work in similar professions. For instance, if you are a computer programmer, the accountants, human resources supervisor, and graphic artists in your company are semi-specialists. These individuals have worked around your company's products or services and, therefore, are familiar with the subject matter.

Your bosses are often semi-specialists because they might no longer work closely with the equipment. Although they might have been actively involved at one time, as they moved more and more into management, they moved further and further away from hands-on knowledge.

Your subordinates might be semi-specialists because of their levels of education or work experience. Finally, colleagues at other companies might be semi-specialists if they are not familiar with your company's procedures or in-house jargon, acronyms, and abbreviations.

Additional information:

See Chapter 10, "Traditional Correspondence: Memos and Letters," and Chapter 7, "Oral Presentations and Nonverbal Communication in the Workplace."

Lay Audience

Customers and clients who neither work for your company nor have any knowledge about your field of expertise are the lay audience. If you work in telecommunications for a telephone company, for example, and you write a letter to a client regarding a problem with the company's phone line, your audience is a lay reader. If your field of expertise is biomedical equipment and you write a procedures manual for the patient who is the end user, you are writing to a lay audience. Although you understand your subject matter, your reader or listener is not an expert in the field. These people are learning about a subject matter which is outside of their daily realm of experience.

Communicating to a lay audience is difficult. It is easy to write or speak to a specialist who thoroughly understands the subject matter you are discussing. However, communicating to a lay audience totally outside your field of expertise is demanding. To communicate successfully to a lay audience, remember that these people share the following characteristics:

- Lay audiences are unfamiliar with your subject matter. They do not understand the product or service. Therefore, you should communicate simply. That is not to say that you should insult the lay audience with a remedial discussion or with a patronizing tone. Explain your topic clearly. Achieve clarity through precise word usage, depth of detail, and simple graphics.

FIGURE 3.1 Specialist Writing to Specialist

Specialists professionally involved in this project would be familiar with the project number. No further background information is needed.

"Sheet C3.1" and "C7.1" are construction documents that only these specialists have access to, not ones readily available to the public.

"Road striping," "stop bars," "signal head #10," "5-section," and "cross slope," common terms in this industry, are clear to specialists.

"ROW," an abbreviation for "right of way," does not need to be defined since it is often used by these specialists.

Missouri Department of Transportation
5117 East 31st Street
Kansas City, MO 64128

April 17, 2009

Mr. Bob Walker
DTR Construction
7120 W. 130th Road
Parkston, MO 76221

Reference: Project No. BRO-451 (17)

Dear Mr. Walker:

We have completed our review of the Highway 135 and Hillside Drive intersection improvements for Parkton Independent School District. The following revisions need to be made to the plans before we can continue our review:

1. On sheet C3.1, road striping revisions are needed. Left turn arrows must be painted 75' behind all new stop bars.
2. On sheet C7.1, signal head #10 needs to be a 5-section head instead of a 3-section head, in accordance with Missouri ordinance 733BR911A. We also need to see the D37 and D38 specifications for these signal designs.
3. Please provide us with a larger typical cross-section of Highway 135. A note on this cross-section should be placed to match the existing cross slope of the highway.
4. We need a legal description of new ROW along the frontage. Also, let me confirm our phone conversation on April 4. The Department of Transportation will honor the gas company's easement for future Highway 135 construction.

Once the plans have been revised according to the itemized points above, please resubmit them. You can fax them to Orlie Vernon, the Department of Transportation's Parkton project contact. His fax number is 816-555-2121. If I can answer any questions, call me at 816-100-9911.

Sincerely,

Dale Askew
Permit Specialist

cc: Orlie Vernon
Parkton Independent School District

- Since lay audiences do not understand your work environment, they will not understand any of your in-house jargon, abbreviations, or acronyms. Define unfamiliar terms.
- A lay audience might need background information. When explaining information to the lay audience, provide sufficient causes, results, or rationale.

FIGURE 3.2 Specialist Writing to Lay Audience

Missouri Department of Transportation
5117 East 31st Street
Kansas City, MO 64128

May 6, 2009

Jamie Wilson
The Bellaire Daily
205 Main
Bellaire, MO

Dear Jamie:

District 6 of the Missouri Department of Transportation is proud to introduce a free feature available for your newspaper. *TransporT* is a bimonthly column from the department personnel who service your newspaper's coverage area. This column delivers *useful* information about transportation issues related to your readers' unique needs. It's a great way to familiarize your readers with the people responsible for the planning, building, and maintaining of roads your city uses every day.

I have enclosed a copy of this month's *TransporT* news column for your review. Please use any of the information you would like in *The Bellaire Daily.* I'll continue to send *TransporT* columns to you twice a month. Later this summer, I'll contact you to see how your readers are enjoying the columns and to find what additional information they might like to read.

Thanks so much for your help, Jamie. If you have any questions, please call me at (813) 555-1212 or e-mail me at joe_b@modot.com.

Sincerely,

Joe Baker
Senior Public Affairs Specialist

Figure 3.1, shown earlier in this chapter, talks about highly technical aspects of transportation, such as slopes, striping, and easements. This letter to a lay audience just mentions "roads."

Positive words, such as "proud," "free," "useful," "great way," and "enjoying," are used to build rapport.

Communicating effectively requires that you recognize the differences among specialists, semi-specialists, and lay audiences. If you incorrectly assume that all readers or listeners are experts in your field, you will create problems for yourself as well as for your readers. Figure 3.2 is a letter written from a specialist (a state department of transportation employee) to a lay reader (a newspaper editor).

Multiple Audiences

Information is not always communicated to just one type of audience. Sometimes your correspondence has multiple audiences. You will write or speak to specialists, semi-specialists, and lay audiences simultaneously. For example, when writing a report, people might assume that the first-line supervisor will be the only reader. This is not the case, however. The first-line supervisor could send a copy of your report to the manager, who could then submit the same report to the executive officer. Similarly, your first-line supervisor might send the report to your colleagues or to your subordinates.

Your report might be sent to other departments. In fact, the report could go outside your company to clients, vendors, newspapers, agency auditors, or eventually to the court system (juries, lawyers, and judges). Figure 3.3 shows what this communication challenge looks like.

Communicating to multiple audiences with different levels of understanding and different reasons for reading creates a challenge. When you add the necessity of using the

FIGURE 3.3 **Multiple Audiences**

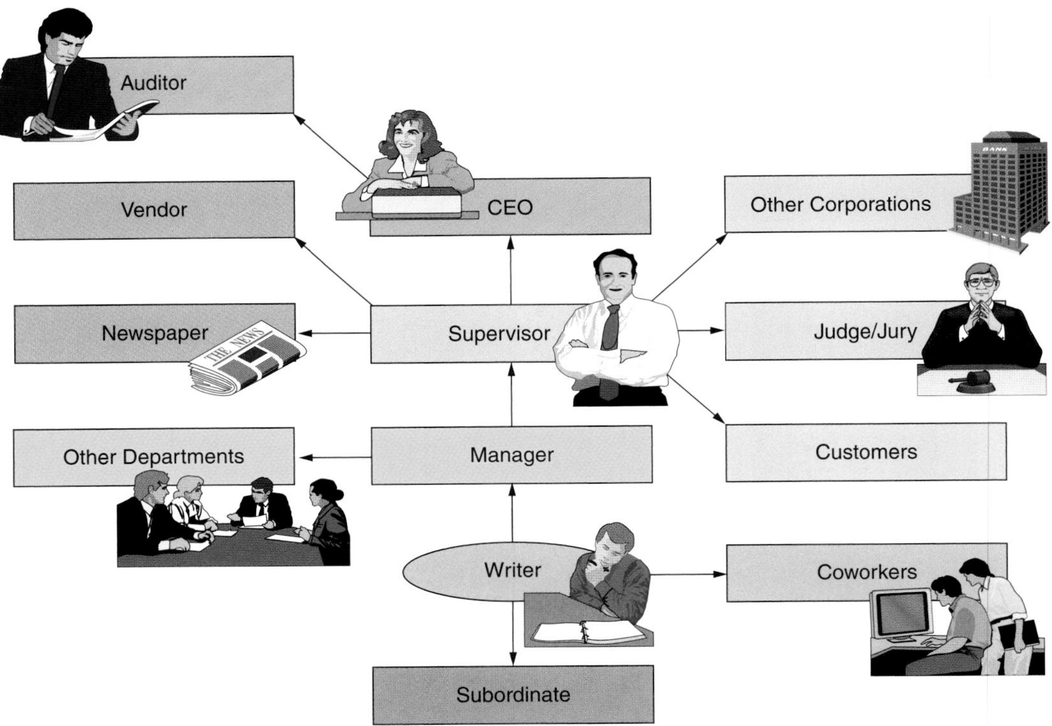

appropriate tone for all of these varied readers, the writing challenge becomes even greater. How do you meet such a challenge? The first key to success is recognizing that multiple audiences exist and that they share the following characteristics:

- Your intended audience will not necessarily be your only readers. Others might receive copies of your correspondence or hear your speech.
- Some of the multiple readers will not be familiar with the subject matter. You will have to provide background data (objectives, overviews) to clarify the history of the report for these readers. In a short letter, memo, report, or e-mail, this background information should be brief. Provide a reference line suggesting where the readers can find out more about the subject matter if they wish—"Reference: Operations Procedure 321 dated 9/21/09." In longer reports, background data will appear in the summary or abstract, as well as in the report's introduction.
- Summaries, abstracts, introductions, and references are especially valuable when you consider one more fact regarding written communication: reports, memos, letters, and e-mail are kept on file. When you write the correspondence initially, you can assume that your reader or readers have knowledge of the subject matter. Months (or years) later when the report is retrieved from the files, will your readers still be familiar with the topic? Will you still have the same readers? Future readers need background information.
- Correspondence for multiple readers must have a matter-of-fact, businesslike tone. You should not be too authoritative, since upper-level management might read the memo, letter, or report. However, you should not be patronizing, since lower-level subordinates might also read the correspondence.

Additional information:

See Chapter 17, "Short, Informal Reports."

Defining Terms for Different Audiences

Multiple audiences often need terms defined. To communicate successfully with multiple audiences, do any or all of the following:

- Define words or phrases either parenthetically or in a glossary. For example, you could define ATM parenthetically: ATM (asynchronous transfer mode). Then,

FIGURE 3.4 Effective Memo for Multiple Audiences

Date: July 7, 2009
To: Distribution
From: Rochelle Kroft
Subject: Revision of Operating Procedure (OP) 354 dated 5/31/09

The reissue of this procedure was the result of extensive changes requested by Engineering, Manufacturing, and Quality Assurance. These procedural changes will be implemented immediately according to Engineering Notice (EN) 185.

Some important changes are as follow:

1. *Substitutions*: An asterisk (*) can no longer be used to identify substitutable items in requirement lists. Engineering will authorize substitutions in work orders (WOs). Substitutions also will be specified by item numbers rather than by generic description names.
2. *Product Quality Requirements* (PQRs): These will be included in WOs either by stating the requirements or by referring to previous PQRs.
3. *Oral Instructions*: When oral instructions for process adjustments are given, the engineer must be present in the department.

All managers will review OP 354 and EN 185. Next, the managers will review the changes with their supervisors to make sure that each supervisor is aware of his or her responsibilities. These reviews will occur immediately.

Distribution:

Rob Harken	Manuel Ramos
Julie Burrton	Jeannie Kort
Hal Lang	Jan Hunt
Sharon Myers	Earl Eddings

Use the designation "Distribution" for multiple readers.

A parenthetical abbreviation following a technical term allows you to use the abbreviation in subsequent references.

Abbreviations can be used after prior parenthetical definition.

Throughout the memo, use an objective tone (passive voice) vs. a directive (active voice) that could offend management or some other reader. The correct tone can ensure that the memo is read.

readers will not misconstrue ATM as the more commonly understood *automatic teller machine*. You could also use a glossary, as follows:

example

HTTPS	Hypertext Transfer Protocol, Secure
TDD	telecommunication device for the deaf
TTY	teletypewriter

- Use a sentence to define terms, either in a glossary or following the unfamiliar terms.

example

Hypertext Transfer Protocol is a computer access code that provides secure communications on the Internet, an intranet, or an extranet.

Effective Memo for Multiple Audiences

The memo in Figure 3.4 is written to a multiple audience. "To: Distribution" is a common way to direct correspondence to numerous readers. The multiple readers affect the memo's content. Terms such as "operating procedure," "engineering notice," and "product quality requirements" are followed by parenthetical abbreviations—(OP), (EN), and (PQR). If these abbreviations alone had been presented, certain specialists would have understood them. Other readers in the distribution list, however, might not have known what OP, EN, or PQR meant. To communicate with all readers in multiple audiences, the writer must correctly use both the written-out terms and the abbreviations.

The tone of the memo is appropriate for readers with different levels of responsibility. Although it contains no direct commands, which may be offensive to management, the memo is still assertive. The matter-of-fact tone simultaneously suggests to management that these actions will be carried out and informs subordinates to do so.

MULTICULTURALISM

Multicultural Communication

Many corporations are doing business internationally. In fact, your company might market its products or services worldwide. International business requires multicultural communication, the sharing of written and oral information between businesspeople from many different countries (Nethery).

The Global Economy

Coca-Cola, which produces 300 different brands in 200 countries, generates 70 percent of its income from outside the United States ("Around the World"). To accommodate its worldwide audience, Microsoft offers the Windows XP Professional operating system in 25 different languages ("Multilingual Features in Windows XP Professional"). Black & Veatch, a global engineering company, has locations in Argentina, Australia, Botswana, Brazil, China, the Czech Republic, Egypt, Germany, Hong Kong, India, Indonesia, Korea, Kuwait, Malaysia, Mexico, Philippines, Poland, Saudi Arabia, Singapore, South Africa, Swaziland, Taiwan, Thailand, Turkey, United Arab Emirates, United Kingdom, United States, Vietnam, Zambia, and Zimbabwe.

FAQs

Q: Why is "multiculturalism" important in workplace communication? Doesn't everyone in the United States speak English?

A: No, not everyone in the United States speaks English. Look at these facts:

- "50 million people in the U.S., 19 percent of the population, . . . speak a language other than English at home and 22 million . . . have limited English proficiency" (Gardner).
- Between 1990 and 2000, the number of Americans speaking a language other than English at home grew by 15.1 million (a 47 percent increase) and the number with limited English proficiency grew by 7.3 million (a 53 percent increase).

Q: OK, so many people in the United States speak a language other than English. But, shouldn't we write for the majority of our audience?

A: Your audience will be diverse—and the "majority" in the United States is changing.

- "America's population has become increasingly diverse as Hispanics, African Americans, Asians and people from many other segments rapidly contribute to the cultural richness of our population."
- Hispanics in 2006 numbered 44.2 million and are expected to surge to 51.4 million by 2011.
- Asian/Pacific Islanders surpassed 13 million in 2006.
- African Americans numbered nearly 36 million in 2006.
- Asians, Blacks & Hispanics accounted for more than 70 percent of population growth, and Hispanics accounted for more than 50 percent since 2000.
- By 2011 the population in the three largest ethnic groups will be more than 107 million and Hispanics will represent nearly half of that population.
- On July 1, 2006, America had 113.5 million households. Of this total, nearly 13 million were Hispanic, nearly 11.5 million African American and nearly 4 million Asian. (Melgoza)

Margaret Keating, Vice President-Operations for Hallmark Cards Inc., oversees Hallmark's North American global operations. Her workforce of 6,000 employees is geographically dispersed. Much of her time is spent writing clear and concise communication to be translated into different languages (Cardarella D19).

The Challenges of Multicultural Communication

The Internet and e-mail affect global communication and global commerce. With these technologies, companies can market their products internationally and communicate with multicultural clients and coworkers at a keystroke. An international market is great for companies since a global economy increases sales opportunities. However, international commerce also creates written and oral communication challenges.

Medtronic, a leading medical technology company, does business in 120 countries. Many of those countries mandate that product documentation be written in the local language. To meet these countries' demands, Medtronic translates its manuals into 11 languages: French, Italian, German, Spanish, Swedish, Dutch, Danish, Greek, Portuguese, Japanese, and Chinese (Walmer 230). Multilingual reports create unique communication challenges as well. Each language version must be identical in content, readability, tone, style, and emphasis. Therefore, external audit firms must read each language version for accuracy (Courtis and Hassan 395).

Multicultural Team Projects

What about international, multilingual project work teams? If, for example, your U.S. company is planning to build a power plant in China, you will work with Chinese engineers, financial planners, and regulatory officials. To do so effectively, you will need to understand that country's verbal and nonverbal communication norms. You also will have to know that country's management styles, decision-making procedures, sense of time and place, and local values, beliefs, and attitudes.

The world's citizenry does not share the same perspectives, beliefs, values, political systems, social orders, languages, or habits. Successful workplace communication takes into consideration language differences, nonverbal communication differences, and cultural differences. Due to the multicultural makeup of your audience, you must ensure that your writing, speaking, and nonverbal communication skills accommodate language barriers and cultural customs. The classic example of one company's failure to recognize the importance of translation concerns a car that was named Nova. In English, "nova" is defined as a star that spectacularly flares up. In contrast, "*no va*" in Spanish is translated as "no go," a poor advertisement for an automobile.

Communicating Globally . . . in Your Neighborhood

Cross-Cultural Workplace Communication

Multiculturalism will not just affect you when you communicate globally. You will be confronted with multicultural communication challenges even in your own city and state. Another term for this challenge is *cross-cultural communication*, writing and speaking between businesspeople of two or more different cultures within the same country (Nethery).

How big a challenge is this? "About 19 million people in the United States are not proficient in English" (Sanchez A1). In addition, look at the statistics regarding America's melting pot shown in Figure 3.5.

These numbers are estimated to change. The U.S. Department of Labor's study of the civilian labor force by age, sex, race, and Hispanic origin projects that in 2014, the White population will fall in percentage to 67 percent, the Black population will fall to 12 percent, and the Asian population will stay the same at 5 percent. The Hispanic population, however, will grow to 16 percent, as shown in Figure 3.6.

One of the challenges presented by the increasingly multicultural nature of our society and workplace is language. Language barriers are an especially challenging situation

FIGURE 3.5 **Occupational Employment in Private Industry by Race/Ethnic Group/Sex and by Industry, United States, 2005.**

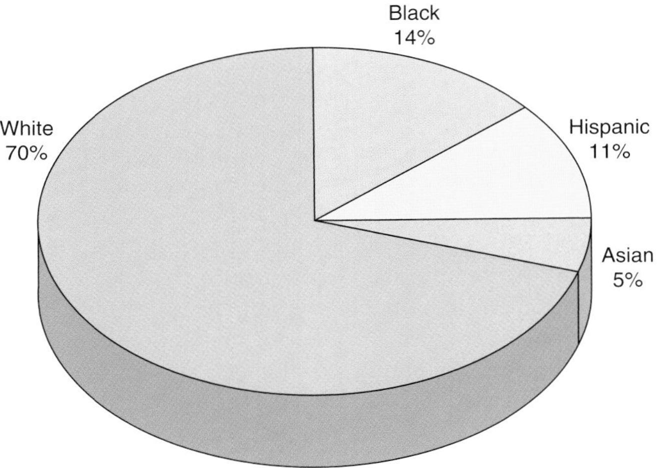

FIGURE 3.6 **Occupational Employment in Private Industry by Race/Ethnic Group/Sex and by Industry, United States, 2014**

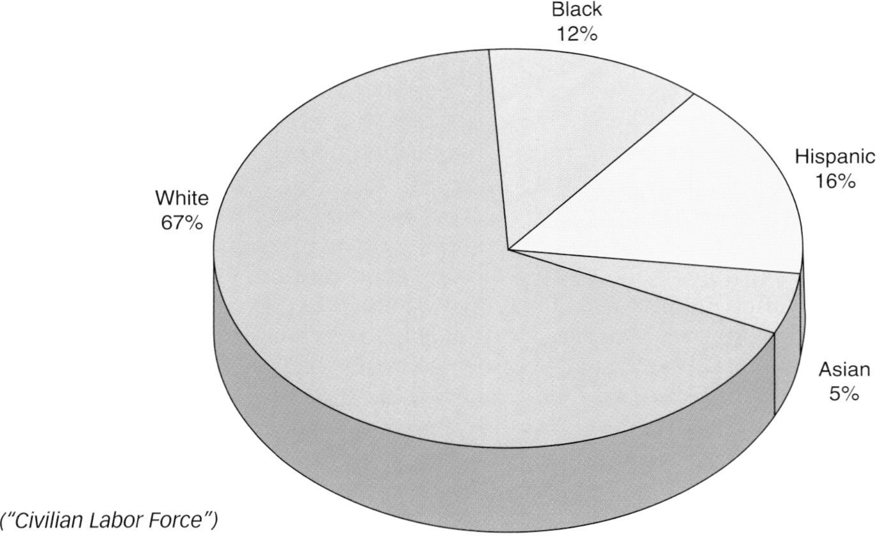

("Civilian Labor Force")

for hospitals, police and fire personnel, and governmental agencies where failure to communicate effectively can have dangerous repercussions.

One hospital reported that it has "13 staff members supplying Spanish, Arabic and Somali translations . . . ? In 2001, nearly 21,000 Spanish interpretations were performed at [the hospital]. The numbers for 2002 . . . exceeded 29,000 interpretations." However, this hospital's successful use of translators to help with doctor-patient communication is rare. "Only about 14 percent of U.S. hospitals provide training for volunteer translators." Most hospitals depend on the patient's relatives. "In one case studied, an 11-year-old sibling translated. The child made 58 mistakes" (Sanchez A4). Imagine how such errors can negatively impact healthcare and medical records.

The communication challenges are not just evident for employees in healthcare and other community infrastructures. Industries as diverse as banking, hospitality (restaurants and hotels), construction, agriculture, meat production and packing, and insurance also face difficulties when communicating with clients and employees for whom English is a second language.

Keys to Successful Communication in a Multicultural Environment

Define Acronyms and Abbreviations

Acronyms and abbreviations cause a problem for most readers. Although you and your immediate colleagues might understand such terminology, many readers won't. This is especially true when your audience is not native to the United States. For example, corporate employees often abbreviate the job title "system manager" as "sysmgr." However, in German, the title "system manager" is called the "system leiter;" in French, it's "le responsable." The abbreviation "sysmgr." would make no sense in either of these countries (Swenson WE-193).

Avoid Jargon and Idioms

The same dilemma applies to *jargon* and *idioms*, words and phrases that are common expressions in English but that could be meaningless outside our borders. Every day in the United States, we use "on the other hand" as a transitional phrase and "in the black" or "in the red" to denote financial status. What do these idioms mean in a global market? Similarly, the computer industry says, "The system crashed so we rebooted." A literal translation of this jargon into Chinese, German, or French might confuse the readers (Swenson WE-194).

Distinguish Between Nouns and Verbs

Many words in English act as both nouns and verbs. This is especially true with computer terms, such as "file," "scroll," "paste," "code," and "help." If your text will be translated, make sure that your reader can tell whether you're using the word as a noun or a verb (Rains 12).

Watch for Cultural Biases or Expectations

Your text will include words and graphics. As a workplace communicator, you need to realize that many colors and images that connote one thing in the United States will have different meanings elsewhere. For example, the idioms "in the red" and "in the black" will not necessarily communicate your intent when they are translated. Even worse, the colors black and red have different meanings in different cultures. Red in the United States connotes danger; therefore, "in the red" suggests a financial problem. In China, however, the word "red" has a positive connotation, which would skew your intended meaning. The word "black" often implies death and danger, yet "in the black" suggests financial stability. Such contradictions could confuse readers in various countries.

Animals represent a multicultural challenge. In the United States, we say you're a "turkey" if you make a mistake, but success will make you "soar like an eagle." The same meanings don't translate in other cultures. Take the friendly piggy bank, for example. It represents a perfect image for savings accounts in the United States, but pork is a negative symbol in the Mideast. If you are "cowed" by your competition in the United States, you lose. Cows, in contrast, represent a positive and sacred image in India (Horton 686–93).

Avoid Humor and Puns

Humor is not universal. In the United States, people talk about regional humor. If a joke is good in the South but not in the North, how could that same joke be effective overseas? Microsoft's software package Excel is promoted by a logo that looks like an "X" superimposed over an "L". This visual pun works in the United States because we pronounce the letters "X" and "L" just as we would the name of the software package. If your readers are not familiar with English, however, they might miss this clever sound-alike image (Horton 686).

Realize That Translations Might Take More or Less Space

If your writing will be conveyed on disk or on the Internet rather than on paper, you must consider software's line-length and screen-length restrictions. For example, a page of hard-copy text in the United States will consist of approximately 55 lines that average 80 characters per line. How could this present a problem? The standard sheet of paper in

Table 3.2 Translations Increase Word Length

English Word	Translations into Other Languages
Print	*Impression* (French)
File	*Archivo* (Spanish)
View	*Visualizzare* (Italian)
Help	*Assistance* (French)
E-mail	*Courriel* (Swiss)

the United States measures 8½″ × 11″. In contrast, the norm in Europe for standard-sized paper is A4—210 × 297 millimeters, or 8.27″ × 11.69″.

Why is the size of paper important? This has nothing to do with language barriers, right? Here is the problem: If you format your text and graphics for an 8½″ × 11″ piece of paper in Atlanta, travel to London, download the files on a computer there, and hit Print, you might find that what you get is not what you hoped for. The line breaks will not be the same. You will not be able to three-hole punch and bind the text. The margins will be off, and so will the spacing on your flowchart or table (Scott 20–21).

An even greater problem occurs when you are writing for the Internet. On a Web site, you will provide a navigation bar and several frames. Why is this a problem? A page of English text translates to the same length in any language, right? The answer is no. The word count of a document written in English will expand more than 30 percent when translated into some European languages. For example, the four-letter word "user" has 12 characters in German (Hussey and Homnack RT-46). In Table 3.2, notice how English words become longer when translated into other languages (Horton 691).

The Swiss government is trying to curb what it defines as "the encroachment of English." To do so, the government's French Linguistics Service is asking its citizens to avoid using the word "spam," instead opting for *courier de masse non sollicite,* meaning "unsolicited bulk mail" ("Swiss fight encroachment of English" A16).

Avoid Figurative Language

Many of us use sports images to figuratively illustrate our points. We "tackle" a chore; in business, a "good defense is the best offense"; we "huddle" to make decisions; if a sale isn't made, you might have "booted" the job; if a sale is made, you "hit a home run." Each of these sports images might mean something to native speakers, but they may not communicate worldwide. Instead, say what you mean, using precise words (Weiss 14).

Be Careful with Numbers, Measurements, Dates, and Times

If your text uses measurements, you are probably using standard American inches, feet, and yards. However, most of the world measures in metrics. Thus, if you write 18 high × 20 wide × 30 deep, what are the measurements? There is a huge difference between 18 × 20 × 30 inches and 18 × 20 × 30 millimeters. In the United States, we tend to abbreviate dates as MM/DD/YY: 05/03/09. In the United Kingdom, however, this could be perceived as March 5, 2009, instead of May 3, 2009. See Table 3.3 for additional examples of interpreting day and time. Time is another challenge. Table 3.4 shows how different countries write times.

In addition to different ways of writing time, you must also remember that even within the United States, 1:00 p.m. does not mean the same thing to everyone. Is that central time, Pacific time, mountain time, or eastern time?

Another challenge with time occurs when we incorrectly assume that everyone everywhere abides by the same work hours. In the United States, the average workweek is 40 hours, and the typical workday is from 8:00 a.m. to 5:00 p.m. However, this is not the norm globally. French laws have reduced the workweek to 35 hours. Many Middle Eastern countries close work "for all or part of Friday," the beginning of Sabbath. Offices in parts of southern Europe shut down for a traditional two-hour lunch (noon to 2 p.m.)

Table 3.3 Different Ways of Understanding and Writing the U.S. Date 05/03/09

Country	Date
United States	May 3, 2009
United Kingdom	March 5, 2009
France	5 mars 2009
Germany	5. Marz 2009
Sweden	09–05–03
Italy	5.3.09

Table 3.4 Different Ways of Writing the U.S. Time 5:15 p.m.

Country	Time
United States	5:15 p.m.
France	17:15
Germany	17.15
Quebec, Canada	17 h 15

(St. Amant "Communication" 28). Therefore, if you write an e-mail telling a coworker in Spain or Jordan that you will call at 2:00 p.m. his or her time, that could be an inappropriate time for your audience.

Finally, even simple words like "today," "yesterday," or "tomorrow" can cause problems. Japan is 14 hours ahead of U.S. Eastern Standard Time. Thus, if you need a report "tomorrow," do you mean tomorrow—the next day *your time*—or tomorrow—two days from your reader's time?

To solve these problems, determine your audience and make changes accordingly. That might mean

- Writing out the date completely (January 12, 2009)
- Telling the reader what standard of measurement you will use ("This document provides all measurements in metrics.")
- Telling the reader what scheme of time presentation you will use ("This document relates time using a 24-hour clock rather than a 12-hour clock.")
- Using multiple formats ("Let's meet at 2:30 p.m./14:30.")
- Avoiding vague words like "today," "tomorrow," or "yesterday"
- Recognizing that people have different work schedules globally

Use Stylized Graphics to Represent People

A photograph or realistic drawing of people will probably offend someone and create a cultural conflict. You want to avoid depicting race, skin color, hairstyles, and even gender. To solve this problem, avoid shades of skin color, choosing instead pure white or black to represent generic skin. Use simple, abstract, even stick figures to represent people. Stylize hands so they are neither male nor female—and show a right hand rather than a left hand, if possible (a left hand is perceived as "unclean" in some countries) (Flint 241).

Recognizing the importance of the global marketplace is smart business and a wise move on the part of the workplace communicator. Figure 3.7 is an example of poor communication in an e-mail message for a multilingual audience. This e-mail fails for many reasons:

1. **Date**—Does "03/07/09" mean March 7, 2009, or July 3, 2009?
2. **Time**—"12:00" not only fails to specify the time zone, but creates the question whether midnight or noon is intended.

BEFORE

FIGURE 3.7 Flawed E-Mail Message for a Multilingual Audience

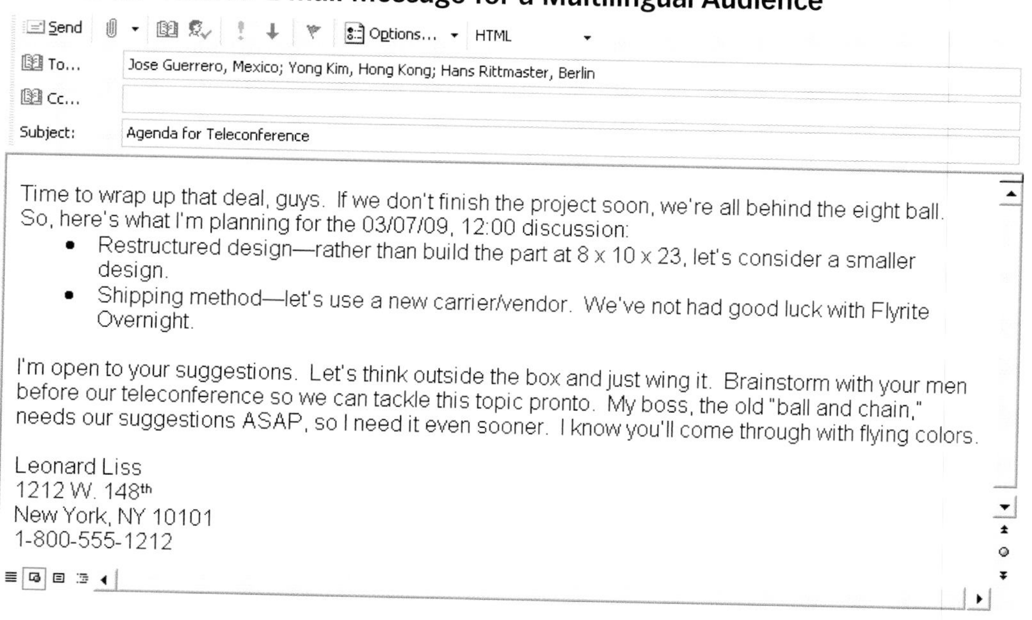

Send | Options... | HTML

To... Jose Guerrero, Mexico; Yong Kim, Hong Kong; Hans Rittmaster, Berlin

Cc...

Subject: Agenda for Teleconference

Time to wrap up that deal, guys. If we don't finish the project soon, we're all behind the eight ball. So, here's what I'm planning for the 03/07/09, 12:00 discussion:

- Restructured design—rather than build the part at 8 x 10 x 23, let's consider a smaller design.
- Shipping method—let's use a new carrier/vendor. We've not had good luck with Flyrite Overnight.

I'm open to your suggestions. Let's think outside the box and just wing it. Brainstorm with your men before our teleconference so we can tackle this topic pronto. My boss, the old "ball and chain," needs our suggestions ASAP, so I need it even sooner. I know you'll come through with flying colors.

Leonard Liss
1212 W. 148th
New York, NY 10101
1-800-555-1212

3. **Figurative language**—"Wrap up that deal," "behind the eight ball," "think outside the box," "just wing it," "brainstorm," "tackle this topic," "ball and chain," and "flying colors" all are idiomatic phrases. Though these phrases might communicate within the United States, they may not translate internationally.
4. **Informal tone**—Informality is common in most stateside businesses, but this is not true in all countries. Germany and Japan, for example, are far more formal in their business communication expectations. Thus, "guys" and the contractions could cause problems creating incorrect tone.
5. **Sexist language**—"Guys" and "men" should be avoided when writing any correspondence. The implication is that only men will be involved in the discussions, which of course is completely erroneous.
6. **Measurements**—"8 × 10 × 23" is confusing. Is the writer discussing inches, feet, or meters?
7. **Slash marks**—"New carrier/vendor" could mean either "new carrier *or* vendor," or "new carrier *and* vendor." Which is it?
8. **Undefined abbreviations or acronyms**—"ASAP" must be defined. Though all readers within the United States might understand this to mean "as soon as possible," you cannot assume the same level of understanding internationally.
9. **Cultural sensitivity**—In the United States, we might see humor in referring to a boss as "the old ball and chain" (though this is highly doubtful). In Japan, however, where saving face is a cultural norm, potentially offending a superior would be a problem.

Finally, using the word "pronto" is cowboy slang. To communicate internationally, you must consider each country's cultural norms—and be careful to avoid offense or confusion.

In contrast, Figure 3.8 corrects these communication problems:

AFTER

FIGURE 3.8 Effective E-Mail Message for a Multilingual Audience

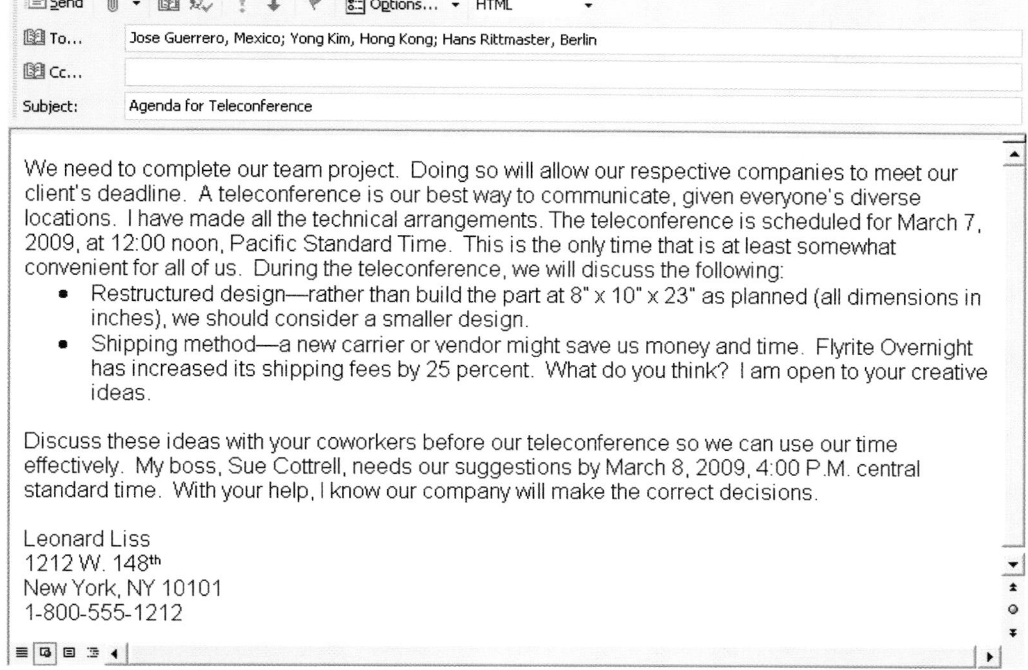

Send | Options... | HTML

To... Jose Guerrero, Mexico; Yong Kim, Hong Kong; Hans Rittmaster, Berlin

Cc...

Subject: Agenda for Teleconference

We need to complete our team project. Doing so will allow our respective companies to meet our client's deadline. A teleconference is our best way to communicate, given everyone's diverse locations. I have made all the technical arrangements. The teleconference is scheduled for March 7, 2009, at 12:00 noon, Pacific Standard Time. This is the only time that is at least somewhat convenient for all of us. During the teleconference, we will discuss the following:

- Restructured design—rather than build the part at 8" x 10" x 23" as planned (all dimensions in inches), we should consider a smaller design.
- Shipping method—a new carrier or vendor might save us money and time. Flyrite Overnight has increased its shipping fees by 25 percent. What do you think? I am open to your creative ideas.

Discuss these ideas with your coworkers before our teleconference so we can use our time effectively. My boss, Sue Cottrell, needs our suggestions by March 8, 2009, 4:00 P.M. central standard time. With your help, I know our company will make the correct decisions.

Leonard Liss
1212 W. 148th
New York, NY 10101
1-800-555-1212

Additional information:

See Chapter 9, "Electronic Communication: E-Mail Messages, Instant Messages, Text Messages, and Blogging."

AVOIDING BIASED LANGUAGE

In addition to recognizing your audience's level of knowledge, roles, and cultural diversity, you also must consider your audience's age, physical limitations, and gender.

Ageist Language

A word like "elderly" could imply feebleness. The words "Old folks" create a negative image. To avoid these biases, write "people over 70" or "retirees." Better yet, avoid reference to age.

BEFORE	AFTER
Professor Jones, an elderly teacher at State University, is publishing a textbook despite his age.	Professor Jones, a State University teacher, is publishing a textbook.

Biased Language About People with Disabilities

The word "handicap" creates a negative image. "Disability" is generally preferred. However, any euphemism can be offensive. You should avoid reference to a person's disabilities. If you need to refer to a physical problem, do so without negative characterizations.

BEFORE	AFTER
Debbie Brown, a blind market researcher, won "Employee of the Month."	Debbie Brown, a market researcher, won "Employee of the Month."
The AIDS victim changed insurance carriers.	The AIDS patient changed insurance carriers.
John suffers from diabetes.	John is diabetic.
Sheila is confined to a wheelchair.	Sheila uses a wheelchair.

Sexist Language

Women constitute half of the workforce in the United States. Thus, you must avoid gender-biased language in your communication. Gender-biased language occurs through

- Omission
- Unequal treatment
- Stereotyping
- Word choice

Omission

When your writing ignores women or refers to them as secondary, you are being biased. The following examples show biased comments and their nonbiased alternatives.

BEFORE	AFTER
The computer information specialists and their wives and children attended the company picnic.	The computer information specialists and their families attended the company picnic.
The congressional legislation on foreign trade agreements was proposed by a woman, Claire McGowan.	The congressional legislation on foreign trade agreements was proposed by Claire McGowan.
When conducting his quarterly review, the auditor must always check for errors.	When conducting a quarterly review, the auditor must always check for errors.
As we acquired scientific knowledge, men began to examine long-held ideas more critically.	As we acquired scientific knowledge, people began to examine long-held ideas more critically.

Unequal Treatment

Modifiers that describe women in physical terms not applied to men are patronizing.

BEFORE	AFTER
The poor women could no longer go on; the exhausted men . . .	The exhausted men and women could no longer go on.
Mrs. Acton, a statuesque blonde, is Joe Granger's assistant.	Jan Acton is Joe Granger's assistant.

Stereotyping

If your writing implies that only men do one kind of job and only women do another kind of job, you are stereotyping. For example, if you suggest that men hold all management jobs and women hold all subordinate positions, this is stereotyping.

BEFORE	AFTER
Current tax regulations allow a head of household to deduct for the support of his children.	Current tax regulations allow a head of household to deduct for child support.
The manager is responsible for the productivity of his department; the foreman is responsible for the work of his linemen.	Management is responsible for departmental productivity. Supervisors are responsible for their personnel.
The secretary brought her boss his coffee.	The secretary brought the boss's coffee.
The teacher must be sure her lesson plans are filed.	The teacher must file all lesson plans.

Pronouns

Biased language disappears when you use pronouns that treat all people equally. Pronouns such as "he," "him," or "his" are masculine. Sometimes you read disclaimers by manufacturers stating that "although these masculine pronouns are used, they are not intended to be biased. They are only used for convenience." This is an unacceptable statement. Use of "he," "him," and "his" exclusively creates a masculine image.

BEFORE	AFTER
Sometimes the doctor calls on his patients in their homes.	Sometimes the doctor calls on patients in their homes.
The typical child does his homework after school.	Most children do their homework after school.
A good lawyer will make sure that his clients are aware of their rights.	A good lawyer will make sure that his or her clients are aware of their rights.

To avoid this bias, avoid masculine pronouns. Instead, use the plural, generic "they" or "their." You also can use "he or she" and "his or her." Sometimes you can solve the problem by omitting all pronouns.

Nouns

Use nouns that are nonbiased. To achieve this, avoid nouns that exclude women and denote that only men *or* women are involved. These are called "gender-tagged words." Most of them have an actual gender attached to them. Usually, this involves the word "man" or "men." However, the suffix "ess" is a feminine ending, also creating a gender tag, just as "princess" equals a woman, while "prince" is a man.

BEFORE	AFTER
mankind	people
manpower	workers, personnel
the common man	the average citizen
wise men	leaders
businessmen	businesspeople
policemen	police officers
firemen	firefighters
foreman	supervisor
chairman	chairperson *or* chair
stewardess	flight attendant
waitress	server

Figures 3.9 and 3.10 illustrate a biased and unbiased letter from a company to a customer.

ACHIEVING AUDIENCE INVOLVEMENT

Recognizing your audience entails knowing their levels of knowledge, roles in relation to your job, multiculturalism, and diversity issues, such as age, physical limitations, and gender. By understanding these audience concerns, you can communicate more effectively. An additional goal is to achieve audience involvement. Effective workplace communication must build rapport to achieve audience buy-in. You want to persuade your audience

FIGURE 3.9 Biased Letter

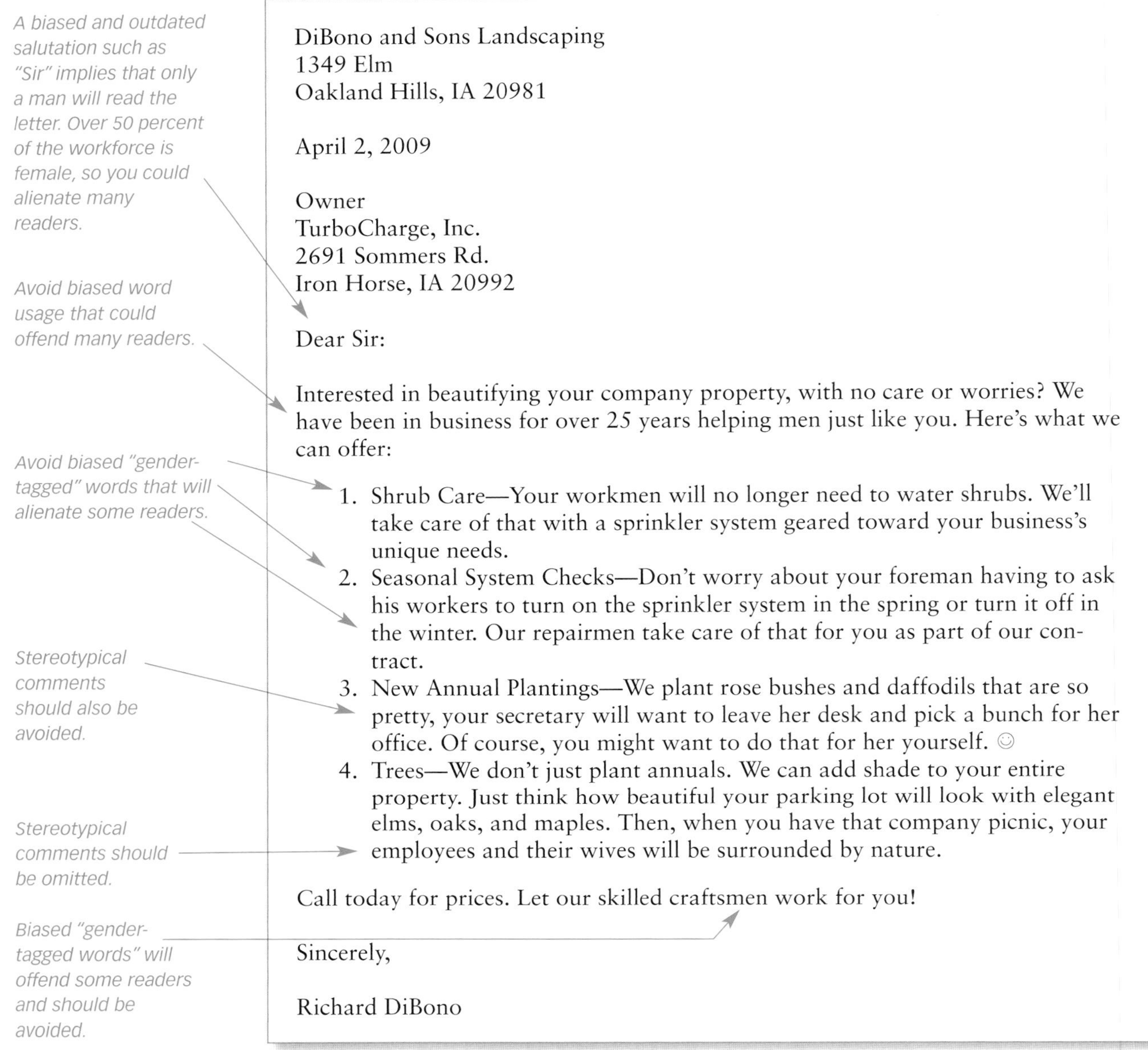

DiBono and Sons Landscaping
1349 Elm
Oakland Hills, IA 20981

April 2, 2009

Owner
TurboCharge, Inc.
2691 Sommers Rd.
Iron Horse, IA 20992

Dear Sir:

Interested in beautifying your company property, with no care or worries? We have been in business for over 25 years helping men just like you. Here's what we can offer:

1. Shrub Care—Your workmen will no longer need to water shrubs. We'll take care of that with a sprinkler system geared toward your business's unique needs.
2. Seasonal System Checks—Don't worry about your foreman having to ask his workers to turn on the sprinkler system in the spring or turn it off in the winter. Our repairmen take care of that for you as part of our contract.
3. New Annual Plantings—We plant rose bushes and daffodils that are so pretty, your secretary will want to leave her desk and pick a bunch for her office. Of course, you might want to do that for her yourself. ☺
4. Trees—We don't just plant annuals. We can add shade to your entire property. Just think how beautiful your parking lot will look with elegant elms, oaks, and maples. Then, when you have that company picnic, your employees and their wives will be surrounded by nature.

Call today for prices. Let our skilled craftsmen work for you!

Sincerely,

Richard DiBono

to accept your point of view—*willingly*. This requires the correct motivational tone, when writing or speaking up to management, down to subordinates, laterally to coworkers, or out to customers. Follow these suggestions to involve your audience.

Avoid Commands

Do not write or say, "You must" or "You should." Write "I recommend that," "Our team suggests that," or "I propose." Words like "recommend," "suggest," and "propose" show that you respect everyone's involvement and right to a say in the matter. By avoiding a commanding tone, you encourage your audience's participation.

Ask Questions

You could write to management, "Would you consider implementing . . . ?" To your subordinates, you could write, "What are your thoughts about increasing . . . ?" To customers, you might say, "What can we do to improve your . . . ?" Questions such as these speak

FIGURE 3.10 Unbiased Letter

DiBono and Sons Landscaping
1349 Elm
Oakland Hills, IA 20981

April 2, 2009

Owner
TurboCharge, Inc.
2691 Sommers Rd.
Iron Horse, IA 20992

Subject: Sales Information

Interested in beautifying your company property, with no care or worries? We have been in business for over 25 years helping **business owners** just like you. Here's what we can offer:

- Shrub Care—Your **employees** will no longer need to water shrubs. We'll take care of that with a sprinkler system geared toward your business's unique needs.
- Seasonal System Checks—Don't worry about your **supervisor** having to **ask his or her** workers to turn on the sprinkler system in the spring or turn it off in the winter. Our **staff** takes care of that for you as part of our contract.
- New Annual Plantings—We plant rose bushes and daffodils that are so pretty, **all of your employees** will want vase-filled flowers on **their** desks.
- Trees—We don't just plant annuals. We can add shade to your entire property. Just think how beautiful your parking lot will look with elegant elms, oaks, and maples. Then, when you have that company picnic, your employees and **their families and friends** will be surrounded by nature.

Call today for prices. Let our skilled **experts** work for you!

Sincerely,

Richard DiBono

Unbiased words are boldfaced for emphasis.

directly to the audience's need to be a participant. You show concern for the audience and involve them in the process.

Use Positive Words or Phrases

"Please" and "thank you" are positive words. Do not write, "I need you to order. . . ." By revising this to read "Please order," you alter the tone and encourage action rather than dictate. Positive words motivate, whether you are writing up or down. Following are examples of positive words which you can use to create motivational tone.

advantage	effective	happy	profitable	successful
asset	efficient	please	satisfied	thank you
benefit	enjoyable	pleased	succeed	value
confident	favorable	pleasure	success	worthwhile

To clarify the importance of positive words, following are examples of negative writing and positive revisions.

FIGURE 3.11 **Audience Focus**

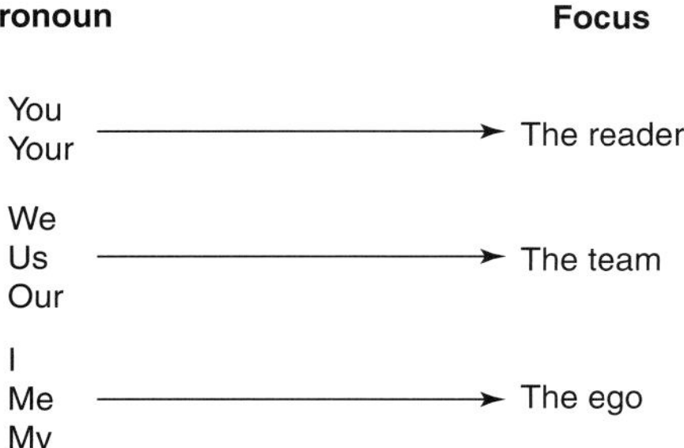

BEFORE	AFTER
I cannot process your request. You failed to follow the instructions.	So that we may process your request more rapidly, please fill in line 6 on the printed form.
The error is your fault. You kept the books incorrectly.	To ensure better customer service, let's review our bookkeeping practices.
I have received your e-mail complaining about the department's quality control practices.	Thank you for suggesting ways to improve the department's quality control procedures.
The invoice you sent was useless by the time it arrived.	To help us maintain our payment schedules, please submit invoices by the 15th of each month.
Your bill is now three weeks overdue. Failure to pay immediately will result in lower credit ratings.	If you're as busy as we are, you've probably misplaced our recent bill (mailed three weeks ago). Please send in your payment this week to maintain your high credit ratings.

Making something positive out of something negative is a challenge, but the rewards for doing so are great. If you attack your audience with negatives, you lose. When you involve your audience through positive words, you motivate them to work with you and for you.

Involve Your Audience with "You Usage"

Limit your use of first-person pronouns like "I," "Me," and "My." Your audience wants to know what is in it for them, not how you will benefit. To achieve a different focus on the audience or the team, emphasize "you usage." Use second-person pronouns like "You" and "Your." Third-person pronouns, such as "We," "Us," and "Our," also are effective to create a sense of team between the writer and audience. Figure 3.11 summarizes these focuses.

When you use "You" or "Your," you are speaking directly to your reader(s). The audience reads the words "You" or "Your" and sees themselves in the pronoun, envisioning that they are being spoken to, focused on, and singled out. By focusing on the pronouns "You" or "Your" in your workplace communication, you involve your audience in the correspondence. "We," "Us," and "Our" are team words to connote group involvement. These pronouns are especially valuable when writing to multiple audiences or when writing to subordinates. In either instance, "We," "Us," and "Our" imply to the readers that, "We're all in this together." Such a team concept helps motivate by making the readers feel an integral part of the whole. "I," "Me," and "My" focus on the writer or speaker. These pronouns can connote egocentricity. To avoid egocentric word usage, emphasize "You" and "Your" and downplay "I," "Me," and "My."

BEFORE	AFTER
Dear Sir: With regard to lost policy 123, enclosed is a lost policy form. Complete this form and return it ASAP. Upon receipt, the company will issue a replacement policy.	Dear Mr. Salinas: Your lost policy 123 can be replaced easily. I've enclosed a form to help us replace it for you. All you need to do is fill it out. As soon as we get the form, we'll send you your replacement policy.
Claims Procedure To obtain service under the Emissions Performance Warranty, take the vehicle to the company dealer as soon as possible after it fails an I/M test along with documentation showing that the vehicle failed an EPA-approved emissions test.	**Claims Procedure** How do you get service under the Emissions Performance Warranty? To get service under this warranty, take your car to the dealer as soon as possible after it has failed an EPA-approved test. Be sure to bring along the document that shows your car failed the test.

Focus on Audience Benefit

To encourage cooperation at all levels (management, coworkers, subordinates, vendors, and customers), show your readers or listeners how everyone will benefit by pursuing your plan. In your letter, memo, report, or e-mail, state the benefit clearly. You can do this anywhere, but here are some options:

- State the benefit early (subject line, first paragraph, or abstract)
- Conclude with the benefit

Placing the benefit early in the communication will interest and involve your audience. For example, you could write, "Congratulations! Your sales force sold 45 percent more maintenance contracts than last quarter." Placing the benefit at the end could provide a motivational close, leaving readers with a positive impression. You could encourage participation at all levels by concluding as follows: "When this action is completed, the company will achieve a 35 percent increase in sales."

Figure 3.12 achieves audience involvement through reader benefit in many ways.

Personalize Your Text with Names

Companies *don't* write to companies; *people write to people*. When you write or speak, you are communicating to people. People want to be treated like individuals. One simple way to personalize is by referring to your audience by name. Within the correspondence, write, "Jim, please report on . . ." or in the face-to-face meeting say, "Mrs. Harken, when did you experience the troubles with your cable?" Doing so personalizes your communication.

Potential Problems

You might encounter the following problems when you personalize with names:

- If you are writing *up to management*, be cautious. If you rarely or never communicate with your company's CEO, for example, using a first name would be inappropriate. If, however, you work closely with your management and regularly communicate on a first-name basis, then continue doing so.
- When you write *outside the company*, using a first name would be inappropriate if you do not have a relationship with your audience. Refer to this person by his or her last name.
- If you are speaking to someone and *you do not know how to pronounce his or her name*, avoid using that person's name.
- If you do not know a person's *marital status* (Mrs. or Miss), use Ms.

Figure 3.13 is impersonal and negative. It does not motivate. Instead, it commands and degrades. Figure 3.14, in contrast, uses pronouns, personalization, and positive words to encourage action and build rapport.

Additional information:

See Chapter 10, "Traditional Correspondence: Memos and Letters."

FIGURE 3.12 Memo Focusing on Audience Benefit

Positive words, such as "Improved," "strengthen," "please," "effective," and "Thank you," focus on audience benefit.

Date: January 2, 2009
To: Mary Ortega
From: Jim Goodwin
Subject: Improved Billing Control Procedures

Three times in the last quarter, one of the billing cycles excluded franchise taxes. This was a result of the CRT input being entered incorrectly.

To strengthen internal controls and reduce the chance of future errors, please implement the following steps, effective January 15, 2009.

1. Input — Ask the accounting clerk to complete the control screen entry form.
2. Entry — Ensure that the CRT entry clerk will enter the control screen input and use local print to produce a hard copy.
3. Verification — Have the junior accountant compare the hard copy to the control screen and initial when it's correct.
4. Notification — Tell the junior accountant to inform data processing when they should start the billing process.

The new procedure will provide more timely and accurate billing to our customers. Customer service will be happier with accounting. We all will have more pride in our contributions. Thank you for your help with this solution.

BEFORE

FIGURE 3.13 Impersonal and Negative Memo

Using a negative tone will alienate and offend many readers.

Using a commanding tone might cause the reader to react negatively. In contrast, creating a "team-oriented" tone will encourage readers to act.

Concluding with a threatening tone will not inspire the reader or encourage participation.

Date: July 22, 2009
To: Martha Collins
From: Bob Kaplan
Subject: Assignment of Contract Administration Project

Work to date on the Contract Administration Project has been too slow. Your staff must make up for other departments' failures to meet our deadline. Here is what you have to do:

1. Analyze other operating companies to see how they create neutrality when administering contracts.
2. Demand that our consultants achieve impartiality.
3. Develop standards for all operating procedures across all departments.
4. Give me a report immediately upon your conclusion of these requirements.

You and your staff must get this right this time. Otherwise, the company will suffer severe setbacks in our competitive bidding process.

AFTER

FIGURE 3.14 Personal and Positive Memo

Date: July 22, 2009
To: Martha Collins
From: Bob Kaplan
Subject: Assignment of Contract Administration Project

Martha, thank you for accepting the Contract Administration Project. To accomplish our goal of competitive bidding, I need your help. Please follow these guidelines with your team:

1. Analyze other operating companies to see how they ensure neutrality when administering contracts.
2. Coordinate our consultants to guarantee that they achieve impartiality.
3. Develop standards for all operating procedures across all departments.
4. Prepare a report for our advisory team. They will need it by the end of the month.

Your successful completion of this project, Martha, will help us improve all future contract administrations. I'm confident that you will do a great job, and I know our company and employees will benefit.

Use the reader's name and positive words such as "thank you" to personalize and promote goodwill.

Showing the reader respect for ability will encourage teamwork.

Positive words and pronoun usage will encourage teamwork and action on the part of the audience.

Ending in a motivational tone will inspire and encourage the reader.

AUDIENCE CHECKLIST

_____ 1. What is the audience's level of understanding regarding the subject matter?
- Specialist
- Semi-specialist
- Lay
- Multiple

_____ 2. Given your audience's level of understanding, have you written accordingly?
- Have you defined acronyms, abbreviations, and jargon?
- Have you supplied enough background data?

_____ 3. What is your role in relation to the audience?
- Do you work for the reader?
- Does the reader work for you?
- Is the reader a peer?
- Is the reader a client?
- Is the reader a vendor?

_____ 4. What response do you want from the audience? Do you want the audience to act, respond, confirm, consider, decide, or file the information for future reference?

_____ 5. Will the audience act according to your wishes? What is the audience's attitude toward the subject (and you)?
- Negative
- Positive
- Noncommittal
- Uninformed

_____ 6. Is the audience in a position of authority to act according to your wishes (can he or she make the final decision)? If not, who will make the decision?

_____ 7. Have you considered diversity?
- Have you avoided biased language that could offend various age groups, people of different sexual orientations, people with physical challenges, or people of different cultures and religions?
- Have you considered that people from different countries and people for whom English is a second language will be involved in the communication? This might mean that you should
  - Clarify time and measurements
  - Define abbreviations and acronyms
  - Avoid figurative language and idiomatic phrases unique to one culture
  - Avoid humor and puns
  - Consider each country's cultural norms

_____ 8. Have you avoided sexist language?
- Have you used "their" or "his or her" to avoid sexist singular pronouns like "his"?
- Have you used generic words such as "police officer" versus sexist words like "policeman"?
- Have you avoided excluding women, writing sentences such as "All present voted to accept the proposal" versus the sexist sentence "All the men voted to accept the proposal"?
- Have you avoided patronizing, writing sentences such as "Mr. Smith and Mrs. Brown wrote the proposal" versus the sexist sentence "Mr. Smith and Judy wrote the proposal"?

_____ 9. Have you motivated the audience to act?
- Have you involved the audience by using pronouns or a person's name?
- Have you shown the audience benefit by using positive words and verbs?

_____10. Have you considered the audience's preferences regarding style?
- Is use of a first name appropriate?
- Should you use a last name?

Phil Met His Challenge

To meet his communication challenge, Phil used the P^3 process.

Planning

To plan his proposal, Phil first considered the following:

- Goals—provide Spanish language skills to non-Spanish-speaking supervisors and basic English skills to Spanish-speaking manufacturing employees. The goal is to improve employer/employee relations, create a safe and more efficient workplace, and ensure quality work.
- Audience—corporate employees and supervisors
- Channels—e-mail messages, face-to-face interviews, cover letters, and a proposal
- Data—drawn from interviews

Planning for the proposal, Phil says, "I need to accomplish the following:

1. **Company needs assessment—**I plan to meet with Arctic management to identify company needs and to establish training priorities. I need to analyze the difficulties limited English-speaking employees were having in terms of interaction with supervisors and other employees, communication breakdowns, or lost work time due to difficulties adapting to American culture.
2. **Participant assessment—**I need to meet with the training participants to assess their language abilities and needs more thoroughly. I can use oral interviews, testing, and job shadowing to gather data.
3. **Methods of instruction—**Based on my findings, I will determine whether to use modeling, intensive drills, oral responses, learning pairs and triads, role-playing, or simulations. When I gather data from research, I then can write the proposal to Arctic."

Packaging

After Phil researched Arctic's language needs, he wrote a rough draft of the proposal. The proposal was prefaced by a cover letter in which Phil hoped to show Arctic how the training sessions would be structured and what benefits the company would derive. He gave the draft of the cover letter (see Figure 3.15) to a colleague for feedback.

Perfecting

According to Phil, "It's my job to propose the most cost-effective solution to my client's problems with multicultural communication. I have to consider how to best customize our course offerings and to determine which of my specialists would work well with the client's employees and management. By assisting companies to communicate better with all of their employees, I am doing a service for our entire community. Good employees benefit not only themselves and the company but also society by being productive on the job. My perfected cover letter (Figure 3.16) offers the specific ways in which we can meet the client's needs to improve multicultural communication."

FIGURE 3.15 **Colleague-Suggested Revisions to Phil's Cover Letter (Shown with Track Changes)**

Comment: delete the "th" Phil. The date alone is sufficient.

Comment: I'd expand this subject line, adding something like "Workplace English" for clarity.

Comment: wordy—how about just writing "We are pleased to propose an on-site . . .?"

Comment: "we developed."

Comment: Throughout you use full block alignment. I've switched to ragged right.

Comment: add "ask"

Comment: Let's boldface the major headings, as shown, and itemize these parts for easier access. The long paragraph is hard to read.

Comment: This is just written "CDs" without the apostrophe.

Comment: The passive voice in this sentence creates what I think English teachers call a dangling modifier. Let's write, "When the training program is complete, I will develop a follow-up plan to help . . ."

Comment: delete

February 24th, 2009

Joan T. Osborn
Manager, Learning and Development Americas
Arctic Cooling Technologies
7401 W 129th Street
Overland Park, KS 66213

SUBJECT: TRAINING PROPOSAL

We are pleased to present to you a proposal to offer an on-site Workplace English program to limited English-speaking Arctic employees. Following a language assessment to determine the appropriate level of instruction, we will develop a comprehensive curriculum to improve each participant's English language skills and knowledge.

The program will offer an easy and quick way to learn limited amounts of everyday workplace English. Instruction will be divided into three components: Speaking in English, Listening in English, and Workplace Application English. Aspects of U.S. culture will also be included in each session to discuss some of the most fundamental aspects of everyday U.S. culture that are often misunderstood.

Speaking in English teaches participants how to say practical, common phrases and questions in American English. We will focus on pronunciation, expressions, and questions for use in everyday interactions. **Listening in English** teaches the participants how to comprehend many basic and common expressions, phrases, and questions used in everyday English. **Workplace-specific English** teaches participants to comprehend and respond to workplace-specific language including phrases and terms that are regularly used in the workplace.

Each participant will receive a manual with CD's that can be used to maintain the language skills they have acquired.

We propose offering on-site classes limited to a maximum of 15 participants for 6 weeks to meet two hours, twice a week, for a total of 24 hours. A once per week meeting format for 12 weeks could also be provided. All instructional material will be provided to participants. The $3,300 estimated budget for this program includes curriculum development, instructional manuals, and on-site training. Following completion of the program a follow-up plan will be developed to help each participant maintain and improve his or her English language skills.

We look forward to finalizing this training program. Thank you for the opportunity to present this proposal to you for your careful consideration. We pride ourselves on providing quality training, language development, career planning, consulting, and economic development services.

FIGURE 3.16 **Perfected Letter for Arctic Training Workshop**

12345 College Boulevard
Overland Park, Kansas 66210
913.469.3845
Fax 913.469.4415
www.centerforbusiness.org

CENTER FOR BUSINESS AND TECHNOLOGY

Phil says, "Focusing on the needs of the audience is an important part of my job. The P³ Communication Process allows me to create the best possible written document to address my audience's needs."

February 24, 2009

Joan T. Osborn
Manager, Learning and Development Americas
Arctic Cooling Technology
7401 W. 129th Street
Overland Park, KS 66213

Subject: Training Proposal for Arctic Workplace English

We are pleased to propose an on-site Workplace English program to limited English-speaking Arctic employees. Following a language assessment to determine the appropriate level of instruction, we developed a comprehensive curriculum to improve each participant's English language skills and knowledge.

The program will offer an easy and quick way to learn limited amounts of everyday workplace English. Instruction will be divided into three components: Speaking in English, Listening in English, and Workplace-Specific English. Aspects of U.S. culture will also be included in each session to discuss some of the most fundamental aspects of everyday U.S. culture that are often misunderstood. Following is an overview of what we can offer Arctic:

> **Speaking in English:** This training component teaches participants how to say practical, common phrases, and to ask questions in American English. We will focus on pronunciation, expressions, and questions for use in everyday interactions.
>
> **Listening in English:** In this training unit, participants learn many common expressions, phrases, and questions used in everyday English.
>
> **Workplace-Specific English:** Here, participants learn how to comprehend and respond to workplace-specific language including phrases and terms that are regularly used in the workplace.

Each participant will receive a manual with CDs that can be used to maintain the language skills they have acquired.

We propose offering on-site classes limited to a maximum of 15 participants for 6 weeks to meet two hours, twice a week, for a total of 24 hours. A once-per-week meeting format for 12 weeks could also be provided. All instructional material will be provided to participants. The $3,300 estimated budget for this program includes curriculum development, instructional manuals, and on-site training. When the training program is complete, I will develop a follow-up plan to help each participant maintain and improve his or her English language skills.

We look forward to finalizing this training program. Thank you for the opportunity to present this proposal. We pride ourselves on providing quality training, language development, career planning, consulting, and economic development services.

Sincerely,

Phil Wegman
Program Director

CHAPTER HIGHLIGHTS

1. To communicate successfully both orally and in writing, you must recognize your audience's level of understanding.
2. Consider whether your audience is a specialist, semi-specialist, lay reader, or multiple when you communicate.
3. You can define terms parenthetically or in a glossary.
4. The tone you express in a communication should be determined by your audience's level of authority.
5. Be positive and avoid commands when you want audience involvement.
6. Use the pronoun "you" frequently to achieve audience involvement.
7. You have to consider the diverse makeup of your audience to communicate effectively.
8. Sexist language creates problems for communicators.
9. To communicate globally, consider your word choices carefully.
10. Be aware that cultural differences exist worldwide, and workplace communication has to adjust to these cultural differences.

WWW *Find additional audience exercises, samples, and interactive activities at http://www.prenhall.com/gerson.*

MEETING WORKPLACE COMMUNICATION CHALLENGES

CASE STUDIES

1. Dove Hill, GA, experienced severe thunderstorms on March 15, 2009. GAI (Goodwin & Associates Insurance Co.) insured much of the territory affected by this storm. Over 1,200 houses insured by GAI had water damage. As written, the homeowner's policy provided full-replacement coverage for water damage. In the homeowner's policies, water damage was limited to situations such as the following:

   - A hailstorm smashes a window, permitting hail and rain to access a home.
   - A heavy rain soaks through a roof, allowing water to drip through a ceiling.
   - A broken water pipe spews water into a home.

In contrast, homes which suffered water damage due to flooding were not covered, unless the homeowners also had taken out a separate flood damage policy from the National Flood Insurance Program. For insurance purposes, "flood" is defined as "the rising of a body of water onto normally dry land." For example, flood damage can include a river overflowing its banks or a heavy rain seeping into a basement. Of the 1,200 homes suffering water damage, only 680 homes qualified for full replacement coverage. The remaining homes suffered flood damage and were thus not covered.

Assignment

Write a letter from GAI to a homeowner denying coverage for the water damage claim. You are writing as a specialist to a lay reader. You want to maintain a good business relationship, despite the bad news. Build rapport while specifying the denial in lay terms.

2. Financial Trust & Annuity (FTA), a bank with over 100 sites throughout the Southwest, is growing at approximately 15 percent per year. To accommodate this growth, the bank needs to hire more tellers. Good business demands that FTA hire diversity: people with disabilities, employees from many races and cultures, men and women, and a variety of age ranges. Doing so will help ensure that the bank's

tellers represent people from all walks of life and that the bank will design its services to appeal to a diverse customer base. In addition, another goal is to hire a percentage of employees who can speak Spanish, Arabic, or Chinese. This will help FTA communicate with a key constituency of its emerging customer base.

Assignment

You are the human resources manager. To meet FTA's hiring needs, write a job advertisement. Include the following information:

- Teller job description
- Pay range
- Fringe benefits
- Work hours
- Qualification for the job (include preferred years of experience, needed skills, and language preferences)

In addition, write a statement addressing the bank's commitment to diversity. To accomplish these goals, research comparable job advertisements, either from the newspaper, online sources, or local bank bulletin boards.

ETHICS IN THE WORKPLACE CASE STUDY

Kimberly Evans works as a claims adjuster for an insurance company. Her office is a cubicle, centrally located in the middle of dozens of other open cubicles. She interacts successfully with her customers and coworkers on the phone and via e-mail.

Because she is a hardworking and dedicated adjuster, Kimberly often eats lunch in her cubicle. During her lunch break, she frequently browses the Internet on her computer. Her two passions are politics and religion, so she visits Web sites dedicated to these two topics. Her computer screen is visible to all of her coworkers, bosses, and customers who visit the office.

Question: Is it ethical for Kimberly to display her religious and political interests on company computers? Justify your answer based on information provided in Chapter 1.

INDIVIDUAL AND TEAM PROJECTS

Achieving Audience Involvement

1. Rewrite the following sentences, focusing on personalization by adding pronouns or names.

   a. The company will require further information before processing this request.

   b. It has been decided that a new procedure must be implemented to avoid further mechanical failures.

   c. The department supervisor wants to extend a heartfelt thanks for the fine efforts expended.

   d. I think you have done a great job. I want you to know that you have surpassed this month's quota by 12 percent. I believe I can speak for the entire department by saying thank you.

   e. If the computer overloads, simultaneously press Reset and Control. Wait for the screen command. If it reads "Data Recovered," continue operations. If it reads "I/O Error," call the computer resource center.

2. Achieve reader benefit in the following sentences by avoiding negative words, using positive words instead.

   a. We cannot lay your cable until you sign the attached waiver.

   b. John, don't purchase the wrong program. If we continue to keep inefficient records, our customers will continue to complain.

c. You have not paid your bill yet. Failure to do so might result in termination of services.

d. If you incorrectly quote and paraphrase, you will receive an F on the assignment.

e. Send me the requested information by January 12.

3. Rewrite the following flawed correspondence. Its tone is too negative and commanding, regardless of the audience. Soften the tone to achieve better audience involvement and motivation.

Date: October 15, 2009
To: Distribution
From: Darryl Kennedy
Subject: Fourth Quarter Goals

Due to a severe lack of discipline, the company failed to meet third quarter goals. To avoid repeating this disaster for the fourth quarter, this is what you all must do—**ASAP.**

1. Demand that the sales department increase cold calls by 15 percent.
2. Require weekly progress reports by all sales staff.
3. Penalize employees when reports are not provided on time.
4. Tell managers to keep on top of their staff, prodding them to meet these goals.

Remember, when one link is weak in the chain, the entire company suffers.
DON'T BE THE WEAK LINK!

Recognizing the Audience

Find examples of writing directed to specialists, semi-specialists, or lay audiences. To do so, read professional journals, find procedures and instructions, look at marketing brochures, read trade magazines, or ask your colleagues and coworkers for memos, letters, or reports.

Once you have found these examples, bring them to class. In small groups, discuss your findings to determine whether they are written for specialists, semi-specialists, or lay readers. Use the following Audience Evaluation Form to record your decisions.

Audience Evaluation Form					
Example #	1	2	3	4	5
Type (Circle One)	Specialist Semi-Specialist Lay	Specialist Semi-Specialist Lay	Specialist Semi-Specialist Lay	Specialist Semi-Specialist Lay	Specialist Semi-Specialist Lay
Criteria *Language* • Abbreviations • Acronyms • Jargon					
Content • General • Specific • Background					
Tone • Formal • Less Formal • Least Formal					

Recognizing Issues of Diversity

1. Rewrite the following sentences for multicultural, cross-cultural audiences.

   a. Let's meet at 8:30 p.m.

   b. The best size for this new component is 16 × 23 × 41.

   c. To keep us out of the red, we need to round up employees who can put their pedal to the metal and get us out of this hole.

   d. We need to produce fliers/brochures to increase business.

   e. The meeting is planned for 07/05/09.

2. Rewrite the following flawed correspondence. Be sure to consider your multicultural/cross-cultural audience's needs.

> To: Andre Castro, Barcelona; Sunyun Wang, Singapore; Nachman Sumani, Tel Aviv
> From: Ron Schaefer, New York
> Subject: Brainstorming
>
> I need to pick your brains, fellows. We've got a big one coming up, a killer deal with a major European player. Before I can make the pitch, however, let's brainstorm solutions. The client needs a proposal by 12/11/09, so I need your input before that date. Give me your ideas about the following:
>
> 1. What should we charge for our product, if the client buys in bulk?
> 2. What's our turnaround time for production?
> 3. What plans should we suggest for international rollout?
> 4. What kinds of training should we implement for the new systems?
>
> E-mail me your feedback by tomorrow, 1:00 p.m. my time, at the latest. Trust me, guys. If we boot this one, everyone's bonuses will be lost. ;)

Avoiding Biased Language

1. Rewrite the following sentences to avoid sexist language.

   a. All the software development specialists and their wives attended the conference.

   b. The foremen met to discuss techniques for handling union grievances.

   c. Every technician must keep accurate records for his monthly activity reports.

   d. The president of the corporation, a woman, met with her sales staff.

   e. Throughout the history of mankind, each scientist has tried to make his mark with a discovery of significant intellectual worth.

2. Rewrite the following flawed correspondence. Be sure to achieve effective audience understanding and involvement. To do so, avoid sexist language and define highly technical terms. Search online to find any abbreviations or acronyms that need defining (or invent definitions). Remember: Though the immediate readers might understand all of the terminology, other readers might not.

West Central Auditors
"Your Technical Engineering and Energy Resource Experts"
1890 River
Pocato, Idaho 89022

March 12, 2009

Marks-McGraw, Inc.
2145 Oceanview
Clackamas, Oregon

Gentleman:

We will visit your plant next week for the TEA. To ensure that our visit goes smoothly, we plan this procedure:

- Your plant foreman will have his engineer assemble his CAD drawings for the power plant at Brush Prairie.
- Our men will review these drawings against TE specs, such as specific cost estimates for energy usage and energy data.
- Our ER group will then help your men plan and prepare your facilities for technology changes mandated by the DOE's office of the EERE.
- We also want to review your engineer's plans for any building envelopes. Make sure he brings all relevant correspondence he has had with his clients.

This audit should take approximately seven man-hours. If we have to go off-site for lunch or breaks, the job will take even longer. Therefore, please ask your secretary to provide drinks and food for our six representatives. Tell her that we have no dietary concerns, so anything she provides will be greatly appreciated.

Sincerely,

Jim Wynn, Team Manager

DEGREE-SPECIFIC ASSIGNMENTS

List ten terms (jargon, acronyms, and/or abbreviations) unique to your degree program. Then, envisioning a lay audience, parenthetically define and briefly explain these terms. To test the success of your communication abilities, orally share these highly technical terms with other students who have different majors. First, state the term to see if they understand it. If they do not, provide the parenthetical definition. How much does this help? Do they understand now? If not, add the third step—the brief explanation. How much information do the readers need before they understand your highly technical terms?

PROBLEM-SOLVING THINK PIECES

1. HBM has suffered several lawsuits recently. A former employee sued the company, contending that it practiced "ageism" by promoting a younger employee over him. In an unrelated case, another employee contended that she was denied a raise due to her ethnicity. To combat these concerns, HBM has instituted new human resources practices. Their goal is to ensure that the company meets all governmental personnel regulations, including those required by the EEOC (Equal Employment Opportunity Commission), the FMLA (Family Medical Leave Act), and the ADA (Americans with Disabilities Act). HBM is committed to achieving diversity in its workplace.

   HBM now needs to hire a new office manager for one of its branch operations. It has three outstanding candidates. Following are their credentials:

   - **Carlos Gutierrez**—Carlos is a 27-year-old recent recipient of an MBA (Masters in Business Administration) from an acclaimed business school. At this university, he learned many modern business applications, including capital budgeting, human resources management, organizational behavior, diversity management, accounting and marketing management, and team management strategies. He has been out of graduate school for only two years, but his work during that time has been outstanding. As an employee at one of HBM's branches, he has already impressed his bosses by increasing the branch's market share by 28 percent through innovative marketing strategies he learned in college. In addition, his colleagues enjoy working with him and praise his team-building skills. Carlos has never managed a staff, but he is filled with promise.

- **Cheryl Huff**—Cheryl is a 37-year-old employee of a rival mortgage company, Farm-Ranch Equity. She has a BGS (Bachelor of General Studies) from a local university, which she acquired while working full time in the mortgage/real estate business. At Farm-Ranch, where she has worked for over 15 years, she has moved up the ladder, acquiring many new levels of responsibility. After working as an executive assistant for five years, she then became a mortgage payoff clerk for two years, a loan processor for two years, a mortgage sales manager for four years, and, most recently, a residential mortgage closing coordinator for the last two years. In the last two positions, she managed a staff of five employees.

  According to her references, she has been dependable on the job. Her references also suggest that she is a forceful taskmaster. Though she has accomplished much on the job, her subordinates have chaffed at her demanding expectations. Her references suggest, perhaps, that she could profit from some improved people skills.
- **Rose Massin**—Rose is 47 years old and has been out of the workforce for 12 years. During that absence, she raised a family of three children. Now, the youngest child is in school, and Rose wants to reenter the job market. She has a bachelor's degree in business, and was the former office manager of this HBM branch. Thus, she has management experience, as well as mortgage experience.

  While she was branch manager, she did an outstanding job. This included increasing business, working well with employees and clients, and maintaining excellent relationships with lending banks and realtors. She was a highly respected employee and was involved in civic activities and community volunteerism. In fact, she is immediate past-president of the local Rotary Club. Though she has been out of the workforce for years, she has kept active in the city and has maintained excellent business contacts. Still, she's a bit rusty on modern business practices.

Assignment

Who will you hire from these three candidates? Based on the information provided, make your hiring decision.

WEB WORKSHOP

1. As director of human resources for your company, your job is to write a non-discrimination policy. To do so, access an online search engine to find other companies' non-discrimination policies. Compare and contrast what you find. Then, based on your research, write your company's policy.
2. You are head of international relations at your corporation. Your company is preparing to go global. To ensure that your company is sensitive to multicultural concerns, research the cultural traits and business practices at five countries of your choice. To do so, access an online search engine and type in "multicultural business practices in ________" (specify the country's name). Report your findings either orally or in a memo.

QUIZ QUESTIONS

1. Why should you consider your audience before you speak or write in business?
2. Distinguish a specialist from a semi-specialist.
3. What is a lay audience?
4. What is a multiple audience?
5. Why is it important to consider your audience's role when you communicate?
6. How can you achieve audience involvement?
7. What do you consider when you recognize that your audience is diverse?
8. What are three things you should do to write effectively for a multicultural audience?
9. Why should you avoid idioms, slang, acronyms, and biased language in your workplace communication?
10. What is "you-usage," and why is it important?

CHAPTER 4

Planning Workplace Communication

▸REAL PEOPLE *in the workplace*

ImageSkill is a marketing firm that deals primarily with governmental and not-for-profit agencies. Their mission is to mold their clients' public relations and to promote an effective overall image. They accomplish this through an array of services including the following:

- Web design
- Desktop publishing
- Editing
- Multimedia production
- Events management
- Image control
- Communications training

ImageSkill recently hired **Nicole Stefani** to be a member of its public relations team. Nicole graduated with honors from the University of Indiana with a degree in journalism and a specialization in public relations. After graduation, she interned in Chicago at a renowned PR company, working with such clients as Nike, Crate & Barrel, and Banana Republic. She brings her expertise as a communications and media specialist to ImageSkill.

In Chicago, she worked as an advocate for businesses and nonprofit associations. She says, "I understand the attitudes and concerns of a community and public interest groups. I work to establish and maintain relationships with my clients, other businesses, and the media." Nicole will be called upon to draft press releases, make promotional films, plan conventions, interact with members of print and broadcast journalism, and prepare annual reports and proposals. ◂

CHAPTER GOALS

When you complete this chapter, you will be able to do the following:

1. Follow the P^3 process of planning, packaging, and perfecting to create effective workplace communication. This chapter discusses planning, the first step in the process. Chapters 5 and 6 discuss packaging and perfecting.
2. Examine your goals by considering audience, determining the channel of communication, and gathering data.
3. Gather data by answering reporter's questions, mind mapping, brainstorming and listing, outlining, storyboarding, creating organization charts, or performing research.

WWW

To learn more about the P^3 communication process, visit our Web Companion at http://www.prenhall.com/gerson.

Nicole's Communication Challenge

Nicole says, "As a new hire at ImageSkill, I need to make my first presentation to a potential client, Greenfield City Management. I have to write a proposal in response to their RFP (request for proposal) and make a PowerPoint presentation to the city council.

Though I am a trained public relations expert, making a professional presentation orally and in writing still is a challenge for me. I am challenged by the scope and importance of the project. My manager says that to accomplish the task, I must do the following:

- **Make an initial client contact**—Through telephone calls, e-mail messages, networking, or a preliminary meeting, I will get a general idea of what the client needs.
- **Meet face to face**—Through personal meetings with upper-level decision makers, I will collect information about the end-user's needs. I will ask questions, such as 'How can we meet your needs?' and 'How will you measure success?'
- **Create a proposal**—Following the initial meeting, I will write a proposal, complete with schedules, project plans, the project's scope, and a description of the deliverables.

This is a large job and an important challenge for me. I want to do good work because I am new to the company and hope to impress my manager. Of greater significance, I now live in Greenfield. I know that my efforts will contribute to Greenfield's growth, and that will be good for our business." Meeting the needs of ImageSkill and the potential client is a communication challenge for newly hired Nicole.

See page 107 to learn how Nicole met her "planning" communication challenge. In addition, see Chapter 5, "Packaging Workplace Communication Through Effective Document Design," pages 131–133, to learn how Nicole met her "packaging" challenge. Then in Chapter 6, "Perfecting Workplace Communication," pages 157–158, follow Nicole through her "perfecting" challenge.

AN INTRODUCTION TO THE P^3 COMMUNICATION PROCESS

Workplace communication is a major part of your daily work experience. A well-written memo, letter, report, or e-mail message or a well-constructed oral presentation gets the job done and makes you and your company look good. In contrast, poorly written correspondence and poorly conceived presentations waste time and create negative images.

However, recognizing the importance of workplace communication does not ensure that your correspondence will be successful. How do you effectively write the memo, letter, or report? How do you successfully produce the finished oral presentation? For some people, producing effective correspondence or presentations simply involves getting out a piece of paper or turning on a computer and leaping in. This is not the case for most business professionals, however. For most writers, the following might occur when they begin to compose correspondence.

1. **Blank page syndrome**—Writers stare at the blank page or screen until beads of sweat form on their foreheads, and they encounter the dreaded blank page syndrome.
2. **Lack of focus**—They write but, in doing so, they wander about aimlessly for several pages, lacking direction and focus.

What writers need is a solution to these and other problems. Workplace communicators need a methodical approach to communicating effectively. The P^3 process to effective communication entails *planning, packaging,* and *perfecting.* In this chapter, you will find a detailed discussion of how to plan your business communication. Chapter 5 discusses the second step of the process, packaging your communication. Chapter 6 shows you how to perfect your business communication, the third step in the process. The three steps are shown in the table below.

The P^3 Communication Process

Planning	Packaging	Perfecting

• Determine your goals
• Consider your audience
• Choose your communication channel
• Gather your data
	• Organize the draft according to some logical sequence
	• Format the content to allow for ease of access
	• Factor in the impact of technology
		• Revise
		  • Add missing details
		  • Delete wordiness
		  • Simplify word usage
		  • Enhance the tone of your communication
		  • Reformat your text for ease of access
		  • Proofread
		  • Correct errors

Planning the Communication

Before you can write your document or give your oral presentation, you must have something to say. By planning, you spend time generating information *prior* to writing the correspondence or composing the notes for your presentation. In the planning stage, you do the following:

- Determine your goals (What do you hope to achieve in the communication?)
- Consider your audience (Who will read your correspondence or hear your comments?)
- Determine the communication channel (Will you write a memo, letter, e-mail, or report? Will you give a formal speech using PowerPoint or an informal presentation?)
- Gather data (What content do you need for your communication?)

Packaging the Communication

Once you have planned, the next step is to package the communication by creating a draft (discussed in Chapter 5). To draft your document, do the following:

- Organize the draft according to some logical sequence
- Format the content to allow for ease of access
- Factor in the impact of technology

Perfecting the Communication

The final step, and one that is essential to successful communication, is to rewrite your draft, making it as perfect as it can be (discussed in Chapter 6). Revision allows you to perfect your memo, letter, e-mail, report, or oral presentation so you can write or speak effectively. Doing so requires that you follow a seven-step procedure:

- Add missing details for clarity
- Delete wordiness for conciseness
- Simplify your word usage
- Enhance the tone of your communication
- Reformat your text for ease of access
- Practice the speech or reread and review your text for content
- Proofread and correct any grammatical or textual errors

HOW TO PLAN YOUR COMMUNICATION

The first stage of the process allows you to *plan* your communication. Through planning, you accomplish many objectives, including

- Determining your goals
- Considering your audience
- Choosing the communication channel
- Gathering your data

Additional information:

See Chapter 1, "Communicating in the Workplace," Chapter 3, "Meeting the Needs of the Audience," and Chapter 16, "Research and Documentation."

Determine Your Goals

The first step is to determine your goals in the correspondence or presentation. You might be communicating to

1. *Inform* an audience of facts, concerns, or questions you might have
2. *Instruct* an audience by directing actions
3. *Persuade* an audience to accept your point of view
4. *Build rapport* by managing work relationships

These goals can overlap. You might want to inform by providing an instruction. You might want to persuade by informing. You might want to build trust by persuading. Following is a detailed discussion of each of these communication goals (Campbell et al.).

Communicating to Inform

Often, you will write letters, reports, and e-mail to inform. Examples of informative communication include the following:

- An e-mail message to inform your staff about the topics to be discussed in an upcoming meeting.
- A trip report to inform your supervisor about the conference you attended.
- A telephone call to inform your boss about a prospective client's needs.
- A follow-up letter to a vendor, answering questions about a product or service.
- A memo informing your coworkers about the corporate picnic, personnel birthdays, or new stock options available to them.

In these situations, your goal is to share information objectively.

Communicating to Instruct

Instructions will play a large role in your communication activities. As a manager, for example, you might need to give employees instructions for correctly following procedures. These could include steps for filling out performance evaluations, travel requests, trip reimbursements, contractor agreements, employee benefits, or payroll forms.

As an employee, you also will provide instructions to help your colleagues complete tasks. Suppose a coworker in the next cubicle asks you how to create file folders in your company's e-mail system. You will give this colleague step-by-step instructions for completing the job. Maybe the toner is low or paper is jammed in your office complex's printer. To correct this problem, you provide the needed instructions to a coworker.

Communicating to Persuade

If your goal in writing or speaking is to change others' opinions or a company's policies, you need to be persuasive. Following are examples of the need for persuasive communication:

- Writing a proposal, a brochure, or a flier to sell a product or a service.
- Preparing your annual job performance report to justify a raise or a promotion.
- Meeting with disgruntled customers to explain why they will not get a rebate or why your company cannot deliver a product on a certain date.
- Convincing a company to repair or replace a damaged product you bought.

Additional information:

See Chapter 13, "Persuasive Workplace Communication."

Communicating to Build Rapport

Building rapport (empathy, trust, and understanding) is an important component of your workplace communication. You need to consider other people's feelings and learn to work well with coworkers and customers.

For example, let's say that a supervisor reads a report from an employee. The supervisor then asks the employee for a follow-up meeting. In this meeting, the supervisor says, "We need more information."

What could be the problem here? This seemingly nonconfrontational comment could lead to hurt feelings. The employee could interpret this statement in several negative ways. The employee might believe that the supervisor is disagreeing with the report's findings. The employee could assume that the supervisor doubts the report's thoroughness. These assumptions might be false, but people are complex, and feelings can be hurt easily.

A supervisor's job, then, is not only to communicate a fact ("we need more information") but also to build rapport—to recognize sensitive issues and avoid threatening comments. The supervisor should say, "Thanks for the information. Now, what do you think our next step should be?" This comment empowers the audience while still asking for more information.

Recognizing the motivation for your correspondence makes a difference. Determining your goals allows you to provide the appropriate tone and scope of detail in your communication. In contrast, failure to assess motivation can cause communication breakdowns. What do you hope to achieve in your writing? What is your goal, your objective, or your purpose? To write effectively, understand your objectives. Decide whether you are informing, instructing, persuading, or building rapport.

Consider Your Audience

What you say and how you say it is greatly determined by your audience. Consider the following questions about audience when you plan your workplace communication:

- Are you writing up to management, down to subordinates, laterally to coworkers, or out to customers and vendors?
- Are you speaking to specialists (experts in your field), semi-specialists (people with some knowledge about your field), or a lay audience (people outside your work environment)?
- Are you speaking to a customer, team member, or boss?
- What issues of diversity should you consider (age, gender, physical limitations, or multiculturalism)?

Additional information:

See Chapter 3, "Meeting the Needs of the Audience."

Decide Which Communication Channel to Use

After you have determined your audience and your goals, decide how best to convey your message. Which communication channel will you use if you write or speak? If you write, should you provide a letter, a memo, a report, an e-mail, a Web site, a proposal, an instructional procedure, a flier, or a brochure? If you speak, will you do so in a teleconference, a videoconference, or a face-to-face meeting?

The scope of your content and your audience might determine your communication channel (written or spoken). If your content is relatively short and aimed at coworkers, perhaps an e-mail is sufficient. However, if your content is relatively long and your audience consists of potential customers, a multi-page proposal might be demanded. If your content is complex, a report is necessary.

If your goal is to receive a brief project update from a team, maybe a short meeting is the best approach. Will you need a teleconference, so individuals can see and hear each other? Wisely selecting the medium for your communication can positively or negatively impact your success. Table 4.1 shows oral and written communication channels.

Additional information:

See Chapter 1, "Communicating in the Workplace."

Gather Your Data

Once you have decided why you are communicating, who your audience is, and which communication channel to use, another goal is to gather data. In this chapter, and throughout the textbook, you will learn numerous ways to gather information, including the following:

- Mind mapping/clustering
- Brainstorming/listing
- Creating organizational ("Org") charts
- Outlining
- Storyboarding
- Researching
- Answering reporter's questions—Who, What, When, Where, Why, and How

See Table 4.2 for explanations and illustrations of gathering information.

Table 4.1 Communication Channels

Written Communication Channels	Oral Communication Channels
• E-mail • Memos • Letters • Reports • Proposals • Fliers • Brochures • Faxes • Internet Web sites • Intranet Web sites • Extranet Web sites • Instant messaging • Blogging • Job information (resumes, letters of application, follow-up letters)	• Leading meetings • Conducting interviews • Making sales calls • Managing others • Participating in teleconferences and videoconferences • Facilitating training sessions • Participating in collaborative team projects • Providing customer service • Making telephone calls • Leaving voice mail messages • Making presentations at conferences or to civic organizations • Participating in interpersonal communication at work • Conducting performance reviews

Table 4.2 Techniques for Gathering Information

Mind Mapping/Clustering

Envision a wheel. At the center is your topic. Radiating from this center, like spokes of the wheel, are different ideas about the topic. Mind mapping allows you to look at your topic from multiple perspectives and then cluster the similar ideas.

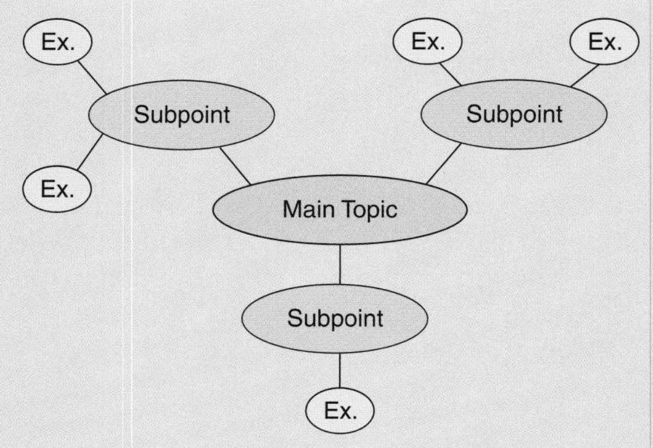

Answering Reporter's Questions

Journalists, both print and broadcast, use reporter's questions to gather data. These include "Who," "What," "When," "Where," "Why," and "How."

Reporter's Questions Checklist	
Who	Joe Kingsberry Sales Representative
What	Need to know • What our discount is if we buy in quantities • What the guarantees are • If service is provided on-site • If the installers are certified and bonded • If Acme provides 24-hour shipping
When	Need the information by July 9 to meet our proposal deadline
Where	Acme Radiators 11245 Armour Blvd. Oklahoma City, Oklahoma 45233 Jkings@acmerad.com
Why	As requested by my boss John, to help us provide more information to prospective customers
How	Either communicate with a letter or an e-mail. I can write an e-mail inquiry to save time, but I must tell Joe to respond in a letter with his signature to verify the information he provides.

Brainstorming/Listing

Either individually or with a group, you can randomly brainstorm ideas about a topic. Listing these ideas helps to organize your thoughts. This method works for almost all kinds of communication. Brainstorming/listing is especially valuable for collaborative activities, such as team projects and meetings.

<u>Brainstorming Notes for Improving Greenfield City Image</u>

- Prepare a questionnaire for city employees soliciting input
- Ask Chamber of Commerce for the types of potential businesses currently interested in moving to Greenfield
- Visit neighborhood home association meetings to learn about community concerns
- Develop city festivals around themes relevant to the city's history
- Meet with the Greenfield City Council to determine if the city will offer tax incentives to induce corporate interest in Greenfield
- Visit other cities (of similar size to Greenfield) which have modeled excellent growth procedures
- Include city residents and officials in the decision-making process
- Perform a S-W-O-T analysis to learn strengths, weaknesses, opportunities, and threats.

Table 4.2 (Continued)

Outlining

This traditional method of gathering and organizing information allows you to break a topic into major and minor components. This is an all-purpose planning tool, useful for almost any communication project—written or oral. For example, a training facilitator was asked to create a new workshop entitled "Dynamic Communication Skills," teaching secretaries, administrative assistants, and support staff how to sell ideas with confidence to colleagues and management. To plan this new workshop, the training facilitator created a topic outline.

1.0. The Communication Process
- 1.1 Planning
  - Planning Techniques
- 1.2 Packaging
  - All-Purpose Organizational Template
  - Organizational Techniques
- 1.3 Perfecting

2.0 Criteria for Effective Workplace Communication
- 2.1 Clarity
- 2.2 Conciseness
- 2.3 Document Design
- 2.4 Audience Recognition
- 2.5 Accuracy
- 2.6 Effective Writing Checklist

3.0 Tips for Effective Oral Presentations
- 3.1 Planning
- 3.2 Organizing
- 3.3 Practicing
- 3.4 Presenting
  - What to Have on Hand
  - How to Speak
  - How to Use Gestures
  - How to Avoid (Minimize) Nervousness
  - How to Handle Conflict

Storyboarding

Storyboarding lets you graphically sketch each page or screen of your text. This allows you to see what your document might look like. Below is an example of a storyboard for a long proposal report.

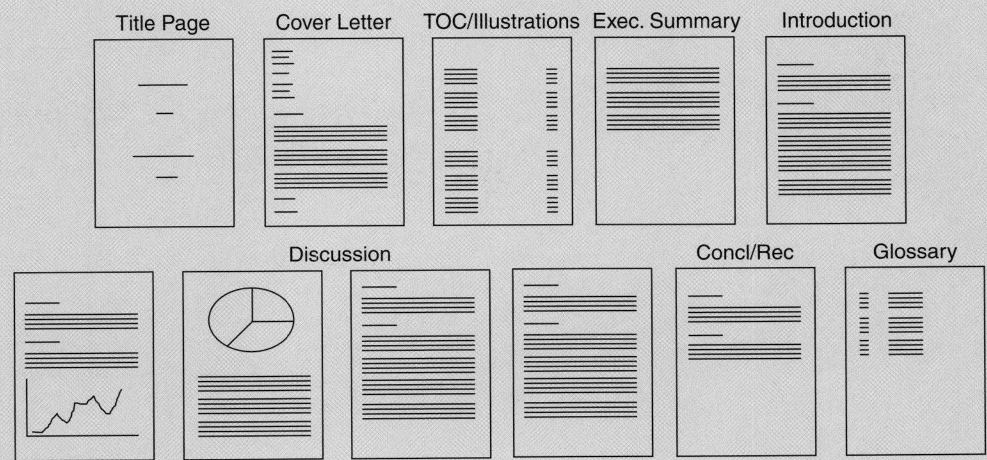

(Continued)

Organizational ("Org") Charts

An "org" chart graphically allows you to see the overall organization of a document as well as the subdivisions within it. You can use this planning method for Web sites, for example.

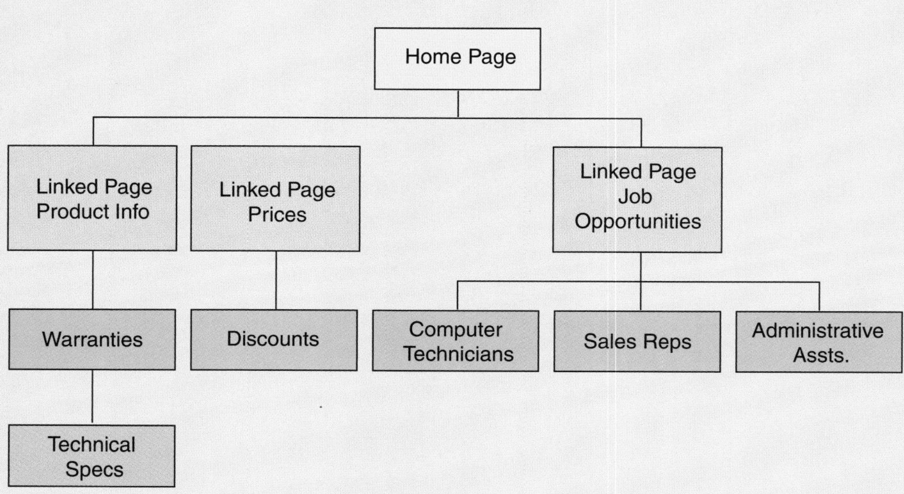

Researching

When you prepare a long report or study a difficult business-related problem, you generally will need to perform some formal research. Detailed information about research techniques and documentation standards can be found in Chapter 16. To the right are some Internet sites to help you research various topics.

Internet Research Sources

Search Engines	Metasearch Engines	Subject Directories
Google	Clusty	Librarians' Index
Yahoo	Dogpile	Infomine
Ask.Com	SurfWax	Academic Info
	Copernic Agent	About.Com
		Google Directory
		Yahoo

PLANNING CHECKLIST

_____ 1. Have you determined your communication goals? Do you want to
   - Inform?
   - Instruct?
   - Persuade?
   - Build rapport?

_____ 2. Did you consider your audience?
   - Are you writing or speaking to a specialist, semi-specialist, or lay audience?
   - What is your audience's role—supervisor, subordinate, colleague, customer, or vendor?
   - What issues of diversity should be considered—age, gender, physical limitations, multicultural, or cross-cultural?

_____ 3. Have you decided on a communication channel—e-mail, letters, memos, reports, videoconferences, teleconferences, telephone calls, group meetings, or face-to-face meetings?

_____ 4. What content (evidence, proof, questions, or comments) will you include in your oral presentation or written correspondence?

_____ 5. Have you chosen a planning technique to help you gather your data—reporter's questions, mind mapping/clustering, brainstorming/listing, outlining, storyboarding, or org charts?

_____ 6. After completing your planning, are you ready to begin packaging your thoughts? If not, which planning technique should you try again?

Nicole Met Her Challenge

To meet her communication challenge, Nicole used the *P*³ process.

Planning

To plan her proposal and presentation, Nicole first considered the following:

- Goals—respond to RFP and produce a winning proposal and oral presentation
- Audience—mayor, city council members, and city of Greenfield citizens
- Channels—e-mail messages, proposal, and PowerPoint presentation
- Data—responses to the RFP and interviews

According to Nicole, "I have to gather data and determine the client's needs. To *plan* my presentation, I will gather data about Greenfield's unique public relations challenges through a series of interviews with Greenfield city employees as well as city residents. When I conclude the interviews, I will collate the responses to find out what Greenfield's perceived strengths are (so the city can capitalize on these values), its weaknesses that need to be overcome, its opportunities for growth and improved service, and whatever demographic threats impede the city's chances for success." To plan her project, Nicole created a brainstorming list (Figure 4.1).

After she brainstormed to ensure that she was focused and maintained a clear direction, Nicole quickly jotted down ideas by answering reporter's questions (Figure 4.2). By answering reporter's questions (Who, What, When, Where, Why, and How), Nicole also considered her audience and decided which communication channel to use.

See Chapter 5 (Packaging Workplace Communication Through Effective Document Design), pages 131–133, to learn how Nicole packaged her rough draft.

FIGURE 4.1 Brainstorming List for Planning

Improving Greenfield City Image

- Prepare a questionnaire for city employees soliciting input
- Ask Chamber of Commerce for the types of potential businesses currently interested in moving to Greenfield
- Visit neighborhood home association meetings to learn about community concerns
- Meet with the Greenfield City Council to determine if the city will offer tax incentives to induce corporate interest in Greenfield
- Visit other cities (of similar size to Greenfield) which have modeled excellent growth procedures
- Include city residents and officials in the decision-making process

FIGURE 4.2 Reporter's Questions for Planning

Who's my audience?	Marc Shabbot, my boss. The city of Greenfield management (for the proposal, PowerPoint presentation, and eventual Web site). The city of Greenfield citizens (they'll see the Web site and be involved in the planning). Prospective business owners and new residents who we hope will move to Greenfield.
What is my goal?	I need to *persuade* and *inform*. The Web site, proposal, and oral presentation will be based on facts. However, I also want to persuade the audience. • We must sell Greenfield on ImageSkill's value. • We must convince new business owners and potential homeowners to move to Greenfield. • We must show current Greenfield residents that the city govt. cares about their health and well-being.
When's it due?	Greenfield needs the information by September 9.
Where?	I have to make the oral presentation in the Greenfield Board Room. I'll need to meet with Greenfield citizens first (here at ImageSkill's conference room?).
Why am I doing this/why is it important?	As requested by my boss Marc, to meet Greenfield's request for proposal. More important—I want to succeed. This is my opportunity to shine, to demonstrate my talents.
How?	The PowerPoint presentation and proposal. How can I get the info. I need? • Invite Greenfield citizens to participate through questionnaires?

Nicole says, "I find planning to be a valuable part of the P^3 process. With adequate planning, I can be sure that I am going in the right direction with my communication. I have to try to understand my audience prior to writing, and planning lets me do that. With an understanding of the audience, I can then collect sufficient details to continue with packaging and perfecting the written communication."

CHAPTER HIGHLIGHTS

1. Planning, packaging, and perfecting are the three steps in the P^3 communication process.
2. Planning allows you to generate quality information prior to writing the correspondence or composing the notes for a presentation.
3. Planning allows you to avoid the blank page syndrome and lack of focus.
4. Packaging lets you organize and format your content.
5. Perfecting is the revision stage in the process.
6. Before you write a document or give a speech, you need to know why you are communicating, so examine your purposes.
7. You need to determine your goals when you plan communication. These could include persuading, instructing, informing, and building rapport.
8. Determine whether your audience is a specialist, semi-specialist, or lay person.
9. When you plan your communication, determine how your communication channel (e-mail, Web sites, PowerPoint, teleconferences, or videoconferences) will affect the presentation of content.
10. Planning techniques help you gather information. You can answer reporter's questions, mind map, brainstorm/list, flowchart, outline, storyboard, or create organizational charts.

WWW *Find additional planning and P^3 exercises, samples, and interactive activities at http.//www.prenhall.com/gerson.*

MEETING WORKPLACE COMMUNICATION CHALLENGES

CASE STUDIES

1. Kate O'Brien has just returned from a business trip, where she traveled to troubleshoot a corporate problem. Her international company SportingStyle, which markets sports apparel, just purchased a smaller sporting goods company. The merger has led to challenges with labor relations, patent laws, licensing, and clashes in management styles.

During her trip, Kate needed to interview the new company's managers and workforce to determine areas needing improvement. After the trip, Kate had to share her findings with her project team and report decisions to SportingStyle management.

Assignment

Review the following list of communication challenges that Kate confronted, and decide which communication channel she needed to use to complete her tasks. For example, will she use written or oral communication? If she needs to write, will she write a memo, an e-mail, a letter, a report, and so on? If she speaks, will it be a professional presentation with a full-range of audio-video aids, or will a casual conversation (face-to-face or even on a telephone) work?

- She needed to make travel arrangements before the trip (hotel and air and ground transportation).
- She had to confer with her project team in the home office while she was traveling.

- She had to interview the new management and workforce during her trip.
- She needed to share her company's management philosophy with the acquired company's entire staff.
- She had to document her expenses once the trip ended.
- She had to report her findings to corporate management in the home office.
- She had to present the information at a stakeholders' meeting.

2. Electronic City is a retailer of DVDs, televisions, CDs, computer systems, cameras, telephones, fax machines, printers, and more. Electronic City needs to create a Web site to market its products and services. The content for this Web site should include the following:

Prices	Store Hours	Warranties	Service Agreements
Job opportunities	Installation fees	Extended holiday hours	Discounts
Technical support	Product information	Special holiday sales	Delivery fees

Assignment

Review the above list of Web site topics. Using an organizational chart, decide how to group these topics. Which will be major links on the Web site's navigation bar? Which will be topics of discussion within each of the major links? Once you have organized the links, sketch the Web site by creating a storyboard.

3. You are the special events planner in the marketing department at Thrill-a-Minute Entertainment Theme Park. You and your project team need to plan the grand opening of the theme park's newest sensation ride—*The Horror*—a wooden roller coaster that boasts a 10 g drop. What activities should your team plan to market and introduce this special event?

Assignment

Using at least three of the planning techniques discussed in this chapter, gather ideas for a day-long event to introduce *The Horror.* Report your findings as follows:

- In the brief report to your marketing department boss, explain why you are writing, give options for the event and clarify which techniques you used to gather ideas, and sum up by recommending what you think are the best marketing approaches.
- Write an e-mail to your teacher providing options for the event and explaining which techniques you used to gather data.
- Give an oral presentation in class providing options for the event and explaining which techniques you used to gather data.

ETHICS IN THE WORKPLACE CASE STUDY

Nicole Stefani is a public relations employee for ImageSkill, a corporate "make-over" company that works to improve a client's image. One of ImageSkill's new clients, the city of Greenfield, has asked for help. Nicole began planning by inerviewing the city's residents and business owners and workers. She used a S-W-O-T analysis, focusing on the city's <u>S</u>trengths, <u>W</u>eaknesses, <u>O</u>pportunities for improvement, and any <u>T</u>hreats to the city's success.

Prior to the interviews, Nicole clarified her goals to the city participants. They all were made aware that their comments eventually would be used for reporting purposes. During the interviews, to ensure accuracy, Nicole either taped what the people said or took very thorough notes. Before writing her rough draft, Nicole always double checks with her sources of information to be sure that their quotes are correct.

Nicole has contacted one individual, Burt Knoblauch, for verification. Upon hearing what Nicole plans to use as his quote, Burt first denies having said it. Nicole, however, has his comment on tape and plays it for him. Burt then tells Nicole that he does not want her to use his comments because his boss and his neighbors might be offended. Nicole reminds Burt that she had told him and all *S-W-O-T* participants that their comments would be used in a follow-up report. She even has signed release forms for Burt and others.

Question: What should Nicole do? She knows for a fact that her quote is accurate. In addition, Burt's quote is essential to her report, since his comments highlight a key problem facing the city. Still, Burt has told her that using the quote could cause him problems at work and in his neighborhood.

INDIVIDUAL AND TEAM PROJECTS

Practicing Planning Techniques. Take one of the following topics. Then, using the suggested planning techniques, gather data.

1. **Reporter's questions.** To gather data for your resume, list answers to the reporter's questions for two recent jobs you have held and for your past and present educational experiences.
2. **Mind mapping/clustering.** Create a mind map of your options for obtaining college financial aid.
3. **Brainstorming/listing.** List five reasons why you have selected your degree program, why you have chosen the college or university you are attending, or why you have (or have not) pledged a fraternity or sorority.
4. **Outlining.** Outline your reasons for liking or disliking a current or previous job.
5. **Storyboarding.** If you have a personal Web site, use storyboarding to graphically depict the various screens. If you don't have such a site, use storyboarding to graphically depict what your site's screens would include.
6. **Creating an org chart.** Use an org chart to show the hierarchical structure of your place of business.

DEGREE-SPECIFIC ASSIGNMENTS

The following degree-specific activities allow you to research your field on the Internet.

1. **Marketing and Public Relations.** Women-owned businesses represent one of the fastest growing segments of the marketplace. Why? Access the *Online Women's Business Center* (http://www.onlinewbc.gov/). Research this site to find unique ways to market a woman-owned business. Share this information with your classmates in a brief oral presentation.
2. **Business Information Systems.** E-business is booming. Almost every company has a Web presence, and every company uses computer information systems for internal and external communication. However, computer systems are prone to security risks. Research the Internet to find information related to computer security. You could focus on acquiring electronic evidence for criminal cases, corporate online security policies, ways to block Internet viruses, or network intrusion devices. Share this information with your classmates in a brief oral presentation.
3. **Accounting.** As an accountant, you need to stay informed about current topics related to your field. Access the American Institute of Certified Public Accountants Web site (http://www.aicpa.org/index.htm). Each month, this site provides a CPA News Center with "News Alerts" and a Spotlight Area with "Special News for CPAs." Research three to five newsworthy articles. In a brief oral presentation to your class, share your findings.

4. **Management.** Are you considering starting your own business? Access the Small Business Help Center Web site (http://www.helpbizowners.com/index.html) to learn more about start-up procedures. This site provides links on buying and selling businesses, finance options, franchising, marketing, taxes, technology, and more. Review the information provided in this site and share your findings in a brief oral presentation.
5. **Finance.** Taxes are a concern for all businesses. To learn more about taxes (filing deadlines, forms, calendars, and estimators), access 1040.com (http://www.1040.com/). Click on the "Newsroom" link for the most recent publications about tax information. Read three to five articles and share your findings in a brief oral presentation.

PROBLEM-SOLVING THINK PIECES

1. **Determining Your Goals for Communicating.** When you communicate, you will do so to *persuade, instruct, inform,* and/or *build rapport*. From the following list, determine the goals for the communication.
   a. You write a sales letter about your home-owned painting and wallpapering company.
   b. As a supervisor, you write an e-mail to motivate a team member or subordinate who is having trouble accomplishing a task.
   c. You are bothered by a departmental policy. You write a proposal to your boss, asking him or her to accept your point of view about changing this departmental policy.
   d. As a sales manager, you rush to an employee's aid by meeting with an angry customer and trying to pacify him or her.
   e. As a mid-level manager, you write an e-mail to an employee, sharing good news you have just received from your supervisor, who agreed to give the employee a raise.
2. **Considering Your Audience.** Your audience determines much of your content in written and oral communication. Is your audience a specialist, semi-specialist, or lay person? This will help you decide whether you need to define terms. Is your audience management, colleagues, subordinates, customers, or vendors? This might affect the tone of your communication. What if your audience is a combination of these roles? How will that change your comments? Finally, is your audience multicultural? This will require that you avoid idioms and regionalisms. From the following topics
   - Determine who your audience is
   - Decide how this audience will affect what you say and how you say it (what tone; what level of detail; what level of word usage)
   - Explain your decisions
     a. You are a sales representative. You are giving a product demonstration to employees from the purchasing department.
     b. As an employee, you are writing your annual Individual Development Plan, explaining your five-year goals.
     c. You need to write a sales brochure for new services your home mortgage company is marketing.
     d. Your company is changing its insurance provider. As benefits coordinator, you must visit all of the company's departments to answer questions about the new benefits package.
     e. Your boss has asked you to create the company Web site. In it, you will list job openings, price changes, an online order form, service information, and technical specifications for your company's product line.

3. **Choosing a Communication Channel.** From the following topics, determine which communication channel you should use and explain your choice.
   a. You need to ask your boss a question about a potential contract with a vendor, and you need the answer now. Your boss is in New York, and you are in Sacramento.
   b. You need your boss's signature on a vendor contract, and you need it now. Your boss is in New York, and you are in Sacramento.
   c. John has received complaints from customers. You need to share this information with John, not only so he knows what has been said but also so you can get his response.
   d. As a public relations employee, you are tasked with promoting an upcoming charity 5K run to support osteoporosis. You need to tell key stakeholders (runners, individuals with osteoporosis, and their families and friends) about the time, place, route, costs, prizes, and so on, regarding this race. The information will be placed in sporting goods stores, hospitals, and health clinics.
   e. Your team is writing a proposal. The team consists of over 20 employees in four locations (Houston, Milwaukee, Boston, and London). The team needs to see topographical maps of a construction site and brainstorm the proposal's contents. Everyone's input is required.

WEB WORKSHOP

USA.gov allows you to research a wide variety of topics, such as education and jobs, benefits and grants, consumer protection, environment and energy, science and technology, and public safety and health. This Web site also provides information on breaking news. Access USA.gov and research a topic relevant to your career goals. Write a memo or report to your instructor summarizing your findings.

QUIZ QUESTIONS

1. How does poor workplace communication affect your job performance?
2. When you try to write workplace correspondence without first planning, what two negative things can happen to you?
3. What are the three parts of the *P*[3] process?
4. When you plan your communication, what four points should you consider?
5. What are different communication channels you can use to convey content to your audience?
6. Why would you write an e-mail message versus a report (and vice versa)?
7. What are the four goals of workplace communication?
8. What types of audience do you consider in workplace communication?
9. What are the different techniques for gathering data?
10. List four Internet metasearch engines.

CHAPTER 5

Packaging Workplace Communication Through Effective Document Design

▸REAL PEOPLE *in the workplace*

Nicole Stefani says, "As a new hire at ImageSkill, I need to write a proposal in response to an RFP (request for proposal) from Greenfield City Management and make a PowerPoint presentation to the city council." Nicole has spent the last three weeks gathering data for her proposal to the city council.

Nicole initially made contact with the client. She visited with them on the telephone and used e-mail correspondence to gather information that the client considered important. Nicole then reviewed their needs with her boss, Marc Shabbot, who suggested some additional questions she could ask in face-to-face meetings. Through two personal meetings with upper-level decision makers, she collected information about the end-user's needs. She asked questions, such as "How can we meet your needs?" and "How will you measure success?"

After Nicole gathered data from the client, she had to write the proposal complete with schedules, project plans, the project's scope, and a description of the deliverables. Marc was reviewing this proposal, so Nicole wanted to do a good job.

Nicole said, "I know the importance of this proposal for ImageSkill, the city council, and for me. This is my first big job packaging a proposal for ImageSkill, and I want the proposal to be well organized, well designed, well written, and persuasive."◂

Nicole's Communication Challenge

Nicole says, "I have so many things to consider as I write my first formal proposal to a potential client. I've always been a good writer in my college classes, but writing for a boss and a client is different. I'm now getting paid for how I communicate on the written page.

As I package the draft of my proposal, I'm going to consider the following aspects of successful workplace writing:

- **Organization of content**—How will I organize the content of the proposal? I know that I have many options, such as analysis, spatial organization, chronology, importance, comparison/contrast, problem/solution, and cause/effect. However, which ones are best for the various parts of my proposal?
- **Page layout and design**—I know that workplace writing should never look like an essay or a research paper. How can I help my audience access the parts of my proposal? What types of visual aids will make my proposal look professional and enjoyable to read? I want to make sure that easy-to-read proposals become a hallmark of my on-the-job communication skills.

CHAPTER GOALS

When you complete this chapter, you will be able to do the following:

1. Package your communication.
2. Organize your content through analysis, spatial organization, chronology, importance, comparison/contrast, problem/solution, and cause/effect.
3. Consider page layout (the document's design) to help you communicate more effectively.
4. Understand the impact of technology on workplace communication.

WWW

To learn more about the *P*³ communication process, visit our Web Companion at http://www.prenhall.com/gerson.

- **Impact of technology on workplace communication**—Technology will be my ally as I attempt to persuade the city council. Which communication channel will work best for this proposal?"

Packaging the draft of her proposal is a challenge for Nicole since this is her first written document for ImageSkill.

See pages 107–108 in Chapter 4, "Planning Workplace Communication," to learn how Nicole met her "planning" challenge. In addition, see pages 131–133 in this chapter to learn how Nicole met her "packaging" challenge. Then in Chapter 6, "Perfecting Workplace Communication," page 142, follow Nicole through her "perfecting" challenge.

PACKAGING YOUR COMMUNICATION

The communication process is composed of three key components: planning, packaging, and perfecting, as shown in the table below. (Planning is discussed in Chapter 4; perfecting is discussed in Chapter 6.)

The P^3 Communication Process

Planning	Packaging	Perfecting

• Determine your goals
• Consider your audience
• Choose your communication channel
• Gather your data
	• Organize the draft according to some logical sequence
	• Format the content to allow for ease of access
	• Factor in the impact of technology
		• Revise
		  • Add missing details
		  • Delete wordiness
		  • Simplify word usage
		  • Enhance the tone of your communication
		  • Reformat your text for ease of access
		  • Proofread
		  • Correct errors

Chapter 4 teaches you how to plan your communication, the first stage of the communication process. After completing the first step in the communication process, you should have sufficient information to begin drafting your text. If it is just a jumble of thoughts, however, will your audience be able to clearly understand your ideas?

Every good product or service must be packaged successfully. That includes automobiles, makeup, clothing, hair conditioning products, telecommunications equipment, food, and many more items. It also includes services such as banking, insurance, consulting, biomedical engineering, and accounting.

When companies package their products and services effectively, they focus on *visual appeal* and *ease of use*. The same holds true for written and oral communication. You can help your audience understand your ideas by packaging them effectively—the second step in the communication process. Packaging lets you put your thoughts into words. In addition, you need to package the content (the draft) on a page or screen, taking into consideration the following:

- **Visual appeal**—If text has visual appeal, it also is easier for an audience to find and use the information you have provided.
- **Ease of use**—When you use various methods of organization in drafting your text, readers can follow your train of thought more easily.

BEFORE DRAFTING YOUR TEXT

Once you have gathered your data, determined your objectives, considered your audience, and chosen your communication channel, perform the following tasks.

Review Your Planning

Additional information:

See Chapter 4, "Planning Workplace Communication."

Before you begin to write, review the information you gathered during the planning stage. By reviewing your mind mapping, outline, reporter's questions, or organization chart, you can see if you have missed any important details. This review also gives you one more reminder of why you are writing (your goals) and what you want to say (your content).

Review the Criteria for Each Type of Communication

This textbook provides easy-to-follow criteria for writing many different workplace documents or for making oral presentations. Before you begin your communication, review the criteria for a memo or letter (Chapter 10), an e-mail message (Chapter 9), a report (Chapters 17 and 18), or other written documents. Review tips for successful PowerPoint presentations (Chapter 7) or the key components of a proposal (Chapter 19). As part of packaging your communication, you should remind yourself of the key components for different types of documentation.

DRAFTING YOUR TEXT

When you are ready to package your content, begin the draft. To help your audience understand and easily access your content, you must do the following:

1. Choose effective *methods of organization*
2. Consider *page layout*
3. Factor in the *impact of technology*

Methods of Organization

Communicators usually know where they are going, but the audience does not have this same insight. When readers pick up your document, they can read only one line at a time. When listeners first hear you speak, they do not know where you are going to lead them. Your audience knows what you are saying at the moment, but they do not know what your goals are. They can only hope that in your written document or oral presentation, you will lead them logically.

To avoid confusing your audience, you need to organize your thoughts. As with planning, you have many organizational options that will let you package your content effectively. These include the following:

- Analysis
- Spatial organization
- Chronology
- Importance
- Problem/Solution
- Cause/Effect
- Comparison/Contrast

Analysis

Large topics can be difficult for audiences to understand. If you break a topic into smaller pieces, however, the content often will make more sense. For example, what does your employee benefits package include? What does your homeowner's insurance cover? What are the benefits of your cell phone calling plan? What courses do you need to take to complete your degree program? When you break any of these topics down into their components, the topics are easier to understand. Analyzing a topic helps you focus on the smaller pieces that make up the whole. For example, Figure 5.1 provides information about cell phone rates found online at one company's Web site.

FIGURE 5.1 Analysis of Cell Phone Rates

Home
Pricing
Services
Contacts

Cell Phone Rates

Look at what AaBb Telecom offers you as part of your cell phone services:

Basic Rates	
Telephone Services	**Fees**
• Basic service (line charge) • Call waiting • Caller ID • Voice mail	This all comes with your basic cell phone charge!

Optional Rates		
Telephone Options	**Fees**	
Long distance calls within the United States	$5.50 unlimited (no roaming fees)	
Long distance calls abroad	When You Call	Our Low Rates (per minute)
	7 a.m.–Midnight	6 cents
	12:01 a.m.–6:59 a.m.	4 cents
	Weekends	4 cents
	Holidays	Same rates apply

Special Fees	
Late payment fees	$3.00 for each day your payment is late.

Notice how the topic of telephone rates is analyzed—broken down into three smaller components: Basic Rates, Optional Rates, and Special Fees. Then each of these topics is further analyzed, broken into smaller pieces. Doing so allows the reader to understand the subject more easily.

Once a topic has been broken into smaller pieces, the next step of analysis is to determine the significance of the content. You should ask yourself questions such as

- "What does this information mean?"
- "What's the impact of this information?"
- "How will this information affect my decisions?"
- "What more information do I need before making a decision?"

Spatial Organization

Additional information:

See Chapter 17, "Short, Informal Reports."

To organize a topic spatially, consider how the topic looks if viewed from left to right, right to left, inside to outside, or bottom to top. This method of organization will not work with all topics. Spatial organization is useful when providing physical descriptions of products or perhaps reporting on work-related accidents or events. For example, an accident report would factor in geographical location—north, east, south, and west. The following example, taken from an engineer's report, spatially describes the condition of a concrete driveway.

example

Italics and boldface indicate the spatial organization of this report.

The poured concrete driveway shows the following flaws:

- The ***top*** wearing slab, which should be a uniform 4″ deep, ranges in thickness from 3½″ to 1¼″.
- The ***middle*** heating units are broken in six places, distributing the heat unevenly.
- The ***bottom*** structural slab consists of badly corroded steel rebars.

In contrast, notice how the home burglary claims report shown in Figure 5.2 uses compass directions for its spatial organization.

Chronology

You might use chronological order to organize many types of communication. If you are a recording secretary for a board, agency, sorority, fraternity, or city council, you could use chronology to report meeting minutes. This would entail noting who said what first, who responded next, and so forth. The minutes also could be organized chronologically by reporting on which topics were discussed first, next, following, and last. Reverse chronology is used for many resumes. In a reverse chronological resume, you discuss your current job or educational status first. Then, you list your prior employment and educational accomplishments.

Additional information:

See Chapter 9, "Electronic Communication: E-Mail Messages, Instant Messages, Text Messages, and Blogging."

Chronology also is a valuable way to organize trip reports. You could itemize and discuss which events or conference presentations you attended, citing them first to last. Chronology is mandatory if you are discussing the steps in a procedure. Finally, agendas follow a chronological order. Figure 5.3 is an e-mail message sent by a company's facilities manager, reporting the chronological steps he followed when dealing with a power outage.

Importance

If you organize your text by importance, you tell readers which parts of the discussion are most important and which are less important. For example, if you are a purchasing agent, you might want to write a memo to a manager, highlighting options for the purchase of a digital camera. You have done the research, and you know which choices are better. In your memo, you can sway your reader's opinion by listing the best choice first and the weaker choices last. Figure 5.4 is a letter written by the chairperson of the board of a publicly traded utility company. In this letter to the other board members, the chair suggests ways in which the organization can meet stockholder concerns regarding changes that are taking place in the industry.

Additional information:

See Chapter 10, "Traditional Correspondence: Memos and Letters."

FIGURE 5.2 **Report Using Spatial Organization**

Claims Report

Date: January 16, 2009
To: Larry Lerner, Regional Manager
From: Susan McGarvey
Subject: Claims Report on Burglary at 1600 Oaklawn

Introduction:

Time/Date When Claim Was Filed: 8:45 p.m./January 15, 2009
Policy Number: 3209-6491
Effective Date: May 15, 2008
Policy Holder(s): Mr. John Stamper and Mrs. Carol Stamper
Mailing Address: 1600 Oaklawn
City/State/Zip: Caligon, MS 34267
Phone: 314-555-2424

Description:

Narrative: The residents (Mr. John Stamper and Mrs. Carol Stamper), returning from an evening out, found their house broken into, vandalism, and missing items.

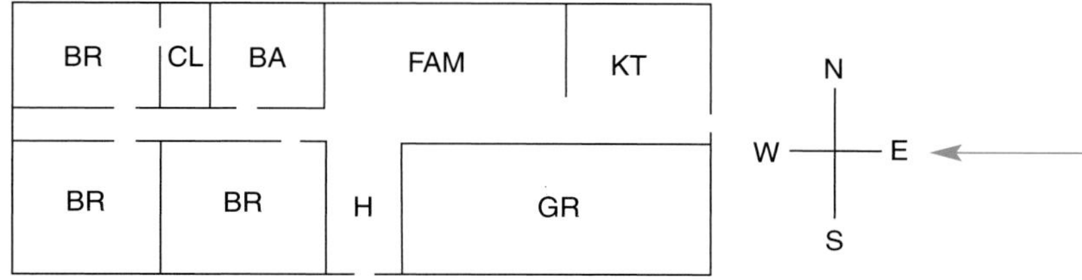

Compass directions (both visual and written) can be used in reports to help you organize your thoughts and help the reader visualize the event or activity.

Exit/Entry: Entry appears to have been made by cutting an L-shaped hole in the northwest bedroom (BR) window. The perpetrators then apparently left the BR and traveled due south to the southwest BR, where vandalism occurred. Then, the culprits walked east down the hall to the family room (FAM). When the Stampers returned home, they found their garage (GR) door open, suggesting that the perpetrators exited south from their home.

Missing Items/Estimated Costs:

- Sony 42" plasma television ($1,400)
- $200 cash that had been laying on the family room desk
- A Sony Play Station ($499)
- A DVD player ($60)

Vandalism/Costs:

- The southwest BR had random spray-paint on the ceiling and walls ($200—materials and labor). Please see the attached photographs.
- Window repair ($125—material and labor)

Status:

This claim has been given to Claims Adjuster Mary Woltkamp for disposition. I have informed Mary that turnaround time on claim clearance must be two weeks maximum to meet our company's new mandates for customer satisfaction.

FIGURE 5.3 **E-Mail Message Using Chronology**

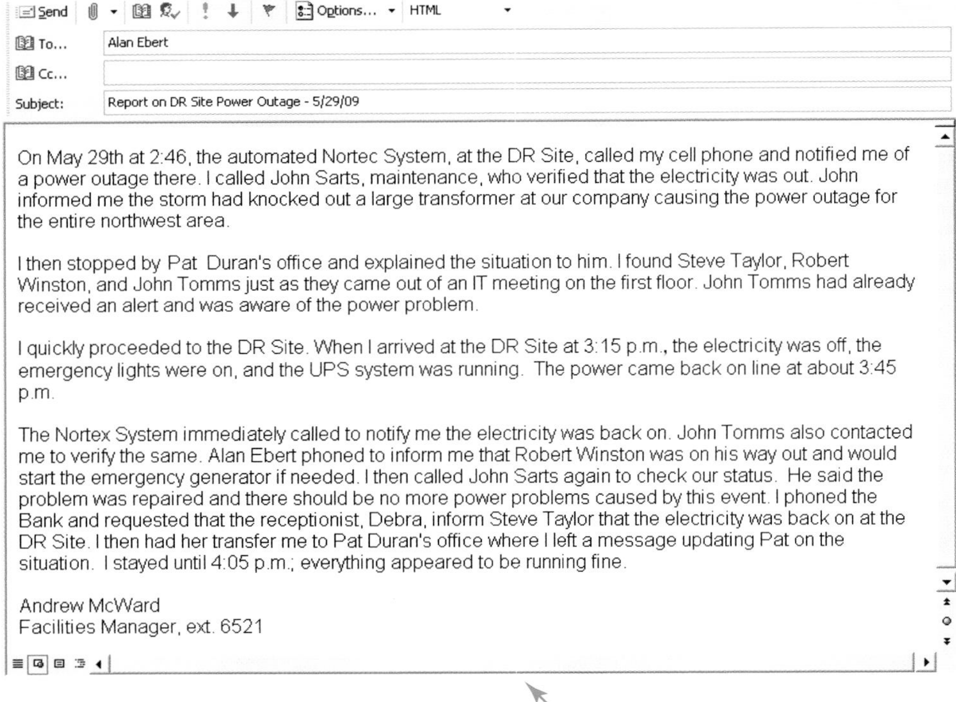

Send | Options... | HTML

To... Alan Ebert

Cc...

Subject: Report on DR Site Power Outage - 5/29/09

On May 29th at 2:46, the automated Nortec System, at the DR Site, called my cell phone and notified me of a power outage there. I called John Sarts, maintenance, who verified that the electricity was out. John informed me the storm had knocked out a large transformer at our company causing the power outage for the entire northwest area.

I then stopped by Pat Duran's office and explained the situation to him. I found Steve Taylor, Robert Winston, and John Tomms just as they came out of an IT meeting on the first floor. John Tomms had already received an alert and was aware of the power problem.

I quickly proceeded to the DR Site. When I arrived at the DR Site at 3:15 p.m., the electricity was off, the emergency lights were on, and the UPS system was running. The power came back on line at about 3:45 p.m.

The Nortex System immediately called to notify me the electricity was back on. John Tomms also contacted me to verify the same. Alan Ebert phoned to inform me that Robert Winston was on his way out and would start the emergency generator if needed. I then called John Sarts again to check our status. He said the problem was repaired and there should be no more power problems caused by this event. I phoned the Bank and requested that the receptionist, Debra, inform Steve Taylor that the electricity was back on at the DR Site. I then had her transfer me to Pat Duran's office where I left a message updating Pat on the situation. I stayed until 4:05 p.m.; everything appeared to be running fine.

Andrew McWard
Facilities Manager, ext. 6521

Explaining an event in a chronological order allows you to create a narrative for the audience. This narrative tells the story of what occurred in an easy-to-understand sequence.

Comparison/Contrast

One way to make decisions is by comparing and contrasting options. Which car should you buy? In which degree program should you enroll? Which job offer should you take? By comparing and contrasting topics or options, you can help your audience see the pros and cons of the choices available. The following example, taken from a company's Web site, uses comparison/contrast to clarify product information and prices for men's and women's outerwear.

example

Men's and Women's Styles/Sizes

	Jackets	Sizes	Prices
Men's	Windbreakers	S-M-L-XL	$35.50
	Trench Coats	S-M-L-XL-XXL	$45.50
	Field Coats	S-M-L-XL	$55.50
	Pea Coats	S-M-L-XL-XXL	$65.50
Women's	Windbreakers	P-S-M-L	$25.50
	Trench Coats	P-S-M-L	$35.50
	Blazers	P-S-M-L	$45.50

Problem/Solution

Another way to organize your content is to focus on problems and solutions. For instance, if you work in customer service, you might have to respond to consumer complaints.

FIGURE 5.4 Letter Organizing Content by Importance

Arrowhead Utilities
1209 Arrowhead Dr.
Lake Washington, IA 39921

September 12, 2009

Ms. Stacy Helgoe
1982 Evening Star Rd.
Lawrence, KS 78721

Dear Ms. Helgoe:

We have experienced rough times lately in the utilities industry. Prices for oil, gas, water, and coal have gone up by over 50 percent, but statewide regulations have disallowed us from raising rates to meet these costs. This has led the Board to consider laying off workers, reducing our geographic area of coverage, limiting our customers' options for service, merging with a utilities competitor, and providing fewer hours of service ("enforced brownouts").

To ensure continued good relations, we should inform our stakeholders of these decisions. How should we proceed? Here are our options:

1. Implement the Board suggestions without notifying the stakeholders directly. A follow-up article in the local newspaper's business section could then report the activity.
2. Implement the Board suggestions and provide a personalized letter to each stockholder detailing the causes and our goals.
3. Present information to the stakeholders at an abbreviated annual meeting, asking for questions and answers, and then taking a vote on which options to pursue.
4. Hold four "small town" meetings prior to any vote or implementation. This will allow stakeholders ample opportunity for discussion.

My suggestion is #4, the best choice to ensure large scale stakeholder buy-in and empowerment. Any other approach, I believe, will create distrust on the part of our primary audience. I want to hear from each of you regarding your thoughts. Please call me at 914-555-7676, ext. 234 by September 21.

Sincerely,

Christy Pieburn
Board Chair

The itemized body in this letter uses analysis to provide the reader options. In the discussion paragraph, the body points are itemized from least important to most important. Notice that point #4 is considered "best choice," thus the most important point.

To do so, you might focus on the problem your customer has identified and then present your company's solutions. This could occur in the following instances:

- **Insurance**—The customer has suffered home or automobile damages (the problem). Your insurance company needs to write a letter to the client, providing solutions to the problem.
- **Banking**—A customer has lost a wallet. Cash and credit cards are all missing. Your job is to tell this client which steps to follow to solve this problem.

- **Hospitality management**—You work at a hotel. A recent visitor to your hotel says the bill was incorrect. This could include telephone services, in-room movies, room service, or being billed for nights not stayed. You must resolve such issues. A problem/solution approach would work well.

Proposals often use a problem/solution method of organization. If you are giving an oral presentation as part of a proposal, for example, you might want to mention the customer's problem and then highlight the many ways in which your company will make improvements. Writing your draft using this method of organization not only helps your audience follow your ideas, but also it allows you to maintain your focus as you write. Following is a proposal's executive summary, which uses problem/solution as a means of organization.

Additional information:

See Chapter 19, "Internal and External Proposals."

example

Executive Summary

Problem

Results from the employee satisfaction survey indicate that the Northwest Group needs to improve current leadership training. Our analysis reveals that managers want to hire staff from outside our Supervision Identification Program's (SIP) pool of "SIP Certified" personnel.

Managers do not currently believe that the SIP pool contains personnel with the skills needed to succeed on the job. They highlighted three areas specifically: diversity management, communication skills, and knowledge of ISO standardization.

Solution

We propose solving this problem as follows:

- Researching supervisory software vendors for improved online and computer-aided individual instruction.
- Implementing improved ongoing and post-assessment techniques.

Cause/Effect

Another method of organizing your draft is cause/effect. In this instance, you would focus on what caused a specific situation or its results—the effects. This method of organization would be useful in writing an incident report; for example, if a customer has complained about a corporate billing error, you can consider cause/effect. As the company's accountant, you cannot blame the problem on a "computer error." After all, errors often must be traced to people—programmers, engineers, or data entry. So who or what caused the problem? The causes could be overcharges, unitemized bills, incorrect interest calculations, or processing charges.

Once you clarify the causes, determine the effects. Maybe the cause for the billing error is a faulty computer code. The effect, then, would be the need to rewrite this code. Perhaps the cause is human—an employee is not doing his or her job correctly. The effect would be to institute more effective employee training.

Analyzing the causes or effects of a situation will help you communicate effectively. Following are instances where a cause/effect report could be used:

- **Retail management**—You work at a regional store for a national corporation. Sales at your store have decreased (the effect). What caused this problem? Your supervisors at the home office want you to analyze the possible causes. Are prices out of line with the competition? Have demographics changed at the location? Are your expenses (rent, salaries, etc.) too high, necessitating a move or layoffs? Using a cause/effect organization will help you plan your report.
- **Information systems**—You are a customer service operator. A customer calls, telling you that his or her computer cursor is frozen on the screen (the effect). What could cause this problem? You analyze the issue and conclude that the causes could be due to conflicts among device drivers, software, and hardware, as well as corrupt elements in specific files. Then, in a follow-up e-mail, you organize your draft by highlighting the effect and discussing the causes.

FIGURE 5.5 Report Organized by Cause/Effect

DATE: March 16, 2009
TO: Edie Kreisler
FROM: Carlos De La Torre
SUBJECT: National Savings and Loan Employee Interview Results

Purpose of Report

In response to your request on February 23, 2009, ExecuMeasure has interviewed the National Savings and Loan's Trust Department employees. Our goal was to determine what might have been causing the department's poor morale and decreasing sales. The following is a report on our procedure and findings.

The Introduction explains why this report was written—to determine the causes of departmental problems.

Analysis of Employee Morale Problems and Decreasing Sales

Procedures. ExecuMeasure's research consultants met with all of the Trust Department's staff and interviewed them on the basis of their job performance and interpersonal communication skills. We used a scale of 1–10, with 1 representing "poor" and 10 representing "outstanding."

The questions regarding job performance are does *name of person* a) represent the company/department in a professional manner, b) return customer calls/e-mail, c) close sales deals efficiently, d) meet his/her quotas.

For interpersonal communication, Carlos and his team asked the employees to assess themselves and other staff, using the same 1–10 point scale. The questions are does *name of person* a) work well with other team members, b) respond in a timely manner to team member calls/e-mail, c) conduct him/herself professionally in departmental meetings, d) act collegially (with honesty and sincerity).

Findings. As the figure below indicates, all six staff received very good ratings on job performance with scores between 7.6 and 8.7. However, two employees scored significantly lower on interpersonal communication skills than the other workers. Barb scored 4 and Dan 3.6. The norm for the other four employees was 8.3.

Figure 1: Overall Employee Rankings

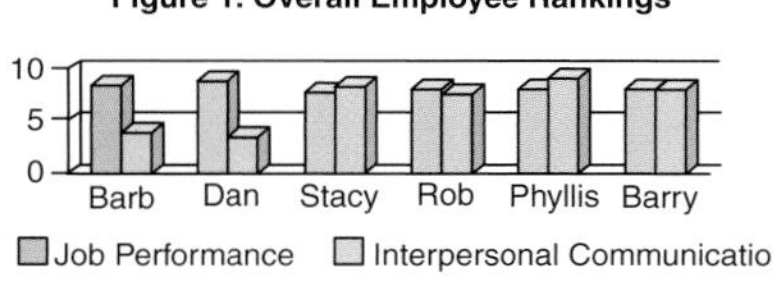

This figure clarifies why Barb and Dan received their low scores, causing low employee morale.

Figure 2: Employee Performance

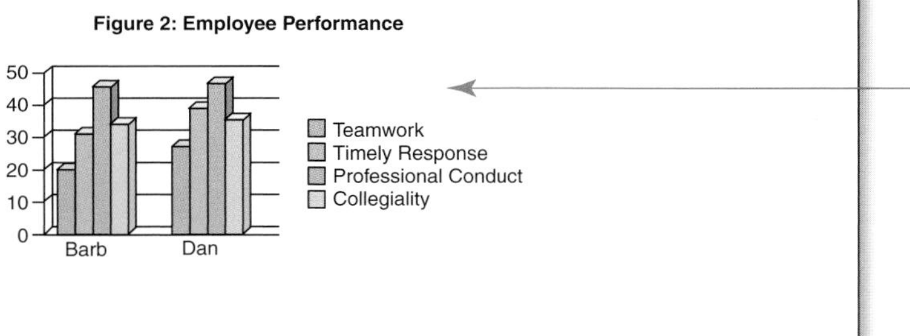

The bar chart specifies the causes for this office's employee problems.

Figure 5.5 shows a report organized by cause/effect.

Page Layout

Your primary goal in packaging, the second stage of the P^3 process, is to put words on the paper or screen. After all, if the words are only inside your head, then no one can understand your thoughts. However, dumping words on a page is not helpful either. Paragraphs filled with wall-to-wall words are not visually appealing to your readers. If your correspondence is visually unappealing, as in the "Before" example on page 124, an audience might not read it.

FIGURE 5.5 Report Organized by Cause/Effect Continued

Carlos De La Torre
Page 2
March 16, 2009

The employee's comments further clarify the causes behind the company's problems.

When we asked the other staff members to explain their scores, they gave us the following causes: "Barb will tell me one thing but then say something else to another employee." "Dan always comes late to meetings and either acts bored or says contrary things." "Barb never returns my e-mail messages or my phone calls." "Dan will mock other coworkers." "Barb thinks that she is the only one who works here." "It's obvious that Dan has no respect for Stacy—or anyone else for that matter."

Summary of Findings

The report's conclusion highlights the effect of the office's poor employee relationships.

As is evident from our survey, Stacy, Rob, Phyllis, and Barry trust and respect each other and their manager. However, they do not have the same trust and respect for Barb and Dan, nor do they believe Barb or Dan treat them professionally. The effect of this office dynamic seems to be poor morale and office inefficiency.

ExecuMeasure specializes not only in determining the causes for office-environment challenges but also in providing solutions to improve office morale and efficiency. If you would like us to work with you and your colleagues on creative solutions, please contact me at 212-555-9856, ext. 234.

BEFORE

Bit, Byte, and Begone (Triple B), your computer training storehouse, realizes that computer training can be expensive. We look for ways to maximize a corporate employee's learning while minimizing the company's costs. Through a combination of benefit-maximizing methods and cost-containment techniques, Triple B students enjoy a tremendous return on their learning investment. To control costs and maximize our client's benefits, Triple B strives for innovative approaches to a client's unique corporate training needs. For workplace communication word processing packages, we suggest the following unit of instruction: "Assist" classes for more than ten people. In an "Assist" class, Triple B trainers provide individual assistance. In fact, we will even "train the trainers." That is, we will teach your staff how to teach others within your company, saving you money down the line. Our classes can be held at your convenience, meeting on weekdays, nights, or weekends. Dual classes can be run simultaneously for more employees. Night classes mean that your employees will not miss valuable work time. Weekend classes mean that participants will avoid distracting phone calls, e-mail messages, or meetings. Lack of productivity caused by employee downtime will hurt a company's profit margin. Our flexible and innovative scheduling options solve that problem. The varied formats provided above let you choose which option works best for your staff and for your company.

This paragraph is well organized. It has an introductory sentence, body information, and a conclusion. Nonetheless, the paragraph might not be easy to read or understand. Visually, it is too demanding on the reader. Because the paragraph is visually dense, it forces the reader to wade through too much clutter. Highlighting techniques will help your audience understand your content, as in the "After" example.

AFTER

Bit, Byte, and Begone (Triple B), your computer training storehouse, realizes that computer training can be expensive. We look for ways to maximize a corporate employee's learning while minimizing the company's costs.

Benefits

Through a combination of benefit-maximizing methods and cost-containment techniques, Triple B students enjoy a tremendous return on their learning investment. To control costs and maximize our client's benefits, Triple B strives for innovative approaches to a client's unique corporate training needs.

AFTER Continued

Options

For workplace communication word processing packages, we suggest "Assist" classes for more than ten people. "Assist" classes provide you these benefits:

- Individual assistance.
- "Train the trainers" classes. We will teach your staff how to teach others within your company, saving you money down the line.
- Convenient meetings on weekdays, nights, or weekends.
- Dual classes run simultaneously for more employees.

Night classes mean that your employees will not miss valuable work time. Weekend classes mean that participants will avoid distracting phone calls, e-mail messages, or meetings. Lack of productivity caused by employee downtime will hurt a company's profit margin. Our flexible and innovative scheduling options solve that problem.

The varied formats provided above let you choose which option works best for your staff and for your company.

"Benefits" and "Options" are examples of headings.

Bullets allow for easy access and readability.

Headings and Talking Headings

To improve your page layout and make content accessible, use headings and talking headings. Headings—words or phrases such as "Introduction," "Discussion," "Conclusion," "Problems with Employees," or "Background Information"—highlight the content in a particular section of a document. When you begin a new section, you should use a new heading. In addition, use subheadings if you have a long section under one heading. This will help you break up a topic into smaller, more readable units of text.

Talking headings, in contrast, are more informative than headings. A heading helps your readers navigate the text by guiding them to key parts of a document. However, headings, such as "Introduction," "Discussion," and "Conclusion," do not tell the readers what content is included in the section. Talking headings, such as "Human Resources Committee Reviews 2009 Benefits Packages," informatively clarify the content that follows.

One way to create a talking heading is to use a subject (someone or something performing the action), a verb (the action), and an object (something acted upon).

example

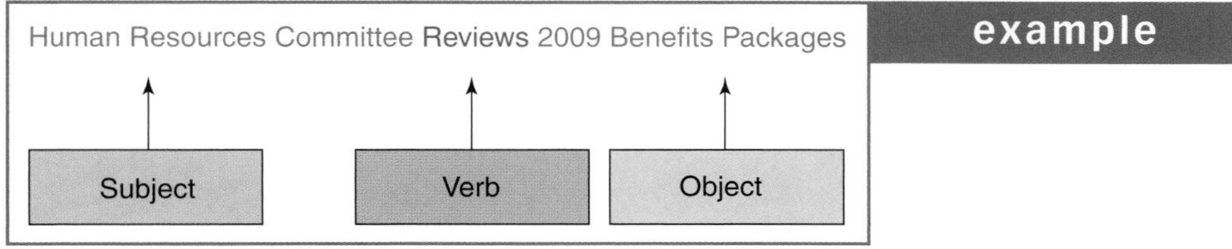

Another way to create talking headings is to use informative phrases, such as "Problems Leading to Employee Dissatisfaction," "Uses of Company Cars for Personal Errands," and "Cost Analysis of Technology Options for the Accounting Department." Table 5.1 provides examples of informative talking headings.

Table 5.1 Talking Headings

Sentences Used as Talking Headings
Rude Customer Service Leads to Sales Losses
Accounting Department Requests Feedback on Benefits Package
Corporate Profit Sharing Decreases to 27 Percent
Parking Lot Congestion Angers Employees
Phrases Used as Talking Headings
Budget Increases Frozen Until 2009
Outsourced Workers Leading to Corporate Layoffs
Harlan Cisneros—New Departmental Supervisor
EEOC: Questions About Company Hiring Practices

Chunking

An easy way to improve your document's design is to break text into smaller chunks of information, a technique called chunking. When you use chunking to separate blocks of text, you help your readers understand the overall organization of your correspondence. They can see which topics go together and which are distinct.

Chunking to organize your text is accomplished by using any of the following techniques:

- White space (horizontal spacing between paragraphs, created by double or triple spacing)
- Rules (horizontal lines typed across the page to separate units of information)
- Section dividers and tabs (used in longer reports to create smaller units)
- Headings

Order

Once a wall of unbroken words has been separated through headings and chunking to help the reader navigate the text, the next benefit a reader wants from your text is a sense of order. What is most important on the page? What is less important? What is least important?

You can help your audience prioritize information by ordering ideas. One way to accomplish this goal is by using a hierarchy of headings set apart from each other through various techniques, such as typeface, type size, capitalization, density, and position.

Typeface Some different typefaces (or fonts) include Times New Roman, Courier, *Monotype Corsiva*, ***Forte***, Curlz MT, Arial, CASTELLAR, **Impact**, Rockwell, and Comic Sans MS.

The typeface you choose will either be a serif or sans serif typeface. Serif type has "feet" or decorative strokes at the edges of each letter. This typeface is commonly used in text because it is easy to read, allowing the reader's eyes to glide across the page. Times New Roman is a serif font, for example.

SERIF ←— decorative feet

Sans serif (as seen in this parenthetical comment) is a block typeface that omits the feet or decorative lines. This typeface is often used for headings. Arial is a sans serif font, for example.

SANS SERIF ←— no decorative feet

Though you have many font typefaces to choose from, all are not appropriate for business documents. Times New Roman and Arial are best to use for letters, memos, reports, resumes, and proposals, because these font types are most professional looking and are easiest to read.

Type Size Another way of prioritizing for your readers is through the size of your type. For example, a first-level heading could be in 18-point type. The second-level heading would then be set in 16-point type, the third-level heading in 14-point type, and the fourth-level heading in 12-point type. Avoid using headings smaller than 12-point type. A smaller font size is hard to read.

Capitalization You can order headings through capitalization. An "all cap" heading (ALL CAP) would take priority over headings typed in "title case." "Title case" includes a mixture of upper- and lowercase characters.

Density The weight of the type also prioritizes your text. Type density is created by **boldfacing.**

Position Your headings can be centered, aligned with the left margin, indented, or outdented (hung heads). No one approach is more valuable or more correct than another. The key is consistency. If you center your first-level heading, for example, and then place subsequent heads at the left margin, this should be your model for all chapters or sections of that report.

Access

A fourth way to assist your audience is by helping them access information rapidly. You can use any of the following highlighting techniques to help readers filter out extraneous or tangential information and focus on key ideas.

White Space In addition to horizontal space, created by double or triple spacing, you can also create vertical space by indenting. This vertical white space breaks up the monotony of wall-to-wall words. White space invites your readers into the text and helps the audience focus on the indented points you want to emphasize.

Bullets Bullets, used to emphasize items within an indented list, are created by using asterisks (*), hyphens (-), bullet characters (●), pictures ( ), and Webdings or Wingdings (✓).

Numbering Whereas bullets create lists of items equal in importance, numbered lists show sequence or importance. Use a numbered list for step-by-step procedures or to show that the first point is more important than subsequently numbered points.

Underlining Use underlining cautiously. If you underline too frequently, none of your information will be emphatic. One underlined word or phrase will call attention to itself and achieve reader access. If you underline too many words or phrases, you lose the emphasis and visually distract your readers.

Italics *Italics*, like underlining, should be used with caution. Italicize to emphasize a key word or phrase. Do not overuse this highlighting technique.

Text Boxes An excellent way to emphasize a key point is by creating a text box. You can add to the emphasis by using shadows as well.

A text box highlights important information. Adding "Note" calls further attention to the point.

example

Note: Be sure to hand-tighten the nuts at this point. Once you have completed the installation, go back and *securely* tighten all nuts.

Color Colors can also provide emphasis. A word typed in red, for example, could be used to emphasize the potential for **Danger. Warnings** are highlighted in orange, and **Cautions** in yellow.

You can use background colors to emphasize ideas, but be careful. Background colors can be distracting, and you must choose text colors that are easy to read against the colored background.

example

Gradients can create an interesting background.	For readability, the best color combination is black text on a white background.

Inverse Type To help readers access information, use inverse type—**printing white on black** versus black on white.

When it comes to highlighting techniques, more is *not* better. A few highlighting techniques help your readers filter out background data and focus on key points. However, too many highlighting techniques are distracting and clutter page layout.

Variety

You might want to use smaller or larger paper; vary the weight of your paper, using 10-pound, 12-pound, or heavier card stock; or even print your text on colored paper. You can vary the document design as follows.

Choose Landscape Orientation Rather than print your text using a vertical, portrait orientation (8½″ × 11″), you could choose to print the text horizontally, using a landscape orientation (11″ × 8½″).

example

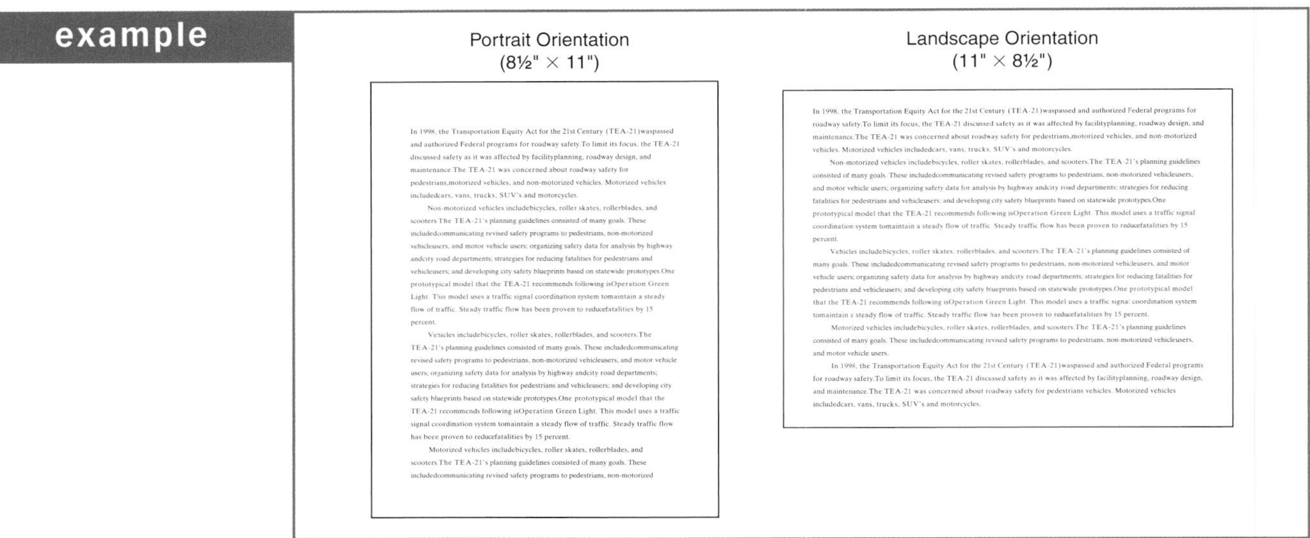

Use More Columns Provide your reader two to five columns of text for variety. Columns are commonly used in brochures, newsletters, fliers, and other marketing publications.

example

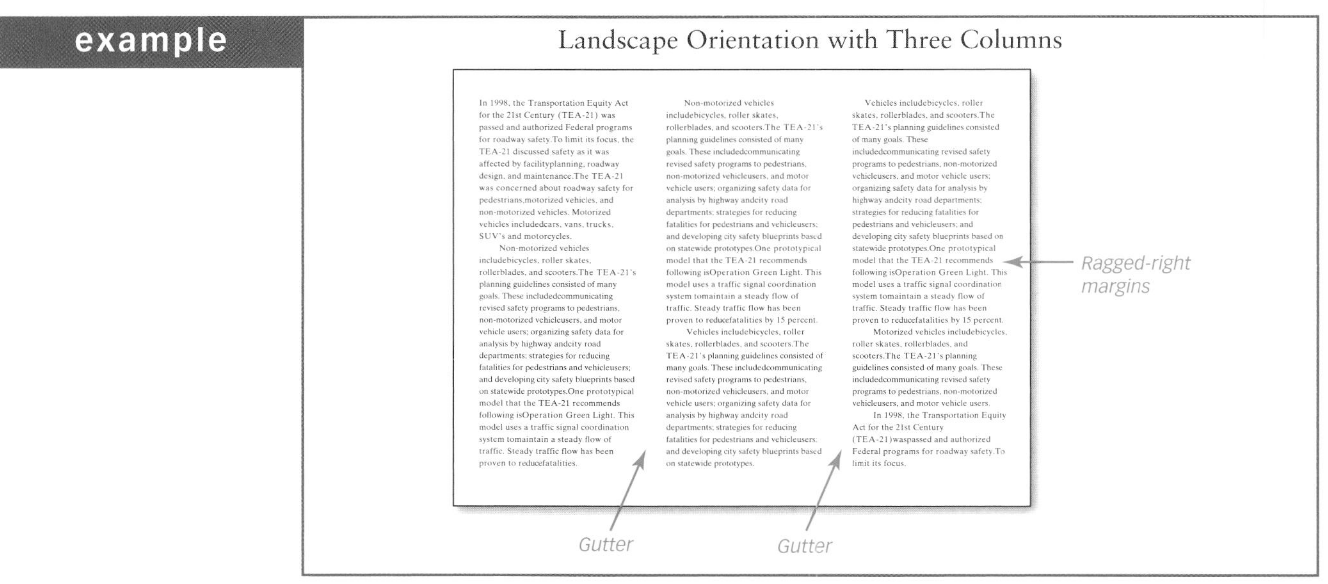

Vary Gutter Width Columns of text are separated by vertical white space called gutters. You can vary these spaces for more visual appeal.

Use Ragged-Right Margins Fully justified text (both right and left margins aligned) was once considered professional, giving the text a clean look. Now, however, studies confirm that right-margin-justified text is harder for the audience to read. It is too rigid. In contrast, ragged-right type (the right margin is not justified) is easier to read and more pleasing to the eye. Ragged-right text is a visually appealing page layout preference.

Technology Tip

Document Design Using Microsoft Word 2007

When it comes to document design, you have a world of options at your fingertips. Not only does your word processing software offer you possibilities to enhance page layout, but also the Internet provides unlimited resources.

Word Processing

In a word processing software program like Microsoft Word 2007, you enhance your document's design in many ways.

From the "**Home**" tab ribbon, you can make changes to **Font, Paragraphs**, and **Styles**.

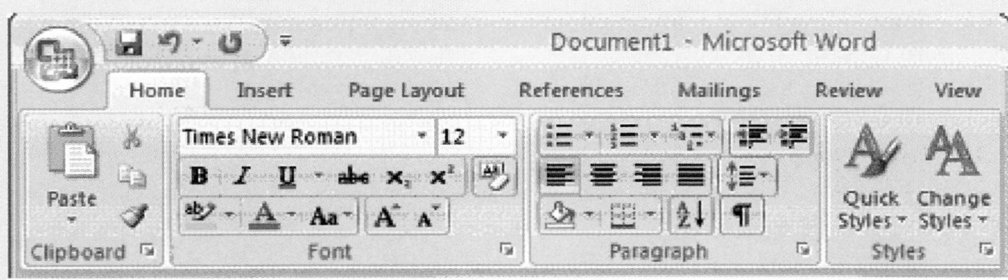

"FONT" CATEGORY

- choose your **FONT TYPE** and **SIZE**

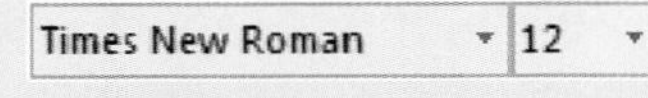

- **bold face,** *italicize,* and underline the text

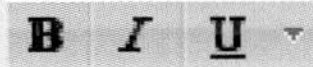

- ~~strikethrough~~, create subscripts and superscripts

- highlight the color of text, change the font color, or change the case of selected text to uppercase or lowercase

- increase or decrease the font size

"PARAGRAPH" CATEGORY

- create bulleted, numbered, or multilevel lists (as with outlines)

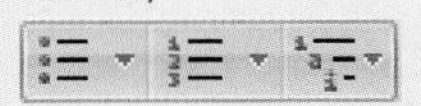

- decrease or increase an indentation

- change the margins (ragged right, centered, block right, or full block)

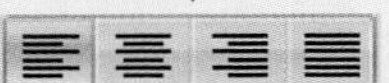

- change the spacing between lines

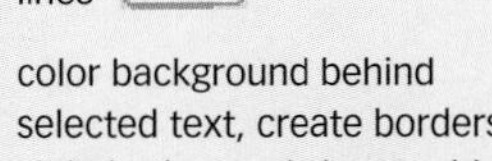

- color background behind selected text, create borders, alphabetize, and show or hide paragraphing

"STYLES" CATEGORY

- format text ("Quick Styles") and change colors and fonts throughout a document ("Change Styles")

FIGURE 5.6 The Shrinking Size of Workplace Communication

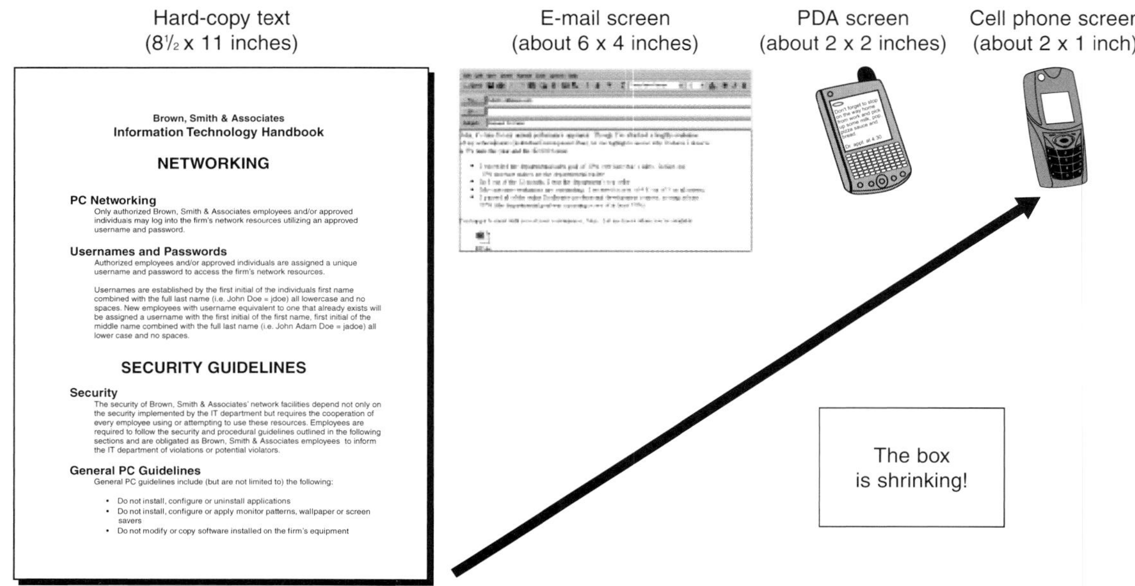

The Impact of Technology

Technology is another consideration when packaging your communication. How will you be presenting your information? Will you send a hard-copy report or an electronic e-mail? Will your e-mail be sent to an audience's desktop computer or to a wireless handheld PDA or cell phone? For an oral presentation, will you supplement your comments with a PowerPoint presentation? Technology is impacting your communication. The size of the screen makes the difference. Screen sizes are shrinking. When you package your comments, you need to consider the way in which technology limits your space. Today, effective workplace communication *must fit in a box*.

Workplace Communication Fits in a Box

How does workplace communication fit in a box? Think about a resume. A resume often fits on one page. Certainly, resumes can be longer, and in some situations must be longer. After all, an employee with many years of work experience will need more than one page to clarify accomplishments. One hard-copy page measures 8½″ by 11″. In contrast, much of today's written communication in the workplace is accomplished through e-mail. E-mail messages are rarely 8½″ by 11″; an e-mail screen tends to be smaller. Therefore, e-mail messages should be limited to about 20 lines of text. In that case, technology impacts how you package your communication. This technological impact is even more dramatic when you consider the screen size for a handheld PDA (about 2″ by 2″) and the screen size for a cell phone (about 1″ by 2″). Figure 5.6 graphically depicts how the size of the box affects communication.

The bottom line is that technology will affect your packaging. If you are filling out an online form, you will be limited by the size of the form's fields. If you are writing correspondence that will be viewed as a PowerPoint presentation or e-mail sent to someone's PDA or cell phone, you need to limit the size of your text. Therefore, as you write your draft for either oral or written communication, consider the impact of technology.

PACKAGING CHECKLIST

_____ 1. Have you decided which method of organization to use?
- Analysis
- Spatial organization
- Chronology
- Importance
- Comparison/contrast
- Problem/solution
- Cause/effect

_____ 2. Have you considered your document's layout to improve visual appeal?
- Shorter paragraphs (four to six typed lines)
- Bullets or numbered points to list key ideas
- Headings or subheadings

_____ 3. Have you factored in the impact of technology, realizing that communication "fits in a box"?
- E-mail screens—approximately 4″ × 6″
- Handheld PDA screens—approximately 2″ × 2″
- Cell phone screens—2″ × 1″
- PowerPoint "autolayouts"

Nicole Met Her Challenge

To meet her communication challenge, Nicole used the *P*³ process.

Packaging

In Chapter 4, you were introduced to Nicole Stefani who had been hired by ImageSkill as a public relations expert. Her first major job as a new hire is to help the city of Greenfield improve their public relations and promote an effective overall image. See Chapter 4, pages 107–108, to learn how Nicole gathered her data and planned her communication. Now she needs to *package* her written communication. She writes a rough draft, organizing her text according to importance.

Nicole says, "To package the proposal, I prioritized the answers from the interviews and focused on which responses were most prevalent. My goal was to organize both written and oral comments according to a 'least important/most important' presentation. That is, I offered the clients options for improving their public relations: the first options were the most cost-effective but the least promising; the latter options were more costly but more rewarding for the city." Figure 5.7 shows how Nicole packaged her rough draft for review by her boss.

To see Marc's revision suggestions, go to Chapter 6 (Perfecting Workplace Communication), pages 157–158.

FIGURE 5.7 Rough Draft (complete with errors) of Nicole's Proposal Submitted to Her Boss, Marc Shabbot, for Review

Date: August 13, 2009
From: Nicole Stefani
To: Greenfield City Council
Subject: Proposal

Introduction:

The city of Greenfield wants to promote the city. Greenfield city management wants to inform new homeowners and prospective businesses to move to Greenfield. ImageSkill can help Greenfield meet this problem.

Discussion:

To determine the best way to promote growth in Greenfield, ImageSkill staff followed this procedure. We met with a representative body of Greenfield citizens to ask questions. From this survey, ImageSkill determined which areas Greenfield citizens considered to be the city's most alluring. Our goal was to decide how best to maximize Greenfield's perceived strengths and opportunities for growth and give us insight into the best area's in which growth could be accomplished.

Following are our findings regarding strengths and growth opportunities:

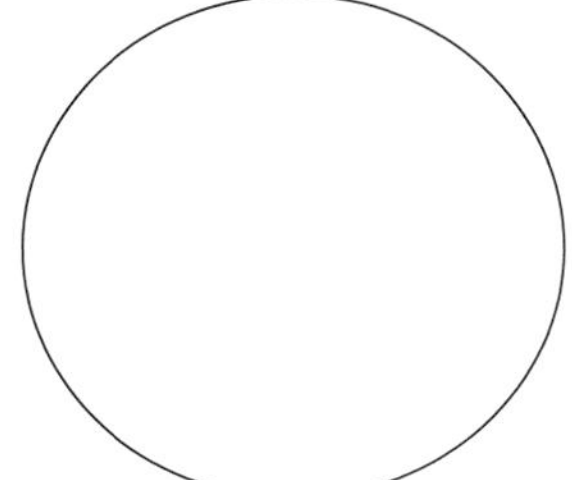

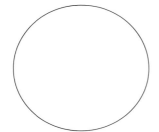

Nicole says, "I told Marc, I plan to insert one or two graphics here for visual impact. Bob in R&D would help me with the number crunching."

On a scale of 1–4, with 4 being best, Greenfield citizens concluded that the city's strengths are it's schools (4), city services (4), infrastructure (2), housing (1), caring people (1), amenities (1), and safety (1). The citizens would like to see growth in the following areas: more jobs/higher pay (7), more entertainment opportunities (6), more greenspace/hiking trails (6), more and improved schools (5), and city promotion (4).

The citizens of Greenfield want to accomplish all of the above goals without hurting the city's strengths. They don't want pollution, traffic, increased taxes, or higher housing costs. They want to maintain Greenfield's "small town friendliness," the city's responsiveness to their needs, and the city's charm. These are the challenges that you are facing.

ImageSkill can help Greenfield solve its problems. As a full service marketing company, we can help you as follows. We can create brochures for your Chamber of Commerce that will market your city's charms and intrinsic values as well as promote growth opportunities for new businesses, educational growth, entertainment, and parks/recreational prospects. We can build a Web site for your city, highlighting all of the values your citizens are so proud of. We can manage image control. As your city grows, citizens will resent change, which is inevitable, so packing this growth is the challenge, and ImageSkill can solve this problem by providing you outstanding radio, tv, and newspaper sound bites as well as periodical articles that promote the city's changes as best for the common good. We can also help you manage your city's events/conferences. As new companies move to your city, the city will experience growth in the convention industry. In addition, one way to ensure that your citizens continue to enjoy their city's "small town charms" is to provide city events (parades, holiday fare, or just occasional activities) that amplify city pride and augment whatever events are currently in place.

Nicole Stefani
Page 2
August 13, 2009

Conclusion:

ImageSkill wants to be your full service marketing agent. We have the talent and the skill to enhance the city of Greenfield's growth opportunities. Here's what our staff of seven employees will provide you:

1. Nicole Stefani—Customer service representative. I am your personal contact at ImageSkill. I have a BS in journalism and public relations from IU and expertise with city/government demographics.
2. Marc Shabbot—CEO, ImageSkill. Marc has over 25 years of experience in not-for-profit marketing and advertising. He is the founder and chairman of our company.
3. Joe Kingsberry—Desktop publishing. When we create your fliers and brochures, Joe will man the helm. He has knowledge of all Microsoft and Adobe Suite applications.
4. Cheryl Huff—Web designer. Cheryl is an expert with all Web design tools, including Flash, FrontPage, Netscape Composer, and IE. Please see our online Web portfolio of her work (http://www.imageskill.com/portfolio).
5. Larry Massin—Customer service rep. Larry and I will partner on your account. He has a BS in marketing and over ten years of experience.
6. Christy McWard—Writer. Christy has a BA in journalism and an MA in marketing. She is our chief wordsmith.
7. Robert Cottrell—R&D. We build your brochures and Web site based on facts. Bob is in charge of all research.

Nicole says, "I told Marc that there would be errors in this rough draft. My primary goal was to get all of the content down in a fairly easy-to-access layout. My plan was to get Marc's feedback and then perfect the text later."

CHAPTER HIGHLIGHTS

1. Packaging your communication makes it accessible and easy to understand.
2. When you organize your content, the audience can logically follow your thoughts.
3. When you analyze a topic, you break it down into smaller pieces.
4. Spatial organization is used when you have physical descriptions of products or work-related accidents or event reporting.
5. Chronological order can be used for meetings, trip reports, procedures, and agendas.
6. When you organize by importance, you tell your audience which part of the material is most important or least important.
7. Comparing/contrasting helps your audience to understand the pros and cons.
8. Organizing content from problem to solution helps your audience focus on unique aspects of a problem and/or optional ways to solve problems.
9. Using highlighting techniques helps you to create visually accessible and readable content in your workplace communication.
10. Technology affects how you package your communication, since so much workplace writing must "fit in a box."

WWW *Find additional packaging and P^3 exercises, samples, and interactive activities at http://www.prenhall.com/gerson*

CASE STUDIES

1. Currently, Carlos De La Torre, CEO of ExecuMeasure, is working with the trust department at National Savings and Loan. The department is having employee problems. Two key employees, both with long-term service to the bank and significant levels of knowledge and performance, are at odds with the rest of the department. These two employees are bickering with other workers, not returning e-mail, failing to work well in team activities, playing workers against each other in power struggles, and creating morale problems in the department. Of course, the simple solution would be to fire the two troublesome employees. However, they are the trust department's key salespeople, representing over 45 percent of all closed sales. In addition, they receive support from the former department manager, who only recently has been promoted to a higher position at the bank. ExecuMeasure's job is to find solutions, other than termination.

Assignment

Brainstorm possible solutions to the trust department's problem. What steps can Edie Kreisler, the department's manager, take to improve morale in her department? After deciding on appropriate solutions, write a short memo or e-mail highlighting the problems and then listing your suggested solutions.

2. You work for the Cove, Oregon, city planning department. Your boss, Carol Haley, has received complaints recently from citizens concerned about a wastewater facility being built in their neighborhood. The homeowners are worried about odors, chemical runoff in nearby Tomahawk Creek, decreases in home values, and a generally diminished quality of life in the neighborhood. The wastewater facility will be built. Despite the citizens' concerns, city planning has decided that the city needs and will profit from the plant. Nonetheless, you must respond to these complaints, acting upon the citizens' issues.

For odor abatement, the wastewater management company plans to control fumes and particulate matter through the use of cross flow and wet scrubbers, thermal oxidizers, absorption materials, and bio-filters. Many of the concerns regarding runoff and home values can be solved through improved land management and ecological restoration. By planting more reeds, bushes, and trees in the green space between the homes and the proposed plant, runoff can be absorbed more efficiently. In addition, green barriers will improve home values. Finally, you have learned that the wastewater company wants to be a good neighbor. To do so, it plans to become actively involved in the community by building more parks, playgrounds, hike/bike trails, and by stocking the nearby pond.

Assignment

In small teams or as individuals, write an e-mail to the boss, Carol Haley, detailing the problems and suggesting solutions. Be sure to consider page layout and space limitations presented by technology.

3. Candice Millard is going on a business-related trip to Japan. Candice must plan her travel agenda. This will include the following requirements:
   - Contacting a translator who will attend meetings with her.
   - Arranging for travel expenses by writing a proposal to her supervisor.
   - Learning the customs of the country and people she will visit.
   - Deciding on appropriate clothing for a businesswoman in Japan.
   - Learning what health and safety precautions are needed for international travel.
   - Acquiring a passport for her travel.

Assignment

In small teams or as individuals, help Candice prioritize her travel arrangements. Rank the previous list in a chronological order, from most important to least important. What must she do first? What, in your opinion, are the lesser important steps? Once you have compiled your list, explain your reasons behind organizing it as you have.

4. Read the following introduction from a proposal submitted by SeeRight Graphics to a prospective customer.

Introduction
In this report, SeeRight Graphics proposes a new logo design for your company, Gulf Drift Homes. With over 50 years of experience, and the recognition of renownd home builders, such as John Ayers, Briant and Jones Inc., and Larson and Sons, we will make your company stand apart from the competition.

Challenges
Our research has found that your current logo faces the following challenges:

1. *Outdated Look*
   Gulf Drift Homes' current logo was designed ten years ago. The Massin Logo Institute (MLI) reports that logos need to be refreshed for new consumer recognition every three years. Doing so gives companies a modern, contemporary look for emerging markets.
2. *Missing Information*
   Since your last logo was developed, you have added two new sales locations. In addition, you have added an e-mail address, a Web site URL, and a fax number. EGR research statistics state that today's customer accesses companies by e-mail 45 percent of the time and by Web 23 percent of the time. You could be losing significant business since your current logo is without these essentials.
3. *No Graphic Representation*
   Your current logo lacks the visual punch needed to attract new customers. A 2005 MLI survey found that 90 percent of customers choose a company based on its eye-catching logo.

Fig 1. Reasons Why People Choose a Company

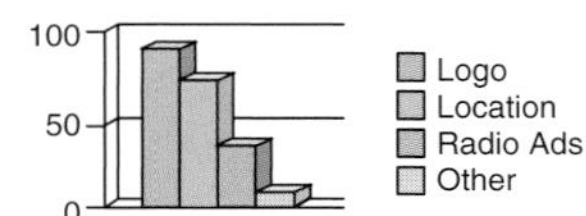

SeeRight researched your competition to define customer recognition of logos. The following figure shows our findings:

Fig 2. Recognition by Customers

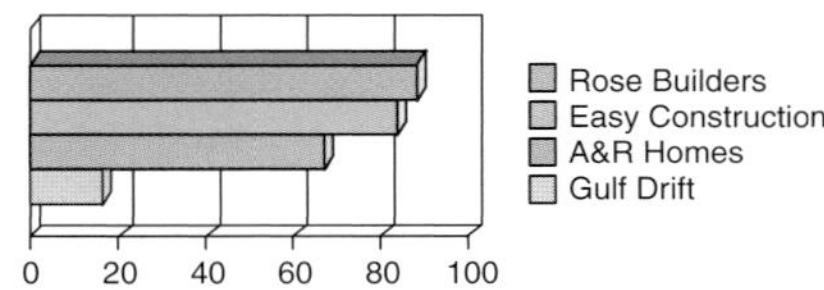

Assignment

What methods of organization are used in this proposal's introduction? Give examples to prove your point.

ETHICS IN THE WORKPLACE CASE STUDY

Ta'isha Roberts contacted her company's computer technician to complain about her company laptop. She told the technician that the laptop wasn't working effectively. When her husband used the laptop to make a PowerPoint presentation at their son's elementary school, the 2007 PowerPoint slides he had created on Ta'isha's laptop were not compatible with the school's 2003 PowerPoint capabilities. Ta'isha wants the technician to fix her laptop. However, the technician realizes that Ta'isha is using her corporate-issued laptop for purposes other than work.

Question: Is it ethical for Ta'isha's husband to use her company laptop? Is it the technician's ethical responsibility to report this use of company equipment to her boss? Justify your answer based on information provided in Chapter 1.

INDIVIDUAL AND TEAM PROJECTS

1. Choosing Methods of Organization

   a. Using spatial organization, write an advertisement describing the interior of a car, truck, or SUV; the exterior of a piece of equipment; a piece of clothing; or an apartment or dormitory room.

   b. Organizing your text chronologically, write a report documenting your drive to school or work, your activities accomplished in class or at work, your discoveries at a conference or vacation, or your activities at a sporting event.

   c. A fashion merchandising retailer asked her buyers to purchase a new line of clothing. In her memo, she provided them the following list to help them accomplish their task. Reorganize the list by importance, and justify your decisions.

DATE: January 15, 2009
TO: Buyers
FROM: Nancy Roetteger
SUBJECT: Clothing Purchases

It's time again for our spring purchases. This year, let's consider a new line of clothing. When you go to the clothing market, focus on the following:

- Colors
- Materials
- Our customers' buying habits
- Price versus markup potential
- Quantity discounts
- Wholesaler delivery schedules

Good luck. Your purchases at the market are what make our annual sales successful.

   d. Visit two or more auto dealerships, clothing stores, computer stores, restaurants, prospective employers, colleges, and so on. Based on your discoveries, write a report using comparison/contrast to make a value judgment. Which of the two cars would you buy, which of the two restaurants would you frequent, at which of the computer stores would you purchase equipment, and so on?

e. Research two or more career fields, job opportunities, or online job search sources (Monster.com, Job-Hunt.org, dice.com, careerbuilder.com, HotJobs.yahoo.com, etc.). Based on your findings, write an e-mail using comparison/contrast to decide which option is best.

2. Practicing Page Layout

a. In small teams or as individuals, reformat the following memo to improve its page layout.

DATE: June 10, 2009
TO: Prakas Patel
FROM: Shin Zhu
SUBJECT: Request for Webpage

Since the launch of our delivery services, business is booming. This is terrific and we have our great staff to thank for the success. To keep our business growing, we need to build a Web site. This is where I need your help. Please construct a Web site focusing on our new delivery services.

The Web site must include information on the following key concerns. Emphasize our delivery services. This issue is crucial to our success, so we need to push it. You will also want a complete list of all service options, including price ranges, delivery choices, on-site pick-up or drop-off, and times. Provide links to our home office Web site and auxiliary companies. These links will be used primarily for intrastate deliveries—note this on our site. Highlight special deals we offer—lower copy rates for small business agents, holiday rates, and frequent shipper programs. You must also include all contact information (phone, e-mail, fax, and street addresses) plus store hours. Finally, graphics are great!

Work your Web design magic on this one, although I know you will! I will need the site up and running by July 10. That will give me time to get word out on the new fabulous Web site.

b. In small teams or as individuals, rewrite the following headings, creating more informative talking headings.

Headings	Talking Headings
Content for Quarterly Report	
Reviewing Corporate Policies	
Laying Off Employees	
Internal Promotions	
Confronted with Angry Customers	

c. Reformat the following case study to improve its document design.

The city of Waluska wants to provide its community a safe and reliable water treatment facility. The goal is to protect Waluska's environmental resources and to ensure community values.

To achieve these goals, the city has issued a request for proposal to update the Loon Lake Water Treatment Plant (LLWTP). The city recognizes that meeting its community's water treatment needs requires overcoming numerous challenges. These challenges include managing changing regulations and protection standards, developing financially responsible treatment services, planning land use of community expansion, and upholding community values.

For all of the above reasons, Hardtack and Sons (H & S) engineering is your best choice. We understand the project scope and recognize your community's needs.

We have worked successfully with your community for a decade, creating feasibility studies for Loon Lake toxic control, developing odor-abatement procedures for your streams and creeks, and assessing your water treatment plant's ability to meet regulatory standards.

H & S personnel are not just engineering experts. We are members of your community. Our dynamic project team has a close working relationship with your community's regulatory agencies. Our Partner in Charge, Julie Schopper, has experience with similar projects worldwide, demonstrated leadership, and the ability to communicate effectively with clients.

H & S offers the city of Waluska an integrated program that addresses all your community's needs.

We believe that H & S is your best choice to ensure that your community receives a water treatment plant ready to meet the challenges of the twenty-first century.

3. Studying the Impact of Technology

   a. Search for online forms. These could include forms that you would fill out for a job search, insurance claims, Social Security information, electronic transfer forms, IRS forms, passport applications, legal agreements, FCC (Federal Communications Commission) forms, and so on. When you find online forms or PDF downloads, give examples of how these forms limit the size of your communication.

   b. Open up PowerPoint and select several of the autolayout options. Type text within these templates and determine the following:

   - How many lines of text are possible on the screens?
   - At what point does the font size change automatically?
   - When does the text run off the screen?

   Based on your discoveries, about how many lines of text should you write for an effective, easy-to-read PowerPoint slide?

   c. Turn on a cell phone. Look at the cell phone's screen and determine

   - How many lines of text fit within that screen.
   - How many words fit on each line.
   - How many characters each line will accommodate.

   Share your findings with your classmates.

   d. With the help of your classmates, look at five or six examples of e-mail messages from different e-mail programs (Outlook, Yahoo, Juno, Eudora, Lotus, etc.). Then determine

   - How many lines of text fit within the e-mail's predefined box, before you need to scroll.
   - How many words fit on each line.
   - How many characters each line will accommodate.

Share your findings with your classmates.

e. Research new, emerging technologies that can be used for business communication purposes. Many of these are hybrid wireless cell phone/PDAs that allow business workers to send and receive faxes and e-mail and access the Internet. These hybrids also include electronic calendars, address books, memopads, and calculators. Others are electronic notepads or tablet PCs. How might the size of these technologies affect communication?

DEGREE-SPECIFIC ASSIGNMENTS

Keys to successful packaging include knowing how to organize information, choosing correct communication channels, and recognizing the impact of technology. Read the following information about a retail clothing store and then answer the assignment questions.

On-the-Go is a clothing store specializing in hip new styles. This hometown store has experienced incredible growth and plans to expand. It will open two new stores in your city. In addition, On-the-Go will offer franchise opportunities in six cities throughout the state and ten stores throughout the region. You have been hired in one of the following capacities:

- Marketing specialist
- Chief information officer, in charge of business information systems
- Head accountant
- Manager of operations
- Chief financial officer

What are your primary responsibilities?

Assignment

Either individually or in small, degree-related groups, complete the following:

1. Using brainstorming, make a list of at least five to ten key concerns facing your department, based on the company's growth plans.
2. Review your list, and organize it according to importance. (Which items in the list must take priority over the other concerns?)
3. Write a brief paragraph (approximately five to ten sentences) explaining why you have included the items on your list, how they relate precisely to your degree area, and why you have prioritized the list as you have.
4. Suppose you will need to communicate these key concerns to management, franchisees, employees, and prospective new hires. Which communication channel would work best for each audience?

PROBLEM-SOLVING THINK PIECES

1. **Choosing Methods of Organization**

   Read the following cases. Then, based on the information you have been provided, choose the method of organization that would package your communication most effectively.

   - Analysis
   - Chronology
   - Problem/Solution
   - Spatial organization

- Importance
- Cause/Effect
- Comparison/Contrast

a. Mary Ramos, a junior at the state university, earned a 2.45 GPA in the fall semester. Unfortunately, to remain in the accounting program, she needed to maintain a 2.5 GPA. She experienced extenuating circumstances in the fall. These included strep throat, theft of her computer, and additional hours spent working at her job. Mary needs to write a letter to her accounting advisor explaining her issues. Which method of organization should she use in her letter?

b. Holbert Lang, auditor for an independent accounting firm, has been hired by a not-for-profit agency to study a current accounting problem. The not-for-profit agency recognizes an impairment of one or more of its long-lived (fixed) assets. The agency needs to decide whether to report this problem or not. Before they make this decision, however, they want to know why the impairment is occurring. Holbert's job is to make this determination. Which method of organization should he use to package his report, once he finds his answers?

c. Lupe Salinas, Webmaster for his company, has to create a corporate Web site. The company has many benefits that it wants to share with potential customers. Before Lupe can create the various screens for his site, however, he has to decide which of the company's many values to focus on. It is a large job. Which method of organization could he use to simplify the challenge and package his findings?

d. Sharon Mitchell, training coordinator, needs to create a PowerPoint presentation for her company's new hires. She needs to teach them the steps they must follow to receive reimbursement for job-related travel. Which method of organization will help her trainees understand this procedure?

e. Mike Moore, hotel manager, is planning an all-staff meeting. The hotel has received numerous complaints lately from hotel patrons. First, Mike needs to explain the nature of these complaints. Then, Mike must suggest to his staff ways in which these challenges can be overcome. Which method of organization should he use to package his oral presentation?

f. Carolyn Jensen, insurance claims adjuster, is writing an e-mail to her boss. The claims office has not done a good job closing open claims files. Carolyn's boss has asked her to study options for improving their close-out times. She has studied the options and is ready to share her findings. She wants to leave the best choice for last. Which method of organization should she use to package her e-mail?

g. Karen Rochlin recently purchased a used car. It came with a 90-day warranty on parts. After only three and a half months, her car's engine has started to stall while idling. She needs to write a letter of complaint to the used car dealer and ask for an extension on her warranty. She knows the car is "officially" out of warranty, but she has extenuating circumstances to share with the dealer. Which method of organization could she use to package her letter?

h. Darryl Kennedy is a home inspection technician. After inspecting a home under consideration for purchase, he now needs to write his follow-up report. The home is in good shape, except for the northeast exterior wall. On this wall, he found evidence of wood rot near the baseboard, fungi in the wall's midsection, and then a damaged gutter where the wall met the roofline. Which method of organization should he use to clarify his findings for the reader?

WEB WORKSHOP

When you communicate in the workplace, you will have numerous communication channels—various ways to package your information. You might choose to write a hard-copy letter, report, or memo. You also might choose to package your content electronically as either a Web site, a PowerPoint presentation, or an e-mail message.

Technology is impacting the shape, size, content, style, and channel of twenty-first century workplace communication. Use an Internet search engine to research ways in which technology is changing workplace communication.

1. In an Internet search engine, type in any of the following key words:
    - whitepaper+BlackBerry
    - whitepaper+e-mail
    - whitepaper+PowerPoint

    Note: A "white paper" is a corporate or educational report made available to the public that focuses on a particular industry issue.

Assignment

After researching issues related to the emerging importance of technology in the workplace, share your findings with your teacher and/or classmates. To do so, you could make a brief oral presentation or write an e-mail summarizing key points.

2. Many articles are now suggesting that instant messaging (IM) might replace e-mail as a primary communication channel in the workplace.

Assignment

Access an Internet search engine and type in phrases such as the following:

- Instant messaging+impact
- Instant messaging+importance
- Instant messaging+significance

You will find many articles about the growing importance of instant messaging as a communication channel. Report your findings either in an oral presentation to your class or in a brief e-mail message to your teacher.

QUIZ QUESTIONS

1. What do you consider when you package workplace communication?
2. List five methods of organization.
3. Define "analysis" and explain its importance.
4. How is spatial organization used?
5. What are three types of workplace communication which depend on chronology?
6. How does organizing by importance help your audience?
7. Explain why you would organize content by comparison/contrast.
8. What are three ways to achieve effective page layout?
9. Why do you consider technology when you package your communication?
10. How do you effectively use problem/solution in proposal writing?

CHAPTER 6

Perfecting Workplace Communication

▸REAL PEOPLE *in the workplace*

ImageSkill has assigned a big job to a new hire, **Nicole Stefani**. She needs to write a proposal in response to Greenfield City Management's RFP (request for proposal) and make a PowerPoint presentation to the city council.

Nicole planned her document by gathering data through telephone conversations, e-mail messages, and face-to-face meetings with the potential client. She also met with her boss, Marc Shabbot, who suggested questions she could ask to help gather information. With the information, Nicole packaged her proposal by considering organization, layout, and design.

She says, "Though I've written the proposal, I know that I'm not finished with it. Success of any written document is dependent on the text being error free. Therefore, I now need to perfect the proposal. Like most writers, I have difficulty finding all my errors. I also can't always determine if the document has the correct tone or addresses all the needs of the audience.

Finally, I know that sometimes my writing is not persuasive. I'm fortunate that Marc is willing to edit my proposal. He has worked for ImageSkill for over ten years and knows how to craft a winning proposal. He's a great team leader, and I look forward to his comments about the proposal. With him and the *P*³ process, I know that I'm going to be a successful communicator in the workplace."◂

CHAPTER GOALS

When you complete this chapter, you will be able to do the following:

1. Perfect your business documents.
2. Use the reporter's questions and focus on specificity of detail to add missing material from your text.
3. Delete dead words and phrases for conciseness and to enhance readability.
4. Simplify words for conciseness and easier understanding.
5. Move information for emphasis.
6. Reformat paragraphs and use highlighting techniques for ease of access.
7. Create a pleasant tone to ensure effective workplace communication.
8. Proofread and correct for accuracy.

WWW

To learn more about perfecting and the P^3 process, visit our Web Companion at http://www.prenhall.com/gerson.

Nicole's Communication Challenge

Nicole says, "Though I am a trained public relations expert, making a professional presentation orally and in writing still is a challenge for me. I am challenged by the scope and importance of this project.

My manager says that to accomplish this task, he and I must ensure the proposal is perfected. Marc says that he will help me review the rough draft for any gaps in content. We'll check to see if I've clarified who does what, when, where, why, and how. While Marc wants our proposal to be complete, he also wants it to be easy to read. Therefore, in the perfecting stage, we'll see where we can delete wordiness and use easy-to-understand words.

Next, I'll work on improving the text's layout and design. With Marc's help, and the help of our graphics staff, I'll probably have to move text around to emphasize key points and add tables and figures. Finally, this proposal must be persuasive. We need to sell our ideas and sell our company's professionalism. To accomplish these goals, in the perfecting stage, I'll try to create a positive tone and avoid all grammar errors.

Marc says that perfecting the proposal will ensure its readability and its persuasiveness. This is a great opportunity for me; I know that if I do a good job, Marc and ImageSkill will be impressed with me and my professionalism. More importantly, I want to meet the needs of my first client. Because I live in Greenfield now, I want to help ensure its continued growth and prosperity."

Meeting the needs of ImageSkill and the potential client with a perfected proposal is a communication challenge for newly hired Nicole.

See pages 107–108 in Chapter 4, "Planning Workplace Communication," to learn how Nicole met her "planning" challenge. In addition, see Chapter 5, "Packaging Workplace Communication Through Effective Document Design," pages 131–133, to learn how Nicole met her "packaging" challenge. Then in this chapter on pages 157–162, follow Nicole through her "perfecting" challenge.

PERFECTING FOR PROFESSIONALISM

The P^3 process is composed of three key components: planning, packaging, and perfecting. (Planning is discussed in Chapter 4; Packaging is discussed in Chapter 5.)

Planning

Chapter 4 taught you how to plan your communication, the first stage of the P^3 process. Your task in that initial stage was to accomplish the following:

- Determine your objectives.
- Consider your audience.
- Choose your communication channel.
- Gather your data.

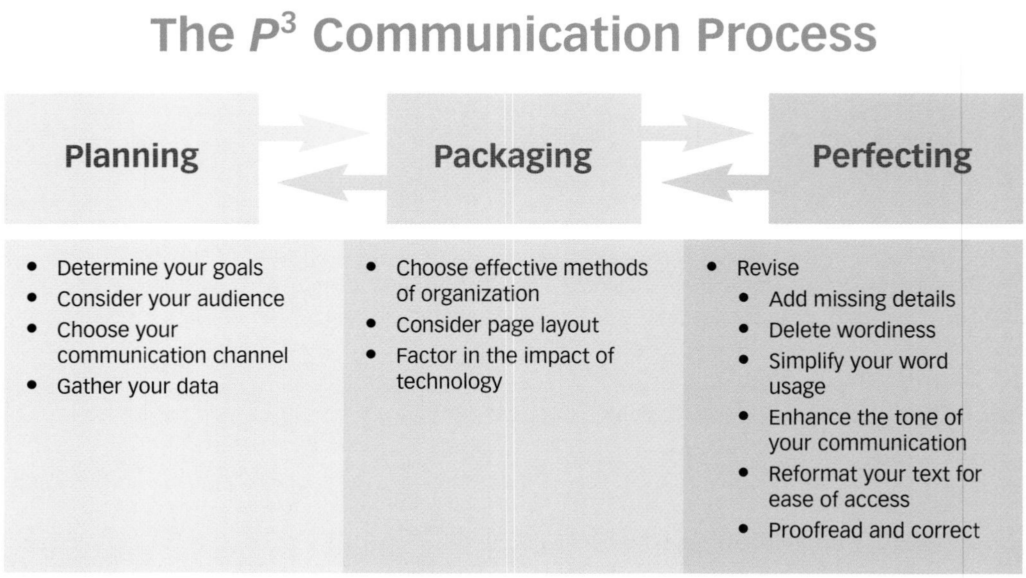

The P^3 Communication Process

Planning	Packaging	Perfecting

• Determine your goals
• Consider your audience
• Choose your communication channel
• Gather your data
	• Choose effective methods of organization
	• Consider page layout
	• Factor in the impact of technology
		• Revise
		  • Add missing details
		  • Delete wordiness
		  • Simplify your word usage
		  • Enhance the tone of your communication
		  • Reformat your text for ease of access
		  • Proofread and correct

Packaging

Chapter 5 taught you how to package your communication. In this second stage of the P^3 process, you write a rough draft of your letter, memo, e-mail, report, or oral presentation. To help your audience understand and easily access your content, you must package your ideas as follows:

- Choose effective methods of organization.
- Consider page layout.
- Factor in the impact of technology.

Perfecting

Once you have written a rough draft, you should revise and perfect your written or oral communication. Revision requires that you do the following:

- *Add* information for clarity.
- *Delete* wordiness for conciseness.
- *Simplify* unnecessarily complex words for easier understanding.
- *Move* information around (cut and paste) to ensure that your most important ideas are emphasized.
- *Reformat* (using highlighting techniques) to ensure reader-friendly ease of access.
- *Enhance* the tone and style of the text.
- *Proofread* and *correct* any errors to ensure accurate grammar and content.

ADDING INFORMATION FOR CLARITY

When you write a first draft of your correspondence or oral presentation, you will not always include every key idea or develop each point thoroughly. Your goal in the first draft, usually, is to get information on the page. Don't worry if you forget to include ideas that the audience will need for clarity. The perfecting stage of the P^3 process is the time to add these missing details. To accomplish this goal, be sure you have answered the *reporter's questions*, and focus on *specificity of detail* (using precise words versus imprecise, vague words).

Additional information:

See Chapter 4, "Planning Workplace Communication."

Reporter's Questions

The reporter's questions, sometimes called journalist's questions, are "Who," "What," "When," "Where," "Why," and "How." Chapter 4 discussed how answering these questions helps you plan your correspondence. Even if you began your communication project by answering the reporter's questions, you still might have omitted some key information. Writers tend to assume that an audience understands the topic and knows the background. Rarely is this true.

Readers are less informed than you might assume for several reasons:

- **If there is a CC (complimentary copy)**—Your primary reader might know what you are talking about. However, the other four or five people who get copies of the memo, e-mail, or report could be less informed. You might be writing primarily for your boss, for example. However, the e-mail also could be "cc'd" to coworkers in your department or in other departments, subordinates, independent auditors from another company, a vendor, or a client. Each of these readers will have different levels of understanding.
- **When many people hear your oral presentation**—If you are speaking to an audience composed of many different people, they might not have the same level of knowledge about your topic. Some will be specialists who have been actively involved in the project. Others will be semi-specialists and less involved. Still others will be lay audiences; the topic might be completely new to them. Your job is to clarify your information so all members of your audience basically have the same level of knowledge.

Additional information:

See Chapter 7, "Oral Presentations and Nonverbal Communication in the Workplace."

- **When time has passed**—Even if your primary reader knows the background and understands the topic, time might have passed between your initial discussion and your actual writing of the correspondence. During that time lapse, he or she can forget why or what you are writing about. In fact, years could have passed. Most reports are filed for future reference. Reports can be read two years later, five years later, or ten years later. It might be read by lawyers in a courtroom or by those who have replaced you when you are promoted to a new position. You need to clarify your information for future readers.
- **Due to distractions**—You might be writing or speaking to a boss. Your boss probably knows what you are talking about. However, he or she is in charge of many employees, has to attend numerous meetings, and must report to several other people up the chain of command. Your boss could have just returned from a vacation, job-related travel, or a doctor's appointment. In other words, your boss is distracted. Again, you should not take too much for granted. Clarify instead by adding missing details.

The "Before" example is vague.

BEFORE

Recently, we attended a conference to learn about HIPAA accountability.

This sentence is uninformative. When was "recently"? Who is "we"? Which "conference" did you attend? What exactly is "HIPAA"? To clarify, you need to add the missing details.

The "After" example provides more clarity by answering reporter's questions.

AFTER

On March 16, 2009, the employee benefits department staff (Pam Garcia, Tamishia Berry, and Karen Pecis) attended the Midwest Health Insurance Conference in Chicago. Our primary goal was to learn about changes in accountability mandated by HIPAA laws (Health Insurance Portability and Accountability Act of 1996).

When
Who
Where
What
Why

Specificity of Detail

A second way to achieve clarity is to add specific details to your communication. In your draft, you might have used connotative words, such as "substantial," "several," "near," "many," "few," "recently," "some," or "soon." These connotative words convey an impression. However, they do not provide quantifiable facts. How "substantial" is "substantial?" How "often" is "often?" In contrast, denotative words provide specificity of detail and quantify information. Whereas "substantial" is connotative, the denotative revision might be "75 percent" or "$100,000."

Look at the following "Before" example of vague writing caused by imprecise, unclear adjectives.

Vague, connotative words are underlined in this activity report.

BEFORE

Activity Report Draft

Our latest attempt at manufacturing plastic backpack clasps has led to some positive results. We spent several hours in Dept. 15 trying different machine settings and techniques. Several good parts were molded using two different sheet thicknesses. Here is a summary of the findings.

First, we tried the thick sheet material. At high heats, this thickness worked well.

Next, we tried the thinner sheet material. The thinner material is less forgiving, but after adjustments we were making good parts. Still, the thin material caused the most handling problems.

The engineer who wrote this report realized that it was unclear. To solve the problem, she rewrote the report, this time replacing the vague words and adding more specific information, as seen in the "After" example.

AFTER

Activity Report Revision

During the week of 10/4/09, we spent approximately 12 hours in Dept. 15 trying different machine settings, techniques, and thicknesses to manufacture plastic backpack clasps. Here is a report on our findings.

0.030" Thick Sheet

At 240°F, this thickness worked well.

0.015" Thick Sheet

This material is less forgiving, but after decreasing the heat to 200°F, we could produce 15 percent more good parts. Still, material at 0.015" causes handling problems.

DELETING DEAD WORDS AND PHRASES FOR CONCISENESS

The sentence in the following "Before" example is clear; it answers reporter's questions and provides specific detail. However, it is not good writing. Though the writer has provided valuable information, the writing is unnecessarily wordy.

BEFORE

On two different occasions, I have made an investigation of your neighbor's adjacent property and have come to the conclusion that your fence has been incorrectly installed, extending beyond the aforementioned property line, and that the said fence must be modified by the end of this month to ensure that you avoid the possibility of legal action.

This same sentence could be written more concisely, as in the "After" example.

The original "Before" sentence is 57 words long. The revision consists of three sentences, totaling 30 words. Each sentence in the "After" example averages about ten words.

AFTER

On June 1 and July 10, I investigated your neighbor's property. Your fence incorrectly extends beyond the property line. To avoid a lawsuit, please move the fence by July 31.

Though you need to be clear in your communication, you also must be concise for at least three reasons:

- **Time**—Other people's time is valuable. Your audience cannot nor should they spend too much time reading your e-mail or listening to you speak. They have their own jobs to perform. Long documents and lengthy oral presentations often waste people's time. Keep it short—for the sake of your audience as well as for your sake as the communicator.
- **Space**—The length of your communication makes a difference. If your writing, for example, is too long, it might not "fit in the box." A PowerPoint screen limits you to only six or seven lines of readable text. An e-mail screen limits you to only 20 or so lines before the reader must scroll. Many people still want you to limit your resume to one page. To ensure that your communication is the appropriate length for your communication channel, delete unnecessary words and phrases.
- **Readability**—"Readability" is the reading level of your document. It defines whether you are writing at a fifth-grade, ninth-grade, or twelfth-grade level, for example. You can gauge your readability level by using many readability formulas, including the Linsear Write Index, the Lazy Word Index, the Flesch Reading Ease Score, and the Flesch-Kincaid Grade Level Score. A popular tool for determining readability is Robert Gunning's Fog Index. Gunning bases his readability findings on the length of your sentence and the length of your word usage.

Readability is important for the following reasons: Only about 30 percent of Americans graduate from college. Therefore, if you are writing at a college level, you might create readability challenges for approximately 70 percent of your audience (U.S. Census Bureau). College graduates *do not* read at a college level. According to the National Center for Education Statistics, "about 41 million Americans," close to 22 percent of the adult population, "read at the lowest literacy level" (Labbe B9).

Given these facts, many businesses ask their employees to write between sixth- to eighth-grade levels. Regardless of your work environment, you will need to communicate

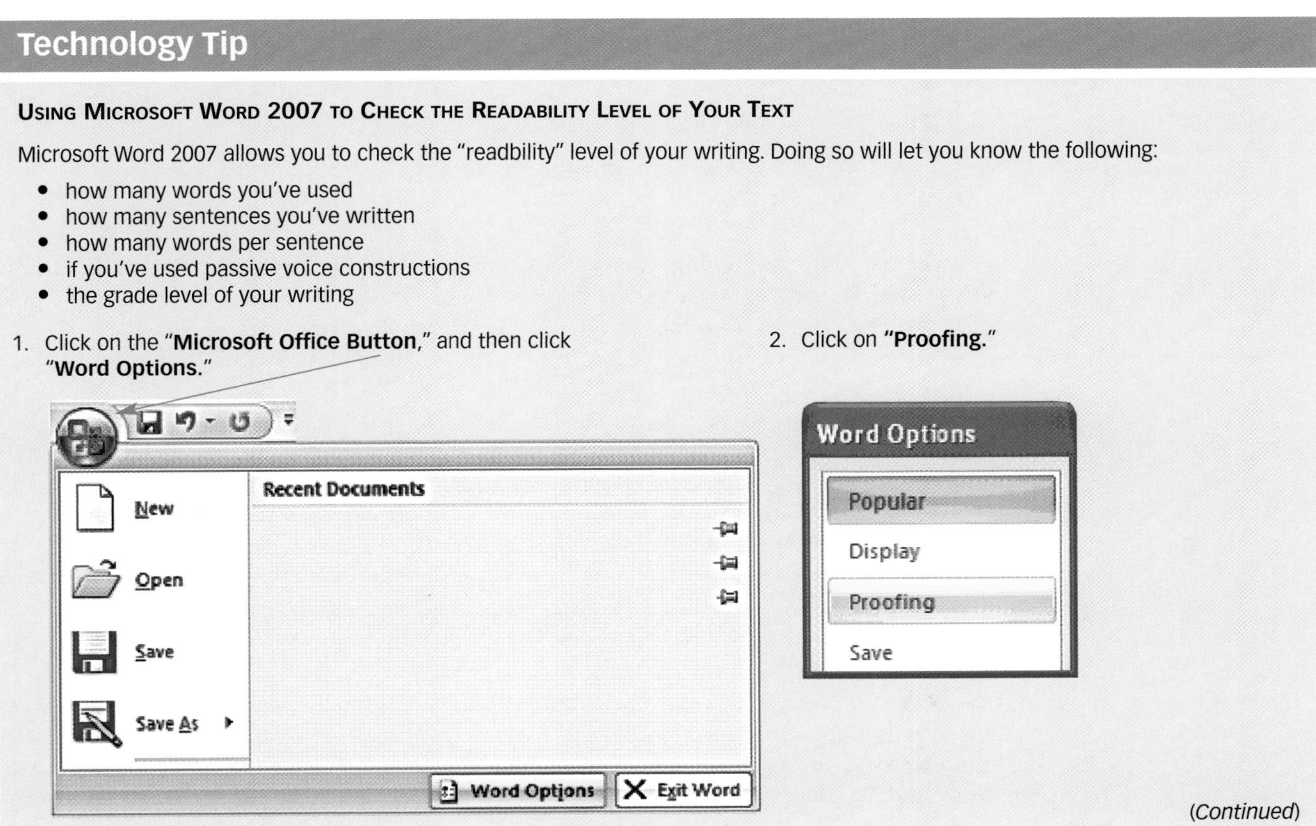

Technology Tip

Using Microsoft Word 2007 to Check the Readability Level of Your Text

Microsoft Word 2007 allows you to check the "readbility" level of your writing. Doing so will let you know the following:

- how many words you've used
- how many sentences you've written
- how many words per sentence
- if you've used passive voice constructions
- the grade level of your writing

1. Click on the "**Microsoft Office Button**," and then click "**Word Options.**"
2. Click on "**Proofing.**"

(Continued)

3. When the following screen pops up, select **"Check grammar with spelling"** and **"Show readability statistics."** Then click **"OK."**

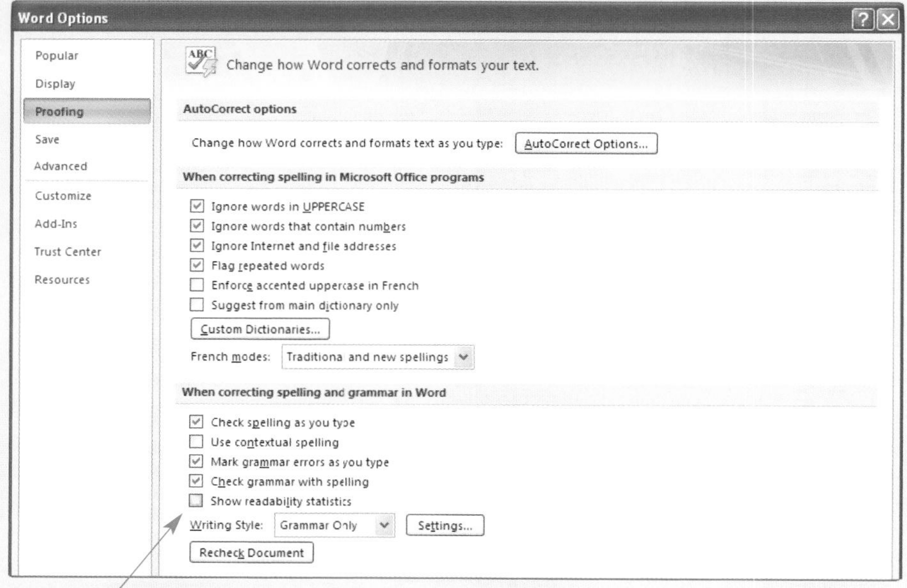

Show readability statistics

Once you have enabled this readability feature, open a file and check the spelling. When Word has finished checking the spelling and grammar, you will see a display similar to the one shown.

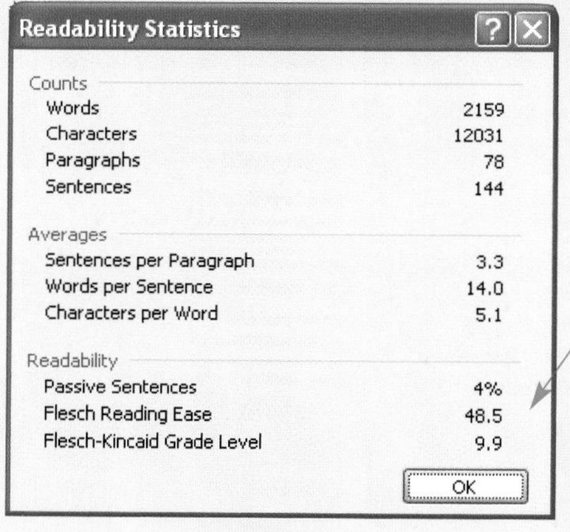

In this example, the text consisted of 2159 words, 78 paragraphs, and 144 sentences. This averaged out to 14 words per sentence, equaling about 10th grade level writing. Four percent of the sentences were written in passive voice.

to the general public. You can help most readers understand your content by writing shorter sentences. Conciseness is a valuable goal in the third stage of the *P*³ process—perfecting.

Sentence Length

To accomplish conciseness and improved readability, limit the length of your sentences. Sentences that are approximately 10 to 12 words long are easier for your audience to read and understand than longer sentences. In contrast, sentences longer than 20 words create readability challenges for your audience. To achieve conciseness, use the following techniques for deleting dead words and phrases.

Deleting "Be" Verbs

"Be" verbs include conjugations of the verb "be": "is," "are," "was," "were," "would," "will," "been," and "am." Often, these verbs create unnecessarily wordy sentences. For example, look at the "Before" and "After" examples.

BEFORE	AFTER
Wordiness Caused by "Be" Verbs	**Deleting for Conciseness**
Bill *is* of the opinion that stock prices will decrease.	Bill thinks stock prices will decrease. (**Note**: The sentence deletes one "be" verb—*is*—but keeps another—*will*.)
I *am* in receipt of your bill for $1,000.	I received your $1,000 bill.
If I can *be* of any assistance to you, please call.	If I can help, please call.
They *are* planning to fax new invoices tomorrow.	They plan to fax new invoices tomorrow.
Barb *had been* hoping to move into her new office complex today.	Barb hoped to move into her new office complex today.
I wonder if you *would be* so kind as to provide discount information about your sales.	Please provide discount information about your sales.
The senior staff accountant *is* responsible for assuring that all liabilities *are* reported on a yearly basis.	The senior staff accountant reports liabilities annually.

Using Active Voice Versus Passive Voice

In active voice sentences, the subject performs an action. In passive voice sentences, the subject is acted upon. Sometimes, "be" verbs create passive voice sentences, as in the "Before" example.

BEFORE
It *has been* decided that Joan Smith *will* head our sales department.

This sentence is written in the passive voice (the primary focus of the sentence, Joan Smith, is acted on rather than initiating the action).

Passive voice causes two problems.

1. Passive constructions are often unclear. Who decided that Joan Smith will head the department? To solve this problem and to achieve clarity, replace the vague indefinite pronoun "it" with a precise noun, as in the "After" example.

AFTER
Larry named Joan Smith head of the sales department.

2. Passive constructions are often wordy. Passive sentences require helping verbs, such as *has been*. The "After" example omits the helping verb *has been* and the "be" verb *will*.

Deleting the Expletive Pattern

When you begin sentences with "there" or "it," you create the expletive pattern of sentence structure. Notice how the expletive pattern again uses "be" verbs. The expletives ("there" and "it") create wordy sentences, as shown in the "Before" examples in contrast to the "After" examples.

BEFORE	AFTER
There are three people who will work for Acme. *It has been* decided that ten engineers will be hired.	Three people will work for Acme. Ten engineers will be hired.

Avoiding *Shun* Words

Another way to write more concisely is to avoid words ending in *-tion* or *-sion*—*shun* sounds. For example, look at the "Before" and "After" examples.

BEFORE	AFTER
Shun Words	**Concise Version**
came to the conclu*sion*	concluded (or decided, ended, stopped)
with the excep*tion* of	except (or but)
make revi*sions*	revise
investiga*tion* of the	investigate (or look at, review, assess)
consider implementa*tion*	implement (or use)
utiliza*tion* of	use

Avoiding Camouflaged Words

Camouflaged words are similar to *shun* words. In both instances, a keyword is buried in the middle of surrounding words (usually helper verbs or unneeded prepositions). For example, in the phrase "make an amendment to," the keyword "amend" is camouflaged behind unnecessary words. Look at the "Before" and "After" examples.

BEFORE	AFTER
Camouflaged Words	**Concise Version**
make an *adjust*ment of	adjust (or revise, alter, change, edit, fix)
have a *meet*ing	meet
*thank*ing *you* in advance	thank you
for the purpose of *discuss*ing	discuss
arrive at an *agree*ment	agree
at a *later* moment	later

Limiting Prepositional Phrases

Prepositions can be important words in your communication. They help you convey information about time and place. Occasionally, however, *prepositional phrases* create wordy sentences. A prepositional phrase includes a *preposition* and a *noun* or *pronoun* that serves as the object of the preposition. For example, "at a later moment" is a prepositional phrase. It includes the preposition "at" and the noun "moment." This prepositional phrase is wordy and can be revised to read "later."

BEFORE	AFTER
Wordy Prepositional Phrases	**Concise Version**
He spoke *at a rapid rate*.	He spoke rapidly (or fast).
She wrote *with regard to* the meeting.	She wrote regarding (or about) the meeting.
I will call *in the near future*.	I will call soon.
On two different occasions, we met.	We met twice.
The manager *of personnel* was hired.	The personnel manager was hired.

SIMPLIFYING WORDS FOR CONCISENESS AND EASIER UNDERSTANDING

To achieve clarity *and* conciseness, you also should simplify your word usage. Use easy-to-understand words. Sometimes, you can accomplish this by limiting the length of your words. You cannot nor should you avoid words like "accountant," "engineer," "telecommunications," "computer," or "nuclear." Though these words have more than two syllables, they are not difficult to understand. When you need longer words, use them. Also, try to avoid old-fashioned, legalistic words, like "pursuant," "accordance," and "aforementioned." Too often, writers and speakers use these words to *impress* their audience. In contrast, communicators should *express* their content clearly and simply. Look at the following, long-winded, "Before" sentence:

BEFORE

I would like you to take into consideration the following points, which I know will assist you in better applying new HIPAA rules and regulations currently burdened by the need to execute all data manually and on paper rather than through standardized, electronic transmissions.

The previous sentence is 44 words long. As noted earlier in this chapter, most people would prefer to read sentences consisting of approximately 10 to 12 words. Also, the sentence contains 10 words with three or more syllables: "consideration," "following," "applying," "regulations," "currently," "execute," "manually," "standardized," "electronic," and "transmissions." None of these words is challenging individually. Still, the mass of syllables makes the sentence hard to understand. In contrast, readers are more comfortable with words under three syllables. To solve the challenges presented by the length of the sentence and the length of the words, simplify as in the "After" example.

AFTER

Please consider the following points. This will help you apply new HIPAA rules by submitting data online instead of having to type all text on separate forms.

In this revision, the 44 words have been reduced to 27 words. Also, the original long sentence has been cut into two smaller sentences. Finally, the remaining two sentences contain only four multisyllabic words ("consider," "following," "submitting," and "separate"). The conciseness saves you and your reader time, and makes the information easier to understand.

Following is a list of long words that can be simplified for conciseness and easier understanding:

BEFORE	AFTER
Long Words	**Concise Version**
Utilize	Use
Anticipate	Await (or expect)
Cooperate	Help
Indicate	Show
Initially	First (or 1.)
Presently	Now
Prohibit	Stop
Inconvenience	Problem
Pursuant	Before
Endeavor	Try
Sufficient	Enough
Subsequent	Next

MOVING INFORMATION FOR EMPHASIS

Another way to perfect your communication is to organize your content effectively. Procedural steps that appear out of chronological order, for example, will confuse your readers. If you are reporting on a meeting, your minutes should reflect the chronology of what was discussed. Using a chronological order also can help your audience follow trends. If you want to organize by importance, are you beginning with less important options and leading to more important ones? If you want to describe events spatially, present your information from north to south, or top to bottom.

Reread your draft to make sure you are organizing your text as you had planned. If you are not following the correct organization, move the information. Word processing lets you move information for emphasis by cutting and pasting. The "Before" example is not organized effectively, while the "After" is reorganized chronologically.

The "Before" example discusses clothing options in the 90s, 50s, 70s, 80s, and 60s. Organizing the topic so randomly makes the text hard to follow.

BEFORE

We have clothing and other things from the 50s to the 90s. Our 90s' clothing includes everything from early 90s' grunge (raggedy cut-off shorts and flannel plaid shirts) to late 90s' hip-hop clothes and add-ons (oversized jewelry, sports jerseys, hooded sweatshirts, and fancy club duds). Our 50s' goodies cover the "beats" to the "blues brothers" clothing. We have black suits, black skinny ties, as well as black berets and turtlenecks. Our 70s' clothing and items include all kinds of disco clothing including leisure suits and everything from Ban-lon to skin-tight pants, shirts, white belts, and boots. We even have disco dance hall balls and posters for your own disco dance floor. Our 80s' clothing focuses on what we call "business-nerd" including pocket protectors, wing-tip shoes, and horned rim glasses (with the nose piece already pre-taped). We also cover "Flash Dance" clothes like leg warmers and off-the-shoulder sweatshirts. Finally, our 60s' hippie stuff is way cool, such as wide, hallucinogenic ties, Nehru jackets, short-short mini-dresses, long-long granny dresses, way-out-there boots, peace symbols, fringe belts, and vests.

Compare the disorganized example with the following "After" example.

The "After" example is improved in at least three ways: the text has been organized chronologically by decade; the bullets make the text easier to access; and the boldface headings highlight each era.

AFTER

We have clothing and other things from the 50s to the 90s.

- The **50s.** Our goodies cover the "beats" to the "blues brothers" clothing. We have black suits, black skinny ties, as well as black berets and turtlenecks.
- The **60s.** Our hippie stuff is way cool. We sell wide, hallucinogenic ties, Nehru jackets, short-short mini-dresses, long-long granny dresses, way-out-there boots, peace symbols, fringe belts, and vests.
- The **70s.** Our merchandise includes all kinds of disco clothing including leisure suits and everything from Ban-lon to skin-tight pants, shirts, white belts, and boots. We even have disco dance hall balls and posters for your own disco dance floor.
- The **80s.** Check out our "business-nerd-wear" including pocket protectors, wing-tip shoes, and horned rim glasses (with the nose piece already pre-taped). We also cover "Flash Dance" clothes like leg warmers and off-the-shoulder sweatshirts.
- The **90s.** We have clothing from early 90s' grunge (raggedy cut-off shorts and flannel plaid shirts) to late 90s' hip-hop clothes and add-ons (oversized bling bling jewelry, sports jerseys, hooded sweatshirts, and fancy club duds).

When you wrote your rough draft in the packaging stage of the P^3 process, your primary concern might have been content—not visual appeal. Now, however, in the perfecting stage, you also must format your text to allow for ease of access. You need to consider how the text will look on the page. If you give your readers a wall of words, they might not read the text. An unbroken page of words is not reader-friendly.

PARAGRAPH LENGTH

The number of lines you write in a paragraph is arbitrary. As the writer, you must decide how much to write. Some paragraphs, due to the complexity of the subject matter, might require more development. Other paragraphs requiring less development can be shorter. Nonetheless, an excessively long paragraph is ineffective. In a long paragraph, you force your reader to wade through many words and digest large amounts of information. This hinders comprehension. In contrast, short, manageable paragraphs invite reading and help your readers understand your content. A paragraph in effective workplace communication should consist of no more than four to six typed lines and no more than 50 words per paragraph.

BEFORE

Our project management approach will provide your city clear deliverables and meet your RFP criteria. Orlin & Sons proposes the following sequence. We will assess the adequacy of your current facilities from a technology perspective, starting on January 13, to be completed by February 1, 2009. Then, beginning on February 5, O&S will meet with residents' focus groups to identify community needs, including health, culture, history, and quality of life issues. This will allow us to identify necessary improvements to meet your current and ongoing requirements. We will complete this project milestone on February 25, 2009. The final step of the process involves setting team goals needed to work with city, county, and state regulatory agencies. We will begin this step on March 8 and conclude by March 15, 2009.

The "Before" paragraph is not visually attractive. To invite your readers into the document, consider the document's design. Make your text open and inviting by using formatting techniques, as shown in the "After" example.

AFTER

Our project management approach will provide your city clear deliverables and meet your RFP criteria. Orlin & Sons proposes the following sequence:

- Assess the adequacy of your current facilities from a technology perspective.
  Begin/End Dates: 1/13–2/1/09.
- Meet with residents' focus groups to identify community needs: health, culture, history, and quality of life issues.
  Begin/End Dates: 2/5–2/25/09.
- Set team goals needed to work with city, county, and state regulatory agencies.
  Begin/End Dates: 3/8–3/15/09.

Break up wall-to-wall words and margin-to-margin text with smaller paragraphs. Use bullets to make each paragraph stand out more effectively. The boldfaced, indented dates emphasize key milestones.

ENHANCING TONE

Another goal for successful communication is tone. As a writer or speaker, your challenge is not just conveying information. You also want to motivate your audience or avoid angering them. For example, when you are communicating with an unhappy customer, you must maintain your professionalism. Remain calm and positive for the following reasons:

- Your words represent the company, your department, your boss, and yourself. Think of it this way—your letter is like a photograph. All the reader might ever know of you and your company is what you have written. You want that photograph to create a pleasant and professional image.
- You want to keep the customer's business. Increasing a customer's anger will turn the customer away.
- You want to build trust.

To communicate with the correct tone, enhance your correspondence through "you usage" and "positive words."

You Usage

People want to be spoken to, not spoken at. Remember, "Companies don't write to companies; people write to people." To achieve a person-to-person feel in your correspondence, personalize your text or speech through pronoun usage. The generally accepted hierarchy of pronoun usage is as follows:

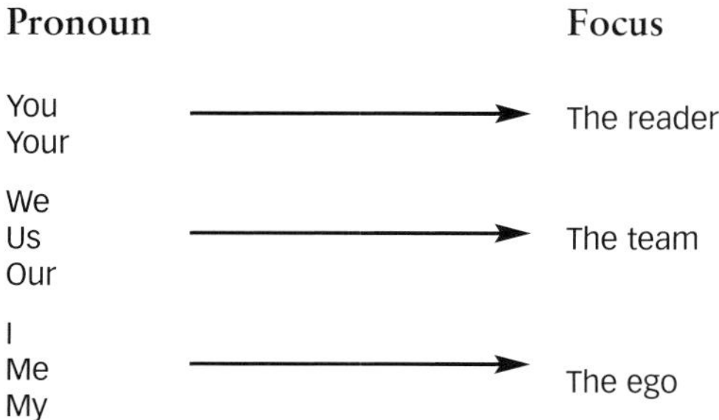

The first group of pronouns, "you/your," is the most preferred. This is called "you usage." When you use "you" or "your," you speak directly to your reader(s) on a one-to-one basis. By focusing on the pronouns "you" or "your" in your workplace communication, you involve your readers in the correspondence. The second and third tiers of pronoun usage ("we," "us," "our," "I," "me," and "my") are good to use also, but are less effective.

Positive Words

You also can enhance the tone of your communication by focusing on positive words. Your goal is to keep your customers, associates, subordinates, and supervisors positively motivated, as well as to minimize negative consequences. To accomplish this challenge, try using any of the following words when writing and speaking:

Positive Verbs	
Accomplish	Improve
Establish	Increase
Help	Initiate
Achieve	Insure (ensure)
Assist	Maintain
Build	Organize
Coordinate	Plan
Create	Produce
Develop	Promote
Implement	Train

The "Before" example fails to speak to the reader personally and uses negative words.

BEFORE

Note the negative words: "demands," "penalized," "failure," "negate," and "denied."

The customer service department policy demands that products must be returned within 30 days of purchase. Customers will be penalized 20 percent of the product cost for shipping and restocking. Failure to do as advised by this policy will negate the department's responsibilities and reimbursement will be denied. Only store credit for other purchases will be allowed for purchases after 30 days. Customers not meeting these policy guidelines must send a written request for return or exchange variances.

This company can convey the same message in a much more positive and personalized way, as in the "After" example.

AFTER

Acme's Easy Return Policy guarantees you satisfaction for every purchase. Just return any item within 30 days of receipt for a refund of your purchase price (less a 20 percent shipping and restocking charge). By meeting these policy standards, you are ensured a timely reimbursement. For returns after our 30-day grace period, simply enjoy store credit for your other purchases. If you'd like additional help, our online returns center will guide you through the exchange variances return process.

In contrast to the "Before," negative version, this "After" text enhances the tone of the communication. "Guarantees," "satisfaction," "ensured," "timely," "enjoy," "additional help," and "guide" create a friendly work atmosphere. Pronouns like "you," "your," and "our" personalize the text, creating a conversational tone.

CORRECTING FOR ACCURACY

A final step in the perfecting process is correctness, for both written and oral communication. Clarity, conciseness, and tone are primary objectives of effective workplace communication. However, if your communication is incorrect—grammatically or textually—then potentially you have wasted your time and destroyed your credibility. To be effective, your workplace communication must be accurate. This could include any of the following concerns:

- Missing works cited information needed to document your sources
- Grammar—spelling and punctuation errors will make you and your company look unprofessional
- Figures/numbers—your numbers must be written and add up correctly. The differences between feet and inches, meters and kilometers could cost your company money.

Accuracy in workplace communication requires that you proofread your text. Look at the example of inaccurate writing on page 156 caused by poor proofreading.

Technology Tips

Using Microsoft Word 2007 for Perfecting

Word processing programs help you perfect your document in many ways:

- **Spell Check**—When you misspell a word, often spell check will underline the error in red (as shown in the following example with "grammer" incorrectly spelled). Spell check, unfortunately, will not catch all errors. If you use a word like "to," instead of "too," spell check will not **no** the difference (of course, that should be "know" but spell check did not mark the error). Microsoft Word 2007's "**Review**" tab also provides you access to proofreading help, and allows you to make comments and track changes.

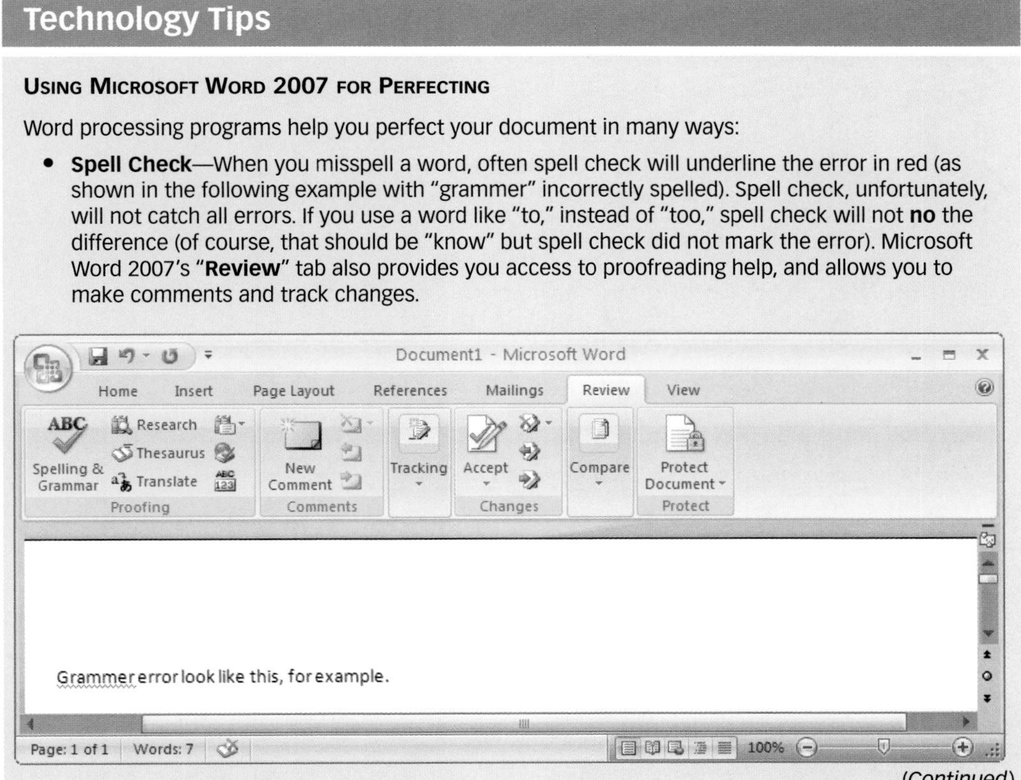

(Continued)

- **Grammar Check**—Word processors also can help you catch grammar errors. Grammar check underlines errors in green. When you right-click on the underlined text, the word processing package will provide an optional correction, such as the example shown.

Grammer error look like this, for example.

error looks
errors look
Ignore Once
Grammar...
About This Sentence
Look Up...
Cut
Copy
Paste

- **Add/delete**—Word processing makes adding new content and deleting unneeded text very easy. All you need to do is place your cursor where you want to add/delete. Then, to add, you type. To delete, you hit the Backspace key or the Delete key.
- **Move**—The Copy, Cut, and Paste features of word processing allow you to move text with ease.
- **Enhance/reformat**—In addition to changing the tone of your text, you also can enhance the visual appeal of your document at a keystroke. From the "**Home**" tab on your toolbar, you can choose from the Word 2007 Ribbon and include bullets, italics, boldface, font changes, numbered lists, and so on.

example

Note that the savings and loan incorrectly typed the customer's street rather than the last name. The underlined words are misspelled.

First City Federal Savings and Loan
1223 Main
Oak Park, Montana

October 12, 2009

Mr. and Mrs. David Harper
2447 N. Purdom
Oak Park, Montana

Dear Mr. and Mrs. Purdom:

In response to your request, your account with us has been close out. We are submitting a check in the amount of $468.72 (your existing balance). If your have any questions, please fill free to contact us. We look forward to hearing form you.

Proofreading Tips

The following list contains some helpful tips for proofreading text.

1. **Let someone else read it.** You miss errors in your own writing for two reasons. First, you might make the error because you do not know any better. Second, you read what you think you wrote, not what you actually wrote. Another reader might catch errors you do not.
2. **Let it sit.** Write your correspondence at 8:00 a.m., and read it at 10:00 a.m. Write it in the morning, and read it again in the afternoon. Write it on Friday, and let it sit until Monday. When you read it again later, you will be more objective.
3. **Read backwards.** You cannot do this for content. You should only read backwards to slow yourself down and to focus on one word at a time to catch typographical errors.
4. **Read one line at a time.** Use a ruler or scroll down your PC to isolate one line of text. This slows you down for proofreading.
5. **Read long words syllable by syllable.** How is the word "responsiblity" misspelled? Look at it again, more closely. Notice that the next to the last "i" is missing. You

might not have caught the error the first time you read the sentence. If you have seen a word thousands of times, you read what it is supposed to be, not what is actually typed. In contrast, if you read the word slowly, one syllable at a time, you see that the word is incorrectly typed "re-spon-si-bli-ty."

6. **Use technology.** Computer spell checks are useful for catching most errors. They might miss proper names, homonyms (their, there; it's, its; to, two, too), or incorrectly used words such as "effect" versus "affect," or "device" to mean "devise." Spell checks do not understand context. For instance, if you mean to write "The fog is heavy tonight" but mistype this and write "The dog is heavy tonight," spell check will not catch this error.
7. **Check figures, scientific and technical equations, and abbreviations.** If you mean $400,000, do not write $40,000. Double-check any number or calculations. If you mean to say HCl (hydrochloric acid), do not write HC (a hydrocarbon).
8. **Read it out loud.** Sometimes you can hear errors that you do not see. For example, you know that *a* outline is incorrect. It sounds wrong. *An* outline sounds better and is correct.
9. **Use a dictionary.** If you are uncertain about how to spell a word, look it up.
10. **Print it out.** Reading a printed copy is easier than reading on a computer screen.

PERFECTING CHECKLIST

____ 1. Have you added information for clarity?
- Answer the reporter's questions (Who, What, When, Where, Why, How).
- Provide specific details.
- Avoid connotative words; use denotative words.

____ 2. Have you achieved conciseness?
- Delete dead words and phrases.
- Limit sentence length.
- Delete "be" verbs.
- Use active voice versus passive voice.
- Delete the expletive pattern ("there" and "it").
- Avoid "shun" words (ending in *-tion* or *-sion*).
- Avoid camouflaged words.
- Limit prepositional phrases.

____ 3. Did you simplify words for conciseness and easier understanding?

____ 4. Have you moved information for emphasis?

____ 5. Did you reformat your text for ease of access?
- Limit paragraph length.
- Use highlighting techniques.
- Insert graphics to clarify complex ideas.

____ 6. Did you enhance tone to build rapport?
- Emphasize "you usage."
- Use positive words.

____ 7. Have you proofread to correct errors?

Nicole Met Her Challenge

To meet her communication challenge, Nicole used the P^3 process.

Perfecting

In Chapter 4, you were introduced to Nicole Stefani, who had been hired by ImageSkill as a public relations expert. Her first major job as a new hire is to help the city of Greenfield improve their public relations and promote an effective overall image. See Chapter 4, pages 107–108, to learn how Nicole gathered her data and planned her communication. In Chapter 5, Nicole packaged her written communication by writing a rough draft of her proposal, organizing the text by importance. See pages 131–133 to learn how Nicole packaged her proposal. Now, Nicole will *perfect* her communication to ensure its professionalism. She will have her boss, Marc Shabbot, review and edit her written text.

Marc makes tracking changes to revise the rough draft as follows:

- Adding details for clarity
- Deleting wordiness for conciseness

(*Continued*)

- Simplifying for easier understanding
- Moving text for emphasis
- Enhancing tone
- Reformatting for access
- Proofreading to correct errors for accuracy

Nicole says, "I know that my documents will be read by decision makers and the public and could result in expenditures of money, time, and people. I want to communicate in the best way possible and help the city of Greenfield to meet the needs of its populace. The *P*³ process lets me feel confident about my writing skills. The suggestions that my boss made will allow me to perfect the proposal before I have to present it to decision makers. I know that I'm not a 'perfect' writer and that there is always room for improvement.

I am pleased with the proposal I wrote for the city of Greenfield. Following the *P*³ process of planning, packaging, and perfecting let me develop the entire proposal and be proud of the result. My boss has already given me some positive strokes. Because of the success of this proposal, I look forward to the oral presentation when I can elaborate on the proposal's content. Some of the fear I usually feel before I make a presentation has dissipated by following a process approach to communication."

Figure 6.1 shows Marc's suggested revisions made using tracking changes.

FIGURE 6.1 Nicole's Rough Draft with Tracking Changes

Comment: Nicole, improve your subject line by adding the focus of this proposal.

Comment: Could you make those headings more informative?

Comment: Delete the semicolon. In fact, you don't need any punctuation after a centered heading.

Comment: Spelling error, Nicole: "areas."

Comment: Be sure to make this visual aid clearly informative by adding a key or headings.

Comment: Spelling error, vs. "its"

Date: August 13, 2009
From: Nicole Stefani
To: Greenfield City Council
Subject: Proposal

Introduction:

The city of Greenfield wants to promote the city. Greenfield city management wants to inform new homeowners and prospective businesses to move to Greenfield. ImageSkill can help Greenfield meet this problem.

Discussion;

To determine the best way to promote growth in Greenfield, ImageSkill staff followed this procedure. We met with a representative body of Greenfield citizens to ask questions. From this survey, ImageSkill determined which areas Greenfield citizens considered to be the city's most alluring. Our goal was to decide how best to maximize Greenfield's perceived strengths and opportunities for growth and give us insight into the best area's in which growth could be accomplished.

Following are our findings regarding strengths and growth opportunities:

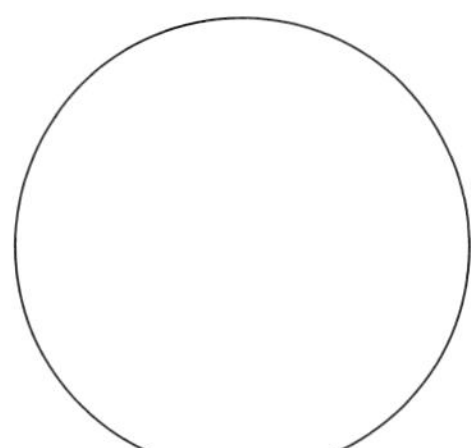

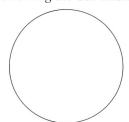

On a scale of 1–4, with 4 being best, Greenfield citizens concluded that the city's strengths are it's schools (4), city services (4), infrastructure (2), housing (1), caring people (1), amenities (1), and safety (1). The citizens would like to see growth in the following areas: more jobs/higher pay (7), more entertainment opportunities (6), more green space/hiking trails (6), more and improved schools (5), and city promotion (4).

FIGURE 6.1 Continued

The citizens of Greenfield want to accomplish all of the above goals without hurting the city's strengths.They don't want pollution, traffic, increased taxes, or higher housing costs.They want to maintain Greenfield's "small town friendliness," the city's responsiveness to their needs, and the city's charm.These are the challenges that you are facing.

ImageSkill can help Greenfield solve its problems. As a full service marketing company, we can help you as follows. We can create brochures for your Chamber of Commerce that will market your city's charms and intrinsic values as well as promote growth opportunities for new businesses, educational growth, entertainment, and parks/recreational prospects. We can build a Web site for your city, highlighting all of the values your citizens are so proud of. We can manage image control. As your city grows, citizens will resent change, which is inevitable, so packing this growth is the challenge, and ImageSkill can solve this problem by providing you outstanding radio, tv, and newspaper sound bites as well as periodical articles that promote the city's changes as best for the common good. We can also help you manage your city's events/conferences. As new companies move to your city, the city will experience growth in the convention industry. In addition, one way to ensure that your citizens continue to enjoy their city's "small town charms" is to provide city events (parades, holiday fare, or just occasional activities) that amplify city pride and augment whatever events are currently in place.

Comment: This is a 50 word sentence, Nicole. Let's cut in half at least.

Comment: Nicole, this entire paragraph is hard to follow. Consider adding subheadings and bulleting key points for easier access.

Conclusion:

ImageSkill wants to be your full service marketing agent. We have the talent and the skill to enhance the city of Greenfield's growth opportunities. Here's what our staff of seven employees will provide you:

1. Nicole Stefani—Customer service representative. I am your personal contact at ImageSkill. I have a BS in journalism and public relations from IU and expertise with city/government demographics.
2. Marc Shabbot—CEO, ImageSkill. Marc has over 25 years of experience in not-for-profit marketing and advertising. He is the founder and chairman of our company.
3. Joe Kingsberry—Desktop publishing. When we create your fliers and brochures, Joe will man the helm. He has knowledge of all Microsoft and Adobe Suite applications.
4. Cheryl Huff—Web designer. Cheryl is an expert with all Web design tools, including Flash, FrontPage, Netscape Composer, and IE. Please see our online Web portfolio of her work (http://www.imageskill.com/portfolio).
5. Larry Massin—Customer service rep. Larry and I will partner on your account. He has a BS in marketing and over ten years of experience.
6. Christy McWard—Writer. Christy has a BA in journalism and an MA in marketing. She is our chief wordsmith.
7. Robert Cottrell—R&D. We build your brochures and Web site based on facts. Bob is in charge of all research.

Comment: Define this abbreviation, please.

Comment: What's "IE"?

Comment: Nicole, I feel like we leave the reader hanging. This proposal needs a conclusion, especially something that ends on an upbeat and positive note. Let's use a conclusion to resell our benefits and to suggest a follow-up action.

Figure 6.2 shows Nicole's perfected proposal.

FIGURE 6.2 Perfected Proposal Presented to the Client

Date: August 13, 2009
From: Nicole Stefani
To: Greenfield City Council
Subject: Proposal to Market the City of Greenfield

How ImageSkill Can Meet Your Needs:
The city of Greenfield wants to promote the city. Greenfield city management wants to inform new homeowners and prospective businesses to move to Greenfield. ImageSkill can help Greenfield meet this problem.

Solutions to Greenfield's Public Relations Challenges:
To determine the best way to promote growth in Greenfield, ImageSkill staff followed this procedure. We met with a representative body of Greenfield citizens who participated in a survey. From this survey, ImageSkill determined which areas Greenfield citizens considered to be the city's most alluring. Our goal was to decide how best to maximize Greenfield's perceived strengths and opportunities for growth. An analysis of the survey gave us insight into the best areas in which growth could be accomplished. Following are our findings regarding strengths and growth opportunities.

Strengths (as defined by the number of responses to the survey)

Figure 1: Greenfield Strengths

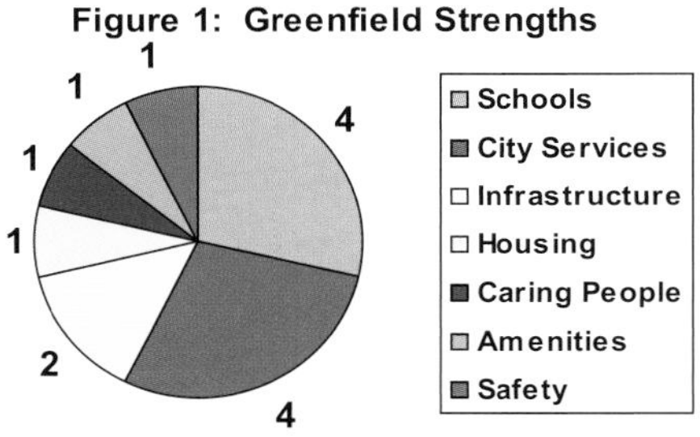

On a scale of 1–4, with 4 being best, Greenfield citizens concluded that the city's strengths are its schools (4), city services (4), infrastructure (2), housing (1), caring people (1), amenities (1), and safety (1).

Growth Opportunities

The citizens would like to see growth in the following areas, as defined by the number of responses to the survey: more jobs/higher pay (7), more entertainment opportunities (6), more green space/hiking trails (6), more and improved schools (5), and city promotion (4).

FIGURE 6.2 **Continued**

Nicole Stefani
Page 2
August 13, 2009

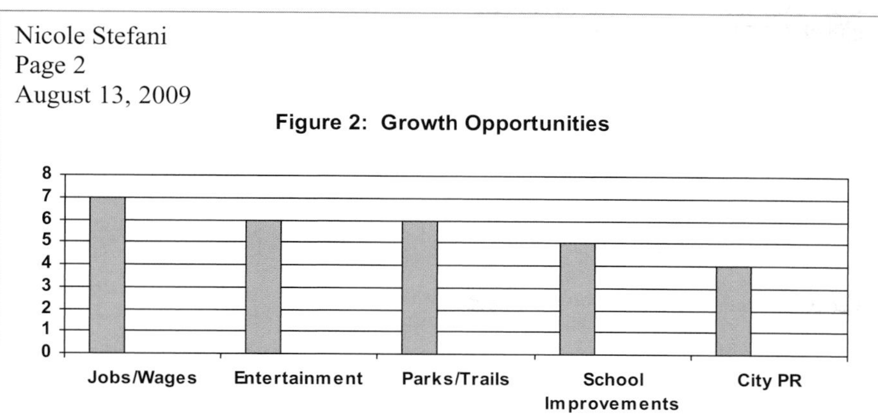

The citizens of Greenfield want to accomplish all of these goals without hurting the city's strengths. They do not want pollution, traffic, increased taxes, or higher housing costs. They want to maintain Greenfield's "small town friendliness," the city's responsiveness to their needs, and the city's charm. These are the challenges that you are facing.

ImageSkill's Value Added

ImageSkill can help Greenfield by providing you the following benefits:

- *Create brochures* for your Chamber of Commerce that
  - ⇨ Market your city's charms and intrinsic values
  - ⇨ Promote growth opportunities for new businesses, educational options, entertainment, and parks/recreational prospects.
- *Build a Web site* for your city, highlighting all of the values which bring pride to your citizens.
- *Manage your city's events/conferences.* For example, we suggest an annual "Homesteader's Day" to honor Greenfield's "small town charms." This festival would amplify city pride and include parades, music, crafts, and sporting events—all under our caring management. Furthermore, as new companies move to your city, your convention industry will grow.
- *Provide image control.* As your city grows, citizens might resent change. This is inevitable. Positively packaging this growth is our challenge and our skill. ImageSkill will achieve this goal as follows:
  - ⇨ Ensure outstanding radio, television, and newspaper coverage.
  - ⇨ Write periodical articles that promote the city's growth, regionally and nationally.

(*Continued*)

Nicole says, "I am confident that the proposal will be well received by my audience. Following the three stages of the P[3] process gave me the time to execute the document to the best of my ability. Using the series of headings allowed me to perfect the layout and design of the proposal."

FIGURE 6.2 Continued

Nicole Stefani
Page 3
August 13, 2009

ImageSkill's Expert Staff

When you hire ImageSkill, you hire the best. Our staff provides you years of experience and diversity of talent.

1. *Nicole Stefani*—Customer service representative. I am your personal contact at ImageSkill. I have a BS in journalism and public relations from Indiana University and expertise with city/government demographics.
2. *Marc Shabbot*—CEO, ImageSkill. Marc has over 25 years of experience in not-for-profit marketing and advertising. He is the founder and chairman of our company.
3. *Joe Kingsberry*—Desktop publishing. To create your fliers and brochures, Joe uses his extensive knowledge of all Microsoft and Adobe Suite applications.
4. *Cheryl Huff*—Web designer. Cheryl is an expert with all Web design tools, including Flash, FrontPage, Netscape Composer, and Internet Explorer. Please see our online web portfolio of her work (http://www.imageskill.com/portfolio).
5. *Larry Massin*—Customer service rep. Larry and I will partner on your account. He has a BS in marketing and over ten years of experience.
6. *Christy McWard*—Writer. Christy has a BA in journalism and an MA in marketing. She brings both print and oral communication skills to your projects.
7. *Robert Cottrell*—R&D. We build your brochures and Web site based on facts. Bob is in charge of all research.

Enhancing Greenfield's Growth:
ImageSkill wants to be your full service marketing agent. We have the talent and the skill to enhance the city of Greenfield's growth opportunities. Please allow me to meet with your city council to make a PowerPoint presentation, showing you examples of ways in which our talents perfectly meet your needs.

CHAPTER HIGHLIGHTS

1. When you perfect a business document, you add information for clarity, delete wordiness for conciseness, simplify complex words, or move information. You can also reformat for ease of access, and enhance the style and tone through positive words and pronouns.
2. Perhaps the most important part of this final stage is correcting errors in your written communication. Accuracy ensures your professionalism.
3. Answering the reporter's questions will help you to add detail for clarity.
4. To achieve conciseness, delete "be" verbs, use active voice verbs, delete expletives, avoid *shun* and camouflaged words, and limit prepositional phrases.
5. Highlighting techniques, such as bullets, numbers, boldface, and italics, will make your text more accessible. However, avoid overusing these methods.

WWW *Find additional perfecting information and exercises, samples, and interactive activities at http://www.prenhall.com/gerson.*

MEETING WORKPLACE COMMUNICATION CHALLENGES

CASE STUDY

Julie Schopper is a human resources supervisor. Her employee, William Huddleston, has exceeded the department's budget for training. Julie's job is to explain the problem to William and to offer solutions. To do so, she wrote a memo to William, but the memo is flawed.

Assignment

Using the techniques discussed in this chapter, revise the flawed memo. Be sure to

- Add details for clarity
- Delete wordiness
- Simplify the word usage
- Move information for emphasis
- Reformat the text
- Enhance the tone
- Correct grammatical errors

Date: April 3, 2009
To: William Huddleston
From: Julie Schopper
Subject: TRAINING CLASSES

Bill, your recent training budget has exceeded our projections. You need to solve this problem are else your department will get the blame. I have discussed this issue with management and we have come up with several suggestions that you need to review and get back to us with your assessments.

You could reduce the number of training classes you have scheduled, fire several trainers, increase the number of participants allowed per class, bill participants separately for training materials, or reduce training material costs by creating online PDF downloads that participants could print out at their own workstations and bring to training sessions. With these suggestions, we would keep the same amount of income from participants, the company would also save a significant amount of money due to the reduction of trainer salaries and training expenditures. The downside might be less effective training, once the trainer to participant ratio is increased. As another option, we could outsource our training. This way we could fire all our in house trainers, which would mean that we would save money on benefits and salaries, as well as offer the same number of training sessions, which would keep our trainer to participant ratio low.

What do you think. We need your feedback before we precede, so even if your busy, get on this right away. E-mail me by Tuesday with your thoughts.

ETHICS IN THE WORKPLACE CASE STUDY

Kim Ngyuen is a manager at a nationwide communications company, XConnect. Because of a declining economy, company revenues have fallen 29 percent since last quarter. On Friday, Kim sent the following press release regarding the company:

"As you all know, our economy has slowed down, but prices for all services and goods are growing rapidly. Plus, the communications industry is changing. Most communications companies are streamlining their operations to meet economic challenges. Many operations are moving offshore. Increasingly, contract workers are being brought in to decrease the rising costs of benefits packages. Investors demand that our company show profitability. Soon, with improved strategies, we will be able to increase our business. However, in the short term, to offset losses, we will cut approximately 1,000 jobs. The company will offer voluntary separation packages followed by involuntary cuts as needed."

Question: Though the information about job loss is in Kim's press release, Kim has placed this information at the end of a long paragraph. Is his placement of content ethical since this is how his employees learned the bad news? Has Kim used the best method of organization to convey his content?

INDIVIDUAL AND TEAM PROJECTS

1. *Adding Information for Clarity*

   a. **Reporter's Questions** The following sentences are vague. Different readers will interpret them differently. Revise these sentences by answering *reporter's questions* (Who, What, When, Where, Why, and/or How).

   1. We need this information for the meeting.
   2. The machinery will replace the broken equipment in our department.
   3. If we fail to meet their request, we will lose the account.
   4. Weather problems in the area resulted in damage to the computer systems.
   5. If we cannot solve this problem, we will not meet the customer's deadline.

   b. **Specificity** The following sentences are vague and need specific details.

   1. We need reports as soon as possible.
   2. Failure to meet the deadline could have a negative impact.
   3. Insufficient personnel caused the most recent occurrences.
   4. Fire in the office led to substantial losses.
   5. By not completing the recent deal, we lost a large percentage of our business.

2. *Deleting Wordiness for Conciseness*

   a. **Readability.** Revise the following paragraph by deleting wordiness and simplifying words.

> Ramifications of yesterday's revised implementation schedule are significant because doing as requested by management could lead to missed deadlines as well as the potential for production malfunctions. I respectively request that management reconsider these suggestions, taking into consideration the short-term longevity of our employees, many of whom are newly hired. Instead, I am of the opinion that any inconveniences our company might experience due to revising the schedule will be offset by the inestimable values we will derive. I am cognizant that changes are challenging to make, but management might consider doing so at this point in time to benefit employee morale.

   b. **Passive voice.** Passive voice often leads to vague, wordy sentences. Revise the following sentences by rewriting them in the active voice.

   1. Installation of the new network-wide software was carried out by the information technology department.
   2. Benefits were derived when George attended the conference.

3. The information was demonstrated and explained in great detail by the training supervisor.
4. Discussions were held with representatives from Allied, who supplied analytical equipment for automatic upgrades.
5. The symposium on HIPAA rules and regulations was attended by the nursing staff.

c. **Sentence length.** The following sentences are unnecessarily wordy. They contain expletives,"be" verb constructions, *shun* words, camouflaged words, and wordy prepositional phrases. Revise the sentences to make them more concise.

1. In regard to the progress reports, they should be absolutely complete by the fifteenth of each month.
2. I wonder if you would be so kind as to answer a few questions about your proposal.
3. I am in receipt of your memo requesting an increase in pay and am of the opinion that it is not merited at this time due to the fact that you have worked here for only one month.
4. In this meeting, our intention is to acquire a familiarization with this equipment so that we might standardize the replacement of obsolete machinery throughout our entire work environment.
5. It is anticipated that these changes will lead to a reduction in cost expense overages.

3. *Simplifying for Easier Understanding* To limit sentence length, limit word length. Find shorter, easier-to-understand words to replace the following words.

1. aforementioned
2. adjacent
3. approximately
4. ascertain
5. attached herewith
6. consequence
7. elucidate
8. demonstrate
9. frequently
10. identify
11. numerous
12. occasion
13. residence
14. subsequent
15. terminate

4. *Reformatting for Access* The following text is not visually appealing. Its denseness makes it hard to read and understand. Reformat the text for access. This can include smaller paragraphs and highlighting techniques, such as headings, itemized lists, boldface and italics, or even graphics (pie chart, bar chart, or table).

The project director is designated as the account manager. The account manager's primary responsibility is to ensure that departmental outgo does not exceed budgeted funds. Biannual and annual operating expenditures are given to each account manager in the prior fiscal year. This way, all budgetary information can be shared with employees during the annual departmental in-service meeting. These budgetary guidelines will be used to assure that all charges on departmental accounts are correct and fit within the budgeted amounts. When reporting budgeted activities, be sure to use the standard code designations and code numbers: Income, 100; Salaries, 200; Benefits, 300; General Supplies, 400; Indirect Costs, 500; Retirement, 600; Travel, 700. For additional information about cost guidelines and for updates, employees and project managers can access the company's intranet budget site: http://www.FWB.com/AccountRule.htm. The password for all interested parties is "AccountUpdate." If you have questions, contact our accounting help desk (1-800-accthelp).

5. ***Moving Information for Emphasis*** Revise the information in the following memo, reorganizing it according to importance. You might want to itemize the body text from most important to least, or from least important to most. Justify your decisions.

Date: April 23, 2009
To: Anita Morales, Bank President
From: Quality Circle Team Members
Subject: Bank Employee Concerns

In our last Quality Circle meeting, our team members, with input from bank employee representatives, discussed the following concerns.

- **Flextime.** Currently, unlike other banks in our region, our employees do not have flextime work scheduling. If we could add this to our benefits package, employees could vary their times of arrival and departure. This would give the bank more coverage, lessen crowding in the parking lot during peak hours, and provide employees more flexibility for childcare, medical appointments, and other personal needs.
- **Employee safety.** We have three safety concerns. First, during winter, sidewalks are not regularly deiced. Last February, for example, one of our employees slipped on the icy sidewalk in front of the bank and broke her arm. Improved cold-weather maintenance would solve this problem. Second, our parking lot is poorly lit. As you know, the bank is in a designated "high-crime area." To solve this problem, we are pairing up to walk to our cars at the close of business. A better solution is necessary. Finally, our security guards are not always at their assigned stations during the day. Supervision should address this issue.
- **Workplace ergonomics.** The interior of the bank is too dark. We need better lighting to improve work conditions. In addition, our desk chairs do not adjust for height, nor do they have flex backs for work comfort.
- **Grievance procedures.** In our current system, if employees have grievances, they are required to report first to their immediate supervisors (the chain of command). If, however, the problem is with a supervisor, employees will not file a complaint for fear of retribution or fear of hurting the supervisor's feelings. A better approach would be to hire either an ombudsperson or a grievance officer. This person would assess complaints more objectively and without the threat of reprisal.
- **Employee evaluations.** Currently, supervisors evaluate employees biannually. These evaluations determine our raises. However, the evaluations consist of little more than casual observations. Our assessments are not based on standardized criteria. Thus, the evaluations are primarily subjective. A committee composed of managers and employees should be formed to create assessment criteria.

Please consider our requests and let us know when we can meet for further discussion. Thank you for your consideration.

6. ***Enhancing Tone*** The following e-mail is demanding, harsh, and impersonal. Revise the flawed e-mail by using positive words and pronouns to promote rather than punish.

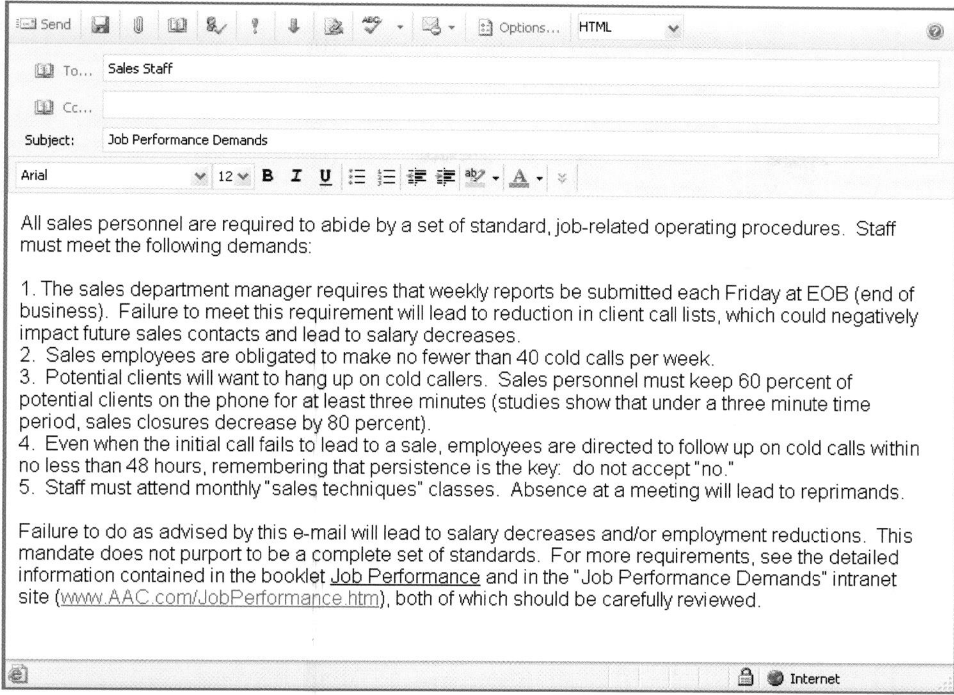

To... Sales Staff

Cc...

Subject: Job Performance Demands

All sales personnel are required to abide by a set of standard, job-related operating procedures. Staff must meet the following demands:

1. The sales department manager requires that weekly reports be submitted each Friday at EOB (end of business). Failure to meet this requirement will lead to reduction in client call lists, which could negatively impact future sales contacts and lead to salary decreases.
2. Sales employees are obligated to make no fewer than 40 cold calls per week.
3. Potential clients will want to hang up on cold callers. Sales personnel must keep 60 percent of potential clients on the phone for at least three minutes (studies show that under a three minute time period, sales closures decrease by 80 percent).
4. Even when the initial call fails to lead to a sale, employees are directed to follow up on cold calls within no less than 48 hours, remembering that persistence is the key: do not accept "no."
5. Staff must attend monthly "sales techniques" classes. Absence at a meeting will lead to reprimands.

Failure to do as advised by this e-mail will lead to salary decreases and/or employment reductions. This mandate does not purport to be a complete set of standards. For more requirements, see the detailed information contained in the booklet Job Performance and in the "Job Performance Demands" intranet site (www.AAC.com/JobPerformance.htm), both of which should be carefully reviewed.

7. ***Correcting Errors for Accuracy*** The following memo contains grammatical errors, including problems with spelling and punctuation, as well as agreement, capitalization, and number usage. Correct the errors to ensure professionalism.

Date: Febuary 12, 2009
To: Martha Collins
From: Richard Davis
Subject: 2009 Digital Carriers

Attached is the supplemental 2009 Digital Carriers reports that is required to support this years growth patterns. As we have discussed in previous phone conversations the January numbers show a decrease in traffic but forecasts still suggest increased traffic therefore we are issuing plans for this contingency.

If the January forecasts prove to be accurate the carriers being placed in the network via these plans will support our future growth accept for areas where growth cannot be predicted. Some areas for example are to densely populated for forecasting. Because the company did not hire enough survey personal to do a thorough job.

Following is an update of our suggested revisions;

Digital Carriers Needed	Capitol Costs
52,304	$3,590,625

If your going to hire anyone to provide follow up forecasts, they should have sufficient lead time. The survey teams, if you want a successful forecast, needs at least three months. 25 team members should be sufficient. If we can provide farther information please let us know.

DEGREE-SPECIFIC ASSIGNMENTS

Following are mission statements from various organizations related to your majors. Individually or in small groups, read these mission statements. Decide where they are successful and how they could be improved. For improvements, focus on this chapter's techniques for perfecting:

- *Add* information for clarity.
- *Delete* wordiness for conciseness.
- *Simplify* unnecessarily complex words for easier understanding.
- *Move* information around (cut and paste) to ensure that your most important ideas are emphasized.
- *Reformat* (using highlighting techniques) to ensure reader-friendly ease of access.
- *Enhance* the tone and style of the text.
- *Proofread* and *correct* any errors to ensure accurate grammar and content.

1. **Marketing and Public Relations.** The following mission statement from *e*Marketing Association can be found online at http://www.emarketingassociation.com/about.htm.

> The eMarketing Association (eMA) is the world's largest international association of emarketing professionals. Members include governments, companies, professionals and students involved with the emarketing arena. The eMA provides marketing resources, services, research, certifications, educational programs and events to its members and the marketing community. The eMA works with a number of organizations, companies and governments on issues related to eCommerce, multi-channel marketing and legislative issues. The eMA has members in over 40 countries around the world, and sponsors or manages over 30 events a year. Over 3,000 marketers have enrolled in eMA online courses and thousands of eMarketing professionals and students have achieved certification status.

2. **Business Information Systems.** The following mission statement from *BT Syntegra USA*, a systems integration and consulting firm, can be found online at http://www.btconsulting.com/.

> BT's Consulting & Systems Integration business helps organizations transform the way that they operate in the digital networked economy by applying business knowledge and technology to make possible new and better ways of working. It provides consultancy and systems integration services including business consultancy, complex program management, and custom systems design, development and operation.
>
> So, why is consulting and systems integration from BT different? Unlike others in the technology world, we don't need to confuse customers to prove our worth—we become part of a collaborative team, working together to develop the most innovative and complete solutions. We pride ourselves not only on the extent of our technical knowledge, but also on the depth of understanding we have about our customers' business and their markets. This expertise ensures that the systems we design and build have an immediate business impact. Impact on efficiency, on performance, on competitiveness. Although every program we work on is different one thing always remains the same. We never start an engagement unless we and our customer are absolutely clear about our role and the benefits that our customer is anticipating.

3. **Accounting.** The following mission statement from the American Institute of Certified Public Accountants can be found online at http://www.aicpa.org/about/mission.htm.

THE AMERICAN INSTITUTE OF CERTIFIED PUBLIC ACCOUNTANTS OBJECTIVES

To achieve its mission, the AICPA:

A. Advocacy
Serves as the national representative of CPAs before governments, regulatory bodies and other organizations in protecting and promoting members' interests.

B. Certification and Licensing
Seeks the highest possible level of uniform certification and licensing standards and promotes and protects the CPA designation.

C. Communications
Promotes public awareness and confidence in the integrity, objectivity, competence and professionalism of CPAs and monitors the needs and views of CPAs.

D. Recruiting and Education
Encourages highly qualified individuals to become CPAs and supports the development of outstanding academic programs.

E. Standards and Performance
Establishes professional standards; assist members in continually improving their professional conduct, performance and expertise; and monitors such performance to enforce current standards and requirements.

4. **Management.** The following mission statement from HumanResources.com can be found online at http://www.humanresources.com/.

HumanResources.com is the source for information, products and services that are related to the human resources field of activity. We cater to human resources professionals, recruiters, career seekers, people on the job, as well as executives. We are committed to providing an environment that cultivates knowledge, insight, and personal development.

We recognize the need for keeping current in your chosen profession and in the world of work in general. At HumanResources.com you will find current news and articles from a variety of well-respected newspapers, magazines, and journals from around the world. You will find interesting perspectives, topical stories, case studies, and helpful advice from the leading minds in the field.

We encourage you to use us as your starting point to link to the various industry associations, government agencies, and suppliers of products and services in order to make your job easier.

Quick Facts:
– Operating since 1999, Major re-design in 2004
– Owned by Sites To See Corporation
– Based in Toronto, Ontario, Canada

We hope you enjoy visiting HumanResources.com. If you have a chance, drop us a line to let us know what you think, and what you would like to see more of on the site. All feedback is highly appreciated!

5. **Finance.** The following mission statement from Financial Executives International can be found online at http://www.fei.org/about/us.cfm.

Our mission: to be the preeminent association for financial executives by:
networking—providing forums for peer networking
knowledge—alerting members to emerging issues
advocacy—advocating the views of financial executives
ethics—promoting ethical conduct

Networking *your key to success*:
15,000 peers–CFOs, VPs of Finance, Treasurers, Controllers, Tax Executives, Academics, Audit Committee members
companies large and small, public and private, cross-industry
conferences, teleconferences, webcasts, publications
chapter meetings, member database, email-based discussion forums

Knowledge *saving you time, helping you work smarter*
emerging issues and market trends
regulatory and legislative updates
practical research, benchmarking and case studies
professional development, online and in-person
valuable career management tools
a one-stop website offering virtual benefits

Advocacy *speaking for your interests*
a widely-respected voice for corporate finance issues
representing you before the SEC/FASB/PCAOB/IASB/Congress

Ethics *raising the bar for the profession*
FEI's code of ethics is an industry model
dedicated to advancing responsible, ethical financial management leadership skills

PROBLEM-SOLVING THINK PIECES

In teams or individually, consider the following issues. Then, provide your answers either by writing a report or by making an oral presentation.

1. This chapter emphasizes the importance of positive word usage to build rapport and trust. However, you will not always be able to use positive words. Think of instances in workplace communication when you would be required to focus on the negative. Find examples from the workplace or online to support your contentions.
2. This chapter emphasizes the importance of "you usage"—pronouns to build rapport and to create a personalized tone. However, in some instances, personalization might not be appropriate in business communication. Decide when correspondence could or should avoid "you usage." Find examples from the workplace or online to prove your point.
3. Many businesses ask their employees to *write between sixth- to eighth-grade levels* to make correspondence easier to read. However, there might be instances where correspondence (memos, letters, e-mail messages, reports, etc.) could be or should be written at a higher grade level. When do you believe that it is acceptable for workplace communication to be written at a "higher" level? Explain your decision, either in writing (memo, e-mail message, or report), or present your findings in an oral presentation. Find examples of workplace communication which, in your opinion, have been appropriately written at a "college" level.

WEB WORKSHOP

1. Access an Internet search engine (Google, Yahoo, Excite, Lycos, AskJeeves, etc.) and type in "readability." Research the sites you find for examples or discussions about the importance of and challenges with readability. Write an e-mail message or memo to your instructor describing your findings.
2. Many newspaper and journal articles are written about the problems caused by poor workplace communication. Access an Internet search engine (Google, Yahoo, Excite, Lycos, AskJeeves, etc.), and research the topic of poor workplace communication. Write an e-mail message or memo to your instructor describing your findings, or share your findings with your class in an oral presentation.

QUIZ QUESTIONS

1. What are at least five ways to perfect your workplace communication?
2. List two ways to add information for clarity.
3. What are reasons for being concise in your workplace communication?
4. List at least five ways to improve readability.
5. List five ways to proofread your text.

CHAPTER 7

Oral Presentations and Nonverbal Communication in the Workplace

▸REAL PEOPLE *in the workplace*

Nurani Singh has just been hired as recruitment specialist in the human resources Department for Monroe, Louisiana. She will be responsible for coordinating the recruitment and hiring of city employees. She also will administer such things as the city's retirement system, unemployment compensation, family medical leave programs, and the city's drug and alcohol abuse policy.

As a recruitment specialist, Nurani will be involved in staffing all merit system positions. This will include ensuring that all employee recruiting, advertising, testing, certification, and applicant record-keeping comply with legal, professional, and merit system standards.

She will meet these responsibilities as follows:

- Advertising and recruiting
- Reviewing approximately 120,000 applications a year to determine whether or not the applicants meet position requirements
- Verifying experience and education
- Establishing, maintaining, and certifying eligibility lists for all vacant city positions

The largest part of her job will be making presentations to city employees and conducting new-employee orientation each month. In this capacity, she will prepare PowerPoint slides and speak in front of small and large groups, in both formal and informal presentations.

Nurani is well educated with a degree in Communications from the University of Texas. However, like most employees, she finds speaking in public to be a challenge.◂

CHAPTER GOALS

When you complete this chapter, you will be able to do the following:

1. Understand the importance of verbal and nonverbal communication in the workplace.
2. Develop effective listening skills.
3. Use the telephone to communicate successfully.
4. Use voice mail effectively.
5. Deliver effective informal oral presentations on the job.
6. Participate in teleconferences.
7. Communicate effectively in a videoconference or Webconference.
8. Deliver formal oral presentations.
9. Use a variety of visual aids to enhance your oral presentations, including techniques for effective PowerPoint presentations.
10. Use the *P³* process—planning, packaging, and perfecting—for informal and formal oral communication.

WWW

To learn more about oral communication, visit our Web Companion at http://www.prenhall.com/gerson.

Nurani's Communication Challenge

When Nurani started working for Monroe, Louisiana, as a specialist for recruitment in human resources, she was not totally familiar with HR rules and regulations. She says, "I had only one night to study and prepare for my first oral presentation, 'Conducting Effective Interviews,' to a group of 25 supervisors. Unfortunately, I mainly read the PowerPoint slides to the audience, and failed to engage them through interaction and activities. After the presentation, for which I was rated only 25 percent satisfaction out of a 100, I asked my supervisor how to improve. She told me to be better prepared and ask questions to engage my listeners.

My challenge was how to make presentations effective so that my preparation was apparent. In addition, I had to figure out how to engage the audience. My effective presentation skills would then enable the participants to gain confidence in me. After the presentation, I knew the attendees had to be comfortable calling me in my position as a human resources recruitment specialist and be assured that I had the answer.

My personal communication philosophy is to try to present my best work every time. With that philosophy in mind, I was challenged to improve my presentation skills to ensure effective communication. Without communication in business, I feel like we're all working in chaos."

See pages 201–203 to learn how Nurani met her communication challenge.

THE ROLE OF VERBAL AND NONVERBAL COMMUNICATION

Here is a fact: Excellent communication does not just depend on the words you say and write. Communication also depends greatly on the tone of your voice, your body language, and other verbal and nonverbal cues you give your audience.

How Important Is Verbal and Nonverbal Communication?

Some researchers suggest that "from 65 to 90 percent of every conversation is interpreted through body language" (Warfield 1). Other researchers state, "Linguists who study nonverbal elements of a conversation—including posture, eye contact, facial expression and gestures—have concluded that these silent elements make up 55% of the message. Tone of voice contributes another 38%. This leaves 7% for the words you use" (Smith 1; Tilton 15).

What Is Verbal Communication?

What you say is either positively or negatively impacted by *how* you say it. In fact, "after your physical appearance, your voice is the first thing people notice about you" (Clarke). Verbal communication includes your pace (rate of speech) and modulation.

Pace

A key concern is pace, the speed with which you speak. Generally, we speak about 150 to 200 words a minute. Speeding up this rate can have several meanings. You could imply that you are happy, fearful, angry, or surprised. In addition, if you talk too fast, not only will your coworkers have trouble following your train of thought, but also they might think you're nervous or impatient. If you are in a meeting, for instance, and you speak too rapidly, your voice might be telling your coworkers it's time to move on to the next topic. You are saying, "I don't have time for you" or "I want to be doing something else (anything else)."

Talking too slowly also causes problems. Slowing down your pace could convey sadness, boredom, or lack of interest. People also might feel you are "talking down" to them. A very slow delivery style could imply condescension—"I assume that you can't keep up with my thoughts, so I'll speak *realllly slowlllly*."

Modulation

Modulation is the loudness, tone, and pitch of your voice as you speak. A very loud voice will be perceived as dominating, while a very quiet voice might make you sound meek. Mumbling can destroy your credibility. A monotone voice, one with no change in pitch, will put your listener to sleep. Modulation, varying your pace and pitch, adds interest to your comments.

What Is Nonverbal Communication?

Nonverbal communication is another phrase for body language. Basically, nonverbals entail "any conscious or unconscious movement of a part or all of the body that communicates an emotional message" (Tilton 15). Nonverbal cues can include any of the following:

Eye Contact

Good eye contact suggests openness, confidence, and interest in your audience. However, too much of anything is potentially bad. Do not overdo eye contact; staring and glaring will make any audience uncomfortable. Many people believe that "duration of the eye contact should last between 3 to 5 seconds" (Tilton 15). Any more than that might be unsettling.

Multicultural Concerns In our multicultural marketplace, you will interact with customers and business professionals from other cultures who might consider direct eye contact offensive (Smith 1). Studies suggest that people from "contact cultures" (Southern Europe, Latin America, and Mexico) "focus directly on the other person's eyes." People from "non-contact cultures" (Asia, India, and Northern Europe) "focus mainly on the other person's head and face" rather than the eyes (Sime).

Facial Expressions

Smiling, frowning, yawning, wide-eyed bewilderment, open-mouthed shock, lip-curling angry glares . . . each of your facial expressions conveys a message. For example, if you are listening to a colleague speak and you are simultaneously wrinkling your nose, arching your eyebrows, scrunching your face, or furrowing your forehead, you are conveying a very negative message. You are showing your coworker your uncertainty or disbelief. In contrast, a pleasant smile shows your involvement and support. Figure 7.1 illustrates facial expressions.

Posture

Imagine you are in a meeting. A coworker is speaking, but you are fidgeting, slouching, placing your hands behind your head, or leaning to the side. You are reflecting boredom and disinterest. However, if you lean forward, you show you are interested and actively

FIGURE 7.1 **Facial Expressions**

Smiling

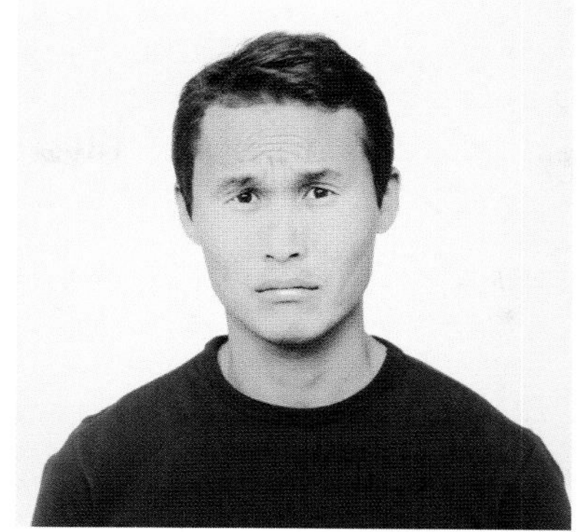

Frowning

Angry

Surprised

Bored

participating in the team meeting. Nodding your head up and down shows your affirmation. Figure 7.2 demonstrates how posture reveals attitude.

Proximity

What does proximity to your audience show? If you stand very close to a spouse or child, you show your love. However, if you stand very close to a coworker, you will encroach upon his or her "personal space." This suggests a threatening or domineering behavior. In *The Silent Language*, Edward T. Hall suggests that we have four main proximity zones of personal space (see Table 7.1).

Figure 7.3 shows the effect of proximity.

Multicultural Concerns Sometimes, proximity is a cultural consideration. Studies suggest that people from many Middle Eastern cultures stand closer while speaking than North Americans are used to. Middle Eastern men might greet each other by grasping both hands and placing kisses on each cheek. There is more physical contact, and conversations often involve touching. These "intimate" and "personal" behaviors in Western culture are acceptable "public" behaviors in the Mideast. In contrast, touching is not common in the Far East (China and Japan) where more formality and distance from the speaker are the norm.

FIGURE 7.2 **Attitude Revealed Through Posture**

Look at the people sitting in the picture. What does each person's posture tell you about his or her attitude?

Table 7.1 Proximity Zones of Personal Space

Zone	Distance	Who Gets Access	What Occurs Within the Zone
Intimate	6–18 inches (15–45 cm)	Spouse, children, and significant others	Intimate relationships, such as touching and hugging
Personal	1 1/2–4 feet (45–125 cm)	Family members and close personal friends	Within an arm's length, less intimate but still very close encounters
Social	4–12 feet (1.25–3.60 m)	Friends and colleagues	Social and business relationships
Public	Over 12 feet (+3.6 m)	Strangers/general public	Little interaction

Gestures

We speak a lot with our hands. A "thumbs up" gesture in the United States signifies that everything is "A-OK." A "thumbs down" gesture denotes a negative opinion. These two gestures are commonly understood in Western culture. Other gestures, however, are subtler. Making a fist shows anger, while rubbing the back of your neck or your forehead shows stress. If you place your hands in front of your mouth, you might be showing secretiveness and deception (Tilton 4). Finger pointing might help you make a point, but it is also very assertive, even threatening. We all know how annoying a person's rhythmic drumming can be. This suggests that the individual is impatient and bored.

Multicultural Concerns Many of our gestures are culturally based. Finger pointing in the Far East is considered to be rude. Worse still are the "thumbs up" and "OK" signs. In Italy and Australia, a "thumbs up" motion can be considered obscene. The "OK" gesture can be considered obscene in Brazil and Germany.

FIGURE 7.3 **Proximity Zones of Personal Space**

Personal

Public

Social

Intimate

Listening Skills

In addition to controlling verbal and nonverbal reactions, you must also learn to listen effectively. If you are not listening while others are speaking, no communication can occur. Successful collaboration demands that all voices be heard. Successful workplaces value everyone's input.

Barriers to Active Listening

What gets in the way of your active listening abilities? Following are barriers to active listening:

Multitasking Envision this scenario: You take your seat at the meeting table while you are still on your cell phone, speaking to a customer. You are also checking your PDA to see when you can meet with this client. Meanwhile, you are thinking about the deadline you have for submitting a proposal. You are wondering if your graphic artist has completed the sketch you need for the upcoming meeting. You are juggling many tasks at the same time. Now, the meeting you are currently in has begun, and your coworker is speaking. How can you also pay attention to this colleague, when you have so many other tasks to complete and so many other conflicting responsibilities?

Preconceived Notions If you believe that you already know what a coworker will say, then you will not hear his or her comments objectively. Assumptions like this are a barrier to successful teamwork. Preconceived notions will diminish the colleague's comments and hinder your contributions to the team effort.

Focusing on Your Response While your colleague is speaking, you do not wait to hear all he or she says. You only focus on your follow-up response. You can barely wait to jump in with a clever retort, a witty comeback, an oppositional point of view, or debunking doubt. This does your colleague and your team a disservice. By thinking only of what you might say in response, you risk not hearing all that is said in the meeting.

Interrupting the Speaker Interrupting a speaker is rude. A person should have the right to his or her say. If you interject your comments abruptly, then your coworker cannot complete his or her thoughts. Waiting to respond is not only good manners; it is also good business.

External Distractions Think of all the possibilities that can be distracting:

- The room is too hot or too cold.
- It is 11:30 a.m. and you are hungry.
- It is Friday and you have exciting plans for the weekend.
- A road crew is repairing the street outside your office complex, complete with the sounds of a jackhammer and the annoying beeping of a truck backing up.

These external distractions might disallow you from focusing on what your coworkers are saying.

Keys to Effective Listening

Effective listening is critical to a company's goal of continuous improvement. Employees can best represent a company by actively listening to vendors, customers, and coworkers. When you hear what others say, you can contribute to discussions more effectively and provide useful feedback. Following are some keys to effective listening (Smith 3):

Stop What You Are Doing; Concentrate on the Task at Hand Turn off your cell phone. Put your PDA away. Leave your unfinished work in your office. Focus on the team's project and on the speaker.

Do Not Talk It is hard to speak and listen at the same time. Sure, you might have valuable things to add, but wait your turn. Let the speaker have his or her say. Once that person has spoken, you can speak.

Make Eye Contact When someone in the group is speaking, look at this person. Do not look out the window, look at your watch, stare at your shoes, or look at your fingernails. Making eye contact with the speaker will help you focus your attention.

Take Notes Another way to stay focused is by taking notes. Write down key ideas for future reference. You can also jot down any thoughts you might have for follow-up discussion. This way, you will not feel compelled to interrupt the speaker. The notes will help you remember what you had wanted to say, once the speaker has completed his or her thoughts.

Be Objective Without a doubt, a colleague might say something with which you disagree. That's OK. We all have different points of view. However, remember that these differences are exactly what make cross-functional teams successful. Be open to alternatives; be willing to accept differences in opinion. Remember to judge the content of the comment instead of the speaker. Often, you will be asked to work with someone you might not like. Even if you do not like his or her politics, work ethic, or personality, that individual still might make valuable comments.

Ask Questions Once someone has spoken, review your notes and ask questions. Be sure your questions are sincere and not oppositional. The goal of a question should be clarity. Then, after the speaker answers your question, repeat what he or she has said to confirm your understanding.

Control Your Reactions If you hear something you do not agree with, avoid just shouting out your opposition. Your goal is to *respond*, not *react*. When you *react*, you tend to do so defensively. Immediate reactions are based on impatience, maybe even aggression. A more effective approach to successful listening is to calmly *respond*, after a brief delay—five to ten seconds. This delay acts as a filter or a buffer. Even a short delay will allow you to digest the issue, consider the person's point of view, organize your thoughts, and then respond more professionally.

THE IMPORTANCE OF ORAL COMMUNICATION

Many people, even the seemingly most confident, are afraid to speak in front of others. A recent Monster.com poll asked, "What is your biggest career-related phobia?" Table 7.2 shows the results.

You may have to communicate orally with your peers, your subordinates, your supervisors, and the public. Oral communication is an important component of your business success because you will be required to speak formally and informally on an everyday basis.

EVERYDAY ORAL COMMUNICATION

"Hi. My name is Bill. How may I help you?" Think about how often you have spoken to someone today or this week at your job. You constantly speak to customers, vendors, and coworkers face-to-face, on the telephone, or by leaving messages on voice mail.

- If you work an 800 hotline, your primary job responsibility is oral communication.
- When you return the dozens of calls you receive or leave voice mail messages, each instance reveals your communication abilities.
- As an employee, you must achieve rapport with your coworkers. Much of your communication to them will be verbal. What you say impacts your working relationships.

Every time you communicate orally, you reflect something about yourself and your company. The goal of effective oral communication is to ensure that your verbal skills make a good impression and communicate your messages effectively.

Table 7.2 Career-Related Phobias
(Monster Career Advice Newsletter)

Percentage	Phobia
42 percent	Giving a speech or presentation
32 percent	Confronting a coworker or boss
15 percent	Networking
11 percent	Writing a report or proposal

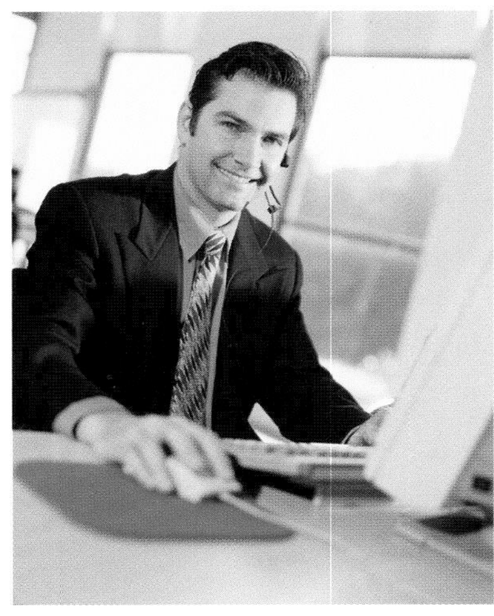

Telephone and Voice Mail

You speak on the telephone dozens of times each week. When speaking on a telephone, make sure that you do not waste either your time or your listener's time.

The following list has ten tips for telephone and voice mail etiquette.

1. Know what you are going to say before you call.
2. Speak clearly and enunciate each syllable.
3. Avoid rambling conversationally.
4. Avoid lengthy pauses.
5. Leave brief messages.
6. Avoid communicating bad news.
7. Repeat your phone number twice, including the area code.
8. Offer your e-mail address as an option.
9. Sound pleasant, friendly, and polite.
10. If a return call is unnecessary, say so.

Informal Oral Presentations

As a team member, manager, supervisor, employee, or job applicant, you often will speak to a coworker, a group of colleagues, or a hiring committee. You may need to communicate orally in an informal setting for any number of reasons such as the following:

- Your boss needs your help preparing a presentation. You conduct research, interview appropriate sources, and prepare reports. When you have concluded your research, you might be asked to share your findings with your boss in a brief, informal oral presentation.
- Your company is planning corporate changes (staff layoffs, mergers, relocations, or increases in personnel). As a supervisor, you want to provide your input in an oral briefing to a corporate decision maker.
- At a departmental meeting, you are asked to report orally on the work you and subordinates have completed and to explain future activities.
- In a team meeting, you participate in oral discussions regarding agenda items.
- Your company is involved in a project with coworkers, contractors, and customers from distant sites. To communicate with these individuals, you participate in a teleconference, orally sharing your ideas.
- You are applying for a job. Your interview, though not a formal, rehearsed presentation, requires that you speak effectively before a hiring committee.
- You are working from a remote location but need to participate in company-sponsored training. You cannot physically attend the workshop, but through a Webconference or Webinar, you and other employees from around the country can be trained simultaneously on your computers.

These informal oral communication instances could be accomplished best through videoconferences, teleconferences, Webconferences, and Webinars.

Videoconferences and Teleconferences

If you are communicating with a group and you want to hear what everyone else is saying simultaneously, videoconferences or teleconferences are an answer. Consider using a videoconference or teleconference when three or more people at separate locations need to talk.

In a videoconference or teleconference, you want all participants to feel as if they are in the same room facing each other. With expensive technology in place, such as cameras, audio components, coders/decoders, display monitors, and user interfaces, you want to avoid wasting time and money with poor communication.

The following list has ten tips for videoconferences and teleconferences.

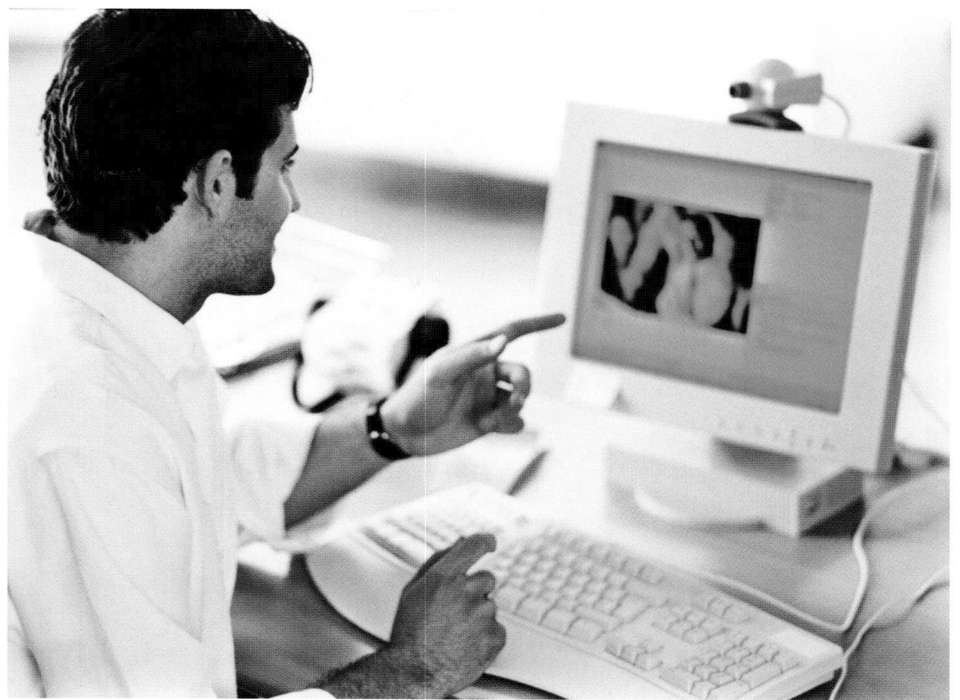

1. Inform participants of the conference date, time, time zone, and expected duration.
2. Make sure participants have printed materials before the teleconference.
3. Ensure that equipment has good audio quality.
4. Choose your room location carefully for quiet and privacy.
5. Consider arrangements for hearing impaired participants. You might need a text telephone (TTY) system or simultaneous transcription in a chat room.
6. Introduce all participants.
7. Direct questions and comments to specific individuals.
8. Do not talk too loudly, too softly, or too rapidly.
9. Turn off cell phones and pagers.
10. Limit side conversations.

Webconferencing

Due to economic hardships, rising costs of airline tickets, time-consuming travel, and the need to complete projects or communicate information quickly, many companies and organizations are using an electronic tool to communicate with their personnel, customers, prospective employees, and vendors—Webconferencing. Webconferencing, sometimes called Webinars or virtual meetings, include Web-based seminars, lectures, presentations, or workshops transmitted over the Internet. A unique feature of a live Webconference or Webinar is interactivity. In a Webconference, the presenter and audience can send, receive, and discuss information ("Webinar"). In contrast to Webconferences, a Webcast or netcast broadcasts either live or delayed audio and video, "more like traditional television or radio" ("Webcast").

Following are instances where a Webconference would be appropriate:

- **Train employees**—Companies are using Webconferences as a training tool not only to save travel expenses but also because of the interactivity a Webconference allows.
- **Make sales presentations**—With Webconferences, a sales team can make numerous sales calls without leaving the office. Through Webconferencing, sales presentations can be enhanced with product demonstrations, question and answer sessions, text messaging, and interactive polling to gather client and customer information (Murray; Van Wagner).
- **Conduct quarterly or annual meetings**—An online conference room allows corporations to "rollout earnings reports and corporate information to employees and investors" (Murray).
- **Hold press conferences**—This allows a company to share corporate information with news agencies at diverse locations.
- **Enhance online collaboration with colleagues at diverse locations**—Along with e-mail, teleconferencing, and videoconferencing, Webconferencing provides companies another option when face-to-face meetings are difficult due to cost and distance.

The following list has ten tips for Webconferencing.

1. Limit Webconferences to 60 to 90 minutes.
2. Limit a Webconference's focus to three or four important ideas.
3. Start fast by limiting introductory comments.
4. Keep it simple. Instead of using too many Internet tools, limit yourself to simple and important features like polling and messaging.
5. Plan ahead. Make sure that all Webconference participants have the correct Internet hardware and software requirements; know the correct date, time, agenda, and Web login information for your Webconference; and have the correct Web URL or password.
6. Before the Webconference, test your equipment, hypertext links, and PowerPoint slide controls.
7. Use both presenter and participant views. One way to ensure that all links and slides work is by setting up two computer stations. Have a computer set on the presenter's view and another computer logged in as a guest. This will allow you to view accurately what the audience sees and how long displays take to load.
8. Involve the audience interactively through questions and/or text messaging.
9. Personalize the presentation. Introduce yourself, other individuals involved in the presentation, and audience members.
10. Archive the presentation.

Formal Presentations

You might make a formal presentation if you

- Attend a civic club meeting and provide an oral presentation to maintain good corporate or community relations.
- Represent your company at a city council meeting. You will give an oral presentation explaining your company's desired course of action or justifying activities already performed.
- Represent your company at a local, regional, national, or international conference by giving a speech.
- Make an oral presentation promoting your service or product to a potential customer.

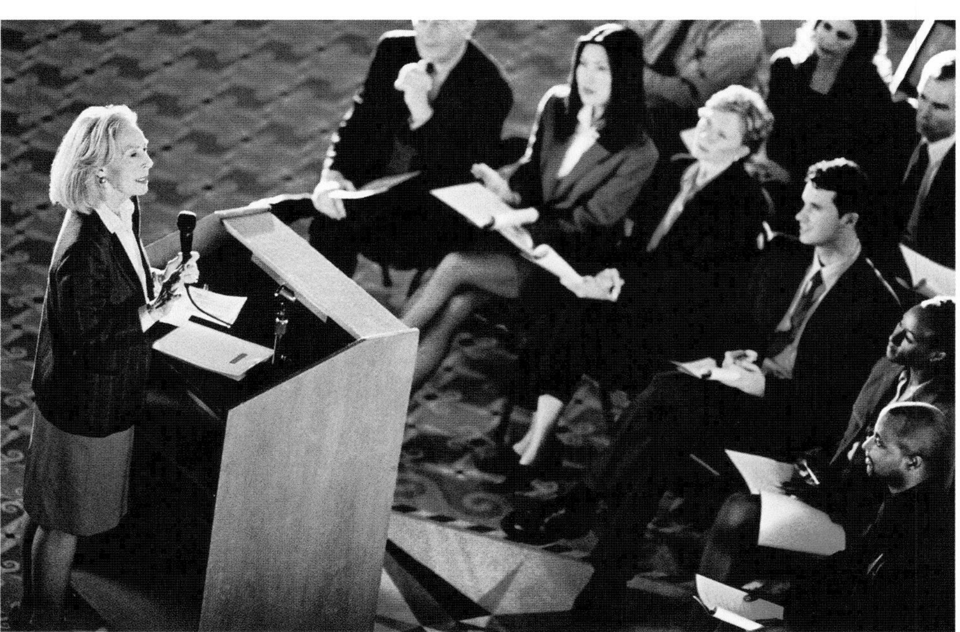

Types of Formal Oral Presentations

Three types of formal oral presentations are as follows:

1. **The memorized speech—**The least effective type of oral presentation is the memorized speech. This is a well-prepared speech that is committed to memory. Although such preparation might make you feel less anxious, too often these speeches sound mechanical and impersonal. They are often stiff and formal, and allow no speaker–audience interaction.
2. **The manuscript speech—**In a manuscript speech, the entire speech is written on paper. You read from a carefully prepared manuscript. This may lessen speaker anxiety and help you to present information accurately, but such a speech can seem monotonous, wooden, and boring to the listeners.
3. **The extemporaneous speech—**Extemporaneous speeches are probably the best and most widely used method of oral communication. You carefully prepare your oral presentation by conducting necessary research and then create a detailed outline. However, you avoid writing out the complete presentation. When you make your presentation, you rely on notes or PowerPoint slides with the major and minor headings for reference. This type of presentation helps you avoid seeming dull and mechanical, allows you to interact with the audience, and still ensures that you correctly present complex information. (See Figure 7.4 for the Oral Presentation Process.)

Parts of a Formal Oral Presentation

A formal oral presentation consists of an introduction, a discussion (or body for development), and a conclusion.

Introduction The introduction should welcome your audience, clarify your intent by providing a verbal "road map," and capture your audience's attention and interest. This is the point in the presentation where you are drawing in your listener, hoping to create enthusiasm and a positive impression.

To create a positive impression, set the table. Address your audience politely by saying "Good morning" or "Good afternoon." Tell the audience your name and the names

FIGURE 7.4 The Oral Presentation Process

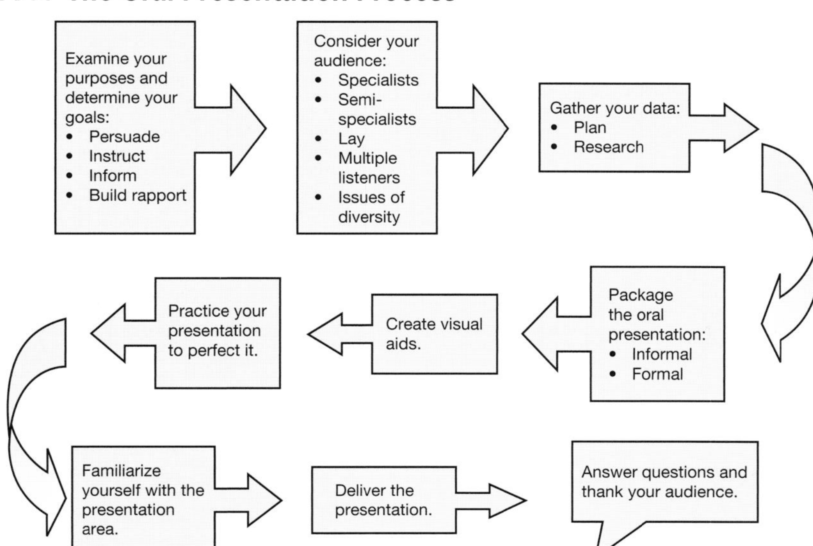

(adapted from "Communications: Oral Presentations")

of others who might be speaking. Welcome the audience and thank them for inviting you to speak.

- **Table Setters for Goodwill**

"Good evening. Thank you for allowing us to speak to you tonight. I'm (name and title). You'll also hear from (name and title)."

Next, provide a "road map," clarifying what points you'll discuss and laying out the order of the topics.

- **Road Map (thesis statement)**

"What we are going to talk about, and in this order, are the following key points:

- The issues that led us to consider road improvements
- Challenges to this construction task
- Optional approaches
- Costs
- A timeframe for your consideration."

You can use a variety of openings to capture your audience's attention, such as the following:

- **Word Pictures**—Anecdotes (specific in *time*, *place*, *person*, and *action*), Quotes, and Data (facts and figures)

Word Pictures to Arouse Your Audience's Interest

"From November 2008 to February 2009, our police department received over 75 reports of problems regarding Elm Street. These ranged from injuries related to hill jumping, cars sliding into the street's ditches, increased rush hour traffic, and tight turn lanes. One stretch of the road, from grid 39 to grid 47, is too narrow for snow removal crews to clean effectively. And these issues promise to get worse with residential growth anticipated to increase by 39 percent. As Sgt. Smith of the police department stated, 'Elm Street is a disaster waiting to happen.'"

- **A Question or a Series of Questions**—Asking questions involves the audience immediately. A training facilitator could begin a workshop as follows:

Questions to Involve Your Audience

"How many reports do you write each week or month? How often do you receive and send e-mail messages to customers and colleagues? How much time do you spend on the telephone? Face it, workplace communication is a larger part of your engineering job than you ever imagined."

These three questions are both personalized and pertinent. Through the use of the pronoun "you," the facilitator speaks directly to each individual in the audience. By focusing on the listeners' job-related activities, the questions directly lead into the topic of conversation—workplace communication.

- **A Quotation from a Famous Person**—The training facilitator in the previous example could have begun his speech with a quote from Warren Buffett, a famous businessperson.

A Quote to Arouse Audience Interest

"How important is effective workplace communication? Just listen to what Warren Buffett has to say on the topic:

> 'For more than forty years, I've studied the documents that public companies file. Too often, I've been unable to decipher just what is being said or, worse yet, had to conclude that nothing was being said. . . . Perhaps the most common problem . . . is that a well-intentioned and informed writer simply fails to get the message across to an intelligent, interested reader. In that case, stilted jargon and complex constructions are usually the villain. . . . When writing Berkshire Hathaway annual reports, I pretend that I'm talking to my sisters. I have no trouble picturing them: Though highly intelligent, they are not experts in accounting or finance. They will understand plain English, but jargon may puzzle them. My goal is simply to give them the information I would wish them to supply me if our positions were reversed. To succeed, I don't need to be Shakespeare; I must, though, have a sincere desire to inform.' (Buffett, 1–2)

That's what I want to impart to you today: good writing is communication that is easy to understand. If simple language is good enough for Mr. Buffett, then that should be your goal."

Discussion (or Body) After you have aroused your listeners' attention and clarified your goals, you have to prove your assertions. In the *discussion* section of your formal oral presentation, provide details to support your thesis statement. You can develop your content in a variety of ways, including the following:

- **Quotes, testimony, anecdotes**—You can find this type of information through primary and secondary research. For example, primary research, such as a survey or questionnaire, will help you substantiate content. By using people's direct comments, you help your audience to relate to the content. You also can find quotes, testimony, or anecdotes in secondary research, such as periodicals, newspapers, books, or online. This type of information validates your comments.
- **Data**—Facts and figures, again found through research or interviews, develop and support your content. Stating exactly how often your company's computer system has been attacked by viruses will support the need for improved firewalls. A statistical analysis of the increase in insurance premiums will clarify the need for a new insurance carrier. Showing the exponential increase in the cost of gasoline over the last ten years will show why your company should consider purchasing hybrids.

To help your audience follow your oral communication, present these details using any of the following modes of organization:

Additional information:

See Chapter 5, "Packaging Workplace Communication Through Effective Document Design."

Comparison/contrast. In your presentation, you could compare different makes of office equipment, employees you are considering for promotion, different locations for a new office site, vendors to supply and maintain your computers, different employee benefit providers, and so forth. Comparison/contrast is a great way to make value judgments and provide your audience options.

Problem/solution. You might develop your formal oral presentation by using a problem-to-solution analysis. For example, you might need to explain to your audience why your division needs to downsize. Your division has faced problems with unhappy customers, increased insurance premiums, decreased revenues, and several early retirements of top producers. In your speech, you can then suggest ways to solve these problems ("We need to downsize to lower outgo and ultimately increase morale"; "Let's create a 24-hour, 1-800 hotline to answer customer concerns"; "We should compare and contrast new employee benefits packages to find creative ways to lower our insurance costs").

Additional information:

See Chapter 13, "Persuasive Workplace Communication."

Argument/persuasion. Almost every oral presentation has an element of argument/persuasion to it—as does all good written communication. You will usually be persuading your audience to do something based on the information you share with them in the presentation.

Importance. Prioritizing the information you present from least to most important (or most important to least) will help your listeners follow your reasoning more easily. To ensure the audience understands that you are prioritizing, provide verbal cues. These include simple words like "first," "next," "more important," and "most important." Do not assume that these cues are remedial or obvious. Remember, sometimes it is hard to follow a speaker's train of thought. Good speakers realize this and give the audience verbal signposts, reminding the listeners exactly where they are in the oral presentation and where the speaker is leading them.

Chronology. A chronological oral presentation can outline for your audience the order of the actions they need to follow. For example, you might need to prepare a yearly evaluation of all sales activities. Provide your audience with target deadlines and with the specific steps they must follow in their reports each quarter.

Maintaining coherence. To maintain coherence, guide your audience through your speech as follows:

- **Use clear topic sentences**—Let your listeners know when you are beginning a new, key point: "Next, let's talk about the importance of conciseness in your workplace writing."
- **Restate your topic often**—Constant restating of the topic is required because listeners have difficulty retaining spoken ideas. Although a reader can refer to a previously discussed point by turning back a page or two, listeners do not have this option.

  Furthermore, a listener is easily distracted from a speech by noises, room temperature, uncomfortable chairs, or movement inside and outside the meeting site. Restating your topic helps your reader maintain focus. "Repeat major points. Reshow visuals, repeat points and ideas several times during your presentation. Put them in your summary, too" (O'Brien).
- **Use transitional words and phrases**—Using transitional words and phrases helps your listeners follow your speech. Transitional words and phrases, like those shown in Table 7.3, aid reader comprehension, emphasize key points, and highlight your speech's organization.

Table 7.3 The Purpose of Transitional Words

Purpose	Examples
Cause and effect	because, since, thus, therefore, for this reason, due to this, as a result of, consequently, in order to
Example	for instance, for example, another
Interpreting jargon	that is, in other words, more commonly called
Sequencing ideas	first, second, next, last, following, finally, above, below
Adding a point	furthermore, next, in addition, besides, not only . . . but also, similarly, likewise
Restating	in other words, that is, again, to clarify
Contrasting	but, instead, yet, however, on the other hand, nevertheless, in contrast, on the contrary, whereas, still
For emphasis	in fact, more importantly, clearly
Summarizing	to summarize, therefore, in summary, to sum up, consequently, therefore

Conclusion Conclude your speech by restating the main points, by recommending a future course of action, or by asking for questions or comments. A polite speaker leaves time for a few follow-up questions from the audience. Gauge your time well, however. You do not want to bore people with a lengthy discussion after a lengthy speech. You also do not want to cut short an important question-and-answer (Q and A) session. If you have given a controversial speech that you know will trouble some members of the audience, you owe them a chance to express their concerns.

VISUAL AIDS

Most speakers find that visual aids enhance their oral communication. "Visuals have the greatest, longest lasting impact—show as much or more visually as what you say. Use pictures; use color. Use diagrams and models" (O'Brien).

Although PowerPoint slide shows, graphs, tables, flip charts, and overhead transparencies are powerful means of communication, you must be the judge of whether visual aids will enhance your presentation. Avoid using them if you think they will distract from your presentation or if you lack confidence in your ability to create them and integrate them effectively. However, with practice, you probably will find that visuals add immeasurably to the success of most presentations.

Table 7.4 lists the advantages, disadvantages, and helpful hints for using visual aids. For all types of visual aids, practice using them before you actually make your presentation. When you practice your speech, incorporate the visual aids you plan to share with the audience.

Table 7.4 Visual Aids—Advantages and Disadvantages

Type	Advantages	Disadvantages	Helpful Hints
Chalkboards	Are inexpensive. Help audiences take notes. Allow you to emphasize a point. Allow audiences to focus on a statement. Help you be spontaneous. Break up monotonous speeches.	Make a mess. Can be noisy. Make you turn your back to the audience. Can be hard to see from a distance. Can be hard to read if your handwriting is poor.	Clean the board well. Have extra chalk. Stand to the side as you write. Print in large letters. Write slowly. Avoid talking with your back to the audience. Don't erase too soon.
Chalkless Whiteboards	Same as above.	Are expensive. Require unique, erasable pens. Can stain clothing. Some pens can be hard to erase if left on the board too long. Pens that run low on ink create light, unreadable impressions.	Use blue, black, or red ink. Cap pens to avoid drying out. Use pens made especially for these boards. Erase soon after use.
Flip Charts	Can be prepared in advance. Are neat and clean. Can be reused. Are inexpensive. Are portable. Help you avoid a nonstop presentation. Allow for spontaneity. Help audiences take notes. Allow you to emphasize key points. Encourage audience participation. Allow easy reference by turning back to prior pages. Allow highlighting with different colors.	Are limited by small size. Require an easel. Require neat handwriting. Won't work well with large groups. You can run out of paper. Markers can run out of ink.	Have two pads. Have numerous markers. Use different colors for effect. Print in large letters. Turn pages when through with an idea so audience will not be distracted. Don't write on the back of pages where print bleeds through.

(Continued)

Table 7.4 Visual Aids—Advantages and Disadvantages (Continued)

Type	Advantages	Disadvantages	Helpful Hints
Overhead Transparencies	Are inexpensive. Can be used with lights on. Can be prepared in advance. Can be reused. Can be used for large audiences. Allow you to return to a prior point. Allow you to face audiences.	Require an overhead projector. Can be hard to focus. Require an electrical outlet and cords. Can become scratched and smudged. Can be too small for viewing. Bulbs burn out.	Use larger print. Protect the transparencies with separating sheets of paper. Frame transparencies for better handling. Turn off the overhead to avoid distractions. Focus the overhead before beginning your speech. Keep spare bulbs. Don't write on transparencies. Face the audience.
Slides (Slide Shows)	Are portable. Slides are easy to protect. Can be used for large groups. Can be prepared in advance. Are entertaining and colorful. Allow for later reference.	Require dark rooms. Hurt speaker–audience interaction (eye contact). Can be expensive. Can be challenging to create. Require screen, machinery, and electrical outlets. Slides can get out of order. Dark room makes taking notes challenging. Machinery can malfunction.	Use a pointer. Use a remote control for freedom of movement. Check slides to make sure none are out of order. Check working condition of the equipment.
Videotapes	Allow instant replay. Can be freeze-framed for emphasis. Can be economically duplicated. Can be rented or leased inexpensively. Are entertaining.	Require costly equipment (monitor and recorder). Are bulky and difficult to move. Can malfunction. Require dark rooms. Require compatible equipment. Deny easy note taking. Deny speaker–audience interaction.	Practice operating the equipment. Avoid long tapes.
Films	Are easy to use. Are entertaining. Have many to choose from. Can be used for large groups.	Require dark rooms. Require equipment and outlets. Make note taking difficult. Deny speaker–audience interaction. Can malfunction and become dated.	Use up-to-date films. Avoid long films. Provide discussion time. Use to supplement the speech, not replace it. Practice with the equipment.
PowerPoint (PPt) Presentations	Are entertaining. Offer flexibility, allowing you to move from topics with a mouse click. Can be customized and updated. Can be used for large groups. Allow for speaker–audience interaction. Can be supplemented with handouts easily generated by PPt. Can be prepared in advance. Can be reused. Allow you to return to a prior slide. Can include animation and hyperlinks.	Require computers, screens, and outlets. Work better with dark rooms. Computers can malfunction. Can be too small for viewing. Can distance the speaker from the audience.	Practice with the equipment. Bring spare computer cables. Be prepared with a backup plan if the system crashes. Have the correct computer equipment (cables, monitors, screens, etc.). Make backup transparencies. Practice your presentation.

PowerPoint Presentations

One of the most powerful oral communication tools is visual—Microsoft PowerPoint (PPt). Whether you are giving an informal or formal oral presentation, your communication will be enhanced by PowerPoint slides.

Today, you will attend very few meetings where the speakers do not use PowerPoint slides. PowerPoint slides are simple to use, economical, and transportable. Even if you have never created a slide show before, you can use the templates in the software and the autolayouts to develop your own slide show easily. An added benefit of PowerPoint slides is that you can print them and create handouts for audience members.

FAQs

Q: Why is the use of PowerPoint so important in the workplace?

A: Widely used by businesspeople, educators, and trainers, PowerPoint is among the most prevalent forms of presentation technology. "According to Microsoft, 30 million PowerPoint presentations take place every day: 1.25 million every hour" (Mahin). Employees in education, business, industry, technology, and government use PowerPoint not only for oral presentations but also as hard-copy text.

Q: Does everyone like PowerPoint? Aren't there any negative attitudes toward this technology?

A: Not everyone likes PowerPoint. Opposition to the use of this technology, however, usually stems from the following problems:

- Dull PowerPoint slides, lacking in variety and interest
- The use of Microsoft's standardized templates
- "Death by Bullet Point," caused by an excessive dependence on bullets

However, these challenges can be overcome easily through techniques discussed in this chapter.

Benefits of PowerPoint

When you become familiar with Microsoft PowerPoint, you will be able to achieve the following benefits:

1. Choose from many different presentation layouts and designs.

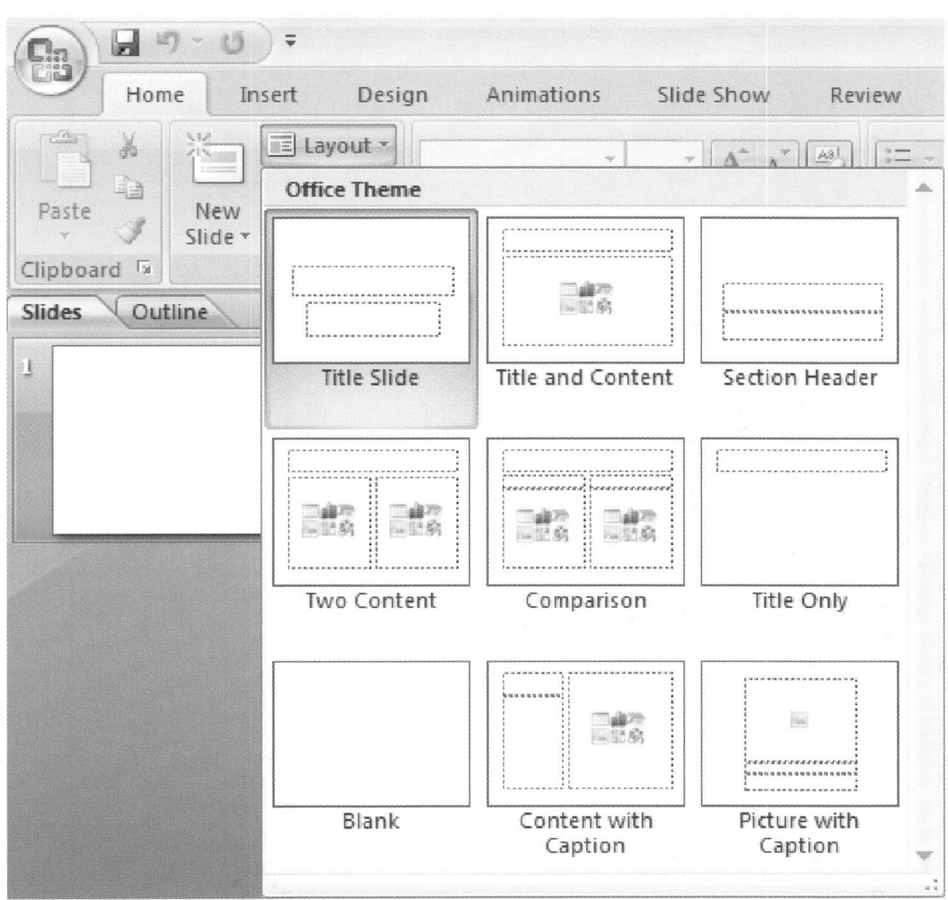

2. Create your own designs and layouts, changing colors and color schemes from preselected designs.

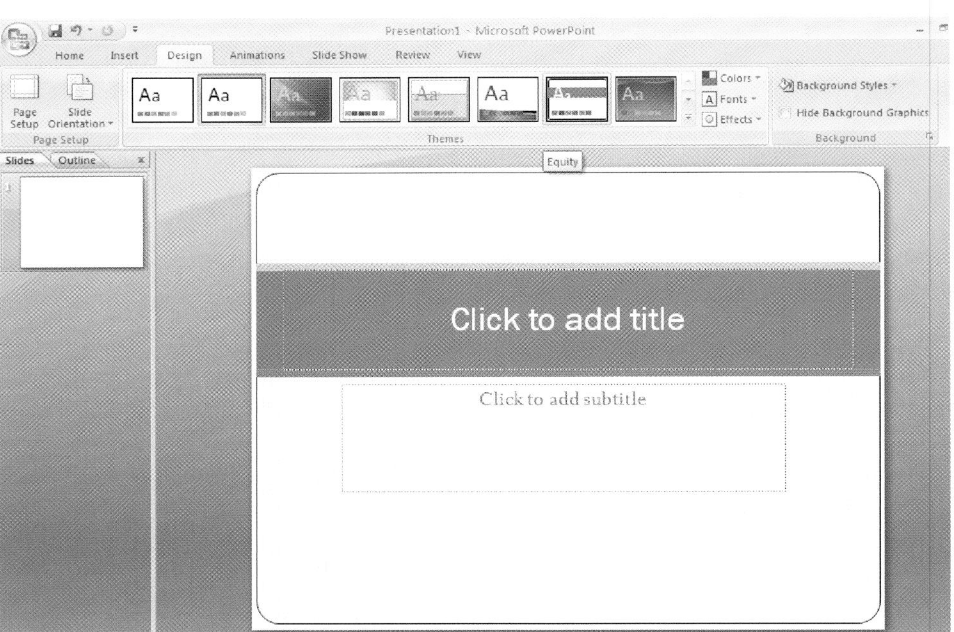

3. Add, delete, or rearrange slides as needed. By left-clicking on any slide, you can copy, paste, or delete it. By left-clicking between any of the slides, you can add a new blank slide for additional information.

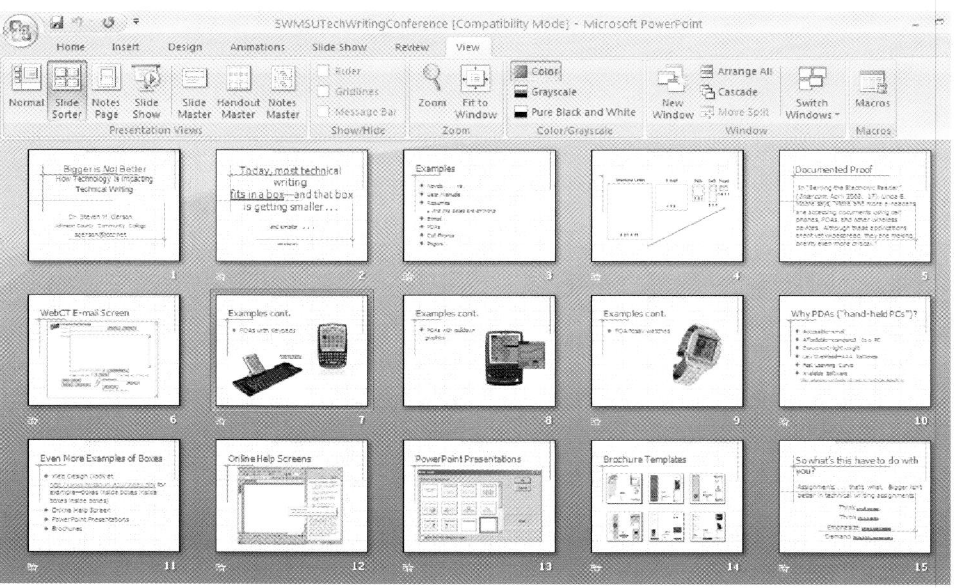

4. Insert art from the Web, add images, or create your own drawings.

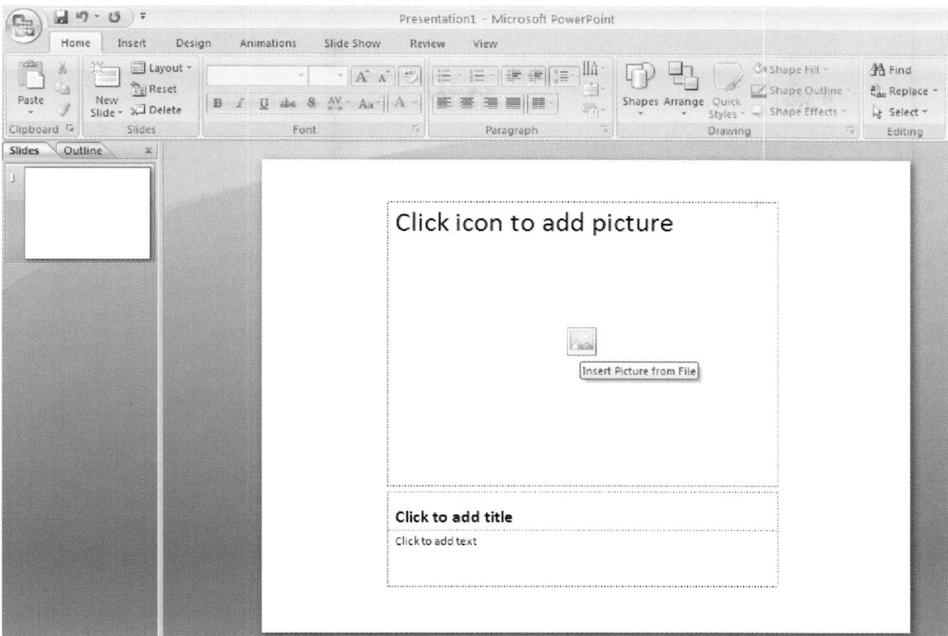

5. Add hyperlinks either to slides within your PowerPoint presentation or to external Web links.

Tips for Using PowerPoint

To make it easy for you to add PowerPoint slides to your presentations, consider the following hints.

1. **Create optimal contrast**—Use dark backgrounds for light text or light backgrounds for dark text. Avoid using red or green text (individuals who are color blind cannot see these colors). Use color for emphasis only.
2. **Choose an easy-to-read font size and style**—Use common fonts, such as Times New Roman, Courier, or Arial. Arial is considered to be the best to use because sans serif fonts (those without feet) show up best in PowerPoint. Use no more than three font sizes per slide. Use at least a 24-point font size for text and 36-point font size for headings.
3. **Limit the text to six or seven lines per slide and six or seven words per line**—Think 6 × 6. Two or more short, simple lines of text are better than one slide with many words. Also, use no more than 40 characters per line (a character is any letter, punctuation mark, or space). You can accomplish these goals by creating a screen for each major point discussed in your oral presentation.
4. **Use headings for readability**—To create a hierarchy of headings, use larger fonts for a first-level heading and smaller fonts for second-level headings. Each slide should have at least one heading to help the audience follow your thoughts.
5. **Use emphasis techniques**—To call attention to a word, phrase, or idea, use color (sparingly), boldface, all caps, or arrows. Use a layout that includes white space. Include figures, graphs, pictures from the Web, or other line drawings.
6. **End with an obvious concluding screen**—Often, if speakers do not have a final screen that *obviously* ends the presentation, the speakers will click to a blank screen and say, "Oh, I didn't realize I was through," or "Oh, I guess that's it." In contrast, an obvious ending screen will let you as the speaker end graciously—and without surprise.

7. **Prepare handouts**—Give every audience member a handout, and leave room on the handouts for note taking.
8. **Avoid reading your screens to your audience**—Remember your audience can read and will become bored if you read slides to them. Speakers lose their dynamism when they resort to reading slides rather than speaking to the audience.
9. **Elaborate on each screen**—PowerPoint should not replace you as the speaker. In contrast, PowerPoint should add visual appeal, while you elaborate on the details. Give examples to explain fully the points in your presentation.
10. **Leave enough time for questions and comments**—Instead of rushing through each slide, leave sufficient time for the audience to consider what they have seen and heard. Both during and after the PowerPoint presentation, give the audience an opportunity to offer input.

See Figures 7.5 to 7.7 for samples of effective PowerPoint slides.

FIGURE 7.5 PowerPoint Slide of Student Headcount

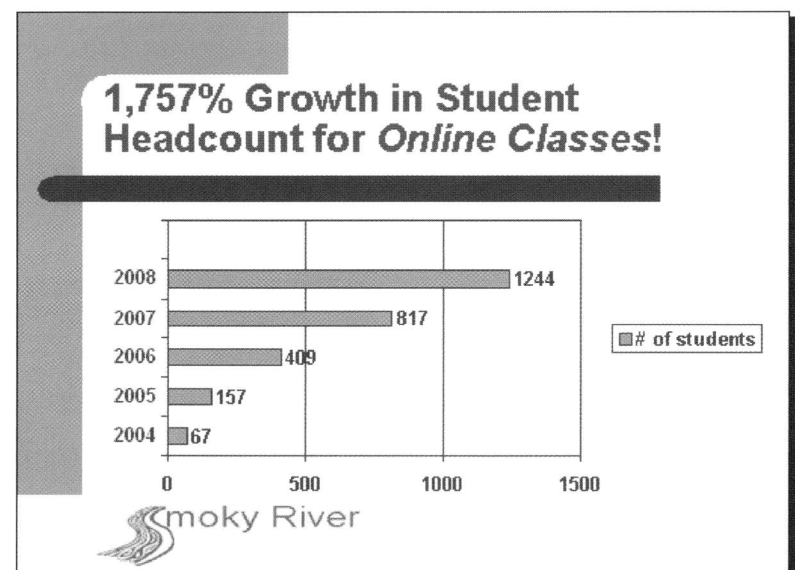

FIGURE 7.6 PowerPoint Slide of Source of Information

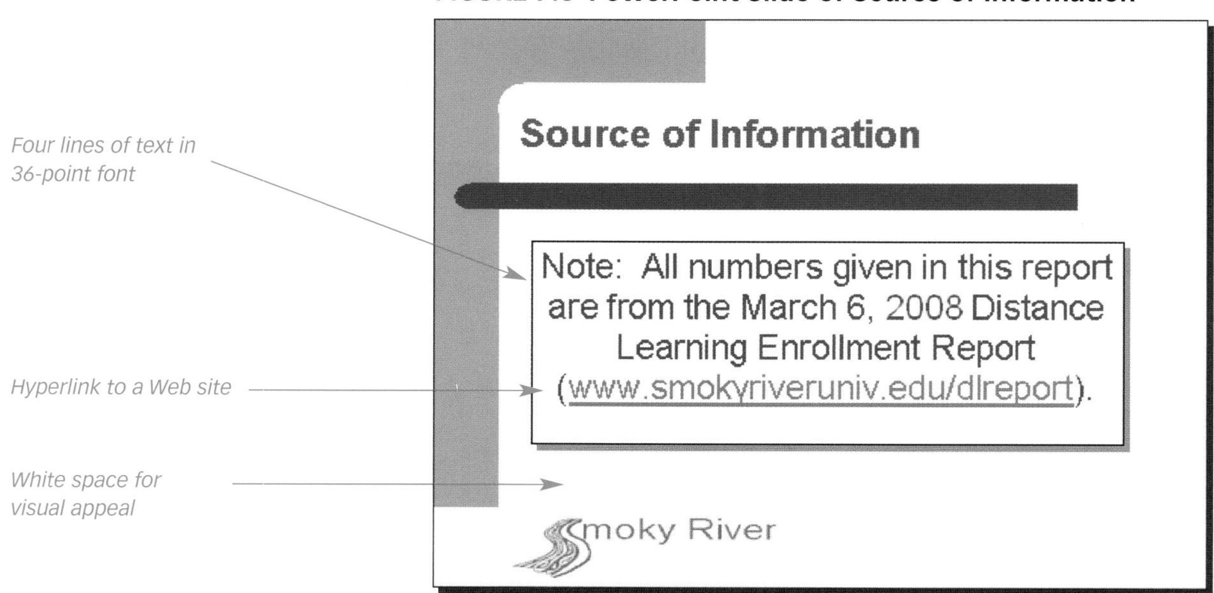

FIGURE 7.7 **PowerPoint Slide of Student Comments**

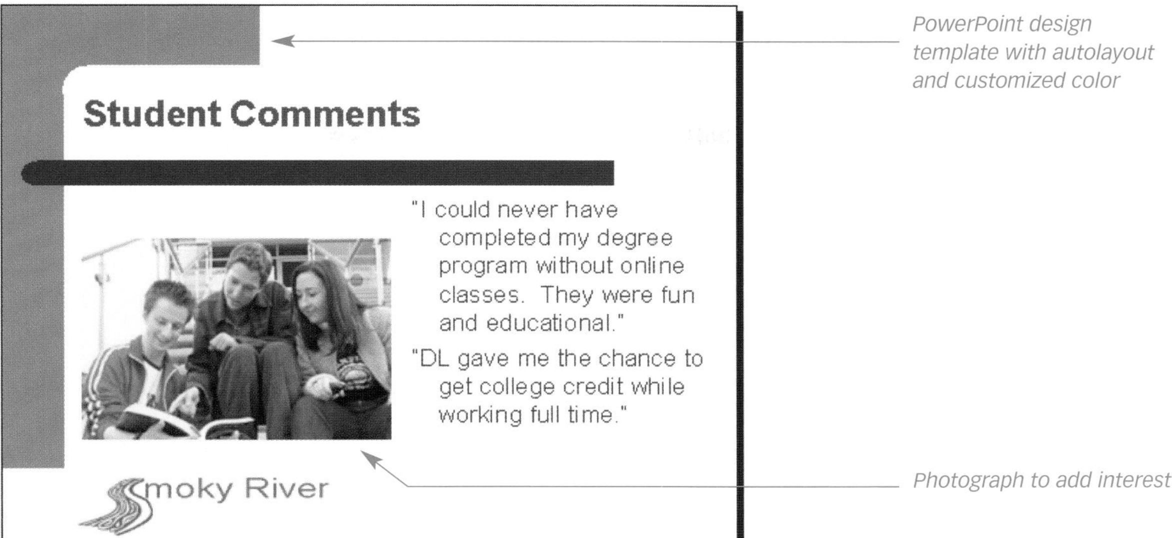

POWERPOINT SLIDES CHECKLIST

_____ 1. Does the presentation include headings for each slide?

_____ 2. Have you used an appropriate font size for readability?

_____ 3. Did you choose an appropriate font type for readability?

_____ 4. Did you limit yourself to no more than three different font sizes per screen?

_____ 5. Has color been used effectively for readability and emphasis, including font color and slide background?

_____ 6. Did you use special effects effectively versus overusing them?

_____ 7. Have you limited text on each screen (remembering the 6 × 6 rule)?

_____ 8. Did you size your graphics correctly for readability, avoiding ones that are too small or too complex?

_____ 9. Have you used highlighting techniques (arrows, color, white space) to emphasize key points?

_____10. Have you edited for spelling and grammatical errors?

THE *P*³ PROCESS AT WORK

As with writing memos, letters, e-mail messages, and reports, approach your oral communication project as a step-by-step process. Doing so will allow you to express yourself with confidence. Follow the *P*³ process to organize your presentation effectively.

Planning

Planning gets you started with your presentation. Similar to planning for written communication, when you plan a speech, you should do the following.

Consider the Purpose

Determine why you are making an oral presentation. Ask yourself questions like

- Are you selling a product or service to a client?
- Do you want to inform your audience of the features in your newly created software?
- Are you persuading your audience to increase corporate spending to enhance a benefits package?
- Are you giving a speech for one of your college classes?

- Has your boss asked for your help in preparing a presentation? After you research the content, will you have to present your information orally?
- Are you a supervisor justifying workforce cuts to your division?
- Did a customer request information on solutions to a problem?
- Are you representing your company at a conference by giving a speech?
- Are you running for an office on campus and giving a speech about your candidacy?
- At the division meeting, are you reporting orally on the work you and your team have completed and the future activities you plan for the project?

Determining the purpose of a presentation will ensure that you choose the appropriate content.

Inform For example, when your speech is to *inform*, you want to update your listeners. Such a speech could be about new tax laws affecting listeners' pay, new management hirings or promotions, or budget constraints or cutbacks. Speeches that inform do not necessarily require any action on the part of your audience. Your listeners cannot alter tax laws, change hiring or promotion practices, or prevent cutbacks. The informative speech keeps your audience up-to-date.

Additional information:

See Chapter 13, "Persuasive Workplace Communication."

Persuade However, some speeches *persuade*. You will speak to motivate listeners. For instance, you might give an oral presentation about the need to hold more regular and constructive meetings. You might tell your audience that teamwork will enhance productivity. Maybe you are giving an oral presentation about the value of quality controls to enhance product development. In each instance, you want your audience to leave the speech ready and inspired to act on your suggestions.

Instruct You might speak to *instruct*. In an instruction, you will teach an audience how to follow procedures. For instance, you could speak about new sales techniques, ways to handle customer complaints, implementation of software, manufacturing procedures, or how to prepare for on-campus interviews. When you instruct, your goal is both to inform and persuade. You will inform your audience how to follow steps in the procedure. In addition, you will motivate them, explaining why the procedure is important.

Build Trust Finally, you might give an oral presentation to *build trust*. Let's say you are speaking at an annual meeting. Your goal not only might be to inform the audience of your company's status, but also to instill the audience with a sense of confidence about the company's practices. You could explain that the company is acting with the audience's best interests in mind. Similarly, in a departmental meeting, you might speak to build rapport. As a supervisor, you will want all employees to feel empowered and valued. Speaking to build trust will accomplish this goal.

Consider Your Audience

When you plan your oral presentation, consider your audience. Ask yourself questions such as the following:

- Are you speaking to a specialist, semi-specialist, or lay audience?
- Are you speaking up to supervisors?
- Are you speaking down to subordinates?
- Are you speaking laterally to peers?
- Are you speaking to the public?
- Are you addressing multiple audience types (supervisors, subordinates, and peers)?
- Is your audience friendly and receptive or hostile?

- Are you speaking to a captive audience (one required to attend your presentation)?
- Is your audience diverse in terms of culture, gender, or age?
- Will you need translators for those with hearing impairments?

Considering your audience's level of knowledge and interest will help you prepare your presentation. You should consider whether or not your audience needs terms defined and what tone you should take. You cannot communicate effectively if your audience fails to understand you or if your tone offends or patronizes them. Plan how you will design your content and style to communicate most effectively with your audience.

Additional information:

See Chapter 3, "Meeting the Needs of the Audience."

Presentation Plan A presentation plan, like the example shown in Figure 7.8, can help you accomplish these goals.

Gather Information

The best delivery by the most polished professional speaker will lack credibility if the speaker has little of value to communicate. As you plan your presentation, you must study and research your topic thoroughly before you package it.

You can rely on numerous sources when you research a topic for an oral presentation. For example, you could use any of the following sources:

- Interviews
- Questionnaires and surveys
- Visits to job sites
- Conversations in meetings or on the telephone

FIGURE 7.8 Oral Presentation Plan

Presentation Plan

Topic: ______________________________

Objectives:

- What do you want your audience to believe or do as a result of your presentation?
- Are you trying to persuade, instruct, inform, build trust, or combinations of the above?

Development: What main points are you going to develop in your presentations?

1.

2.

3.

Organization: Will you organize your presentation using *analysis, comparison/contrast, chronology, importance,* or *problem/solution?*

Visuals: Which visual aids will you use?

- Company reports
- Internet research
- Library research including periodicals and books
- Market research

Using information from company reports or other sources such as the Internet, books, or periodicals requires that you read and document your research (discussed in Chapter 16). Gathering information through interviews, questionnaires, surveys, or conversations, however, requires help from other people. To ensure that you get this assistance, consider doing the following:

- Ask politely for their assistance.
- Explain why you need the interview and information.
- Explain how you will use the information.
- Make a convenient appointment for the interview or to fill in the survey.
- Come prepared. Research the subject matter so you will be prepared to ask appropriate questions. Write your interview questions or the survey before you meet with the person. Take the necessary paper, forms, pencils, pens, laptop, electronic notepads, handheld PDAs, or recording devices you will need for the meeting.
- When the interview ends or the individuals complete the survey, thank them for their assistance.

Figure 7.9 shows a sample questionnaire used by a team researching the feasibility of adding a day-care center to their university campus for use by students and staff.

Packaging the Presentation

After you obtain your information, your next step is to write a draft and consider visual aids for the presentation. The writing step in the communication process lets you use the research you gathered in the planning stage. When you organize your information, you will determine whether or not additional material is needed or if you can delete some of the material you have gathered.

Don't write out the complete text of the presentation. Too often when people have a complete text in front of them, they rely too heavily on the written words. They end up reading most of the paper to the audience rather than speaking more conversationally. Instead of writing out a complete copy of the presentation, use an outline or note cards to present your speech.

Outline

You may want to write a more detailed outline focusing on your speech's major units of discussion and supporting information. A skeleton speech outline (Figure 7.10) provides a template for your presentation.

Note Cards

If you decide that presenting your speech from the outline will not work for you, consider writing highlights on 3″ × 5″ note cards. Avoid writing complete sentences or filling in the cards from side to side. If you write complete sentences, you will be tempted to read the notes rather than speak to the audience. If you fill in the cards from side to side, you will have trouble finding key ideas. Write short notes (keywords or short phrases) that will aid your memory when you make your presentation, such as in Figure 7.11.

Perfecting the Presentation

In the perfecting step of the communication process, consider all aspects of style, delivery, appearance, and body language and gestures. Then, most importantly, practice. Even if you have excellent visual aids and well-organized content, if you fail to deliver effectively, your audience could miss your intended message.

FIGURE 7.9 Sample Research Questionnaire

Student and Staff Questionnaire
for Proposed Child Care Center

1. Are you male or female?
2. Are you in a single- or double-income family?
3. Are you a student or staff?
4. How many children do you have?
5. What are the ages?
6. Would you be interested in having a child care center on campus?
7. How much would you be willing to pay per hour for child care at this center?
8. Do you think a child care center would increase enrollment at this university?
9. How many hours per week would you enroll your child/children?
10. What hours of operation should the child care center cover?
11. What credentials should the child care providers have?
12. What should be the number of children per classroom?

If student:

13. Are you enrolled full time or part time at the university?
14. Do you attend mornings, afternoons, evenings, weekends, or a combination of the above (please specify)?
15. Would you be willing to work in the child-care center part time?

If staff:

16. What hours do you work at the university?
17. What hours would you need to use for child care at the center?

Additional comments (if any):

Thank you for your assistance.

Style

As with good writing, effective oral communication demands clarity and conciseness. To achieve clarity, stick to the point. Your audience does not want to hear about your personal life or other irrelevant bits of information. You need to maintain focus on the topic. Concise oral presentations depend on the same skills evident in concise writing—word and sentence length. Trim your sentences of excess words (12 to 15 words per sentence is still the preferred length).

Remember to speak so that your audience can understand you and your level of vocabulary. You should speak to communicate rather than to impress your listeners.

Delivery

Effective oral communicators interact with and establish a dynamic relationship with their audiences. The most thorough research will be wasted if you are unable to create rapport and sustain your audience's interest. Although smaller audiences are usually easier to connect with, you can also establish a connection with much larger audiences through a variety of delivery techniques.

FIGURE 7.10 Skeleton Speech Outline

Skeleton Outline

Title:
Purpose:

I. Introduction
   A. Attention getter:

   B. Focus statement:

II. Body
   A. First main point:
      1. Documentation/subpoint:
         a. Documentation/subpoint:
         b. Documentation/subpoint:
      2. Documentation/subpoint:
      3. Documentation/subpoint:
   B. Second main point:
      1. Documentation/subpoint:
      2. Documentation/subpoint:
         a. Documentation/subpoint:
         b. Documentation/subpoint:
      3. Documentation/subpoint:
   C. Third main point:
      1. Documentation/subpoint:
      2. Documentation/subpoint:
      3. Documentation/subpoint:
         a. Documentation/subpoint:
         b. Documentation/subpoint:

III. Conclusion
   A. Summary of main points:

   B. Recommended future course of action:

Eye Contact Avoid keeping your eyes glued to your notes. You will find it easy to speak to one individual because you will naturally look him or her in the eye. The person will respond by looking back at you.

With a larger audience, whether the audience has 20 or 200-plus people, keeping eye contact is more difficult. Try looking into different people's eyes as you move through your

FIGURE 7.11 **Sample Note Card**

Sample 3 × 5 Note Cards

Need for Parking Lot Expansion

- Safety
- Accessibility
- Potential growth

Procedure for Parking Lot Expansion

- Stakeholders' meeting and vote
- City council approval
- Arrangement with contractors

presentation (or look slightly above their heads if that makes you more comfortable). Most of the audience has been in your position before and can sympathize.

Rate Because your audience wants to listen and learn, you need to speak at a rate slow enough to achieve those two goals. Determine your normal rate of speech, and cut it in half. *Slow* is the best rate to follow in any oral presentation. You could speed up your delivery when you reach a section of less interesting facts. Slow down for the most important and most interesting parts. Match your rate of speaking to the content of your speech, just as actors vary their speech rate to reflect emotion and changes in content.

Enunciation Speak each syllable of every word clearly and distinctly. Rarely will an audience ask you to repeat something even if they could not understand you the first time. It is up to you to avoid mumbling. Remember to speak more clearly than you might in a more conversational setting. Slowing down your delivery will help you enunciate clearly.

Pitch When you speak, your voice creates high and low sounds. That's *pitch*. In your presentation, capitalize on this fact. Vary your pitch by using even more high and low sounds than you do in your normal, day-to-day conversations. Modulate to stress certain keywords or major points in your oral presentations.

Pauses One way to achieve a successful pace is to pause within the oral presentation. Pause to ask for and to answer questions, to allow ideas to sink in, and to use visual aids or give the audience handouts. These pauses will not lengthen your speech; they will only improve it.

A well-prepared speech will allow for pauses and will have budgeted time effectively. Know in advance if your speech is to be 5 minutes, 10 minutes, or an hour long. Then plan your speech according to time constraints, building pauses into your presentation. Practice the speech beforehand so you can determine when to pause and how often.

Emphasis You will not be able to underline or boldface comments you make in oral presentations. However, just as in written communication, you will want to emphasize key ideas. Your body language, pitch, gestures, and enunciation will enable you to highlight words, phrases, or even entire sentences.

Interaction with Listeners You might need your audience to be active participants at some point in your oral presentation, so you will want to encourage this response. Your attitude and the tone of your delivery are key elements contributing to an encouraging atmosphere.

Conflict Resolution You might be confronted with a hostile listener who either disagrees with you or does not want to be in attendance. You need to be prepared to deal with such a person. If someone disagrees with you or takes issue with a comment you make, try these responses:

- "That is an interesting perspective."
- "Thanks for your input."
- "Let me think about that some more and get back to you."
- "I have got several more ideas to share. We could talk about that point later, during a break."

If you are confronted with a challenge, consider the following:

- Put it off until later so it does not distract from your presentation.
- Let the situation diffuse.
- Give yourself some more time to think about it.
- Give the person time to cool off.

The important point to remember is to not allow a challenging person to take charge of your presentation. Be pleasant but firm and maintain control of the situation. You will be unable to please all of your listeners all of the time. However, you should not let one unhappy listener destroy the effect of your presentation for the rest of the audience.

Appearance

When you speak to an audience, they see you as well as hear you. Therefore, avoid physical distractions. For example, avoid wearing clothes or jewelry that might distract the audience. You might be representing your company or trying to make a good impression for yourself when you speak, so dress appropriately.

Body Language and Gestures

During an oral presentation, nonverbal communication can be as important as verbal communication. Your appearance and attitude are important. In addition, the way you present your speech through your movements and tone of voice will affect your listeners. If you are enthusiastic about your topic, your listeners will respond enthusiastically. If you are bored or ambivalent, your tone and mannerisms will reflect your attitude. If you are negative, your tone will communicate negativity to your audience.

To communicate effectively, be aware of your body language and your gestures:

- Avoid standing woodenly. Move around somewhat, scanning the room with your eyes, stopping occasionally to look at one person. Remember to look at all parts of the room as you make the oral presentation.
- If you are nervous and your hands shake, try holding onto a chair back, lectern, the top of the table, or a paper clip.
- Use hand motions to emphasize ideas and provide transitions. For instance, you could put one finger up for a first point, two fingers up for a second point, and so forth.
- Avoid folding your arms stiffly across your chest. This projects a negative, defensive attitude.

Post-Speech Questions and Answers

After your presentation, be prepared for a question-and-answer session. Politely invite your audience to participate by saying, "If you have any questions, I am happy to answer them."

When an audience member asks a question, make sure everyone hears it. If not, you can repeat the question. If you fail to understand the question, ask the audience member to repeat it and clarify it.

If you have no answer, tell the audience. You could say that you will research the matter and get back with them. Faking an answer will only harm your credibility and detract from the overall effect of your presentation. Another valid option is to ask the audience what they think regarding the question or if they have any possible solutions or answers to the question. This is not only a good way to answer the question but also to encourage audience interaction.

Practice

Practice your speech—including manipulation of your visual aids—so you use them at appropriate times and places during the presentation. As you practice, you will grow more comfortable and less dependent on your note cards or outline. Use the Effective Oral Presentation Checklist to determine if you are sufficiently prepared for your oral presentation.

You will find that the more you practice, the more comfortable you feel. Practicing will help you achieve the following:

- Decrease your fear.
- Process your thoughts.
- Become more comfortable with the topic.
- Pronounce troublesome words.
- Decide what to emphasize and how to emphasize it.
- Enhance verbal and nonverbal cues.
- Rearrange your content.
- Add further details.
- Know when and how to use your visual aids.

EFFECTIVE ORAL PRESENTATION CHECKLIST

_____ 1. Does your speech have an introduction:
- Arousing the audience's attention?
- Clearly stating the topic of the presentation?

_____ 2. Does your speech have a body:
- Explaining what exactly you want to say?
- Developing your points thoroughly?

_____ 3. Does your speech have a conclusion:
- Suggesting what is next?
- Explaining when (due date) a follow-up should occur?
- Stating why that date is important?

_____ 4. Does your presentation provide visual aids to help you make and explain your points?

_____ 5. Do you modulate your pace and pitch?

_____ 6. Do you enunciate clearly so the audience will understand you?

_____ 7. Have you used body language effectively:
- Maintaining eye contact?
- Using hand gestures?
- Moving appropriately?
- Avoiding fidgeting with your hair or clothing?

_____ 8. Have you prepared for possible conflicts?

_____ 9. Do you speak slowly and remember to pause so the audience can think?

_____ 10. Have you practiced with any equipment you might use?

Nurani Met Her Challenge

Nurani met her communication challenge by using the *P*^{3} process.

Planning

To plan her presentation, Nurani considered the following:

- Goal—to communicate human resources information and regulations
- Audience—mid-level managers
- Channels—oral presentation, PowerPoint slides, and handouts
- Data—details about governmental regulations illustrated by real-life experiences

(*Continued*)

Because Nurani gives a variety of oral presentations ("Conducting Effective Interviews," "Recruitment and Selection," and "Coaching and Counseling"), she has to be organized and consider her audience. She says, "The audience is easily distracted by many things such as the following:

- Other seminar participants
- The temperature in the room
- Lack of visibility
- Hunger
- Work they're not doing

My content has to be 'on point' expressing my expertise and explaining the new information unfamiliar to the audience. I also have to determine which content to place on PowerPoint slides so that the presentation is outlined visually for the audience. Listing the highlights of my presentation is a great way for me to get started." Figure 7.12 shows how Nurani used brainstorming to plan her communication.

Packaging

Because Nurani was unfamiliar with the content the first time she presented it, she felt she would be more comfortable by writing everything down on PowerPoint slides. This detailed overview would help Nurani organize the content and determine the sections of the topics to be discussed. To package her presentation, Nurani drafted PowerPoint slides, such as the one shown in Figure 7.13.

Perfecting

Perfecting Nurani's presentation goes hand-in-hand with perfecting her PowerPoint slides. To be an effective oral communicator, Nurani says, "I had to learn to speak slowly and avoid reading excessively long PowerPoint slides." Nurani also had to do the following:

- Give an explanation of an overall policy.
- Provide a real-life situation so her listeners could apply the rule or policy.

  For example, if she was demonstrating the importance of asking only bonafide occupational qualities (BFOQ), she would use specific illustrations of questions interviewers *cannot ask*: "Would your husband mind if you travel?"; "Are you planning to get pregnant?"; "What is your favorite sports team?"
- Move around the room so everyone remained alert and remember to make eye contact with the listeners.

Perfecting Nurani's PowerPoint slides entailed the following:

- Using just bullet points.
- Having short points on slides that she elaborates on verbally.
- Making sure the slides are an outline for her listeners so they can follow the presentation.
- Including an appealing graphic for audience involvement.
- Proofreading to ensure correct grammar, punctuation, and spelling.
- Checking that her slides are updated with the constantly changing HR rules and regulations.
- Including only the highlights of the topics being discussed.

According to Nurani, "Since my first presentation, I have learned much about the process of effective oral and written communication. I have also learned that people don't always take information presented and apply it—and that is another challenge."

She now consistently scores in the 90 percent range on her evaluations, attesting to her success as a communicator. Even more important, through Nurani's presentations, she believes she is demonstrating that effective communication is extremely important in business, for without it "there is no business." Figure 7.14 shows Nurani's perfected PowerPoint slide.

FIGURE 7.12 **Nurani's Brainstorming List**

Before the interview: Be prepared, find the best place and time for the interview, make up the interview questions, invite the panel of interviewers, get water and glasses, paper and pens.

During the interview: Create a comfortable environment, ask questions correctly (check with our HR legal staff to find out what we can and can't ask), use correct body language when speaking to the applicant, and be sure to manage time (don't go too long, so we have enough time for the next interview).

At the end of the interview: Thank the applicant, walk her or him out of the building (or to meet management), meet with the panel to discuss the interview, rate the applicant, and prepare documentation for records.

Nurani says, "After my first unsuccessful presentation, I realized I was too 'slide dependent.' My slides had too much content and were way too boring and I read them to my audience! No wonder I scored poorly on the effectiveness of the presentation."

FIGURE 7.13 **Rough Draft of a PowerPoint Slide for Nurani's Presentation on "Conducting Effective Interviews"**

Create a Comfortable Environment

Not every applicant will be comfortable in an interview. It's your job to make them comfortable. Conduct the interview in a pleasant, informal, and conversational manner. Here's what to do:

Welcome applicants and make them feel comfortable by creating a climate that shows trust. Introduce any panel members and explain how the interview will be conducted. Tell the applicant how many questions will be asked, who will ask them, the sequence of the questions, and let them know about note-taking resources available during the interview, like pen, paper, or reference materials. Let applicants know that they will be allowed to ask questions at the end of the interview if there's time and encourage them to organize their thoughts before responding to questions.

Create a positive body language by making eye contact and facing the applicants so that you seem caring and interested. Avoid negative body language. Don't frown, cross your arms, look at your watch, take phone calls, or yawn during the interview.

Nurani says, "I love this PowerPoint slide! I can't believe how different it is from my original cluttered approach to creating a PowerPoint slide. Without considering layout and how it would affect the audience's ability to read the slide during the presentation, I would never have been able to create such a simple, easy-to-read slide. My audiences are obviously enjoying my presentations, and several have asked me how to prepare slides for their presentations—this is a sure indicator of my success as a communicator."

FIGURE 7.14 **Perfected PowerPoint Slide**

Create a Comfortable Environment

In the interview, be pleasant, informal, and conversational.

- Welcome applicants by using their name.
- Introduce the interviewers.
- Allow applicants to ask questions.
- Avoid negative body language.

CHAPTER HIGHLIGHTS

1. Effective oral communication ensures that your message is conveyed successfully and that your verbal, nonverbal, and listening skills make a good impression.
2. You will communicate verbally and nonverbally on an everyday basis, informally and formally.
3. Some points to consider when you use the telephone are to script your telephone conversation, speak clearly and slowly, and avoid rambling.
4. To prepare for a teleconference, plan ahead, check equipment for audio quality, and choose a quiet and private location.
5. A Webinar is a way to offer online training for employees at remote locations.
6. An effective oral presentation introduction might include an anecdote, question, quote, facts, or a "table setter."
7. In the discussion section of a formal oral presentation, organize your content according to comparison/contrast, problem/solution, argument/persuasion, importance, and chronology.
8. Conclude your formal speech by restating your main points, recommending a future course of action, or asking for questions and comments.
9. One of the most powerful oral communication tools is Microsoft PowerPoint.
10. Some benefits of PowerPoint slides include different autolayouts and design templates, the ability to add and delete slides easily, hypertext links, and the incorporation of sound and graphics.
11. When you use PowerPoint slides, ensure contrast between the font color and background, choose an easy-to-read font size and type, use few words and lines per slide, use headings for navigation, and prepare print handouts for your audience.
12. During a PowerPoint presentation, avoid merely reading the slides, choosing instead to elaborate on each slide with additional details.
13. The P^3 process of planning, packaging, and perfecting allows you to create effective oral communication.
14. When you plan your presentation, consider purpose, recognize audience, and gather information.
15. To package your presentation, use a presentation plan, outline, note cards, and visual aids.
16. Many visual aids can enhance your oral presentation. Choose from chalkboards, chalkless whiteboards, flip charts, overhead projections, PowerPoint, videos, and film.
17. To perfect your oral presentation (the third part of the communication process), work on your delivery, including eye contact, rate of speech, enunciation, pitch, pauses and emphasis, interaction with the audience, and conflict resolution.
18. Appearance, body language, and gestures will impact your oral presentation.
19. Restate main points frequently in an oral presentation.
20. Be prepared for questions and discussion after you conclude your oral presentation.

WWW *Find additional oral communication exercises, samples, and interactive activities at http://www.prenhall.com/gerson.*

MEETING WORKPLACE COMMUNICATION CHALLENGES

CASE STUDIES

1. A customer is treated poorly by a salesperson at TechStop. Vice President of Customer Service Shuan Wang has received reports of other problems with customer–sales staff interaction. Shuan needs to make a formal oral presentation to all sales personnel to improve customer service.

Assignment

Based on this scenario, outline the content for Shuan's oral presentation as follows:

- Determine what kind of introduction should be used to arouse the audience's interest.
- Provide a "road map" in the introduction.
- Develop information to teach employees sales etiquette and customer interaction, store policy regarding customer satisfaction, the impact on poor customer relations, and the consequences of failing to handle customers correctly. Research these topics (either online, in a library, or through interviews) to find your content.
- Organize the speech's content with appropriate transitions to aid coherence.
- Determine which types of visual aids would work best for this presentation.

2. Halfmoon Outdoor Equipment is an online wholesaler of hiking, biking, and boating gear. Their staff is located in four northeastern states. Halfmoon's CEO, Montana Wildhack, wants his employees to excel in communication skills, both oral and written. He has hired a consultant to train the employees. To accomplish this training, the consultant will provide a teleconference. Fifty employees will meet with the consultant in a training room, while another 150 Halfmoon staff members view the training session from their different work locations.

During the presentation, the consultant highlights his comments by writing on a chalkless whiteboard, using green and yellow markers. He attempts humor by suggesting that Halfmoon's thermal stocking caps would be a perfect gift for one of the seminar's balding participants, Joe. In addition, the consultant has participants break into small groups to role-play. They take mock customer orders, field complaints, and interact with vendors, using a prepared checklist of do's and don'ts.

At one point, when a seminar participant calls in a question, the consultant says, "No, that would be wrong, wrong, wrong! Why would you ever respond to a customer like that? Common sense would dictate a different response." During a break, the consultant does not mute his clip-on microphone. To conclude the session, the consultant asks for questions and comments. When he cannot provide a good answer, he asks the participants to suggest options, and he says he will research the question and place an answer on his corporate Web site.

Assignment

Using the guidelines in the chapter, discuss the speaker's oral communication skills. Explain what went wrong and why. Explain what was effective.

ETHICS IN THE WORKPLACE CASE STUDY

Pakash Patel is an outgoing employee at a large real estate corporation with offices in several locations. His company created a listserv for its 750 employees to ensure collegiality and to disseminate corporate news. Pakash consistently uses the company listserv for the following reasons: to share jokes, to advertise his daughter's Girl Scout cookie sales, to sell his son's trash bags for soccer team travel, to sell tickets to the Easter pageant at his church, to distribute chain letters, and to provide daily words of wisdom from various religious denominations.

Question: Is it ethical for Pakash to use his company's listserv in these ways? Justify your answer based on information provided in Chapter 1.

INDIVIDUAL AND TEAM PROJECTS

After reading the following scenarios, answer these questions:

- Would your oral presentation be everyday, informal, or formal?
- Would you use a videoconference, teleconference, or face-to-face meeting?
- Which visual aids would work best for your presentation?
- Is your oral presentation goal to persuade, instruct, inform, or build trust?

1. You work at FlashCom Electric, a company that has created a new interface for modems. Your boss has asked you to make a presentation to sales representatives from 20 potential vendors in the city. The oral presentation will explain to the vendors the benefits of your product, the sales breaks you will offer, and how the vendors can increase their sales.
2. After working for Friendly's, a major discount computer hardware and software retailer, for over a year, you have created a new organizational plan for their vast inventory. In an oral presentation, you plan to show the CEO and board of directors why your plan is cost effective and efficient.
3. You are the manager of an automotive parts supply company, Plugs, Lugs, & More. Your staff of 10 in-store employees lacks knowledge of the store's new merchandise, has not been meeting sales goals, and does not always treat customers with respect and care. It's time to address these concerns.
4. As CEO of your engineering/architectural firm (Levin, Lisk, and Lamb), you must downsize. Business is decreasing and costs are rising. To ensure third-quarter profitability, 10 percent of the staff must be laid off. That will amount to a layoff of over 500 employees. You now must make an oral presentation to your stockholders and the entire workforce (located in three states) at your company's annual meeting to report the situation.
5. Your quality circle team is composed of account representatives, management, engineers, and technicians. You have been asked to research new ways to improve product delivery. Now that you have concluded your research, you need to share your suggestions with your 15 team members.

DEGREE-SPECIFIC ASSIGNMENTS

1. **Marketing and Public Relations** Your company, PR&U, has been hired by a client to provide advertising options for their line of office products. The client has an advertising budget of $35,000 and wants whatever advertising campaign you create to be cost effective, easy to update, allow for customer access, and visually appealing (the company wants to present an attractive and youthful look).

Assignment

Brainstorm advertising possibilities for the client, create a skeleton outline for your company's oral presentation to the client, and then give the required speech.

2. **Business Information Systems** Your BIS department has been asked to develop and implement a company-wide information system. Various software applications, accessed via the company's intranet, will provide services for the following departments:
   - The accounting department will use software for risk management assessments, business valuations, comparative analyses, and evaluation of debt structure and receivables.

- The marketing department will use software to create highly targeted prospect lists, sales plans, and marketing evaluations.
- Manufacturing will use software to optimize internal productivity and regulate shipping.
- Human resources will use optical character recognition (OCR) software for assessing job candidates and benefits software for compensation and insurance calculations.
- All departments will use in-house developed communication software (e-mail, instant messaging, listservs, and chat rooms).

Assignment

To share this information with each department, use PowerPoint to create an outline and then give the required oral presentation.

3. **Accounting** As an accountant for Liss, Levin, and Myers Accounting, you have been asked to audit a client's businesses. Reggie King is CEO of King Associates, a parent company for a dozen wholly owned subsidiaries. Mr. King's subsidiaries include three construction firms, six building supply companies (manufacturing and marketing wall board, lumber, hardware, stucco, paint, and carpeting), two plumbing companies, and one land development firm.

You must travel to these companies, located in three different cities, to perform your audits. Primarily, you will focus on the following risk assessments:

- An assessment of the company's financial records and account balance levels
- An evaluation of the company's internal control structure
- An assessment of the company's procedures for controlling and detecting risk

Your audit found the following problems. At three of the visited sites, management was not communicating effectively with Mr. King about the perceived risks (offshore production, limited workforce, slow delivery of goods, and increased costs of services). The problem was due to what you have defined as an overly complex organizational structure, involving unusual managerial lines of authority. At four sites, you found material misstatements in the financial statements. These were due to high turnover rates in the companies' accounting and information technology staffs. At two sites, you found that accounting practices were not uniform. Most troubling, at three sites, you found inadequate or missing controls related to the approval of payroll transactions. This was due to inadequate record keeping of assets, an inadequate system for authorizing transactions, and an inadequate system for approving transactions.

Assignment

Brainstorm possible solutions for this company's accounting problems, create a skeleton outline, and then make an informal oral presentation reporting on your findings.

4. **Management** You are the events planning manager for the Lewis Clark Convention Center in Galveston, Texas. The convention center is attached to the Surf Winds Hotel located on Seawall Boulevard. Convention sales were down last quarter. You need to give an oral presentation, instructing the sales employees how to increase their convention sales.

Assignment

Brainstorm ways to increase convention sales, create a skeleton outline for your company's oral presentation to employees, and then give the required oral presentation.

5. **Finance** You work for a mortgage company as a loan officer. A client asks you to recommend the best loan option for his planned home purchase. The client, Randall Bridge, hopes to purchase a $250,000 home in Naples, Florida. Randall,

a first-time buyer, earns $60,000 a year as a biomedical engineer. He does not have any money for a down payment, so he hopes to finance 100 percent of the loan.

Assignment

Brainstorm mortgaging options for Randall Bridge, create a skeleton outline for your oral presentation to him, and then give the required speech.

PROBLEM-SOLVING THINK PIECES

Evaluating an Oral Presentation

Listen to a speech. You could do so on television; at a student union; at a church, synagogue, or mosque; at a civic event, city hall meeting, or community organization; at a company activity; or in your classroom. Answer the following questions:

1. What was the speaker's goal? Did the speaker try to persuade, instruct, inform, or build trust? Explain your answer by giving examples.
2. What type of introduction did the speaker use to arouse listener interest (anecdote, question, quote, or facts)? Give examples to support your answer.
3. What visual aids were used in the presentation? Were these visual aids effective? Explain your answers.
4. How did the speaker develop the assertions? Did the speaker use analysis, comparison/contrast, argument/persuasion, problem/solution, or chronology? Give examples to prove your point.
5. Was the speaker's delivery effective? Use the Presentation Delivery Rating Sheet in Table 7.5 to assess the speaker's performance by placing check marks in the appropriate columns.

Giving an Oral Presentation

1. Find an advertised job opening in your field of expertise or degree program. You could look either online, in the newspaper, at your school's career placement center, or at a company's human resources office. Perform a mock interview for this position. To do so, follow this procedure:
   - Designate one person as the applicant.
   - He or she should prepare for the interview by making a list of potential questions (ones that the applicant believes he or she will be asked, as well as questions to ask the search committee).
   - Designate others in the class to represent the search committee.
   - The mock search committee should prepare a list of questions to ask the applicant.

Table 7.5 Presentation Delivery Rating Sheet

Delivery Techniques	Good	Bad	Explanation
Eye Contact			
Rate of Speech			
Enunciation			
Pitch of Voice			
Use of Pauses			
Emphasis			
Interaction with Listeners			
Conflict Resolution			

2. Research a topic in your field of expertise or degree program. The topic could include a legal issue, a governmental regulation, a news item, an innovation in the field, or a published article in a professional journal or public magazine. Make an oral presentation about your findings.

Creating a PowerPoint Slide Presentation

For assignments 1 and 2 in the section "Giving an Oral Presentation," create PowerPoint slides to enhance your oral communication. To do so, follow the guidelines provided in this chapter.

Assessing PowerPoint Slides

After reviewing the Microsoft PowerPoint slides below (pages 209–210), determine which are successful, which are unsuccessful, and explain your answers, based on the guidelines provided in this chapter.

WEB WORKSHOP

For more information about PowerPoint, both techniques for using this technology effectively and potential problems, research the following online articles. Compare and contrast the various points of view and make an informal oral presentation to your class.

Sites Related to Effective PowerPoint Usage

Finklestein, Ellen. "PowerPoint Tips Blog." http://www.ellenfinkelstein.com/powerpoint_tip.html.

Paradi, Dave. "How to Avoid Death by PowerPoint." http://www.communicateusingtechnology.com/articles/avoid_death_by_ppt.htm.

"The PowerPoint FAQ." http://www.pptfaq.com/.

Russell, Wendy. "10 PowerPoint Tutorials for Beginners—How to Use PowerPoint." http://presentationsoft.about.com/od/powerpoint101/a/begin_guide.htm.

Wuorio, Jeff. "Presenting with PowerPoint: 10 Do's and Don'ts." http://www.microsoft.com/smallbusiness/resources/technology/business_software/presenting_with_powerpoint_10_dos_and_donts.mspx#EDD.

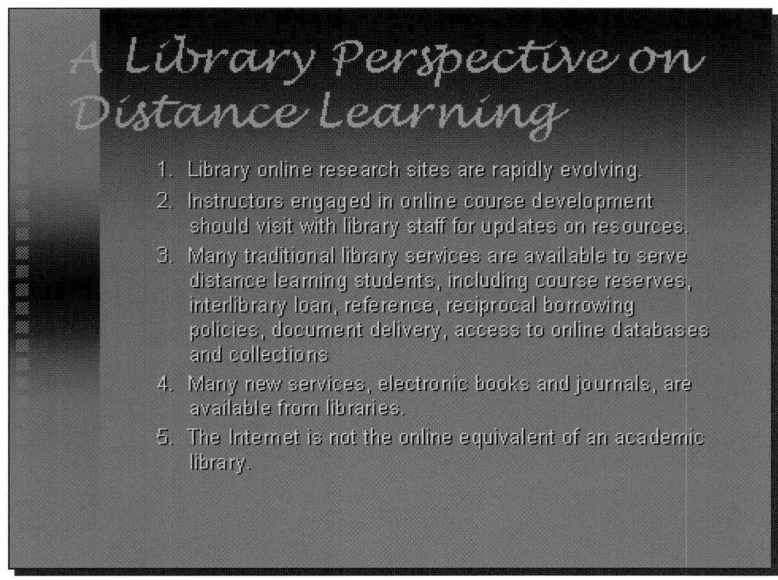

A Technology Perspective on Online Education

Students—24-hour Call Centers answer student hardware/software needs

Faculty—The Tech Center helps faculty with course creations and tech resources

As DISTANCE LEARNING classes have risen in popularity, many *diverse areas* of interest across campuses have converged.

These include counseling, library services, technology, faculty, administration, continuing education, etc.

To ensure student success, *Distance Learning Coordinating Committees* need to draw from these multiple disciplines and collaborate on solutions to arising technology challenges.

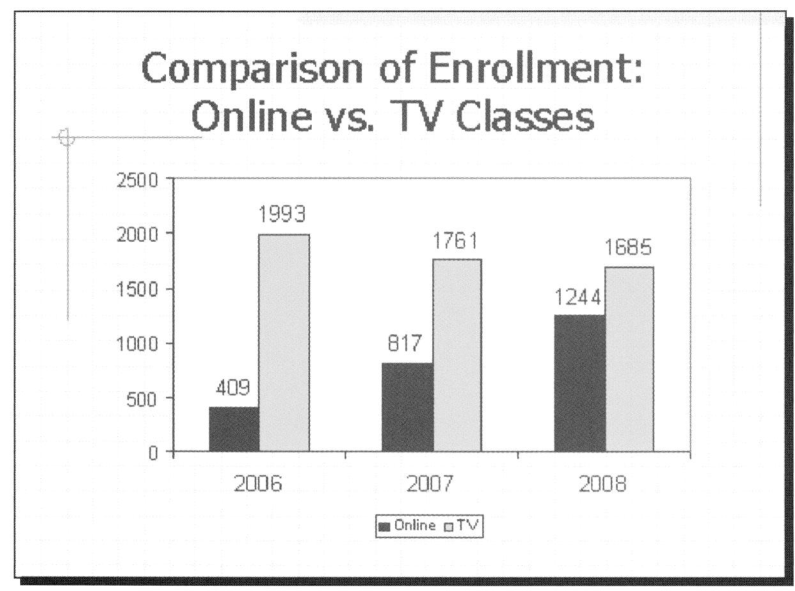

Opposing Points of View Regarding PowerPoint

Garber, Angela. "Death by PowerPoint." http://www.smallbusinesscomputing.com/biztools/article.php/684871. April 1, 2001.

Tufte, Edward. "PowerPoint Is Evil. Power Corrupts. PowerPoint Corrupts Absolutely." http://www.wired.com/wired/archive/11.09/ppt2.html.

QUIZ QUESTIONS

1. What are five elements of nonverbal communication?
2. What are five barriers to active listening?
3. What are four keys to effective listening?
4. How can you create an effective teleconference or videoconference?
5. In what ways can you arouse audience attention in an oral presentation?
6. How can you conclude a formal oral presentation?
7. What are the benefits of using Microsoft PowerPoint slide presentations?
8. List three ways to create an effective PowerPoint slide presentation.
9. Instead of writing out an entire formal oral presentation, how can you prepare your speech?
10. In addition to PowerPoint, what are four other types of visual aids?

CHAPTER 8

Visual Aids in Workplace Communication

▸REAL PEOPLE *in the workplace*

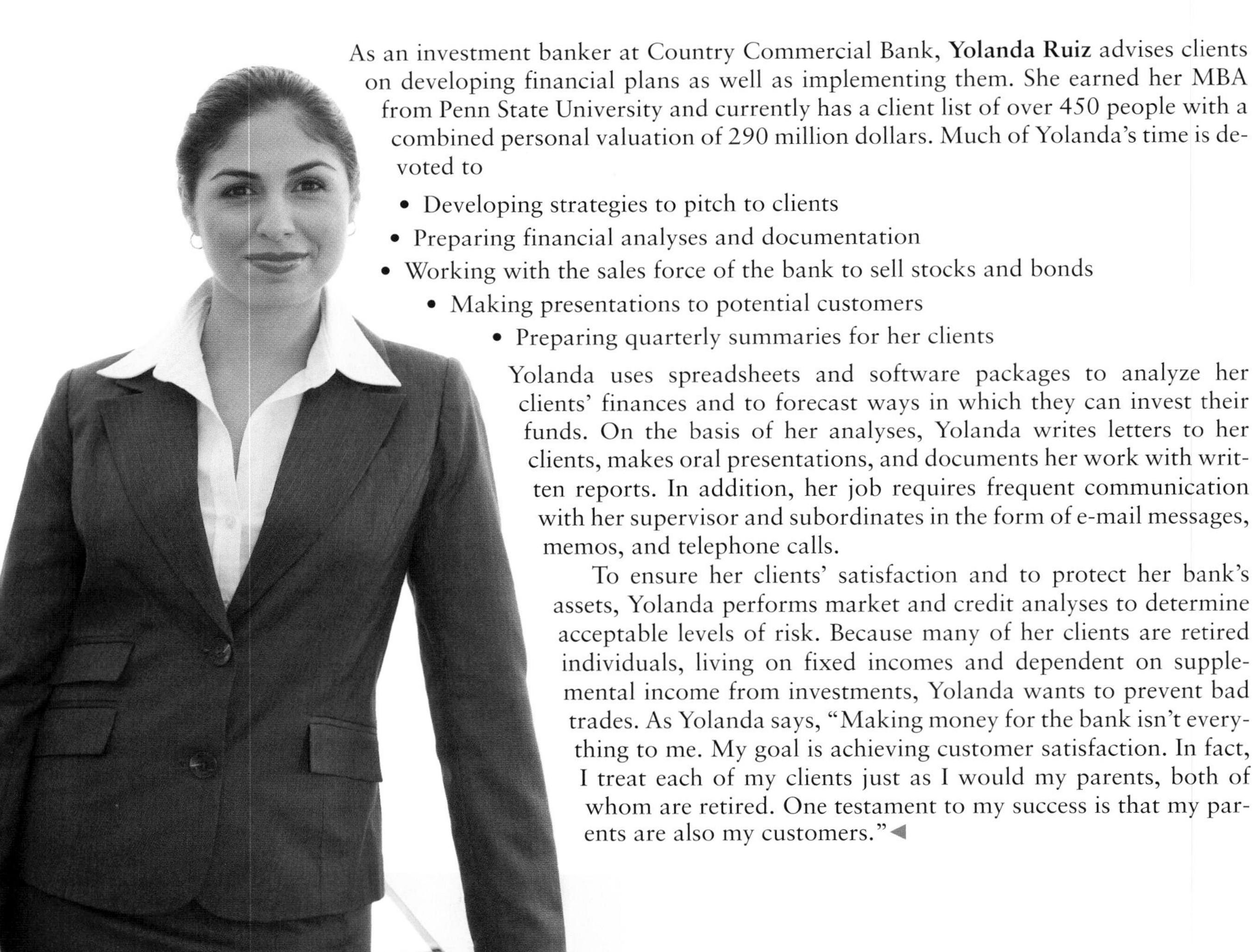

As an investment banker at Country Commercial Bank, **Yolanda Ruiz** advises clients on developing financial plans as well as implementing them. She earned her MBA from Penn State University and currently has a client list of over 450 people with a combined personal valuation of 290 million dollars. Much of Yolanda's time is devoted to

- Developing strategies to pitch to clients
- Preparing financial analyses and documentation
- Working with the sales force of the bank to sell stocks and bonds
- Making presentations to potential customers
- Preparing quarterly summaries for her clients

Yolanda uses spreadsheets and software packages to analyze her clients' finances and to forecast ways in which they can invest their funds. On the basis of her analyses, Yolanda writes letters to her clients, makes oral presentations, and documents her work with written reports. In addition, her job requires frequent communication with her supervisor and subordinates in the form of e-mail messages, memos, and telephone calls.

To ensure her clients' satisfaction and to protect her bank's assets, Yolanda performs market and credit analyses to determine acceptable levels of risk. Because many of her clients are retired individuals, living on fixed incomes and dependent on supplemental income from investments, Yolanda wants to prevent bad trades. As Yolanda says, "Making money for the bank isn't everything to me. My goal is achieving customer satisfaction. In fact, I treat each of my clients just as I would my parents, both of whom are retired. One testament to my success is that my parents are also my customers."◂

CHAPTER GOALS

When you complete this chapter, you will be able to do the following:

1. Recognize the benefits of visual aids in your workplace communication.
2. Consider the use of color or three-dimensional graphics to enhance workplace communication.
3. Understand the criteria for creating effective tables and figures.
4. Know how to use a variety of visual aids.
5. Evaluate your visual aids with the checklist.

WWW

To learn more about visual aids in workplace communication, visit our Web Companion at http://www.prenhall.com/gerson.

Yolanda's Communication Challenge

Yolanda says, "I have to write a proposal to a potential client, Sylvia Light, a retired public health nurse. Sylvia is 68 years old and worked for the Texas Public Health Department for 36 years. She has earned her State of Texas retirement and Social Security benefits. She also has $315,500 in savings allocated as follows: $78,000 in an individual retirement account (IRA), $234,000 in a low-earning certificate of deposit (CD), and $3,500 in her checking account.

Sylvia contacted me for help organizing a portfolio for a comfortable retirement. I have studied Sylvia's various accounts and considered lifestyle and expenditures. Now, I am ready to write the proposal. In this proposal, I want to show Sylvia how to reallocate funds. Sylvia should keep some ready money available and invest a portion of capital for long-term returns. Currently, too much of her money is tied up in a CD earning 3.1 percent.

Like most people, Sylvia is unfamiliar with financial planning. Though she was an expert in her health field—tuberculosis and AIDS treatment—money matters confuse her. Numbers alone will not explain my vision for money management. My challenge is to communicate complex financial information to Sylvia, a lay person with no financial expertise."

See pages 231–232 to learn how Yolanda met her communication challenge.

THE BENEFITS OF VISUAL AIDS

Although your writing may have no grammatical or mechanical errors and you may present valuable information, you won't communicate effectively if your information is inaccessible. Consider the following paragraph.

example

According to an article in *Business Week*, WiFi usage worldwide is growing. For example, in the Asia Pacific Rim region, in 2004, 32,937,000 people were WiFi users. This number grew to 55,341,000 in 2005. In 2006, the article said that the number grew to 81,048,000 and then to 168,193,000 in 2009. In Western Europe, in 2004, WiFi users numbered 16,681,000. By 2005, 24,877,000 people were WiFi users. This number grew to 33,546,000 in 2006 and then to 63,746,000 in 2009. Central and Eastern Europe had fewer WiFi users, numbering 2,109,000 in 2004, 3,172,000 in 2005, 4,383,000 in 2006, and 9,875,000 in 2009. Latin American WiFi users were

(Continued)

example

similar to those in Europe, with the following numbers: 2,386,000 in 2004, 3,401,000 in 2005, 4,528,000 in 2006, and 8,331,000 in 2009. The lowest number of WiFi users exist in Africa and the Middle East, with only 287,000 in 2004, 664,000 in 2005, 1,096,000 in 2006, and 2,747,000 in 2009. North American WiFi users are second only to Asia and the Pacific Rim. In North America, WiFi users numbered 20,570,000 in 2004, 30,235,000 in 2005, 40,454,000 in 2006, and 74,174,000 in 2009. The total number of WiFi users worldwide were 74,970,000 in 2004, 117,690,000 in 2005, 165,055,000 in 2006, and 327,066,000 in 2009 ("WiFi Gaining Traction").

If you read the preceding paragraph in its entirety, you are an unusually dedicated reader. Such wall-to-wall words mixed with statistics do not create easily readable writing. The goal of effective workplace communication is accessible information. The example paragraph fails to meet this goal. Readers cannot digest the data easily or see clearly the comparative changes annually from one region of the world to another.

To present large blocks of data or reveal comparisons, you can supplement, if not replace, your text with graphics. In workplace communication, visual aids accomplish several goals. Graphics (whether hand drawn, photographed, or computer generated) will help you achieve conciseness, clarity, and cosmetic appeal.

Conciseness

Visual aids allow you to provide large amounts of information in a small space. Words used to convey data (such as in the example paragraph above) double, triple, or even quadruple the space needed to report information. By using graphics, you can also delete many unnecessary words and phrases.

Clarity

Visual aids can clarify complex information. Graphics help readers see the following:

- **Trends**—Certain trends, such as increasing or decreasing sales figures, are most evident in line graphs.
- **Comparisons between like components**—Comparisons such as actual monthly versus average WiFi usage can be seen in grouped bar charts.
- **Percentages**—Pie charts help readers discern these.
- **Facts and figures**—A table states statistics/numbers more clearly than a wordy paragraph.

Additional information:

See Chapter 6, "Perfecting Workplace Communication."

Cosmetic Appeal

Visual aids help you break up the monotony of wall-to-wall words. If you only give unbroken text, your reader will tire, lose interest, and overlook key concerns. Graphics help you sustain your reader's interest. Let's face it; readers like to look at pictures. The two types of graphics important for workplace communication are tables and figures. This chapter helps you correctly use both.

COLOR

All graphics look best in color . . . don't they? Not necessarily. Without a doubt, a graphic depicted in vivid colors will attract your reader's attention. However, the colors might not aid communication. For example, using colored graphics could have these drawbacks (Reynolds and Marchetta 5–7):

1. The colors might be distracting (glaring orange, red, and yellow combinations on a bar chart would do more harm than good).
2. Colors that look good today might go out of style in time.
3. Colored graphics increase production costs.
4. Colored graphics consume more disk space and computer memory than black-and-white graphics.
5. The colors you use might not look the same to all readers.

Let's expand on this last point. Just because you see the colors one way does not mean your readers will see them the same. We are talking about what happens to your color graphics when someone reproduces them as black-and-white copies. We are also talking about computer monitor variations. The color on a computer monitor depends on its resolution (the number of pixels displayed) and the monitor's RGB values (how much red, green, and blue light is displayed). Because all monitors do not display these same values, what you see on your monitor will not necessarily be the same as what your reader sees. To solve this problem, test your graphics on several monitors. Also, limit your choices to primary colors instead of the infinite array of other color possibilities. More important, use patterns to distinguish your information so that the color becomes secondary to the design.

THREE-DIMENSIONAL GRAPHICS

Many people are attracted to three-dimensional (3-D) graphics. After all, they have obvious appeal. Three-dimensional graphics are more interesting and vivid than flat, one-dimensional graphics. However, 3-D graphics have drawbacks. A 3-D graphic is visually appealing, but it does not convey information quantifiably. A word of caution: Use 3-D graphics sparingly. Better yet, use the 3-D graphic to create an impression; then include a table to quantify your data.

CRITERIA FOR EFFECTIVE GRAPHICS

Figure 8.1 is an example of a cosmetically appealing, clear, and concise graphic. At a glance, the reader can pinpoint the comparative prices per barrel of crude oil between 2005 and 2009. Thus, the line graph is clear and concise. In addition, the writer has included an interesting artistic touch. The oil gushing out of the tower shades parts of the graph to emphasize the dollar amounts. Envision this graph without the shading. Only the lines would exist. The shading provides the right touch of artistry to enhance the information communicated.

The graph shown in Figure 8.1 includes the traits common to effective visual aids. Whether hand drawn or computer generated, successful tables and figures

1. Are integrated with the text (i.e., the graphic complements the text; the text explains the graphic).
2. Are appropriately located (preferably immediately following the text referring to the graphic and not a page or several pages later).
3. Add to the material explained in the text (without being redundant).

FIGURE 8.1 Line Graph with Shading

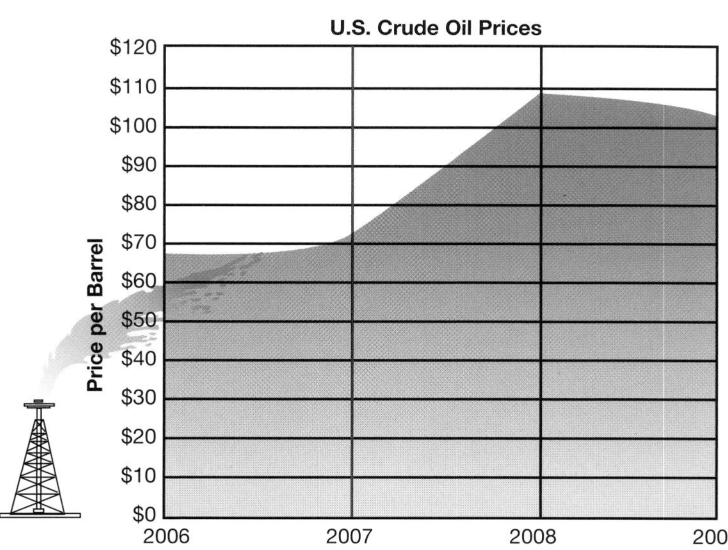

4. Communicate important information that could not be conveyed easily in a paragraph or longer text.
5. Do not contain details that detract from rather than enhance the information.
6. Are an effective size (not too small or too large).
7. Are neatly printed to be readable.
8. Are correctly labeled (with legends, headings, and titles).
9. Follow the style of other figures or tables in the text.
10. Are well conceived and carefully executed.

TYPES OF GRAPHICS

Graphics can be broken down into two basic types: tables and figures. Tables provide columns and rows of information. You should use a table to make factual information—such as numbers, percentages, and monetary amounts—easily accessible and understandable. Figures, in contrast, are varied and include bar charts, line graphs, photographs, pie charts, schematics, line drawings, and many more.

Tables

Let's tabulate the information about worldwide WiFi usage presented earlier on pages 213–214. Because effective workplace communication integrates text and graphics, you will want to provide an introductory sentence prefacing Table 8.1, as follows: "Table 8.1 shows WiFi usage by regions in the world."

This table has advantages for both the writer and the reader. First, the headings eliminate needless repetition of words, thereby making the text more readable. Second, the audience easily can see the comparison between yearly WiFi usage worldwide. Thus, the table highlights the content's significant differences. Finally, if this information is included in a report, the writer will reference the table in the Table of Contents' List of Illustrations. This creates ease of access for the reader.

Criteria for Effective Tables

To construct tables correctly, do the following:

1. Number tables in order of presentation (i.e., Table 1, Table 2, Table 3, etc.).
2. Title every table. In your writing, refer to the table by its number, not its title. Simply say, "Table 1 shows . . . ," "As seen in Table 1," or "The information in Table 1 reveals. . . ."
3. Present the table as soon as possible after you have mentioned it in your text. Preferably, place the table on the same page as the appropriate text, not on a subsequent, unrelated page, or in an appendix.
4. Don't present the table until you have mentioned it.
5. Use an introductory sentence or two to lead into the table.

Table 8.1 WiFi Usage by Worldwide Regions
("Wi-Fi Gaining Traction")

	2004	2005	2006	2009
Asia Pacific	32,937,000	55,341,000	81,048,000	168,193,000
Western Europe	16,681,000	24,877,000	33,546,000	63,746,000
Central and Eastern Europe	2,109,000	3,172,000	4,383,000	9,875,000
Latin America	2,386,000	3,401,000	4,528,000	8,331,000
Africa/Middle East	287,000	664,000	1,096,000	2,747,000
North America	20,570,000	30,235,000	40,454,000	74,174,000
Total	**74,970,000**	**117,690,000**	**165,055,000**	**327,066,000**

6. After you have presented the table, explain its significance. You might write, "Thus, WiFi usage has grown exponentially since 2004. The peak year for WiFi usage by region worldwide is 2009."
7. Write headings for each column. Choose terms that summarize the information in the columns. For example, you could write "Percent (%) of Error," "Length in Ft.," or "Amount in $."
8. Because the size of columns is determined by the width of the data or headings, you may want to abbreviate terms (as shown in item 7). If you use abbreviations, however, be sure your audience understands your terminology.
9. Center tables between right and left margins. Don't crowd them on the page.
10. Separate columns with ample white space, vertical lines, or dashes.
11. Show that you have omitted information by printing two or three periods or a hyphen or dash in an empty column.
12. Be consistent when using numbers. Use either decimals or numerators and denominators for fractions. You could write 3¼ and 3¾ or 3.25 and 3.75. If you use decimal points for some numbers but other numbers are whole, include zeroes on the whole numbers. For example, write 9.00 for 9.
13. If you do not conclude a table on one page, write *Continued* in parentheses.

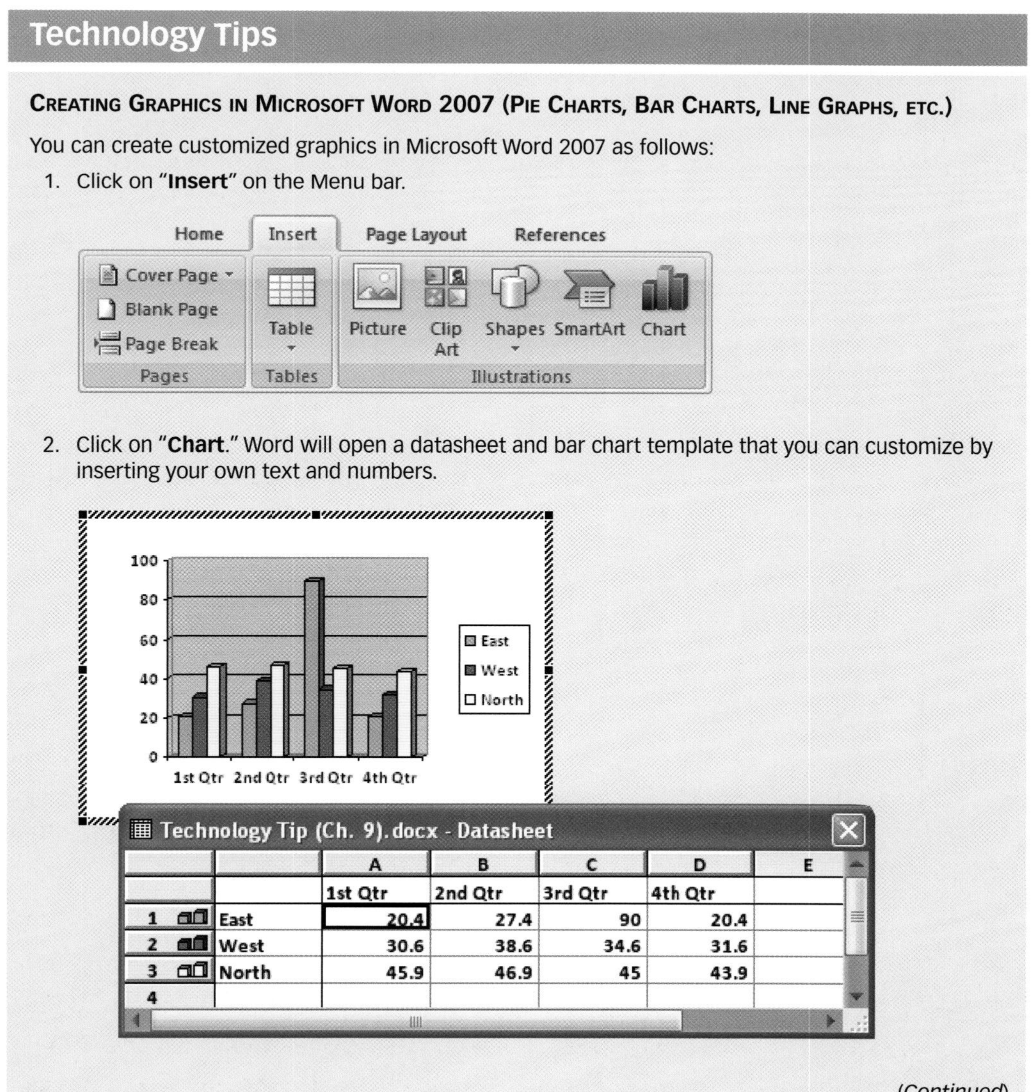

Technology Tips

CREATING GRAPHICS IN MICROSOFT WORD 2007 (PIE CHARTS, BAR CHARTS, LINE GRAPHS, ETC.)

You can create customized graphics in Microsoft Word 2007 as follows:

1. Click on "**Insert**" on the Menu bar.

2. Click on "**Chart**." Word will open a datasheet and bar chart template that you can customize by inserting your own text and numbers.

		A	B	C	D	E
		1st Qtr	2nd Qtr	3rd Qtr	4th Qtr	
1	East	20.4	27.4	90	20.4	
2	West	30.6	38.6	34.6	31.6	
3	North	45.9	46.9	45	43.9	
4						

(*Continued*)

Once you have opened Word's graphic datasheet and template, you can customize the graphic further as follows:

3. Choose the type of graphic you want by clicking on "**Chart**" on the Menu bar and scrolling to and selecting "**Chart Type**." The "**Chart Type**" dialog box displays, allowing you to select a chart type.

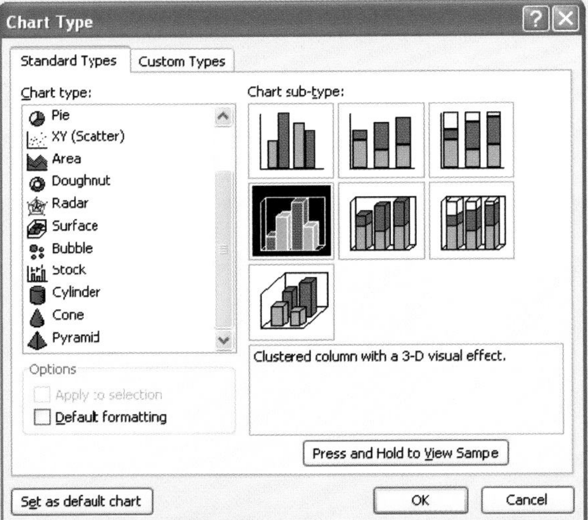

4. To add figure numbers, figure titles, legends, gridlines, data labels, and data tables, click on "**Chart**" and scroll to and select "**Chart Options**."

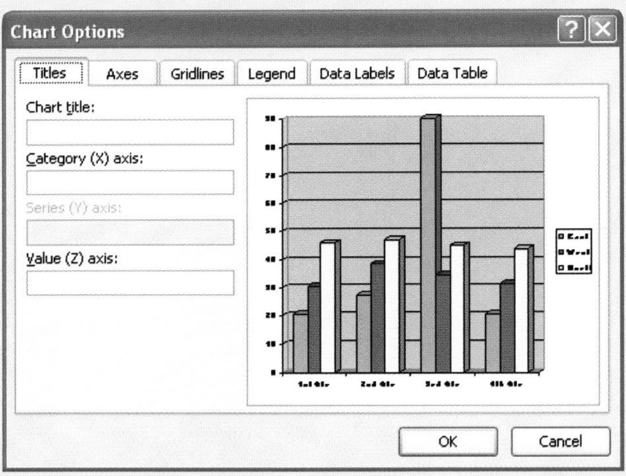

Figures

Another way to enhance your workplace communication is to use figures. Whereas tables eliminate needless repetition of words, figures highlight and supplement important points in your writing. Like tables, figures help you communicate with your reader.

Types of figures include the following:

- Bar charts
  - Grouped bar charts
  - 3-D (tower) bar charts
  - Pictographs
  - Gantt charts
  - Pie charts

- Line charts
  - Broken line charts
  - Curved line charts
- Combination charts
- Flowcharts
- Organizational charts
- Line drawings
  - Exploded views
  - Cutaway views
  - Renderings
  - Virtual reality drawings
- Photographs
- Icons
- Internet graphics

All of these types of figures can be computer generated using an assortment of computer programs. The program you use depends on your preference and hardware.

Criteria for Effective Figures

To construct figures correctly, do the following:

1. Number figures in order of presentation (i.e., Figure 1, Figure 2, Figure 3, etc.).
2. Title each figure. When you refer to the figure, use its number rather than its title: for example, "Figure 1 shows the relation between the average price for houses and the actual sales prices."
3. Preface each figure with an introductory sentence.
4. Don't use a figure until you have mentioned it in the text.
5. Present the figure as soon as possible after mentioning it instead of several paragraphs or pages later.
6. After you have presented the figure, explain its significance. Don't let the figure speak for itself. Remind the reader of the important facts you want to highlight.
7. Label the figure's components. For example, if you are using a bar or line chart, label the x- and y-axes clearly. If you're using line drawings, pie charts, or photographs, use clear *call-outs* (names or numbers that indicate particular parts) to label each component.
8. When necessary, provide a legend or key at the bottom of the figure to explain information. For example, a key in a bar or line chart will explain what each differently colored line or bar means. In line drawings and photographs, you can use numbered call-outs in place of names. If you do so, you will need a legend at the bottom of the figure explaining what each number means.
9. If you abbreviate any labels, define these in a footnote. Place an asterisk (*) or a superscript number (^{1}, ^{2}, ^{3}) after the term and then at the bottom of the figure where you explain your terminology.
10. If you have drawn information from another source, note this at the bottom of the figure.
11. Frame the figure. Center it between the left and right margins or place it in a text box.
12. Size figures appropriately. Don't make them too small or too large.

Bar Charts

Bar charts show either vertical bars (as in Figure 8.2) or horizontal bars (as in Figure 8.3). These bars are scaled to reveal quantities and comparative values. You can shade, color, or crosshatch the bars to emphasize the contrasts. If you do so, include a key explaining what each bar represents, as in Figure 8.3. *Pictographs* (as in Figure 8.4) use picture

FIGURE 8.2 **Vertical Bar Chart**

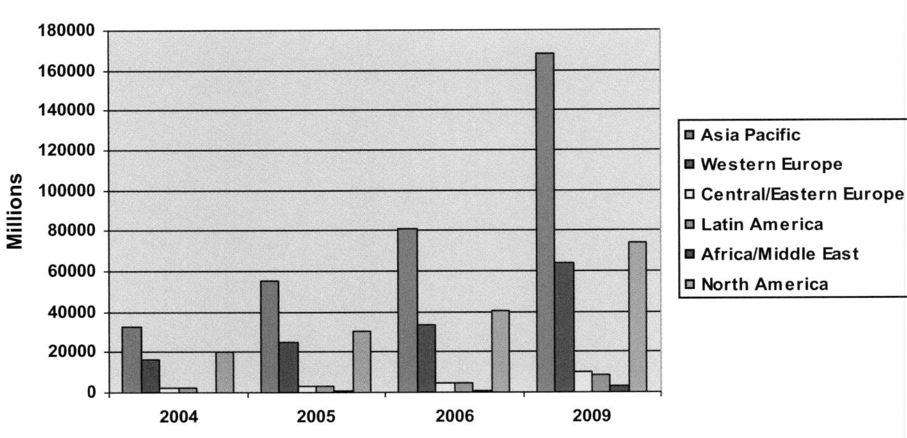

("WiFi Gaining Traction")

FIGURE 8.3 **Horizontal Bar Chart for High-Tech Readers**

Comparative Sales Revenue of Office Products Dealers

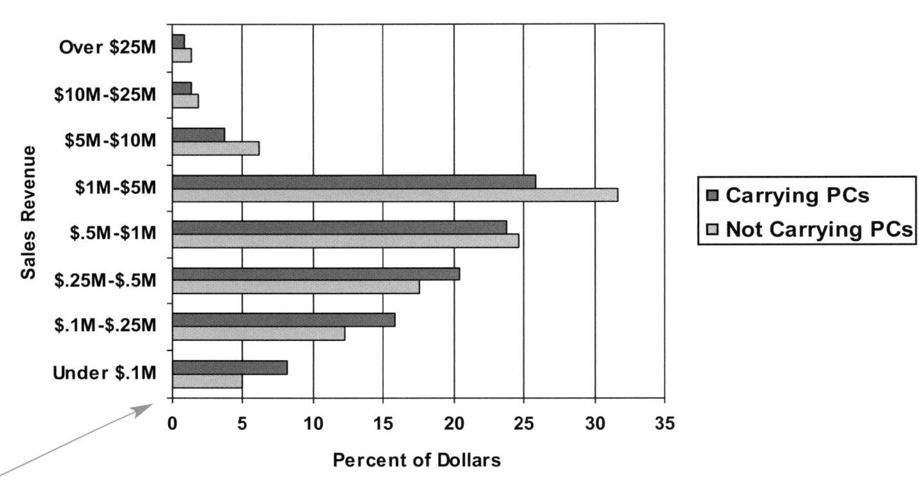

This bar chart, geared toward office supply sales managers, is as factual as the pictograph in Figure 8.4. However, Figure 8.3, which omits the drawings of the PCs, keeps the graphic more businesslike for the intended audience of specialists.

symbols instead of bars to show quantities. To create effective pictographs, keep the following in mind:

1. The picture should be representative of the topic discussed.
2. Each symbol equals a unit of measurement. The size of the units depends on your value selection as noted in the key or on the x- and y-axes.
3. Use more symbols of the same size to indicate a higher quantity; do not use larger symbols.

FIGURE 8.4 Pictograph for Lay Audience

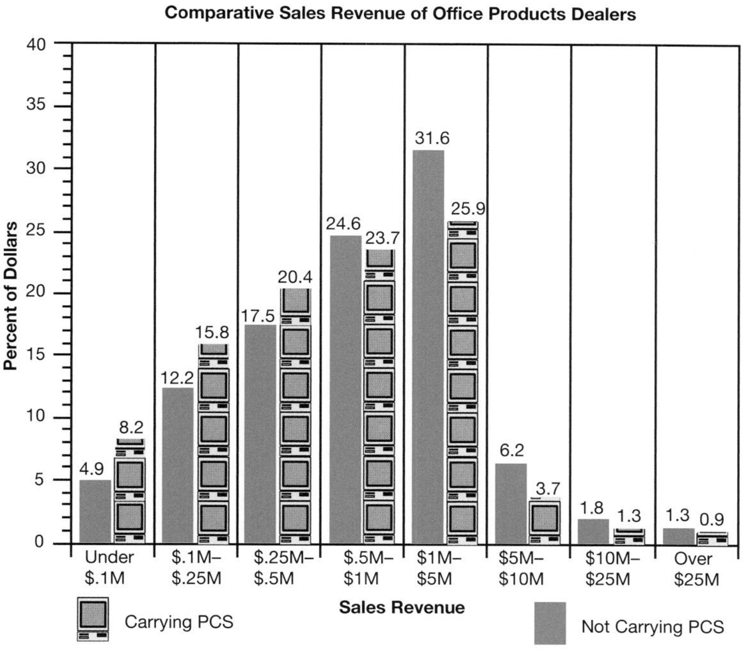

Comparative Sales Revenue of Office Products Dealers

The pictograph is as factual as the bar chart in Figure 8.3. However, since the pictograph is designed for lay readers, it uses symbols of PCs to enhance the visual appeal of this topic. In addition, the PCs make the subject matter more interesting for the audience.

Gantt Charts

Gantt charts, or *schedule charts* (as in Figure 8.5), use bars to show chronological activities. For example, your goal might be to show a client phases of a project. This could include planned start dates, planned reporting milestones, planned completion dates, names of team members along with the assigned duties, actual progress made toward completing the project, and work remaining. Gantt charts are an excellent way to represent these

FIGURE 8.5 Gantt Chart

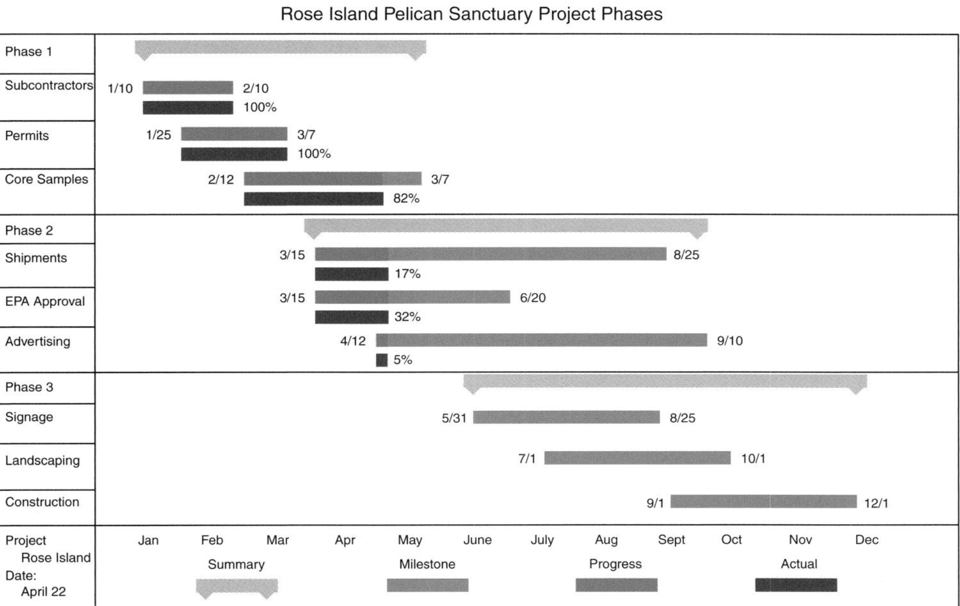

activities visually. They are often included in proposals to project schedules or in reports to show work completed. To create successful Gantt charts, do the following:

1. Label your x- and y-axes. For example, if the y-axis represents the various activities scheduled, then the x-axis represents time (either days, weeks, months, or years).
2. Provide gridlines (either horizontal or vertical) to help your readers pinpoint the time accurately.
3. Label your bars with exact dates for start or completion.
4. Quantify the percentages of work accomplished and work remaining.
5. Provide a legend or key to differentiate between planned activities and actual progress.

Pie Charts

Use pie charts (as in Figure 8.6) to illustrate portions of a whole. The pie chart represents information as pie-shaped parts of a circle. The entire circle equals 100 percent or 360 degrees. The pie pieces (the wedges) show the various divisions of the whole.

To create effective pie charts, do the following:

1. Be sure that the complete circle equals 100 percent or 360 degrees.
2. Begin spacing wedges at the 12 o'clock position.
3. Use shading, color, or crosshatching to emphasize wedge distributions.
4. Use horizontal writing to label wedges.
5. If you don't have enough room for a label within each wedge, provide a key defining what each shade, color, or crosshatching symbolizes.
6. Provide percentages for wedges when possible.
7. Do not use too many wedges—this would crowd the chart and confuse readers.
8. Make sure that different sizes of wedges are fairly large and dramatic.

Line Charts

Line charts reveal relationships between sets of numbers. To make a line chart, plot sets of numbers and connect the sets with lines. These lines create a picture showing the upward and downward movement of quantities. Line charts of more than one line (see Figure 8.7) are useful in showing comparisons between two sets of values. However, avoid creating line charts with too many lines, as these will confuse your readers.

FIGURE 8.6 Pie Chart

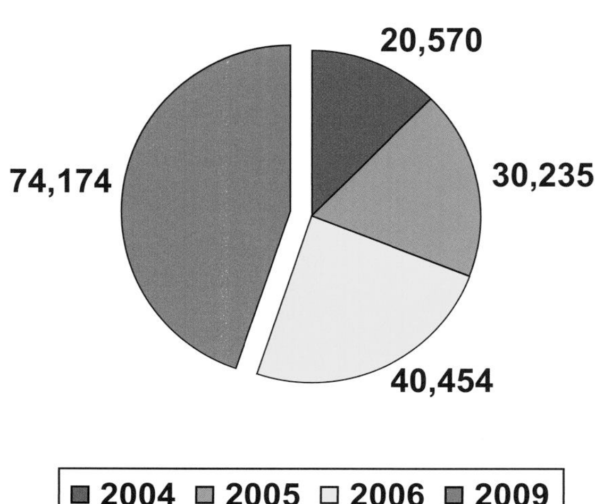

FIGURE 8.7 Line Chart

WiFi Usage by Worldwide Regions

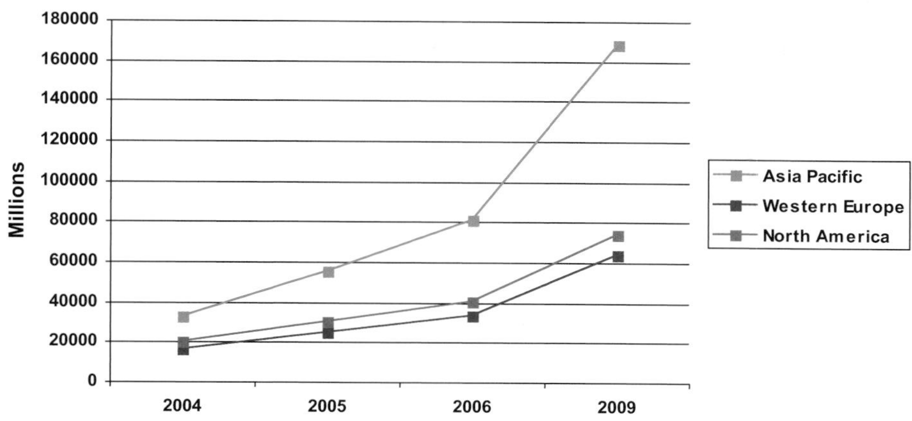

Combination Charts

A combination chart reveals relationships between two sets of figures. To do so, it uses a combination of figure styles, such as a bar chart and a line chart (as shown in Figure 8.8). The value of a combination chart is that it adds interest and distinguishes the two sets of figures by depicting them differently.

Flowcharts

You can show the chronological sequence of activities using a flowchart. Flowcharts are especially useful for writing technical instructions. When using a flowchart, remember that ovals represent starts and stops, rectangles represent steps, and diamonds equal decisions (see Figure 8.9).

Organizational Charts

Organizational charts (as in Figure 8.10) show the chain of command in an organization. You can use boxes around the information or use white space to distinguish among levels

FIGURE 8.8 Combination Chart (Bar and Line)

Comparison of Customer Purchases for Krenshaw & Bailey, Inc., 2004-2009

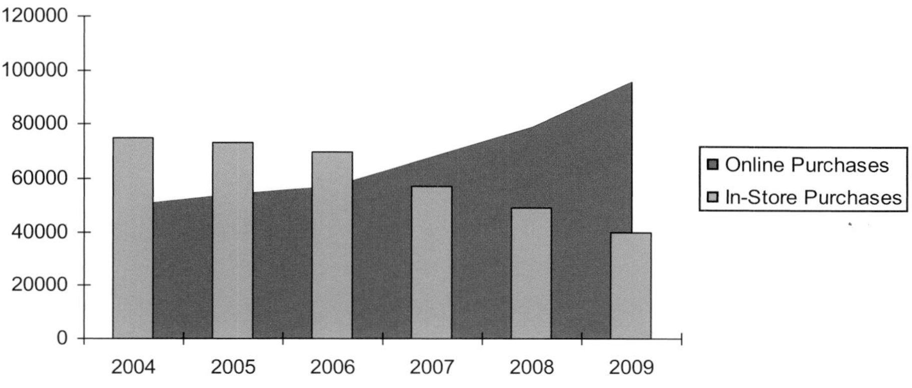

FIGURE 8.9 **Flowchart**

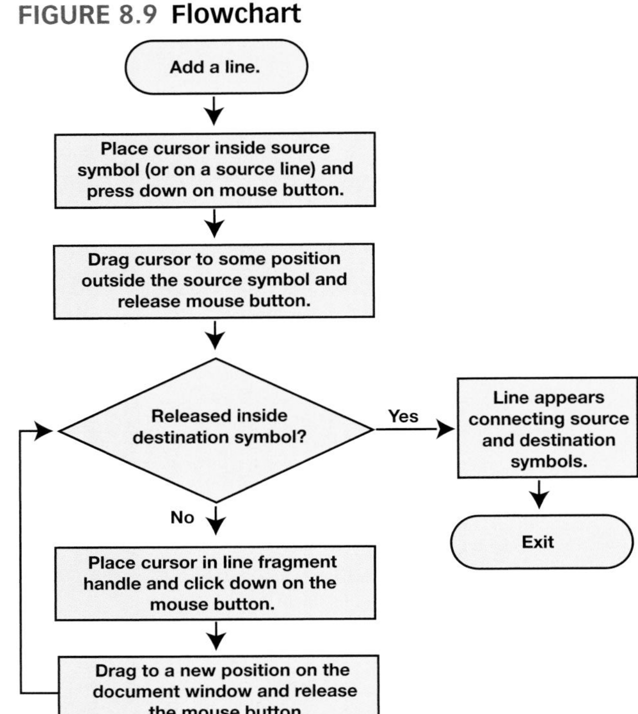

in the chart. An organizational chart helps your readers see where individuals work within a business and their relation to other workers.

Line Drawings

Use line drawings to show the important parts of a mechanism or to enhance your text cosmetically. To create line drawings, do the following:

1. Maintain correct proportions in relation to each part of the object drawn.
2. If a sequence of drawings illustrates steps in a process, place the drawings in left-to-right or top-to-bottom order.
3. Using call-outs to name parts, label the components of the object drawn (see Figure 8.11).
4. If there are numerous components, use a letter or number to refer to each part. Then reference this letter or number in a key (see Figure 8.12).
5. Use exploded views (Figure 8.12) or cutaways (Figures 8.13) to highlight a particular part of the drawing.

Renderings and Virtual Reality Drawings

Two different types of line drawings are renderings and virtual reality views. Both offer 3-D representations of buildings, sites, or things. Often used in the architectural/engineering industry, these 3-D drawings (as shown in Figures 8.14 and 8.15) help clients get a visual idea of what services your company can provide. Renderings and virtual reality drawings add lighting, materials, and shadow and reflection mapping to mimic the real world and allow customers to see what a building or site will look like in a photorealistic setting.

Photographs

A photograph can illustrate your text effectively. Like a line drawing, a photograph can show the components of a mechanism. If you use a photo for this purpose, you will need to label (name), number, or letter parts and provide a key. Photographs are excellent visual

FIGURE 8.10 Organizational Chart

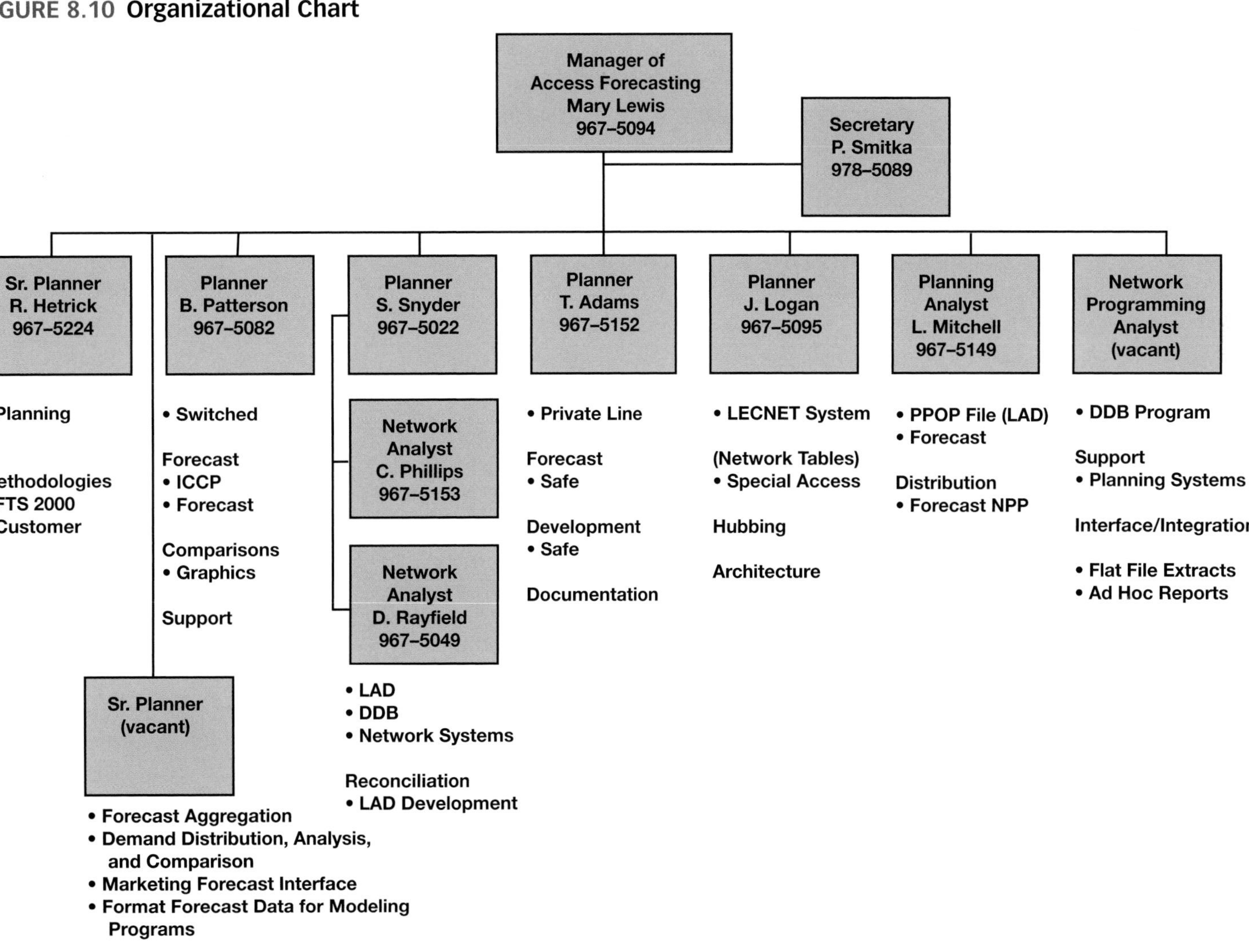

FIGURE 8.11 **Line Drawing (Exploded View with Callouts)**

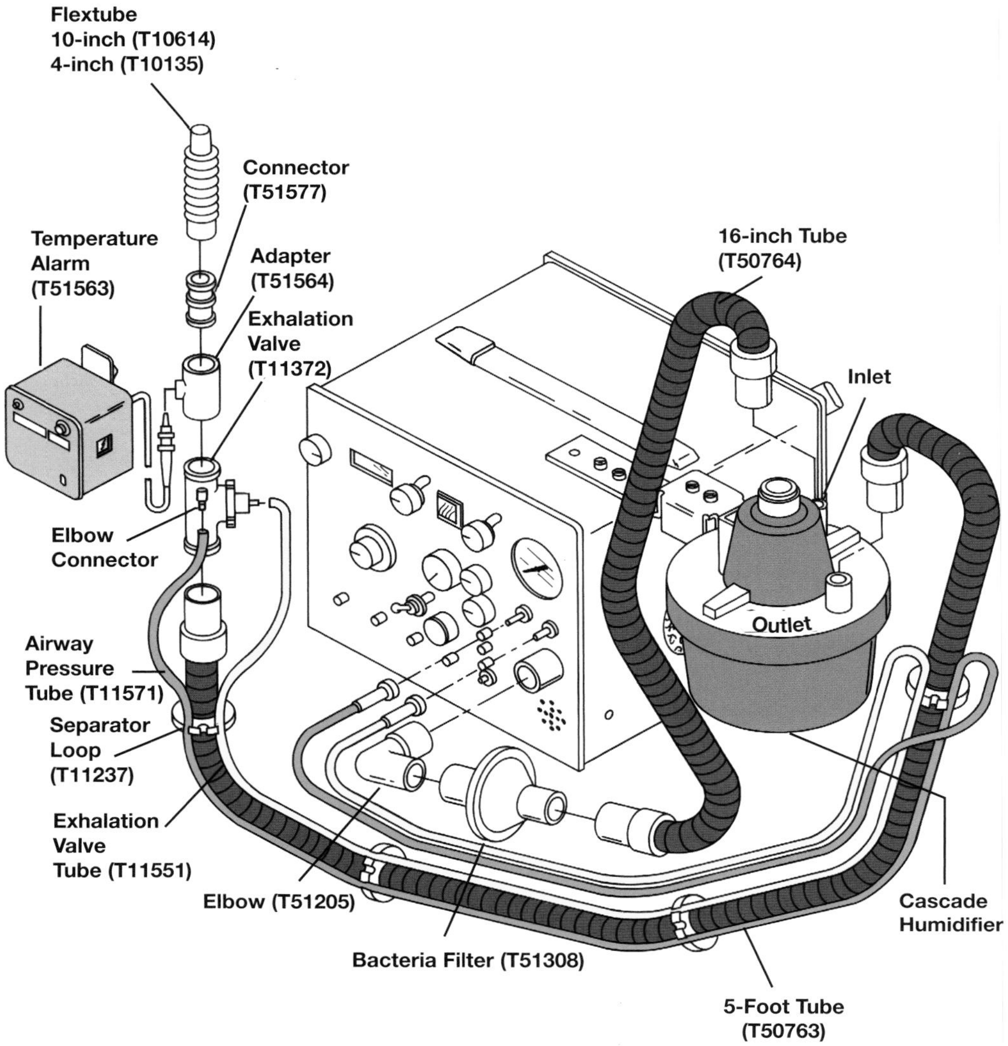

aids because they emphasize all parts equally. Their primary advantage is that they show something as it truly is.

Photographs have one disadvantage, however. They are difficult to reproduce. Whereas line drawings photocopy well, photographs do not. See Figure 8.16.

Icons

Approximately 23 percent of America's population is functionally illiterate. In today's global economy, consumers speak diverse languages. Given these two facts, how can workplace writers communicate to people who cannot read and to people who speak different languages? Icons offer one solution. Icons (as in Figures 8.17 to 8.20) are visual representations of a capability, a danger, a direction, an acceptable behavior, or an unacceptable behavior.

For example, the computer industry uses icons—open manila folders—to represent computer files. In manuals, a jagged lightning stroke iconically represents the danger of electrocution. On streets, an arrow represents the direction we should travel; on computers, the arrow shows us which direction to scroll. Universally depicted stick figures of men and women greet us on restroom doors to show us which rooms we can enter and which rooms we must avoid.

FIGURE 8.12 Line Drawing (Exploded View with Key)

Exhalation Valve Parts List		
Item	Part Number	Description
1	000723	Nut
2	003248	Cap
3	T50924	Diaphragm
4	Reference	Valve Body
5	Reference	Elbow Connector
—	T11372	Exhalation Valve

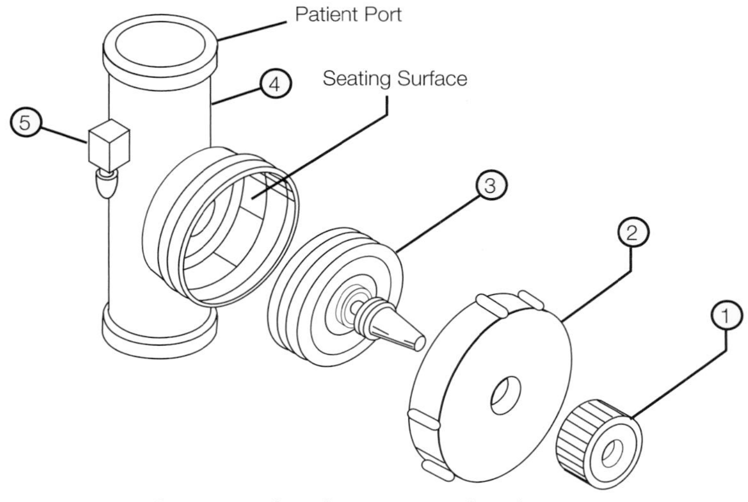

FIGURE 8.13 Line Drawing (Cutaway View)

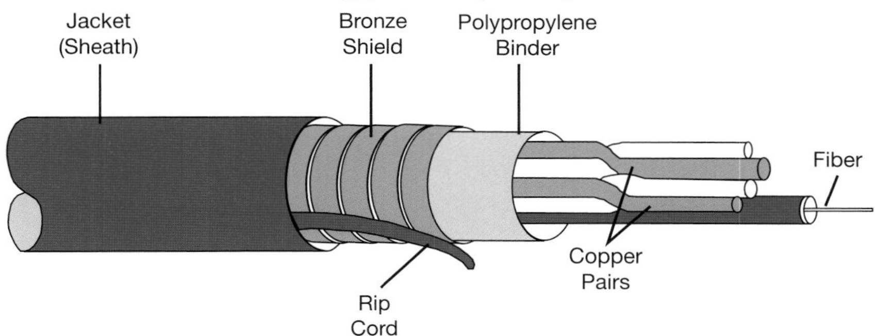

FIGURE 8.14 Architectural Rendering

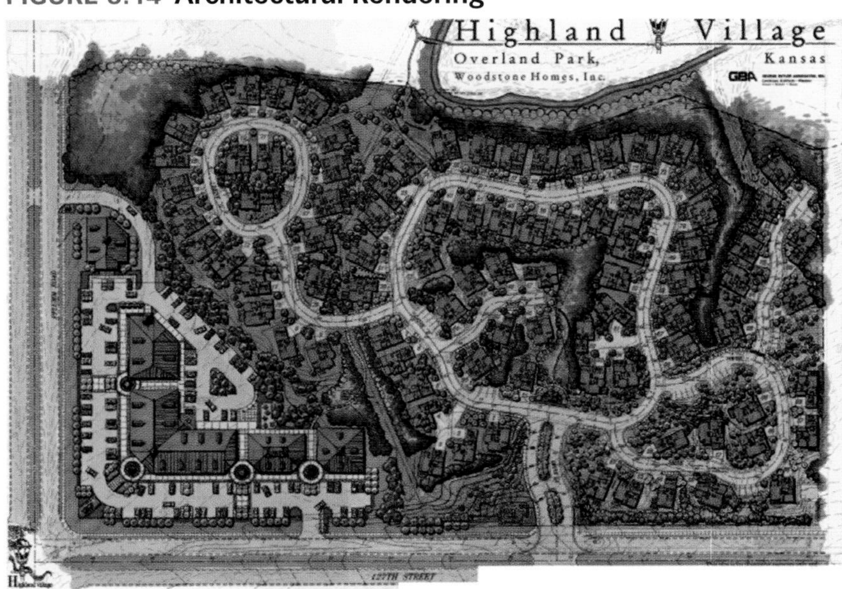

(Courtesy of George Butler Associates)

FIGURE 8.15 3-D Drawing (Virtual Reality)

(Courtesy of Johnson County Community College)

When used correctly, icons can save space, communicate rapidly, and help readers with language problems understand the writer's intent.

To create effective icons, follow these suggestions:

1. **Keep it simple**—You should try to communicate a single idea. Icons are not appropriate for long discourse.
2. **Create a realistic image**—This could be accomplished by representing the idea as a photograph, drawing, caricature, outline, or silhouette.
3. **Make the image recognizable**—A top view of a cell phone or computer terminal is confusing. A side view of a playing card is completely unrecognizable. Select the view of the object that best communicates your intent.
4. **Avoid cultural and gender stereotyping**—For example, if you are drawing a hand, you should avoid showing any skin color, and you should stylize the hand so it is neither clearly male nor female.
5. **Strive for universality**—Stick figures of men and women are recognizable worldwide. In contrast, letters—such as *P* for *parking*—will mean very little in China, Africa, or Europe. Even colors can cause trouble. In North America, red represents danger, but red is a joyous color in China. Yellow calls for caution in North America, but this color equals happiness and prosperity in the Arab culture (Horton 682–93).

Download Existing Online Graphics

The Internet contains thousands of Web sites that contain graphics, including photographs, line drawings, cartoons, icons, animated images, arrows, buttons, horizontal lines, balls, letters, bullets, hazard signs, and more. You can download images from any Web site.

To download graphics, place the cursor on the graphic and right-click on the mouse. A pop-up menu will appear. Either click on "Copy" or click on "Save Picture." You can save your image in the file of your choice. The images from the Internet will already be GIF (graphics interchange format) or JPEG (joint photographic experts group) files.

However, when downloading images, you must abide by copyright laws and ethical considerations. "Every element of a Web page—text, graphics, and HTML code—is protected by U.S. and International copyright laws, whether or not a formal copyright application has been submitted" (LeVie 20-21). If you and your company "borrow" from an

FIGURE 8.16 **Photograph of Mechanical Piping**

(Courtesy of George Butler Associates)

FIGURE 8.17 **An Icon of Explosives**

FIGURE 8.18 **An Icon of Dangerous Machinery**

FIGURE 8.19 **An Icon of Electric Shock**

FIGURE 8.20 **An Icon of Corrosive Material**

existing Internet site, thus infringing upon that site's copyright, you can be assessed actual or statutory damages. To meet ethical considerations,

- assume that any information on the Internet is covered under copyright protection laws unless proven otherwise.
- obtain permission for use from the original creator of graphics or text.
- copyright any information you create.
- include a copyright notice.
- modify existing images.
- create your own graphics and text (Johnson 17).

Can you ethically make changes to a graphic taken from your own company's archives or from boilerplate documents? Do these graphics need a source citation? If your company owns these images, you can modify them and use the changed graphics without citing the source. See figures 8.21 and 8.22 for examples of original and modified photographs taken from a company's archives.

Modifying or Creating New Graphics

To modify graphics, you can download them in two ways. First, you can print the screen by pressing the Print Screen key (usually found on the upper right of your keyboard). This captures the entire screen image in a clipboard. Then you can open a graphics program and paste the captured image.

FIGURE 8.21 Original Photograph with Two Men and Pipe

This photograph includes two men working on a pipe. Note the writing on the pipe.

FIGURE 8.22 Modified Photograph (with the two men and *PLO5* removed)

In this computer-altered photograph, the two men and the writing have been digitally erased.

(Courtesy of George Butler Associates)

Second, you can save the image in a file and then open the graphic in a graphics package. Most graphics programs will allow you to customize a graphic. Popular programs include Paint, Paint Shop Pro, PhotoShop, Corel Draw, and Adobe Illustrator. In these graphics programs, you can manipulate the images by changing colors, adding text, reversing the images, cropping, resizing, redimensioning, rotating, retouching, deleting or erasing parts of the images, overlaying multiple images, joining multiple images, and so forth.

Another option is to create your own graphic. If you are artistic, draw your graphic in a graphics program. This option might be more challenging and time consuming. However, creating your own graphic gives you more control over the finished product, provides a graphic precisely suited to your company's needs, and helps avoid infringement of copyright laws.

VISUAL AID CHECKLIST

_____ 1. Will a visual aid add to your workplace communication and make the document concise, make it clear, and add cosmetic appeal?

_____ 2. Should you use color or three-dimensional graphics in your visual aids?

_____ 3. Have you considered the following criteria when you created your tables or figures:

- a. Are visuals integrated with the text?
- b. Do the visuals add to the text and enhance it without being redundant?
- c. Do they communicate information visually that could not easily be conveyed in text?
- d. Are the visuals the correct size, labeled, readable, and similar in style?

_____ 4. Did you use a table when you presented factual information, such as numbers, percentages, or monetary amounts?

_____ 5. Have you used a figure to highlight and supplement important points in your writing?

Yolanda Met Her Challenge

To meet her communication challenge, Yolanda used the *P³* process.

Planning

To plan her proposal, Yolanda first considered the following:

- Goal—document financial information and recommend actions to be taken by clients
- Audience—specialists (bank employees) and lay reader (client, Sylvia Light)
- Channels—financial proposal to be reviewed by both the client and the bank
- Data—specific client financial information, including assets

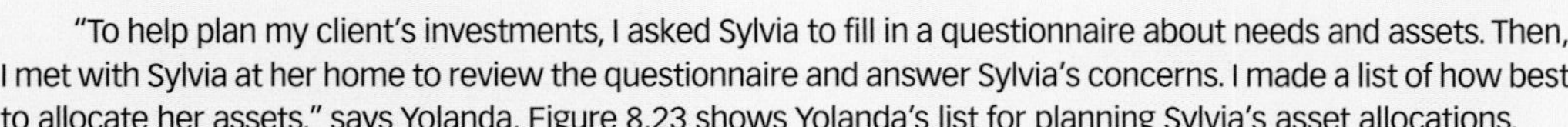

"To help plan my client's investments, I asked Sylvia to fill in a questionnaire about needs and assets. Then, I met with Sylvia at her home to review the questionnaire and answer Sylvia's concerns. I made a list of how best to allocate her assets," says Yolanda. Figure 8.23 shows Yolanda's list for planning Sylvia's asset allocations.

Packaging

Next, Yolanda had to decide how to present her information to Sylvia. Yolanda wrote a rough draft of the asset allocation in paragraph form (Figure 8.24):

Perfecting

Yolanda met her communication challenge by revising her rough draft. Yolanda says, "I knew that raw numbers and lots of words would not communicate clearly to my customer. When I propose investment plans, it's just too hard for most people to visualize where their money will be located. To solve this problem, I created a pie chart for Sylvia. I also emphasized where I planned to invest the largest percentage of her money (annuity allocations) by pulling a wedge apart from the pie. Finally, I made sure that I quantified the allocations by providing the percentages and using a legend to identify each wedge. A picture really is worth a thousand words!" Figure 8.25 shows Yolanda's perfected pie chart of asset allocation.

FIGURE 8.23 Planning List of Asset Allocations

Sylvia could reallocate funds as follows:

- $110,000 in an annuity
- $78,000 in an individual retirement account (IRA)
- $45,000 in municipal bonds
- $37,000 in a stock fund
- $42,000 in certificates of deposit (CD)
- $3,500 in a checking account

FIGURE 8.24 Rough Draft of Asset Allocation

"Sylvia, after reviewing your financial situation and considering your lifestyle, you should consider reallocating your assets in the following way. Put $110,000 in an annuity, $78,000 in an individual retirement account (IRA), $45,000 in municipal bonds, $37,000 in a stock fund, $42,000 in certificates of deposit (CD), and $3,500 in a checking account. This will allow you to have a higher overall rate of return, yet remain almost risk free in regards to your capital. I know that preservation of capital is of paramount importance at this time in your life."

FIGURE 8.25 Perfected Pie Chart of Asset Allocation

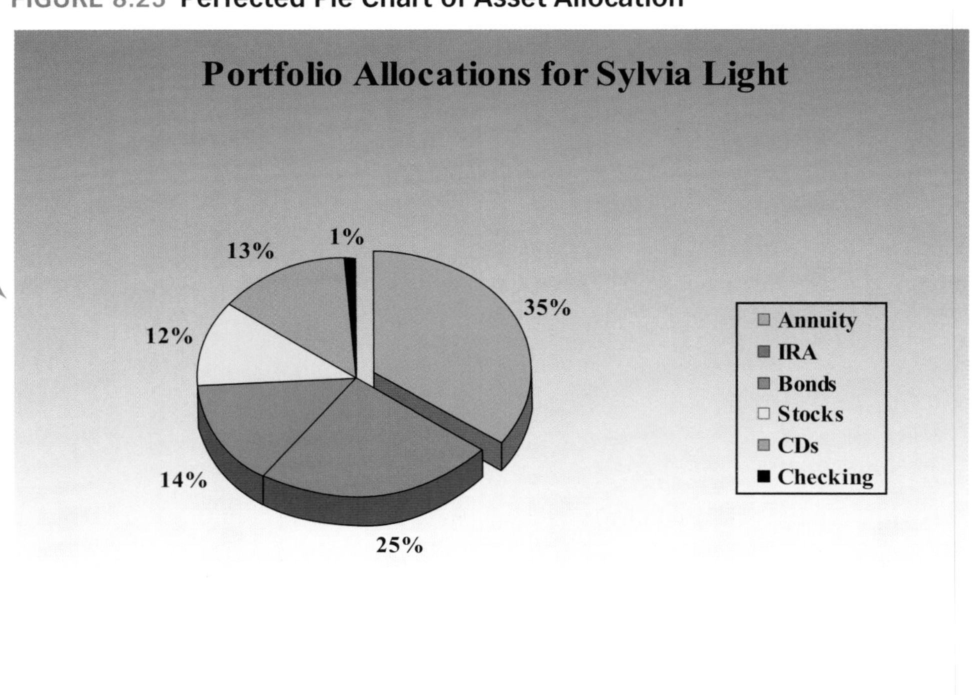

Yolanda says, "I approached my potential client the way I discuss money with my parents. Like Sylvia, they, too, are very successful in their fields. However, money confuses them. This confusion makes it difficult for them to understand how to make their money both safe and wisely invested. Using a visual aid in my presentation to Sylvia allowed her to 'see' how I could manage her money and keep her capital safe but growing. We are now moving forward with reallocation of all her finds, and she told me that she is comfortable with my plans. I am thrilled to be working with her. Planning, packaging, and perfecting allowed me to choose the most successful method of communication—a visual aid."

CHAPTER HIGHLIGHTS

1. Using graphics can allow you to create a more concise document.
2. Graphics add variety to your text, breaking up wall-to-wall words.
3. Color and 3-D graphics can be effective. However, these two design elements can also cause problems. Your color choices might not be reproduced exactly as you planned, and a 3-D graphic could be misleading rather than informative.
4. Tables are effective for presenting numbers, dates, and columns of figures.
5. You can often communicate more easily with your audience when you use figures, such as bar charts, pie charts, line charts, flowcharts, and organizational charts, to highlight and supplement important parts of your text.

WWW *Find additional visual aid exercises, samples, and interactive activities at* http://www.prenhall.com/gerson.

MEETING WORKPLACE COMMUNICATION CHALLENGES

CASE STUDIES

1. Yolanda Ruiz is an investment banker at Country Commercial Bank, as noted in this chapter's beginning scenario. She is writing a proposal to a potential client, Sylvia Light, a retired public health nurse. She now has $315,500 in savings, allocated as follows: $78,000 in an individual retirement account (IRA), $234,000 in a low-earning certificate of deposit (CD), and $3,500 in her checking account.

Sylvia contacted Yolanda, asking her to help organize a portfolio for a comfortable retirement. Yolanda has studied Sylvia's various accounts and considered her lifestyle and expenditures. Now, she is ready to write the proposal. Yolanda plans to propose that Sylvia could reallocate her funds as follows:

- $110,000 in an annuity
- $78,000 in an individual retirement account (IRA)
- $45,000 in municipal bonds
- $37,000 in a stock fund
- $42,000 in certificates of deposit (CD)
- $3,500 in a checking account

To make this proposal visually appealing and more readily understandable to Sylvia, Yolanda will use visual aids.

Assignment

- Create a table to show how Yolanda wants to invest Sylvia's money.
- Create a bar chart comparing current allocations with the proposed allocations.

2. Jim Goodwin owns an insurance agency, Goodwin and Associates Insurance. Letters are a major part of his workplace communication with vendors, clients, and his insurance company's home office. However, many of Jim's employees fail to recognize the importance of business communication. To prove how important their communication is to the company, Jim plans to summarize the amount

of writing they do on the job. He has determined that they write the following each week:

- **57 inquiry letters.** Insurance coverage changes constantly. To clarify these changes for customers, employees write inquiries, asking the home office questions about new insurance laws, levels of coverage, coverage options, and rate changes.
- **253 response letters.** A key to GAI's success is acquiring new customers. When potential customers call, e-mail, or write letters asking for insurance quotes, Jim and his employees write response letters.
- **43 cover (transmittal) letters.** Once customers purchase insurance policies, Jim and his coworkers mail these policies, prefaced by cover letters. The transmittal letters clarify key points within the documents.
- **12 good news letters.** Jim's clients receive discounts when they add new cars to an existing policy or include both car and home coverage. GAI writes good news letters to convey this information.
- **25 adjustment letters.** When accidents occur or when losses take place, GAI writes adjustment letters stating that the claim is covered.
- **2 bad news letters.** Like any business, GAI needs to communicate bad news to clients or vendors.
- **7,000 e-mail messages.** Jim and his 20 employees send and receive a minimum of 50 e-mail messages each day.

Assignment

To show his employees the importance of workplace communication, Jim wants to create a visual aid depicting this data. Create the appropriate graphic for GAI's employees.

ETHICS IN THE WORKPLACE CASE STUDY

Flow Inc., a water purification company that has been in business for over 40 years, is known throughout the Midwest for their logo and slogan, "Go with the Flow."

Nurani Rama grew up in the Midwest and is familiar with Flow Inc. She now works for a software development company named GO—Gaming Online. Nurani is in charge of marketing and public relations and was asked to create a logo and slogan for her company. She fondly remembers Flow Inc.'s slogan and logo and used it as her inspiration. At a board meeting, Nurani promoted the phrase "Flow with the GO" and showed her creation.

Question: Though Nurani's company is in a different city and industry than Flow Inc., is it ethical for Nurani to pattern her slogan and logo after Flow Inc.'s? Justify your answer based on information provided in Chapter 1.

INDIVIDUAL AND TEAM PROJECTS

1. Present the following information in a pie chart, a bar chart, and a table.
   In 2009, the Interstate Telephone Company bought and installed 100,000 relays. It used these for long-range testing programs that assessed failure rates. It purchased 40,000 Nestor 221s; 20,000 VanCourt 1200s; 20,000 Macro R40s; 10,000 Camrose Series 8s; and 10,000 Hardy SP6s.
2. Using the information presented in activity 1 and the following revised data, show the comparison between 2009 and 2010 purchases through two pie charts, a grouped bar chart, and a table.
   In 2010, after assessing the success and failure of the relays, the Interstate Telephone Company made new purchases of 200,000 relays. It bought 90,000 VanCourt 1200s; 50,000 Macro R40s; 30,000 Camrose Series 8s; and 30,000 Hardy SP6s. No Nestors were purchased.
3. Create a line chart. To do so, select any topic you like. The subject matter, however, must include varying values. For example, present a line chart of your grades in one class, your salary increases (or decreases) at work, the week's temperature ranges, your weight gain or loss throughout the year, the miles you've run during the week or month, amounts of money you've spent on junk food during the week, and so forth.
4. Analyze the following graphic and explain why it succeeds or fails. How could you improve it?

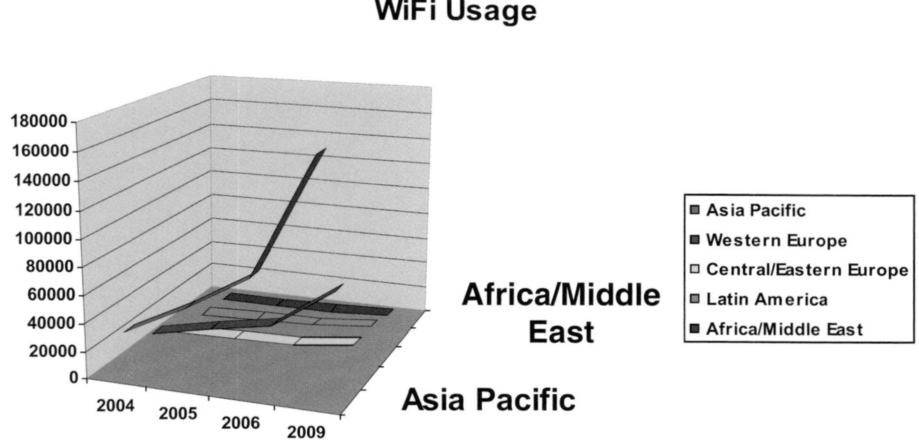

DEGREE-SPECIFIC ASSIGNMENTS

1. In your workplace communication classroom, what are the students' majors? Tabulate your findings, and create a pie chart or bar chart to show these figures.
2. In teams, research average salaries for the following career fields: management, business information systems, accounting, marketing, and finance. Using any of the visual aids discussed in the chapter, depict the information you have found.
3. Go to *Homefair's* "Salary Calculator" at www.homefair.com/real-estate/salary-calculator.asp. This site will allow you to enter a job field, city, and state to show you what you might earn in this locale. Pick five cities you might want to work and live in, research the medium earnings, and tabulate your results. Create a bar chart, line graph, or pie chart to illustrate these earning figures.

PROBLEM-SOLVING THINK PIECES

1. Angel Guerrero, computer information systems technologist at HeartHome Insurance, was responsible for making an inventory of his company's hardware. He learned the following: The company had 75 laptops, 159 PCs, 27 printers, 10 scanners, 59 handheld computers (PDAs), 238 cell phones with e-mail capabilities, and 46 digital cameras.

   To write his inventory report, Angel needs to chart the previous data. Which type of visual aid should Angel use? Explain your answer, based on the information provided in this chapter. Create the appropriate visual aid.

2. Minh Tran works in the marketing department at Thrill-a-Minute Entertainment Theme Park (TET). Minh and her project team need to study entry prices, ride prices, food and beverage prices, and attendance of their park versus their primary competitor, Carnival Towne (CT).

   Minh and her team have found that TET charges $16.50 admission, while CT charges $24.95. Most of TET's rides are included in the entry price, but special rides (the Horror, the Bomber, the Avenger, and Peter Pan's Train) cost $2.50. At CT, the entry fee covers many rides, excluding Alice's Teacup, Top-of-the-World Ferris Wheel, and the Zinger, which cost $2.00 each. Food and beverages at TET cost $1.95 for a hot dog, $2.50 for a burger, $3.95 for nachos, and $1.50–2.50 for drinks. At CT, food and beverages cost $1.75 for hot dogs, $2.75 for burgers, $2.50 for nachos, and $1.50–2.50 for drinks. Attendance at TET last year was 250,000, while attendance at CT was 272,000.

   What type of visual aid should Minh and her team use to convey this information? Explain your decision, based on the criteria for graphics provided in this chapter. Create the appropriate visual aid.

3. Toby Hebert is human resource manager at Crab Bayou Industries (Crab Bayou, LA), the world's largest wholesaler of frozen Cajun food. Toby and her management are concerned about the company's hiring trends, as prospective employees complained about discriminatory hiring practices at Crab Bayou.

   To prove that the company has not practiced discriminatory hiring practices, Toby has studied the last ten years' hires by age. She found that in 1999, the average age per employee was 48; in 2000, the average age was 51; in 2001, the average age was 47; in 2002, the average age was 52; in 2003, the average age was 45; in 2004, the average age was 47; in 2005, the average age was 42; in 2006, the average age was 39; in 2007, the average age rose to 42; in 2008 and 2009, the average age fell to 29 and 30, respectively (due to a large number of early retirements).

   What type of visual aid should Toby use to convey this information? Explain your decision, based on the criteria for graphics provided in this chapter. Create the appropriate visual aid.

4. Yasser El-Akiba is a member of his college's International Students' Club. Yasser is a native of Israel. Other members of the club are from other countries: 3 from Australia, 2 from Ecuador, 8 from Mexico, 5 from Africa, 2 from England, 3 from Canada, 4 from the Dominican Republic, and 9 from China.

   What kind of visual aid could Yasser create to show his club members' homelands? Defend your decision based on criteria for graphics in this chapter. Create the appropriate visual aid.

WEB WORKSHOP

Using an Internet search engine, type in phrases such as "automobile sales+line graph," "population distribution by age+pie chart," or "California+ organizational chart." Create similar phrases for bar charts, pictographs, flowcharts, tables, and so forth. Open several links from your Web search and study the examples you have found. Which examples of graphics are successful and why? Which examples of graphics are unsuccessful and why? Explain your reasoning, based on this chapter's criteria.

QUIZ QUESTIONS

1. How can you effectively present large blocks of data or reveal comparisons?
2. What are four things graphics help readers to see?
3. What can happen when you use only unbroken text in a document?
4. When should you avoid using color in a graphic?
5. What are drawbacks to 3-D graphics?
6. When you use graphics, what are five criteria to follow to achieve successful graphics?
7. What are five types of figures?
8. What are two types of bar charts?
9. What is a pictograph?
10. When would you use a Gantt chart?

CHAPTER 9

Electronic Communication: E-Mail Messages, Instant Messages, Text Messages, and Blogging

▸REAL PEOPLE *in the workplace*

Robert Jordan, PE, is president of Robert Jordan Associates, Inc. (RJA), an architectural–engineering consulting firm. He supervises 270 employees, located in Kansas, Missouri, and Illinois. These include architects, engineers, scientists, and administrative support staff such as human resources, marketing, information systems, and accounting.

Robert receives about 40 "meaningful e-mail" a day—and about 60 spam e-mail, attempting to tell him how to lose weight, buy gifts for his loved ones, make cost-effective travel arrangements, and so on. Robert deletes the spam, which is obvious at a glance in his e-mail in-box, and responds to the business-related correspondence.

He says that e-mail has been essential at RJA for at least three reasons.

The Good News:

1. **E-mail is fast and efficient.** For example, Robert finds that e-mail is ideal for setting up meetings with groups of individuals, including coworkers and clients. Before e-mail, he had to make multiple calls to his staff or customers, usually missing them while they were off-site. Or, he'd have to walk down the halls and up the stairs to leave messages on desks. Instead, with e-mail, Robert knows that the message will arrive at its intended location and that the recipients will respond.
2. **E-mail provides an "electronic record."** When RJA used to rely more heavily on the telephone, people had trouble remembering to document conversations. E-mail provides a communication trail: multiple e-mail messages reporting all communication that has occurred regarding a topic.
3. **E-mail is great for attaching larger files.** Though e-mail messages shouldn't contain excessive amounts of information, you can attach longer documents to e-mail, using the e-mail as a cover message explaining what the longer document contains.

Unfortunately, Robert tells us that e-mail also has a few downsides.

The Bad News:

1. **E-mail lacks privacy.** Robert says that though phone calls made record keeping hard, they ensured privacy. All you needed to do was shut your office door or speak quietly. E-mail, in contrast, inadvertently can be sent to many people.

CHAPTER GOALS

When you complete this chapter, you will be able to do the following:

1. Recognize the importance of electronic communication.
2. Understand the characteristics of online communication.
3. Understand the components of successful e-mail messages.
4. Use e-mail samples as guidelines for effective e-mail components, organization, writing style, and tone.
5. Recognize techniques for successfully using instant messages in the workplace.
6. Understand the use of instant messages in the workplace.
7. Understand the importance of text messaging.
8. Understand the purpose of blogging.
9. Follow criteria to create effective corporate blogging.
10. Follow the *P*³ process—planning, packaging, and perfecting—to write effective electronic communication.

WWW

To learn more about electronic communication, visit our Web Companion at http://www.prenhall.com/gerson.

2. **E-mail can lead to inaccurate communication.** "Didn't I read somewhere that 90 percent of communication is body language?" Robert asks. This leads to a second problem. With e-mail, the people involved can't see each other's eyes, hand motions, shrugs, smiles, or frowns, nor can they hear the grunts, groans, or laughs. Much of this body language is lost with electronic communication.
3. **E-mail can be depersonalized.** Body language not only helps communication, but also it personalizes. E-mail, Robert notes, can be impersonal. In some instances, especially those regarding "contentious situations," Robert knows that a face-to-face discussion is the best way to solve a problem. In fact, Robert concludes that some people who are "conflict resistant" use e-mail as a way of avoiding person-to-person communication. To maintain good business relationships (with personnel as well as with customers), face-to-face talks often are needed.

Still, for speed and efficiency, e-mail is hard to beat.◄

Robert's Communication Challenge

According to Robert, "One of RJA's clients has experienced roofing failure before the end of a 20-year warranty. The building's roof was constructed in 1995 but showed signs of failing in 2008. My company has been hired as a consultant to assess the roofing problem, review three roofing bids, and recommend a solution to the client's problem.

RJA has concluded their research into the roofing failure. Now, I have to send detailed information about the roofing problem to an external insurance auditor, and I have to send it quickly because the roof is failing. The situation must be resolved before further interior damage occurs.

I am facing the following communication challenges:

1. A letter will take too long to mail.
2. The information is too complex to be conveyed in a short e-mail message.
3. Some of the report's material, such as a table, cannot be sent in an e-mail message.
4. The report is dry and formal, but I always like to convey a more personalized and friendly tone. This is one of my 'best-business practices.'

My challenge is to figure out how to best communicate complex information quickly to my audience."

See page 255 to learn how Robert met his communication challenge.

THE IMPORTANCE OF ELECTRONIC COMMUNICATION

The written word has undergone significant changes, leaping from the printed page into cyberspace. Correspondence, once limited to hard-copy letters, reports, and memos, is now often available electronically as e-mail, Web sites, Web logs (blogs), instant messages, text messages, and online help. Corporate brochures and newsletters, once paperbound, now are online. Product and service manuals, once paperbound, are also online. Resumes are online. Research is online. Workplace communication in the twenty-first century is increasingly electronic.

The National Commission on Writing reports about the changing role of electronic communication ("Writing: A Ticket to Work" 11). This report states that e-mail messages are the most prevalent type of communication written on the job. The National Commission on Writing follow-up report further emphasizes the importance of electronic communication. In this report, survey results state that 83.7 percent of respondents "almost always" use e-mail; 100 percent "frequently or almost always" use e-mail. These numbers more than doubled all other communication channels used in the workplace ("Writing: A Powerful Message" 17).

Additional information:

See Chapter 1, "Communicating in the Workplace."

This trend toward electronic communication is increasing with the growing presence of "WiFi" (wireless fidelity) hotspots. At WiFi hotspots, people can connect to the Internet by way of laptops, Tablet PCs, cell phones, and handheld computers. This allows people to work off-site, anywhere and anytime.

THE CHARACTERISTICS OF ONLINE COMMUNICATION

Electronic communication is an entirely different mode of communication than hard-copy text. What are the differences between paper text and online communication?

Online Readers Are Topic Specific

In libraries or book stores, we wander up and down aisles, looking for any book that interests us. In contrast, online readers tend to access the Web or blogs with specific goals in mind. You go online to search for specific information found in specific Web sites: CD prices, automobile loan rates, hotel room availability, restaurant menus, technical specifications for laser printers, the start date for your college's spring semester, and so forth.

Online Readers Want Information Quickly

Online readers often scan, skim, and skip over text, looking just for the information they want and ignoring the rest of the text. In fact, readers want to find information in "ten seconds. Your web site visitors love skim-readable pages. They're not lazy. They're just in a hurry. They don't want to waste time and money reading the wrong web page, when there are millions of other pages to choose from" (McAlpine).

Electronic Communication Platforms Are Diverse

Another difference between an online reader and the same person who reads a hard-copy novel is the way in which he or she accesses the document. Most hard-copy text is printed on either book-sized pages or on 8½" × 11" pieces of paper. In contrast, online readers can access electronic communication using cell phones, Blackberries, iPods, WiFi compatible laptops, and other wireless devices. Thus, screen resolution and the size of type font are key elements for online readability. Furthermore, online readers use FireFox, Internet Explorer, AOL, Facebook, MySpace, and a host of other electronic platforms to view text. Each of these electronic platforms differs in subtle ways.

Electronic Communication Encourages Random Access

When you write a blog or a Web site, you must reconsider hypertext links versus chronological reading. We read books from beginning to end, sequentially. Web sites, however, allow us, even encourage us, to leap randomly from screen to screen, from Web site to Web site, or from blog to blog.

Electronic Communication Is Often More Casual Than Other Forms of Written Communication

Another unique distinction of electronic communication is its casual tone. Letters and reports have formal formats which elevate the tone of your correspondence. In contrast, instant messages, blogs, text messages, the Internet, and e-mail messages tend to be more informal, even conversational.

WHY IS E-MAIL IMPORTANT?

Many companies rely on e-mail extensively for the following reasons.

Time

The primary driving force behind e-mail's prominence is time. E-mail is quick. Whereas a posted letter might take several days to deliver, e-mail messages can be delivered within seconds.

Convenience

With wireless communication, you can send e-mail from notebooks to handhelds. Current communication systems combine a voice phone, a personal digital assistant, and e-mail into a package that you can slip into a pocket or purse. Then, you can access your e-mail messages anywhere, anytime.

Internal/External

E-mail allows you to communicate internally to coworkers and externally to customers and vendors. Traditional communication channels, like letters and memos, have more limited uses. Generally, letters are external correspondence written from one company to another company; memos are internal correspondence transmitted within a company.

Cost

E-mail is cost effective because it is paper-free. With an ability to attach files, you can send many kinds of documentation without paying shipping fees. This is especially valuable when considering international business.

Documentation

E-mail provides an additional value when it comes to documentation. Because so many writers merely respond to earlier e-mail messages, what you end up with is a series of e-mail messages conveying comments and earlier responses. When e-mail is printed out, often the printout will contain dozens of e-mail messages, representing an entire string of dialogue. This provides a company with an extensive record for future reference. In addition, most companies archive e-mail messages in backup files. Even though an individual employee's computer will keep copies of sent messages and deleted messages, periodically the employee's computer will reach maximum capacity. The employee will then need to delete files to open space for new correspondence. In this case, an employee might want to make hard copies of key documents for future reference.

TECHNIQUES FOR WRITING EFFECTIVE E-MAIL MESSAGES

To convey your messages effectively and to ensure that your e-mail messages reflect professionalism, follow these tips for writing e-mail.

FIGURE 9.1 Flawed E-Mail Message

This e-mail is flawed due to its use of instant messaging abbreviations, lowercase text, ellipses (...), slang, and emoticons. Though these techniques might be appropriate for personal communication, overly casual e-mail messages are not appropriate for workplace communication.

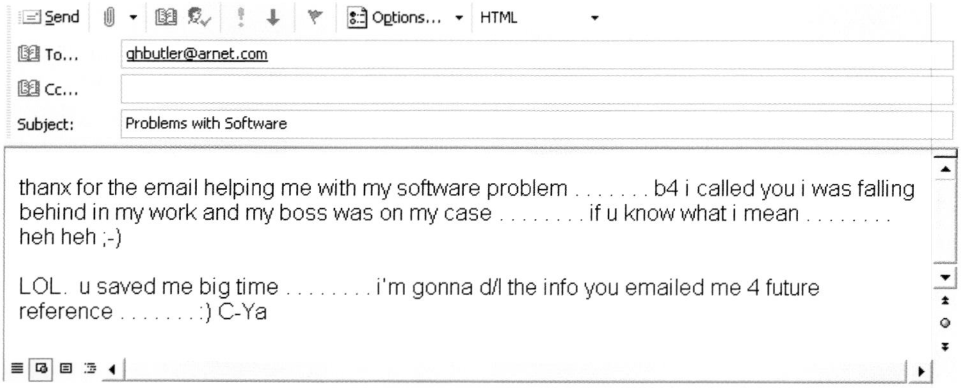

Send | Options... | HTML

To... ghbutler@arnet.com

Cc...

Subject: Problems with Software

thanx for the email helping me with my software problem . . . . b4 i called you i was falling behind in my work and my boss was on my case . . . . if u know what i mean . . . . heh heh ;-)

LOL. u saved me big time . . . . i'm gonna d/l the info you emailed me 4 future reference . . . . :) C-Ya

Recognize E-Mail's Lack of Privacy and Corporate Ownership

E-mail messages written in the workplace are owned by the corporation. As such, you must be aware of the following:

- Your e-mail is not confidential.
- Employers have the right to keep track of how their employees spend their work hours. The 1986 Electronic Communications Privacy Act asserts that the company, *not the writer*, owns an employee's e-mail.
- A company's systems operator is allowed to intercept messages during transmission.
- Even deleting e-mail doesn't ensure confidentiality. Deleted e-mail can be stored in backup systems.

Avoid Casual, Unprofessional Tone

E-mail lets you "talk" back and forth with your e-mail reader. Thus, e-mail tends to become friendlier than memos or letters. This can lead to unprofessional-sounding communication, as shown in Figure 9.1.

Recognize Your Audience

Additional information:

See Chapter 3, "Meeting the Needs of the Audience."

E-mail messages can be sent to managers, coworkers, subordinates, vendors, and customers, among other audiences. Your e-mail readers will be specialists, semi-specialists, and lay people. Thus, you must factor in levels of knowledge.

If an e-mail message is sent internationally, you also might have to consider your readers' language. Remember that abbreviations and acronyms are not universal. Dates, times, measurements, and monetary figures differ from country to country. In addition, your readers' e-mail system might not have the same features or capabilities that you have. Hard-copy text will look the same to all readers. E-mail platforms, in AOL, Outlook, Juno, HotMail, Yahoo, and so on, display text differently. To communicate effectively, recognize your audience's level of knowledge, unique language, and technology needs.

Identify Yourself

Identify yourself by name, affiliation, or title. You can accomplish this either in the "from" line of your e-mail or by creating a signature file or ".sig file." This .sig file acts like an online business card. You should also include contact information including phone number and fax number. Once this identification is complete, readers will be able to open your e-mail without fear of corrupting their computer systems.

Provide an Effective Subject Line

Readers are unwilling to open unsolicited or unknown e-mail, due to fear of spam and viruses. In addition, corporate employees receive approximately 50 e-mail messages each day. They might not want to read every message sent to them. To ensure that your e-mail messages are read, avoid uninformative subject lines, such as "Hi," "What's New," or "Important Message." In contrast, successful subject lines summarize the memo's content. One-word subject lines do not communicate effectively, as in the following flawed subject line. The "Before" sample has a *topic* (a what) but is missing a *focus* (a what about the what).

BEFORE	AFTER
Subject: Sales	Subject: Report on Quarterly Sales

Following are examples of successful subject lines:

example

Subject: Meeting Dates for Business Training Workshop
Subject: Information about Health Insurance Plan for 2009–2010
Subject: Upgrades for Desktop Hardware in the Accounting Department
Subject: Installation of Metal Detectors in the Visitor's Center

Keep Your E-Mail Message Brief

To help readers access information quickly, "Apply the 'top of the screen' test. Assume that your readers will look at the first screen of your message only" (Munter et al. 31). Limit your message to one screen (if possible).

Organize Your E-Mail Message

Successful writing usually contains an introductory paragraph, a discussion paragraph or paragraphs, and a conclusion. Although many e-mail messages are brief (only a few sentences), you can use the introductory sentences to tell the reader why you are writing and what you are writing about. In the discussion, clarify your points thoroughly. Use the concluding sentences to tell the reader what is next. For example, explain when a follow-up is required and why that date is important.

Use Highlighting Techniques Sparingly

Many e-mail packages will let you use highlighting techniques, such as boldface, italics, underlining, computer-generated bullets and numbers, centering, font color highlighting, and font type changes. Many other e-mail platforms will not display such visual enhancements. To avoid having parts of the message distorted, limit your highlighting to asterisks (*), numbers, and double spacing.

Proofread Your E-Mail Message

Errors will undermine your professionalism and your company's credibility. Recheck your facts, dates, addresses, and numerical information before you send the message. Although you won't have the time to employ the following tips for all your e-mail messages, you might find these tips helpful in certain instances (particularly if you are writing about a sensitive subject, and you want to be sure you have correctly worded the message):

- Type your text first in a word processing package, like Microsoft Word.
- Print it out. Sometimes it is easier to read hard-copy text than text online. Also, your word processing package, with its spell check and/or grammar check, will help you proofread your writing.

Additional information:

See Chapter 6, "Perfecting Workplace Communication."

- Once you have completed these two steps (writing in Word or Works and printing out the hard-copy text), copy and paste the text from your word processing file into your e-mail.

Make Hard Copies for Future Reference

Making hard copies of all e-mail messages is not necessary because most companies archive e-mail. However, in some instances, you might want to keep a hard copy for future reference. These instances could include transmissions of good news. For example, if you have received compliments about your work, you might want to save this record for your annual job review. You also might save a hard copy of an e-mail message regarding flight, hotel, car rental, or conference arrangements for business-related travel.

Be Careful When Sending Attachments

When you send attachments, tell your reader within the body of the e-mail message that you have attached a file; specify the file name of your attachment and the software application that you have used (HTML, PowerPoint, PDF, RTF [rich text format], Word, or Works); and use compression (Zip) files to limit your attachment size. Zip files are necessary only if an attachment is quite large.

Practice Netiquette

When you write your e-mail messages, observe the rules of "netiquette."

- **Be courteous**—Do not let the instantaneous quality of e-mail negate your need to be calm, cool, deliberate, and professional.
- **Be professional**—Occasionally, e-mail writers compose excessively casual e-mail messages. They will lowercase a pronoun like "i," use ellipses (. . .) or dashes instead of more traditional punctuation, use instant messaging shorthand language such as "LOL" or "BRB," and depend on emoticons (☺ ☹). These e-mail techniques might not be appropriate in all instances. Don't forget that your e-mail messages represent your company's professionalism. Write according to the audience and communication goal.
- **Avoid abusive, angry e-mail messages**—Because of its quick turnaround abilities, e-mail can lead to negative correspondence called flaming. Flaming is sending angry e-mail, often TYPED IN ALL CAPS. Readers can perceive the all caps as yelling, such as in Figure 9.2.

FIGURE 9.2 FLAMING E-Mail

Subject: EXCESSIVE PRINTER PAPER

I HAVE TALKED WITH SEVERAL PEOPLE AND THIS SEEMS TO BE A PROBLEM IN YOUR DEPARTMENT. SOMEONE PRINTS INFORMATION AND WON'T PICK IT UP AT THE PRINTER. THEN THE NEXT PERSON HAS TO SORT THROUGH THE PRINTED MATERIAL TO FIND WHAT HE OR SHE WANTS. SOME PRINTOUTS ARE NEVER USED; THEY JUST SIT THERE FOR DAYS, GETTING IN OTHER PEOPLE'S WAY. PEOPLE SHOULD JUST PICK UP THEIR PRINTING AND GET IT OUT OF EVERYONE ELSE'S WAY. THAT'S ONLY COMMON COURTESY. BUT THE PEOPLE IN YOUR DEPARTMENT AREN'T EVEN REMOTELY CONSIDERATE OF OTHERS. IF YOU MANAGEMENT PEOPLE WOULD JUST DO YOUR JOBS, NONE OF THIS WOULD HAPPEN.

ORGANIZING AND WRITING E-MAIL MESSAGES

You can write e-mail messages to accomplish many purposes in business and industry. E-mail messages can be directive, telling employees to complete a task. An e-mail message can preface an attachment, document expenses or incidents, confirm attendance at a meeting, or provide step-by-step procedures. You might want to write an e-mail message that recommends action or studies the feasibility of a project. An e-mail message might be written to ask questions or provide updates.

Many e-mail messages only require a sentence or two. If you need to convey more information than can be accomplished in only a few sentences, try the easy-to-use template shown in Figure 9.3.

SAMPLES OF E-MAIL MESSAGES

See Figures 9.4, 9.5, and 9.6 for examples of well-written e-mail messages.

Figure 9.5 shows an e-mail written by an information technology supervisor that will be distributed companywide to many different employees with varied levels of technical expertise. Note how the writer uses a simple style of writing, carefully omitting any technical terms. Only the information necessary to assure audience comfort with the new identification system is included.

In contrast to Figure 9.5, which is designed for an audience with multiple levels of expertise, Figure 9.6 is written for an audience of specialists. Since this specialized audience is familiar with the technical terms included in the e-mail, the writer does not have to define such terminology. In addition, the writer includes sophisticated details and content which the technically proficient audience can understand easily.

FIGURE 9.3 E-Mail Template

Subject: *Topic* and *Focus* of the E-Mail

Opening sentence or paragraph. Briefly tell the reader *why* you are writing and *what* you are writing about.

Discussing the topic. Develop your ideas in one or two short paragraphs (three to four sentences per paragraph) or in a list. Use this discussion section to explain *what exactly* you need to communicate. This can include meeting dates, steps to follow, a list of questions, or actions required.

Ending the e-mail. Tell the reader *what's next*. This could include a follow-up action and a due date. You should also thank readers if they are assisting you in some way.

.sig file
Name
Company
Address
Phone number

FIGURE 9.4 Successful E-Mail Message About the Status of a Company's Web Site

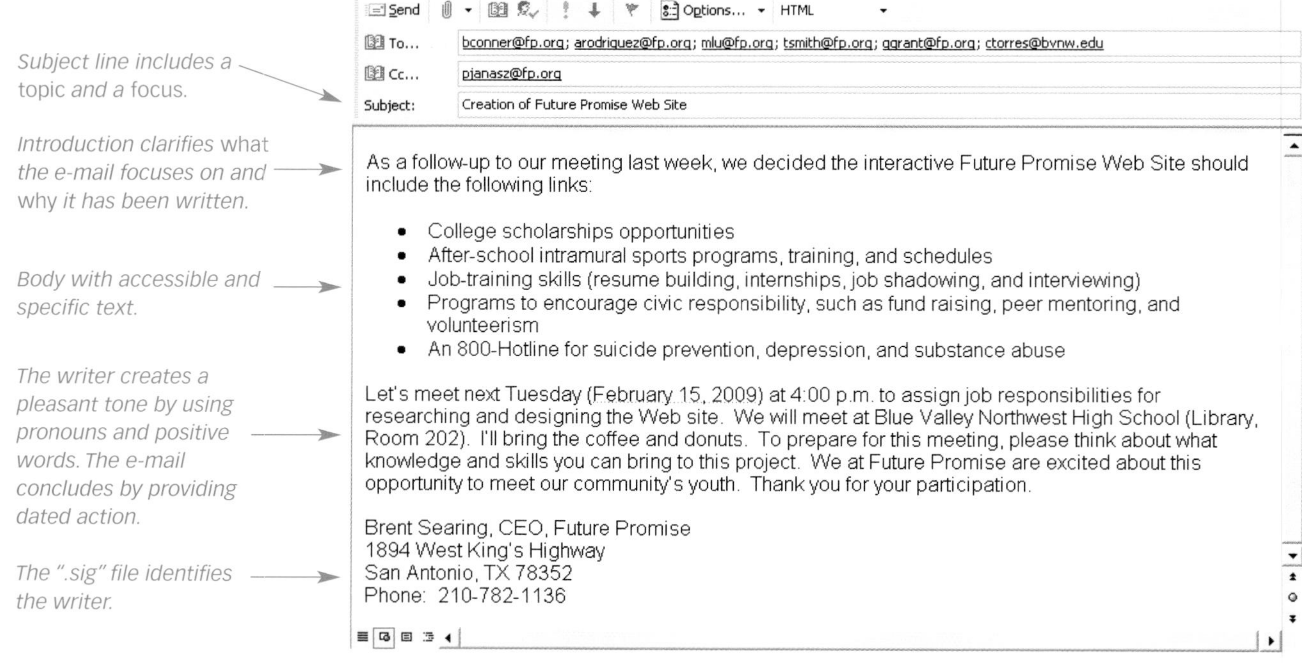
Subject line includes a topic and a focus.
Introduction clarifies what the e-mail focuses on and why it has been written.
Body with accessible and specific text.
The writer creates a pleasant tone by using pronouns and positive words. The e-mail concludes by providing dated action.
The ".sig" file identifies the writer.
Send
Options...
HTML

To... bconner@fp.org; arodriguez@fp.org; mlu@fp.org; tsmith@fp.org; ggrant@fp.org; ctorres@bvnw.edu
Cc... pjanasz@fp.org
Subject: Creation of Future Promise Web Site

As a follow-up to our meeting last week, we decided the interactive Future Promise Web Site should include the following links:

- College scholarships opportunities
- After-school intramural sports programs, training, and schedules
- Job-training skills (resume building, internships, job shadowing, and interviewing)
- Programs to encourage civic responsibility, such as fund raising, peer mentoring, and volunteerism
- An 800-Hotline for suicide prevention, depression, and substance abuse

Let's meet next Tuesday (February 15, 2009) at 4:00 p.m. to assign job responsibilities for researching and designing the Web site. We will meet at Blue Valley Northwest High School (Library, Room 202). I'll bring the coffee and donuts. To prepare for this meeting, please think about what knowledge and skills you can bring to this project. We at Future Promise are excited about this opportunity to meet our community's youth. Thank you for your participation.

Brent Searing, CEO, Future Promise
1894 West King's Highway
San Antonio, TX 78352
Phone: 210-782-1136

FIGURE 9.5 E-Mail for Multiple Audience Levels

Date: May 31, 2009
To: ncomlistserv
Cc: lhughes@ncomlistserv
From: vpatel@ncom.net, Information Technology Supervisor

Subject: Rollout of New Computer Security System

Our company is rolling out a new computer security system next week. This is an important step for the company for at least three reasons: to deter hackers from breaking into our company's computers, to minimize spam and viruses, and to ensure that our company's intellectual assets are secure from terrorists. This week, our information technology (IT) specialists will install the appropriate hardware and software needed to achieve this important security. Your workflow, however, will not be affected. Following are the key steps to this procedure:

1. To authenticate each computer user, we'll ask you to type in your 5-digit PIN (Personal Identification Number). If you have forgotten this number, just give us a call at 1-help-055.
2. Once the computer system verifies your identity, our new firewall system kicks in. This software will give you access to company data while denying the same access to others outside the company.
3. Finally comes protection from spam and viruses—"about time," you're probably saying. You'll never have to worry about this step. It's all taken care of for you by our network administrators.

Though our new system security should be seamless for end users, we're offering quick and easy hands-on demonstrations for interested employees. To make this convenient for you, look for our computer specialists in the company cafeteria from 11:00 a.m.–1:00 p.m. The staff wearing yellow shirts and sitting at the

(Continued)

FIGURE 9.5 Continued

"FAQ" tables will answer your questions and walk you through the new procedures. As always, if you have any questions, just e-mail me (vpatel@ncom.net). We're here to help.

Veejay Patel
Supervisor, Information Technology
nCom
2134 Industrial Way
Sacramento, CA 22109
vpatel@ncom.net

To communicate effectively to multiple audiences, this writer defines technical terms like "IT" and "PIN." The e-mail uses a conversational tone and simple instructions to make this challenging topic more friendly and informative. The emphasis is on helpfulness vs. depth of detail.

FIGURE 9.6 E-Mail for Specialists

Date: May 31, 2009
To: lhughes@ncom.net, Director, Network Systems
From: vpatel@ncom.net, Supervisor, Information Technology
Subject: Rollout of New Computer Security System

Lorette, as a follow-up to our meeting last week, below is an overview of the new computer security system we plan to implement:

1. nCom's network security begins with user authentifications. Starting next week, we will ask each company employee to use a PIN. These 5-digit numbers, ranging from 00000-99999, will give our employees 100,000 possible numbers to choose from, thus essentially negating chances of duplication. In addition, the nCom PIN verification system will allow employees only three attempts at logging in. This safeguards our system from intrusion, since hackers have only a 1/30,000 chance at guessing the correct PIN before access is blocked.
2. Authentification through PINs protects access, but it does not deter malware. That's where our firewall system kicks in. Our new Aponex firewall uses a two-pronged, doubleback security system to deflect adware, spyware, viruses, worms, and trojan horses. First, packet filtering will allow the passage of predefined data to pass through the system while discarding suspicious data. Next, as a backup filter, our software also uses stateful inspection. This allows us to match data against predefined characteristics, set by our network administrators. Together, these IPS systems detect anomalies to deny access and protect corporate privacy.
3. Finally, we have installed surveillance honeypots as network-accessible decoys. The honeypots, essentially, create a shadow network which we can monitor against unauthorized access attempts.

My IT staff is working this weekend to install all required hardware and software. In doing so, we will have the system up and running by Monday of next week, without interfering with regular workplace operations. IT staff will also be accessible to employees next week to help in the transition. I have arranged for FAQ tabling during lunch hours. I want to take this opportunity to thank you, Lorette, for your support and my staff for their expertise during this network security implementation.

Veejay Patel
Supervisor, Information Technology
nCom
2134 Industrial Way
Sacramento, CA 22109
vpatel@ncom.net

For specialists, this e-mail does not define terms, such as "IPS" and "PIN." The e-mail contains detailed content about technical information regarding system security that the audience both needs and will understand. The tone of this e-mail is more businesslike, since the writer is a subordinate writing to management.

The E-mail Checklist will give you the opportunity for self-assessment and peer evaluation of your writing.

E-MAIL CHECKLIST

_____ 1. **Does the e-mail use the correct address?**

_____ 2. **Have you identified yourself?** Provide a "Sig" (signature) line. Include contact information such as a phone number and fax number.

_____ 3. **Did you provide an effective subject line?** Include a *topic* and a *focus*.

_____ 4. **Have you effectively organized your e-mail?** Consider including the following:
- an opening sentence(s) telling *why* you are writing and *what* you are writing about.
- a discussion unit with itemized points telling *what exactly* the e-mail is discussing.
- a concluding sentence(s), *summing up* your e-mail message or telling your audience what to do next.

_____ 5. **Have you used highlighting techniques sparingly?**
- Avoid boldface, italics, color, or underlining.
- Use asterisks (*) for bullets, numbers, and double spacing between paragraphs for access.

_____ 6. **Did you practice "netiquette"?**
- Be polite, courteous, and professional.
- Don't **FLAME**.

_____ 7. Is the e-mail **concise**?

_____ 8. **Did you identify and limit the size of attachments?**
- Tell your reader(s) if you have attached files and what types of files are attached (PPt, PDF, RTF, Word, etc.).
- Limit the files to 750K.

_____ 9. Does the memo recognize **audience?**
- Define acronyms or abbreviations where necessary.
- Consider a diverse audience (factoring in issues, such as multiculturalism or gender).

_____10. Did you avoid **grammatical errors**?

INSTANT MESSAGING

Fast as it is, e-mail may be too slow for today's fast-paced workplace. As a result, instant messaging (IM) might replace e-mail in the workplace. IM pop-ups are already providing businesses with many benefits.

Benefits of Instant Messaging

The following list contains benefits of instant messaging:

- Increased speed of communication.
- Improved efficiency for geographically dispersed workgroups.
- Collaboration by multiple users in different locations.
- Communication with colleagues and customers at a distance in real time, like the telephone.
- The avoidance of costly long distance telephone rates. (Note: Voice over IP (VoIP) services, which allow companies to use the Internet for telephone calls, could be more cost-efficient than IM.)
- A more "personal" link than e-mail.
- A communication channel that is less intrusive than telephone calls.
- A communication channel that allows for multitasking. With IM, you can speak to a customer on the telephone or via an e-mail message *and simultaneously* receive product updates from a colleague via IM.

(Hoffman; Shinder)

Perhaps due to the benefits of IM, the Radicati Group, a technology market research firm, states that IM usage in business will grow from "867 million accounts in 2005, to approximately 1.2 billion accounts in 2009" ("Generation Y" 2).

Challenges of Instant Messaging

Using IM in the workplace could lead to problems, including security, archiving, monitoring, and employee misuse (Hoffman; Shinder):

- **Lost productivity**—Use of IM on the job can lead to job downtime. First, we tend to type more slowly than we talk. Next, the conversational nature of IM leads to "chattiness." If employees are not careful, or monitored, a brief IM conversation can lead to hours of lost productivity.
- **Employee abuse**—IM can lead to personal messages rather than job-related communication with coworkers or customers.
- **Distraction**—With IM, a bored colleague easily can distract you with personal messages, online chats, and unimportant updates.
- **Netiquette**—As with e-mail, due to the casual nature of IM, people tend to relax their professionalism and forget about the rules of polite communication. IM can lead to rudeness or just pointless conversations.
- **Spim**—IM lends itself to "spim," instant messaging spam—unwanted advertisements, pornography, pop-ups, and viruses.
- **Security issues**—This is the biggest concern. IM users are vulnerable to hackers, electronic identity theft, and uncontrolled transfer of documents. With unsecured IM, a company could lose confidential documents, internal users could download copyrighted software, or external users could send virus-infected files.

Techniques for Successful Instant Messaging

To solve potential problems, consider these suggestions:

- **Choose the correct communication channel**—Use IM for speed and convenience. If you need length and detail, other options—e-mail messages, memos, reports, letters—are better choices. In addition, sensitive topics or bad news should never be handled through IM. These deserve the personal attention provided by telephone calls or face-to-face meetings.
- **Document important information**—Most IM programs archive information and let users search for text by keywords or phrases. These systems are perfect for future reference (Hoffman; Shinder).
- **Summarize decisions**—IM is great for collaboration. However, all team members might not be online when decisions are made. Once conclusions have been reached that affect the entire team, the designated team leader should e-mail everyone involved. In this e-mail, the team leader can summarize the key points, editorial decisions, timetables, and responsibilities.
- **Tune in, or turn off**—The moment you log on, IM software tells everyone who is active online. Immediately, your IM buddies can start sending messages. IM pop-ups can be distracting. Sometimes, in order to get your work done, you might need to turn off your IM system. Your IM product might give you status options, such as "on the phone," "away from my desk," or "busy." Turning on IM could infringe upon your privacy and time. Therefore, turning off might be the answer.
- **Limit personal use**—Your company owns the instant messaging in the workplace. IM should be used for business purposes only.
- **Create "buddy" lists**—Create limited lists of IM users, including legitimate business contacts (colleagues, customers, and vendors).
- **Avoid public directories**—This will help ensure that your IM contacts are secure and business related.
- **Disallow corporate IM users from installing their own IM software**—A company should require standardized IM software for safety and control.

FIGURE 9.7 **Instant Message**

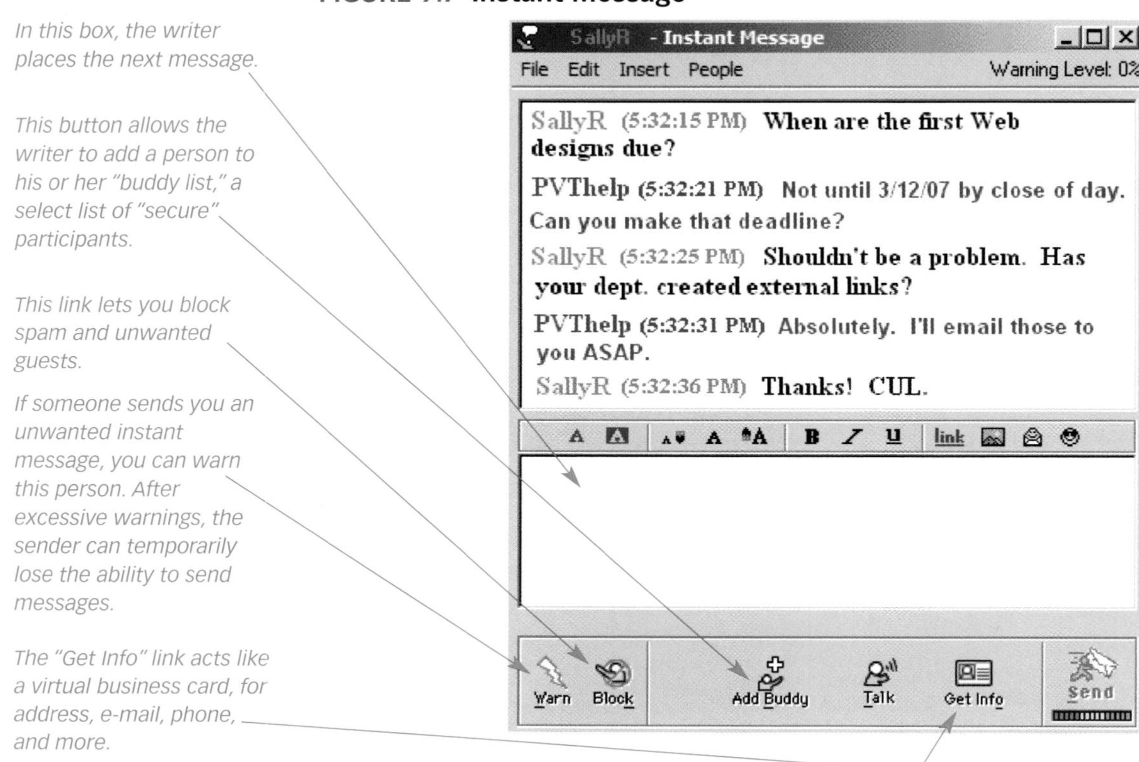
In this box, the writer places the next message.
This button allows the writer to add a person to his or her "buddy list," a select list of "secure" participants.
This link lets you block spam and unwanted guests.
If someone sends you an unwanted instant message, you can warn this person. After excessive warnings, the sender can temporarily lose the ability to send messages.
The "Get Info" link acts like a virtual business card, for address, e-mail, phone, and more.
SallyR - Instant Message
File Edit Insert People
Warning Level: 0%
SallyR (5:32:15 PM) When are the first Web designs due?
PVThelp (5:32:21 PM) Not until 3/12/07 by close of day. Can you make that deadline?
SallyR (5:32:25 PM) Shouldn't be a problem. Has your dept. created external links?
PVThelp (5:32:31 PM) Absolutely. I'll email those to you ASAP.
SallyR (5:32:36 PM) Thanks! CUL.
link
Warn
Block
Add Buddy
Talk
Get Info
Send

- **Never use IM for confidential communication**—Use another communication channel if your content requires security.

See Figure 9.7 for a sample instant message.

TEXT MESSAGING

Text messaging, also known as SMS (short message service), is a growing presence in workplace communication. "About 7.3 billion text messages are sent within the United States every month, up from 2.9 billion a month a year ago, according to CTIA, the wireless industry's trade group" (Noguchi). The Mobile Marketing Association and Pew Internet & American Life Project Surveys state that close to 40 percent of mobile phone users send text messages ("Text Messaging"). Verizon reported an industry record in June 2007 when customers sent and received over 10 billion text messages (Maddock).

A unique characteristic of TM is its demand for conciseness. Text messages over 160 characters are delivered in multiple segments. Each segment is billed as a separate message. That's why users try to limit messages to 160 characters—to save expenses.

"Pretty soon, you're going to have to teach text messaging in your classes," Robert Clark said. "Why's that?" we asked. He responded, "30 percent of my interaction with staff is through text messaging (TM)" (Clark). Robert, a facilities manager for a real estate company, drives between eight apartment complexes to supervise his staff at each location. Robert says that he depends on text messaging (TM) instead of e-mail for the following reasons.

Reasons for Using TM

- **Cost**—Though Robert's cell phone is supplied by his employer, the staff whom he supervises pay for their own cell phones. It's cheaper for them to use text messaging if their cell phone plans allow for unlimited TM.

- **Technological access**—Staff members who work under Robert do not have computers at their work sites, nor does Robert have a computer in his car as he drives around town. However, all of Robert's employees have access to a cell phone.
- **Speed**—TM is a quick and easy way to communicate short messages like "John will be late for work today. Is it OK if I stay overtime?" That's 63 characters (counting the letters, spaces, and punctuation marks). Since TM limits text to 160 characters, Robert and his staff must use other means of communication for longer correspondence.
- **Multitasking**—TM is a great way for Robert to multitask. Robert might be in a meeting, for example. He says, "If three of my staff need to attend training on a certain day, I can text all three employees right then and there and not disrupt the meeting. They'll respond quickly, allowing me to tell the meeting facilitator which of my employees can or can't attend. Then we can reschedule accordingly."
- **Decreases the *"intimidation factor"***—Many people don't like to write letters, memos, reports, or even e-mail messages. Many people don't like face-to-face communication, either. They're intimidated by writing or by bosses. Robert says that TM decreases this "intimidation factor" for his employees. With TM, they don't have to actually speak to Robert, they don't have to face him person-to-person, and they don't have to worry about grammar.
- **Documentation**—Finally, TM allows Robert to document his conversations, something that's not always possible with phone calls. TM is an instant record of a dialogue. A TM account saves incoming and outgoing calls for a few days until it's full. This allows an employee to review them if needed to clarify any later misconceptions.

IM/TM Corporate Usage Policy

To clearly explain the role of IM and TM (text messaging) in the workplace, a company should establish a corporate IM and TM usage policy. Many industries might already have such a policy in place, in relation to existing restrictions established by Sarbanes-Oxley or HIPAA. Consider a policy as follows:

- Train employees to use IM and TM effectively for business.
- Teach employees IM and TM security concerns.
- Explain which services are allowable in the workplace. For example, conversation between coworkers is acceptable, but chat between employees and outside individuals might not be advisable. File transfers of proprietary information need to be restricted and monitored.
- Do not allow employees to store IM or TM passwords on computer desktops or other sites that are easily accessed.
- Install appropriate security measures, including automated encryption, to protect against external threats, such as spam, spim, and viruses.
- Log and archive instant messages and text messages for compliance to your company policy (Bradley; "How To"; Ollman).

BLOGGING FOR BUSINESS

Jonathan Schwartz, president and chief operating officer of Sun Microsystems, says that blogging is a "must-have tool for every executive." "It'll be no more mandatory that they have blogs than that they have a phone and an e-mail account," says Schwartz (Kharif). Bill Gates, Microsoft's CEO, says that blogs could be a better way for firms to communicate with customers, staff, and partners than e-mail and Web sites. "More than 700 Microsoft employees are already using blogs to keep people up to date with their projects" ("Gates Backs Blogs for Business"). Mark Cuban, owner of the Dallas Mavericks basketball team, has a blog (Ray). IBM created "the largest corporate blogging initiative" by encouraging its 320,000 employees to become active bloggers with the goal of achieving "thought leadership" in the global information technology market (Foremski). Nike launched an "adverblog" to market its products. Technorati, a blog search engine, currently tracks "112.8 million blogs and over 250 million pieces of tagged social media" ("Welcome to Technorati").

What is "blogging," and why are so many influential companies and business leaders becoming involved in this new communication channel?

Blogging—a Definition

A blog, the shortened version of the words Web log, is created online and can include text, graphics, links to other Web sites, and video. Blogs, a form of social media, not only allow but also encourage input from many readers. Daily, millions of people use blogs to publish their thoughts. Though many blogs feature text, blogs take different forms, including

- **Vlogs**—The posting of videos through RSS feeds ("Really Simple Syndication" or "Rich Site Summary") to create miniprograms, such as seen on YouTube.
- **MP3 blogs**—The creation of music files available for downloading. These are also known as "musicblogs" or "audioblogs."
- **Podcasting**—News, market information, music, product announcements, public service announcements, how-to videos, photographs, audio files, and more can be conveyed through podcast blogs. The term "podcast" is a combination of the words "iPod" and "broadcast."
- **Micro-blogging**—Brief text, usually less than 200 characters. Facebook and MySpace, two popular blog sites, include micro-blogging features.
- **Twitter**—A free micro-blogging service. Users can send short "tweets" or updates limited to 140 characters to cell phones or other blog sites.
- **Flickr**—A site for sharing photographs.

Ten Reasons for Blogging

The rapid growth of blogging might be reason enough to consider using this online communication channel. However, following are ten reasons why companies blog:

1. **Communicate with colleagues**—Chris Winfield, president of an online marketing company, says that he uses blogging to improve "communication flow to his employees" (Ray). Many companies encourage their employees to use blogging for project updates, issue resolutions, and company announcements. The engineering department at Disney ABC Cable Networks Group uses blogs "to log help desk inquiries" (Li).
2. **Communicate with customers**—In contrast to private, intranet-based blogs used for internal corporate communication, many companies have public blogs. Through public blogs, a company can initiate question/answer forums, respond to customer concerns, allow customers to communicate with each other, create interactive newsletters, and build rapport with customers, vendors, and stakeholders.
3. **Introduce new product information**—Because blogs are quick and current, they allow companies to provide up-to-date information about new products and services.
4. **Reach an influential audience segment**—The Pew Institute's 2006 "Internet & American Life Project" reported the following facts (Lenhart and Fox):
   - 8 percent of Internet users age 18 and older, or about 12 million American adults, keep a blog.
   - 39 percent of Internet users age 18 and older, or about 57 million American adults, read blogs.
   - 19 percent of Internet users age 12 to 17 keep a blog.
   - 38 percent of online teens read blogs.
   - 54 percent of bloggers are under age 30.
5. **Improve search engine rankings**—Marketing is a key attraction of corporate blogs. A blog post using keywords, allowing for comments and responses, and providing

references and links to other sites tends to rank in the top 10 to 20 listings in Internet search engines (Ray).

6. **Network through "syndication"**—To access a Web site, a reader must know the URL or use a search engine, such as Google. In contrast, blogs can be distributed directly to the end users through a "Web feed." The writers of a blog can syndicate their content. Then, through the Web feed, readers can subscribe to the blog. Syndication makes Web feeds available for people to access. By using feed programs, such as RSS, Atom, or Current Awareness, bloggers can syndicate their blogs or be notified when topics of interest are published. Thus, blogs can be very personalized—essentially delivered to your door on a moment's notice.
7. **Facilitate online publishing**—A key value to blogging is its ease and affordability. Blogging services and software, such as Blogger, Movable Type, TypePad, Xanga, Live Journal, and Weblogger, give bloggers access to easy-to-use Web-based forms.
8. **Encourage "bubble-up" communication**—Blogging promotes brainstorming and public forums. A corporate blog builds networks, allows people to dialogue, and encourages conversation.
9. **Track public opinion**—"Trackback" features, available from many blog services, let companies track blog usage. Tracking lets companies monitor their brand impact and learn what customers are saying (good news or bad news) about their products or services.
10. **Personalize your company**—Personal responses to customer comments help personalize a company. Blogging offers a refreshing option that reaches out to customers, offering a human dimension.

Ten Guidelines for Effective Corporate Blogging

If you and your company decide to enter the "blogosphere," the world of blogging, follow these guidelines.

1. **Protect company information**—Do not disseminate private, confidential, or proprietary materials. Seek permission from management prior to communicating company information on your blog.
2. **Identify your audience**—As with all workplace communication, audience recognition and involvement is crucial. Before blogging, decide what topics you want to focus on, what your unique spin will be, what your goals are in using a blog, and who your blog might appeal to.
3. **Achieve customer contact**—Blogs are innately personal. Take advantage of this feature. Make your blogs fun and informal. You can give your blog personality and encourage customer outreach by including "interesting news of the day, jokes," personnel biographies, question/answer forms, updates, an opportunity to add comments, as well as information about products and services (Ray). In addition, make sure the blog is interactive, allowing for readers to comment, check out new links, or add links.
4. **Determine where to locate the blog**—You can locate and market blogs in many different places: Add a blog link to your existing Web site; create a totally independent blog site; distribute your blog through RSS aggregators; add a blog link to your e-mail signature; or mention your blog in marketing collateral (Wuorio "Blogging"; Li).
5. **Start "blogrolling"**—Once you have determined audience and blog location, it's time to start blogrolling. Start talking. Not only do you need to start the dialogue by adding content to your blog, but also you want to link your blog to other sites (Wuorio "Blogging").
6. **Emphasize keywords**—By mentioning keywords in your blog titles and text, you can increase the number of hits your blog receives from Internet search engines.
7. **Keep it fresh**—Avoid a stale blog. By posting frequently (daily or weekly), you encourage bloggers to access your site and return to it often.

8. **Respond quickly to criticism**—In a communication like the blogosphere where dialogue is the desired end, companies inevitably will receive criticism about products or services. Blogs allow companies to respond quickly to and manage bad news.
9. **Build trust**—Don't use the blog only to promote new corporate products. Transparent marketing tends to backfire on bloggers who want sincere content, an opportunity to learn more about a company's culture, and a chance to engage in dialogue.
10. **Develop guidelines for corporate blog usage**—Because blogs encourage openness from customers as well as employees, a company must install guidelines for corporate blog usage. One potential problem associated with blogging, for example, is divulging company information, such as financial information or trademark secrets. Figure 9.8 offers a sample code of ethics created by Forrester Research (Li).

See Figure 9.9 for a sample blog site.

FIGURE 9.8 Blogging Code of Ethics

Blogging Code of Ethics

1. I will tell the truth.
2. I will write deliberately and with accuracy.
3. I will acknowledge and correct mistakes promptly.
4. I will preserve the original post, using notations to show where I have made changes so as to maintain the integrity of my publishing.
5. I will never delete a post.
6. I will not delete comments unless they are spam or off-topic.
7. I will reply to e-mails and comments when appropriate and will do so promptly.
8. I will strive for high quality with every post—including basic spell-checking.
9. I will stay on topic.
10. I will disagree with other opinions respectfully.
11. I will link to online references and original source materials directly.
12. I will disclose conflicts of interest.
13. I will keep private issues and topics private because discussing private issues would jeopardize my personal and work relationships.

FIGURE 9.9 Sample Blog "Technotipster"

(Courtesy of Jonathan Bacon)

Robert Met His Challenge

To meet his communication challenge, Robert used the P^3 process.

Planning

To plan his communication, Robert considered the following:

- Goal—to communicate complex information quickly about roofing damage
- Audience—external auditors (specialists)
- Channels—e-mail cover message and attached short report
- Data—notes taken from site visits and three roofing subcontractor bids

Robert says, "Planning this communication entails gathering data for the report. To do so, my team and I visit the building site to evaluate the extent of the roofing damage. The team takes notes, measurements, and photographs. These will be included in a request for proposal (RFP), which is sent to roofing companies. Eventually, the notes and photographs will be submitted to the insurance auditor.

After the initial site visit, I received three bids in response to the RFP. I reviewed these responses with my team to analyze which roofing company best meets the client's needs." Figure 9.10 shows Robert's researched material in a table.

In addition to using a table for planning, Robert also organized his notes about the roof failure by listing, as shown in Figure 9.11.

Packaging

Robert reviewed his researched material and determined it was far too complex for an e-mail message. Robert says, "A letter report would work, but mailing a letter will take too long. Since I want to get the report to the client quickly, I'm going to use a different communication channel, an e-mail cover message. Then, I will attach the more formal, short, detailed report with a table."

Robert quickly wrote an e-mail message and saved it as a draft to be reviewed after talking to the insurance auditor about the roof as shown in Figure 9.12.

Perfecting

When Robert reviewed his draft, he realized that he needed to add more information for clarity, including the following:

- client's street and city address
- brief summary of the bid totals

Robert says, "I want to improve the e-mail's tone to avoid words like 'low-balled' which could create a negative image of a company. Then I need to alter the e-mail's format for easier access to ensure effective communication." Robert's perfected e-mail cover message in Figure 9.13 meets his communication challenge.

FIGURE 9.10 Planning Table

Table Analyzing the Three Bids					
Bidder	**Base Bid**	**Painting ($/Sq. Ft.)**	**Deck Overlay ($/Sq. Ft.)**	**Deck Removal/ Replacement ($/Sq. Ft.)**	**Interior Protection ($/Sq. Ft.)**
Elbert Roofing Co.	$88,128	0.60	3.00	5.50	0.50
FCP Roofing Corp.	$86,631	0.86	2.30	5.00	0.26
Van Horn Roofing	$120,959	0.65	3.00	5.50	0.90

FIGURE 9.11 **Listing to Plan Notes about Roof Failure**

1. In reviewing both the base bid and the unit prices, Elbert Roofing Company becomes the apparent low bidder. The two low bids are extremely close. As previously indicated, the original roofer is affiliated with Elbert Roofing Company.
2. As the apparent low bidder, Elbert's unit prices for painting, deck overlayment, and deck removal/replacement are within the unit price guidelines established by the RFP.
3. The total roof area is approximately 14,200 square feet. The apparent low bid is $88,128, which yields a $6.21 sq. ft. unit price for the new roofing system. The general unit cost for a 4-ply BUR roof of this size is $5.00+/−$0.50.
4. The original roof system contained 1.5″ of Coopers Rx. The low bidder is utilizing hardboard and ISO-Board with a comparable "R-Value" of 12.5.
5. Elbert has included a unit price of $0.50/sq. ft. for interior protection. The test cuts revealed a large 4′ by 10′ SRA with holes through the 22 ga. prime painted roof deck. Our inspection revealed approximately 75 percent visible deck.
6. The issue of depreciation must be addressed. The original warranty period began in January 1995. The owner filed their claim in November 2008. Therefore, the owner has had beneficial usage of the original roof system for 168 months.

FIGURE 9.12 **E-Mail Message Saved in Draft Mode for Future Revision**

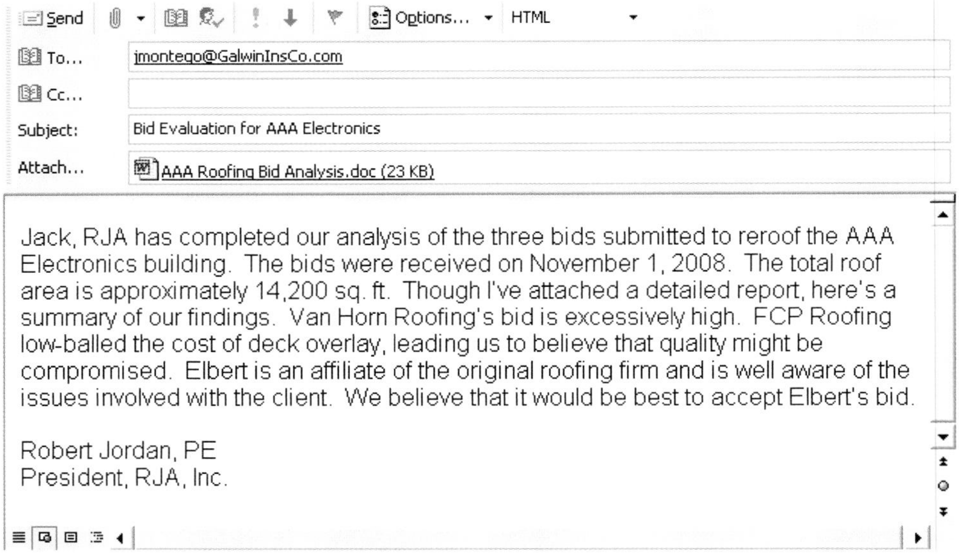

Send | Options... | HTML

To... jmontego@GalwinInsCo.com

Cc...

Subject: Bid Evaluation for AAA Electronics

Attach... AAA Roofing Bid Analysis.doc (23 KB)

Jack, RJA has completed our analysis of the three bids submitted to reroof the AAA Electronics building. The bids were received on November 1, 2008. The total roof area is approximately 14,200 sq. ft. Though I've attached a detailed report, here's a summary of our findings. Van Horn Roofing's bid is excessively high. FCP Roofing low-balled the cost of deck overlay, leading us to believe that quality might be compromised. Elbert is an affiliate of the original roofing firm and is well aware of the issues involved with the client. We believe that it would be best to accept Elbert's bid.

Robert Jordan, PE
President, RJA, Inc.

CHAPTER HIGHLIGHTS

1. To be an effective communicator in business and industry today, you have to communicate electronically.
2. A blog, the shortened version of the words "Web log," is created online and can include text, graphics, links to other Web sites, and video.
3. Text messaging, also known as SMS (short message service), is a growing presence in workplace communication.
4. Following are benefits of instant messaging: increased speed of communication, improved efficiency for geographically dispersed workgroups, collaboration by

FIGURE 9.13 Perfected E-Mail Cover Message

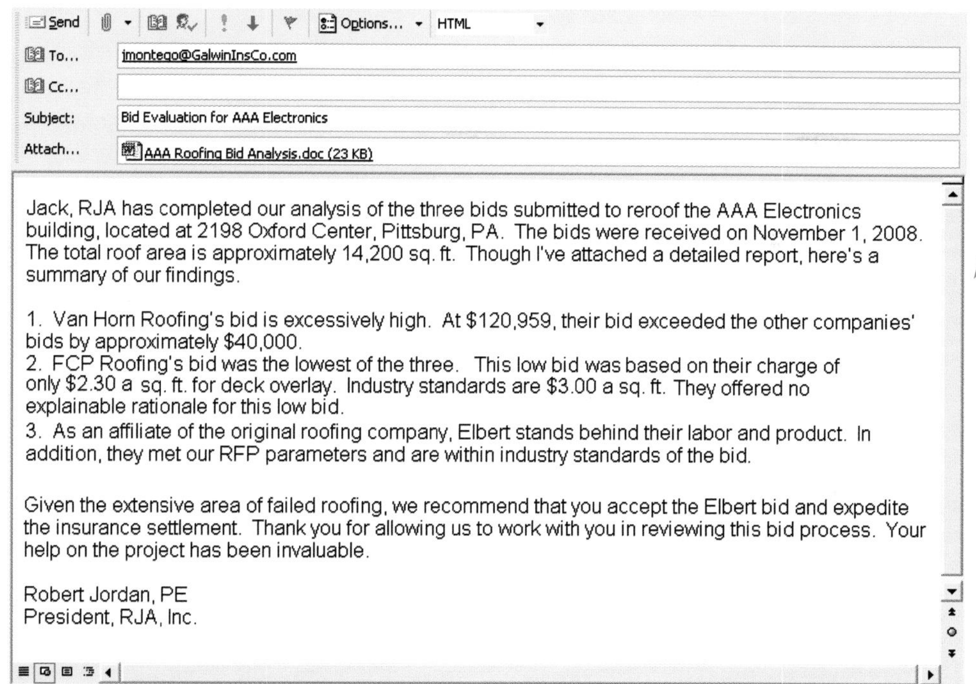

To... jmontego@GalwinInsCo.com

Cc...

Subject: Bid Evaluation for AAA Electronics

Attach... AAA Roofing Bid Analysis.doc (23 KB)

Jack, RJA has completed our analysis of the three bids submitted to reroof the AAA Electronics building, located at 2198 Oxford Center, Pittsburg, PA. The bids were received on November 1, 2008. The total roof area is approximately 14,200 sq. ft. Though I've attached a detailed report, here's a summary of our findings.

1. Van Horn Roofing's bid is excessively high. At $120,959, their bid exceeded the other companies' bids by approximately $40,000.
2. FCP Roofing's bid was the lowest of the three. This low bid was based on their charge of only $2.30 a sq. ft. for deck overlay. Industry standards are $3.00 a sq. ft. They offered no explainable rationale for this low bid.
3. As an affiliate of the original roofing company, Elbert stands behind their labor and product. In addition, they met our RFP parameters and are within industry standards of the bid.

Given the extensive area of failed roofing, we recommend that you accept the Elbert bid and expedite the insurance settlement. Thank you for allowing us to work with you in reviewing this bid process. Your help on the project has been invaluable.

Robert Jordan, PE
President, RJA, Inc.

Robert says, "Short, simple, numbered paragraphs are the best way for me to communicate in an e-mail message. Too many people rely on one, long paragraph in their e-mails. It's often difficult for people to read such poorly designed e-mails. I like to have an introduction, a discussion section, and a concluding paragraph in an e-mail. Then I feel that my e-mails are 'perfected.'"

multiple users in different locations, communication with colleagues and customers at a distance in real time, like using the telephone.

5. E-mail allows you to communicate internally to coworkers and externally to customers and vendors.
6. Because so many writers merely respond to earlier e-mail messages, what you end up with is a virtual paper trail, allowing for effective archiving.
7. Practice "netiquette"—online etiquette—when writing e-mail and instant messages.
8. Developing guidelines for corporate blogging can ensure ethical and appropriate use.
9. Use effective subject lines in e-mail to make sure your audience opens the message rather than deleting it.
10. Use IM for speed and convenience. If you need length and detail, other options—e-mail messages, memos, reports, letters—are better choices.

WWW *Find additional electronic communication exercises, samples, and interactive activities at http://www.prenhall.com/gerson.*

MEETING WORKPLACE COMMUNICATION CHALLENGES

CASE STUDIES

1. Barney Allis Stores (BAS) home based in Seattle, Washington, has department stores in all 50 states and in the following international cities: Toronto, Vancouver, London, Paris, Berlin, Barcelona, Rome, Beijing, Hong Kong, Sidney, Mexico City, and Rio de Janeiro.

It's time for BAS's marketing department to inform all BAS stores about which products they should "push" for the Christmas season sales. This year, BAS has created a new product line, called "BASK"—Barney Allis Stores–Kids. The focus will be on winter and spring children's wear, for ages Pre-K through fifth grade. Marketing needs to highlight three item lines:

- Corduroy pants
- Reversible jackets
- Long-sleeve T-shirts. These will be customized to each country with slogans (in the country's language) or cartoon characters decorating the shirtfronts. The cartoon characters will also be customized, stemming from each country's Saturday morning cartoon television shows.

Assignments

a. As BAS director of marketing, your job is to write an e-mail message to all stateside and international marketing managers. The e-mail not only will provide the previous information but also act as a cover correspondence, transmitting an attached product catalogue. The e-mail must direct the marketing managers to create print, radio, and television advertisements for their local markets. Direct them to the BAS intranet site http://www.bas.com/BASK/salestalk.html. In writing this e-mail, follow the criteria provided in this chapter.

b. As BAS director of marketing, your job is to introduce your company's new line of products to customers. One way to do so is through a blog. Review several blog sites from companies such as Harley Davidson, Hewlett-Packard, Microsoft, and Honda. Using blog comments from these sites as samples, start blogrolling. Write a blog entry to customers to introduce BAS's new line of children's clothing.

2. Suburban Day Surgery (SDS) publicized an RFP, requesting proposals from architectural/engineering companies. SDS wanted to expand its facility, adding new surgical rooms, recovery rooms, offices, waiting rooms, and a cafeteria.

After reviewing several proposals, SDS has narrowed its choices. One company that SDS is especially interested in is Lamb & Sons Construction Co. This company has a wonderful reputation, as shown in their proposal's client references. In addition, SDS has seen Lamb's work through visiting other sites the company has built.

However, SDS management has a few remaining questions. The proposal did not specify when Lamb & Sons could begin work, what their timeframe was for completing the project, and the credentials of their subcontractors.

Assignment

Write an e-mail to Bill Lamb (blamb@l&sco.com). In the e-mail, explain why you are writing, specify your requests, and give a deadline for the response. Follow the criteria for e-mail provided in this chapter.

3. Yasser El-Akiba is president of his college's FBLA club (Future Business Leaders of America). Riverbend Community College (RCC), Riverbend, Idaho, has sponsored the FBLA team's travel to a regional convention.

During the trip, another FBLA team member, Harry Tinker, was caught drinking in his hotel room, failed to abide by the FBLA's mandatory curfew, and was unprepared for his part in the team's "New Business Idea" competition.

As president of the club, Yasser must report Harry's infractions to the team sponsor, Dr. Stapleton (gstaple@rcc.edu).

Assignment

Write an e-mail for Yasser, following the criteria provided in this chapter. Explain why the e-mail is being written, develop the assertions, and conclude either by recommending action or by asking for follow-up instructions.

4. Bob Ward, an account manager at HomeCare Health Equipment, has not gotten the raise that he thinks he deserves. When Bob met with his boss, Alicia Green, last Thursday for his annual evaluation, she told him that he had missed too many days

of work (eight days during the year); was unwilling to work beyond his 40-hour work week to complete rush jobs; and had not attended two mandatory training sessions on the company's new computerized inventory system.

Bob agrees that he missed the training sessions, but he was out of town on a job-related assignment for one of those sessions. He missed eight days of work, but he was allowed five days of sick leave as part of his contract. The other three days missed were due to his having to stay home to take care of his children when they were sick. He believed that these absences were covered by the company's parental leave policy. Finally, he does not agree that employees should be required to work beyond their contractual 40 hours.

Assignment

Write an e-mail to Alicia Green, stating Bob's case. In this e-mail, begin by reminding Alicia why this message is being sent. Then, develop Bob's assertions. Finally, ask for follow-up action. Follow the criteria for e-mail presented in this chapter, and be careful with tone when writing this message.

5. You are the manager of WhiteOut, a store that sells snowboarding equipment and clothing. You have seven employees. Since you believed that today would be a light day for sales, you asked only one employee to work. However, more customers than you had expected suddenly showed up, and you and your store employee are overwhelmed. You need help—fast. E-mailing your other employees might not work, since they aren't necessarily near a computer. In contrast, you know they all have cell phones. The fastest and most trustworthy way to communicate with them is through text messaging.

Assignment

Write a brief text message to the employees. Tell them the circumstances at your store and ask them for assistance.

ETHICS IN THE WORKPLACE CASE STUDY

Carlos Delgado is a facilities maintenance employee for the city. His job requires that he travel throughout the city to various locations but to always be available for emergency calls. Therefore, the city issued him a cell phone for use during his 40-hour-a-week job.

As Carlos's manager, you are responsible for checking cell phone bills. When you review Carlos's bill, you find that he sent 2,248 text messages last month. Most facilities' maintenance employees send an average of 500 text messages a month.

When you questioned Carlos about this excessive cell phone use, Carlos said that he had frequent TM conversations with coworkers at different job sites. However, Carlos also said that his son had been ill that month. Carlos had text messaged his home to check on his son's health.

Question: Are personal calls acceptable on company equipment? Is it ethical for Carlos to have made so many text messages on his corporate-issued cell phone? Justify your answer based on information about ethics in the workplace provided in Chapter 1.

INDIVIDUAL AND TEAM PROJECTS

1. E-mail is used to convey many types of information in business and industry. Write an e-mail message to accomplish any of the following purposes (pick any topic you would like to complete the assignment).
   - Tell a subordinate or a team of employees to complete a task.
   - Tell a reader or readers that you have attached a document and list the key points that are included in the attachment.
   - Report on expenses, incidents, accidents, problems encountered, projected costs, study findings, hirings, firings, and reallocations of staff or equipment.

    - Tell a reader about a meeting agenda, date, time, and location; decisions to purchase or sell; topics for discussion at upcoming teleconferences; conclusions arrived at; fees, costs, or expenditures, and so on.
    - Explain how to set up accounts, research on the company intranet, operate new machinery, use new software, apply online for job opportunities through the company intranet, create a new company Web site, or solve a problem.

2. Visit five to ten local companies. When meeting with these companies' employees, ask them how much time they spend writing e-mail, sending instant messages, and/or text messaging. Ask them what they generally write about, who their audiences are, and whether their e-mail, instant messages, and/or text messages are important parts of their jobs. Then, report your findings in an e-mail message to your instructor.

DEGREE-SPECIFIC ASSIGNMENTS

1. **Marketing and Public Relations.** Aunt Rose's Bakery has been in business for 75 years. Home-based in Houston, this company markets products such as brownie, cookie, cake, cupcake, and biscuit mixes. It also markets cake decoration items such as sprinkles, frostings, and cupcake tins. Business has been good throughout the company's history. In fact, business has been so good that the company never saw a great need to market itself. Aunt Rose's Bakery assumed that people would continue to buy their tried and true products.

   However, with market proliferations and increased competition, Aunt Rose's profits have declined 15 percent every year for the past three years. Research into this problem has suggested that changes are necessary in the marketing of Aunt Rose's products. For example, one focus group concluded that Aunt Rose's Web site was a key contributor to the company's declining profits. The Web site was perceived as dull, uninviting, and "too corporate." The site did not encourage audience interaction, online purchases, personalized features about employee profiles, or insight into the company's history and culture. All the Web site provided was nutrition facts, corporate earnings, and shareholder dividends.

   Aunt Rose's marketing team wants to make the Web site more user-friendly.

Assignment

You are the marketing manager at Aunt Rose's Bakery. You need to communicate the problem with the company's current marketing strategy and suggest ways to improve this problem. Write an e-mail message to your staff. This correspondence, written from a manager's perspective down to subordinates, must accomplish four goals: inform the readers of the problem, instruct them about ways to resolve this issue, persuade them of the importance of this task, and build rapport.

2. **Business Information System.** United Packaging has provided overnight delivery of items for business and industry for 35 years. This company is now expanding globally and changing its name to International Packaging. This expansion created a need for radical changes in the company's business information systems. For example, International Packaging's expansion demanded the hiring of over 1,000 new employees, the purchase of over 100 new facilities around the world, contracts with dozens of new suppliers, and interconnected relations with 16 different countries. These business developments not only affected customers and employees, but also created legal issues, the need for dissemination of news to stakeholders, and complex auditing challenges.

   For International Packaging to handle its expansion, a complex computer software platform was mandatory, including an Internet, extranet, and intranet presence.

Assignment

As manager of the information technology department at International Packaging, you have been asked by the company's CEO to train all company employees on the use of a new software platform. Write an e-mail to personnel representatives from purchasing, accounting, legal, delivery, and human resources, inviting them to a training session. This e-mail not only should clarify the purpose of the training but also persuade them of this training session's importance. In addition, the correspondence should provide a training timetable and topics for discussion (use your own knowledge and experience to develop discussion topics). Finally, build rapport to ensure that the training sessions are productive.

3. **Accounting.** In 2006, employers gained access to a new retirement savings option: Roth contributions to 401(k) plans. Corporate employees can designate all or part of their contributions to their 401(k)s on an after-tax basis. The change makes most distributions tax free. Although the change means more complexity for employers, most will offer the option because of the big benefit that workers can enjoy from tax-free buildup in their 401(k)s.

   As manager of AccountingHelp, you need to inform your colleagues of this new tax option and the laws regulating it, so your accountants can pass the news on to their customers. Basically, the general characteristics of a Roth contribution include the following: Employees must make an irrevocable designation of a Roth contribution when they make the 401(k) election; employers must treat the contributions as included in the employee's gross income for the purposes of all taxes, as if the employee had received cash instead; contributions must be maintained in a separate account; designated Roth contributions are limited to elective contributions; employer-matching contributions and nonelective contributions are not eligible; finally, an employee's contributions to a Roth 401(k) account were limited—for 2006, the maximum for 401(k) contributions is $15,000 (the figure will be adjusted for inflation for future years).

Assignment

Write an e-mail message informing your colleagues at AccountingHelp of this new law. This correspondence must be informative, persuasive, and instructional. Follow the criteria provided in this chapter for successful e-mail messages.

4. **Management.** EleckTek, a small but profitable telecommunications company, is involved in a hostile takeover. The much larger, national telecommunications company WireNet is absorbing EleckTek. This will result in an upheaval of employee relations in an environment of change. Due to redundancies, substantial job losses will occur, the pay structure will alter, benefits will change, retirement packages will alter, profit sharing will be affected, and jobs must be restructured.

Assignment

As human resources manager, you have been asked to send an e-mail message to EleckTek employees. In this correspondence, you must inform the audience of changes occurring due to the hostile takeover. In addition, this correspondence will inform the audience of EleckTek's next steps: training seminars about job transitioning (interview techniques, resume preparation, and job search skills); state-of-the-company video sessions; and the initiation of an intranet feedback loop. This correspondence must be informative, persuasive, instructional, and positive so that rapport is reestablished.

5. **Finance.** You hope to open your own online mortgage company. To accomplish this goal, you will need to write a business plan for submission to a local bank, who will underwrite your financing needs. Before you write your business plan, you have a number of questions. You need to know what licensing is required for your city, county, and state and what bonding is needed to operate your business. You want to know the kinds and amounts of references the bank requires. You also will need loans for start-up office equipment, furniture, and hardware and software. Finally, you know that marketing will be a key to your success (hard-copy fliers and brochures, as well as Internet-based marketing—pop-ups, for example). You hope that the bank's small business division will provide you information about what they consider to be approximate costs for office equipment and a marketing budget.

Assignment

You have been working with a loan officer at the bank, Larry Raymond. Write an e-mail to Larry, asking him the questions you have in preparation for writing your business plan. Abide by the criteria provided in this chapter. Include any other details that you believe are necessary to accomplish your goals.

PROBLEM-SOLVING THINK PIECES

1. You are the manager of a human resources department. You are planning a quarterly meeting with your staff (training facilitators, benefits employees, personnel directors, and company counselors). The staff works in three different cities and 12 different offices. You need to accomplish four goals: get their input regarding agenda items, find out what progress they have made on various projects, invite them to the meeting, and provide the final agenda.

How should you communicate to them? Should you write an e-mail, an instant message, a text message, or a blog? Write an e-mail message to your instructor explaining your answer. Be sure to give reasons for and against each option.

2. Check out Stonyfield Farm's blog site at http://www.stonyfield.com/weblog/index.cfm. This site says, "Welcome to the Stonyfield Farm blog 'Cow'munities" where they provide "a chance for you to look inside Stonyfield and get to know us, and us to know you." Read entries from their two blog sites: "Baby Babble" and "The Bovine Bugle." Then answer these questions:
   - What are Stonyfield's goals in creating these two blogs, seemingly unrelated to their product?
   - How do these blogs relate to corporate business?
   - If you were hired by a company of your choice, what storylines would your blogs feature and why?

WEB WORKSHOP

1. Most city, county, and state governments have written guidelines for their employees' e-mail. For example, see "Electronic and Voice Mail: Connecticut's Management and Retention Guide for State and Municipal Government Agencies" (http://www.cslib.org/email.htm). In addition, see the e-mail policy provided by Hennepin County, Minnesota (http://www.co.hennepin.mn.us/wemail.html). Similarly, most companies and college/universities have written guidelines for their employees' and students' e-mail use. For example, see what Florida Gulf Coast University says about e-mail use (http://admin.fgcu.edu/compservices/policies3.htm).

Research your city, county, and/or state e-mail policies. Research your city or college/university's e-mail policies. Determine what they say, what they have in common, and how they differ.

Write an e-mail to your instructor reporting your findings.

2. Bloggers provide up-to-date information on newsbreaking events and ideas.
   - For e-commerce news, visit http://blog.clickz.com/ (ClickZ Network—Solution for Marketers).
   - For business blogging, visit http://www.businessweek.com/the_thread/blogspotting/ (Business Week Online) to learn where the "worlds of business, media and blogs collide."
   - For technology news (information technology, computer information systems, biomedical informatics, and more), visit http://blogs.zdnet.com/. Once in this site, use their search engine to find a technology topic that interests you.
   - For information about criminal justice, visit http://lawprofessors.typepad.com/crimprof_blog/criminal_justice_policy/ (Law Professor Blog Network).

- For information about science, visit http://www.scienceblog.com/cms/index.php (ScienceBlog), which has links to blogs about aerospace, medicine, anthropology, geoscience, and computers.
- The Car Blog (http://thecarblog.com/) has news items about automotives.
- Every news agency, such as ABC, NBC, CBS, CNBC, has a news blog.

To see what's new in your field of interest, check out a blog. Then, report key findings to your instructor in an e-mail message.

3. Research information about blogging by checking out the following sites. Then, write an e-mail message to your professor, summarizing your findings:
   - Foremski, Tom. "IBM Is Preparing." *SiliconValleyWatcher.* 13 May 2005. http://www.siliconvalleywatcher.com/mt/archives/2005/05/can_blogging_bo.php.
   - "Gates Backs Blogs for Business." *BBCNews.* 21 May 2005. http://news.bbc.co.uk/2/hi/technology/3734981.stm.
   - Jackson, Lorrie. "Blogging Basics: Creating Student Journals on the Web." 18 July 2006. http://www.education-world.com/a_tech/techtorial/techtorial037print.shtml.
   - Kharif, Olga. "Blogging for Business." *BusinessWeekOnline.* 9 Aug. 2004. http://www.businessweek.com/technology/content/aug2004/tc2004089_3601_tc024.htm.
   - Li, Charlene. "Blogging: Bubble or Big Deal?" *Forrester.* Nov. 2004. http://www.forrester.com/Research/Print/Document/0,7211,35000,00.html.

4. Research information about text messaging by checking out the following sites. Then, write an e-mail message to your professor, summarizing your findings:
   - Calvey, Mark. "RU Ready? Wells Initiates Banking by Text Message." *Kansas City Business Journal.* 22 Oct. 2007. 3 Nov. 2007. http://kansascity.bizjournals.com/kansascity/othercities/sanfrancisco/stories/2007/10/22/daily9.html?b=1193025600%5E1539015.
   - Kedrosky, Paul. "Why We Don't Get the (Text) Message." *CNN.Money.* Oct. 2006. 3 Nov. 2007. http://money.cnn.com/magazines/business2/business2_archive/2006/08/01/8382255/index.htm.2.
   - Noguchi, Yuki. "Life and Romance in 160 Characters or Less: Brevity Gains New Meaning as Popularity of Cell Phone Text Messaging Soars." *Washington Post.* 29 Dec. 2005. 2 Nov. 2007. http://www.washingtonpost.com/wp-dyn/content/article/2005/12/28/AR2005122801430.html.
   - Tedeschi, Bob. "Reaching More Customers with a Simple Text Message." *The New York Times.* 3 Nov. 2007. http://www.nytimes.com/2007/07/16/business/media/16ecom.html 16. Jul.
   - "Text Messaging: Frequently Asked Questions." *AT&T.* 2 Nov. 2007. http://www.wireless.att.com/learn/messaging-internet/messaging/faq-text-messaging.jsp.

QUIZ QUESTIONS

1. What are the characteristics of online readers?
2. What is blogging?
3. Why should a corporation consider blogging?
4. What are three benefits of text messaging in the workplace?
5. What are three benefits of using instant messaging in the workplace?
6. What are three challenges of using instant messaging in the workplace?
7. What is "spim"?
8. Why is a ".sig" file important in e-mail?
9. Which highlighting techniques should not be used in e-mail?
10. What is "netiquette"?

CHAPTER 10

Traditional Correspondence: Memos and Letters

▸REAL PEOPLE *in the workplace*

Kim Suzaki is a financial planner with GMM&J, a firm that offers comprehensive financial planning to high net-worth clients. GMM&J's financial plans address a complete review and analysis of their clients' cash flow, tax planning, retirement planning (saving for retirement and taking distributions once retired), estate planning, adequacy of insurance (life, medical, long-term disability, property), and education planning for clients with children. Kim's ongoing challenge is to communicate effectively with a variety of clients, principals in his firm who review his work, and other professionals such as the clients' CPAs and attorneys.

Mr. Suzaki's communications with his clients are both written and verbal. When he needs information or has a question, he will call the client, other professionals, or people with whom he already has letters of authorization (LOAs) to communicate directly. In addition to the phone calls, he and his firm prepare written communication and correspondence with these same audiences. The shelf-life of these documents is forever, so much care and attention are paid to every word, every number, every nuance in the communication. For example, the memo that details the topics discussed during the meeting between Mr. Suzaki, his firm's principals, and the client is a perpetual record of that meeting. The written memo may serve as a helpful reference for both the client and Kim, but that same document may also take on a legal quality if there is ever any question as to what financial advice was given or acted upon.

A sampling of the various documents prepared by GMM&J includes the following:

- Direct communication with clients. Such documents include letters and memos, meeting reminders, requests for information, memos documenting a meeting, and newsletters
- Letters to clients' accountants before tax season (requesting information used for the tax return)
- Letters to brokers, insurance agents, attorneys, and other professionals
- Internal documentation and reports ◂

CHAPTER GOALS

When you complete this chapter, you will be able to do the following:

1. Understand the differences between memos and letters.
2. Follow all-purpose templates to write memos and letters.
3. Use memo samples as guidelines for memo components, organization, writing style, and tone.
4. Evaluate your memos and letters with checklists.
5. Write different types of routine messages, including the following:
   - Inquiry
   - Response
   - Cover (transmittal)
   - 100 percent yes adjustment
   - Order
   - Confirmation
   - Recommendation
   - Thank you
6. Follow the P^3 process—planning, packaging, and perfecting—to create memos and letters.

WWW

To learn more about routine memos and letters, visit our Web Companion at http://www.prenhall.com/gerson.

Kim's Communication Challenge

Kim says, "As a financial planner, I analyze and document clients' financial status on a quarterly basis. I focus on their investment performances, income taxes, cash flow, and financial projections. After assessing these figures, I must write a memo that documents my findings and recommendations. My company writes memos rather than e-mail messages for two reasons: (1) the hard-copy memos will be inserted into three-ring binders (one for the client and one for the company) for future documentation; (2) a memo has a more formal air about it than a casual e-mail.

I can use boilerplate templates for much of the memo. However, communicating effectively still creates challenges, especially related to audience. My memo must communicate clearly to the client, to the firm's principals, and to future audiences.

- The client wants key information. This could include a variety of possibilities. One client might want to know his or her net worth, while another client might want to know 'what's next on the to-do list.' One client might want to know if she is doing OK overall financially, while another client just needs to know what his next tax payment is. Some clients only want to know if they 'have enough cash to redecorate the kitchen.' My challenge is communicating to meet the client's needs. Adding to this challenge is the fact that the client often will be a lay reader with little knowledge about financial planning.
- The firm's principals have different interests. The principals, specialists in the field, care primarily about accuracy and completeness of the financial plan. My job for this audience is to make sure that all of the numbers and the analyses are correct and that tax and financial projections are valid.
- The final audience is unseen and unknown—future readers. The memo will be archived for future reference. Future audiences could include lawyers who are involved in legal issues or a future financial planner who takes over the client's account. In the latter instance, the memo must be clear enough to provide the future financial planner enough information to understand the client's history.

Writing one memo to three or more different audiences with varied levels of expertise creates communication challenges for me."

See page 295 to learn how Kim met his communication challenge.

THE DIFFERENCES BETWEEN MEMOS AND LETTERS

This chapter focuses on traditional correspondence—memos and letters. To give you an overview of the differences and similarities between memos and letters, look at Table 10.1.

Table 10.1 Memos versus Letters

Characteristics	Memos	Letters
Destination	Internal: correspondence written to colleagues within a company.	External: correspondence written outside the business.
Format	Identification lines include "Date," "To," "From," and "Subject." The message follows.	Includes letterhead address, date, reader's address, salutation, text, complimentary close, and signatures.
Audience	Generally specialists or semi-specialists, mostly business colleagues.	Generally semi-specialists and lay readers, such as vendors, clients, stakeholders, and stockholders.
Topic	Generally topics related to internal corporate decisions; abbreviations and acronyms often allowed.	Generally topics related to vendor, client, stakeholder, and stockholder interests; abbreviations and acronyms usually defined.
Tone	Informal due to peer audience.	More formal due to audience of vendors, clients, stakeholders, and stockholders.
Attachments or Enclosures	Hard-copy attachments can be stapled to the memo. Complimentary copies (cc) can be sent to other readers.	Additional information can be enclosed within the envelope. Complimentary copies (cc) can be sent to other readers.
Delivery Time	Determined by a company's in-house mail procedure.	Determined by the destination (within the city, state, or country). Letters could be delivered within 3 days but may take more than a week.
Security	If a company's mail delivery system is reliable, the memo will be placed in the reader's mailbox. Then, what the reader sees on the hard-copy page will be exactly what the writer wrote. Security depends on the ethics of coworkers and whether the memo was sent in an envelope.	The U.S. Postal Service is very reliable. Once the reader opens the envelope, he or she sees exactly what the writer wrote. Privacy laws protect the letter's content.

MEMOS

Reasons for Writing Memos

Memos are an important means by which employees communicate with each other. Memos, hard-copy correspondence written within your company, are important for several reasons. The National Commission on Writing substantiated the importance of correspondence (memos and letters). Their survey of 120 major companies employing approximately 8 million workers found that 70 percent of the companies require employees to write memos and letters ("Writing: A Ticket to Work").

You will write memos to a wide range of readers, including your supervisors, coworkers, subordinates, and multiple combinations of these audiences. Since memos usually are copied (cc: complimentary copies) to many readers, a memo sent to your boss could be read by an entire department, the boss's boss, and colleagues in other departments.

Because of their frequent use and widespread audiences, memos may represent a major component of your interpersonal communication skills within your work environment. Furthermore, memos are very flexible and can be written for many different purposes. Consider these options:

- **Documentation**—Expenses, incidents, accidents, problems encountered, projected costs, study findings, hirings, firings, reallocations of staff or equipment, and so on.
- **Confirmation**—A meeting agenda, date, time, and location; decisions to purchase or sell; topics for discussion at upcoming teleconferences; conclusions arrived at; fees, costs, or expenditures; and so on.
- **Procedures**—How to set up accounts, research on the company intranet, operate new machinery, use new software, apply online for job opportunities through the company intranet, create a new company Web site, solve a problem, and so on.

FAQs

Q: Why write a memo? Haven't memos been replaced by e-mail?

A: E-mail is rapidly overtaking memos in the workplace, but employees still write memos for the following reasons:

1. Not all employees work in offices or have access to computers. Many employees who work in warehouses or in the field cannot easily access an e-mail account. They must depend on hard-copy documentation like memos.
2. Not all companies have e-mail. This may be hard to believe in the twenty-first century, but still it's a fact. These companies depend on hard-copy documentation like memos.
3. Many unions demand that hard-copy memos be posted on walls, in break rooms, in offices, and so on, to ensure that all employees have access to important information. Sometimes, unions even demand that employees initial the posted memos, thus acknowledging that the memos have been read.
4. Some information cannot be transmitted electronically via e-mail. A bank we've worked with, for example, sends hard-copy cancelled checks as attachments to memos. They cannot send the actual cancelled check via e-mail.
5. E-mail messages are very easy to disregard. We get so many e-mail messages (many of them spam) that we tend to quickly delete them. Memos, in contrast, make more of an official statement. People might take hard-copy memos more seriously than e-mail messages.

- **Recommendations**—Reasons to purchase new equipment, fire or hire personnel, contract with new providers, merge with other companies, revise current practices, renew contracts, and so on.
- **Feasibility**—Studying the possibility of changes in the workplace (practices, procedures, locations, staffing, equipment, missions/visions), and so on.
- **Status**—Daily, weekly, monthly, quarterly, biannually, or yearly statements about where you, the department, or the company is regarding many topics (sales, staffing, travel, practices, procedures, finances, etc.).
- **Directive (delegation of responsibilities)**—Informing subordinates of their designated tasks.
- **Inquiry**—Asking questions about upcoming processes or procedures.
- **Cover**—Prefacing an internal proposal, long report, or other attachments.

Criteria for Writing Memos

Memos contain the following key components:

- Memo identification lines—Date, To, From, and Subject
- Introduction
- Discussion
- Conclusion
- Audience recognition
- Appropriate memo style and tone

Figure 10.1 shows an ideal, all-purpose organizational template that works well for memos.

Subject Line

The subject line summarizes the memo's content. One-word subject lines do not communicate effectively, as in the following flawed subject line. The "Before" sample has a *topic* (a what) but is missing a *focus* (a what about the what).

BEFORE	AFTER
Subject: Sales	Subject: Report on Quarterly Sales

FIGURE 10.1 All-Purpose Memo Template

DATE:
TO:
FROM:
SUBJECT: Focus + Topic

Introduction: A lead-in or overview stating *why* you are writing and *what* you are writing about.

Discussion: Detailed development, made accessible through highlighting techniques, explaining *exactly what* you want to say.

-
-
-

Conclusion: State *what* is next, *when* this will occur, and *why* the date is important.

Introduction

Once you have communicated your intent in the subject line, get to the point in the introductory sentence(s). Write one or two clear introductory sentences which tell your readers *what* topic you are writing about and *why* you are writing. The following example invites the reader to a meeting, thereby communicating *what* the writer's intentions are. It also tells the reader that the meeting is one of a series of meetings, thus communicating *why* the meeting is being called.

example In the third of our series of sales quota meetings this quarter, I'd like to review our productivity.

Discussion

The discussion section allows you to develop your content specifically. Readers might not read every line of your memo (tending instead to skip and skim). Thus, traditional blocks of data (paragraphing) are not necessarily effective. The longer the paragraph, the more likely your audience is to avoid reading. Make your text more reader-friendly by itemizing, using white space, boldfacing, creating headings, or inserting graphics.

BEFORE

Example—Unfriendly Text

This year began with an increase, as we sold 4.5 million units in January compared to 3.7 for January 2008. In February we continued to improve with 4.6, compared with 3.6 for the same time in 2008. March was not quite so good, as we sold 4.3 against the March 2008 figure of 3.9. April was about the same with 4.2, compared to 3.8 for April 2008.

AFTER

Example—Reader-Friendly Text

Comparative Quarterly Sales (in Millions)

	2008	2009	Increase/Decrease
Jan.	3.7	4.5	0.8+
Feb.	3.6	4.6	1.0+
Mar.	3.9	4.3	0.4+
Apr.	3.8	4.2	0.4+

Conclusion

Conclude your memo with "thanks" and/or directive action. A pleasant conclusion could motivate your readers, as in the following example. A directive close tells your readers exactly what you want them to do next or what your plans are (and provides dated action).

example

> If our quarterly sales continue to improve at the current rate, we will double our sales expectations by 2009. Congratulations! Next Wednesday (12/22/09), please provide next quarter's sales projections and a summary of your sales team's accomplishments.

Audience Recognition

Since letters go outside your company, your audience is usually a semi-specialist or lay reader, demanding that you define your terms specifically. In a memo, your in-house audience is easy to address (usually a specialist or semi-specialist). You often can use more acronyms and internal abbreviations in memos than you can in letters.

You will write the message to "Distribution" (listing a group of readers) or send the memo to one reader but "cc" (send a "carbon copy" or "complimentary copy") to other readers. Thus, you might be writing simultaneously to your immediate supervisor (specialists), to his or her boss (semi-specialist), to your colleagues (specialists), and to a CEO (semi-specialist). To accommodate multiple audiences, use parenthetical definitions, such as Cash in Advance (CIA) or Continuing Property Records (CPR).

Additional information:
See Chapter 3, "Meeting the Needs of the Audience."

Style and Tone

Because memos are usually only one page long, use simple words, short sentences, specific detail, and highlighting techniques. In addition, strive for an informal, friendly tone. Memos are part of your interpersonal communication abilities, so a friendly tone will help build rapport with colleagues.

In memos, audience determines tone. For example, you cannot write directive correspondence to supervisors mandating action on their part. It might seem obvious that you can write directives to subordinates, but you should not use a dictatorial tone. Though the subordinates are under your authority, they must still be treated with respect. You will determine the tone of your memo by deciding if you are writing vertically (up to management or down to subordinates) or laterally (to coworkers), as shown in Figure 10.2.

FIGURE 10.2 Vertical and Lateral Communication Within a Company

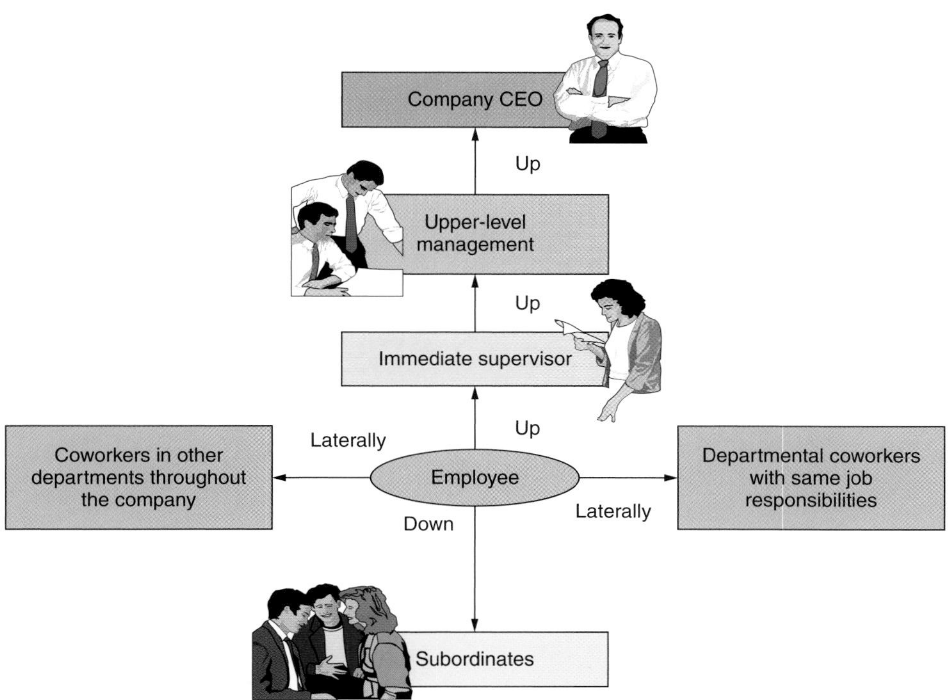

The following examples show the differences between using a friendly tone versus an unfriendly one.

BEFORE	AFTER
Example 1—Unfriendly, Demanding Style We will have a meeting next Tuesday, Jan. 11, 2009. Exert every effort to attend this meeting. Plan to make intelligent comments regarding the new quarter projections.	**Example 2—Friendly, Personal Style** Let's meet next Tuesday (January 11, 2009). Even if you're late, I'd appreciate your attending. By doing so, you can have an opportunity to make an impact on the new quarterly projections. I'm looking forward to hearing your comments.

SAMPLE MEMOS

See Figures 10.3 and 10.4 for sample memos.

FIGURE 10.3 Comparison/Contrast Feasibility Memo

"Distribution" indicates that the memo is being sent to a number of employees. Notice a distribution list is provided at the bottom of the memo.

The introduction states the purpose of this memo, providing dates, personnel, options, and intended action.

The memo's discussion analyzes the criteria used to decide which radio to purchase. Tables are used to ensure reader-friendly ease of access.

MEMORANDUM

DATE: December 12, 2009
TO: Distribution
FROM: Luann Brunson
SUBJECT: Replacement of Maintenance Radios

On December 5, the manufacturing department supervisor informed the purchasing department that our company's maintenance radios were malfunctioning. Purchasing was asked to evaluate three radio options (the RPAD, XPO 1690, and MX16 radios). Based on my findings, I have issued a purchase order for 12 RPAD radios.

The following points summarize my findings.

1. *Performance*
During a one-week test period, I found that the RPAD outperformed our current XPO's reception. The RPAD could send and receive within a range of 5 miles with minimal interference. The XPO's range was limited to 2 miles, and transmissions from distant parts of our building broke up due to electrical interference.

2. *Specifications*
Both the RPAD and the MX16 were easier to carry, because of their reduced weight and size, than our current XPO 1690s.

	RPAD	XPO 1690	MX16
Weight	1 lb.	2 lbs.	1 lb.
Size	5″ × 2″	8″ × 4″	6″ × 1″

3. *Cost*
The RPAD is our most cost-effective option because of quantity cost breaks and maintenance guarantees.

	RPAD	XPO 1690	MX16
Cost per unit	$70.00	$215.00	$100.00
Cost per doz.	$750.00	$2,580.00	$1,100.00
Guarantees	1 year	6 months	1 year

FIGURE 10.3 Continued

Purchase of the RPAD will give us improved performance and comfort. In addition, we can buy 12 RPAD radios for approximately the cost of 4 XPOs. If I can provide you with additional information, please call. I'd be happy to meet with you at your convenience.

Distribution:	M. Ellis	M. Rhinehart	T. Schroeder
	P. Michelson	R. Travers	R. Xidis

The conclusion summarizes the importance of the action. It also ends in a personalized and positive tone to ensure reader involvement and to build rapport.

FIGURE 10.4 Cover Memo Prefacing Attachments

Memo — CompuMed

DATE: November 11, 2009
TO: CompuMed Management
FROM: Bill Baker, Human Resources Director
SUBJECT: Information about Proposed Changes to Employee Benefits Package

As of January 1, 2010, CompuMed will change insurance carriers. This will affect all 5,000 employees' benefits packages. I have attached a proposal, including the following:

1. Reasons for changing from our current carrier. — page 2
2. Criteria for our selection of a new insurance company. — pages 3–4
3. Monthly cost for each employee. — pages 5–6
4. Overall cost to CompuMed. — page 7
5. Benefits derived from the new healthcare plan. — page 8

Please review the proposal, survey your employees' responses to our suggestions, and provide your feedback. We need your input by December 1, 2009. This will give the human resources department time to consider your suggestions and work with insurance companies to meet employee needs.

Enclosure: Proposal

A cover memo directs the reader to the content in an attachment, such as a proposal. You can summarize the most important parts of the proposal, providing page numbers. In addition, your memo can include dated action so that the reader knows what to do next.

The Memo Checklist will give you the opportunity for self-assessment and peer evaluation of your writing.

MEMO CHECKLIST

____ 1. Does the memo contain identification lines (Date, To, From, and Subject)?

____ 2. Does the subject line contain a topic and a focus?

____ 3. Does the introduction clearly state
- Why this memo has been written?
- What topic the memo is discussing?

____ 4. Does the body explain exactly what you want to say?

____ 5. Does the conclusion
- Tell when you plan a follow-up or when you want a response?
- Explain why this dated action is important?

____ 6. Are highlighting techniques used effectively for document design?

____ 7. Is the memo concise?

____ 8. Is the memo clear
- Achieving specificity of detail?
- Answering reporter's questions?

____ 9. Does the memo recognize audience
- Defining acronyms or abbreviations where necessary for various levels of readers (specialists, semi-specialists, and lay)?

____ 10. Does the memo avoid grammatical errors? Errors will hurt your professionalism.

Technology Tip

Using Memos and Letter Templates in Microsoft Word 2007

1. Click on the **"Office Button"** located on the top left of your toolbar and scroll to **"New."**

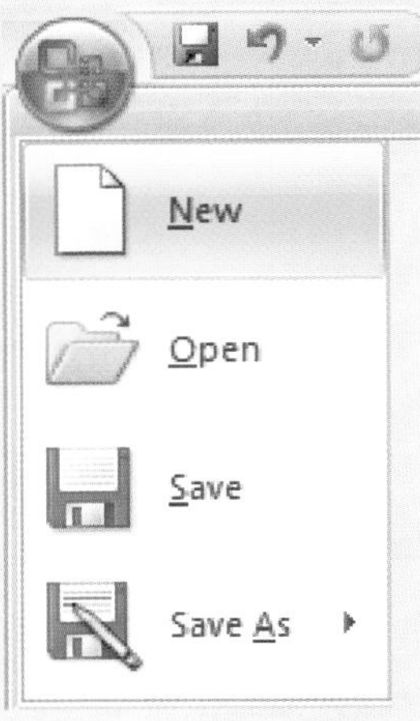

This screen will pop up.

2. Click on the type of document you want to write, such as **"Letters"** or **"Memos."**

   When you choose the communication channel, the window will pop up.

Templates for letters and memos

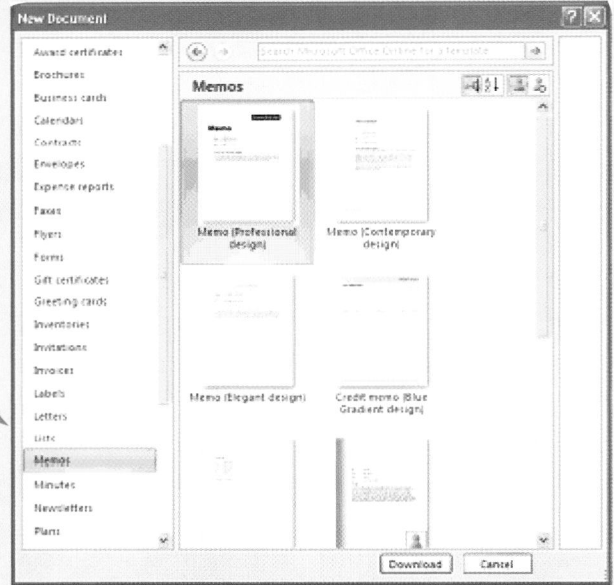

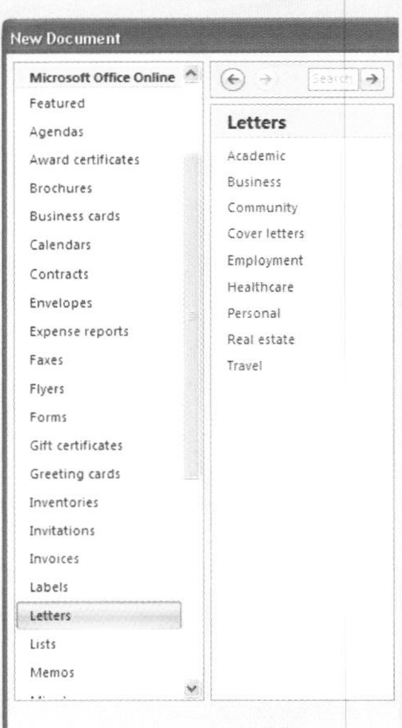

You can choose from one of ten memo templates or one of literally hundreds of letter templates. Each of these templates gives you an already-designed format, complete with spacing, font selection, and layout. In addition, these templates provide fields in which you merely need to type the appropriate information (address, company name, date, salutation, complimentary close, your name and title, etc.).

These templates and wizards are both good and bad. They remind you which components can be included in a memo or letter, they make it easy for you to include these components, and they let you choose ready-made formats.

However, the templates also can create some problems. First, they are somewhat limiting in that they dictate what information you should include and where you should put this information. The content and placement of this information might contradict your teacher's or boss's requirements. Second, the templates are prescriptive, limiting your choice of font sizes and types. Our advice would be to use these templates and wizards with caution.

LETTERS

Reasons for Writing Letters

Letters are external correspondence that you send from your company to a colleague working at another company, to a vendor, to a customer, to a prospective employee, and to stakeholders and stockholders. Letters usually leave your work site (as opposed to memos, which stay within the company).

Because letters are sent to readers in other locations, your letters not only reflect your communication abilities but also are a reflection of your company. This chapter provides letter criteria and examples to help you write the following correspondence: inquiry, cover (transmittal), response, 100 percent yes adjustment, order, confirmation, thank-you, and recommendation letters.

See Chapter 11 for information on employment letters, such as letters of application, follow-up letters, and job acceptance letters.

Essential Components of Letters

Your letter should be typed or printed on 8½″ × 11″ paper. Leave 1″ to 1½″ margins at the top and on both sides. Choose an appropriately business-like font (size and style), such as Times New Roman or Arial (12 point). Though "designer fonts," such as Comic Sans and Shelley Volante, are interesting, they tend to be harder to read and less professional. (See Chapter 8 for more information on font selection and readability.)

Your letter should contain the essential components shown in Figure 10.5.

FAQs

Q: Why write a letter? Haven't letters been replaced by e-mail?

A: Though e-mail is quick, it might not be the best communication channel, for the following reasons:

1. E-mail might be too quick. In the workplace, you will write about topics that require a lot of thought. Because e-mail messages can be written and sent quickly, people too often write hurriedly and neglect to consider the impact of the message.
2. E-mail messages tend to be casual, conversational, and informal. Not all correspondence, however, lends itself to this level of informality. Correspondence related to contracts, for example, requires the more formal communication channel of a letter. The same applies to audience. You might want to write a casual e-mail to a coworker, but if you were writing to the president of a company, the mayor of a city, or a foreign dignitary, a letter would be a better, more formal choice of communication channel.
3. E-mail messages tend to be short. For content requiring more detail, a longer letter would be a better choice.
4. We get so many e-mail messages a day that they are easy to disregard—they are even easy to delete. Letters carry more significance. If you want to ensure that your correspondence is read and perceived as important, you might want to write a letter instead of an e-mail.
5. Letters allow for a "greater paper trail" than e-mail. Most employees' e-mail inboxes fill up quickly. To clean these inboxes up, people tend to delete messages that they don't consider important. In contrast, hard-copy letters are wonderful documentation.

FIGURE 10.5 Essential Letter Components

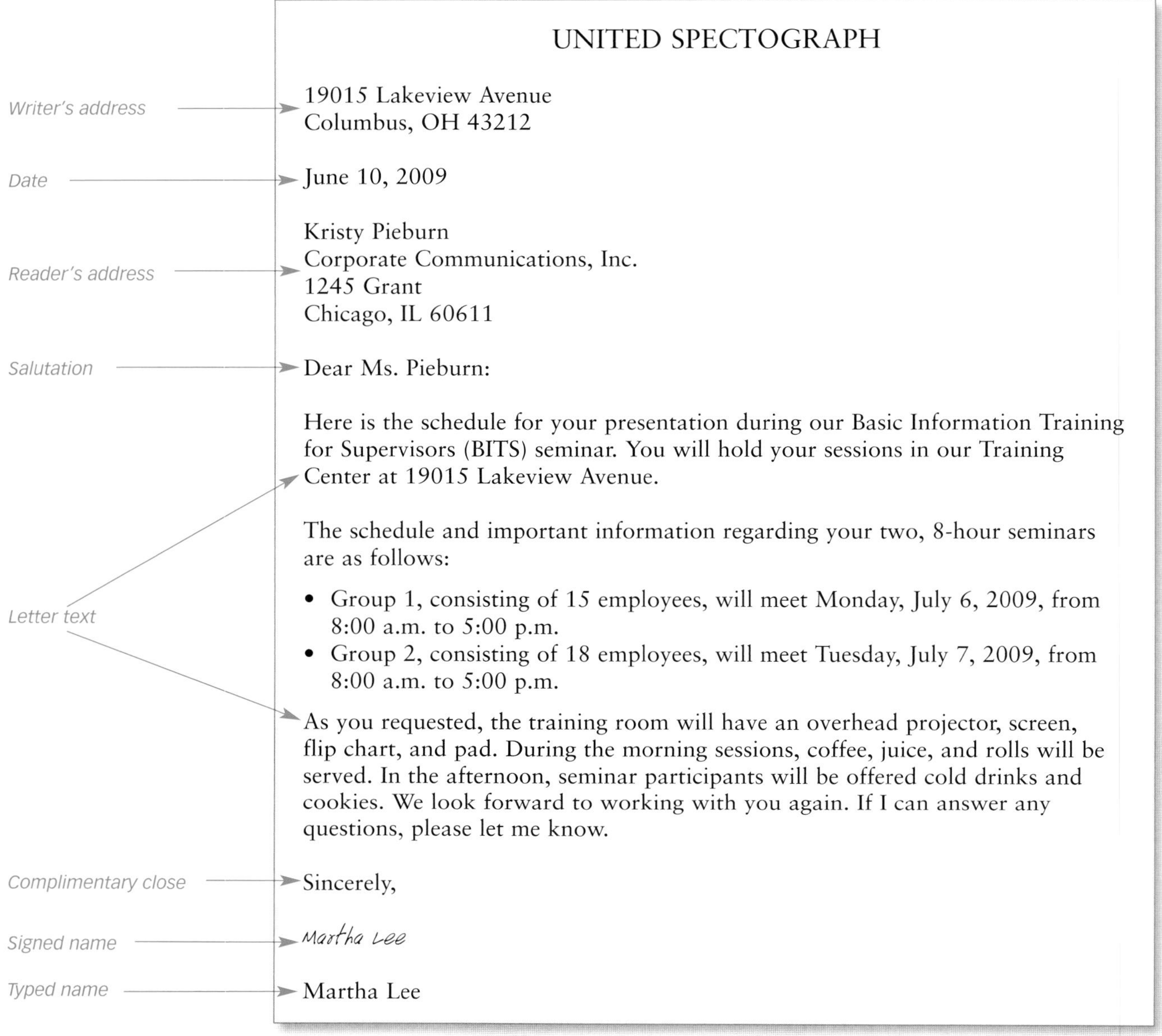

UNITED SPECTOGRAPH

19015 Lakeview Avenue
Columbus, OH 43212

June 10, 2009

Kristy Pieburn
Corporate Communications, Inc.
1245 Grant
Chicago, IL 60611

Dear Ms. Pieburn:

Here is the schedule for your presentation during our Basic Information Training for Supervisors (BITS) seminar. You will hold your sessions in our Training Center at 19015 Lakeview Avenue.

The schedule and important information regarding your two, 8-hour seminars are as follows:

- Group 1, consisting of 15 employees, will meet Monday, July 6, 2009, from 8:00 a.m. to 5:00 p.m.
- Group 2, consisting of 18 employees, will meet Tuesday, July 7, 2009, from 8:00 a.m. to 5:00 p.m.

As you requested, the training room will have an overhead projector, screen, flip chart, and pad. During the morning sessions, coffee, juice, and rolls will be served. In the afternoon, seminar participants will be offered cold drinks and cookies. We look forward to working with you again. If I can answer any questions, please let me know.

Sincerely,

Martha Lee

Martha Lee

Writer's Address

This section contains either your personal address or your company's address. If the heading consists of your address, you will include your street address; the city, state, and zip code. The state may be abbreviated with the appropriate two-letter abbreviation.

If the heading consists of your company's address, you will include the company's name; street address; and city, state, and zip code.

Date

Document the month, day, and year when you write your letter. You can write your date in one of two ways: May 31, 2009, or 31 May 2009. Place the date one or two spaces below the writer's address.

Reader's Address

Place the reader's address two lines below the date.

- Reader's name (If you do not know the name of this person, begin the reader's address with a job title or the name of the department.)
- Reader's title (optional)
- Company name
- Street address
- City, state, and zip code

Salutation

The traditional salutation, placed two spaces beneath the reader's address, is *Dear*, a title, and your reader's last name, followed by a colon (Dear Mr. Smith:).

You can also address your reader by his or her first name if you are on a first-name basis with this person (Dear John:). If you are writing to a woman and are unfamiliar with her marital status, address the letter Dear Ms. Jones. However, if you know the woman's marital status, you can address the letter accordingly (Dear Miss Jones *or* Dear Mrs. Jones).

Letter Body

Begin the body of the letter two spaces below the salutation. The body includes your introductory paragraph, discussion paragraph(s), and concluding paragraph. The body should be single spaced with double spacing between paragraphs. Whether you indent the beginning of paragraphs or leave them flush with the left margin is determined by the letter format you employ.

Complimentary Close

Place the complimentary close, followed by a comma, two spaces below the concluding paragraph. Typical complimentary closes include "Sincerely," "Yours truly," and "Sincerely yours."

Signed Name

Sign your name legibly beneath the complimentary close.

Typed Name

Type your name four spaces below the complimentary close. You can type your title one space beneath your typed name. You also can include your title on the same line as your typed name, with a comma after your name.

Optional Components of Letters

In addition to the letter essentials, you can include the following optional components.

Subject Line

Place a subject line two spaces below the inside address and two spaces above the salutation.

example

Dr. Ron Schaefer
Linguistics Department
Southern Illinois University
Edwardsville, IL 66205

Subject: Linguistics Conference Registration Payment

Dear Dr. Schaefer:

You also could use a subject line instead of a salutation.

example

Linguistics Department
Southern Illinois University
Edwardsville, IL 66205

Subject: Linguistics Conference Registration Payment

A subject line not only helps readers understand the letter's intent but also (if you are uncertain of your reader's name) helps you avoid such awkward salutations as "To Whom It May Concern," "Dear Sirs," and "Ladies and Gentlemen." In the simplified format, both the salutation and the complimentary close are omitted, and a subject line is included.

New-Page Notations

If your letter is longer than one page, cite your name, the page number, and the date on all pages after page 1. Place this notation either flush with the left margin at the top of subsequent pages or across the top of subsequent pages. (You must have at least two lines of text on the next page to justify another page.)

example

Left Margin, Subsequent Page Notation

Mabel Tinjaca
Page 2
May 31, 2009

example

Across Top of Subsequent Pages

Mabel Tinjaca   2   May 31, 2009

Writer's and Typist's Initials

If the letter was typed by someone other than the writer, include both the writer's and the typist's initials two spaces below the typed signature. The writer's initials are capitalized, the typist's initials are typed in lowercase, and the two sets of initials are separated by a colon. If the typist and the writer are the same person, this notation is not necessary.

example

Sincerely,

W. T. Winnery

WTW:mm

Enclosure Notation

If your letter prefaces enclosed information, such as an invoice or report, mention this enclosure in the letter and then type an enclosure notation two spaces below the typed signature (or two spaces below the writer and typist initials). The enclosure notation can be abbreviated "Enc."; written out as "Enclosure"; show the number of enclosures, such as "Enclosures (2)"; or specify what has been enclosed—"Enclosure: January Invoice."

Copy Notation

If you have sent a copy of your letter to other readers, show this in a copy notation. A complimentary copy is designated by a lowercase "cc." List the other readers' names following the copy notation. Type the copy notation two spaces below the typed signature or two spaces below either the writer's and typist's initials or the enclosure notation.

example

> Sincerely,
>
> Brian Altman
>
> Enclosure: August Status Report
>
> cc: Marcia Rittmaster and Larry Rochelle

Formatting Letters

There are three common types of letter formats: Figure 10.6 shows a **full block.** Figure 10.7 shows **a full block with subject line,** and Figure 10.8 shows a **simplified.** Two popular and professional formats used in business are full block and full block with subject line. With both formats, you type all information at the left margin without indenting paragraphs, the date, the complimentary close, or your signature. The full block with subject line differs only with the inclusion of a subject line.

Another option is the simplified format. This type of letter layout is similar to the full block format in that all text is typed margin left. The two significant omissions include no salutation ("Dear ________:") and no complimentary close ("Sincerely,"). Omitting a salutation is useful in the following instances:

- You do not know your reader's name (NOTE: avoid the trite salutation, "To Whom It May Concern:").
- You are writing to someone with a non-gender specific name (Jesse, Terry, Stacy, Chris, etc.) and you do not know whether to use "Mr.," "Mrs.," or "Ms."

The Administrative Management Society (AMS) suggests that if you omit the salutation, you also should omit the complimentary close. Some people feel that omitting the salutation and the complimentary close will make the letter cold and unfriendly. However, the AMS says that if your letter is warm and friendly, these omissions will not be missed. More importantly, if your letter's content is negative, beginning with "Dear" and ending with "Sincerely" will not improve the letter's tone or your reader's attitude toward your comments.

The simplified format includes a subject line to aid the letter's clarity.

Criteria for Different Types of Letters

Though you might write many different types of letters, consider using the all-purpose letter template shown in Figure 10.9.

Letter of Inquiry

If you want information about degree requirements, equipment costs, performance records, turnaround time, employee credentials, or any other matter of interest to you or your company, you write a letter requesting that data. Letters of inquiry require that you be specific. For example, if you write "Please send me any information you have on your computer systems," you are in trouble. You will either receive any information the reader chooses to give you or none at all. Look at the following flawed letter of inquiry from a biochemical waste disposal company.

BEFORE

> Dear Mr. Jernigan:
>
> Please send us information about the following filter pools:
> 1. East Lime Pool
> 2. West Sulphate Pool
> 3. East Aggregate Pool
>
> Thank you.

FIGURE 10.6 **Full Block Format**

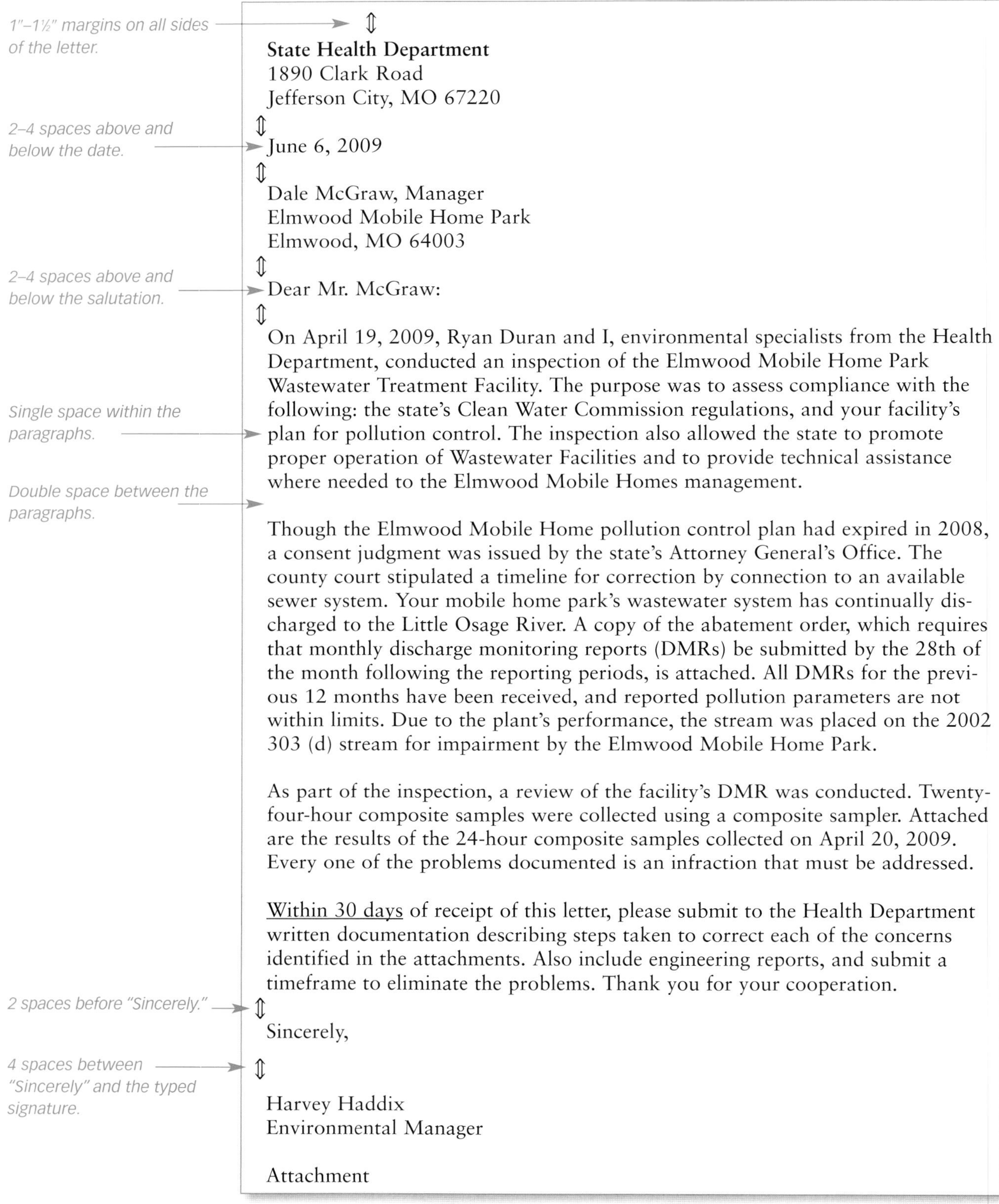

State Health Department
1890 Clark Road
Jefferson City, MO 67220

June 6, 2009

Dale McGraw, Manager
Elmwood Mobile Home Park
Elmwood, MO 64003

Dear Mr. McGraw:

On April 19, 2009, Ryan Duran and I, environmental specialists from the Health Department, conducted an inspection of the Elmwood Mobile Home Park Wastewater Treatment Facility. The purpose was to assess compliance with the following: the state's Clean Water Commission regulations, and your facility's plan for pollution control. The inspection also allowed the state to promote proper operation of Wastewater Facilities and to provide technical assistance where needed to the Elmwood Mobile Homes management.

Though the Elmwood Mobile Home pollution control plan had expired in 2008, a consent judgment was issued by the state's Attorney General's Office. The county court stipulated a timeline for correction by connection to an available sewer system. Your mobile home park's wastewater system has continually discharged to the Little Osage River. A copy of the abatement order, which requires that monthly discharge monitoring reports (DMRs) be submitted by the 28th of the month following the reporting periods, is attached. All DMRs for the previous 12 months have been received, and reported pollution parameters are not within limits. Due to the plant's performance, the stream was placed on the 2002 303 (d) stream for impairment by the Elmwood Mobile Home Park.

As part of the inspection, a review of the facility's DMR was conducted. Twenty-four-hour composite samples were collected using a composite sampler. Attached are the results of the 24-hour composite samples collected on April 20, 2009. Every one of the problems documented is an infraction that must be addressed.

Within 30 days of receipt of this letter, please submit to the Health Department written documentation describing steps taken to correct each of the concerns identified in the attachments. Also include engineering reports, and submit a timeframe to eliminate the problems. Thank you for your cooperation.

Sincerely,

Harvey Haddix
Environmental Manager

Attachment

FIGURE 10.7 **Full Block Format with Subject Line**

State Health Department
1890 Clark Road Jefferson City, MO 67220

June 6, 2009

Dale McGraw, Manager
Elmwood Mobile Home Park
Elmwood, MO 64003

Subject: Pollution Control Inspection

Dear Mr. McGraw:

On April 19, 2006, Ryan Duran and I, environmental specialists from the Health Department, conducted an inspection of the Elmwood Mobile Home Park Wastewater Treatment Facility. The purpose was to assess compliance with the following: the state's Clean Water Law, Clean Water Commission regulations, and your facility's plan for pollution control. The inspection also allowed the state to promote proper operation of Wastewater Facilities and to provide technical assistance where needed to the Elmwood Mobile Homes management.

Though the Elmwood Mobile Home pollution control plan had expired in 2008, a consent judgment was issued by the state's Attorney General's Office. The county court stipulated a timeline for correction by connection to an available sewer system. Your mobile home park's wastewater system has continually discharged to the Little Osage River. A copy of the abatement order, which requires that monthly discharge monitoring reports (DMRs) be submitted by the 28th of the month following the reporting periods, is attached. All DMRs for the previous 12 months have been received, and reported pollution parameters are not within limits. Due to the plant's performance, the stream was placed on the 2002 303 (d) stream for impairment by the Elmwood Mobile Home Park.

As part of the inspection, a review of the facility's DMR was conducted. Twenty-four-hour composite samples were collected using a composite sampler. Attached are the results of the 24-hour composite samples collected on April 20, 2009. Every one of the problems documented is an infraction that must be addressed.

Within 30 days of receipt of this letter, please submit to the Health Department written documentation describing steps taken to correct each of the concerns identified in the attachments. Also include engineering reports, and submit a timeframe to eliminate the problems. Thank you for your cooperation.

Sincerely,

Harvey Haddix
Environmental Manager

Attachment

FIGURE 10.8 **Simplified Format Omitting "Dear . . ." and "Sincerely"**

State Health Department
1890 Clark Road
Jefferson City, MO 67220

June 6, 2009

Dale McGraw, Manager
Elmwood Mobile Home Park
Elmwood, MO 64003

Subject: Pollution Control Inspection

On April 19, 2009, Ryan Duran and I, environmental specialists from the Health Department, conducted an inspection of the Elmwood Mobile Home Park Wastewater Treatment Facility. The purpose was to assess compliance with the following: the state's Clean Water Law, Clean Water Commission regulations, and your facility's plan for pollution control. The inspection also allowed the state to promote proper operation of Wastewater Facilities and to provide technical assistance where needed to the Elmwood Mobile Homes management.

Though the Elmwood Mobile Home pollution control plan had expired in 2008, a consent judgment was issued by the state's Attorney General's Office. The county court stipulated a timeline for correction by connection to an available sewer system. Your mobile home park's wastewater system has continually discharged to the Little Osage River. A copy of the abatement order, which requires that monthly discharge monitoring reports (DMRs) be submitted by the 28th of the month following the reporting periods, is attached. All DMRs for the previous 12 months have been received, and reported pollution parameters are not within limits. Due to the plant's performance, the stream was placed on the 2002 303 (d) stream for impairment by the Elmwood Mobile Home Park.

As part of the inspection, a review of the facility's DMR was conducted. Twenty-four-hour composite samples were collected using a composite sampler. Attached are the results of the 24-hour composite samples collected on April 20, 2009. Every one of the problems documented is an infraction that must be addressed.

Within 30 days of receipt of this letter, please submit to the Health Department written documentation describing steps taken to correct each of the concerns identified in the attachments. Also include engineering reports, and submit a timeframe to eliminate the problems. Thank you for your cooperation.

Harvey Haddix
Environmental Manager

Attachment

FIGURE 10.9 All-Purpose Letter Template

Writer's Address

Date

Reader's Address

Salutation:

A lead-in or overview stating *why* you are writing and *what* you are writing about.

Detailed development, made accessible through highlighting techniques, explaining *exactly what* you want to say.
-
-
-

State *what* is next, *when* this will occur, and *why* the date is important.

Complimentary close,

Signed Name

Typed Name

The reader replied as follows:

BEFORE

Dear Mr. Kimmel:

I would be happy to provide you with any information you would like. However, you need to tell me what information you require about the pools.

I look forward to your response.

The first writer, recognizing the error, rewrote the letter as follows:

AFTER

Dear Mr. Jernigan:

My company, Jackson County Hazardous Waste Disposal, Inc., needs to purchase new waste receptacles. One of our clients used your products in the past and recommended you. Please send us information about the following:

1. Lime Pool—costs, warranties, time of installation, and dimensions
2. Sulphate Pool—costs, material, and levels of acidity
3. Aggregate Pool—costs, flammability, maintenance, and discoloration

We plan to install our pools by March 12. We would appreciate your response by February 20. Thank you.

Providing specific details makes your letter of inquiry effective. You will save your reader's time by quantifying your request.

To compose your letter of inquiry, include the following items.

Introduction Clarify your intent in the introduction. Until you tell your readers why you are writing, they do not know. It is your responsibility to clarify your intent and explain your rationale for writing. Also tell your reader immediately what you are writing about (the subject matter of your inquiry). You can state your intent and subject matter in one to three sentences.

Discussion Specify your needs in the discussion. To ensure that you get the response you want, ask precise questions or list specific topics of inquiry. You must quantify. For example, rather than vaguely asking about machinery specifications, you should ask more precisely about "specifications for the 12R403B Copier." Rather than asking, "Will the roofing material cover a large surface?" you need to quantify—"Will the roofing material cover 150′× 180′?"

Conclusion Conclude precisely. First, explain when you need a response. Do not write, "Please respond as soon as possible." Provide dated action and tell the reader exactly when you need your answers. Second, to sell your readers on the importance of this date, explain why you need answers by the date given.

Figure 10.10 is a sample letter of inquiry.

Cover (Transmittal) Letter

A *cover* or *transmittal letter* precedes attached or enclosed documents, informing the reader by giving an overview of the material that follows.

In business, you are often required to send information to a client, vendor, or colleague. You might send attachments or enclosures prefaced by a cover letter. These attachments may include

Reports	Invoices	Drawings
Maps	Contracts	Specifications
Instructions	Questionnaires	Proposals

A cover letter accomplishes two goals: It tells readers up front what they are receiving and focuses the readers' attention on key points within the enclosures.

Introduction What if the reader has asked you to send the documentation? Do you still need to explain why you are writing? The answer is yes. Although the reader requested the information, time has passed, other correspondence has been written, and your reader might have forgotten the initial request. Introductory sentences provide the reader information about why you are writing and what you are sending.

Discussion In the body of the letter, you want to accomplish two things. You either want to tell your reader exactly what you have enclosed or exactly what of value is within the enclosures. In both instances, you should provide an itemized list or easily accessible, short paragraphs. Page numbers are a friendly gesture toward your audience. You are helping the reader locate the important information, and you are achieving audience recognition and involvement.

However, including page numbers has a greater benefit than audience involvement. These page numbers also allow you to focus your reader's attention on what you want to emphasize. In other words, if you provide the enclosure without a cover letter, you leave it up to readers to sift through the information and decide what is important. In contrast, by providing an itemized list with page numbers, you direct the reader's attention. You can emphasize, for reader benefit, the points in the enclosure which you consider to be important.

FIGURE 10.10 Letter of Inquiry

CompuMed

8713 Hillview Reno, NV 32901 1-800-551-9000 Fax: 1-816-555-0000

September 12, 2009

Sales Manager
OfficeToGo
7622 Raintree
St. Louis, MO 66772

Subject: Request for Product Pricing and Shipping Schedules

My medical technology company has worked well with OfficeToGo for the past five years. However, in August I received a letter informing me that OTG had been purchased by a larger corporation. I need to determine if OTG remains competitive with other major office equipment suppliers in the Reno area.

In the introduction, briefly explaining why you are writing establishes the context of the inquiry for the audience.

Please provide the following information:

1. What discounts will be offered for bulk purchases?
2. Which freight company will OTG now be using?
3. Who will pay to insure the items ordered?
4. What is the turnaround time from order placement to delivery?
5. Will OTG be able to deliver to all my satellite sites?
6. Will OTG technicians set up the equipment delivered, including desks, file cabinets, bookshelves, and chairs?
7. Will OTG be able to personalize office stationery onsite, or will it have to be outsourced?

A detailed and itemized list informs the audience of the exact questions you need answered. If you write precise questions, the audience will be able to provide the information you are requesting.

Please respond to these questions by September 30 so I can prepare my quarterly orders in a timely manner. I continue to expand my company and want assurances that you can fill my growing office supply needs. You can contact me at the phone number provided above or by e-mail (jgood@CompuMed.com). Thank you for your help.

Pleasantly state when and why you need a response by a specific date to encourage your audience. In the conclusion, you can also include detailed contact information not provided in the letterhead.

Jim Goodwin

Owner and CEO

Conclusion Your conclusion should tell your readers what you want to happen next, when you want this to happen, and why the date is important.

Figure 10.11 is an example of a successful cover letter.

Response Letter

In a *response letter*, you provide information, details, or answers to a request. For example, you or your company might need to answer questions about quotes on equipment

FIGURE 10.11 **Cover Letter**

Positive tone and introduction build rapport and inform the reader why this letter is being written: in response to a request.

A lead-in informing the reader that enclosures follow.

An itemized body clarifying what is in the enclosure and where the information can be found.

A conclusion providing dated action, noting when a follow-up call will be made.

Positive word usage and pronouns to build rapport: "at your convenience," "Thank you," and "exceptional."

AMERICAN
HEALTHCARE
1401 Laurel Drive
San Mateo, CA 91404
November 11, 2009

Jan Pascal
Director of Outpatient Care
St. Michael's Hospital
Westlake Village, CA 91362

Dear Ms. Pascal:

Thank you for your recent request for information about our specialized outpatient care equipment. American Healthcare's stair lifts, bath lifts, and vertical wheelchair lifts can help your patients. To show how we can serve you, we have enclosed a brochure including the following information:

- Maintenance, warranty, and guarantee information 1–3
- Technical specifications for our products, including sizes, weight limitations, colors, and installation instructions 4–6
- Visuals and price lists for our products 7–8
- An American Healthcare order form 9
- Our 24-hour hotline for immediate service 10

Early next month, I will call to make an appointment at your convenience. Then we can discuss any questions you might have. Thank you for considering American Healthcare, a company that has provided exceptional outpatient care for over 30 years.

Sincerely,

Toby Sommers

Enclosure

costs, maintenance fees, delivery options, and technical specifications on makes and models from various vendors. You might need to write a letter of response with information about room availability at your hotel, food arrangements, technology and presentation equipment rental fees, presentation room setups, entertainment options, and transportation from the airport. If a client needs to change its insurance carrier, your insurance company responds to their inquiry with quotes for insurance premium costs, levels of coverage, deductibles, and accessibility of claims adjusters.

Introduction Begin with a pleasant reminder of when you spoke with a person or heard from the audience via e-mail or through a letter of inquiry. This explains *why* you are writing the letter. Then, specifically state *what* topic you are writing about.

Discussion Organize your discussion section into as many paragraphs as you need. When possible, remember to use bulleted or numbered lists for easier access. Include in this section the details or explanations needed to answer the inquiry. Consider including any of the following:

Times	Types of activities	Enrollment periods
Dates	Discounts	Enclosures
Amounts of people	Costs	Technology or equipment

Conclusion End your letter of response in an upbeat and friendly tone. You can also include your contact information (e-mail, phone number, address), if not printed on the letterhead. See Figure 10.12 for a sample response letter.

100 Percent Yes Adjustment Letter

A *100 percent yes adjustment letter* agrees with all aspects of a customer's complaint and provides that customer 100 percent satisfaction.

Maybe you purchased a computer desk at a local office supply company. When the product was delivered to your office, you noticed that the shipping box was torn and dented. Upon opening the box, you discovered that the desk was scratched. After complaining, you received a 100 percent yes adjustment letter from the delivery company upholding your claim. They promise that the damaged product will be picked up at their expense and that a new product will be delivered.

Additional information:

See Chapter 12, "Communicating Bad News in the Workplace."

Introduction Since you have good news for the reader, there is no reason to delay the message. In the introductory paragraph, state that you agree with your reader's complaint and will honor the request for adjustment.

Discussion In the discussion paragraph(s), explain *what* happened, *why* the problem occurred, and *how* the problem will be avoided in the future. You want to reassure the customer that the problem will not reoccur.

Conclusion In your concluding paragraph, resell to maintain customer satisfaction and to achieve reader rapport. End the letter upbeat. Should you apologize in the letter? If an apology infers a legal obligation on the part of your company, then be careful. For example, if an injury has occurred due to faulty equipment, your apology might be actionable in court. Check with your company's legal staff or your boss to determine culpability.

See Figure 10.13 for a sample 100 percent yes adjustment letter.

Order Letter

In business, you will need to place orders. Though these often can be placed by fax, telephone, or online, you might order by letter. An *order letter* provides you documentation for your order. When you write your letter, be concise and clear to assure the correctness of your order.

Introduction The introduction of an order letter contains the following:

- **Reason for placing the order**—To meet holiday demands, upgrade office equipment, or maintain inventories, for example.
- **Authorization for placing the order**—Your position as purchasing manager, office manager, or supervisor gives you the authority to place orders for your corporation. However, your boss might have asked you to place an order.
- **Method of delivery**—FedEx, UPS, overnight, mail, express delivery, and so on.
- **Source of item information**—This may be a catalog or sales brochure.

FIGURE 10.12 Response Letter

Beginning the letter with a pleasant, personalized tone builds rapport. This introduction also reminds the reader why the letter has been written.

Anticipate questions the reader might have by including as much detail as possible. Using a list rather than sentences or paragraphs helps highlight key points.

Ending the letter with contact information (phone number and e-mail address) encourages the reader to respond.

City Training Institute
2187 Broadway
San Antonio, TX 78213

May 12, 2009

Gary Morton
City Manager
Round Rock, TX 77463

Dear Gary:

I enjoyed speaking to you this morning, Gary. As promised, I am sending you materials regarding the Regional Council's City Training Institute.

The City Training Institute (CTI) meets the training and organizational development needs of public organizations. The CTI strives to be the training provider of choice for local governments in the greater Hill Country area. Through open enrollment seminars and customized training, the CTI provides a comprehensive array of professional development opportunities.

Though we offer over 100 training workshops, following is a sample list of options:

- Customer Service
- Serving a Diverse Population
- Resolving Conflicts
- Effective Supervisory Skills
- Proactive Listening
- Giving and Receiving Constructive Feedback
- Improving Your Business Writing
- Navigating Change
- Conducting Performance Reviews
- Dynamic Communication Skills
- Managing Time

Gary, thank you for your interest in the CTI. Please let me know how we can help provide you either on-site, custom-designed training or further information about any of our workshops. You can reach me at 1-800-555-1212 or e-mail me at jstaples@cti.org

Sincerely,

John Staples, Training Manager

Discussion In the order letter's body, provide

- *A sentence lead-in.*
- *An itemized listing of the order.*
- *Precise details*, including costs, sizes, shapes, colors, materials, descriptions, or titles. Though serial numbers are useful, not all catalogs (particularly those

FIGURE 10.13 **100 Percent Yes Adjustment Letter**

1101 21st Street
Galveston, TX 77001
(712) 451-1010
May 31, 2009

Mr. Carlos De La Torre
1234 18th Street
Galveston, TX 77001

Dear Mr. De La Torre:

Thank you for your recent letter. I am pleased to inform you that Gulfstream will happily replace your defective shock absorber according to the warranty agreement.

The Trailhandler Performance XT shock absorbers that you purchased were discontinued in October 2009. Mr. Blanton, the mechanic to whom you spoke, incorrectly assumed that Gulfstream was no longer honoring the warranty on that product. Because we no longer carry that product, we either will replace it with a comparable model or refund the purchase price. Ask for Mrs. Cottrell at the automotive desk on your next visit to our store. She is expecting you and will handle the exchange.

We appreciate your business, Mr. De La Torre. I'm glad you brought this problem to my attention. If I can help you in the future, please contact me.

Sincerely,

Holbert Lang,
Sales Manager

cc: Julie Cottrell, Supervisor
Jim Gaspar, CEO

Positive word usage ("Thank you") achieves audience rapport.

The introduction uses positive words ("pleased" and "happily") and immediately states the good news.

The discussion explains what created the problem and provides an instruction telling the customer what to do next.

The conclusion ("we appreciate your business") resells to maintain customer satisfaction.

online) provide this information. The key is being as specific as you can to ensure receipt of the correct merchandise. By writing precisely, you will avoid phone calls or e-mail messages asking for clarity.

Conclusion You need to include the following in your conclusion:

- *Date needed by.*
- *Method of payment.*
- *Contact information*, such as telephone number, e-mail address, or fax number.
- *Positive close*, such as "Thank you for your help." See Figure 10.14 for a sample.

FIGURE 10.14 Order Letter

In the body, include as many details as possible to ensure that the order is correctly filled. You should include exact dollar amounts and amounts of units being ordered. Also provide the serial or catalog number of the item.

Prime Publishing, Inc.
9516 W. 148th St.
Phoenix, AZ 45612
September 5, 2009

A&L Sound Systems
11517 Grant
Tucson, AZ 45510

Subject: Order of Sound and Video Equipment

Our production company is upgrading the multimedia capabilities of employees' desktop PCs. As director of purchasing, I learned of your products from your June 2009 sales catalog. Please ship my order either by FedEx or UPS.

Please send the following items in the amounts indicated:

Quantity	Catalog Number	Description	Cost	Total
10	Wireless	Headphone	$23	$230
10	APA2P1G8	Subwoofer	6	60
10	EDE497HX	Speaker Dolby	7	70
10	RADEON9800Pro	Video Card	223	2,230
10	P4ROPro	Video Capture	136	1,360
10	ATA47294	Digital Video Creator	57	570
10	Wireless	Sound Cards	58	580
10	Live Platinum 5.1	Sound Blaster	9	90
10	MIDI	Music Hardware	47	470
10	249	Keyboard	47	470
10	X7295Y	Microphone	2	20
Total				**$6,150** 375 (tax) 250 (shipping) **$6,775**

I need delivery within 14 days. Enclosed is a check for $6,775. If you have questions, you can reach me at 913-555-2121 or e-mail me at tmix@unicorn records.com. Thank you for helping with this order.

Natalia Johnson
Information Technology

Confirmation Letter

In business, letters represent an official contract. Often, when clients and vendors make arrangements for the purchase of services or products, they must write a *letter of confirmation*. This confirmation letter verifies the details of the agreement.

For example, as head of corporate training, suppose you met with a consulting firm to discuss services they could provide your company. After the meeting, you might write a letter to summarize the outcome of the discussion, confirming payment, dates, and training content.

Maybe you are an accountant. A client has contracted with you to prepare documents for a complex divorce case. This will include reports of joint property, taxes paid, and cash and investments on hand. When you return to the office, you will write a letter confirming your responsibilities, fees, and turnaround time.

Introduction In your confirmation letter's introduction, state the context of the letter so that the reader can relate to the discussion.

Discussion The letter's body clarifies the details of the agreement. Since this constitutes a legally binding document, you must specify anything agreed upon. Using highlighting techniques to make your content accessible, consider including any of the following items.

Dates	Makes/models/serial numbers	Locations	Retainer fees
Times	Personnel and certification requirements	Audiovisual equipment	Length of agreement
Costs	Menus and decorations	Parking	Room setup

Conclusion The conclusion of a confirmation letter tells the reader what to do next. You might include a request for signature, payment due dates, or method of payment.

See Figure 10.15 for a sample confirmation letter.

Recommendation Letter

You might write a letter of recommendation for many reasons:

- An employee deserves a promotion.
- An ex-employee asks for a reference for a new job.
- A colleague is nominated for an award.
- An acquaintance is applying for an education scholarship.
- A governmental agency is checking references.
- A consultant requests a reference for a new client.

When someone asks you to write a letter of recommendation, consider the *Do's* and *Don'ts* in Table 10.2.

Introduction In the introduction, include any of the following:

- Your position (or title)
- Your relationship with the person
- The length of your relationship
- The applicant's name
- The position, scholarship, or award

FIGURE 10.15 Confirmation Letter

The introduction explains why the letter is being written ("in response to your request"), specifies the type of letter (confirmation), and clarifies the topic ("training services" and "fees contracted").

The letter's discussion provides specific details regarding the agreement: costs, locations, and contact.

The conclusion ends positively and instructs the reader, stating what must be done next to confirm the agreement.

Metro Consulting

600 Broadway Albuquerque, NM 23006
510-234-1818 www.metrocon.com

January 23, 2009

Mr. Carl Meyers
ProfCom
1999 Saguaro Dr.
Santa Fe, NM 23012

Dear Mr. Meyers:

In response to your request, this letter confirms our discussion from last week. Below I summarize the agreement we reached regarding the training services your company will provide and the fees contracted.

ProfCom is scheduled to offer the following workshops in 2009:

- Fifteen days of "Customer Service" training to Albuquerque municipal employees. The workshops will be held throughout the city at locations to be determined later. Each location will provide audiovisual equipment, per your specifications. ProfCom will be paid $500 for each half-day workshop and $700 for each full-day workshop.
- Thirty days of "Managing Diversity" training to Albuquerque municipal employees. These workshops will be held in the Albuquerque Civic Center, 1800 Mountainview Dr. ProfCom will be paid $900 a day for each of these workshops. Please contact Mr. Silvio Hernandez, 1-800-ALCIVIC, to request your audiovisual needs.
- Ten days of "Supervisor/Management" training, leading to a "Supervisor/Management Certificate." These workshops will be held in the Albuquerque City Hall, Conference Room A. ProfCom will be paid $1,000 a day for each workshop. Please contact Mary O'Sullivan, 1-510-222-5150, to request your audiovisual needs.
- Undecided additional training workshops, including "Ethical Decision Making," "Accounting for Non-Accountants," and "Dynamic Presentation Skills." We will work with you to schedule these as enrollment figures are calculated. MetroConsulting will pay ProfCom a monthly retainer fee of $3,000 to ensure your availability for these workshops. The $3,000 retainer will be adjusted against complete payment for services rendered.

Thank you for agreeing to provide us these services, Mr. Meyers. MetroConsulting is excited about the prospect of working with your firm. We hope to continue offering these workshops annually if both parties agree upon the success of the training. By signing and dating this letter of confirmation below, you indicate your agreement with the assigned services and compensation.

Rob Harken, Director
MetroConsulting

_______________, Date: _______ _______________, Date: _______

Table 10.2 Dos and Don'ts of Recommendation Letters

Dos	Don'ts
• Agree to write the letter only if you can be supportive • Request a current resume • Obtain information about the position, scholarship, or award the person is applying for • Write the letter with that specific position in mind • Study any information about the person to avoid omissions • Keep your letter to a reasonable length of one to two pages	• Avoid writing if you feel the candidate is weak • Avoid writing if you cannot be positive • Avoid writing letters longer than one page • Avoid writing these letters if you have only vague memories of the person's work • Avoid writing a reference letter if you cannot also talk with confidence about the person to the interviewer

Discussion In the body, include the specific details of the applicant's skills. Consider including

- Examples of the applicant's job performance
- Illustrations proving how and why the person will benefit the company
- Evaluations of the person's chances of success in the company or program
- Differences from other people
- Examples of projects the applicant worked on
- Projects supervised
- Team skills
- Communication abilities
- Names of classes attended or certifications acquired to enhance the applicant's skills
- Honors earned at work, school, or military

Conclusion In the conclusion, sum up why this person is deserving of consideration for the job, award, or scholarship. Be sure to include contact information, such as your telephone number or e-mail address. This will help the recipient of the recommendation letter reach you for a follow-up discussion.

See Figure 10.16 for a sample recommendation letter.

Thank-You Letter

When an employee, customer, vendor, supervisor, coworker, or any business professional does you a favor, you should write a follow-up thank-you letter. Doing so is not only courteous, but also it is good business. By writing a thank-you letter, you show your appreciation and build continued rapport. Create a pleasant tone and a positive attitude in your thank-you letter. Combined with a clear writing style and a well-designed letter, you can show your gratitude. In the introduction, remind the reader why you are writing. You can include the date of the reader's letter or kind words. You also should mention the topic of the reader's comments. In the letter's body, explain how the reader's actions or words helped you. In the conclusion, thank the reader for his or her time and kindness.

See Figures 10.17, 10.18, and 10.19 for the following series of letters: inquiry, response, and thank you.

FIGURE 10.16 **Recommendation Letter**

Subject line to clarify the letter's intent.

Introduction giving the applicant's name, how the writer knows her, and the topic for this letter.

Discussion specifying ways in which the applicant excels. These include grades, team projects, interpersonal communication abilities, and personality traits.

Conclusion sums up the candidate's assets, focusing on ways in which this person meets the needs of a specific program. The conclusion also provides contact information for follow up.

The complimentary close omits "Sincerely." This letter follows the "Simplified Letter Format" with a subject line instead of a salutation.

Note how the recommendation letter uses positive words throughout to highlight the applicant's abilities. Some of these words include "succeed," "happy," "deserving," "pleased," "impressed," and "excel."

Midwest Technological College

15431 College Blvd. Milwaukee, WI 32556 451-987-0101

March 15, 2009

Dr. Anne Cohen
University of Wisconsin Medical Center
1900 E. 39th Street
Madison, WI 35567

Subject: Letter of Recommendation for the University of Wisconsin Medical Center, Department of Health Information Management

I am pleased to write this letter recommending Pekkahm Shoumavong for admission to the WU Department of Health Information Management. Pekkahm was a student in my Business Communication class last semester (Fall 2008). She impressed me not only with her scholarship, but also with her team skills and her conscientious desire to excel.

Pekkahm earned an A in my class due to her excellent written and oral communication skills (letters, memos, reports, e-mail, instructions, and speeches accompanied by PowerPoint presentations). More importantly, I was impressed with her ability to work well with others and to take on leadership roles. My Business Communication class emphasizes collaboration. Pekkahm's team was one of my more successful groups due to their ability to work in concert toward a team goal and to communicate professionally with each other—valuable skills in any environment. This was an especially impressive achievement since Pekkahm's team was multicultural and comprised of students with vastly different abilities.

In addition, I remember that her team encountered a problem with lost computer files. Rather than become distressed, Pekkahm remained calm and worked through the problem professionally. Her even-keeled and congenial personality, interpersonal communication talents, and ability to get along with others while demanding high-quality work helped her team succeed. Finally, Pekkahm attended all classes and turned in all assignments on time, evincing her discipline, conscientiousness, and dedication to excellence.

Self-motivation, professionalism, problem-solving skills, and outstanding communication abilities suggest a high level of potential. I believe these traits distinguish Pekkahm and indicate her ability to succeed in the WU Department of Health Information Management. Pekkahm is deserving of your consideration for admission to your program. Thus, I am happy to be able to recommend her and serve as a reference. If you would like to discuss her attributes further, please call me at 451-987-0101, ext. 59, or e-mail me at cescobar@mtc.edu.

Dr. Carmen Escobar
Communication Professor

FIGURE 10.17 **Letter of Inquiry**

1408 N. Hawker
Independence, MO 64050
March 24, 2009

R4 Technologies
1579 W. Pacific Highway
San Diego, CA 92447

Subject: Request for Game Designer Qualifications

I am interested in becoming a video game designer and would like to know the requirements for the field.

Please send me the following information:

1. What type of education do I need?
2. Which universities offer this degree program?
3. How long does it take to acquire sufficient education or training for the field?
4. What is the starting pay for a new game designer?
5. What are the requirements for getting hired in your firm?

Your response by April 5, 2009, will help me plan for enrollment at a university. Thank you for your time.

Daniel Black

FIGURE 10.18 **Response to the Letter of Inquiry**

R4 Technologies
1579 W. Pacific Highway
San Diego, CA 92447
April 2, 2009

Daniel Black
1408 N. Hawker
Independence, MO 64050

Dear Daniel:

I recently received your request for becoming a game designer. Unfortunately, I am unable to help you because R4 is a publishing company, and we outsource all design work.

However, I pulled some articles that I think you will find useful. They are enclosed. Also, several sources on the Internet can help you find what you are looking for (also enclosed).

Best wishes,

Susan Cardez
Director of Human Resources
Enc.

FIGURE 10.19 Thank-You Letter

1408 N. Hawker
Independence, MO 64050
April 10, 2009

Susan Cardez
R4 Technologies
1579 W. Pacific Highway
San Diego, CA 92447

Dear Ms. Cardez:

The articles you sent on April 2, "How Do I Become a Game Designer?" from GameDesignX, and "I Really Like Games—How Do I Get a Job as a Game Designer?" from Obscure Productions, are informative.

The information is very helpful for the following reasons:
1. Both explain types of designers.
2. The GameDesignX article tells about education needed.
3. The Obscure Productions' article has an excellent list of references.

Thank you for your time and for sending me the articles. Your kindness in responding has helped provide me focus in my career search.

Sincerely,

Daniel Black

The Letters Checklist will give you the opportunity for self-assessment and peer evaluation of your writing.

LETTERS CHECKLIST

____ 1. **Letter essentials.** Does your letter include the eight essential components (writer's address, date, recipient's address, salutation, text, complimentary close, writer's signed name, and writer's typed name)?

____ 2. **Introduction.** Does the introduction state *what* you are writing about and *why* you are writing?

____ 3. **Discussion.** Does your discussion clearly state the details of your topic depending on the type of letter?

____ 4. **Highlighting/page layout.** Is your text accessible? To achieve reader-friendly ease of access, use headings, boldface, italics, bullets, numbers, underlining, or graphics (tables and figures). These add interest and help your readers navigate your letter.

____ 5. **Organization.** Have you helped your readers follow your train of thought by using appropriate modes of organization? These include chronology, importance, problem/solution, or comparison/contrast.

____ 6. **Conclusion.** Does your conclusion give directive action (tell what you want the reader to do next and when) and end positively?

____ 7. **Clarity.** Is your letter clear, answering reporter's questions and providing specific details that inform, instruct, or persuade?

____ 8. **Conciseness.** Have you limited the length of your words, sentences, and paragraphs?

____ 9. **Audience recognition.** Have you written appropriately to your audience? This includes avoiding biased language, considering the multicultural/cross-cultural nature of your readers, and your audience's role (supervisors, subordinates, coworkers, customers, or vendors). Have you created a positive tone to build rapport?

____ 10. **Correctness.** Is your text grammatically correct? Errors will hurt your professionalism. See the Appendix for grammar rules and exercises.

Kim Met His Challenge

To meet his communication challenge, Kim used the P^3 process.

Planning

To plan his communication, Kim considered the following:

- Goal—document financial information and recommend actions to be taken by clients
- Audience—specialists (financial planners) and lay readers (clients)
- Channels—memo to be reviewed by both the client and the company
- Data—specific client financial information, including assets, liabilities, taxes, and net worth

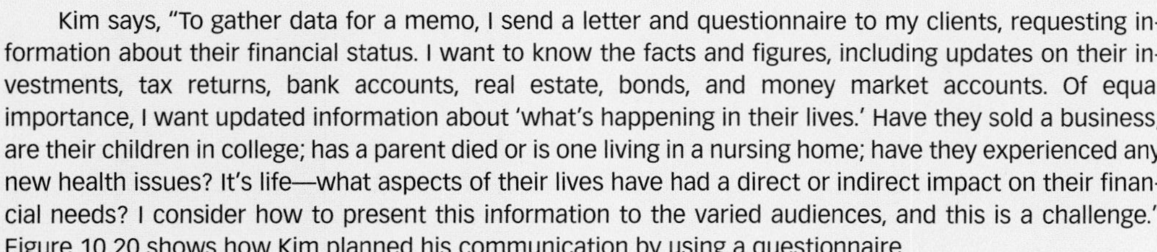

Kim says, "To gather data for a memo, I send a letter and questionnaire to my clients, requesting information about their financial status. I want to know the facts and figures, including updates on their investments, tax returns, bank accounts, real estate, bonds, and money market accounts. Of equal importance, I want updated information about 'what's happening in their lives.' Have they sold a business; are their children in college; has a parent died or is one living in a nursing home; have they experienced any new health issues? It's life—what aspects of their lives have had a direct or indirect impact on their financial needs? I consider how to present this information to the varied audiences, and this is a challenge." Figure 10.20 shows how Kim planned his communication by using a questionnaire.

Packaging

According to Kim, "In my financial planning company, the financial planner gathers data about the client and then packages the information by writing a memo draft that serves two purposes:

1. The memo draft, given to the principal, is used to update the principal on his or her knowledge of the client's financial situation, needs, and suggested actions to be taken. The memo draft ensures that the principal will be better informed prior to meeting with the client.
2. While meeting with the client, the principal will add notes to the memo draft. Later, based on these notes, the financial planner will edit and revise the text for the final memo which will be placed into a pair of three-ring binders (one for the client's files and one for the company's files)." Figure 10.21 shows how Kim packaged his memo rough draft.

Perfecting

Kim met his communication challenge by perfecting the memo draft. He considered his boss's comments by reformatting the text, personalizing the tone, and making the memo more appropriate for his lay audience. Most importantly, Kim corrected the mathematical error that his boss pointed out. As Kim says, "If what you communicate is unclear or, worse, inaccurate, this will confuse the client. They might make the wrong decision which could lead to financial penalties. Planning, packaging, and perfecting my documents ensure that I will create the best possible written communication." See Figure 10.22 for Kim's perfected memo.

FIGURE 10.20 Planning Questionnaire Kim Sends to Clients

To help us prepare for your upcoming meeting, please provide as much of the information and documentation listed below.

1. The 4-30-08 balances for the following accounts:
   - Joint Bank of America Checking Account
   - Mary's CitiBank Checking Account
   - John's business Checking Account
2. Copies of your most recent paycheck stubs.
3. Your 4-30-08 home equity line of credit balance, if any.
4. Any information from State U. regarding Annette's financial aid application.
5. Has anything important happened to you since our last meeting that we should know about that may affect your present or future financial situation?

FIGURE 10.21 Memo Rough Draft with Comments from the Firm's Principal

Kim, be sure to include the memo identification lines (date, to, from, and subject).

Memo Draft
Clients: SMITH, JOHN AND MARY

We met with the Smiths for their quarterly meeting. We reviewed the preliminary December 31, 2008, Balance Sheet, which reflects the following data:

Total Assets	$2,000,000
Liabilities	− 200,000
Deferred Taxes	− 100,000
After-Tax Net Worth	$1,700,000

Note the mathematical error: $1,700,000 here vs. $1,750,000 in the paragraph below.

After John provides us with the year-end K-1s from his business interests, we will finalize this Balance Sheet and update the Smiths' Net Worth Analysis. We also reviewed the December 31, 2008, Balance Sheet, which reflects an increased after-tax net worth of $1,750,000. We reviewed the performance of the Smiths' composite investment portfolio, as well as each of its individual holdings. The time-weighted return earned by the composite portfolio during 2008 was +12.00 percent. Nearly every position in the composite portfolio experienced positive returns in 2008.

To help our clients understand the text, let's add headings and subheadings where possible.

At the time of our meeting, there was a balance of $10,000 in the money market fund of Mary's living trust account. This account supports the Smiths' $3,000 automatic monthly distributions. We will now raise more cash by depositing the Smiths' 2007 IRA Required Minimum Distributions into Mary's trust account. We reviewed our final income tax projection for 2008, which results in a federal income tax liability of $14,000 and a California tax liability of $8,000. We project that the Smiths will owe about $800 to the IRS and $2,000 to California when they file their tax returns.

Let's consider reformatting the text. Maybe a table above would summarize the information. We can add this boilerplate content to an appendix.

We took our first look at our income tax projection for 2009. We estimate that the Smiths will have a federal tax liability of about $15,000 and a California tax liability of about $2,000. Since the projected federal tax liability is higher than our estimate of the Smiths' 2008 tax liability, we recommend that they follow a tax payment strategy that targets 100 percent of their actual 2008 tax liability. We estimate that John will have federal tax withholding of $3,600 from his pension. Therefore, we estimate that the Smiths will need to make quarterly estimated tax payments of $2,600 federally and $1,800 to California. We will determine the actual quarterly payment amounts after we receive copies of the Smiths' 2008 income tax returns.

The Smiths provided us with their year-end cash balances before our meeting, so we were able to compare them to their cash balances at the beginning of the year. This comparison indicated that the Smiths spent about $27,000 less on their living expenses than we estimated. Therefore, we will reduce our estimate of their living expenses to $88,000. After this reduction is incorporated, our initial cash flow projection for 2009 indicates that the Smiths will have a total cash flow shortfall of about $70,000 after accounting for John's pension and both of their Social Security benefits. The Smiths have already withdrawn $10,000 for their home repairs. They will also take a total of $35,000 of automatic monthly distributions over the course of the year. Therefore, we estimate that they will need to take an additional $25,000 from their cash and investment accounts over

FIGURE 10.21 **(Continued)**

the rest of the year to meet their cash flow needs. After we received the transfer of John's IRA from his former employer, we reviewed the research on each of the stocks in the portfolio. We retained the positions in the Acme, Business Growth, and C-Company stocks because they have solid business fundamentals, strong earnings growth, and projected price appreciation over the next year that exceeds the market's potential. We also retained the Growth Fund. We liquidated the rest of the positions and integrated the proceeds into the Smiths' composite portfolio by adding to the following allocations:

$140,000	Large Growth Stocks
$100,000	Large Value Stocks
$ 30,000	Real Estate
$ 30,000	Corporate Bonds
$ 30,000	Government Bonds/CDs

We will contact the Smiths to schedule the next appointment. In the meantime, if they have any questions or concerns, they should give us a call.

Kim, since this memo will eventually go to the Smiths, let's make this conclusion more personal by using the second-person pronoun "you" and more positive by thanking them for their business.

FIGURE 10.22 **Perfected Memo for Client and Company Files**

Memo

Clients: John and Mary Smith
Date: March 15, 2009
Subject: Quarterly Review of Smith Account #12004
Financial Planner: Kim Suzaki

Thank you, Mr. and Mrs. Smith, for allowing GMM&J Financials to help you meet your financial goals. After meeting with you last Tuesday, we have reviewed your December 31, 2008, Balance Sheet and have prepared your 2009 financial projections.

Balance Report Summary:
Your Balance Sheet reflects the following data:

Total Assets	$2,000,000
Liabilities	– 200,000
Deferred Taxes	– 100,000
After-Tax Net Worth	**$1,700,000**

Investment Performance:
We reviewed the performance of your composite investment portfolio, as well as each of its individual holdings. The time-weighted return earned by the composite portfolio during 2006 was +12.00 percent. Nearly every position in the composite portfolio experienced positive returns in 2008.

Cash Account Balances:
At the time of our meeting, there was a balance of $10,000 in the money market fund of Mary's living trust account. This account supports your $3,000 automatic monthly distributions. We will now raise more cash by depositing your 2007 IRA Required Minimum Distributions into Mary's trust account.

(*Continued*)

FIGURE 10.22 Perfected Memo for Client and Company Files (Continued)

Page 2 March 15, 2009

Income Tax Projection:
We reviewed your final income tax projections for 2008 and 2009.

Year	Federal Taxes	California Taxes
2008	$14,000	$8,000
2009	$15,000	$2,000

For 2009, John will have federal tax withholding of $3,600 from his pension. Therefore, we estimate that you will need to make quarterly estimated tax payments of $2,600 to the federal government and $1,800 to California.

Cash Flow Planning:
In 2008, you spent about $27,000 less on your living expenses than we estimated. We will reduce our estimate of your 2009 living expenses to $88,000. After this reduction is incorporated, our initial cash flow projection for 2009 indicates that you will have a total cash flow shortfall of about $70,000 after accounting for John's pension and both of your Social Security benefits. You have already withdrawn $10,000 for home repairs. You will also take $35,000 of automatic monthly distributions over the year. You will need to take an additional $25,000 from your cash and investment accounts over the rest of the year to meet cash flow needs.

Current and Ongoing Issues:
After we received the transfer of John's IRA from his former employer, we reviewed the research on each of the stocks in the portfolio. We retained the positions in the Acme, Business Growth, and C-Company stocks because they have solid business fundamentals, strong earnings growth, and projected price appreciation over the next year that exceeds the market's potential. We also retained the Growth Fund. We liquidated the rest of the positions and integrated the proceeds into your composite portfolio by adding to the following allocations:

Amounts	Investments
$140,000	Large Growth Stocks
$100,000	Large Value Stocks
$ 30,000	Real Estate
$ 30,000	Corporate Bonds
$ 30,000	Government Bonds/CDs

After John provides us with the year-end K-1s from his business interests, we will finalize this Balance Sheet and update your Net Worth Analysis. Our next regularly scheduled meeting will be in June. We will contact you to schedule the appointment. In the meantime, if you have any questions or concerns, please call.

Kim says, "I'm pleased with how the memo turned out. All the errors were corrected, and both my boss and the clients will have an easy time accessing the information. Both the tables and the headings will help them focus on specific areas of concern. I think both the tone and the writing style are correct for the intended audience of lay people."

CHAPTER HIGHLIGHTS

1. Memos and letters are an important part of your interpersonal communication on the job.
2. Memos and letters differ in destination, purpose, format, audience, tone, delivery time, and security.
3. Use an effective subject line including a topic and a focus.
4. Follow the all-purpose template for memos and letters including an introduction, a body or discussion section, and a conclusion.
5. Consider the type of audience whether you are writing a memo or letter.

WWW *Find additional memos and letters exercises, samples, and interactive activities at http://www.prenhall.com/gerson.*

MEETING WORKPLACE COMMUNICATION CHALLENGES

CASE STUDIES

After reading the following case studies, write the appropriate correspondence required for each assignment.

1. As director of human resources at CompuMed biotechnology company, Andrew McWard helps employees create and implement their Individual Development Plans (IDPs). Employees attend 360-Degree Assessment Workshops where they learn how to get feedback on their job performance from their supervisors, coworkers, and subordinates. They also provide self-evaluations. Once the 360-Degree Assessments are complete, employees submit them to Andrew, who, with the help of his staff, develops IDPs.

Andrew sends the IDPs to the employees, prefaced by a cover letter. In this cover letter, he tells them why he is writing and what he is writing about. In the letter's body, he focuses their attention on the attachment's contents: supervisor's development profile, the schedule of activities which helps employees implement their plans, the courses designed to increase their productivity, the costs of each program, and guides to long-term professional development.

In the cover letter's conclusion, Andrew ends upbeat by emphasizing how the employees' IDP help them resolve conflicts and make better decisions.

Assignment

Based on the information provided, write a cover letter for Andrew McWard. He is sending the letter to Sharon Baker, Account Executive,1092 Turtle Hill Road, Evening Star, GA 20091.

2. Mark Shabbot works for Apex, Inc., at 1919 W. 23rd Street, Denver, CO 80204. Apex, a retailer of computer hardware, wants to purchase 125 new flat-screen monitors from a vendor, Omnico, located at 30467 Sheraton, Phoenix, AZ 85023. The monitors will be sold to Northwest Hills Educational Cooperative. However, before Apex purchases these monitors, Mark needs information regarding bulk rates, shipping schedules, maintenance agreements, equipment specifications, and technician certifications. Northwest Hills needs this equipment before the new term (August 15).

Assignment

Write a letter of inquiry for Mr. Shabbot based on the preceding information.

3. Gregory Peña (121 Mockingbird Lane, San Marcos, TX 77037) has written a letter of complaint to Donya Kahlili, the manager of TechnoRad (4236 Silicon Dr., San Marcos, TX 77044). Mr. Peña purchased a computer from a TechnoRad outlet in San Marcos. The *San Marcos Tattler* advertised that the computer "came loaded

with all the software you'll need to write effective letters and perform basic accounting functions." (Mr. Peña has a copy of this advertisement.) When Mr. Peña booted up his computer, he expected to access word processing software, multiple fonts, a graphics package, a grammar check, and a spreadsheet. All he got was a word processing package and a spreadsheet. Mr. Peña wants Ms. Kahlili to upgrade his software to include fonts, graphics, and a grammar check; wants a computer technician from TechnoRad to load the software on his computer; and wants TechnoRad to reimburse him $400 (the full price of the software) for his trouble.

Ms. Kahlili agrees that the advertisement is misleading and will provide Mr. Peña software including the fonts, graphics, and grammar check (complete with instructions for loading the software).

Assignment

Write Ms. Kahlili's 100 percent yes adjustment to Mr. Peña based on the information provided.

4. Your company is planning a picnic to celebrate July 4. Connie Dow is in charge of organizing the picnic. She must write a memo to the company's 200 employees informing them of the picnic's logistics. This memo will be placed on the company's cafeteria bulletin board, in break rooms, and in department offices.

She needs to tell the employees that the picnic will begin at 11:00 a.m. and conclude around 3:00 p.m. Food will be provided (hamburgers, hot dogs, chips, watermelon, and brownies). No alcohol is permitted, but lemonade and soft drinks will be provided. Employees should bring their own recreational equipment for ball games. The picnic will be held at McArthur Park (1231 Oceanview Drive). All employees are encouraged to bring their families.

Assignment

Write Connie's memo, based on the information provided and following the criteria in the book.

ETHICS IN THE WORKPLACE CASE STUDY

George Penrose works for a hospital in their health information management department. He loves his job and has enjoyed the hospital's work environment for ten years.

Recently, the hospital hired a new supervisor. The supervisor has treated all employees fairly, and George is impressed with the supervisor's knowledge of health information management. However, last week, the supervisor wrote a memo to all of the department's employees regarding a change in departmental policy. Starting next month, the supervisor will host non-denominational Bible study sessions. Regardless of an employee's religious beliefs, all employees are encouraged to sign up for these Bible classes. The classes are voluntary, but employees will be given time off their jobs to attend the sessions.

Question: Is it ethical for the company to offer religious study classes? Justify your answer based on information provided in Chapter 1.

INDIVIDUAL AND TEAM PROJECTS

1. Write a letter of inquiry. You might write to a college or university requesting information about a degree program, or to a manufacturer for information about a product or service. Whatever the subject matter, be specific in your request.
2. Write a cover letter. Perhaps your cover letter will preface a report you are working on in school, a report you are writing at work, or documentation you will need to send to a client.
3. Write an adjustment letter. Envision that a client has written a complaint letter about a problem he or she has encountered with your product or service. Write a 100 percent yes adjustment letter in response to the complaint.

4. Write a memo requesting office equipment. Your company plans to purchase new office equipment. Your memo will explain your office's needs. Specifically state what equipment and furniture you want and why these purchases are important.

5. Write a memo reporting on a project's progress. Draw from your experiences in one of your classes. How are you progressing on an assignment? What work have you accomplished? What problems have you encountered? What work remains on this project? In a memo to your instructor, detail this status.

DEGREE-SPECIFIC ASSIGNMENTS

1. **Marketing and Public Relations.** Your marketing and public relations firm, PlaceSetting, specializes in event planning. The company focuses on wedding receptions, post bar and bat mitzvah events, quinceañera celebrations, high school and college graduation parties, and college fraternity and sorority events. One sorority has written PlaceSetting a letter of inquiry, requesting information about entertainment, food, and pricing for an upcoming parent's weekend.

   Write a letter of response. To do so, brainstorm or research possible entertainment, food options, and pricing appropriate for a parents' weekend. Based on your findings, write the required letter. Your letter, which will provide information, persuade the sorority that your company is their best choice, and build rapport for future positive relationships, should follow the criteria provided in this chapter.
2. **Information Technology.** Your company, IT Systems, has developed a new software package to aid communications within organizations. The system includes intranet and extranet capabilities which allow companies to implement business solutions; plan and manage developmental applications; create capital improvement projects; manage transition of vendor services; and manage scheduling. To help your client use the software, you have created an instructional manual that highlights the steps for loading the software, accessing the various system components, performing the appropriate steps, using the online tutorials, and troubleshooting problems.

   Write the cover letter that will preface the user manual. Remember to tell the readers why you are writing and what you are writing about; itemize key points in the manual which you consider to be of special audience interest, and conclude by providing contact information for any follow-up communication.
3. **Accounting.** Your accounting firms—Nelson, Jones, Brownstone—received the following complaint letter from an unhappy customer:

> Subject: 2009 Tax Audit
>
> Ms. Brownstone has prepared my taxes for the last six years. Every year she also prepares my estimated taxes for the next year. I am being audited this year, and I owe $8,900 on my 2008 taxes.
>
> I trust my accountant to calculate my current and estimated taxes so that I can avoid a tax audit and a large tax bill, including interest penalties.
>
> I believe that your firm should guarantee its work and provide an accountant to accompany me when I am audited. In addition, I believe that your company should be responsible for the interest penalties since your accountant underestimated my taxes.
>
> Sincerely,
>
> Ruth Schneider

Respond to this letter of complaint with a 100 percent yes adjustment letter. This letter, following the criteria provided in the chapter, should inform and build rapport.

4. **Management.** As the director of human resources at White Swan Shoes, you recently invited an insurance company, Employee Care, to make presentations to your 150 employees. The presentations covered new insurance premiums, benefits packages, a list of physician providers, life insurance options for employees and family members, and disability insurance costs. After the presentations, the presenters worked individually with employees, answering questions and providing consultations geared toward employees' unique needs. The feedback you received was positive. Comments included, "most informative presentation I've attended," "pleasant and personable presenters," "helpful and knowledgeable staff," and "useful and accessible handouts and brochures."

   Write a thank-you letter to Employee Care Insurance Company. Follow the criteria in this chapter, and build rapport to maintain a positive working relationship.

5. **Finance.** You want to open a small business to develop gaming software and hardware. You will need start-up capital to purchase office equipment, to hire part-time employees, to rent office space, and to market your products. Your potential market includes the United Kingdom, Canada, and Japan, in addition to the United States. You can advertise your gaming software and hardware online, so you also will need an Internet service provider.

   To finance your business, you need a short-term loan. Your banker suggests a microloan from a 7(a) Loan Program. This would provide you a short-term loan up to $35,000 for working capital or the purchase of inventory, supplies, furniture, fixtures, machinery and/or equipment. Before you acquire this loan, you need to find out if the loan can be used to pay your existing education debts or to purchase property. You also want to find out if the loan agency provides management and technical assistance, who guarantees the loan, and whether microloan programs are available in your state.

   To find answers to your questions, write a letter of inquiry. Follow the criteria provided in this chapter. If you need more information about 7(a) microloans, research this topic.

PROBLEM-SOLVING THINK PIECES

Northwest Regional Governmental Training Consortium (GTC) provides educational workshops for elected and appointed officials, as well as employees of city and state governmental offices.

One seminar participant, Mary Bloom, supervisor of the North Platte County Planning and Zoning Department, attended a GTC seminar entitled "Developing Leadership Skills" on February 12, 2009. Unfortunately, she was disappointed in the workshop and the facilitator. On February 16, Mary wrote a complaint letter to GTC's director, Sue Randall, stating her dissatisfaction. Ms. Bloom said that the training facilitator's presentation skills were poor. According to Mary, Doug Aaron, the trainer, exhibited the following problems:

- Late arrival at the workshop
- Too few handouts for the participants
- Incorrect cables for his computer, so he could not use his planned PowerPoint presentation
- Old, smudged transparencies as a backup to the PowerPoint slides

Mary also noted that the seminar did not meet the majority of the seminar participants' expectations. She and the other government employees had expected a hands-on workshop with breakout sessions. Instead, Mr. Aaron lectured the entire time. In addition, his information seemed dated and ignored the cross-cultural challenges facing today's supervisors.

Neither Sue nor her employees had ever attended this workshop. They offered the seminar based on the seemingly reliable recommendation of another state agency, the State

Data Collection Department. From Doug Aaron's course objectives and resume, he appeared to be qualified and current in his field.

However, Mary Bloom deserves consideration. Not only are her complaints justified by others' comments, but she is a valued constituent. The GTC wants to ensure her continued involvement in their training program.

Assignments

1. Sue needs to write a 100 percent yes adjustment letter. In this letter, Sue wants to accommodate the dissatisfied client. How should she recognize Mary's concerns, explain what might have gone wrong, and offer satisfaction?
2. Sue needs to write an internal memo to her staff. This memo will provide standards for hiring future trainers. What should the standards include?

WEB WORKSHOP

Many state governments provide guidelines for writing letters either to government officials or government employees. In addition, you can go online, type something like "how to write state government letters" in a search engine, and find sites with instructions and sample letters for writing to state officials or state agencies.

Research the Internet and find letter samples and guidelines. Once you have done so

- Review sample letters to or from governmental agencies.
- Determine whether these letters are successful, based on the criteria provided in this chapter.
- Write a letter or memo to your instructor explaining how and why the letters succeeded.
- If you find letters that can be improved, rewrite them, using the guidelines provided in this chapter.

QUIZ QUESTIONS

1. What are four differences between letters and memos?
2. What are four reasons for writing a memo?
3. What are the components of a successful subject line?
4. Why would you write a letter of inquiry?
5. What are the goals of a cover letter?
6. List at least four things you could include in the body of a letter of recommendation.
7. In an order letter, what types of details should you include to ensure the correctness of your order?
8. What should you say in the introduction of a 100 percent yes adjustment letter?
9. When should you avoid apologizing in a 100 percent yes adjustment letter?
10. List at least four things to include in the body of a confirmation letter.

CHAPTER 11

Employment Communication: Resumes, Application Letters, Interviewing, and Follow-Up Letters

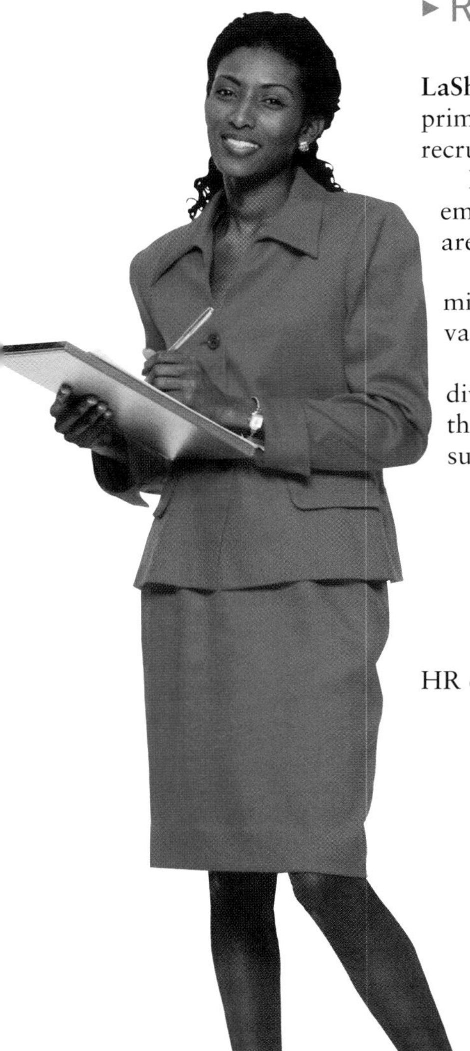

▸REAL PEOPLE *in the workplace*

LaShanda Brown is a human resources specialist with Oklahoma City, Oklahoma. Her primary job responsibility is recruiting. However, her role differs from that of a traditional recruiter who might visit high schools or colleges to recruit applicants.

In contrast, as LaShanda says, her "department serves as the alpha and omega for city employees. They have to start here when they apply, and they end here when they resign, are terminated, or retire."

In the recruitment division, LaShanda reviews applications and resumes to determine if the applicants meet the minimum qualifications for education and experience for various positions.

According to the OKC human resources Web site, "The recruitment and selection division is responsible for staffing all merit system positions." This includes ensuring that all of the following comply with legal, professional, and merit system standards, such as

- Employee recruiting
- Advertising
- Testing
- Certification
- Applicant record-keeping

HR employees do the following:

- Advertise and recruit, including specialized recruiting for technical, professional, or other positions that are difficult to fill
- Review thousands of applications to determine whether or not the applicants meet position requirements
- Verify experience and education when either is required for new or promoted employees
- Establish, maintain, and certify eligibility lists for all city departments that have vacant positions to fill
- Provide employment assistance to both current city employees and to applicants seeking employment in our city government ("Human Resources Department")

In the course of a year, LaShanda and her colleagues review approximately 120,000 applications. She also performs background checks for potential

CHAPTER GOALS

When you complete this chapter, you will be able to do the following:

1. Search for job openings applicable to your interests, education, and experience.
2. Compose effective letters of application that gain attention and are persuasive.
3. Choose either to write a functional or reverse chronological resume.
4. Write effective resumes consisting of your objectives, summary of qualifications, work experience, education, and professional skills.
5. Decide on the correct method of delivery of your resume, either through the mail, as an e-mail attachment, or scannable.
6. Understand effective interview techniques that demonstrate your professionalism.
7. Write appropriate follow-up correspondence to restate how you can benefit the company.
8. Write a job acceptance letter.
9. Use the job search checklists to evaluate your resume, letter of application, interview, follow-up letter, and job acceptance letter.
10. Follow the *P*³ process to write effective employment communication.

WWW
To learn more about employment communication, visit our Web Companion at http://www.prenhall.com/gerson.

employees, presents new employee orientation training, and presents other classes for city employees. LaShanda attends career fairs in addition to conducting the city's own annual career fair. She also coaches employees and citizens on the application process with the city and gives help as needed with resumes and any other career-related questions.◄

LaShanda's Communication Challenge

How did LaShanda find her job in human resources? When she began her job search, LaShanda didn't just need a job; she needed a career change. Not only was she relocating to a new city, but also she was looking to find employment in her chosen field of interest, after having worked in other fields for several years. LaShanda says, "I had a degree in Communications with a specialization in public relations and an interest in human resources. Unfortunately, I had been working in a completely different field, lease acquisitions.

To find a new job in a new city, I used most of the traditional methods:

- Surfing online search engines (Monster, HotJobs, CareerBuilder, and USAJobs).
- Looking at companies prevalent in the city, such as Devon Energy, AT&T, and Boeing.
- Asking family and friends questions like, 'Where do you work? What kinds of jobs do you have there? Know anyone whom I should speak to?'
- Attending job fairs that I found advertised in the local newspaper.
- Working a temp job.

When I found openings, I submitted my resume. And then, I heard nothing—no responses . . . period." LaShanda knew she was doing something wrong. Her challenge was to figure out why she wasn't getting her foot in the door in the department of human resources and how she could improve her job search.

See pages 330–333 to learn how LaShanda met her communication challenge.

HOW TO FIND JOB OPENINGS

When it's time for you to look for a job, how will you begin your search? You know you can't just wander up and down the street, knocking on doors randomly. That would be time consuming, exhausting, and counterproductive. Instead, you must approach the job search more systematically.

Visit Your College or University Job Placement Center

Your college job placement center is an excellent place to begin a job search, for several reasons.

- Your school's job placement service might have job counselors who will counsel you regarding your skills and job options.
- Your job placement center can give you helpful hints on preparing resumes, letters of application, and follow-up letters.
- The center will post job openings within your community and possibly in other cities.
- The center will be able to tell you when companies will visit the campus for job recruiting.
- The service can keep on file your letters of recommendation or portfolio. The job placement center will send these out to interested companies upon your request.

Attend a Job Fair

Many colleges and universities host job fairs. A job fair will allow you to research job openings, make contacts for internships, or submit resumes for job openings. If you attend a job fair, treat it like an interview. Dress professionally and take copies of your resume and letters of recommendation.

Talk to Your Instructors

Whether in your major field or not, instructors can be excellent job sources. They will have contacts in business, industry, and education. They may know of job openings or people who might be helpful in your job search. Furthermore, because your instructors obviously know a great deal about you (having spent a semester or more working closely with you), they will know which types of jobs or work environments might best suit you.

Network with Friends and Past Employers

A *Smart Money* magazine article reports that 62 percent of job searchers find employment through "face-to-face networking" (Bloch 12). A study performed by Drake Beam Morin confirms this, stating that "64 percent of . . . almost 7,500 people surveyed said they found their jobs through socializing and meeting people" (Drakeley 5). Tell friends and acquaintances that you are looking for a job. They might know someone for you to call.

Why is networking so important? It is simple math. Your friends and past employers know people. Those people know people. By visiting acquaintances, you can network with numerous individuals. The more people you talk to about your job quest, the more job opportunities you will discover.

Get Involved in Your Community

There are many ways to network. In addition to talking to your family or past employers, you can network by getting involved in your community. Consider volunteering for a community committee, pursuing religious affiliations, joining community clubs, or participating in fund-raising events. Join Toastmasters. Take classes in accounting, HTML, RoboHelp, or Flash at your local community college. Each networking activity you pursue not only teaches you a new skill but also provides an opportunity to meet new people. Then, "when a job comes open, you'll be on their radar screen" (Drakeley 6).

Check Your Professional Affiliations and Publications

If you belong to a professional organization, this could be a source of employment in three ways. First, your organization's sponsors or board members might be aware of job openings. Second, your organization might publish a listing of job openings. Finally, you can gain professional experience by serving on boards or being an officer in an organization.

These experiences can be included in your resume under professional skills, work experience, and affiliations.

Read the Want Ads

Check the classified sections of your local newspapers or newspapers in cities you might like to live and work in. These want ads list job openings, requirements, and salary ranges. In addition, by reading want ads, you can learn which "keywords" your industry is focusing on and incorporate these terms in your resume.

Read Newspapers or Business Journals to Find Growing Businesses and Business Sectors

Which companies are receiving grants, building new sites, winning awards, or creating new service or product lines? Which companies have just gained new clients or received expanded contracts? Newspapers and journals report this kind of news, and a growing company or business sector might be good news for you. If a company is expanding, this means more job opportunities for you to pursue.

Take a "Temp" Job

Temporary ("temp") jobs, accessed through staffing agencies, pay you while you look for a job, help you acquire new skills, allow you to network, and can lead to full-time employment.

Get an Internship

Internships provide you outstanding job preparedness skills, help you meet new people for networking, and improve your resume. An unpaid internship in your preferred work area might lead to full-time employment. An internship "gives you the opportunity to show your skills, work ethic, positive attitude, and passion for your work." By interning, you can prove that you should be "the next employee the company hires" (Drakeley 7). In fact, the National Association of Colleges and Employers lists internships as one of the top ten skills employers want ("Top Ten Qualities").

Job Shadow

Job shadowing allows you to visit a work site and follow employees through work activities for a few hours or days. This allows you to learn about job responsibilities in a certain field or work environment. In addition, job shadowing also helps you find out if a company or industry is hiring, allows you to make new contacts, and places you in a favorable position for future employment at that company.

Set Up an Informational Interview

In an informational interview, you talk with people currently working in a field, asking them questions about career opportunities and contacts. Informational interviews allow you to

- explore careers and clarify your career goal
- expand your professional network
- build confidence for your job interviews
- access the most up-to-date career information
- identify your professional strengths and weaknesses

Figure 11.1 shows some tips to prepare you for an informational interview.

Research the Internet

In the mid-to-late 1990s, the Internet was the preferred means by which job seekers found employment. That has changed drastically. *Newsweek* magazine calls the Internet "a time waster." Quoting the head of a Chicago outplacement firm, *Newsweek* writes that some

FIGURE 11.1 **Preparation Tips for Informational Interviews**

Informational Interview *Do's*

☑ Prepare ahead of time by calling or writing to ask for an informational interview.
☑ Call before your interview to confirm the date and time.
☑ Arrive on time.
☑ Dress appropriately.
☑ Come prepared to take notes.
☑ Bring your resume.
☑ Ask for referrals.
☑ Always send a thank-you note.
☑ *Never* ask for a job.

Great Questions to Ask

☑ What's your typical day like on the job?
☑ What job experiences have led to your job?
☑ How could a student acquire the needed experience?
☑ What are various jobs in this field?
☑ How is the economy affecting this profession's job outlook?
☑ Are job opportunities and opportunities for advancement good?
☑ What are the key skills needed for jobs in this field?
☑ What levels of education are needed for jobs in this field?
☑ What other types of certifications or training are needed for jobs in this area?
☑ What special advice would you have for a student seeking employment in this field?

of the Internet's popular job sites are "big black holes"—your resume goes in, but you never hear from anyone again (Stern 67).

Others are equally pessimistic about the Internet's value as a source for jobs. Only 10 percent of technical or computer-related jobs are found from electronic job searches, "about 13 percent of interviews for managerial-level jobs result from responding to an online posting," and a mere 4 percent of jobs overall are found through the Internet (Bloch 12).

Nonetheless, you should make the Internet part of your job search strategy. If it is not the best place to find a job, the Internet still provides numerous benefits. Internet job search engines provide excellent job search resources, such as the following:

- **Resumes**—These sites explain the difference between resumes and curriculum vitae (CV), note how to address gaps in your career history, identify how to avoid resume mistakes, and provide tips for writing winning resumes.
- **Interviews**—These sites cover interviewing to get the job and handling illegal questions.
- **Cover letters and thank-you letters**—These sites highlight sample cover letter techniques and ways to write a better thank-you letter.
- **Job search tips**—Topics include leaping to a twenty-first-century technology career and employing the correct netiquette.

Table 11.1 provides some online job search links that may be useful.

Table 11.1 Online Job Search Links

- http://online.onetcenter.org/—O*Net Online bills itself as "the nation's primary source of occupational information, providing comprehensive information on key attributes and characteristics of workers and occupations."
- http://www.Monster.com—lets you post resumes and search for jobs, and provides career advice.
- http://www.CareerBuilder.com—lets you search for jobs by company, industry, and job type.
- http://www.CollegeRecruiter.com/—lists the latest job postings, "coolest career resources, and most helpful employment information."
- http://www.CareerJournal.com—the *Wall Street Journal*'s career search site; provides salary and hiring information, a resume database, and job hunting advice.
- http://www.FlipDog.com—provides national job listings in diverse fields by job title, location, and date of listing.
- http://www.HotJobs.yahoo.com—lets you search for jobs by keyword, city, and state.
- http://www.WantedJobs.com—lists jobs in the United States and Canada.
- http://www.WetFeet.com—gives company and industry profiles, resume help, city profiles, and international job sites.
- http://www.CareerLab.com—provides a cover letter library.
- http://www.CareerCity.com—offers a detailed discussion and samples of functional versus chronological resumes.
- http://www.JobOptions.com—lists job opportunities in the fields of accounting, customer service, engineering, human resources, sales, and technology.
- http://www.CareerMag.com—allows for online job searches by keyword.
- http://www.CareerShop.com—helps you post and edit resumes, offers career advice, and lists hiring employers.
- http://www.ajb.dni.us/—America's Job Bank posts new jobs and provides resources to help you develop credentials, acquire educational financial aid, upgrade and measure your job-related skills, and help you start your own business.
- http://www.dice.com/—Dice, a Web site focusing on technology careers.
- http://www.JobWeb.com/—helps college students, seniors, new college graduates, and alumni with career development and the job search.
- http://www.Job-Hunt.org/—called by *PC Magazine* and *Forbes* the Internet's Best Web site for job hunting and resources.
- http://www.FedWorld.gov/jobs/jobsearch.html—a site dedicated to government job searches and advice.

THE THREE R'S OF JOB SEARCHING

Before you write your resume, consider the three R's of job searching: research yourself, research the company, and research the position. By performing this research, you will improve the focus of your job search. Learning more about your goals, the company's culture, and the specifics of the position will help you to develop your resume.

Research Yourself

You are not right for every job, nor is every job right for you. To ensure that you are applying for the correct position, ask yourself the following questions:

- What are your skills, attributes, and accomplishments?
- What do you have to offer this company?
- What can you bring to the company that is unique?
- What are your selling points?
- How do your qualifications coincide with the position's requirements?
- What interests you about this company and the position?

Research the Company

By researching the company, you educate yourself about the company's culture, values, products, and services. To research the company, view the employer's Web site, read annual reports, or speak with an employee. Consider these questions:

- How does the employer's future look?
- What is the employer's product or service?
- How do your values coincide with the company's values?
- What is the company's vision or mission statement?
- What is the company's culture or philosophy?

JOB SEARCH CHECKLIST

_____ 1. Did you visit your college or university job placement center?

_____ 2. Did you talk to your professors about job openings?

_____ 3. Have you networked with friends or past employers?

_____ 4. Have you checked with your professional affiliations or looked for job openings in trade journals?

_____ 5. Did you read the want ads in the newspapers?

_____ 6. Did you search the Internet for job openings?

_____ 7. Did you consider taking a "temp" job or applying for an internship?

_____ 8. Have you considered an informational interview?

- What are the needs or problems of the employer?
- What role does the employer play in the community?
- Is the employer expanding?

Research the Position

When you consider the job opening, review its requirements and duties. Doing so will help you to determine if this job is right for you. Ask yourself these questions:

- What skills and talents are needed for the position?
- What freedoms does the position allow?
- What is the structure of the department?
- Who does the position report to?

("Resume Guidelines")

Once you have researched these topics and answered these questions, you can decide whether to apply for the position. If you choose to apply, the next step is to create a resume.

CRITERIA FOR EFFECTIVE RESUMES

Once you have found a job that interests you, it is time to apply. Your job application will start when you send the prospective employer your resume. Resumes are usually the first impression you make on a prospective employer. If your resume is effective, you have opened the door to possible employment—you have given yourself the opportunity to sell your skills during an interview. If, in contrast, you write an ineffective resume, you have closed the door to opportunity.

Your resume should present an objective, easily accessible, detailed biographical sketch. However, do not try to include your entire history. The primary goal of your resume, together with your letter of application, is to get an interview. You can then use your interview to explain in more detail any pertinent information that does not appear on your resume.

When writing a resume, you have two optional approaches. You can write either a reverse chronological resume or a functional resume.

Reverse Chronological Resume

Write a reverse chronological resume if you

- Are a traditional job applicant (a recent high school or college graduate, ages 18 to 25)
- Hope to enter the profession in which you have received college training or certification
- Have made steady progress in one profession (promotions or salary increases)
- Plan to stay in your present profession

Functional Resume

Write a functional resume if you

- Are a nontraditional job applicant (are returning to the workforce after a lengthy absence, are older, are not a recent high school or college graduate)
- Plan to enter a profession in which you have not received formal college training or certification
- Have changed jobs frequently
- Plan to enter a new profession

Key Resume Components

Whether you write a reverse chronological or a functional resume, include the following key components.

Identification

Begin your resume with the following:

- Your name (full first name, middle initial, and last name). Your name can be in boldface and printed in a larger type size (14-point, 16-point, etc.).
- Contact information. Include your street address, your city, state (use the correct two-letter abbreviation), and zip code. If you are attending college or serving in the armed forces, you might also want to include a permanent address. By including alternative addresses, you will enable your prospective employer to contact you more easily.
- Your area code and phone numbers. As with addresses, you can provide alternative phone numbers, such as cell phone numbers. However, limit yourself to two phone numbers, and don't provide a work phone. Having prospective employers call you at your present job is not wise. First, your current employer will not appreciate your receiving this sort of personal call; second, your future employer might believe that you often receive personal calls at work—and will continue to do so if he or she hires you.
- Your e-mail or Web site address and fax number. Be sure that your e-mail address is professional sounding. An e-mail address, such as "ILuvDaBears," "Hotrodder," or "HeavyMetalDude," is not likely to inspire a company to interview you.

Career Objectives

A career objectives line is not mandatory. Some employers like you to state your career objectives; others do not. You will have to decide whether or not you want to include this section in your resume.

The career objectives line is like a subject line in a memo or report. Just as the subject line clarifies your memo or report's intent, your career objective informs the reader of your resume's focus. Be sure your career objective is precise. Too often, career objectives are so generic that their vagueness does more harm than good.

BEFORE

Flawed Career Objective

Career Objective: Seeking employment in a business environment offering an opportunity for professional growth.

This poorly constructed career objective provides no focus. What kind of business? What kind of opportunities for professional growth? Employers don't want to hire people who have only vague notions about their skills and objectives.

This improved career objective not only specifies which job the applicant is seeking but also how he or she will benefit the company.

AFTER

Better Career Objective

Career Objectives: To market financial planning programs and provide financial counseling to ensure positive client relations.

Summary of Qualifications

As an option to beginning your resume with career objectives, you might want to consider starting with a summary of qualifications. According to Monster.Com, "resumes normally get less than a 15-second glance at the first screening" (Isaacs). A summary of qualifications allows the employer an immediate opportunity to see how you can add value to the company.

A summary of qualifications should include the following:

- An overview of your skills, abilities, accomplishments, and attributes
- Your strengths in relation to the position you are applying for
- How you will meet the employer's goals

To write an effective summary of qualifications, list your top three to seven most marketable credentials.

example

Summary of Qualifications

- Over four years combined experience in marketing and business
- Developed winning bid package for promotional brochures
- Promoted to Shift Manager in less than two years
- Maintained database of customers, special ordered merchandise, and tracked inventory

Employment

The employment section lists the jobs you've held. This information must be presented in reverse chronological order (your current job listed first, previous jobs listed next). This section must include the following:

- Your job title (if you have or had one)
- The name of the company you worked for

FAQs

Q: Do I really need to limit my resume to one page?

A: Do not worry about limiting yourself to the traditional one-page resume. Conciseness is important in all workplace communication, but if your education, work experience, and professional skills merit more than a page, you must show your accomplishments. In addition, if you submit your resume as part of an e-mail message, readers will scroll. However, don't pad the resume. Limit yourself to jobs within the last ten years and skills that fit the job you're seeking.

Q: In my resume, which should I list first, my work experience or my education?

A: It's all about location, location, location. You should present your most important section first. If education is your strength and will help you get the job, lead with education. If, in contrast, your work experience is stronger, begin your resume with work.

Q: Can I omit jobs that I didn't like?

A: Yes and no. You cannot have any large gaps in your resume, such as a missing year or more. If you have any large gaps, you must either explain the gap or fill it with other activities (education, volunteerism, or childrearing, for instance). However, a missing month or so is not a problem. If you worked a job for a month, left that job, and then found other employment, you do not need to list the short-term job.

- The location of this company (either street address, city, and state or just the city and state)
- The time period during which you worked at this job
- Your job duties, responsibilities, and accomplishments

This last consideration is important. This is your chance to sell yourself. Merely stating where you worked and when you worked there will not get you a job. Instead, what did you achieve on the job? In this part of the resume, you should detail how you met deadlines, trained employees, cut expenses, exceeded sales expectations, decreased overage, and so forth. Plus, you want to quantify your accomplishments.

example

Assistant Manager

McConnel Oil Change, Beauxdroit, LA 2007 to present

- Tracked and maintained over $25,000 inventory
- Trained a minimum of four new employees quarterly
- Achieved a 10 percent growth in customer car count for three consecutive years
- Developed a written manual for hazardous waste disposal, earning a "Citizen's Recognition Award" from the Beauxdroit City Council

Listing your job title, company name, location, and dates of employment merely shows where you were in a given period of time. To prove your contributions to the company, provide specific details highlighting achievements.

Education

In addition to work experience, you must include your education. Document your educational experiences in reverse chronological order (most recent education first; previous schools, colleges, universities, military courses, and training seminars next). When listing your education, provide the following information:

- Degree. If you have not yet received your degree, you can write "Anticipated date of graduation June 2009" or "Degree expected in 2009."
- Area of specialization.
- School attended. Do not abbreviate. Although you might assume that everyone knows what *UT* means, your readers won't necessarily understand this abbreviation. Is UT the University of Texas, the University of Tennessee, the University of Tulsa, or the University of Toledo?
- Location. Include the city and state.
- Year of graduation or years attended.

As you can see, this information is just the facts and nothing else. Many people might have the same educational history as you. For instance, just imagine how many of your current classmates will graduate from your school, in the same year, with the same degree. Why should a company hire you over them? The only way you can differentiate yourself from other job candidates with similar degrees is by highlighting your unique educational accomplishments.

These might include any or all of the following:

Grade point average	Academic honors, scholarships, and awards
Academic club memberships and leadership offices held	Fraternity or sorority leadership offices held
Unique coursework	Percentage of your college education costs you paid for while attending school
Special class projects	Technical equipment you can operate

Please note a key concern regarding your work experience and education. You should have no chronological gaps when all of your work and education are listed. You can't omit a year without a very good explanation. (A missing month or so is not a problem.)

Professional Skills

If you are changing professions or reentering the workforce after a long absence, you will want to write a functional resume. Therefore, rather than beginning with education or work experience, which won't necessarily help you get a job, focus your reader's attention on your unique skills. These could include

Proficiency with computer hardware and software	New techniques you have invented or implemented
Procedures you can perform	Numbers of and types of people you have managed
Special accomplishments and awards you have earned	Publications you have created or worked on
On-the-job training you have received	Certifications you have earned
Training you have provided	Languages you speak, read, and write

These professional skills are important because they help show how you are different from all other applicants. In addition, they show that although you have not been trained in the job for which you are applying, you can still be a valuable employee.

example

Highlight your professional skills that will set you apart from other potential employees.

Professional Skills

- Proficient in Microsoft Word, Excel, PowerPoint, and FrontPage
- Knowledge of HTML, Java, Visual Basic, and C++
- Certified OSHA Hazardous Management Safety Trainer
- Fluent in Spanish and English
- Completed Second Shift Administration Certificate

Military Experience

If you served for several years in the military, you might want to describe this service in a separate section. You would state your

Rank	Discharge status
Service branch	Special clearances
Location (city, state, country, ship, etc.)	Achievements and professional skills
Years in service	Training seminars attended and education received

Professional Affiliations

If you belong to regional, national, or international clubs or have professional affiliations, you might want to mention these. Such memberships might include the Rotarians, Lions Club, Big Brothers, Campfire Girls, or Junior League. Maybe you belong to the Society for Technical Communication, the Institute of Electrical and Electronic Engineers, the National Office Machine Dealers Association, or the American Helicopter Society. Listing such associations emphasizes your social consciousness and your professional sincerity.

Optional Resume Components

Portfolios

As an optional component for your job search, consider using a portfolio. If you are in fashion merchandising, heating/ventilation/air conditioning (HVAC), engineering, drafting, architecture, or graphic design, for example, you might want to provide samples of your work. These samples could include schematics, drawings, photographs, certifications, or other samples of your work.

However, avoid sending an unsolicited portfolio electronically. Many companies are reluctant to open unknown attachments. Make hard copies of your work or a CD-ROM. When you have an interview with a prospective employer, you can give the employer the hard copy or CD-ROM.

References

Avoid a reference line that reads "Supplied on request," "Available on request," or "Furnished on request." Every employer knows that you will provide references if asked. Instead of wasting valuable space on your resume with unnecessary text, use this space to develop your summary of qualifications, education, work experience, or professional skills more thoroughly. Create a second page for references, and bring this reference page to your interview. On the reference page, list three or four colleagues, supervisors, teachers, or community individuals who will recommend you for employment. Provide their names, titles, addresses, and phone numbers. By bringing the reference page to your interview, you will show your prospective employer that you are proactive and organized.

Personal Data

Do not include any of the following personal information on a resume: birth date, race, gender, religion, height, weight, religious affiliations, marital status, or pictures of yourself. Equal opportunity laws disallow employers from making decisions based on these factors.

Effective Resume Style

The preceding information suggests *what* you should include in your resume. Your next consideration is *how* this information should be presented. As mentioned throughout this textbook, format or page layout is essential for effective workplace communication. The same holds true for your resume.

Choose Appropriate Font Types and Sizes

As with most workplace communication, the best font types to use are Times New Roman and Arial. These are readable and professional looking. Avoid "designer fonts," such as Comic Sans, and cursive fonts, such as Shelley Vollante. In addition, use a 10- to 12-point font for your text. Smaller font sizes are hard to read; larger font sizes look unprofessional. Headings can be boldface and 12- to 14-point font size. Limit your resume to no more than two font types.

Avoid Sentences

Sentences create three problems in a resume. First, if you use sentences, the majority of them will begin with the first-person pronoun *I*. You'll write, "I have . . . ," "I graduated . . .," or "I worked . . . ?" Such sentences are repetitious and egocentric. Second, if you choose to use sentences, you'll run the risk of committing grammatical errors: run-ons, dangling modifiers, agreement errors, and so forth. Third, sentences will take up room in your resume, making it longer than necessary.

Format Your Resume for Reader-Friendly Ease of Access

Instead of sentences, highlight your resume with easily accessible lists. Set apart your achievements by adding bullets to your accomplishments, awards, unique skills, and so on. In addition to bullets, make your resume accessible by capitalizing and boldfacing headings, indenting subheadings to create white space, and italicizing subheadings and major achievements. Avoid underlining headings. Studies show that most people find underlined text hard to read (Vogt).

Begin Your Lists with Verbs

To convey a positive, assertive tone, use verbs when describing your achievements. Table 11.2 contains a list of verbs you might use.

Table 11.2 Active Verbs to Highlight Achievements

Accomplished	Designed	Initiated	Planned
Achieved	Developed	Installed	Prepared
Analyzed	Diagnosed	Led	Presented
Awarded	Directed	Made	Programmed
Built	Earned	Maintained	Reduced
Completed	Established	Managed	Resolved
Conducted	Expanded	Manufactured	Reviewed
Coordinated	Gained	Negotiated	Sold
Created	Implemented	Ordered	Supervised
Customized	Improved	Organized	Trained

Quantify Your Achievements

Your resume should not tell your readers how great you are; it should prove your worth. To do so, quantify by precisely explaining your achievements.

BEFORE	AFTER
Maintained positive customer relations with numerous clients.	Maintained positive customer relations with 5,000 retail and 90 wholesale clients.
Improved field representative efficiency through effective training.	Improved field representative efficiency by writing corporate manuals for policies and procedures.
Achieved production goals.	Achieved 95 percent production, surpassing the company's desired goal of 90 percent.

Practice Resume Ethics

When you write your resume, you must be honest for at least two reasons. First, it is unethical to lie. Second, failure to represent yourself honestly can lead to the following:

- Misrepresentation by an employee during recruitment may result in termination.
- An employer can use your resume misrepresentations as a defense in a wrongful dismissal case.
- An employer could sue an employee for damages due to a misrepresentation.

("To Lie or Not to Lie")

How would an employer know if you are misrepresenting yourself? Phony degrees are easy to catch. All a recruiter needs to do is call or fax a college or university. Other resume falsehoods are harder to catch. Recruiters look for the following red flags:

- Gaps and overlaps in work experience
- Unrealistic and vague qualifications
- Lack or misuse of industry terminology

Make It Perfect

You cannot afford to make an error in your resume. Remember, your resume is the first impression you'll make on your prospective employer. You don't want to make a poor first impression by having errors in your resume.

METHODS OF DELIVERY

Whether you choose to create a reverse chronological or a functional resume, you can deliver your document in several ways.

Mail Version

The traditional way to deliver a resume is to insert it into an envelope and mail it. This resume can be highly designed, using bullets, boldface, horizontal rules, indentations, and different font sizes. Because this document will be a hard copy, what the reader sees will be exactly what you mail.

Technology Tip

Using Resume Templates in Microsoft Word 2007

Microsoft Word 2007 provides you a resume template if you want help getting started.

1. Click on the "**Office Button**" located on the top left of your toolbar and scroll to "**New**."

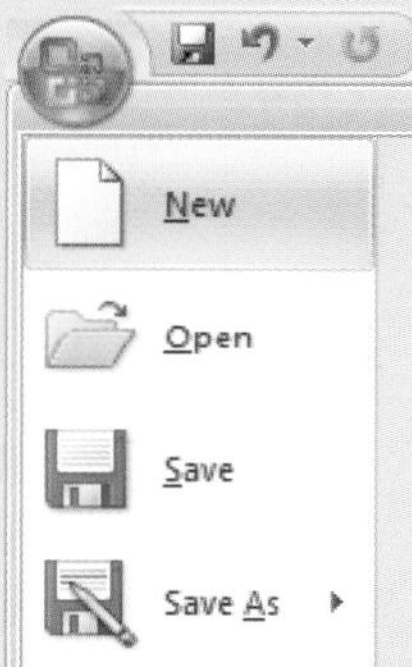

The following "**New Document**" window will pop up.

2. Click on the **"Resumes and CVs"** tab to find 24 optional resume layouts ("Traditional" chronological resume design, "Modern" chronological resume design, "Minimalist" chronological resume design, etc.).

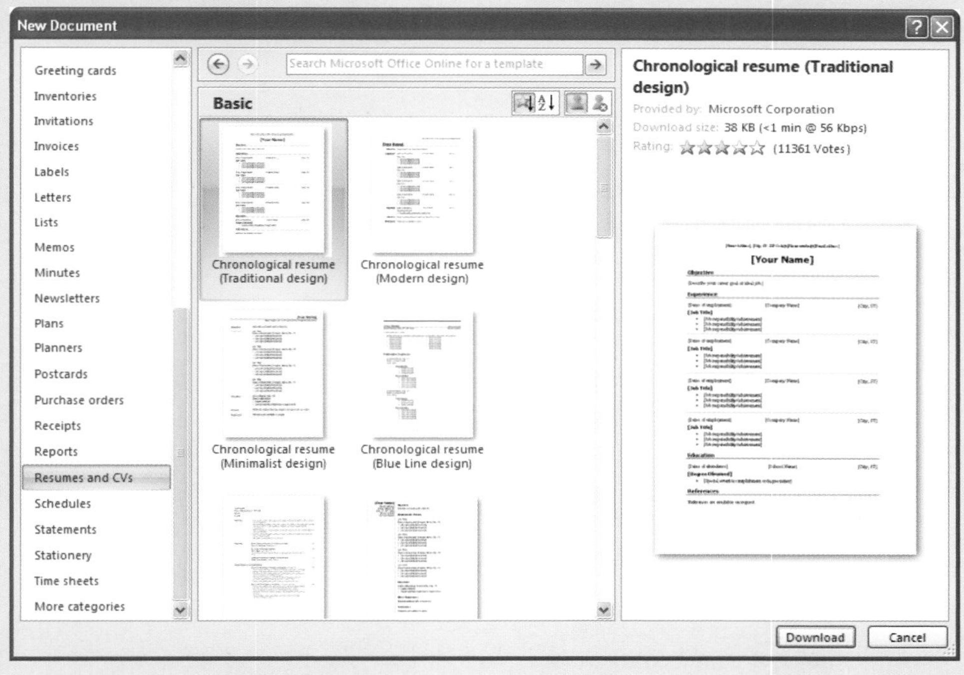

Do not be tempted to overdesign your resume, however. For example, avoid decorative fonts, clip art, borders, or photos. Do not print your resume on unusual colors, like salmon, baby blue, tangerine, or yellow. Instead, stick to heavy white paper and standard fonts, like Times New Roman and Arial.

Resume templates provide benefits as well as create a few problems. On the positive side, the templates are great reminders of what to include in your resume, such as objectives, work experience, education, and skills. In contrast, the templates also limit you and perhaps suggest that you include information that isn't needed. First, the templates mandate font sizes and page layout. Second, a few of the templates suggest that you include information about "Interests" and "References." Rarely should you include "Interests," such as hobbies. Furthermore, most experts suggest that you omit the "References" line, saving valuable resume space for more important information. You can include references on a separate page, especially for interviews. More important, if you use the same templates that everyone else does, then how will your resume stand out as unique?

A good compromise is to review the templates for ideas and then create your own resume with your unique layout.

Figures 11.2 and 11.3 are excellent examples of traditional resumes, ready to be mailed.

FIGURE 11.2 Reverse Chronological Resume

You could replace the "Objective" line with a "Summary of Qualifications," as follows:

Summary of Qualifications
- *Over five years customer service experience*
- *Work experience in public relations*
- *Proven record of written and interpersonal communication abilities*
- *Outstanding leadership skills*
- *Fluent in Spanish*

Sharon J. Barenblatt

1901 Rosebud Avenue
Boston, MA 12987
Cell phone: 202-555-2121
E-mail: sharonbb@juno.com

OBJECTIVE:

Employment as an Account Manager in public relations, using my education, work experience, and interpersonal communication skills to generate business.

EDUCATION:

BS, Business. Boston College. Boston, MA (2009)
- 3.2 GPA
- Social Justice Chair, Sigma Delta Tau (2008)
- Study Abroad Program, Madrid, Spain (2007)
- Internship, Ace Public Relations, Boston, MA (2006)

Frederick Douglas High School. Newcastle, MA (2005)
- 3.5 GPA
- Member, Honor Society
- Captain, Frederick Douglas High School tennis team

List your education and work experience in reverse chronological order.

WORK EXPERIENCE:

Salesperson/assistant department manager. Jessica McClintock Clothing Store. Boston, MA (2008 to present)
- Prepare nightly deposits, input daily receipts
- Open and close the store
- Provide customer service
- Trained six new employees

List current jobs using present tense verbs and past jobs using past tense verbs.

Salesperson. GAP Clothing. Newcastle, MA (2004 to 2005)
- Assisted customers
- Stocked shelves

Do not only list where you worked and when you worked there. Also include your job responsibilities.

PROFESSIONAL SKILLS:
- Made oral presentations to the Pan-Hellenic Council to advertise sorority philanthropic activities
- Helped plan community-wide "Paul Revere's Ride Day"
- Created advertising brochures and fliers

E-Mail Resume

Delivering your resume by U.S. mail can take several days. In contrast, the quickest way to get your resume to the prospective employer is as an e-mail attachment. Speed isn't the only issue. "Hiring managers and recruiters have become as addicted to e-mail as everyone else. More than one-third of human-resource professionals reported a preference for e-mailed resumes, according to . . . the Society for Human Resource Management" (Dixson).

If you choose to submit your resume as an attachment, be sure to write a brief e-mail cover message (we discuss this later in the chapter, under letters of application). In addition, you must clarify what software you have used for this attachment—Microsoft Word, Works, WordPerfect, or Rich Text Format (RTF), for example.

FIGURE 11.3 Functional Resume

JODY R. SEACREST
1944 W. 112th Street
Salem, OR 64925
(513) 451-4978 jseacrest12@hotmail.com

PROFESSIONAL SKILLS

- Operated sporting goods/sportswear mail-order house. Business began as home-based but experienced 125 percent growth and was purchased by a national retail sporting goods chain.
- Managed a retail design studio producing over $500,000 annually.
- Hired, trained, and supervised an administrative staff of 15 employees for a financial planning institution.
- Sold copiers through on-site demonstrations. Exceeded corporate sales goals by 10 percent annually.
- Provided purchaser training for office equipment, reducing labor costs by 25 percent.
- Acquired modern management skills through continuing education courses.

In a functional resume, emphasize skills you have acquired which relate to the advertised position. Also quantify your accomplishments.

WORK EXPERIENCE

Office manager, Simcoe Designs, Salem, OR (2008 to present).
Sales representative, Hi-Tech Office Systems, Salem, OR (2007 to 2008).
Office manager, Lueck Finances, Portland, OR (2005 to 2007).
President, Good Sports, Inc., Portland, OR (2002 to 2004).

In a functional resume, you still should list education and work experience in a reverse chronological order.

COMPUTER PROFICIENCY

Microsoft Office XP, Visual Basic 6, C++, Oracle, Microsoft SQL Server, Network Administration

MILITARY EXPERIENCE

Corporal, U.S. Army, Fort Lewis, WA (1997 to 2001). Honorably discharged.

- Served as company network administrator.
- Planned and budgeted all IT purchases.

A functional resume is organized by importance. Begin with the skills or accomplishments that will get you the job. Place less important information lower in the resume.

EDUCATION

BA, Communication Studies, Portland State University, Portland, OR (1996).

Scannable Resume

Many companies use computers to screen resumes with a technique called electronic applicant tracking. The company's computer program scans resumes as raster (or bitmap) images. Next, the software uses artificial intelligence to read the text, scanning for keywords. If your resume contains a sufficient number of these keywords, the resume will then be given to someone in the human resources department for follow-up.

A scannable resume can be e-mailed or sent through the U.S. mail. To create a scannable resume, type your text using Notepad for Windows, Simpletext for Macintosh, or Note Tab, which is available as freeware. You also can type your resume using Microsoft Word and save the document as a text file, with a *.txt* extension (Dikel 3).

To create a successful scannable resume, try these techniques:

- Use high-quality paper (send an original), optimum contrast (black print on white paper), and paper without wrinkles or folds (if you are mailing the resume).
- Use Courier, Helvetica, or Arial typeface (10- to14-point type).
- Place your name at the top of the page. "Scanners assume that whatever is at the top is your name. If your resume has two pages, place your name and a 'page two' designation on the second page, and attach with a paper clip—no staples" (Kendall).
- Avoid italics, underlining, colors, horizontal and vertical bars, and iconic bullets.
- White space is still important, but do not use your Tab key for spacing. Tabs will be interpreted differently in different computer environments. Use your space bar instead.
- Avoid organizing information in columns.
- Do not center text.
- Limit your screen length to approximately 60 characters per line.
- Hit a hard return at the end of each line instead of using word wrap.
- Use all-cap headings and place your text below the headings, spacing for visual appeal.
- Create bullets using an asterisk (*) or a hyphen (-).
- Create horizontal rules by using the equal sign (=) or underline (_).
- Use *keywords* in your summary of qualifications, work experience, and professional skills.

Keywords are the most important feature of scannable resumes. OCR (optical character recognition) searches focus on keywords and phrases specifically related to the job opening. The keywords include job titles, skills and responsibilities, corporate buzzwords, acronyms and abbreviations related to hardware and software, academic degrees, and certifications.

You can find which keywords to focus on by carefully reading the following:

- job advertisements
- your prospective employer's Web site
- government job descriptions
- industry-specific Web sites
- the *Occupational Outlook Handbook* (found online at http://www.bls.gov/oco/)
- career-related discussion groups or blogs
- sample resumes found online

When using keywords, be specific; avoid vague words and phrases.

BEFORE	AFTER
Knowledge of various software products	Can create online help using AuthorIT and have expertise with PageMaker and Quark

Table 11.3 Hardware and Software Keywords

NT	FrontPage
DOS	AutoCAD
Microsoft Office	LAN/WAN
Corel Office Suite	IBM Client Access
Lotus Notes	CAD/CAM
Novell GroupWise	Windows XP
Internet Explorer	Visual Basic
Netscape Communicator	IBM AS/400
Flow Chart	C++
Excel	Dreamweaver
PowerPoint	Flash

Table 11.4 Employment Skills Keywords

Oral presentations	Customer service
Oral communication	Telemarketing
Effective writing skills	Marketing
Interpersonal communication	Product information
Teamwork	Self-motivated
Flexibility	Organization
Time management	Innovation
Management	Ethical
Web design	Quality assurance
Project management	Training
Supervision	Problem solving

BEFORE	AFTER
Familiar with computer technology	Proficient in multimedia, HTML, and Windows and Macintosh platforms

Other industry keywords are shown in Table 11.3.

In addition to hardware and software skills, OCR also scans for "soft skills," as shown in Table 11.4.

Figure 11.4 shows an excellent example of a scannable resume.

RESUME CHECKLIST

_____ 1. Have you decided to write a reverse chronological resume or a functional resume?

_____ 2. Have you chosen which method of delivery to use—mail, e-mail, or scannable?

_____ 3. Have you included contact information (name, address, phone numbers, e-mail address, Web site URL, and/or fax number)?

_____ 4. Have you provided an objective line that is specific and focuses on company benefit?

_____ 5. In the work experience section, have you included your job title, company name, city, state, and date of employment?

_____ 6. In the education section of the resume, have you included your degree, school, city, state, and date of graduation?

_____ 7. Have you itemized specific achievements at work and school, beginning with verbs?

_____ 8. Have you avoided using sentences, passive voice, and the word "I"?

_____ 9. Does your resume use highlighting techniques to make it reader-friendly?

_____ 10. Have you proofread your resume to find grammatical and mechanical errors?

FIGURE 11.4 Scannable Resume

Place your name at the top of a scannable resume and avoid centering text.

Rochelle J. Kroft
1101 Ave. L
Tuscaloosa, AL 89403
Home: (313) 690-4530
Cell: (313) 900-6767
E-mail: rkroft90@aol.com

OBJECTIVES
==
To use HAZARDOUS WASTE MANAGEMENT experience and knowledge to ensure company compliance and employee safety.

SUMMARY OF QUALIFICATIONS
=======================================
* HAZARDOUS WASTE MANAGEMENT with skills in teamwork, end-user support, quality assurance, problem solving, and written documentation.
* Five years' experience working with international and national businesses and regulatory agencies.
* Skilled in assessing environmental needs and implementing hazardous waste improvement projects.
* Able to communicate effectively with multinational, cross-cultural teams, consisting of clients, vendors, coworkers, and local and regional stakeholders.
* Excellent customer service, using strong problem-solving techniques.
* Effective project management skills, able to multitask.

Use keywords to summarize your accomplishments.

COMPUTER PROFICIENCY
======================================
Microsoft Windows XP, FrontPage, PowerPoint, C++, Visual Basic, Java, CAD/CAM

EXPERIENCE
======================================
Hazardous Waste Manager
Shallenberger Industries, Tuscaloosa, AL (2007 to present)
* Assess client needs for root cause analysis and recommend strategic actions.
* Oversee waste management improvements, using project management skills.
* Conduct and document follow-up quality assurance testing on all newly developed applications to ensure compliance.
* Develop training manuals to ensure team and stakeholder safety. Shallenberger has had NO injuries throughout my management.
* Manage a staff of 25 employees.
* Achieved "Citizen's Recognition" Award from Tuscaloosa City Council for safety compliance record.

Use equal signs for design elements such as line dividers and asterisks (*) for bullets.

Hazardous Waste Technician
CleanAir, Montgomery, AL (2005–2007)
* Developed innovative solutions to improve community safety.
* Created new procedure manuals to ensure regulatory compliance.

Type your scannable resume in Courier, Arial, Verdana, or Helvetica. Avoid "designer fonts" like Comic Sans, Lucida, or Corsiva.

EDUCATION
====================================
B.S. Biological Sciences, University of Alabama, Tuscaloosa, AL (2005)
* Biotechnology Honor Society, President (2004)
* Golden Key National Honor Society

AFFILIATIONS
==================================
Member, Hazardous Waste Society International

CRITERIA FOR EFFECTIVE LETTERS OF APPLICATION

Your resume, whether hard copy or electronic, will be prefaced by a letter of application (or cover e-mail if you submit the resume as an e-mail attachment). These two components of your job package application serve different purposes.

The resume is generic. You'll write one resume and use it over and over again when applying for numerous jobs. In contrast, each letter of application is different, customized specifically for each job. Criteria for an effective letter of application include the following.

Additional information:

See Chapter 10, "Traditional Correspondence: Memos and Letters."

Letter Essentials

Letters contain certain mandatory components: your address, the date, your reader's address, a salutation, the letter's body, a complimentary close, your signed name, your typed name, and an enclosure notation if applicable. If you are submitting an electronic resume along with an e-mail cover message, you will not need these letter essentials. We discuss the e-mail cover message later in this chapter.

Introduction

In your introductory paragraph, include the following:

- Tell where you discovered the job opening. You might write, "In response to your advertisement in the May 31, 2009, *Lubbock Avalanche Journal* . . ." or "Bob Ward, manager of human resources, informed me that. . . ."
- State which specific job you are applying for. Often, the classified section of your newspaper will advertise several jobs at one company. You must clarify which of those jobs you're interested in. For example, you could write, "Your advertisement for a computer maintenance technician is just what I have been looking for."
- Sum up your best credentials. "My BS in chemistry and five years of experience working in a hazardous materials lab qualify me for the position."

Discussion

In the discussion paragraph(s), sell your skills. To do so, describe your work experience, your education, and your professional skills. This section of your letter of application, however, is not meant to be merely a replication of your resume. The resume is generic; the letter of application is specifically geared toward your reader's needs. Therefore, in the discussion, follow these guidelines:

- Focus on your assets uniquely applicable to the advertised position. Select only those skills from your resume that relate to the advertisement and which will benefit the prospective employer.
- Don't explain how the job will make you happy: "I will benefit from this job because it will teach me valuable skills." Instead, using the pronouns "you" and "your," show reader benefit: "My work with governmental agencies has provided me a wide variety of skills from which your company will benefit."
- Quantify your abilities. Don't just say you're great: "I am always looking for ways to improve my job performance." Instead, prove your assertions with quantifiable facts: "I won the 2009 award for new ideas that saved the company money."

Conclusion

Your final paragraph should be a call to action. You could say, "I hope to hear from you soon" or "I am looking forward to discussing my application with you in greater detail." You could tell your reader how to get in touch with you: "I can be reached at 913-469-8500." If you are more daring, you could write, "I will contact you within the next two weeks to make an appointment at your convenience. At that time, I would be happy to discuss my credentials more thoroughly."

In addition to these suggestions, you should mention that you have enclosed a resume. You can do this either in the introduction, discussion, or conclusion. Select the place that best lends itself to doing so.

E-Mail Cover Message

If you submit your resume as an attachment to an e-mail, you should write a brief e-mail cover message. This e-mail message will serve the same purpose as a hard-copy letter of application. Therefore, you want to include an introductory paragraph, a body, and a conclusion.

- **Introduction**—Tell the reader which job you are applying for and where you learned of this position.
- **Body**—State that you have attached a resume. Tell the reader which software you have used to write your resume.

example

> I have attached a resume for your review. To open this document, you will need Microsoft Word.
>
> I have saved the resume as an RTF (rich text file).

Briefly explain why you are the right person for the job. You can do this in a short paragraph or by briefly listing three to five key assets.

- **Conclusion**—Sum up your e-mail message pleasantly. Tell your reader that you would enjoy meeting him or her and that you look forward to an interview.

Online Application Etiquette

If you send your resume as an attachment to an e-mail message, be sure to follow online etiquette:

- **Do not use your current employer's e-mail system**—That clearly will tell your prospective employer that you misuse company equipment and company time.
- **Avoid unprofessional e-mail addresses**—Addresses such as Mustang65@aol.com, Hangglider@yahoo.com, or HotWheels@juno.com are inappropriate for business use. When you use e-mail to apply for a job, it is time to change your old e-mail address and become more professional. Use your initials or your name instead.
- **Send one e-mail at a time to one prospective employer**—Do not mass mail resumes. No employer wants to believe that he or she is just one of hundreds to whom you are writing.
- **Include a clear subject line**—Announce your intentions or the contents of the e-mail: "Resume—Vanessa Diaz" or "Response to Accountant Job Opening."

Additional information:

See Chapter 9, "Electronic Communication: E-mail Messages, Instant Messages, Text Messages, and Blogging."

Figure 11.5 shows an effective cover e-mail message prefacing an attached resume. Figure 11.6 is an example of an effective letter of application.

TECHNIQUES FOR INTERVIEWING EFFECTIVELY

The goal of writing an effective resume and letter of application is to get an interview. The resume and letter of application open the door; however, only a successful interview will win you the job. In fact, some sources suggest that the interview is *the most important stage of your job search*. The Society for Human Resource Management states that 95 percent of respondents to a survey ranked "interview performance" as a "very influential" factor when deciding to hire an employee. "Interview performance [is] more influential than 17 other criteria, including years of relevant work experience, resume quality, education levels, test scores or references" (Stafford L1).

Dress Professionally

Professionalism starts with your appearance. The key to successful dressing is to wear clean, conservative clothing. No one expects you to spend money on high-fashion, stylish

FIGURE 11.5 E-Mail Cover Message

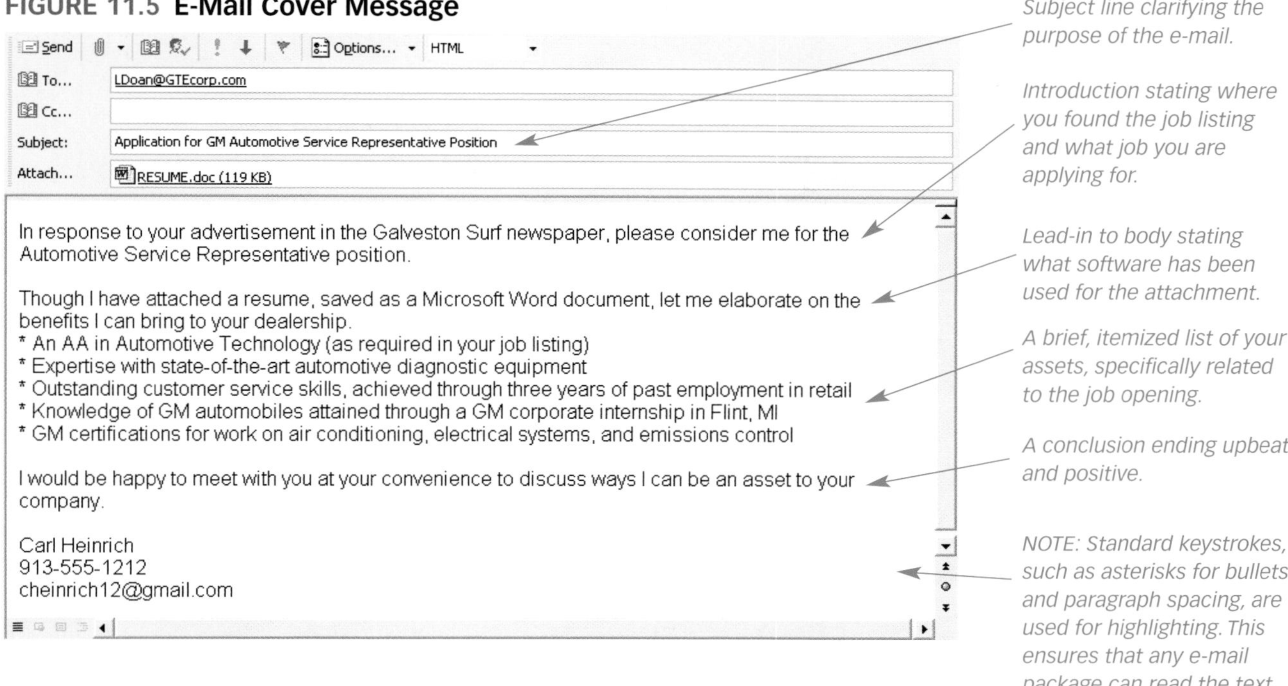
Send | Options... | HTML

To... LDoan@GTEcorp.com

Cc...

Subject: Application for GM Automotive Service Representative Position

Attach... RESUME.doc (119 KB)

In response to your advertisement in the Galveston Surf newspaper, please consider me for the Automotive Service Representative position.

Though I have attached a resume, saved as a Microsoft Word document, let me elaborate on the benefits I can bring to your dealership.
* An AA in Automotive Technology (as required in your job listing)
* Expertise with state-of-the-art automotive diagnostic equipment
* Outstanding customer service skills, achieved through three years of past employment in retail
* Knowledge of GM automobiles attained through a GM corporate internship in Flint, MI
* GM certifications for work on air conditioning, electrical systems, and emissions control

I would be happy to meet with you at your convenience to discuss ways I can be an asset to your company.

Carl Heinrich
913-555-1212
cheinrich12@gmail.com

FAQs

Q: Do prospective employers really care about how well I write? Aren't they more concerned about my area of expertise?

A: According to a report from The National Commission on Writing, employers care deeply about the quality of your writing.

- Writing is a basic consideration for hiring and promotion. More than 75 percent of respondents report that they take writing into consideration in hiring and promoting professional employees.
- State agencies frequently require writing samples from job applicants. Fully 91 percent of respondents that "almost always" take writing into account also require a writing sample from prospective "professional" employees.
- Poorly written applications are likely to doom candidates' chances of employment. Four of five respondents agree that poorly written materials would count against "professional" job applicants either "frequently" or "almost always."

("Writing: A Powerful Message from State Government")

clothes, but everyone expects you to look neat and acceptable. Business suits are still the best option for both men and women.

Be on Time

Plan to arrive at your interview at least 20 to 30 minutes ahead of schedule.

Watch Your Body Language

To make the best impression, don't slouch, chew your fingernails, play with your hair or jewelry, text, or check your watch. These actions will make you look edgy and impatient. Sit straight in your chair, even leaning forward a little to show your enthusiasm and energy. Look your interviewer in the eye. Smile and firmly shake your interviewer's hand.

FIGURE 11.6 Letter of Application

The introduction mentions where the writer found the job, which job he is applying for, and key qualifications for the position.

The letter's body quantifies the writer's abilities. The body also shows how the writer meets the job requirement.

The conclusion ends positively, using words and phrases like "convenience," "please," and "appreciate." It also provides follow-up information.

11944 West 112th Street
St. Louis, MO 66221

December 11, 2009

Mr. Harold Irving
CEO
DiskServe
9659 W. 157th St.
St. Louis, MO 78580

Dear Mr. Irving:

I am responding to your advertisement in the November 24, 2009, issue of *The St. Louis Courier* for a job in your computer technology department. Because of my two years experience in computer information systems, I believe I have the skills you require.

Although I have enclosed a resume, let me elaborate on my achievements. While working as a computer technician for Radio Shack, I troubleshot motherboards, worked the help desk, and made service calls to businesses and residential customers. Your advertisement listed the importance of customer service and technical skills. I have expertise in both areas.

Your job description also mentioned the importance of working in a team environment. At Radio Shack, I frequently made service calls with a team of technicians. Together, we provided quality service.

I would like to meet with you and discuss employment possibilities at your company. Please call me at (913) 469-8500 so that we can set up an interview at your convenience. I appreciate your consideration.

Sincerely,

Macy G. Heart

Enclosure (Resume)

Don't Chew Gum, Smoke, or Drink Beverages During the Interview

The gum might distort your speech; the cigarette will probably offend the interviewer, particularly if he or she is a non-smoker; and you might spill the beverage.

Turn Off Your Cell Phone

Today, cell phones are commonplace. However, the interview room is one place where cell phones must be avoided. Taking a call or texting while you are being interviewed is rude and will ensure that you will not be hired.

Watch What You Say and How You Say It

Speak slowly, focus on the conversation, and don't ramble. Once you have answered the questions satisfactorily, stop.

LETTER OF APPLICATION CHECKLIST

Letter of Application

_____ 1. Have you included all of the letter essentials?

_____ 2. Does your introductory paragraph state where you learned of the job, which job you are applying for, and your interest in the position?

_____ 3. Does your letter's discussion unit pinpoint the ways in which you will benefit the company?

_____ 4. Does your letter's concluding paragraph end cordially and explain what you will do next or what you hope your reader will do next?

_____ 5. Is your letter free of all errors?

E-Mail Cover Message

_____ 1. Have you avoided using an unprofessional e-mail address?

_____ 2. Do you have a precise subject line?

_____ 3. Did you include an introduction, body, and conclusion?

_____ 4. Did you tell the reader which software you have used for your attached resume?

_____ 5. Have you kept your e-mail message short (20 to 25 lines)?

_____ 6. Have you considered online e-mail application etiquette?

Bring Supporting Documents to the Interview

Supporting documents can include extra copies of your resume, a list of references, letters of recommendation, employer performance appraisals, a portfolio (paper copies and electronic version) of technical documents you have written, and transcripts.

Research the Company

Show the interviewer that you are sincerely interested in and knowledgeable about the company. Dr. Judith Evans, vice president of Right Management Consultants of New York, says that the most successful job candidates show interviewers that they "know the company inside and out" (Kallick D1).

Be Familiar with Typical Interview Questions

You want to anticipate questions you will be asked and be ready with answers. Some typical questions include the following:

What are your strengths and weaknesses?	Can you travel?
Why do you want to work for this company?	Will you relocate?
Why are you leaving your present employment?	What do you want to be doing in five years; in ten years?
What did you like least about your last job?	How would you handle this (hypothetical) situation?
What computer hardware are you familiar with, and what computer languages do you know?	What was your biggest accomplishment in your last job or while in college?
What machines can you use?	What about this job appealed to you?
What special techniques do you know, or what special skills do you have?	What starting salary would you expect?
What did you like most about your last job?	How do you get along with colleagues and with management?

When Answering Questions, Focus on the Company's Specific Need

For example, if the interviewer asks if you have experience using FrontPage, explain your expertise in that area, focusing on recent experiences or achievements. Be specific. In fact, you might want to tell a brief story to explain your knowledge. This is called "behavioral description interviewing" (Ralston et al. 9). It allows an interviewer to learn about your speaking abilities, organization, and relevant job skills. To respond to a behavioral description interview question, answer questions as follows (Ralston et al. 11):

- Organize your story chronologically.
- Tell who did what, when, why, and how.

- Explain what came of your actions (the result of the activity).
- Depict scenes, people, and actions.
- Make sure your story relates exactly to the interviewer's needs.
- Stop when you are through—do not ramble on. Get to the point, develop it, and conclude.

If, however, you do not have the knowledge required, then "explain how you can apply the experience you *do* have" (Hartman 24). You could say, "Although I've never used FrontPage, I have built Web sites using Netscape Composer and HTML coding. Plus, I'm a quick learner. I was able to learn RoboHelp well enough to create online help screens in only a week. Our customer was very happy with the results." This will show that you understand the job and can adapt to any task you might be given.

INTERVIEW CHECKLIST

_____ 1. Will you dress appropriately?
_____ 2. Will you arrive ahead of time?
_____ 3. Will you avoid gum, cigarettes, and beverages?
_____ 4. Have you practiced answering potential questions?
_____ 5. Have you researched the company so you can ask informed questions?
_____ 6. Will you bring to the interview additional examples of your work or copies of your resume?

CRITERIA FOR EFFECTIVE FOLLOW-UP CORRESPONDENCE

Once you have interviewed, don't just sit back and wait, hoping that you will be offered the job. Write a follow-up letter or e-mail message. This follow-up accomplishes three primary goals: it thanks your interviewers for their time, keeps your name fresh in their memories, and gives you an opportunity to introduce new reasons for hiring you.

A follow-up letter or e-mail message contains an introduction, discussion, and conclusion.

Additional information:

See Chapter 9, "Electronic Communication: E-mail Messages, Instant Messages, Text Messages, and Blogging," and Chapter 10, "Traditional Correspondence: Memos and Letters."

1. **Introduction**—Tell the readers how much you appreciated meeting them. Be sure to state the date on which you met and the job for which you applied.
2. **Discussion**—In this paragraph, emphasize or add important information concerning your suitability for the job. Add details that you forgot to mention during the interview, clarify details that you covered insufficiently, and highlight your skills that match the job requirements. In any case, sell yourself one last time.
3. **Conclusion**—Thank the readers for their consideration, or remind them how they can get in touch with you for further information. Don't, however, give them any deadlines for making a decision.

See Figure 11.7 for a sample of follow-up correspondence.

FIGURE 11.7 Follow-Up Correspondence

For an e-mail follow-up, you would include your reader's e-mail address and a subject line, such as "Thank You for the July 8 Interview" or "Follow-Up to July 8 Interview."

For a hard-copy follow-up letter, you would include all letter components: writer's address, date, reader's address, salutation, complimentary close, and signature.

Thank you for allowing me to interview with Acme Corporation on July 8. I enjoyed meeting you and the other members of the team to discuss the position of Account Representative.

You stated in the interview that Acme is planning to expand into international marketing. With my Spanish speaking ability and my study-abroad experience, I would welcome the opportunity to become involved in this exciting expansion.

Again, thank you for your time and consideration. I look forward to hearing from you. Please e-mail me at gfiefer21@aol.com.

The correspondence in Figure 11.7 succeeds for several reasons. First, it is short, merely reminding the reader of the writer's interest, instead of overwhelming him or her with too much new information. Second, it is positive, using words such as "enjoyed," "ability," "welcome," "opportunity," "exciting," and "thank you." Finally, the correspondence provides the reader an e-mail address for easy follow-up.

FOLLOW-UP CORRESPONDENCE CHECKLIST

_____ 1. Have you included all the letter essentials or written your e-mail to the correct address?

_____ 2. Does your introductory paragraph remind the readers when you interviewed and what position you interviewed for?

_____ 3. Does the discussion unit highlight additional ways in which you might benefit the company?

_____ 4. Does the concluding paragraph thank the readers for their time and consideration?

_____ 5. Does your correspondence avoid all errors?

JOB ACCEPTANCE LETTER

Great news! After working hard to find a job, your efforts have paid off. You've just received a job offer. Now what? Sometimes, accepting the offer over the phone isn't enough. Your new employer might want you to write and sign an official acceptance letter. In this brief letter, you will want to accomplish the following goals:

- Thank the company for the job opportunity
- Officially accept the job offer
- Restate the terms of employment (salary, benefits, location, position, job responsibilities, and/or start date)

Address the letter to whomever offered you the position for a more personalized touch. Be sure to include your phone number, e-mail address, or mailing address, just in case the company needs to contact you.

This acceptance letter actually could be seen as your first day on the job. Therefore, take as much care in writing this letter as you did in applying for the job. Make sure your letter is grammatically correct, well organized, and conveys a positive tone. Show your new boss or colleagues that you are a professional asset to the company. Figure 11.8 provides a sample Job Acceptance Letter.

JOB ACCEPTANCE LETTER CHECKTLIST

_____ 1. Have you included all the letter essentials?

_____ 2. Does your introductory paragraph explain *why* you are writing (in response to a job offer) and *what* you are writing about (accepting the job)?

_____ 3. Does the discussion unit confirm the particulars of the offer (salary, benefits, job duties, location, start date, and so on)?

_____ 4. Does the concluding paragraph thank the reader for the job opportunity?

_____ 5. Does your letter avoid errors and show your professionalism?

FIGURE 11.8 Job Acceptance Letter

Amy Zhang
9103 Stonefield Rd.
Georgetown, TX 77829

May 15, 2009

George Smithson
Dell
2134 Silicon Way
Round Rock, TX 77112

Dear Mr. Smithson:

In response to your phone call yesterday (5/14/09), I am very happy to accept the position as a Technical Sales Representative II at Dell Computers. I know that with my education, experience, and energy, I will be an asset to your workforce.

As we discussed, my salary will be $35,000 with health and life insurance benefits provided after 90 days of employment. I will work in your Round Rock, Texas, facility, where my job will entail working with a team to meet the schedules, budgets, product costs, and production ramp rates for projects assigned to me and my team. In addition, I will manage small to medium platform/peripheral development programs.

I look forward to joining the Dell team on June 10, 2009. If you have any questions or need additional information, please let me know. Thank you, Mr. Smithson, for this outstanding opportunity.

Sincerely,

Amy Zhang
713-555-0112
azhang@gt.edu

LaShanda Met Her Challenge

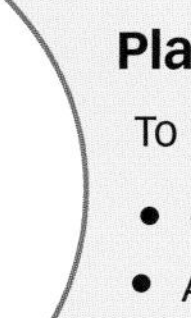

To meet her communication challenge, LaShanda used the *P*³ process.

Planning

To plan her job search, LaShanda considered the following:

- Goal—to find a job in the field of human resources
- Audience—directors of human resources
- Channels—resume and letter of application
- Data—information drawn from her experiences

According to LaShanda, "Since my job search for a position in human resources had not been successful, I needed to refine my search technique. Before I wrote a letter of application and resume, I planned the correspondence by considering the reporter's questions:

- Who should I send my resume and cover letter to?
- What should I include about my experience in human resources?

- When is the application due?
- Where should I send the application (what's the city/state address, e-mail address, or URL)?
- Why am I interested in this job posting?
- How should I send the resume and application—by e-mail, online form, or hard copy?

Next, I called human resources departments to ask the following questions:

- What is your hiring practice?
- Does your company want resumes and cover letters to be submitted by fax, e-mail, or online through the company's Web site?
- What's your company's timeframe for decision making?
- Once I apply, will I be notified that you've received my application or if I've been selected for an interview?"

Packaging

LaShanda went to her husband's military career center for help writing her resume. They told her to use a functional resume because that's what they tended to advocate for people leaving the military. Functional resumes are especially valuable in the military to help service personnel transition into the civilian workplace. When she applied for jobs, she says, "I just downloaded my resume into a company's Web site." See Figure 11.9 for LaShanda's functional resume draft.

Perfecting

LaShanda's functional resume didn't work, especially for a city government job. The functional resume listed jobs in one part of the resume and then responsibilities in another part of the text. This format didn't allow the director of human resources for the city to match dates and duties. Her functional resume left too many holes. To meet the city's needs for more specificity, LaShanda changed strategies. She says, "I restructured my functional resume into a chronological resume. Plus, rather than merely downloading the resume into the city's Web site, I also filled in their customized field requirements. This way, I could make sure that my achievements were clear and quantified. This allowed the city to see that I met the minimum requirements for the job openings in human resources. The result? I got a job in human resources, and now I hire for the city!" Figure 11.10 is LaShanda's perfected chronological resume.

CHAPTER HIGHLIGHTS

1. Use many different resources to locate possible jobs, such as college placement centers, instructors, friends, professional affiliations, want ads, and the Internet.
2. Use either a reverse chronological resume or a functional resume.
3. Write either a traditional hard-copy resume, an e-mail resume, or a scannable resume.
4. Indicate a specific career objective on your resume.
5. Write your letter of application so that it targets a specific job.
6. Prepare before your interview so you can anticipate possible questions.
7. Use a follow-up letter or e-mail message to impress the interviewer and remind him or her of your strengths.
8. Over 60 percent of jobs are found through networking.
9. Relying only on the Internet for your job search is a mistake. Less than 4 percent of new jobs are obtained this way.
10. On your resume, place education first if that is your strongest asset, or begin with work experience if this will help you get the job.

WWW *Find additional employment communication exercises, samples, and interactive activities at http://www.prenhall.com/gerson.*

FIGURE 11.9 Example of LaShanda's Functional Resume

LaShanda did not change her career objective's line to match her new search for a job in human resources.

The functional resume does not show when and where she acquired her expertise.

The resume's content does not clearly relate to the field of human resources. Since this resume was submitted online, LaShanda needed to include keywords which OCR (optical character recognition) software would recognize.

LaShanda E. Brown
833 Hampton Rd., Apt 1
Virginia Beach, VA 23460
757-555-5555
labrown10@yahoo.com

Objective: Retail Management

HIGHLIGHTS OF QUALIFICATIONS

- 3 years retail management and recruiting experience.
- Expertise in oral and written communications. Able to present ideas and goals clearly.
- Motivated team player that thrives in a fast-paced, multifaceted environment.
- Proficient Microsoft Office (Word, Excel, PowerPoint, Outlook, Access), as well as basic office hardware (fax, copiers, cash registers).
- Familiar with most video game consoles (PS1, PS2, Xbox, NES, etc.).

EXPERIENCE

Management

- Supervised five management trainees and taught them all aspects of the rental car industry to include accounts receivable, marketing accounts as well as competition, managing a rental fleet, etc.
- Oversee four telemarketers and two sales representatives. Managed training and quality customer service when representing the company via telephone.
- Improved loss prevention programs within a retail setting by breaking the store up into teams and zones so that all areas were covered at all times. Loss prevention decreased 6 percent in 3 months.
- Earned all-stars sales award four months in a row. Average of 91 percent optional coverage sales on rentals within that four-month frame.

Customer Service

- Implemented new strategies for customer satisfaction, such as interactive games to give customers an opportunity to win coupons on current and future rentals. Customer satisfaction increased 15 percent.
- Greeted customers enthusiastically and efficiently to provide a positive company image and impression.
- Researched and provided solutions to customer service issues, ensuring customer needs were met and they were satisfied.
- Coordinated customer follow-ups ensuring they had a good rental experience with us and would return if they needed our rental services again.

EMPLOYMENT

August 2003–Present	Office Manager	Best Value Remodeling
January 2003–July 2003	Insurance Sales Agent	Geico Direct
May 2001–December 2002	Management Assistant	Enterprise Rent-A-Car

EDUCATION AND TRAINING

- Bachelors in Communication, emphasis in Public Relations
  Missouri Western State College
- Elite Sales Training, Enterprise Rent-A-Car
- Effective Management Training, Best Value Remodeling

FIGURE 11.10 Example of LaShanda's Perfected Reverse Chronological Resume

LaShanda E. Brown
833 Hampton Rd., Apt 1
Virginia Beach, VA 23460
757-555-5555
labrown10@yahoo.com

EDUCATION
Bachelor of Science in Communication, Public Relations emphasis, 5/2001
Missouri Western State College, St. Joseph, MO

- Related courses: small group communication, presentational communication, consumer marketing, persuasive speech, media in communications, public relations communication analysis, human resources management, nonverbal communication, advertising, desktop publishing.

EMPLOYMENT
Office Manager
Best Value Remodeling   Virginia Beach, VA   7/2003–2/2005

- Screen and hire new employees, as well as doing reviews for employees for raises
- Make spreadsheets using MS Excel to track sales by outside sales team and manage commissions
- Update Microsoft Access database to track results from sales calls
- Create PowerPoint presentations for general manager to use when going on corporate sale leads
- Oversee four telemarketers and two sales representatives

Insurance Sales Agent
Geico Direct   Virginia Beach, VA   1/2003–6/2003

- Provided excellent customer service on the phone
- Performed data entry by setting up insurance policies

Management Assistant
Enterprise Rent-A-Car   Oklahoma City, OK   5/2001–12/2002

- Recruited at career fairs
- Won award as "Best New Hire 2001"
- Supervised five management trainees
- Created advertisements locally for the branch
- Improved loss prevention programs by 6 percent in three months
- Earned All-stars sales award four months in a row. Average of 91 percent optional coverage sales on rentals within that four-month frame

Customer Service Supervisor
Office Max   St. Joseph, MO   9/1998–5/2001

- Trained new hires
- Worked front desk
- Provided customer service

Data Entry Clerk
Internal Revenue Service   Oklahoma City, OK   12/1998–5/1999

- Input data entry of tax forms from scanned images
- Downloaded data entry from scanned microfilm into IRS database

SOFTWARE KNOWLEDGE

- Proficient with Microsoft Office programs, including MS Word, MS Excel, MS Access, MS PowerPoint, MS Outlook, and MS Project
- Proficient with PeopleSoft program (HR module)
- Proficient with Lotus programs, including Lotus Notes, Lotus 1-2-3

LaShanda says, "By leading with education, I'm emphasizing my skills as they relate to a job search in human resources."

"My 'employment' section highlights human resources experience, allowing the city's human resources department to see that I meet the requirements for a position."

"The chronological resume better suits my needs in that it matches job responsibilities with specific jobs and dates."

CASE STUDIES

Macy Heart is interested in changing jobs. However, his resume has some problems that need corrections.

Assignment

1. DiskServe, a St. Louis-based company, is hoping to hire a customer service representative for their computer technology department. In addition to DiskServe's CEO, Harold Irving, the hiring committee will consist of two managers from other DiskServe departments and two coworkers in the computer technology department.
2. Rewrite the flawed resume on page 335. Revise the errors and create three types of resumes for Macy G. Heart—a chronological resume, a functional resume, and a scannable resume.

The position requires a bachelor's degree in Information Technology (or a comparable degree) and/or four years' experience working with computer technology. Candidates must have knowledge of C++, Visual Basic (VB), SQL, Oracle, and Microsoft Office applications. In addition, customer service skills are mandatory.

Four candidates were invited to DiskServe's work site for personal interviews—Brian McHenry, Aaron Brown, Rosemary Lopez, and Robin Scott.

Brian has a bachelor's degree in Computer Information Systems. He has worked two years part time in his college's technology lab helping faculty and students with computer hardware and software applications, including Microsoft Office and VB. He worked for two years at a computer hardware/software store as a salesperson. Brian's supervisor considers him to be an outstanding young man who works hard to please his supervisors and to meet customer needs. According to the supervisor, Brian's greatest strength is customer service, since Brian is patient, knowledgeable, and respectful.

Aaron has an Information Technology certificate from Microsoft, where he has worked for five years. Aaron began his career at Microsoft as a temporary office support assistant, but progressed to a full-time salesperson. When asked where he saw himself in five years, Aaron stated, "The sky's the limit." References proved Aaron's lofty goals by calling him "a self-starter, very motivated, hardworking, and someone with excellent customer service skills." He is taking programming courses at night from the local community college, focusing on C++, Visual Basic, and SQL.

Rosemary has an associate's degree in Information Technology. She has five years experience as the supervisor of Oracle application. Prior to that, Rosemary worked with C++, VB, and SQL. She also has extensive knowledge of Microsoft Office. Rosemary was asked, "How have you handled customer complaints in the past?" She responded, "I rarely handle customer complaints. In my past job, I assigned that work to my subordinates."

Robin has a bachelor's degree in Information Technology. To complete her degree, Robin took courses in C++, VB, Oracle, and SQL. She is very familiar with Microsoft Office. Since Robin just graduated from college, she has no full-time experience in the computer industry. However, she worked in various retail jobs (food services, clothing stores, and book stores) during high school, summers, and in her senior year. She excelled in customer service, winning the "Red Dragon Employee of the Month Award" from her last job as a server in a Chinese restaurant.

Assignment

Who would you hire? Give an oral presentation or write an e-mail or memo to Harold Irving explaining which of the candidates he should hire. Explain your choice.

1890 Arrowhead Dr.
Utica, MO 51246
710-235-9999

Resume of Macy G. Heart

Objective: Seeking a position in Computer Technology Customer Service where I can use my many technology and people-person skills. I want to help troubleshoot software and hardware problems and work in a progressive company which will give me an opportunity for advancement and personal growth.

Work Experience

Jan. 2006 to now   Aramco.net   St. Louis, Missouri   Tech Support Specialists
Primarily I provide customer support for customer problems with C ++, Visual Basic, Java, Networking, and Databases (Access, Oracle, SQL). I provide solutions to software and hardware problems and respond to e-mail queries in a timely manner.

Oct. 2004 to Dec. 22, 2005   DocuHelp   Chesterfield, MO   Computer Consultant
Provide technical support for PC's and Mac's. I also trained new PC and Mac users in hardware applications. When business was slow, I repaired computer problems, using my many technology skills.

May 2002 to Oct. 2003   Ram-on-the-Run   East St. Louis, Illinois   Computer Salesman
Sold laptops, PCs, printers, and other computer accessories to men and women. Answered customer questions. Won "Salesman of the Month Award" three months in a row due to exceeding sales quotas.

Jan. 2002 to May 2002   Carbondale High School   Carbondale, IL   Lab Tech
Worked in the school's computer lab, helping Mr. Jones with computer-related classwork. This included fixing computer problems and tutoring new students having trouble with assignments.

Education

Aug. 2006   Bachelor's Degree, Information Technology, Carbondale Institute of Technology, Carbondale, IL
Concentration: Database/Programming Applications
Relevant classes: Business Information Systems, Hardware Maintenance, Database Management, Visual Basic, Systems Analysis and Design, C++, Web Design

May 2002   Graduate Carbondale High School
Member of the Computer Technology Club
Member of FFA and DECA
Principal's Honor Roll, senior year

Computer Expertise

Cisco Certified, knowledge of C++, VB, Microsoft Office Suite XP, HTML, Java, SML

References

Mr. Oscar Jones, Computer Applications Teacher, Carbondale High School
Mr. Renaldo Gomez, Manager, Aramco.net
Mr. Ted Harriot, Technical Support Supervisor, DocuHelp

Additional Information

A good team player, who works well with others
Made all A's in my college major classes
Built my own computer from scratch in high school
Starting football player in high school, Junior Varsity tight end

ETHICS IN THE WORKPLACE CASE STUDY

Shelly Gurwell just graduated from college with a degree in business. She has begun interviewing for jobs at brokerage firms. During one interview, Shelly was asked the following questions:

1. Do you plan on getting married?
2. Do you want to have children?
3. When you have children, who will take care of them if they get sick?
4. Would you move to another city if your husband had to relocate?
5. What's your undergraduate GPA?
6. Do you require days off for religious observation?
7. What are your salary requirements?
8. Do you plan to participate in our company's wellness and weight loss program?
9. Do you have any tattoos, and if so, where are they?
10. Were you in a sorority?

Question: Which of these questions are ethical? Justify your answer based on information provided in Chapter 1.

INDIVIDUAL AND TEAM PROJECTS

1. Practice a job search. To do so, find examples of job openings in newspapers, professional journals, at your college or university's job placement service, and online. Bring these job possibilities to class for group discussions. From this job search, you and your peers will get a better understanding of what employers want in new hires.
2. An informational interview can help you learn about the realities of a specific job or work environment. Interview a person currently working in your field of interest. You can find such employees as follows:

   - **Alumni office**—Ask your college or university for a list of alums who are willing to speak to students about careers.
   - **Career center**—Your school's career center might give you names of people to contact.
   - **Human resources**—Visit a company in your field of interest and ask the human resources staff for help.
   - **Networking**—Do you have friends or family who know of employees in your field of interest?

   Once you find an employee willing to help, visit with him or her and ascertain the following:

   - What job opportunities exist in your field?
   - What does a job in your field require, in terms of writing, education, interpersonal communication skills, teamwork, and so on, as well as the primary job responsibilities?

   After gathering this information, write a thank-you letter to the employee who helped you. Then, write a report documenting your findings and give an oral presentation to your classmates.

3. Write a resume. To do so, follow the suggestions provided in this chapter. Once you have constructed this resume, bring it to class for peer review. In small groups, discuss each resume's successes and areas needing improvement. If you prefer, make transparencies of each resume and have the entire class review them for suggested improvements.
4. Write a letter of application. To do so, find a job advertisement in your newspaper's classified section, in your school's career planning and placement office, at your

work site's personnel office, or in a trade journal. Then construct the letter according to the suggestions provided in this chapter. Next, in small groups review your letter of application for suggested improvements.

5. Practice a job interview in small groups, designating one student as the job applicant and other students as the interview committee. Ask the applicant the sample interview questions provided in this chapter or any others you consider valid. This will give you and your peers a feel for the interviewing process.

DEGREE-SPECIFIC ASSIGNMENTS

1. **Research Your Field**—Whether you are majoring in Marketing and Public Relations, Information Technology, Accounting, Management, or Finance, research your field by going to the *Occupational Outlook Handbook* (http://www.bls.gov/oco/). In this site, you can find the following:
   - Nature of the work
   - Working conditions
   - Training and education needed
   - Advancement potential
   - Employment outlook
   - Earnings

   Either orally or in an e-mail message, report your findings regarding your job outlook.
2. **Research Yourself**—You are not right for every job, nor is every job right for you. Based on the information discovered in activity 1, ask yourself the following questions:
   - What are your skills, attributes, and accomplishments?
   - What do you have to offer this company?
   - What can you bring to the company that is unique?
   - What are your selling points?
   - How do your qualifications coincide with the position's requirements?
   - What interests you about this company and the position?

   Either orally or in an e-mail message, report your findings regarding your job outlook.

PROBLEM-SOLVING THINK PIECES

1. You need to submit a resume for a job opening. However, you have problems with your work history, such as one of the following:
   - You have had no jobs.
   - You have worked as a baby sitter, or you have cut grass in your neighborhood.
   - You have been fired from a job.
   - You have had five (or more) jobs in one year.

   Consider how you would meet the challenges of your job history.
2. You have found a job that you want to apply for. The job requires a bachelor's degree in a specific field. Though you were enrolled in that specific degree program for three years, you never completed the degree.

   What should you say in your resume and letter of application to apply for this position, even though you do not meet the degree requirement? Which type of resume (reverse chronological or functional) should you write? Explain your answer.
3. You are in the middle of an interview. Though the interview was scheduled for 1:00 p.m. to 1:45 p.m., it is running late. You had planned to pick up your son from daycare at 2:00 p.m. How do you handle this problem?
4. You have just completed the interview process for a job in your field. You have impeccable credentials, meeting and exceeding all of the requirements. Your interview went very well. At the close of the interview, one of the interviewers says, "Thank you for interviewing with us. You did a great job answering our questions,

and your credentials are truly excellent. However, I think you would find the job unchallenging, maybe even boring. You are overqualified." How should you handle this situation?

5. During an interview, you are asked to "describe a problem you encountered at work and explain how you handled that challenge." How do you answer this typical question, but avoid giving an answer that paints a negative picture of a boss, coworker, or your work environment?

WEB WORKSHOP

1. Access any Internet search engine to find information about a job search. You can go online to research resumes, cover letters, and follow-up thank-you notes. For example, check out JobsMart.org and Monster.com for resume guidelines, samples, resume makeovers, and do's and don'ts. Research CareerJournal.com for up-to-date articles about resume writing (tips, samples, whether you should pay to have someone write a resume for you, and case studies about "red-flag" resumes). CareerPerfect.com provides guidelines for electronic resume submission and techniques for creating scannable versions of your resume. Check out JobSearchTech to learn why and how to write follow-up thank-you letters, and to see samples.

   Once you research any of these sites, analyze your findings. How do the letters and resumes compare to those discussed in this textbook? What new information have you learned from the online articles?

   a. Report your findings, either in an oral presentation or in writing (e-mail, memo, letter, or report).
   b. If you find sample letters or resumes that you dislike, rewrite them according to the criteria provided in this textbook.

2. Using an Internet search engine, find job openings in your area of interest. Which companies are hiring, what skills do they want from prospective employees, and what keywords are used to describe preferred skills in this work field? Report your findings either to your instructor by writing a memo or e-mail, or give an oral presentation to your class about the job market in your field.

QUIZ QUESTIONS

1. What are four places you can search for a job?
2. What are two types of resumes?
3. What are some optional inclusions in your education section?
4. What types of personal data should you exclude?
5. How can you design a resume effectively?
6. What is the main difference between a resume and a letter of application?
7. What are two ways to conclude a letter of application?
8. What are five things you can do to interview effectively?
9. What are three things you can accomplish in a follow-up letter?
10. How can networking help you find a job?

CHAPTER 12

Communicating Bad News in the Workplace

▸REAL PEOPLE *in the workplace*

David King is a vice president of First National, a savings and loan company with over 30 branches. He oversees dozens of employees and deals with thousands of customers. He occasionally must convey bad news internally to employees and externally to customers. He says, "We have to open our drive-through windows at 8:30 a.m., so if an employee isn't on the job, we lose money and fail to meet our customers' expectations. Tellers have to be here when the windows open." If an employee is late or fails to show up, that issue must be addressed.

For rare absences, David pulls employees aside and pleasantly reminds them of job expectations. If problems become habitual, then David communicates the bad news, called **"corrective reviews."**

Customers also occasionally receive bad news from the bank. For instance, if a client's loan application is denied, the customer receives an **adverse action letter.** The letter can be a partial approval or 100 percent denial. The customer might have asked for a $100,000 home loan but only qualifies for $80,000. In these instances, David responds with a partial approval letter communicating the bad news. He will write, "We are sorry we are unable to grant the loan you requested, but are pleased to offer you a loan with the following terms."

On other occasions, David must deny the customer's loan request completely. A 100 percent denial could be based on the customer's insufficient employment or poor credit history. The customer might have had insufficient collateral. If David and his bank cannot verify employment, income, residence, or credit rating, this also could lead to an adverse action letter. In these instances, David writes, "We are sorry but we are unable to grant the loan you requested at this time."

David, like all good managers, wants employees and customers to be happy with the work environment. However, occasionally bad news must be communicated. When David and his bank document bad news, they must be precise, meeting national banking and employment regulations. Dollars and "sense" demand directness and factual documentation.◂

CHAPTER GOALS

When you complete this chapter, you will be able to do the following:

1. Communicate bad news by
    - Selecting the appropriate communication channel.
    - Using positive words versus negative words.
    - Establishing rapport with the audience by using pronouns.
    - Explaining the reasons behind the bad news thoroughly.
    - Choosing the direct or indirect method of organization.
2. Write different types of bad news letters, e-mail messages, and memos, including
    - Complaint
    - 100 percent negative response
    - Partial adjustment
    - Bad news from a company to a customer or vendor
    - Bad news from a company to an employee
3. Use the P^3 process to write effective bad news messages.

WWW

To learn more about communicating bad news, visit our Web Companion at http://www.prenhall.com/gerson.

David's Communication Challenge

A highly ranking manager at First National is causing problems. David and other employees have reported that the bank's teller supervisor, Joan O'Brien, consistently has made sarcastic comments to subordinates and has yelled at both subordinates and customers. In addition to these interpersonal communication concerns, Joan also has performance-related issues. She has arrived late to work, her division has decreasing productivity, and on numerous occasions tellers under her supervision have left unbalanced budget records. All of this has led to poor reviews from her subordinates, colleagues, customers, and management.

So far, David has approached the situation according to the following bank protocol.

- David says, "I gave Joan a 'first verbal counseling.' In this meeting, I pleasantly but firmly told Joan what was expected on the job, documented where rules had not been followed, and explained what would happen next if her behavior continued.
- When Joan's work habits did not improve, I provided what the bank calls 'second verbal counseling.' In this second, less pleasant meeting, I asked Joan why she was still not following codes of conduct and sternly explained what would happen if her behavior persisted."
- When Joan's performance problems continued beyond the first two counseling meetings, Joan received a "job in jeopardy" letter from the bank's human resources department documenting the problems and placing her on 30- to 90-day probation.

Despite all of David's counseling and the bank's official letter, Joan's poor work behavior has continued. Now, David is challenged by his next step, delivering the bad news that the bank is terminating her as teller supervisor.

See pages 360–362 to learn how David met his communication challenge.

REASONS FOR COMMUNICATING BAD NEWS

Communicating bad news is difficult. No one wants to give or receive bad news. When you state bad news in business, you often have to do so and still retain that person as a customer, vendor, or employee. Anytime you need to share bad news, you have a communication challenge.

You might need to communicate bad news for the following reasons:

- Rejecting a job applicant
- Telling a vendor that a proposal has not been accepted

- Rejecting a customer's request for a refund
- Firing an employee
- Telling stockholders that your company's quarterly earnings are down
- Informing staff of an impending merger which could result in company-wide changes

TIPS FOR COMMUNICATING BAD NEWS

Though communicating bad news is challenging, you still must handle the job. Try these tips to be an effective communicator of bad news.

- **Do not blame or criticize your reader or listener**—If you have to discipline an employee, avoid attacking the person. When people are attacked, they are more likely to fight back. Address the behavior or issue instead.
- **State the bad news clearly and concisely**—Once you have the audience's attention, you must deliver the bad news clearly. Do not equivocate or convey bad news in vague, abstract terms. If you do so, the audience will not understand the problem.
- **Avoid conveying bad news angrily**—Balance your emotions when communicating bad news. When you convey bad news, do so calmly and professionally. Similarly, if your audience responds with emotion, you have to maintain your composure. Reacting in anger is unprofessional.
- **Be polite when you state the bad news**—Use positive words and pronouns to motivate your audience rather than negative words to demean.
- **Empathize with the recipient of the bad news**—Put yourself in the audience's place. What would *you* need to read or hear to accept bad news?
- **Be careful about apologizing**—If a customer complains about late deliveries, you can apologize for this oversight without fear of reprisal. However, if one of your delivery trucks has been in a wreck and the blame has not yet been determined, refrain from a quick apology. Apologizing might suggest your company is at fault.

Companies often refrain from admitting wrongdoing, fearing litigation. At times, however, an apology can mitigate the need for court action. Courtroom litigation is costly; therefore, settlements are preferred by attorneys, judges, and corporate executives. Sometimes, a simple apology can contribute to an out-of-court settlement (Patel and Reinsch 10–12).

CRISIS COMMUNICATION

One illustration of extremely bad news is a crisis situation. Today, workplace communicators must be prepared to communicate information about crises that occur on the job. A crisis could occur for any number of reasons including an armed intruder on the premises, severe weather, a bomb threat, receipt of potentially hazardous material through the mail system, a terrorism threat, and more. According to Gerard Braud, award-winning journalist and specialist in crisis management, one of the most important aspects of crisis management is to have an established "crisis communication plan" ("Crafting a Crisis Communication Plan"). The following are the three components of an effective plan:

- **Create an emergency operations plan**—This includes coordinating with police, fire, and emergency medical services. In addition, a company needs to focus on safe evacuation of buildings, use off-premise meeting sites during the crisis, realize the importance of backing up critical computer files—on and off site, determine the location of temporary offices, and establish chains of authority and responsibility.
- **Develop a means of communication**—This plan establishes how to communicate in a crisis situation. Include who does what and which communication channels will be used. These channels could include telephone calls, e-mail messages, text messages, or loudspeaker announcements. Additional means of electronic communication include social media, such as Web postings, blogs, podcasts, Flickr, and Twitter (Griffin).

- **Institute a business continuity plan**—This plan explains how to restart the business after the crisis has subsided. Such a plan can calm the concerns of stakeholders, keep stock values from falling, ensure that company revenues continue to flow, and maintain the integrity of the company.

CRITERIA FOR COMMUNICATING BAD NEWS

If the situation is not a crisis, when you communicate bad news, do the following to avoid angering or offending your reader.

- Select the appropriate communication channel.
- Use positive versus negative words.
- Establish rapport with the audience by using pronouns.
- Explain the reasons behind the bad news thoroughly.
- Choose the direct or the indirect method of organization.

Select the Appropriate Communication Channel

When you communicate bad news, choose the appropriate communication channel. This could include the telephone, e-mail, your corporate Web site, newspapers, radio, television, newsletters, memos, or letters. The communication channel you choose will be affected by your audience, your familiarity with this person or these individuals, and the subject matter.

Table 12.1 shows the pros and cons of using various communication channels.

Additional information:

See Chapter 1, "Communicating in the Workplace."

Use Positive Words versus Negative Words

Audiences respond favorably to positive words. If you use negative words, you could offend your reader. In contrast, positive words will help you control your readers' reactions, build goodwill, and persuade your audience to accept your point of view.

Choose your words carefully. Even when an audience expects bad news, they still need a polite and positive response. The words in Table 12.2 can cause a negative audience reaction.

When you can, use positive words, like those in Table 12.3, to cushion the news and encourage your reader to accept your point of view.

Consider the following examples of negative versus positive words.

BEFORE	AFTER
1. Do not send responses to RFPs until after April 15.	1. Please send your responses to RFPs after April 15.
2. We are unable to process your request. You failed to follow the instructions.	2. We will be happy to process your request once you provide the information we have highlighted on the attached form.
3. The error is your fault. You scheduled incorrectly and cannot complain about our deliveries. If you would cooperate with us, we would work with you to solve this problem.	3. To improve deliveries, let's work together on our companies' scheduling practices.
4. I cannot grant your request for a company car on July 7.	4. We will be able to grant your request for a company car after July 7. The corporate fleet is being serviced until that time.
5. I regret to inform you that we will not replace the motor in your dryer unless we have proof of purchase.	5. When you provide us proof of purchase, we will be happy to replace the motor in your dryer.

Establish Rapport with the Audience by Using Pronouns

It's bad enough when people receive bad news. It's worse, however, if your audience members feel as if they are being treated like numbers instead of valued employees or customers. One way to achieve rapport with the audience is through using pronouns, which personalize your communication.

Table 12.1 Communication Channels for Delivering Bad News

Communication Channels	Pros	Cons
Social media (text messages, instant messages, and blogs—Flickr, Twitter, Blogger, Facebook, and MySpace)	Many people can be reached instantaneously. These types of communication are especially important for disseminating information and managing crises.	People could panic. People have to be at their computers or have their cell phones turned on.
E-mail	E-mail saves time, ensures an accurate message, and might be less uncomfortable for the messenger, according to the Institute for Operations Research and the Management Sciences.	Using e-mail to convey bad news could be seen as passive–aggressive. It is a way of avoiding speaking face to face.
Web sites	Web sites allow you to provide instantaneous news to an infinite number of people at geographically diverse locations.	Web sites disallow privacy and are impersonal.
Newsletters	Through a corporate newsletter, you can communicate bad news to employees and shareholders. You also can provide a great deal of detail.	Newsletters are not private and are impersonal.
Media (television, newspaper, radio, press releases)	Use the media to communicate bad news which impacts the public.	The media is not private. In addition, when using the media to communicate bad news, your company spokesperson must be careful about what is conveyed about the company. Public disclosure of bad news about a corporation could affect its stock value.
Memos	Memos, as internal correspondence, allow for the private dissemination of bad news. A memo can be directed to one person. Memos also allow you to document the bad news and develop it thoroughly.	As with e-mail, a memo could be seen as passive–aggressive. Face-to-face communication might be a more appropriate channel.
Letters	Letters document your bad news, allow you to develop the content clearly, and help build rapport through proper word usage.	Letters could be perceived as impersonal, even "corporate."
Face-to-face meetings	Use face-to-face meetings for at least two reasons: a large group of stakeholders can hear the bad news at the same time; an individual meeting provides a more personal touch.	People in a large audience can misinterpret your comments. You also might not have clear documentation of what has been discussed.
Telephone	The telephone is effective if you are communicating bad news and cannot meet the person face to face.	When speaking on the phone, neither the speaker nor the listener can see facial expressions or body language. Telephone calls disallow documentation.

Table 12.2 Negative Words

Blunder	Erroneous	Miscalculation	Reject
Cannot	Error	Misconstrue	Setback
Contradict	Failure	Mistake	Trouble
Damage	Fault	No	Unable
Decline	Harm	Predicament	Unfortunately
Deny	Inaccurate	Problem	Veto
Dilemma	Incorrect	Refuse	Will not
Do not	Irresponsible	Regret	Wrong

Table 12.3 Positive Verbs

Accomplish	Improve
Achieve	Increase
Assist	Initiate
Assure	Insure (ensure)
Build	Maintain
Coordinate	Organize
Create	Plan
Develop	Produce
Encourage	Promote
Establish	Satisfy
Help	Train
Implement	Value

Look at the following insurance claims procedure, which omits pronouns and reads impersonally.

Additional information:

See Chapter 3, "Meeting the Needs of the Audience."

BEFORE

Claims Procedure

To make a homeowner's policy claim, submit a minimum of three bids from licensed and bonded roofing contractors. The company will process the claim when all paperwork has been submitted and a claims adjuster inspects the roof.

Compare this flat, dry, impersonal example with the improved After *sample.*

AFTER

Claims Procedure

How do you make a homeowner's policy claim? Please submit a minimum of three bids from the licensed and bonded roofing contractors of your choice. Our company will process your claim as soon as you provide us the paperwork. Then, we will send one of our claims adjusters to inspect your roof.

Note the use of personal pronouns in this improved version.

Thoroughly Explain the Reasons for the Bad News

When communicating, you want to build rapport, and you want to focus on the positive. However, when you are providing bad news, your audience still needs proof. Why is the company merging with a competitor? Why are you changing vendors? Why have you chosen another person's proposal? Why are raises denied?

After you state the bad news, you need to explain the decision. Be as analytical as possible and provide sufficient details to cover any possible questions the reader might have.

In your explanation, quantify by providing any of the following:

Dates	Fees	Regulations	Decision makers' names	Warranties
Contracts	Time limits	Qualifications required	Technical specifications	Job performance appraisals

See Figure 12.1 for an example of delivering bad news.

FIGURE 12.1 Detailed Explanation of Bad News to an Employee

Positive word usage and pronouns build rapport.

Specific details (dates, contractual agreements, timelines, and amounts) develop the bad news.

Positive words and a personalized tone ("Tom") help maintain a pleasant exchange of information.

Date: April 12, 2009
To: Thomas Hartfort, Director of Information Technology
From: Susan Massin, Human Resources Manager
Subject: Response to Request for Early Retirement Buyout

Thank you for your recent inquiry about our Early Retirement Buyout Program (ERBP). We have reviewed your request and determined that we cannot grant early retirement at this time.

Though you meet a number of the criteria, the following prevent you from benefiting from the ERBP:

- We have reached our buyout quota for fiscal year 2009.
- Your position, designated as "hard-to-fill" or "hard-to-staff" by your union contract, is excluded from ERBP coverage.
- Three of the large projects you are currently chairing will not be completed until 2010.
- You had a 16-week leave of absence in 2008. Before you can be considered for early retirement, you must contribute to your banked leave hours.

We would be happy to review this request next year, at which time our buyout quotas should increase. Please stop by, Tom, if you have any questions.

METHODS OF ORGANIZATION FOR BAD NEWS MESSAGES

When you write a bad news message, consider your audience and the topic. What is the best way to share this news with your readers or listeners? Should they hear the news immediately, or should you cushion the blow? You have to decide on the **direct** or the **indirect** method of organization.

The Direct Method of Organization

In the direct method, you state the bad news *immediately*. This could be in a memo or e-mail subject line, the first sentence of a letter, flashing or scrolling text on your company's Web site, a newsletter article's headline, or the first words you say to the audience.

Bad news can range from somewhat minor information to significantly troubling problems. Consider using the direct method of conveying bad news in the following instances.

- **The bad news is relatively unimportant**—Your department has been planning a bowling party for family and friends. Having this party would be a nice way to build rapport and to celebrate departmental achievements. However, the party must be cancelled or delayed due to unexpected renovations at the neighborhood bowling alley. It would have been nice to have this party. Nonetheless, postponing the event will not affect anyone seriously. In this situation, your audience simply needs to hear that plans have changed. Just tell them directly.

  In an e-mail, you could write, "We'll need to cancel this weekend's bowling party. Roll-a-Bowl just told me they're closing tomorrow to renovate their party room. No problem. We'll reschedule as soon as possible."
- **The bad news is easily explained**—Sometimes precedence, fiscal realities, corporate culture, or common sense lead the reader or listener to accept bad news. For example, one of your employees wants to attend a conference in another city. However, your home office recently e-mailed all managers that travel is discontinued for the current fiscal year because of profit margin declines. Your

employee deserves and will accept a direct response from you confirming that travel is impossible in the current business climate.

You could write in an e-mail message, "Because of a home office decision based on lower profits, all travel plans are cancelled for the remainder of fiscal year 2009."

- **The bad news is expected**—Imagine this scenario: For weeks, rumors have swirled around the office complex. Employees have known that their beloved CEO's health was failing. Though he had founded the company and considered all employees his "family members," the pace of the company had become too much for him. In addition, his only child, recently married, just moved to another state with a warmer climate. Thus, the CEO decided to retire sooner than planned. He needed to take better care of himself, and his daughter's move gave him an opportunity to retire to the Sun Belt.

  How should he communicate this corporate bad news? He does so directly. He calls a meeting for the entire staff. To avoid drama, he walks to the microphone and states, "I'm retiring; I'm moving, but you people will always be close to my heart. I look forward to hearing about your continued success."
- **The audience and writer have a friendly relationship and communicate frequently**—Luckily, in many instances, supervisors and subordinates have friendly, respectful, and trusted relationships. Maybe you and your boss have worked together for years. You value each other's skills. You respect each other's decisions. You communicate often, openly, and honestly. In situations like this, even when bad news must be conveyed, the best approach is directness.

  For example, you have asked your supervisor for a new computer, a change in office location, or a new team member for a corporate project. The boss needs to say no for valid reasons. If he or she is used to speaking to you directly, that should be the approach taken in this situation as well. "I know that you need a new computer (*or office or team member*), but I can't do this now. We're a bit short on office funds (*or space or available workers for the project*). I'll make it up to you later."
- **People's health or safety is at risk**—When people are endangered, you cannot delay the message. The bad news must be emphasized immediately. For instance, your company has manufactured a product that is defective. The company must recall the product. First, as a good citizen, your company does not want the public—its clientele and neighbors—to suffer injuries. Second, governmental laws mandate this recall.

  In a situation like this, the company should make public service announcements on television and radio, tell newspapers, and mail letters to customers who have purchased this product. In a letter, the company can provide a boldfaced, subject line, **"Safety Recalls of 2009 Waterspout Washing Machines."**
- **The bad news will affect many stakeholders**—Your company is going out of business, merging with another company, moving to another state, or planning to "offshore" jobs as a cost-saving strategy. Any of these decisions could have an effect on hundreds or thousands of employees, as well as their families and communities. Lost jobs do not just impact those suddenly unemployed. Businesses (restaurants, department stores, dry cleaning establishments, filling stations, grocery stores) that cater to the employees suffer a trickledown effect. School enrollments change. Doctors' offices and hospital staff feel the loss. City taxes might be affected, which impacts police, fire personnel, and street maintenance.

  If a company's status changes significantly, the company must communicate this fact quickly and directly to the public. The local media must be contacted immediately, and the bad news must be stated explicitly.

Formatting Direct Bad News

For an effective approach to communicating bad news directly, consider using an introduction in which you get to the point, a discussion to develop your ideas, and a conclusion to build rapport.

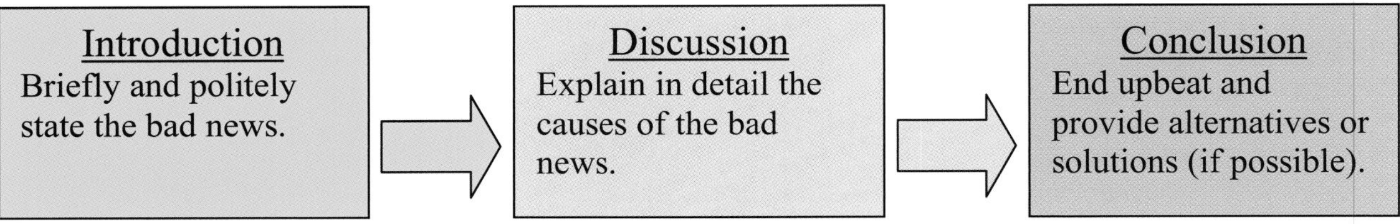

Introduction

Whether you are writing or speaking, begin with an introduction. In a memo, letter, or e-mail message, you could preface your first paragraph with a subject line. For the direct method of organization, this subject could state the bad news. The introductory paragraph in written communication should be brief, polite, and to the point. One to two sentences are sufficient to state the bad news.

Discussion

Include in the discussion paragraph(s) reasons for the bad news. Be sure to provide specific information to develop your ideas. The more details you provide, the less chance there is for questions, renewed discussions, or hurt feelings. Remember to keep the tone pleasant and the document design visually accessible.

Conclusion

In the conclusion, end upbeat. You want to encourage the reader and maintain good relations. Even though you have communicated bad news, you might need to work with this person in the future. In addition, your conclusion could provide alternatives or solutions.

Figure 12.2 is a bad news letter using the direct method of organization.

The Indirect Method of Organization

The direct method of organization states the bad news *immediately*. In contrast, sometimes you need to cushion the blow without conveying bad news too abruptly. Doing so might offend, anger, or cause hurt feelings. If you are rejecting a job applicant, denying a raise, telling a vendor that a proposal has not been accepted, or rejecting a customer's complaint, tact is required. In these instances, delaying bad news in written or oral communication gives you an opportunity to explain your position and justify the bad news. Consider using the indirect method of conveying bad news in the following instances.

- **The bad news is not easily explained**—A new position has been created at your company. One employee believes he deserves the job due to his longevity of service. However, management is thinking about hiring in another direction. Your bosses want to hire someone from another company, thinking he or she could contribute new ideas. In addition, management wants to hire someone with a master's degree in business. The internal candidate only has a bachelor's degree. How do you explain to your internal candidate that he's not "entitled" to the promotion?
- **The bad news is unexpected**—You have to decrease budget in your division by 10 percent due to revenue shortfalls. The company had been doing well financially prior to this news but faced an unexpected market downturn. Your employees believed that the company's stock was secure. How did this sudden change occur?
- **The audience and writer do not communicate frequently**—You are a new manager in charge of ten employees. Although you are located at the home office in Memphis, your employees work throughout the United States. Despite your position of authority in the business, you have had little interaction with your subordinates. Now you have to communicate bad news. How do you do this effectively?

FIGURE 12.2 Recall Letter Using the Direct Method of Communicating Bad News

Aero Home Appliances Corporation
9179 Uvalde San Antonio, TX 78213 805-766-3030

April 1, 2009

Steve Miller
28652 Springhill Dr.
Ogden, UT 56322

Subject: Safety Recall of Model 1299J Dryer

Dear Mr. Miller:

Thank you for purchasing our Aero Model 1299J Dryer. We are proud of its increased capacity, low energy consumption, and outstanding safety record. Our manufacturing processes are highly demanding. However, as you can imagine, defects occasionally occur. Aero's safety engineers have discovered a defect relating to the dryer vent in **some 2009 Aero Model 1299J Dryers.**

In accordance with the National Home Product Safety Act, we would like to inform you of the problem and your follow-up options:

The problem is . . .	A wire has been found to be chafed in the dryer vent. This could lead to fires.
What Aero and your dealer will do to help . . .	Aero will repair your product **free of charge.** Your dealer will install a protective cover on the wiring, a job that only takes about 30 minutes of your time. If you would like, you could pick up a repair kit at your local Aero dealership and repair this problem at your own convenience.
What you can do to ensure your safety . . .	Contact your dealer to schedule repairs. Refer to this letter when you call. If you choose to pick up the repair kit, please bring this letter with you. It identifies the required service.
If you need help . . .	For any follow-up questions, please contact Aero's Customer Service Department (1-800-AERODYN). A representative will be happy to help.

If your dealer is unable to correct this problem, you may submit a written complaint to the National Home Product Safety Administration, NHPSA.com. We are sorry for any inconvenience and are truly concerned about you and your family's safety. Thank you for your patronage and your prompt attention to this issue.

Sincerely,

Harold Lawrence
Supervisor, Customer Service

Due to safety concerns and in response to a national safety act, this letter directly states the bad news.

The introduction begins with positive words of customer appreciation and pride in the company's reputation.

The introduction continues with a statement that reasonable readers can accept. The introduction's final sentence directly states the bad news.

The body uses italicized headings with pronouns to appeal to the reader. The text specifies the bad news and the solution to this problem.

The conclusion provides the reader options for follow-up. It conveys a concern for the reader and ends upbeat and positive, using words such as "Thank you."

Buffers to Cushion the Blow

Typically, letters begin with an introduction explaining why you are writing and what you are writing about. However, using such conciseness and clarity in an indirect bad news letter would be too abrupt. Therefore, preface your bad news with a ***buffer***.

- **Make your buffer concise, one to two sentences**—"Thank you for writing. Customer comments give us an opportunity to improve service."
- **Be sure your buffer leads logically to the explanation that follows. Consider mentioning the topic, as in the following example about billing practices**—"Several of our clients have noted changes in our corporate billing policies. Your letter was one of several addressing this issue."
- **Avoid placing blame or offending the reader**—Rather than stating, "Your bookkeeping error cost us $9,890.00," write, "Mistakes happen in business. We are refining our bookkeeping policies to ensure accuracy."
- **Establish rapport with the audience through positive words to create a pleasant tone**—Instead of writing, "We received your complaint," be positive and say, "We always appreciate hearing from customers."
- **Sway your reader to accept the bad news to come with persuasive facts**—"In the last quarter, our productivity has decreased by 16 percent, necessitating cost-cutting measures."
- **Provide information that both you and your audience can agree upon**—"With the decline of dotcom jobs, many information technology positions have been lost."
- **Compliment your reader or show appreciation**—"Thank you for your June 9 letter commenting on fiscal year 2009."
- **Connect with your reader by expressing understanding or your personal relationship**—Avoid accusing your audience. Don't write, "Your proposal is late, and this could negatively impact our bottom line." Empathize with the audience by writing, "I know how hard your team worked on the proposal. However, for future projects, we need to meet our deadlines."

Formatting Indirect Bad News

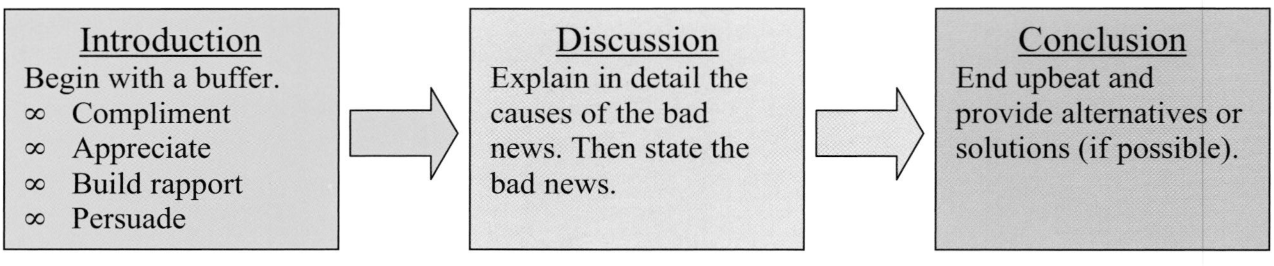

Introduction

Preface your bad news with a buffer. Doing so helps you maintain good relationships with your reader and avoids offending someone blatantly. For example, though you might not hire a job applicant today, he or she might be a viable candidate at a later date. You might need to reject a customer's complaint, but you want to keep that customer's business. Though you might not choose a contractor for one project, you might want to work with this firm in the future. Stating bad news too abruptly could shut the door. In contrast, using a more indirect approach by cushioning bad news leaves the door open for future business relationships.

Discussion

The body paragraph(s) accomplishes two goals: (1) It presents reasons and explanations to develop your proof and lead to the bad news, and (2) It states the bad news.

FIGURE 12.3 **Inverted Journalist's Pyramid**

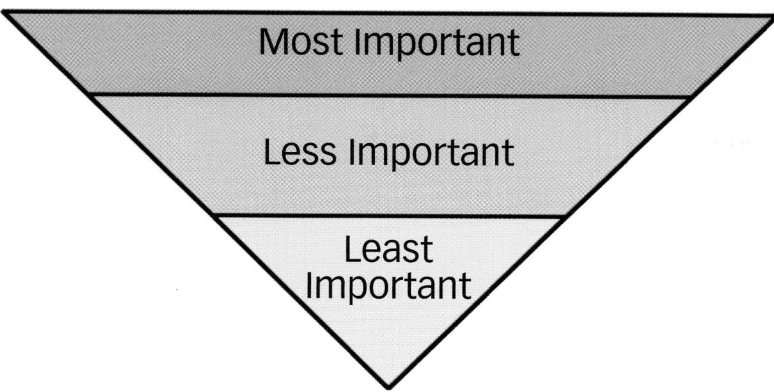

Develop Your Proof If you begin with the bad news, readers might stop reading before you have proven your assertions. Furthermore, a blunt statement of bad news could anger your readers. Preface the bad news with detailed explanations. To develop your reasons, provide your reader quantifiable information, including any of the following:

Dates	Company policy
Costs	Timelines
Percentages	Required qualifications
Warranties	Specifications

Doing so will ensure that your reader understands your point of view. In addition, complete development will lessen the chance for misunderstandings or the need for follow-up discussions.

Organize your details by presenting the most persuasive facts first. Otherwise, your reader might miss this key information. Consider using the *Inverted Journalist's Pyramid* in Figure 12.3. Using this guide, list the most important information first, followed by less important details.

State the Bad News Don't hedge, and don't apologize. Just state the bad news. You have already led the reader to this point in the correspondence through detailed development. To hesitate now will do you and your company damage. In addition, avoid prefacing the bad news with "warning markers," like "however," "but," "nonetheless," or "unfortunately."

Conclusion

If you end your bad news letter with the bad news, then you leave your reader feeling defeated. You want to maintain a good customer–client, supervisor–subordinate, or employer–employee relationship. Therefore, conclude your letter by giving your readers an opportunity for future success. Provide your readers alternatives which allow them to get back in your good graces, seek employment in the future, or reapply for the refund you have denied. Then, end upbeat and positively. See Figure 12.4 for a sample of a letter using the indirect method of communicating bad news.

TYPES OF BAD NEWS MESSAGES

Letter of Complaint

In a *letter of complaint*, you tell an individual or business that you are dissatisfied with a product or service and that you want action taken on your complaint. For example, you or your company might complain as follows:

- **Biotechnology**—You are the purchasing director of a hospital. The radiology department's new CT scan was not calibrated correctly by the installing technician.

Delivering bad news to a job applicant is a difficult task for human resources employees. To do so effectively, begin with a buffer statement, and then in the body, explain why the applicant is being refused. Conclude by giving the applicant options for future contact. Maintain a pleasant, positive tone.

FIGURE 12.4 Job Applicant Rejection Letter Using the Indirect Method of Communicating Bad News

Home and Farm Insurance, Inc.
1234 Elm Street
Chico, CA 22221

March 12, 2009

Stewart Eiband
3199 Guadalupe
Sacramento, CA 22199

Dear Mr. Eiband:

Thank you for your recent letter of application. We appreciate your interest in Home and Farm Insurance. As you can imagine, we received many letters from highly qualified applicants.

The advertisement specifically required that all applicants have an M.S. in computer science and at least five years of experience in information technology. We also suggested that knowledge of insurance regulations would be preferred. Your years of experience fell below our requirements, and your resume did not mention insurance regulation expertise. Therefore, we must reject your application.

If you have regulatory knowledge or have acquired additional job experiences which pertain to our work requirements, we would be happy to reconsider your application. In any case, we will keep your letter on file. When new positions open up, your letter will be reassessed. Good luck in your job search.

Sincerely,

Sue Cottrell
Director, Human Resources

Due to this, several patients had to delay their treatment by at least 24 hours. In-house technicians could not repair the CT scans, and the manufacturing company did not answer their 24-hour hotline or respond to your e-mail messages for help. You plan to write a letter of complaint to the manufacturing company to request immediate repair or replacement of the CT scan.

- **Public relations**—You are the events coordinator for your company. The company hosted a conference at a convention center. Services provided by the convention center did not meet contract stipulations. For example, Internet connections never worked in the display rooms; audiovisual equipment malfunctioned during keynote speeches; convention center workers set up equipment noisily, disrupting presenters; and special menus—vegetarian and kosher—were either not served as requested or were served up to 30 minutes after other patrons had eaten. Your job is to write a complaint letter to the convention center management, requesting future discounts and immediate refunds.
- **Human resources**—Your company has changed its insurance carrier. The new carrier sought to increase insurance premiums after only three months, challenged claims which were supposedly covered, and did not provide accessible claims

adjusters as had been advertised. With a letter of complaint, you can seek solutions to these problems.

To improve your relations with customers or vendors and receive the services you want, write a letter of complaint following our criteria.

Introduction

In the introduction, politely state the problem. Although you might be angry over the service you have received, you want to suppress that anger. Blatantly negative comments do not lead to communication; they lead to combat. Because angry readers will not necessarily go out of their way to help you, your best approach is diplomacy. To strengthen your assertions, include supporting details, in the introduction, such as the following:

Serial numbers	Invoice numbers	Names of salespeople involved in the purchase
Dates of purchase	Check numbers	Receipts

Discussion

In the discussion paragraph(s), explain in detail the problems experienced. These could include

Dates	Information about shipping	An itemized listing of defects
Contact names	Breakage information	Poor service

Be specific. Generalized information will not sway your readers to accept your point of view. In a complaint letter, you suffer the burden of proof. Help your audience understand the extent of the problem. After documenting your claims, state what you want done and why.

Conclusion

End your letter positively. Remember, you want to ensure cooperation with the vendor or customer. You also want to be courteous, reflecting your company's professionalism. Your goal should be to achieve a continued rapport with your reader. In this concluding paragraph, include your contact information and the times you can best be reached. See Figure 12.5 for a sample letter of complaint.

Partial Adjustment

Life is complex. Sometimes you can agree with part of a writer's complaint but disagree with other parts. If you are not going to agree completely or disagree completely, then you will want to write a *partial adjustment letter*.

Introduction

In the introductory paragraph, state the good news first. As always, you want to be diplomatic, bringing the reader to your point of view.

Discussion

In the discussion paragraph(s), precisely explain what happened, and then state the bad news.

Conclusion

In the conclusion, resell. Your reader will not be completely happy with you. You have provided some relief but not all that was requested. The last paragraph is your opportunity to win back a bit of the reader's good faith in your company. See Figure 12.6 for a sample partial adjustment letter.

Additional information:

See Chapter 10, "Traditional Correspondence: Memos and Letters."

FIGURE 12.5 Letter of Complaint

The introduction includes the date of purchase and the problem encountered.

In the body, explain what happened, state what you want done, and justify your demand. This letter develops its claim with warranty information.

Conclude your letter by providing contact information and an upbeat, pleasant tone.

To ensure that the reader is aware of your documentation, list supporting enclosures.

11517 Montgall Ave. #210
Kansas City, MO 64122

April 4, 2009

ABT Phone Co.
4901 Town Center Dr.
Leawood, KS 66297

Subject: Incorrect Phone Replacement

I purchased a Sabra 99 phone from your store on December 21, 2008. On March 14, 2009, the headset jack on my phone stopped working. Enclosed are copies of my original receipt and the receipt for my replacement phone.

I came into the store on March 14, and I was told that my phone could not be repaired in the store. Therefore, a replacement phone would need to be ordered. The representative informed me I would receive the same phone as my original phone, and it would arrive at my home in three to five business days. Instead, the following happened:

1. The replacement phone was shipped to your store.
2. I was not informed you received the phone until April 3, 14 business days after it had been ordered.
3. The replacement phone that was sent was a Sabra 410.

When I came back into your store to pick up the replacement phone, I was told by the manager that the new phone was not an equal replacement. Since the phone I received as a replacement was an older, less expensive model, I believe I should be given another replacement phone that is equivalent to the phone I originally purchased. I also would like credit for the 14 days I had to wait.

I can be reached from 9 a.m. to 5 p.m. any day of the week at 785-823-6504. I look forward to hearing from you soon. Thank you for your time and help.

Brandi R. Dixon

Enclosures (2)

100 Percent Negative Response Message

Unfortunately, sometimes you have to say no to a letter of complaint. If the customer is wrong, you have no alternative but to write a *100 percent negative response.* However, you still want to maintain positive customer relationships. Although you will deny the customer's request for adjustment, you want to keep that customer for future business.

Introduction

In the introductory paragraph, begin with a buffer. This should include a positive statement and facts that all readers can accept, regardless of the situation.

FIGURE 12.6 Partial Adjustment Letter from a Vendor to a Customer

ABT Phone Co.
4901 Town Center Dr.
Leawood, KS 66211

April 10, 2007

Brandi Dixon
11517 Montgall Ave. #210
Kansas City, MO 64122

Dear Ms. Dixon:

Thank you for your letter regarding your replacement phone. I apologize for the mishap and the inconvenience it has caused you.

After receiving your letter, I reviewed your account and found that it was our fault the phone was shipped to the store. However, the phone model was ordered correctly. The warehouse accidentally filled the request wrong. In order to compensate you for our mistake, the phone we are giving you is a Sabra 719. This model of phone is the upgraded model of the original phone you purchased. Your new phone will include these added features and accessories:

1. An MP3 player and sync cable to download and listen to music
2. A micro SD card slot for memory expansion and a 64-megabyte SD card
3. ABT power vision (high-speed Internet) access

Your new phone will be at the store this Friday, April 13. I will make sure the correct phone arrives, and I will activate it while you are in the store. However, I am unable to refund you for the 14 days of service because your phone was still usable.

Thank you for your professionalism in bringing this issue to my attention. If you have any questions or concerns about this new replacement phone, please call me. We appreciate your continued business and thank you for using ABT for your wireless needs.

Sincerely,

Alexander Spezak

Sometimes when you respond to customer complaints, you are unable to satisfy their complete request. However, you can grant them part of their request. In these instances, you can write a partial adjustment. To do so, explain the good news and steps the customer must follow, and then state any bad news. Maintain a pleasant and helpful tone.

Discussion

In the discussion paragraph(s), precisely explain what happened, providing facts, figures, and incontestable detail. Then state the bad news.

Conclusion

In the conclusion, end on an upbeat note. To do so, use positive words and provide the reader with an alternative. See Figure 12.7 for a 100 percent negative response e-mail.

FIGURE 12.7 100 Percent Negative Response E-Mail from a Vendor to a Client

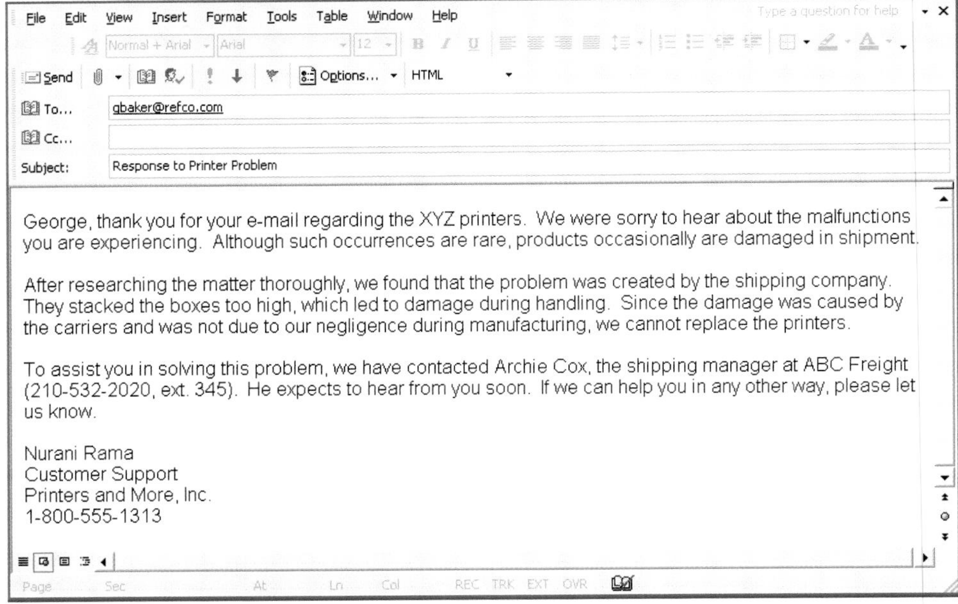
To... gbaker@refco.com

Cc...

Subject: Response to Printer Problem

George, thank you for your e-mail regarding the XYZ printers. We were sorry to hear about the malfunctions you are experiencing. Although such occurrences are rare, products occasionally are damaged in shipment.

After researching the matter thoroughly, we found that the problem was created by the shipping company. They stacked the boxes too high, which led to damage during handling. Since the damage was caused by the carriers and was not due to our negligence during manufacturing, we cannot replace the printers.

To assist you in solving this problem, we have contacted Archie Cox, the shipping manager at ABC Freight (210-532-2020, ext. 345). He expects to hear from you soon. If we can help you in any other way, please let us know.

Nurani Rama
Customer Support
Printers and More, Inc.
1-800-555-1313

Bad News from a Company to a Customer or Vendor

In business, at times, you will need to write a *bad news letter to customers or vendors*. For example, you or your company might convey bad news as follows:

- **Commercial banking**—One of your customers has been overdrawn four times in the last 60 days. This customer also was late on the bank-held mortgage payment. You must write a bad news letter telling this customer that his or her checking account privileges have been cancelled.
- **Insurance**—You are the supervisor at an insurance company. One of your customers has had numerous water-related claims in the last 24 months. You need to write a bad news letter to this customer canceling the insurance due to excessive claims.
- **Wholesale office supplies**—As the owner of a business, you are writing a bad news letter to a long-term vendor. The vendor has supplied paper goods, office furniture, and computer hardware. However, you are canceling your business with them because a new office supply warehouse is offering a 15 percent discount on the same products.

Introduction

In the introductory paragraph, begin with a buffer. This should include a positive statement and facts that all readers can accept, regardless of the situation.

Discussion

In the discussion paragraph(s), precisely explain what happened, providing facts, figures, and incontestable detail. Then state the bad news.

Conclusion

In the conclusion, end on an upbeat note. To do so, use positive words and provide the reader an alternative if possible. See Figure 12.8 for a sample of a bad news letter from a claims company to a customer.

FIGURE 12.8 Bad News Letter from a Claims Company to a Customer

EPCS

Eagle Pass Claims Services
700 Peachtree
Atlanta, GA 30091
678-555-2121

February 22, 2009

Stephani Ortega
1880 Roundrock
Saguaro, AZ 21110

Subject: Computer Loss in Harold Hall, Univ. of Northeast Arizona
File #: EPCS1009112022

Thank you, Ms. Ortega, for submitting a claim about your missing computer system (keyboard, monitor, and hard drive), dated January 5, 2009. We have carefully reviewed the reports from the Saguaro, AZ, police officers assigned to the case, as well as reports from your dormitory's management.

The introduction begins with a buffer, a mutually agreeable statement, which documents the reasons for writing this letter.

Following are our findings:

- Your renter's lease specifically says that "any personal property kept on the premises shall be at the Tenant's sole risk and the Landlord shall not be liable for any damages or loss of property due to theft" (Section A.1). We have enclosed a copy of this lease agreement for your records.
- Section A.2 of your lease states, "The renter bears the responsibility for obtaining renter's insurance to cover various losses, including losses due to theft or vandalism."
- The Saguaro, AZ, police report cannot verify any wrongdoing on the part of Harold Hall management.

The body itemizes the facts which support the denial of the claim.

Thus, since our investigation reveals no negligence or breach of duty on the part of the residence hall, we must deny your claim.

The bad news is stated without hedging.

We suggest that you file a claim with your renter's or homeowner's insurance policy for this loss of personal property. If we can answer any questions, please contact us.

The conclusion provides the reader an alternative solution.

Janet Miller
Sr. Claims Specialist

Enclosure

Bad News from a Company to an Employee

In business, at times, you will need to write an internal *bad news memo, letter, or e-mail to an employee or employees.* At other times, you might need to have face-to-face meetings (individually or in groups) to communicate bad news.

When should you write bad news versus communicating in person? If bad news is relatively unimportant (the department's party has been postponed, computer replacements are delayed, or travel logs will be audited), you could convey such a message in an e-mail. However, in certain instances, it is inappropriate to communicate bad news through a memo, e-mail message, instant messaging, or a telephone call. Even if doing so makes you uncomfortable, you must communicate bad news in person. For example, suppose you have worked with an employee (Joe) for ten years. Recent revenue and client losses demand a reduction in the workforce. You have to fire Joe, whose job has become redundant. You cannot fire Joe in an e-mail message or memo. This would be unkind and a startling way for anyone to learn about bad news. The more professional approach is to convey this bad news to Joe in a face-to-face meeting. After the meeting, you will document the bad news in written form.

Following are illustrations when you or your company might convey written bad news:

- **Writing negative performance appraisals**—One of your employees is frequently late for work, misses work often, has decreased productivity, is rude to customers and coworkers, receives personal phone calls at work, or uses the company's e-mail system for personal correspondence. You must document the poor job performance.

  As a supervisor, your job is to document a subordinate's failure to meet company expectations. Your job also requires you to provide the necessary follow-up steps to correct the problem.
- **Denying promotions**—A subordinate has applied for the night supervisor position that is being vacated in March. According to the new corporate policy, vacated supervisory positions will not be filled. The company plans to increase the job responsibilities of the other supervisors. You need to share this bad news with the applicant.
- **Communicating company-wide bad news**—This may include company restructuring, decrease in profits, layoffs, hostile takeovers, mergers, falling stock prices, or legal problems. Employees need to be informed about any bad news that affects their jobs.

Communicating bad news to coworkers is a challenge. Building or sustaining rapport is essential. Remember to use positive words and to develop your ideas thoroughly. Putting yourself in the audience's place will help you achieve empathy and correct word usage.

Introduction

In the introductory paragraph of a bad news communication, use either the direct or the indirect method of organization. Begin with a positive statement when possible. If you need to start with a buffer, provide facts that all readers can accept, regardless of the situation. If the bad news is expected or minimal, directly state it without a buffer.

Discussion

In the discussion paragraph(s), precisely explain what happened, providing facts, figures, and incontestable detail. If you have used the indirect method of organization, state the bad news after your explanations.

Conclusion

In the conclusion, end on an upbeat note. To do so, use positive words and provide the reader an alternative if possible.

The human resources director at Valley Vale Community College (V^2C^2) occasionally needs to communicate bad news to the entire faculty and staff. The director relies on e-mail messages to communicate quickly with her 1,000-person internal audience. See Figure 12.9.

Additional information:

See Chapter 9, "Electronic Communication: E-Mail Messages, Instant Messages, Text Messages, and Blogging."

FIGURE 12.9 Bad News E-Mail to Employees

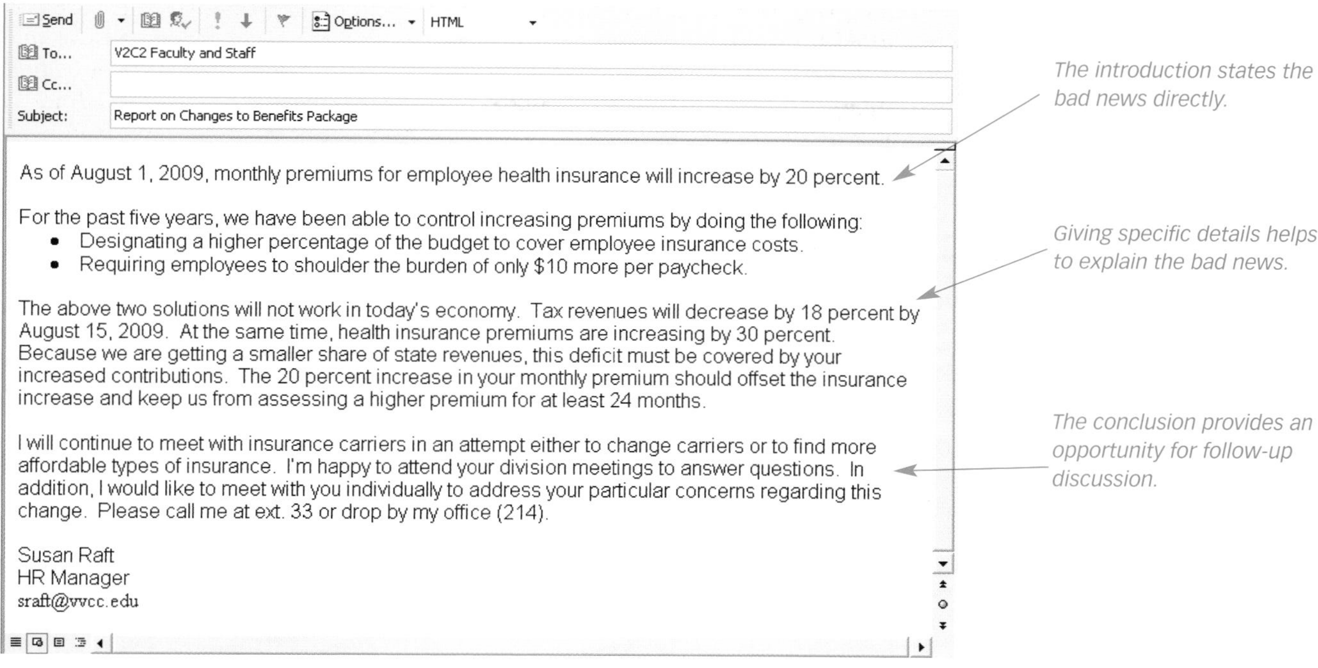

To... V2C2 Faculty and Staff

Cc...

Subject: Report on Changes to Benefits Package

As of August 1, 2009, monthly premiums for employee health insurance will increase by 20 percent.

For the past five years, we have been able to control increasing premiums by doing the following:

- Designating a higher percentage of the budget to cover employee insurance costs.
- Requiring employees to shoulder the burden of only $10 more per paycheck.

The above two solutions will not work in today's economy. Tax revenues will decrease by 18 percent by August 15, 2009. At the same time, health insurance premiums are increasing by 30 percent. Because we are getting a smaller share of state revenues, this deficit must be covered by your increased contributions. The 20 percent increase in your monthly premium should offset the insurance increase and keep us from assessing a higher premium for at least 24 months.

I will continue to meet with insurance carriers in an attempt either to change carriers or to find more affordable types of insurance. I'm happy to attend your division meetings to answer questions. In addition, I would like to meet with you individually to address your particular concerns regarding this change. Please call me at ext. 33 or drop by my office (214).

Susan Raft
HR Manager
sraft@vvcc.edu

Send
Options...
HTML
The introduction states the bad news directly.
Giving specific details helps to explain the bad news.
The conclusion provides an opportunity for follow-up discussion.

BAD NEWS MESSAGES CHECKLIST

_____ 1. **Communication channel.** Have you decided whether to communicate the bad news via social media, telephone, e-mail, Web site, newsletter, media, memo, letter, or face-to-face meeting?

_____ 2. **Organization.** Have you decided whether to use the *direct* or the *indirect* method of organization?

_____ 3. **Introduction.**
- If you have used the *direct* method of organization, did you briefly and politely state the bad news?
- If you have used the *indirect* method of organization, did you begin with a *buffer?*

_____ 4. **Discussion.** Does your discussion clearly explain the details of your bad news?
- This could include specific dates, fees, regulations, company policy, warranties, contracts, time limits, and more.
- In the *indirect* method of organization, did you state the bad news after the above details?

_____ 5. **Highlighting/page layout.** Is your text accessible? To achieve reader-friendly ease of access, you need to use headings, boldface, italics, bullets, numbers, underlining, or graphics (tables and figures). These add interest and help your readers navigate your message.

_____ 6. **Conclusion.** Does your conclusion end upbeat and give alternatives if possible?

_____ 7. **Clarity.** Is your message clear, answering reporter's questions (Who, What, When, Why, Where, and How) and providing specific details?

_____ 8. **Conciseness.** Have you limited the length of your sentences, words, and paragraphs?

_____ 9. **Audience recognition.** Have you written appropriately to your audience? Use positive words versus negative words, as well as personal pronouns to achieve audience rapport.

_____ 10. **Correctness.** Is your text grammatically correct?

David Met His Challenge

To meet his communication challenge, David used the P^3 process.

Planning

To plan his bad news message, David considered the following:

- Goals—document a notice of dismissal
- Audience—employee and human resources director
- Channels—e-mail
- Data—drawn from six months of documented "corrective reviews"

David says, "After repeated discussions and interaction with Joan about her on-the-job behavior, I had to take the last and most difficult step—firing her. To handle this challenging situation and document my actions, I used the reporter's questions to plan the content of my written communication." David used the checklist in Figure 12.10 to plan his bad news message.

Packaging

David gathered information and wrote a draft of his communication to Joan, the employee. He chose to write an e-mail message because that seemed to be both quick and simple. However, he wasn't completely satisfied with the draft e-mail, so he asked a colleague to review it for him and provide some feedback. Figure 12.11 shows how David packaged his rough draft.

In response to David's rough draft, one of his colleagues made the following comments: "David, the overall tone is far too mean. You must change this tone and be less negative. Here are some things I think you could do to change this message: 1. Don't send an e-mail. You should fire someone face-to-face and then write a follow-up memo documenting the dismissal. 2. Don't ever use all caps—that's the same as screaming! 3. Avoid vague words like 'poor performance,' 'decreased productivity,' 'most absenteeism,' and 'highest daily miscounts.' These need to be clarified with details. 4. Definitely drop the last sentence, which is harsh and angry sounding. 5. Itemize your content when you prove your points. 6. I found some grammar errors, too. 'Eventhough' is two words, not one. 'Worse of all' should be 'Worst of all,' but I wouldn't use that phrase—again, too negative. 7. Most importantly, try using the 'indirect method.' Telling Joan that she has been fired in the first sentence is too abrupt. Lead up to this point with a buffer."

Perfecting

Once David read the feedback, he realized that he had to make numerous revisions. Even more than revisions, he reconsidered his communication channel. David says, "Like many people in business today, first, I used an e-mail message without considering the audience or the consequences because e-mail seemed easy. After my coworker's review, I knew that I needed to use a more appropriate communication channel. I decided to meet with the employee and then document the meeting and dismissal by writing a hard-copy, follow-up short report. In this perfecting stage of the communication process, I adjusted my tone and worked to create a positive message in a difficult situation—firing an employee. I am now confident that I met my communication challenge." See Figure 12.12 for David's perfected bad news correspondence.

FIGURE 12.10 Reporter's Questions Checklist for a Bad News Message

_____ 1. **Who is my audience?**

Joan O'Brien, teller supervisor

_____ 2. **Why am I writing?**

To inform Joan that she is being fired

_____ 3. **What is the general topic of my message?**

Firing an employee

_____ 4. **What exactly are the reasons for this bad news?**

- Late to work
- Sarcastic attitude toward subordinates
- Yelling at subordinates
- Decreased productivity in division
- Consistent poor reviews over a long period of time
- Unbalanced budget record

_____ 5. **What alternatives can I offer?**

None remaining—already probationary

_____ 6. **When will the bad news take effect (dated action)?**

June 10

_____ 7. **Why is this date important?**

End 30- to 90-day probationary period

FIGURE 12.11 Rough Draft Bad News Message

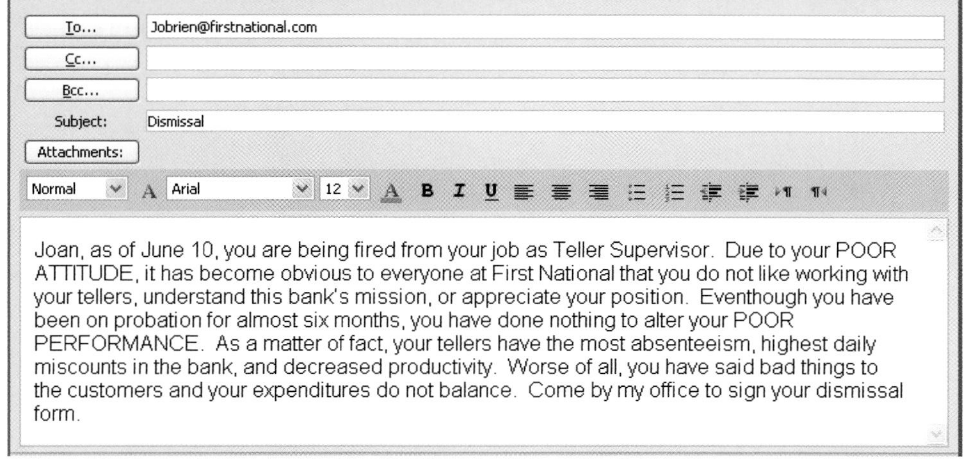

To...
Jobrien@firstnational.com
Cc...
Bcc...
Subject:
Dismissal
Attachments:
Normal
Arial
12
Joan, as of June 10, you are being fired from your job as Teller Supervisor. Due to your POOR ATTITUDE, it has become obvious to everyone at First National that you do not like working with your tellers, understand this bank's mission, or appreciate your position. Eventhough you have been on probation for almost six months, you have done nothing to alter your POOR PERFORMANCE. As a matter of fact, your tellers have the most absenteeism, highest daily miscounts in the bank, and decreased productivity. Worse of all, you have said bad things to the customers and your expenditures do not balance. Come by my office to sign your dismissal form.

FIGURE 12.12 Perfected Bad News Report

David says, "A confidential report is a better communication channel than less secure e-mail messages."

"I chose to use the direct method for conveying bad news since the news is expected by my audience."

"My report quantifies the rationale behind the dismissal, providing specific details leading up to Joan's dismissal."

"I'm using headings and bullets to make the content more accessible."

According to David, "First National offers an alternative to help Joan accept the dismissal and leave the company without resentment. Documenting such bad news is always difficult to write and difficult to receive."

Date: June 9, 2009
To: Mary Arnett, Director of Human Resources
CC: Joan O'Brien, Teller Supervisor
From: David King, Vice President
Subject: REPORT ON EMPLOYEE DISMISSAL

As a follow-up to my meeting with Joan O'Brien, teller supervisor, on June 7, 2009, this report documents the reasons for her dismissal.

Joan had been on 30- to 90-day probation. However, Joan continued to have the following problems with the tellers in her department:

- *Poor attitude toward subordinates.* On seven occasions, her subordinates reported that Joan yelled at them in front of customers and treated them rudely and sarcastically in meetings. The seven instances are documented in Joan's file, with statements from the affected employees.
- *Negative attitude toward the bank's mission of customer service.* Rather than respect the bank's reputation and standards, Joan spoke negatively about the bank to three long-time customers. They reported that Joan said, "First National never offers the best rates on CDs or listens to customer complaints about service."
- *Poor performance.* Her last three quarterly reports revealed that her teller division had the most absenteeism and highest daily miscounts in the bank. Our bank's goal for each teller is less than 2 percent absenteeism and 0 miscounts. Joan's department shows a 3 percent decrease in quarterly productivity. In addition, her department expenditures do not balance. Internal audits reveal a $7,500 total of unaccounted for expenditures.

During the past six months, we worked with Joan to improve her interpersonal communication skills with employees and customers. She attended sensitivity training seminars at the bank's expense. Though she promised to pay more attention to her job responsibilities, including departmental performance and daily counts, we see no evidence of improvement. Thus, I informed Joan that she would be terminated from her job at First National as of June 10. We have offered Joan access to First National's Job Transition Assistance (JTA). Our JTA consultants will help her with resume and job search skills.

CHAPTER HIGHLIGHTS

1. Communicating bad news in business is difficult because you still hope to retain the audience as a customer, vendor, or employee.
2. Choosing the appropriate communication channel will help you to express bad news.
3. Positive words in a bad news message will help you control your reader's reactions, build goodwill, and persuade your audience to accept your point of view.
4. Positive words can cushion the bad news.
5. In a bad news message, pronouns achieve audience rapport and personalize the communication.
6. To explain bad news, consider including fees, data, regulations, warranties, time limits, and contracts.
7. In the direct method of organization, you state the bad news immediately.
8. Expected or easily explained bad news can be stated directly.

9. In the indirect method, begin your bad news with a buffer.
10. A buffer avoids blame, establishes audience rapport, and is persuasive.

WWW *Find additional communicating bad news exercises, samples, and interactive activities at http://www.prenhall.com/gerson.*

MEETING WORKPLACE COMMUNICATION CHALLENGES

CASE STUDIES

After reading the following case studies, write the appropriate letters required for each assignment.

1. Jim Goodwin, owner of Goodwin Insurance Agency, has a client—Mountain Range Engineering (MRE), located in Las Vegas, NV, 2387 Desert Road. MRE provides engineering services for a six-state region. Their employees travel extensively. Therefore, MRE has a fleet of over 1,000 cars. Jim and his company have insured MRE's fleet of cars for three years.

   Jim has bad news to share with MRE. In a bad news letter, Jim must tell his contact at MRE, Rob Harken, that premiums are going up 25 percent due to excessive claims filed by MRE this year. In addition, because MRE's fleet of cars is aging, Jim no longer can provide full-replacement coverage. Finally, he also wants MRE to pay higher deductibles (currently $500, increasing to $1,000).

Assignment

Write the bad news letter for Jim. Use the information provided and any additional details you need. Choose whether to use the direct or indirect method of organization.

2. Mark Shabbot works for Apex, Inc., at 1919 W. 23rd Street, Denver, CO 80204. Apex, a retailer of computer hardware, purchased 125 new flat-screen monitors from a vendor, Omnico, located at 30467 Sheraton, Phoenix, AZ 85023. Thirty of the flat-screen monitors proved to be defective (customers returned them complaining about poor resolution, distorted images, and flickering screens). Mark needs to write a letter of complaint to the vendor, Omnico. Mark wants the 30 monitors replaced and the remaining products tested at no expense to Apex. Apex and Omnico have worked together successfully for ten years.

Assignment

Decide whether to use the direct or indirect method of organization, and then write the complaint letter for Mr. Shabbot.

3. You work at Omnico, in public relations. You have just received a letter of complaint from Mark Shabbot, purchasing director for Apex, Inc. Mark has asked for replacement of 30 monitors and free troubleshooting of an additional 95 units.

Assignment

Write a partial adjustment letter to Mr. Shabbot. Follow the criteria provided in this chapter.

4. Sharon Baker works in corporate communication for Prismatic Consulting Engineering, 123 Park, Boston, MA 01755. In response to an RFP (request for proposal), she has written a proposal to the Oceanview City Council, 457 E. Cypress Street, Oceanview, MA 01759. The proposal suggests ways in which Oceanview can improve its flood control. She provides costs, timeframes, problems which

could occur if the proposed suggestions are not implemented, ways in which the proposal will solve these problems, and Prismatic's credentials. The Oceanview City Council must reject this proposal due to Prismatic's excessively high costs, its limited credentials and client base, and its lengthy timeframe.

Assignment

Write the bad news letter to the vendor.

5. Suburban Day Surgery, 3678 Park Lane, Austin, TX 77877, had planned on expanding. This outpatient facility hoped to enlarge by adding four new surgical rooms, six new recovery rooms, two new doctors' offices, a cafeteria, a three-story parking garage, and increased exterior lighting. Suburban needed approval from the city council for this expansion. However, the Blue River Neighborhood Association fought this expansion successfully. The association convinced the city council that the garage and exterior lighting would negatively impact home privacy and valuations.

Assignment

Write a bad news letter from the Austin City Council to Suburban. Decide whether to use the direct or indirect method, and follow the criteria provided in the chapter.

ETHICS IN THE WORKPLACE CASE STUDY

All corporations adhere to a non-disclosure policy about stock valuation and company profits and losses. Employees are always reminded not to talk about changes in the corporate environment that might affect stock values.

Two colleagues, Rebecca Bacero and Tom Suttles, are eating lunch in a restaurant. During lunch, they discuss a new software product that has not met the company's expectations. Problems have occurred during usability testing which could negatively impact company revenues. In their disappointment, they loudly talk about how the company's stock is expected to decrease by at least 35 percent as a result of this new product's failure.

At a nearby table, a fellow employee, Daniel Solomon, eating with some friends, overhears Rebecca and Tom's loud conversation. One of Daniel's friends says, "Isn't that your company they're talking about? I have money invested in that stock. Should I sell all of your company's stock?"

Question: What should Daniel do? Is it ethical for Daniel to respond to his friend's question? Is Daniel required to report the conversation to his boss? Justify your answer based on information provided in Chapter 1.

INDIVIDUAL AND TEAM PROJECTS

1. Write a letter of complaint. You might want to write to a retail store, a manufacturing company, a restaurant, or a governmental organization. Whatever the subject matter, be specific in your complaint. Follow the criteria for writing letters of complaint.

2. Write a 100 percent negative response letter. You might work for a manufacturing plant, a retail or wholesale operation, a hospital, a training facility, and so on. Whatever your work area, customers will potentially complain about your product or service. Your job is to respond to their complaints. Follow the criteria for writing a 100 percent negative response letter.

3. Write a partial adjustment letter. You might work for a manufacturing plant, a retail or wholesale operation, a hospital, a training facility, and so on. Whatever your work area, customers will potentially complain about your product or service. Your job is to respond to their complaints. Follow the criteria for writing a partial adjustment letter.

4. Write a bad news e-mail from a company to an employee. Deny an employee's request for a letter of recommendation, deny a promotion, refuse a request for job-related travel to a conference, or refuse a request for upgraded office equipment. Follow the criteria for communicating bad news.
5. Write a bad news letter from a company to a vendor. You have received a bad order from an office supply vendor. Maybe you received bad service from your landscaper. The asphalt in your company's parking lot cracked. The windows at your company have not been cleaned as contracted. The snow removal company is not keeping your corporation's parking lot or sidewalks cleared as agreed. Communicate this or any other bad news to your company's vendor.

DEGREE-SPECIFIC ASSIGNMENTS

1. **Marketing and Public Relations.** Judy Huxoll is the events coordinator for S&S Consulting. The consulting company hosted a conference at the Deer Creek Convention Center (1287 Doe Rd, Riverbend, OK 86210). Services provided by the convention center did not meet contract stipulations. Internet connections never worked in the display rooms; audiovisual equipment malfunctioned during keynote speeches; convention center workers set up equipment noisily, disrupting presenters; and special menus—vegetarian and kosher—were either not served as requested or were served up to 30 minutes after other patrons had eaten.

Assignment

Judy needs to write a complaint letter to the convention center management, attention Candice Millard, requesting apologies, future discounts, and immediate refunds. Write the letter of complaint for Judy.

2. **Business Information Systems.** You are the business information systems manager at Blue Valley Independent School District, serving three high schools, seven middle schools, and fifteen elementary schools. Your school district recently purchased a system-wide software platform, costing 17 million dollars.

   This software will be used as follows: student registration, records, grade keeping, and attendance; employee benefits and personnel records; intranet and Internet Web site hosting; e-mail platform; and faculty and staff online training.

   Because of this new software system's widespread impact, all employees must be trained in its use. This will entail a series of online tutorials, requiring approximately 50 hours of graded training. These online tutorials must be passed with a score of 70 percent, and records of performance will be kept on file. Because of the expense of this software platform, the school district will need to delay replacement of the employees' aging hardware and software.

Assignment

As BIS manager, you must communicate the implications of the school district's purchase. The bad news includes the many hours of mandatory training and deferred replacement of hardware and software.

How should you convey this bad news message? Which communication channel would be most effective? Explain your answers. Then, write the memo, letter, e-mail, or report (even if you conclude that a face-to-face meeting would be the best communication channel, the written correspondence could be used for follow-up documentation).

3. **Accounting.** Oak Valley Realty & Associates' (OVR) in-house accounting staff discovered an accounting error in the spreadsheet calculations used by the company's construction division for its "percentage of completion accounting." The net effect is a cumulative reduction in revenue of $2,450,890 over the three-year reporting period 2006 to 2009 with an equivalent charge to pre-tax net income.

The error had a favorable effect in the current year, 2009, because it added approximately $226,000 to OVR's 2009 income before taxes for the nine months ended December 1, 2009.

Assignment

Your in-house accounting staff needs to report the accounting error to independent auditors to achieve Sarbanes-Oxley compliance. Write the bad news report. Include any additional information required to clarify the content.

4. **Management.** Grocers' Warehouse, Inc., supplies wholesale canned goods to grocery stores throughout your state. They pride themselves on timely deliveries, polite delivery associates, and merchandise delivered in quality condition (their motto is "You can't find a dented can, can you?"). One client, Sam's Corner Stores, reported problems with seven recent deliveries. Two of Sam's stores received partial deliveries of their canned vegetable orders. They had ordered 14 cases of mixed vegetables, stewed tomatoes, corn, carrots, and beans. However, they only received eight of the ordered cases. This negatively impacted their primary client, Heartland Food Bank, which provides hot meals for the homeless. Four of Sam's stores have complained about their delivery driver, who was rude to customers and Sam's employees. Another Sam's store was promised a delivery of nine cases of soup on Wednesday by 10:00 a.m. The soup arrived three days later.

Assignment

As Sam's Corner Store's purchasing manager, write the letter of complaint to Grocer's Warehouse, Inc. (GW). Remember that GW is a long-time vendor, with whom you have had a positive relationship. You want to achieve rapport and continue to work with GW. Document your complaint in a positive and professional manner.

5. **Finance.** You are the owner of an online mortgage company, FirstFinancingOrg. You have been working with one client, Rachel Tau, who wants to refinance her house. Unfortunately, she overbought and is having trouble keeping up with her bills. To make up her deficits, she wants to pull $40,000 equity from her home to consolidate her debt.

   However, Ms. Tau has fallen behind on her mortgage payments and her car loans, and she has exceeded the limit of her credit cards. Due to these problems, her credit rating has fallen below 500. Thus, she does not qualify for the loan she is requesting.

Assignment

Write the bad news message informing Ms. Tau that she does not qualify for a loan. Abide by the criteria provided in this chapter. Remember, you want to maintain good relations with Ms. Tau, who might be a repeat customer in the future.

PROBLEM-SOLVING THINK PIECES

After reading the following case studies

- Determine which communication channel would work best.
- Decide whether to use the direct or indirect method of organization.

If you decide that the indirect method of organization would work best, what kind of buffer should you use? Explain your answers either in writing or orally.

1. Café Ole, a chain of 150 coffeehouses located throughout the Southwest, needs to close 20 of its stores. Business is good overall, but these 20 stores are in poor locations and are suffering business losses. Café Ole's CEO needs to tell her

franchise-wide employees of these closures, as well as communicate the bad news to the public.

2. Chez Chic, an upscale women's boutique, has two locations in the Omaha, Nebraska, area. One is located downtown; one is in a suburban mall. Chez Chic's owner, Cindy Snyder, plans to close the downtown store and concentrate her business in the suburban site. She needs to tell her ten, downtown employees, the owner of the downtown building from whom she rents space, and her downtown customers.
3. Joe McDonald, an assistant manager at GreenSpace Grocery Store, has not been doing his job. He has over-ordered fresh produce, meat, and dairy products. Doing so has led to waste. Joe has under-ordered seasonal items, such as hamburger and hotdog buns. This has led to lost profits. In addition, he has failed to alter pricing on sales goods. Though items were marked down on the shelves, the goods rang up higher than patrons expected at the checkout lanes. This has led to customer complaints. These problems have gone on for four months despite warnings from upper management. You need to fire Joe.
4. Mr. Smith took his car to A&A Auto for an oil change. Though the company advertised oil changes in only 30 minutes, Mr. Smith had to wait over an hour. When he got home, he discovered oil leaking in his garage. An oil cap had not been screwed on tightly enough. Mr. Smith drove the car back to A&A. On the way, his car overheated and the oil light went on. Subsequent repairs to his engine cost over $500. Mr. Smith needs to complain.
5. Jane has been working part time at Reader's Roost Bookstore for over four years while going to college. Now that it is summer, she wants to work more hours, be given a raise, and start receiving benefits. Since a major book wholesaler has opened in the area, Jane's boss, Sarah, cannot grant any of these requests. How should Sarah convey this bad news to Jane?

WEB WORKSHOP

By going to any Internet search engine, you can find many articles about the challenges of communicating bad news. Access an online search engine and type "communicating corporate bad news," for example. You will find articles about how to report corporate restructuring, plant closures, layoffs, downsizing, quarterly losses, and poor job performance.

Assignment

1. Find articles which not only explain the challenges encountered when you have to communicate bad news but also provide helpful hints.
2. Report your findings either in writing or orally.

QUIZ QUESTIONS

1. List four reasons why you might need to communicate bad news.
2. How will positive words help you to communicate bad news?
3. When you communicate bad news, how can you avoid angering or offending your reader?
4. What are three components of a crisis communication plan?
5. Why should you use pronouns when you communicate bad news?
6. What are two ways to organize a bad news message?
7. What is a buffer?
8. What is one good way to organize your proof or explanation of bad news?
9. In which type of letter do you agree with part of a complaint but deny another part?
10. Why should you end upbeat in the conclusion of a bad news message?

CHAPTER 13

Persuasive Workplace Communication

▸ REAL PEOPLE *in the workplace*

Dr. Georgia Nesselrode is director of government training for Mid-America Regional Council's (MARC) Government Training Institute (GTI). MARC is the metropolitan planning organization for local governments, and GTI offers a comprehensive selection of professional development opportunities to municipal and county officials (elected and appointed) and their employees. Dr. Nesselrode's ongoing challenge is to communicate effectively with this widely diverse metropolitan audience.

There are nine counties and 120 cities that make up the membership of MARC. These organizations vary in size from delivering services using contract staff to those governments that have thousands of employees. It is important that GTI offers training programs that can satisfy all MARC's members. Georgia will use the full gamut of communication tools to reach her audience, including

- Cover letters with enclosures
- Marketing fliers
- E-mail messages with Web site links to fliers with more details
- Brochures

The marketing flier is an integral part of Dr. Nesselrode's writing because it is a means to convey concise information about specific training programs. Fliers contain the "Who," "What," "Where," "When," and "How," along with the course description and learning objectives. This provides the pertinent information that a local government official needs to make a decision on whether to participate. It is also a format that is easily saved and can be forwarded to colleagues or distributed organization-wide.

Her audiences include her local government human resources/training liaisons; elected and appointed officials such as mayors, city managers, police chiefs, and fire captains; and municipal employees who work in their government's health, accounting, tax, water, public works, and parks and recreation departments.

Because she writes to governmental agencies, officials, and employees, Dr. Nesselrode ensures the correspondence is complete, informative, and presents a professional image. The success of GTI can be attributed, in part, to its strong communications strategy. ◂

CHAPTER GOALS

When you complete this chapter, you will be able to do the following:

1. Understand the importance of argument and persuasion in workplace communication.
2. Recognize the traditional methods of argumentation.
3. Use the *ARGU* technique to organize persuasive workplace communication.
4. Avoid logical fallacies in persuasive communication.
5. Follow the *P³* process to write persuasive workplace communication.
6. Evaluate persuasive documents using the checklist.

WWW

To learn more about persuasive workplace communication, visit our Web Companion at http://www.prenhall.com/gerson.

Georgia's Communication Challenge

A persuasive flier to municipal and county officials is a great communication tool for MARC. In a well-written flier, MARC provides alternative approaches to city problems, gives the communities problem-solving ideas, helps them make sound decisions, and gives them ways to examine their decisions. However, creating a successful flier is challenging for many reasons:

- Georgia says, "To draft a flier, I must have all of the elements of a training program in place. Planning an effective flier starts with confronting a problem faced by a community, choosing a topic of value to the community, selecting the right training facilitators, and picking the correct time and place to hold the training.
- Once all of the information has been gathered, the next challenge is meeting deadlines. This entails getting a draft copy to public affairs, sending the draft to the printer, mailing or posting the flier on MARC's Web site, and making sure that there is plenty of time to reach the target audience before the first program. I need anywhere from eight weeks to three months lead time to successfully market the workshops.

We cannot succeed without proper communication skills—proper in that our communication channel relays purpose and specifies what we have to offer. We couldn't do our business without good communication focused on meeting the needs of our audience, and that is a constant challenge."

See pages 388–390 to learn how Georgia met her communication challenge.

THE IMPORTANCE OF ARGUMENT AND PERSUASION IN WORKPLACE COMMUNICATION

When communicating workplace information, you will write and speak for many reasons. You might write a memo, letter, or e-mail message to *inform* about an upcoming meeting, a job opportunity, a new product release, or a facilities change. You might give an oral presentation to local businesses, government, or educational organizations in which you hope to *build rapport*. In writing a user manual, your goal will be to *instruct*. When you write a proposal, your goal is to *recommend* changes. If you write a report, you might *analyze* options to a client's problems.

In addition to informing, building rapport, instructing, recommending, and analyzing, you also will need to communicate persuasively. Let's say you are a customer who has purchased a faulty product. You might want to write a letter of complaint to the manufacturer of this product. To make your case strongly, you will need to convince your audience, clarifying how the product failed. If your argument is effective, then you will persuade the company to give you a refund or new product.

Additional information:

See Chapter 7, "Oral Presentations and Nonverbal Communication in the Workplace," Chapter 9, "Electronic Communication: E-mail Messages, Instant Messages, Text Messages, and Blogging," Chapter 10, "Traditional Correspondence: Memos and Letters," Chapter 17, "Short, Informal Reports," and Chapter 19, "Internal and External Proposals."

Professionally, you will need to use argument and persuasion daily. As a manager, you might need to argue the merits of a company policy to an unhappy customer. If your colleagues have decided that the department should pursue a course of action, you might need to persuade your boss to act accordingly. Maybe you are asking your boss for a raise or promotion, for office improvements, or for changes to the work schedule. Your task is to persuade the boss to accept your suggestions. In these instances, you will communicate persuasively using any of the following communication channels: memos, e-mail messages and instant messages, letters, reports, proposals, or oral presentations.

Figure 13.1 is a persuasive memo from a subordinate to a boss, documenting a problem and suggesting a course of action. Frequently, you will write e-mail messages to supervisors and colleagues, persuading these readers to accept your point of view. Topics could range from requests for promotion, equipment needs, days off from work, assistance with projects, or financial assistance for job-related travel.

The e-mail messages on page 371 seek to persuade a colleague to attend a work-related conference instead of the e-mail writer. The "Before" sample is not effectively persuasive. It reads more like a command and fails to consider the audience's reaction. The "After" sample is more persuasive.

FIGURE 13.1 Persuasive Memo

DATE:	March 22, 2009
FROM:	Bob Ward
TO:	Lynn Richards
SUBJECT:	Maintenance of Photo Lab Machine

On March 21, our one-hour photo employee Brian Syoung reported that the photo lab machine was malfunctioning. It was cropping photos out of specification and producing blurred images. A service technician examined the machine and estimated that repairs would cost $1,000. Just as a reminder, Lynn, in one week, we are scheduled to have this machine replaced with an upgrade.

The problem is whether to repair the machine or wait a week for the replacement. Here are some facts that I've collected to help us decide.

The body objectively recognizes both sides of an issue and alternative approaches. The first two bullets focus on delaying the repair. The last two bullets explain why repair delays are bad for business.

- The lab makes $850 in profits each week. If we spend $1,000 to repair the machine, we will lose money.
- Closing the lab will save the company around $200 in supplies and electricity.
- If the photo lab is closed, then our higher-paid photo specialists will have to be assigned other jobs. We will lose approximately $2.00 an hour paying them to do work usually assigned to lower-paid employees.
- Closing the photo lab for a week to save repair expenses will inconvenience our repeat customers. In fact, they might shop elsewhere, which would impact us negatively long term.

The body develops the argument persuasively by providing facts and figures. In doing so, the writer reveals his knowledge of the subject matter.

The conclusion emphasizes persuasively the urgency of action.

Lynn, though repairing the machine costs $1,000, we will more than make up for this cost in customer satisfaction. We should repair the machine for the sake of our customers. Since this is a difficult decision, let's meet later this afternoon at 3:00 p.m. in your office. We need to resolve this issue today before we lose more money.

BEFORE

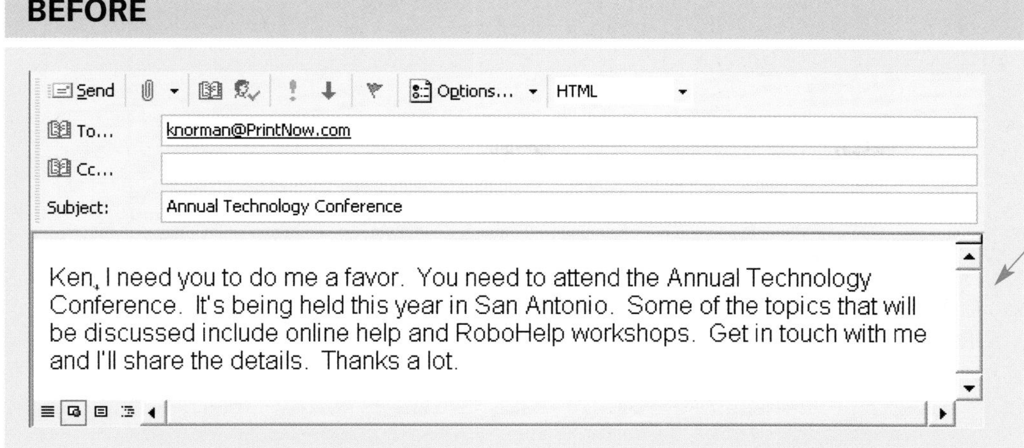

To... knorman@PrintNow.com

Cc...

Subject: Annual Technology Conference

Ken, I need you to do me a favor. You need to attend the Annual Technology Conference. It's being held this year in San Antonio. Some of the topics that will be discussed include online help and RoboHelp workshops. Get in touch with me and I'll share the details. Thanks a lot.

This e-mail message is not persuasive for a number of reasons: the verb "need" is too commanding; the e-mail fails to arouse the reader's interest or provide sufficient detail to convince the reader of the conference's worth; it also fails to urge the reader to action.

AFTER

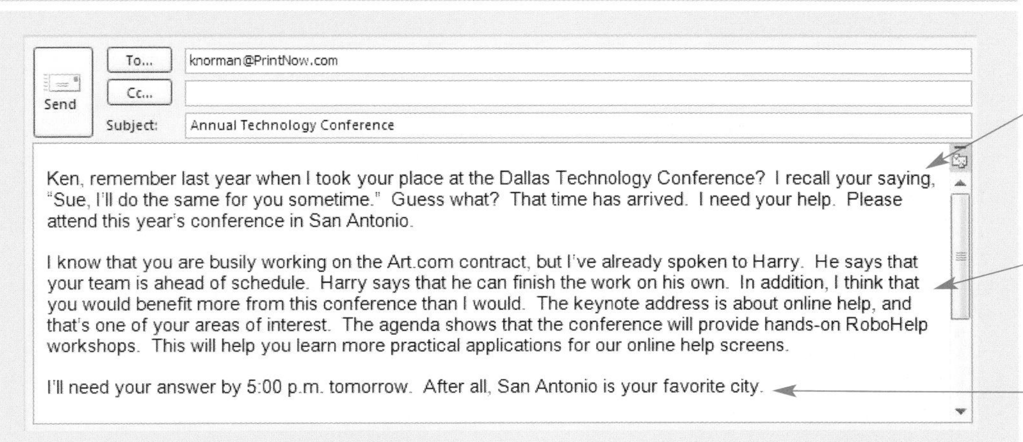

To... knorman@PrintNow.com

Cc...

Subject: Annual Technology Conference

Ken, remember last year when I took your place at the Dallas Technology Conference? I recall your saying, "Sue, I'll do the same for you sometime." Guess what? That time has arrived. I need your help. Please attend this year's conference in San Antonio.

I know that you are busily working on the Art.com contract, but I've already spoken to Harry. He says that your team is ahead of schedule. Harry says that he can finish the work on his own. In addition, I think that you would benefit more from this conference than I would. The keynote address is about online help, and that's one of your areas of interest. The agenda shows that the conference will provide hands-on RoboHelp workshops. This will help you learn more practical applications for our online help screens.

I'll need your answer by 5:00 p.m. tomorrow. After all, San Antonio is your favorite city.

The introductory paragraph arouses reader interest with an anecdote and a question.

The e-mail body begins by refuting any objections. Then, the following sentences show audience benefit.

The conclusion urges action by giving a due date and highlighting reader enjoyment.

FAQs

Q: How does marketing fit in with persuasive workplace communication?

A: Workplace communication is more than hard, cold memos, letters, reports, and user manuals. Workplace communication also has a soft side—*marketing*. Bottom line—every company is in business to make money. Thus, every employee should perceive himself or herself as marketing personnel. Employees need to understand how to be persuasive communicators so they can effectively market products and services.

George Butler Associates (GBA), an architectural and engineering company in Kansas, Missouri, and Illinois, asks every one of its employees to take marketing classes on site. The goal is to help GBA employees work well with clients, vendors, and partners. GBA employees have to persuade potential customers of the value of working with their company. These employees have to understand how to market persuasively and, thus, be a more effective workplace communicator.

TRADITIONAL METHODS OF ARGUMENT AND PERSUASION

To argue a point persuasively, you can use any of the traditional methods of argumentation: ethical (*ethos*), emotional (*pathos*), and logical (*logos*) appeals. These three appeals to an audience are called "the rhetorical triangle," as shown in Figure 13.2.

FIGURE 13.2 **The Rhetorical Triangle**

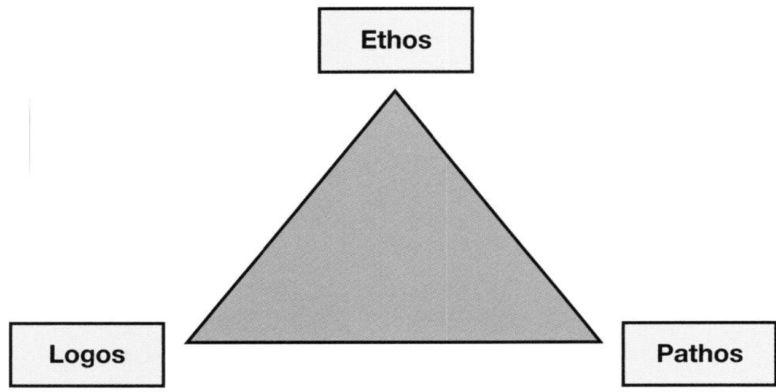

This equilateral triangle suggests that each part of a persuasive appeal is as important as any other part. In addition, it emphasizes the need for balance of all three appeals or types of proof. Excess emotion, for example, might detract from the logical appeal of your argument. Cold, hard facts might fail to persuade your audience.

Ethical (*Ethos*)

Ethos translates as "ethics." Arguments based on ethics depend on your character. If you make arguments based on personal experiences, you must appear to be trustworthy and credible as a writer or speaker. To accomplish this goal, you need to present information that is unbiased, reliable, and evenhanded.

Tiger Woods and Mia Hamm are classic examples of corporate spokespeople hired for their character. When Tiger promotes golf equipment, you trust him. He is a reliable voice for these products due to his successes on the links. The same holds true for Mia. When she speaks about soccer, you know she is an expert. In fact, these athletes are so trustworthy in the eyes of the public that Tiger and Mia also can sell products outside of their primary areas of expertise, such as cars and healthcare products.

Former New York City Mayor Rudy Guiliani wrote a book entitled *Leadership*. Oprah Winfrey writes and speaks about health, education, entertainment, books, money, philanthropy, and more. When these individuals make arguments drawn from personal experiences, audiences respond favorably. Given their reputations for excellence and success, both are trustworthy and reliable experts.

Emotional (*Pathos*)

Pathos translates as "emotion." Arguments based on emotion seek to change an audience's attitudes and actions by focusing on feelings. If you want to move an audience emotionally, you would appeal to passion. You can do this either positively or negatively.

To sway an audience positively, focus on positive concepts like joy, hope, honor, pleasure, happiness, success, and achievement. In addition, use positive words to create an appealing message. In contrast, you also can appeal to emotions negatively. Fear, horror, anger, and unhappiness can be powerful tools in an argument.

Notice how the National Center for Environmental Health warns about the dangers of carbon monoxide: "Carbon monoxide is a silent killer. This colorless, odorless, poisonous gas kills nearly 500 U.S. residents each year, five times as many as West Nile virus" ("CDC and CPSC Warn of Winter Home Heating Hazards"). To highlight the dangers associated with carbon monoxide, the Center for Disease Control uses emotional words such as "killer" and "poisonous," and compares this problem to the frightening "West Nile virus."

Logical (*Logos*)

Logos translates as "logic." Argumentation based on logic depends on rationality, reason, and proof. You can persuade people logically when you provide them the following:

- Facts—statistics, evidence, data, and research.
- Testimony—citing customer or colleague comments, citing expert authorities, and sharing results of interviews.

FIGURE 13.3 Persuasive E-Mail Effectively Using *Ethos*, *Pathos*, and *Logos*

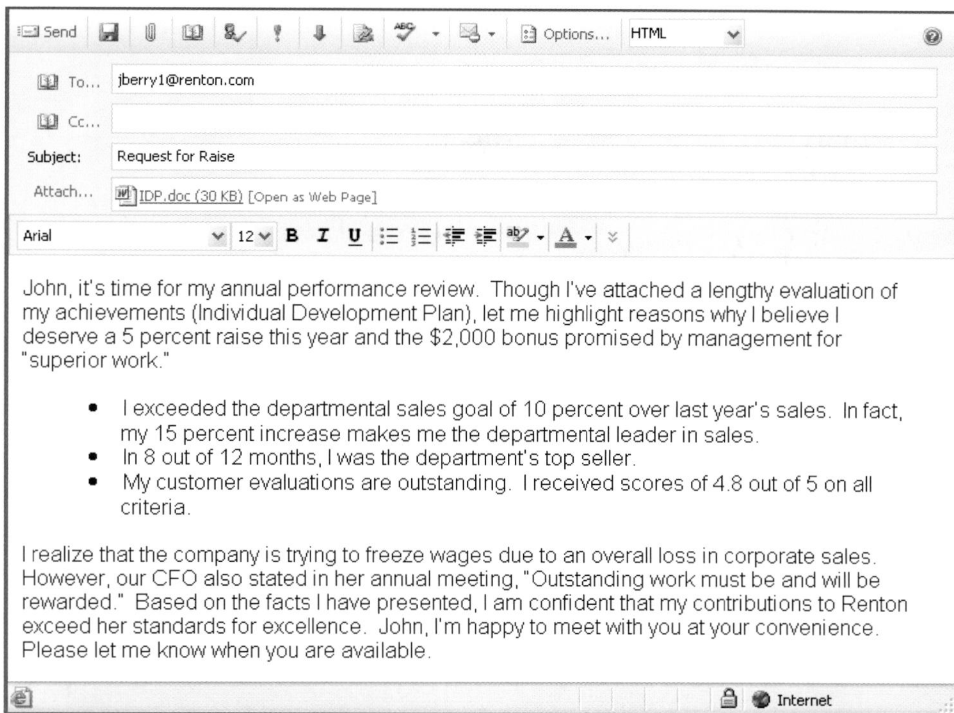

To... jberry1@renton.com

Cc...

Subject: Request for Raise

Attach... IDP.doc (30 KB) [Open as Web Page]

John, it's time for my annual performance review. Though I've attached a lengthy evaluation of my achievements (Individual Development Plan), let me highlight reasons why I believe I deserve a 5 percent raise this year and the $2,000 bonus promised by management for "superior work."

- I exceeded the departmental sales goal of 10 percent over last year's sales. In fact, my 15 percent increase makes me the departmental leader in sales.
- In 8 out of 12 months, I was the department's top seller.
- My customer evaluations are outstanding. I received scores of 4.8 out of 5 on all criteria.

I realize that the company is trying to freeze wages due to an overall loss in corporate sales. However, our CFO also stated in her annual meeting, "Outstanding work must be and will be rewarded." Based on the facts I have presented, I am confident that my contributions to Renton exceed her standards for excellence. John, I'm happy to meet with you at your convenience. Please let me know when you are available.

Ethics (ethos) *is shown when the writer references his achievements at work.*

Facts ("4.8 out of 5") and testimony from customer evaluations support the writer's case logically (logos).

The e-mail addresses the opposing point of view ("freeze wages") to present a "balanced" argument.

Note the use of positive words, such as "achievements," "highlight," "exceeded," and "outstanding." These are examples of pathos—*emotional words to sway the reader.*

- Examples—anecdotes, instances, and personal experiences.
- Strong, clear claims—including warranties and guarantees.
- Acknowledgement of the opposing points of view to ensure that information is "balanced."

The e-mail message in Figure 13.3 argues for a raise by focusing on ethics (*ethos*). The writer refers to his work-related achievements to support his credibility. He uses logic (*logos*)—facts and testimony. Also note how the e-mail factors in an opposing point of view. In addition, the e-mail uses emotion (*pathos*) to sway the reader through positive words.

ARGU TO ORGANIZE YOUR PERSUASION

Effective persuasive communication entails ethical, logical, and emotional appeals. Understanding the importance of this rhetorical triangle is only part of your challenge as a persuasive communicator. The next step is deciding how best to present your argument. Using our *ARGU* approach will help you organize your argument:

- *A*rouse reader interest—grab the audience's attention in the introduction of your communication.
- *R*efute opposing points of view in the body of your communication.
- *G*ive proof to develop your thoughts in the body of your communication.
- *U*rge action—motivate your audience in the conclusion.

*A*rouse Reader Interest

You have only about five to eight seconds to grab your audience's attention in a sales letter, persuasive e-mail message, marketing brochure, speech, or any persuasive communication. You must arouse the audience's interest imaginatively in the first few sentences of your document or oral presentation. Try any of the following attention grabbers in the **introduction** of your persuasive message.

- **Use an anecdote—a brief, dramatic story relating to the topic**—Stories engage your audience. You can involve your readers or listeners by recounting an interesting story that they can relate to. The story should be specific in time, place, person, and action. The drama should highlight an event. However, this must be a short scenario, since your time is limited. If you do not capture their interest quickly, the audience might lose interest.

example

This anecdote is specific in time, place, person, and action.

E-Mail Message Persuading a Manager to Take Action

Sam, last week, a customer fell down in our parking lot, cutting her knee, tearing her slacks, and requiring medical attention. We can't let this happen again. Please consider hiring a maintenance crew to salt and sand our lot during icy weather.

- **Start with a question to interest your audience**—By asking a question, you involve your audience. Questions imply the need for an answer. A question can make readers or listeners ask themselves, "How would I answer that?" This encourages their participation.

example

By asking questions, you involve your audience.

Sales Letter from a Financial Planner

"Where will I get money for my kid's college education?" "How can I afford to retire?" "Will my insurance cover all medical bills?" You have asked yourself these questions. Our estate planning video has the answers.

- **Begin with a quotation to give your communication the credibility of authority**—By quoting specialists in a field or famous people, you enhance your credibility and support your assertions. For example, a quote about economics from Warren Buffett, world-renowned investor and businessman, gives credence to a financial topic. Similarly quotes from Bill Gates, founder of Microsoft, about the computer industry are trustworthy.

example

Use a quote from an authority to make your argument believable.

An Oral Presentation About Stock Performance to Shareholders

As Warren Buffett says, "Our favorite holding period is forever." Though our stock prices are down now, don't panic and sell. The company will rebound.

- **Let facts and figures enhance your credibility**—Factual information (percentages, sales figures, amounts, or dates) catches the audience's attention and lays the groundwork for your persuasion.

example

Facts and figures help persuade your reader to your point of view.

Sales Letter About Computer Technology

Eighty-seven percent of all college students own a computer. Don't go off to college unprepared. Buy your laptop or PC at *CompuRam* today.

- **Appeal to the senses**—You can involve your audience by letting them hear, taste, smell, feel, or visualize a product.

example

An appeal to the senses can involve your reader.

Marketing Flier Advertising a New Restaurant

Hickory-smoked goodness and fire-flamed Grade A beef—let Roscoe's BBQ bring the taste of the South to your neighborhood.

- **Use comparison/contrast to highlight your message**—Comparison/contrast lets you make your point more persuasively. For example, you might want to show a client that your computer software is superior to the competition. You might want to show a boss that following one plan for corporate restructuring is better than an alternative plan. You could justify why an employee has not received a raise by comparing/contrasting his or her performance to the departmental norm. Comparison/contrast is useful in many different arguments for a variety of audiences.

E-Mail from Management to Subordinates Proposing Changes in Procedures

Last year our department fell short of corporate sales goals. This year we must surpass expectations. To do so, I propose a seven-step procedure.

example

Comparing one year to another provides a benchmark for the writer's argument.

- **Begin with poetic devices**—Advertising has long used alliteration (the repetition of sounds), similes (comparisons), and metaphors to create audience interest. Poetic devices are memorable, clever, fun, and catchy.

Sales Letter from a Bank

Looking for a low-loan **lease**? Let USB Bank take care of your car leases. Our 2 percent loans **beat the best**.

example

Use alliteration, the repetition of sounds, to appeal to an audience.

- **Create a feeling of comfort, ease, or well-being**—To welcome your audience and make them feel calm and peaceful, invoke nostalgia or good times.

Slogan from a Mortgage Company Brochure

Come home again to Countryside Mortgage. We treat our customers like neighbors.

example

Interest your audience by making them feel comfortable and welcomed.

- **Create a feeling of discomfort, fear, or anxiety**—Another way to involve your audience is through stress. Beginning communication by highlighting a problem allows you to persuade the audience that you offer the solution.

Sales Letter from an Apartment Complex

Why pay too much? If you paid over $300 a month for your apartment last semester, you got robbed. We'll charge 20 percent less—guaranteed.

example

A negative example can incite an audience to action.

Refute Opposing Points of View

You can strengthen your argument by considering opposing opinions. Doing so shows your audience that you have considered your topic thoroughly. Rather than looking at the subject from only one perspective, you have considered alternatives and discarded them as lacking in merit. In addition, by refuting opposing points of view, you anticipate negative comments an audience might make and defuse their argument.

To refute opposing points of view in the **discussion** (**body**) of your communication, follow these steps:

- Recognize and admit conflicting views.
- Let the audience know that you understand their concerns.

- Provide evidence.
- Allow for alternatives.

Figure 13.4 shows a letter of application that successfully uses refutation as part of its persuasion.

Give Proof to Develop Your Thoughts

In the **body** of your communication, develop your argument with proof. Arousing an audience's interest and refuting opposing points of view will not necessarily persuade the audience. Most people require details and supporting evidence before making decisions. You can provide specific details to support your argument using any of the techniques in Table 13.1.

FIGURE 13.4 Letter of Application Using Refutation

1901 West King's Highway
San Antonio, TX 77910

February 27, 2009

Melissa Flores
Manager, Human Resources
PMBR Marketing
20944 Wildrose Dr.
San Antonio, TX 78213

Dear Ms. Flores:

In response to your advertisement in the *San Antonio Daily Register,* please consider me for the position of Marketing Accounts Associate. I have enclosed a resume elaborating on ways I can benefit your company.

Admit opposing points of view such as the BA requirement.

Your advertisement requires someone with a BA in marketing. Though I do not have this degree, my experience and skills prepare me for this position:

The first five bullets provide evidence to show that the writer has skills appropriate for the job even without the desired degree.

- Ten years experience in marketing.
- Prepared press releases and created public service announcements.
- Created 30-second spots for local radio and television stations.
- Participated in press conferences providing corporate information about declining stock values and company layoffs.
- Created PowerPoint presentations for local governmental agencies.
- Currently enrolled in a marketing class at the university.

The current enrollment provides an alternative to the degree requirements.

My experience and abilities will make me an asset in your marketing department. I would be happy to meet with you to discuss ways in which I can benefit PMBR. Please contact me at jobrien22@hotmail.com

Sincerely,

Joan O'Brien

Attachment: Resume

The writer's refutation anticipates negative comments the audience might have and diffuses the argument through proof and alternatives.

*U*rge Action—Motivate Your Audience

Throughout your communication, you have worked to persuade the audience to accept your point of view. In the **conclusion**, you need to motivate the audience to action. This could include any of the following: attend a meeting, purchase new equipment, invite you to interview for a position, vote on a proposition, promote you, give you a raise, allow you to work a flexible schedule, or change a company policy.

To urge the audience to action, consider the techniques in Table 13.2.

Table 13.1 Techniques for Supporting an Argument

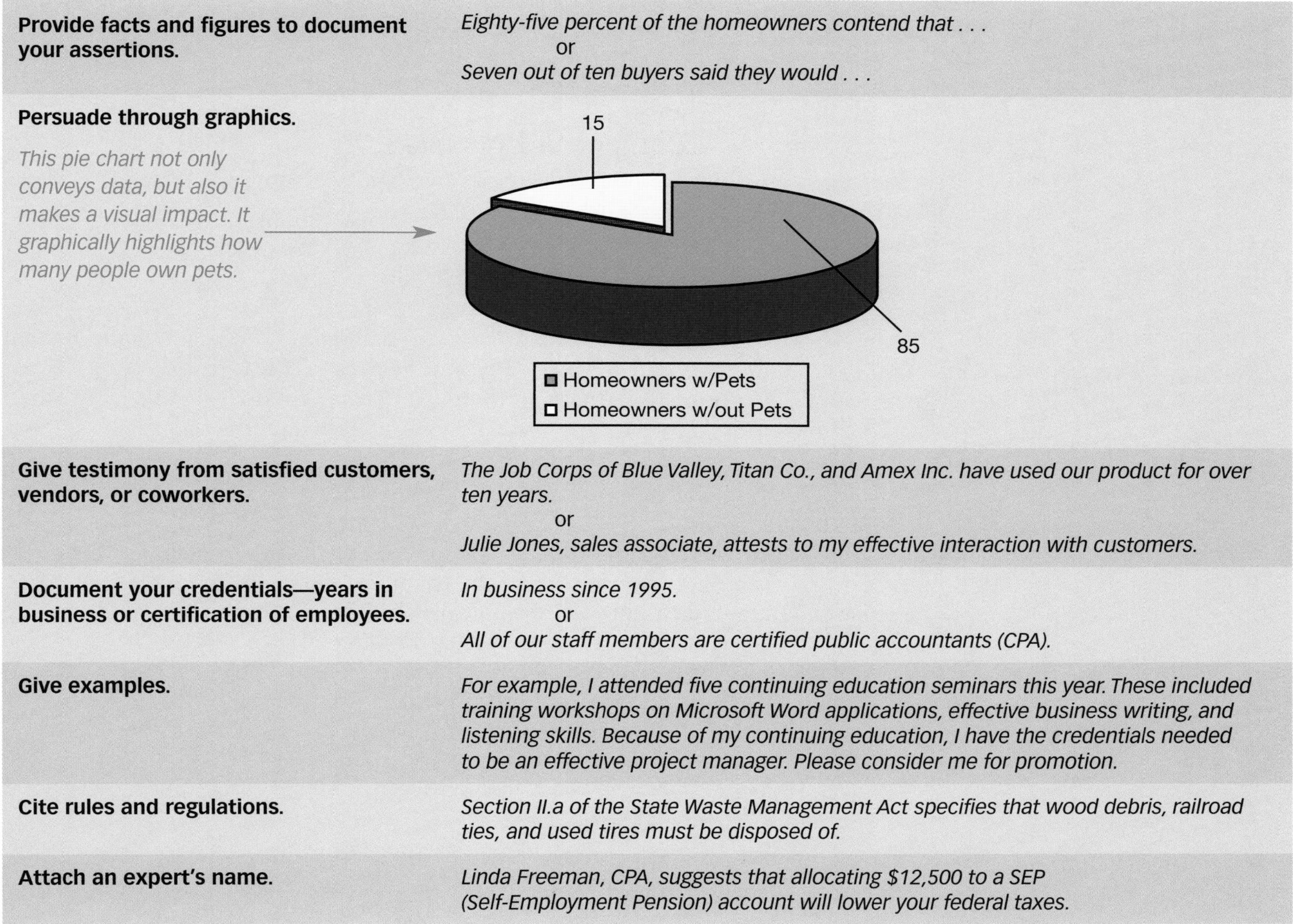

Technique	Example
Provide facts and figures to document your assertions.	*Eighty-five percent of the homeowners contend that . . .* or *Seven out of ten buyers said they would . . .*
Persuade through graphics. *This pie chart not only conveys data, but also it makes a visual impact. It graphically highlights how many people own pets.*	15 85 Homeowners w/Pets Homeowners w/out Pets
Give testimony from satisfied customers, vendors, or coworkers.	*The Job Corps of Blue Valley, Titan Co., and Amex Inc. have used our product for over ten years.* or *Julie Jones, sales associate, attests to my effective interaction with customers.*
Document your credentials—years in business or certification of employees.	*In business since 1995.* or *All of our staff members are certified public accountants (CPA).*
Give examples.	*For example, I attended five continuing education seminars this year. These included training workshops on Microsoft Word applications, effective business writing, and listening skills. Because of my continuing education, I have the credentials needed to be an effective project manager. Please consider me for promotion.*
Cite rules and regulations.	*Section II.a of the State Waste Management Act specifies that wood debris, railroad ties, and used tires must be disposed of.*
Attach an expert's name.	*Linda Freeman, CPA, suggests that allocating $12,500 to a SEP (Self-Employment Pension) account will lower your federal taxes.*

Table 13.2 Techniques for Motivating an Audience to Action

Technique	Example
Give due dates.	*Please respond by January 15.*
Explain why a date is important.	*Your response by January 15 will give me time to prepare a quarterly review and meet with you if I have additional questions.*
Provide contact information for follow-up.	*Please submit your proposal to Hank Green, Project Director. You can e-mail him at hgreen@modernco.com.*
Suggest the next course of action.	*We need to plan our presentation before the next city council meeting, so please attend Tuesday's meeting at 9:00 a.m.*
Show negative consequences.	*You must repair your sidewalk within 30 days to comply with city laws regarding pedestrian safety. Failure to do so will result in a $150 fine.*
Reward people for following through.	*Following these ten simple steps will help you load the software easily and effectively.*

The following paragraph concludes a persuasive letter from a governmental agency to a business owner.

example

This conclusion provides a due date, explains why the date is important, shows negative consequences, and tells the reader what to do next. Until you tell the reader what to do, he or she may not be inspired to act.

> Your business has not been in compliance for over 16 months. We have written three letters to give you guidance and information necessary to get the facility into compliance. You have not responded to our efforts, with the exception of last week's phone call. The deadline for the pending abatement order is 30 days. You must reply to this letter in a timely manner to avoid fines. We are open to having a meeting at our regional office to discuss the corrective measures you can take to address the pending order. Thank you for considering your options.

AVOIDING LOGICAL FALLACIES

In a corporate environment, you must persuade your audience not only logically but also ethically. Your persuasive communication must be honest and reasonable. Honesty demands that you avoid logical fallacies, such as the following.

Inaccurate Information

Additional information:

See Chapter 8, "Visual Aids in Workplace Communication."

Facts and figures must be accurate. One way in which dishonest appeals are made in communication is through inaccurate graphics. Look at Figures 13.5 and 13.6.

Figure 13.5 shows a $25,000 deficit in the third quarter. So does Figure 13.6. However, the size of the bar in Figure 13.6 is misleading, visually suggesting that the loss is less. This is an inaccurate depiction of information.

Unreliable Sources

Being an expert in one field does not mean you are an expert in all fields. For example, quoting a certified public accountant to support an important healthcare issue is illogical. Not all specialists are reliable in all situations.

FIGURE 13.5 **Annual Income**

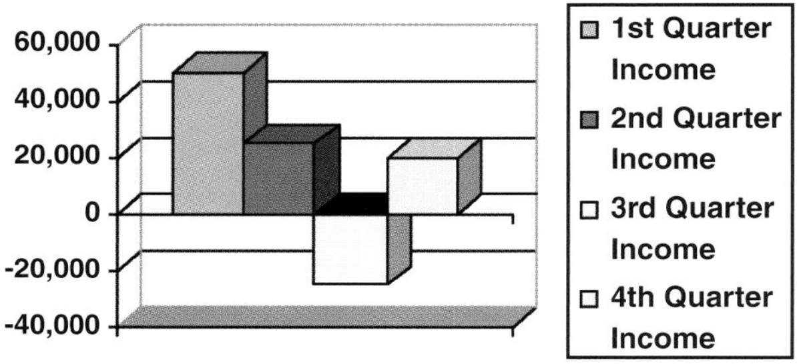

FIGURE 13.6 **Annual Income**

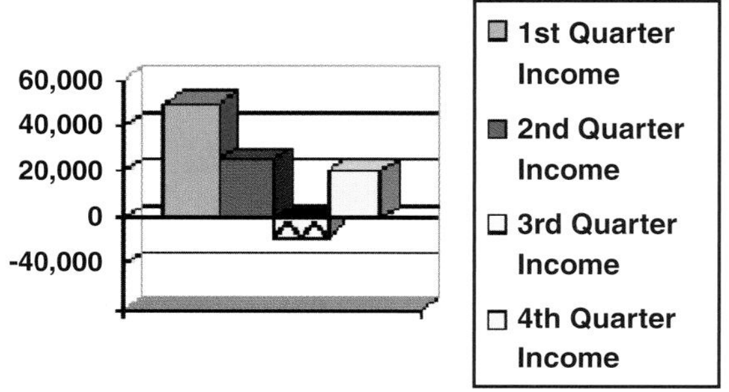

Sweeping Generalizations

Avoid exaggerating. Allow for exceptions. For instance, it is illogical to say that "*All* marketing experts believe that newsletters are effective." Either qualify this with a word like "some" or quantify with specific percentages.

Either . . . Or

Suggesting that a reader has only two options is deceitful if other options exist. Allow for other possibilities. It is wrong to write, "Either all employees must come to work on time, or they will be fired." This blanket statement excludes alternatives or exceptions. Is the "either . . . or" statement true if an employee has a car accident, if an employee's child is sick, or if the employee is caught in heavy traffic due to a snowstorm?

Circular Reasoning (Begging the Question)

"Some accountants are ambitious because they wish to succeed." This statement is illogical and uses circular reasoning because it states the same thing twice. "Ambitious" and "succeed" are essentially synonyms. The writer fails to prove the assertion.

Inaccurate Conclusions

When communicating persuasively, consider all possible causes and effects. Exact causes of events often are difficult to determine.

- A condition that *precedes* another is not necessarily the *cause* of it. This error is called *post hoc, ergo propter hoc.*

  > "The contractor lost the bid, so he cannot expect to have increased revenues this fiscal year." **example**

  Yes, the contractor lost a bid, but this does not necessarily mean that the contractor won't increase revenues in some other way. To say so is to make a hasty conclusion based on too little information.

- A condition that *follows* is not necessarily the *effect* of another. This is called a *non sequitur.*

  > "Because the manager is inexperienced, the report will be badly written." **example**

Again, this is a hasty conclusion. Lack of experience in one area does not necessarily lead to a lack of ability in another area.

Red Herrings

An irrelevant issue used to draw attention from a central issue is called a red herring. For instance, you have failed to pay fines following citations for the mishandling of hazardous wastes. You contact the state environmental agency and complain about state taxes being too high. This is an irrelevant issue. By focusing on high taxes, you are merely avoiding the central issue.

To write effective persuasive documents, you must develop your assertions correctly, avoiding logical fallacies.

TYPES OF PERSUASIVE DOCUMENTS

You have manufactured a new product (a mini digital camera, a paper-thin plasma computer monitor, a fiber optic cable, or a microfiber rain jacket). Perhaps you have just created a new service (home visitation health care, a mobile accounting business, a Web design consulting firm, or a gourmet food preparation and delivery service). Congratulations!

However, if your product sits in your basement gathering dust or your service exists only in your imagination, what have you accomplished? To benefit from your labors, you must market your product or service.

To convince potential customers to purchase your merchandise, you could write any of the following persuasive documents:

- Sales letters
- Fliers
- Brochures

Sales Letters

To write your sales letter, follow the format for letters discussed in Chapter 10 (Traditional Correspondence). You need to include the letter essentials (letterhead address, date, reader's address, salutation, text, complimentary close, and signatures). Your sales letter should accomplish the following objectives, relating to effective persuasion.

*A*rouse Reader Interest

The introductory paragraph of your sales letter tells your readers why you are writing (you want to increase their happiness or reduce their anxieties, for example). Your introduction should highlight a reader problem, need, or desire. If the readers do not need your services, then they will not be motivated to purchase your merchandise. The introductory sentences also should mention the product or service you are marketing, stating that this is the solution to their problems. Arouse your readers' interest with anecdotes, questions, quotations, or facts.

*R*efute Opposing Points of View

By mentioning competitors and undercutting their worth, you can emphasize your product's value.

*G*ive Proof to Develop Your Thoughts

In the discussion paragraph(s), specify exactly what you offer to benefit your audience or how you will solve your readers' problems. You can do this in a traditional paragraph. In contrast, you might want to itemize your suggestions in a numbered or bulleted list. Whichever option you choose, the discussion should provide data to document your assertions, give testimony from satisfied customers, or document your credentials.

*U*rge Action

Make readers act. If your conclusion says, "We hope to hear from you soon," you have made a mistake. The concluding paragraph of a sales letter should motivate the reader to act.

Conclude your sales letter in any of the following ways:

- Give directions (with a map) to your business location.
- Provide a tear-out to send back for further information.
- Supply a self-addressed, stamped envelope for customer response.
- Offer a discount if the customer responds within a given period of time.
- Give your name or a customer-contact name and a phone number (toll free if possible).

Figure 13.7 provides a sample sales letter using the *ARGU* method of persuasion.

Additional information:

See Chapter 10, "Traditional Correspondence: Memos and Letters" for a discussion of "Letter Formats."

Fliers

Corporations and companies; educational institutions; religious organizations; museums, zoos, and amusement parks; cities and states—they all need to communicate with their constituencies (clients, citizens, members). An effective way for these organizations to

FIGURE 13.7 Sales Letter Using the *ARGU* Method

4520 Shawnee Dr. Tulsa, OK 86221 721-555-2121

November 12, 2009

Bill Schneider
Office Manager
REM Technologies
2198 Silicon Way
Tulsa, OK 86112

Dear Mr. Schneider:

Are hardware and software upgrades making your profits plummet? Would you like to reduce your company's computer purchase and maintenance costs? Do computer breakdowns hurt your business productivity? Don't let technology breakdowns harm your bottom line. Many companies have taken advantage of **Office Station's** computer prices, service guarantees, and certified technicians.

The introduction arouses reader interest by using alliteration ("profits plummet") and asking questions. The questions highlight reader problems: profits, costs, productivity, and breakdowns.

The last sentence shows how Office Station will solve the problems.

Office Station, located in your neighborhood, offers you the following benefits:

- Purchase prices at least 10 percent lower than our competitors.
- IBM-trained technicians, available on a yearly contract or per-call basis.
- An average response time to service calls of under two hours.
- Repair loaners to keep your business up and running.
- Over 5,000 satisfied customers, like IBM, Ford, Chevrolet, and Boeing.
- All-inclusive agreements that cover travel, expenses, parts, and shop work.
-  State-of-the-art technologies, featuring the latest hardware and software.

The letter uses positive words to persuade: "advantage," "guarantees," "certified," "benefits," "satisfied," "prompt," and "courteous."

The body provides specific proof to sway the reader: 10 percent lower, IBM-trained technicians, two-hour response, and renowned satisfied customers.

Our service is prompt, our technicians are courteous, and our prices are unbeatable. For further information and a written proposal, please call us at **721-555-2121** or e-mail your sales contact, Steve Hudson (shudson@os.com). He's waiting to hear from you. Take advantage of our *Holiday Season Discounts!*

The conclusion urges action by giving contact names and numbers and seasonal discounts.

Sincerely,

Rachel Adams
Sales Manager

Office Station
Authorized Sales and Service for
Gateway 3M Microsoft HP Apple Dell Swingline

communicate persuasively is with a one- or two-page flier. Fliers can be short and sales-oriented or longer and more informative.

Fliers provide the following benefits:

- **Cost effective**—A flier costs less than an expensive advertising campaign and can be produced in-house.

- **Time efficient**—Creating a flier can take only a few hours of work or less by the company's employees.
- **Responsive to immediate needs**—Different fliers can be created for different audiences and purposes to meet unique, emerging needs.
- **Personalized**—Fliers can be created with a specific market or client in mind. Then, these fliers either can be mailed to that client or hand-delivered for more personalization.
- **Persuasive**—In a compact format, fliers concisely communicate audience benefit.

Criteria for Writing Fliers

When writing your flier, follow these criteria:

- **Determine the length of the flier**—Though one page might be preferable, you could create a two-page flier, using the front and back of an 8½" × 11" piece of paper. If you keep your flier to one page (front only), then you can save money by folding the flier in thirds, stapling it, and using the blank side for mailing purposes (addresses and stamp). A flier even could be smaller, the size of a postcard, for example. Many companies create electronic fliers, transmitted via their Web sites. This is even more cost effective than a hard-copy flier, since no postage is needed.
- **Focus on one idea, topic, or theme per flier**—A flier should make one key point. This is how you make the flier's content relevant to your audience, fulfilling that audience's unique needs. For example, if your company's focus is automotive parts, avoid writing a flier covering every car accessory. Write the flier with one accessory in mind, such as windshield wipers, batteries, or custom rims.
- **Use a title at the top of your flier to identify this flier's theme**—The title can be one or two words long, you could use a phrase, or you might want to write an entire sentence at the top of the flier. An effective persuasive approach is to begin your flier with a question to immediately arouse reader interest: "Software giving you a headache?" or "How usable is your Web site?"
- **Limit your text**—Using few words, provide reader benefit, involve the audience, and motivate them to act. The action could be to purchase a product, attend an event, or contact you for additional information. By limiting your text, you avoid overwhelming either the flier's appearance or the audience's attention span. Getting to the point in a flier is a key concern. Limiting your text helps you achieve this goal.
- **Increase font size**—In a flier you can use a 16-point font and up for text, and a 20-point font and up for titles. This will make the text more readable and more dramatic. Your heading must be eye catching. To accomplish this goal, make sure the heading's font size and style are emphatic—at a glance, even from across a room.
- **Use graphics**—One graphic, at least, will emphasize your theme and visually make your point memorable. Another graphic could include your company logo (for corporate identity and namesake recognition). In addition, the logo should be accompanied by a street address, e-mail address, Web site URL, fax number, or phone number so clients can contact you or visit your site. You also can use text boxes to highlight text.
- **Use color for audience appeal**—Pick one or two dominant colors to emphasize key points. Use a color in your company logo to remind your reader of your company's identity. However, don't overuse color. Excess will distract your reader.
- **Use highlighting techniques**—Bullets, white space, tables, boldface, italics, headings, or subheadings will help your reader access information. A little highlighting goes a long way, especially on a one-page flier. Too much makes a jumble of your text and distracts from the message.

FIGURE 13.8 A Sample Persuasive Flier

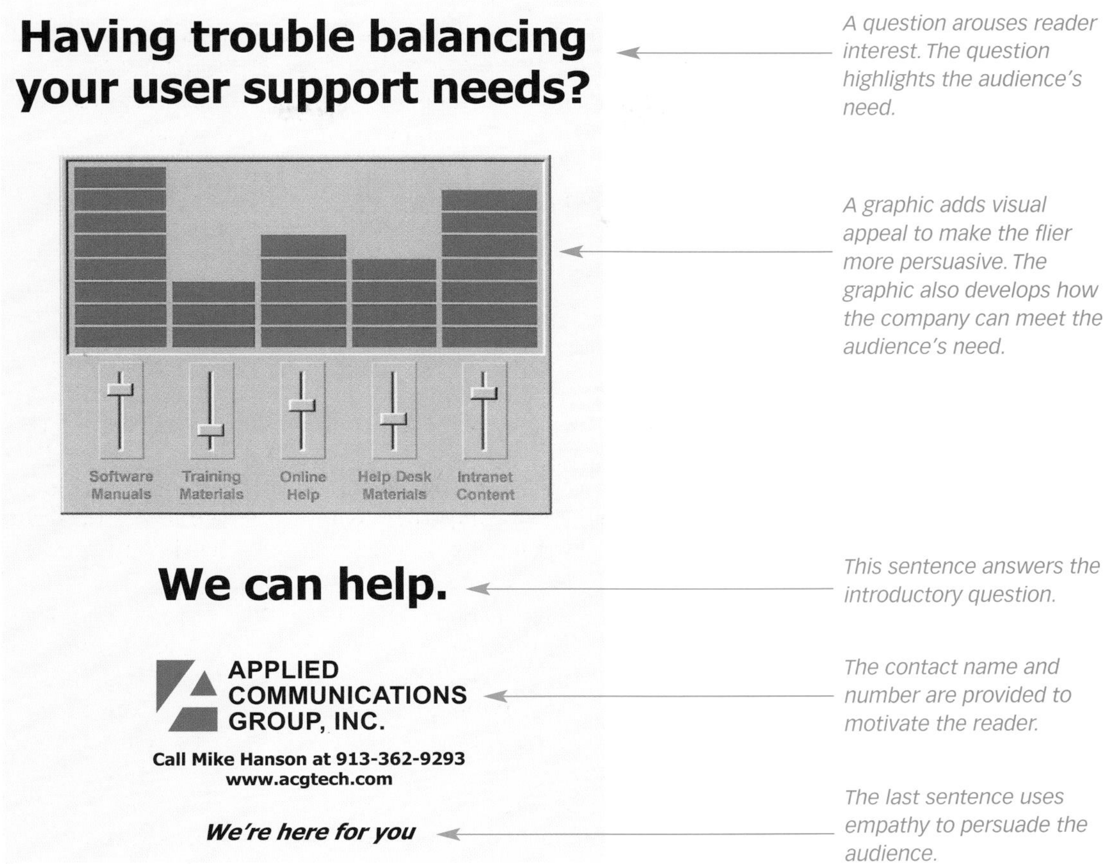

- **Recognize your audience**—You want to show the readers how your product or service will benefit them. Understand your audience's needs and direct the flier to meet those concerns. In addition, you want to engage the reader. To do this, use *pronouns* which speak to the reader on a personal level and *positive words* which motivate the reader to action. Remember to speak at the reader's level of understanding, defining terms as necessary.

Figure 13.8 provides a sample persuasive flier.

Brochures

A flier *must* be short—one or two pages. If you have more information to convey than can fit on a one-page flier, a brochure might be a good option. Brochures offer you a detailed overview of products, services, options, and opportunities, complete with photographs, maps, or charts. Brochures are persuasive because they

- Create awareness of your company, product, or service
- Increase understanding of a product, service, or your company's mission
- Advertise new aspects about your company, product, or service
- Change negative attitudes
- Show ways in which your company, product, or service surpasses your competition
- Increase frequency of use, visit, or purchase
- Increase market share

Technology Tip

USING WORD 2007 FOR FLIER AND BROCHURE DESIGN TEMPLATES

You can access many design templates for fliers and brochures using your word processing software. For example, in Microsoft Word 2007, try this approach:

1. On the Menu bar, click on the **Office Button** and scroll to ***New***. You will see **New Document/Templates** pop up.

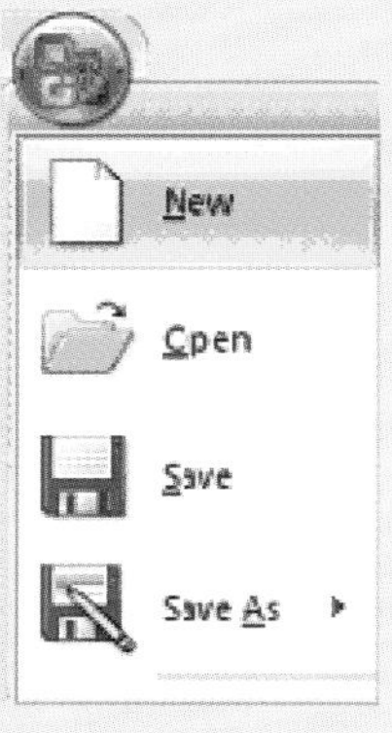

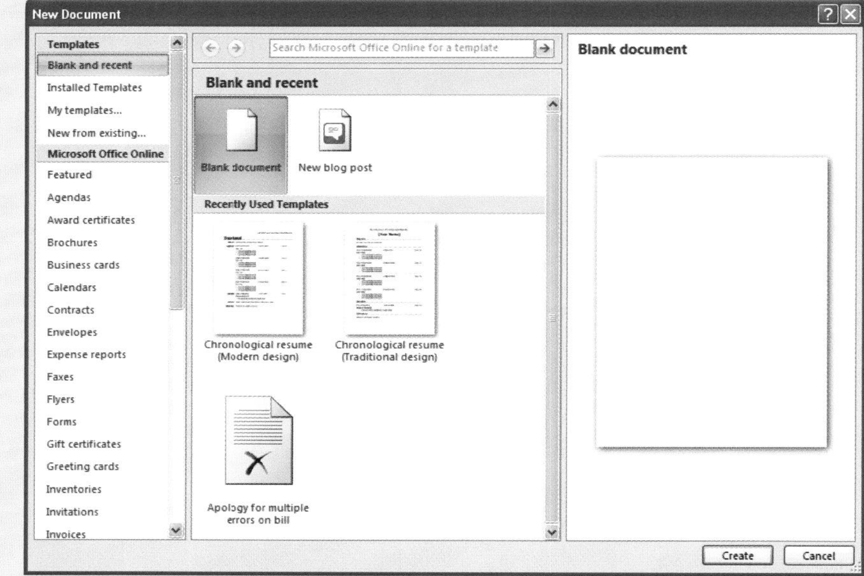

2. Click on the document of your choice (**Fliers** or **Brochures**). Word 2007 gives you 14 different brochure options for events, travel, recruiting, professional services, real estate, and financial. These include templates for 8½" × 11" landscape with three folds, 8½" × 14" landscape with three and four folds, and 8½" × 11" portrait letter folds. When you click on fliers, you get templates for 25 different "Events" fliers, 13 "Marketing" options, 3 "Real Estate" choices, and 5 "Other Fliers."

Note: The templates will not tell you what to write in each panel. Instead, Word 2007 provides you a design layout, a shell for you to fill with your content.

Creating Your Own Brochure Layout

Instead of using a predetermined template, you can create your own as follows:

1. Change from portrait to landscape by clicking on the **Page Layout** tab and then **Orientation**.

2. Create three panels for your brochure by clicking on the **Columns** icon.

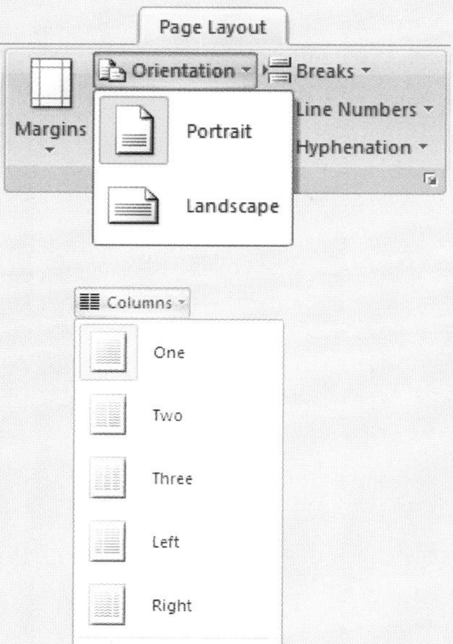

Criteria for Writing Brochures

Brochures come in many shapes and sizes. They can range from a simple front and back, four-panel, 8½" × 5½" brochure (one landscape 8½" × 11" page folded in half vertically) to 6-, 8-, or even 12-panel brochures printed on any size paper you choose. Figure 13.9 shows some examples. Brochures, like fliers, also can be transmitted electronically as Web pages. Your topic and the amount of information you are delivering will determine your brochure's size and means of transmission. To determine what you will write in each panel of your brochure and how the brochure should look, follow these criteria for writing an effective persuasive brochure.

- **Title page (front panel)**—Usually, the title page includes the following components:
  - **Topic**—Name the topic. This includes a product name, a service, a location, or the subject of your brochure.
  - **Graphic**—Include a graphic to appeal to your reader's need for a visual representation of your topic. The graphic will sell the value of your subject (its beauty, its usefulness, its location, or its significance) or visually represent the focus of your brochure.
- **Back panel**—The back panel could include the following:
  - **Conclusion**—Summarize your brochure's content. Restate the highlights of your topic or suggest a next step for your readers to pursue.
  - **Mailing**—The back panel can be used like the face of an envelope. On this panel, when left blank, you can provide your address, a place for a stamp or paid postage, and your reader's address.
  - **Coupons**—As a tear-out, this panel can be an incentive for your readers to visit your site or use your service. Here you can urge action by providing discounts or complimentary tickets.
  - **Location**—Another consideration is to provide your reader with your address, hours of operation, phone numbers, e-mail, and a map to help them locate you.
  - **Contact information**—Include your name, your company's name, street address, city, state, zip code, telephone number, fax number, Web URL, and e-mail address.
- **Body panels (fold-in and inside)**—The following are some suggestions for creating the brochure's text:
  - **Provide headings and subheadings**—These act as navigational tools to guide your readers, direct their attention, and help them find the information they need. The headings and subheadings should follow a consistent pattern of font type and size. First-level headings should be larger and more emphatic than your second-level subheadings. The headings must be parallel to each other grammatically.

    For example, if your first heading is entitled "Introduction," a noun, all subsequent headings must be nouns, like "Location," "Times," "Payment Options," and "Technical Specifications." If your first heading is a complete sentence, like "This is where it all began," then your subsequent headings must

FIGURE 13.9 Four- and Six-Panel Brochures

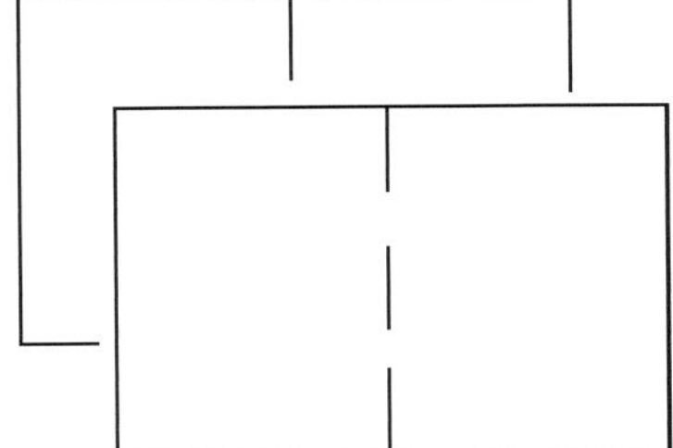

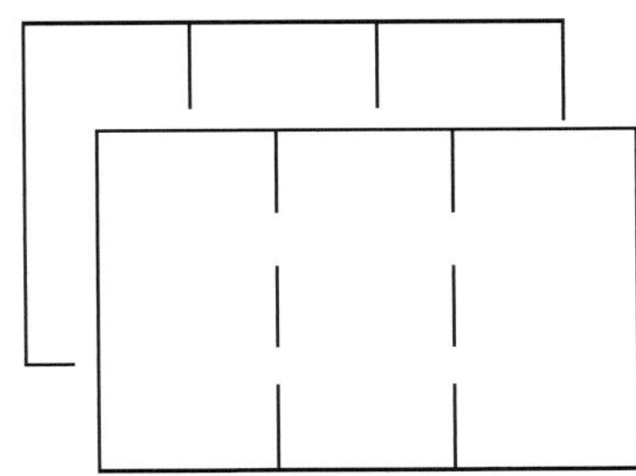

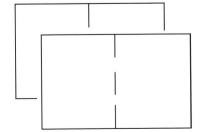

also be complete sentences: "It's still beautiful," "Here's how to find us," and "Prices are affordable."
    - **Use graphics (photographs, maps, line drawings, tables, or figures)**—These vary the page layout, add visual appeal, and enhance your text.
    - **Develop your ideas**—Consider including locations, options, prices, credentials, company history, personnel biographies, employment opportunities, testimonials from satisfied customers, specifications, features, uses of the product or service, payment schedules, or payment plans.
    - **Persuade your audience**—Review the tips provided in this chapter for persuasive arguments. Use ethics, logic, and emotion to sway your reader.
- **Document design**—Visual appeal helps to interest and persuade an audience. Compelling graphics, for example, can help to convince an audience. Use pie charts, bar charts, tables, or photographs to highlight key concerns. In addition to graphics, make your brochure visually appealing by doing the following:
    - Limit sentence length to 10 to 12 words and paragraph length to 4 to 6 lines. When you divide paper into panels, text can become cramped very easily. Long sentences and long paragraphs then become difficult to read. By limiting the length of your text, you will help your readers access the information.
    - Use white space instead of wall-to-wall words. Indent and itemize information so readers won't have to wade through too much detail.
    - Use color for interest, variety, and emphasis. For example, you can use a consistent color for your headings and subheadings.
    - Bulletize key points.
    - Boldface or underline key ideas.
    - Do not trap yourself within one panel. For variety and visual appeal, let text and graphics overlap two or more of the panels.
    - Place graphics at angles (occasionally) or alternate their placement at either the center or the right or left margin of a panel. Panels can become very rigid if all text and graphics are square. Find creative ways to achieve variety.

See Figure 13.10 for a sample brochure.

FIGURE 13.10 Sample Brochure

Title page, back panel, and fold-in body panel.

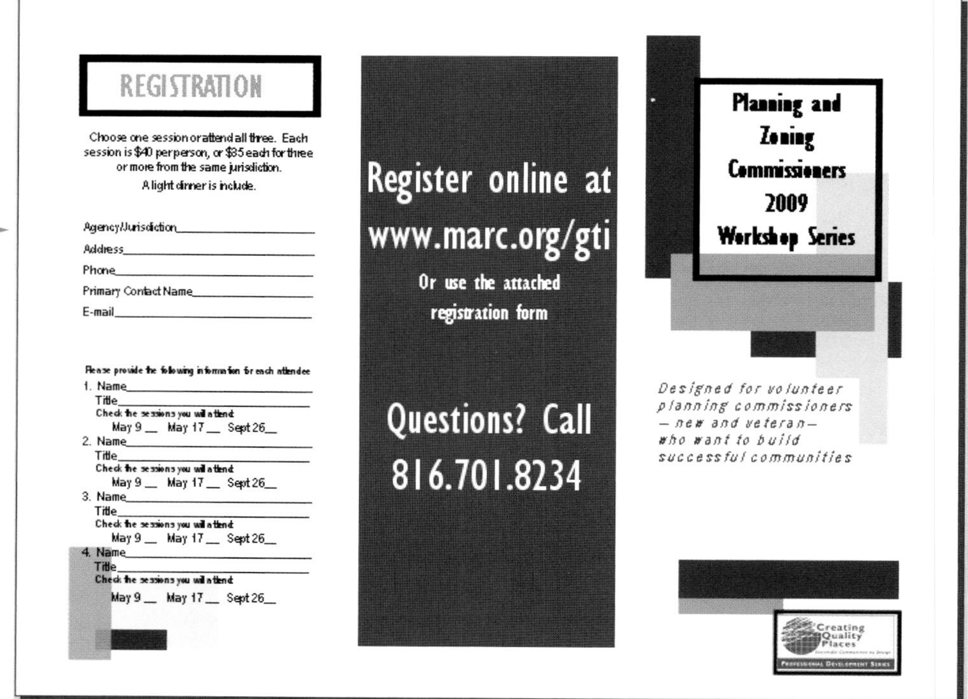

FIGURE 13.10 Continued

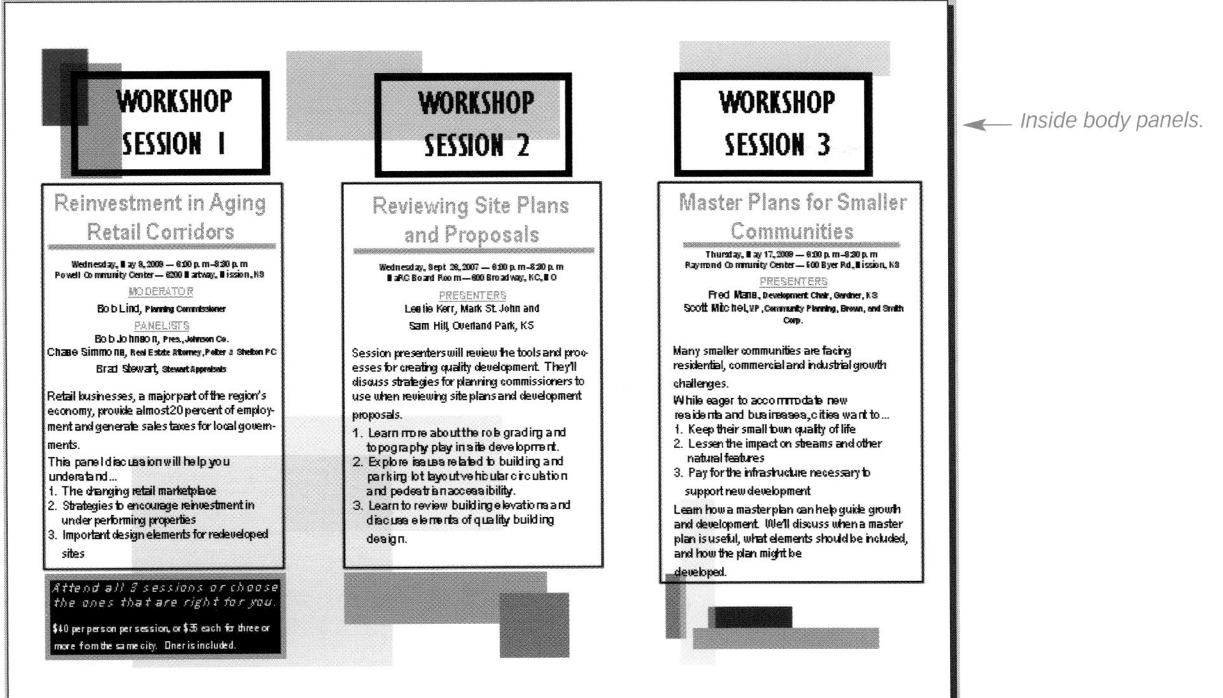

Inside body panels.

EFFECTIVE BROCHURE CHECKLIST

Title Panel

_____ 1. Does the first panel name the product, company, or service?

_____ 2. Does the first panel provide a graphic to attract the reader's attention and pictorially represent the topic?

Back Panel

_____ 1. Does the back panel summarize your brochure's content?

_____ 2. Does the back panel provide your address, a place for a stamp or paid postage, and your reader's address?

_____ 3. Does the back panel provide a coupon, discounts, or complimentary tickets?

_____ 4. Does the back panel give a map of your location?

_____ 5. Does the back panel include your name, your company's name, street address, city, state, zip code, telephone number, fax number, Web URL, or e-mail address?

Body Panels

_____ 1. Are headings presented similarly, maintaining parallelism?

_____ 2. Are graphics effectively used for interest and to clarify ideas?

_____ 3. Does the brochure vary font sizes and type to emphasize key points and add visual appeal?

_____ 4. Are bullets and numbers used to itemize ideas for better access?

_____ 5. Has ample white space been used to help the reader access information and to make reading easier?

_____ 6. Has color been used effectively for visual appeal?

Content

_____ 1. Are all unfamiliar terms defined?

_____ 2. Is the content correct, verified by peer review?

_____ 3. Is the brochure grammatically correct?

_____ 4. Is the text clear, answering reporter's questions (*"who," "what," "when," "where," "why,"* and *"how"*)?

_____ 5. Is the text concise, using short words, short sentences, and short paragraphs?

_____ 6. Does the brochure meet the writer's goals: to create awareness of the company, product, or service; to increase understanding; to advertise new aspects; to change negative attitudes; to show ways in which the topic surpasses the competition?

_____ 7. Are maps used to help the reader find the writer's location?

_____ 8. Are discounts or promotional incentives offered to ensure reader participation?

Audience

_____ 1. Is the level of writing appropriate for the audience (specialist, semi-specialist, lay, multiple readers)?

_____ 2. Has the appropriate tone been achieved through positive words and personalized pronouns?

PERSUASIVE COMMUNICATION CHECKLIST

_____ 1. **Ethical argumentation.** Have you made an ethical argument based on character? You must be trustworthy and credible as a writer or speaker.

_____ 2. **Emotional argumentation.** Have you used emotion to change an audience's attitudes? You can appeal to an audience's emotions either positively or negatively.

_____ 3. **Logical argumentation.** Have you developed your persuasion by depending on rationality, reason, and proof? You can persuade people logically by providing facts, testimony, and examples.

_____ 4. **Arouse reader interest.** Have you used questions, quotes, anecdotes, comparison/contrast, poetic language, or an appeal to senses to interest your audience?

_____ 5. **Refute opposing points of view.** Have you presented a balanced argument? To do so, recognize and admit conflicting views, let the audience know that you understand their concerns, and allow for alternatives.

_____ 6. **Give proof.** Have you provided evidence to prove your point?

_____ 7. **Urge to action.** Have you motivated your audience to act? To do so, provide incentives, give discounts, mention warranties, provide contact information, or suggest a follow-up action.

_____ 8. **Highlighting/page layout.** Is your text accessible? To achieve reader-friendly ease of access, use headings, boldface, italics, bullets, numbers, underlining, or graphics (tables and figures). These add interest and help your readers navigate your text.

_____ 9. **Conciseness.** Have you limited the length of your sentences, words, and paragraphs?

_____ 10. **Audience recognition.** Have you written appropriately to your audience? This includes avoiding biased language, considering the multicultural/cross-cultural nature of your readers, and your audience's role (supervisors, subordinates, coworkers, customers, or vendors).

_____ 11. **Correctness.** Is your text grammatically correct? Errors will hurt your professionalism.

Georgia Met Her Challenge

To meet her communication challenge, Georgia used the P^3 process.

Planning

To plan her communication, Georgia considered the following:

- Goal—help communities meet their infrastructure needs
- Audience—small-town city councils
- Channels—informative and persuasive flier, mailed as hard-copy text and posted on MARC Web site
- Data—specific workshop topics of discussion

The challenge at MARC is communicating to a diverse audience in different-sized cities. Georgia says, "Some cities have professional, experienced staffs, and some staffs are less experienced. Some cities have billion-dollar budgets, while other cities have far more limited financial resources. MARC needs to communicate its role to meet the needs of the different audiences from the different-sized cities. MARC works with communities to address the particular needs of these cities whether it's about the environment or ways to protect the 'small-town way of life.' My challenge is to create fliers that highlight what MARC can do for the different audiences. I start by brainstorming a list of things to do." To plan her communication, Georgia brainstormed and created the list in Figure 13.11.

Packaging

To package the flier, Georgia wrote the rough draft in Figure 13.12. The draft included all of the information she needed to share with her intended audience. One goal was to provide answers to reporter's questions: "Who," "What," "When," "Where," "Why," and "How." "My primary goal, however, was to persuade the audience to enroll in the workshop," says Georgia.

Perfecting

Georgia says, "I worked to perfect my draft by adding persuasive techniques, including *ethos, logos,* and *pathos.* For instance, the moderator and panelists, experts in their fields, represent *ethos* for they are considered trustworthy, credible, unbiased, and reliable. By stating that 'Retail businesses are a major part of a region's economy, providing almost 20 percent of the employment base and generating sales taxes for local governments,' I demonstrate *logos,* logical statistical support of the topic. I used the emotional appeal of *pathos* by referring to my audience's 'small-town quality of life' and the importance of 'streams and other natural features.' Most importantly, I relied on visually appealing layout and design to perfect the flier. We're getting positive feedback from the flier." See Georgia's perfected flier in Figure 13.13.

FIGURE 13.11 **Georgia's Brainstorming List**

1. Brainstorm topics with internal staff and advisory committee (made up of planners and community development representatives)
2. Narrow the topics down to three or four workshops
3. Draft course descriptions
4. Recruit instructors
5. Determine dates and times
6. Secure locations
7. Provide draft information to public affairs to create flier
8. Proof and finalize marketing materials
9. Send e-mail announcements with flier attachment to targeted audience
10. Hand out fliers at MARC-sponsored event
11. Post flier on Web site

FIGURE 13.12 **Georgia's Rough Draft for the Flier**

This draft focuses on the content. Georgia's major concern was getting all of the details down on paper. She knew that she would improve layout, design, and tone in the perfecting stage.

Master Plans for Smaller Communities
Thursday, May 17, 2009
6:00–8:30 p.m.
Grandview Community Center—The View
13500 Byars Road
Grandview, MO

Moderator: Bob Linder, Planning Commissioner

Panelists: Chase Simpson, Real Estate Attorney
William Shane, Shane Appraisals, Inc.
Scott Mitchell, Vice President, Mitchell Engineering

Smaller communities throughout the Kansas City metropolitan area are facing growth challenges—residential, commercial, and industrial. Our smaller cities are eager to accommodate new residents and businesses, but are concerned about losing the small town quality of life, impacting streams and other natural features, and paying for infrastructure to support new development. The preparation and adoption of a master plan or land use plan provides communities with an important citizen engagement opportunity and new tools to guide growth and development.

The two presenters will describe the conditions under which a master plan is useful or necessary; the elements that might be included in a master plan; the various ways that master plans might be developed, including ways to involve residents and other stakeholders; the adoption process; and how the planning commission uses the adopted plan in making decisions on development applications.

Georgia says, "I am pleased with the success of the flier. By going through the three stages of the P^3 process, I was able to create a flier that has been well received by my audience. Not only have I received some complimentary verbal feedback, but also our enrollment in this program is 37 percent higher than I anticipated. My staff and I have decided that this flier will be the prototype for all of our fliers. It would be great if all of our training seminars and panel discussions saw a significant increase in enrollment."

FIGURE 13.13 Georgia's Perfected Flier

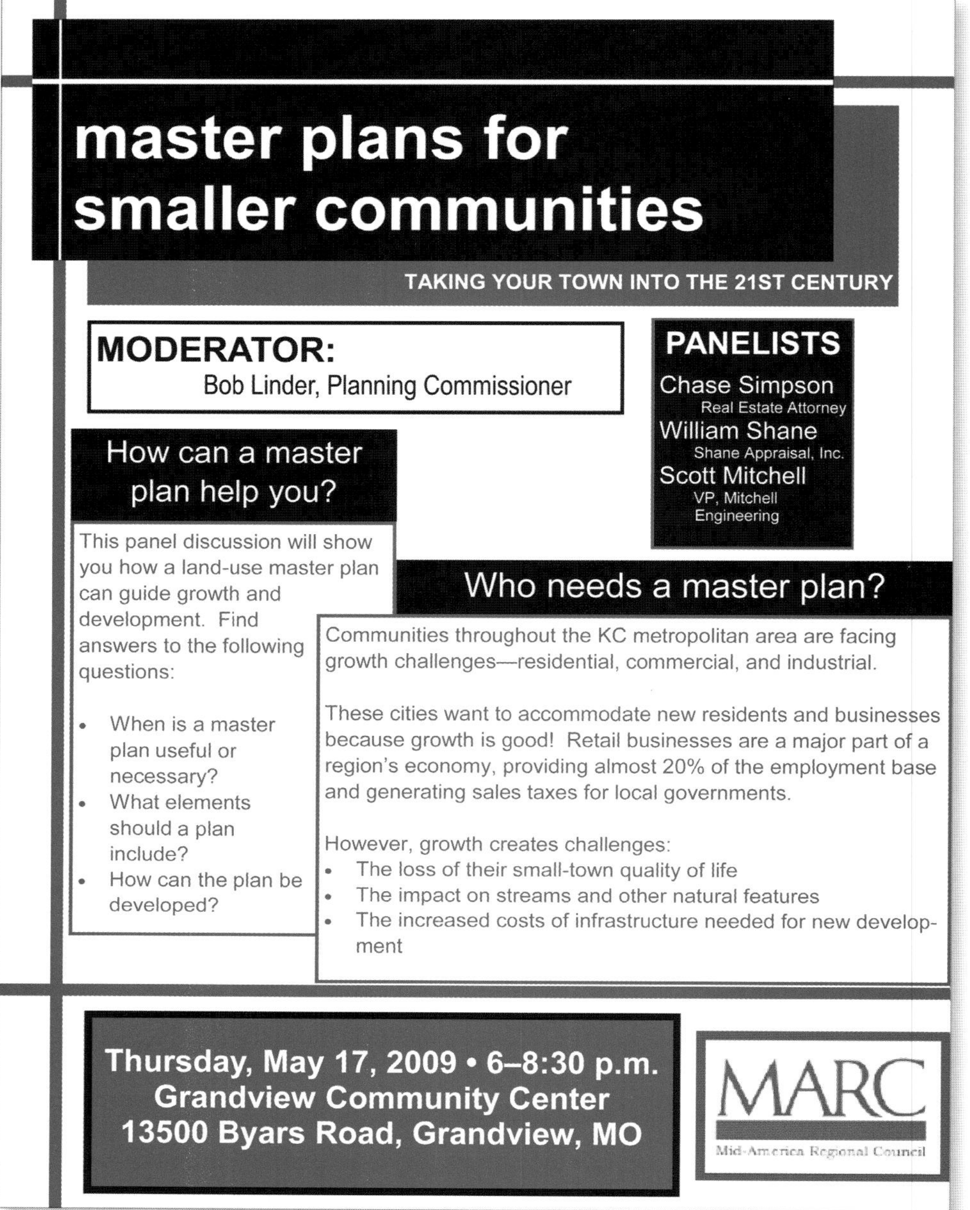

CHAPTER HIGHLIGHTS

1. Persuasive workplace communication consists of a combination of *ethos*, *logos*, and *pathos*.
2. Use the *ARGU* technique to create persuasive fliers, brochures, or sales letters.
3. Avoid logical fallacies when communicating persuasively.
4. Sales letters market services and products.
5. Fliers should be one to two pages long, focused on a topic, titled, have limited text, and be visually appealing.
6. Recognize your audience and their needs when you write a flier, brochure, or sales letter.
7. The front panel of a brochure contains a title and a graphic that depicts the topic.

8. The interior panels of a brochure contain information about the topic.
9. The back panel of a brochure can be used for mailing purposes.
10. Document design is important when you create a flier, brochure, or sales letter.

WWW *Find additional persuasive communication exercises, samples, and interactive activities at http://www.prenhall.com/gerson.*

MEETING WORKPLACE COMMUNICATION CHALLENGES

CASE STUDIES

1. Creative Seminars sends persuasive sales letters, on page 392, to prospective customers. Read the letter and accomplish the following assignments.

Assignments

- Determine where *ethos, logos,* and *pathos* are used persuasively. Give examples to prove your point.
- Based on the criteria provided in this chapter, decide how the sales letter is successful and where it could be improved. Rewrite the letter according to your suggestions for improvement.

2. CompuRam's business profits are falling. This company, which manufactures and maintains computer hardware and software, has studied the problem and found the following reasons for their profit losses.

Increased production in the late 1990s led to the hiring of 50 highly paid information technologists. The average annual salary for these employees is $55,000, plus benefits. In addition, the company had to hire another 50 employees to work the 1-800 help line, answering customer questions about software and hardware concerns. These employees earn approximately $32,000 annually. Each of the employees hired needed office space and office equipment. This led to facility expansion and additional costs of approximately $1,500,000. Though business is good, the increased salaries, benefits packages, and office equipment costs have hurt CompuRam's bottom line.

CompuRam's CEO, Tom Lisk, has concluded that two actions will save the company: a number of employees must be let go, and jobs must be outsourced. These measures, Tom believes, will save the company approximately $3,000,000 annually. Outsourcing and firing employees is not good news for some community members, but keeping the company financially sound benefits the majority of the community.

Assignment

Tom needs to communicate his decision both to employees and to the community.

- Write a letter to the city newspaper persuasively explaining Tom's decision.
- Write an e-mail to the company's employees persuasively explaining Tom's decision.

In both cases, use the ***ARGU*** approach discussed in this chapter. ***A***rouse reader interest, ***R***efute opposing arguments, ***G***ive proof to support your argument, and ***U***rge the readers to accept this planned action.

ETHICS IN THE WORKPLACE CASE STUDY

Alyssa Adams is a sales representative for a biomedical software company. She frequently travels throughout the country to meet with potential clients. In the past year, her company

May 22, 2009

Tom Johnson
Director of Human Resources
ABC Trucking
1234 S. Main St.
Kansas City, MO 64111

Dear Tom:

I am pleased to have made contact with you. As promised, I am sending the packet of information about Creative Seminars. Since 1984, we have been making it easier for many Kansas Citians and others in the region to do their jobs, ultimately making them more productive and their work more consistent.

Our specialty is customized training programs and material that focus on what workers need to *do*, not just what they need to *know*. Our tried and true process is applicable to any industry, any content area, and any delivery medium. A few of our projects include

- Performance management training for trucking company supervisors.
- Orientation training for over-the-road drivers.
- Operation training for tire retreading equipment.
- Safety and facilities management training for federal prison personnel.

We are a local company with strong ties to the community. Our clients include American Century, Bandag, Inc., H&R Block, Helzberg Diamonds, Payless ShoeSource, Sprint, Teva Neuroscience, and Yellow Freight. We are a valuable resource right in your backyard. I would appreciate the opportunity to visit with you about the match between your needs and our capabilities. I will call next week to set up an appointment.

Sincerely,

Connie Smith
President

Enclosure

2782 State Line, Suite 2 ■ Kansas City, MO 64182 ■ 816-555-5555
Fax: 816-555-5511

www.creativeseminars.com

created a new policy for dealing with clients: sales representatives are limited to a maximum entertainment expenditure reimbursement of $25 a year per client.

For the past ten years, Alyssa has been a successful sales representative for the company with a high level of productivity. She likes to do the following for her clients and potential clients: buy them tickets to baseball games, take them out to play golf, buy them lunches, and give them fruit baskets at Christmas.

She abided by the company policy limiting her to $25 per client, but her productivity decreased by 22 percent this year.

Question: Is it ethical for Alyssa to supplement company funds with her own money to do favors for her clients? Justify your answer based on information provided in Chapter 1.

INDIVIDUAL AND TEAM PROJECTS

ARGU

For the ***ARGU*** method of organization, you ***A***rouse reader interest, ***R***efute opposing arguments, ***G***ive proof to support your argument, and ***U***rge reader action. Read the following situations and complete the assignments.

- You plan to sell a flash drive that is small enough to fit on a key chain. For a sales letter, write five different introductions to *arouse* reader interest. Use any of the options provided in the chapter for arousing reader interest.
- Write a body that *refutes* opposing arguments (too expensive, easily damaged, easily lost, etc.) and that *gives proof* to support your product claims.
- In the sales letter for the portable computer flash drive, conclude by *urging reader action*. Use at least two of the methods discussed in this chapter.

Analysis of Persuasive Writing

Find examples of persuasive writing (sales letters, fliers, or brochures). Bring these to class and, in small groups or individually, accomplish the following:

1. Decide which methods of persuasion have been used. Where in the documents do the writers appeal to logic, emotion, and ethics? Give examples and explain your reasoning, either in writing or orally.
2. Have the writers aroused reader interest? Give examples and explain your reasoning, either in writing or orally. If the writers have not aroused reader interest, should they have done so? Explain why. Rewrite the introductions using any of the techniques discussed in this chapter to arouse reader interest.
3. Have the writers refuted opposing points of view? Give examples and explain your reasoning, either in writing or orally. If the writers have not negated opposing points of view, should they have? Explain why. Rewrite the text using any of the techniques discussed in this chapter to negate opposing points of view.
4. Have the writers developed their arguments persuasively? Give examples and explain your reasoning, either in writing or orally. If the writers have not provided persuasive proof, rewrite the text using any of the techniques discussed in this chapter to improve the arguments.
5. Do any of the examples you have found use logical fallacies to persuade the readers? Give examples and explain your reasoning, either in writing or orally. If the writers have used logical fallacies, rewrite the text using any of the techniques discussed in this chapter to improve the arguments.

DEGREE-SPECIFIC ASSIGNMENTS

1. **Marketing and Public Relations**—Write a persuasive brochure about a topic, such as
   - A school organization or club
   - A church activity or club
   - A city organization, agency, or company (the local computer store, animal shelter, nursing home, community center, Rotary club, or city government)
   - Your business
   - Your place of employment
   - A vacation spot
   - A site or location (park, museum, zoo, amusement park, historical site, athletic field or stadium, etc.)
   - A product or line of products (tools, automotive parts, computer accessories, software, etc.)
   - A service (home improvement, hobbies, consulting, animal services, lawn and garden care, installation, automotive care, etc.)

2. **Business Information Systems**—Write a persuasive flier about any topic, such as a
   - New computer software application
   - Computer hardware innovation
   - Web site addition
   - Method for solving a company's problem with online billing, accounting, shipping, delivery, etc.
   - New telecommunications technology
3. **Accounting**—Write a persuasive sales letter. To do so, imagine what product or service you could provide to a potential client. Select one from your area of expertise or academic major. Write your letter according to the criteria for sales letters and the techniques discussed in this chapter.
4. **Management**—Write a persuasive e-mail or memo about any of the following:
   - A request for vacation time
   - A promotion
   - A raise
   - A new office
   - Improved office equipment
   - Additional team members
   - More funding for a project
   - More time to complete a project
   - Financial assistance for a graduate degree
   - Travel authorization for job-related training
5. **Finance**—Write a persuasive complaint letter about any of the following:
   - Defective merchandise
   - Breakage during shipment
   - Late deliveries
   - Overcharges from a business
   - Failure to complete a job to satisfaction
   - Duplicate billings

PROBLEM-SOLVING THINK PIECES

Logical, Emotional, and Ethical Appeals

Read the following situations and determine whether the persuasive arguments appeal to logic, emotion, or ethics. These argumentation techniques can overlap. Explain your answers.

1. If you purchase this product, you can benefit from a healthier, happier, and longer life!
2. Seventy-two percent of SUV car owners say that high gas prices will influence their next car purchases.
3. CEO Jim Snyder, an expert in the field of sports management, says, "Building a downtown sports arena enhances a city's image."
4. Style-tone Hair Gel improves your hair quality by preventing split ends, generating new hair growth, and inhibiting "frizzies."
5. Failure to recycle will cause 52 percent more dangerous hydrofluoric carbons to be released into the atmosphere, leading to harmful decreases in the ozone layer.

Logical Fallacies

Read the following logical fallacies and revise them, ensuring that the sentences provide logical, ethical, and correct argumentation.

1. All marketing experts believe that newsletters are effective.
2. Either all employees must come to work on time, or they will be fired.
3. Some accountants are ambitious because they wish to succeed.

4. The contractor lost the bid, so he cannot expect to have increased revenues this fiscal year.
5. Because the manager is inexperienced, the report will be badly written.

WEB WORKSHOP

You can find electronic brochures and fliers on the Internet. Using a search engine of your choice, type in phrases like "online brochure," "e-brochure," "online flier," "e-flier," "electronic brochure," or "electronic flier." Once you find examples of these online persuasive documents, do the following:

- Compare and contrast the electronic documents with hard-copy brochures and fliers.
- Compare and contrast the electronic documents with the criteria provided in this chapter.
- Decide how the online communication is similar to and different from the criteria and from hard-copy versions.
- If you decide that the online versions can be improved, print them out and revise them.

QUIZ QUESTIONS

1. What are the traditional methods of argumentation?
2. Which type of argumentation depends on your character?
3. Which type of argumentation focuses on feelings?
4. Which type of argumentation depends on rationality, reason, and proof?
5. What are four ways to arouse reader interest?
6. Which steps do you follow to refute opposing points of view in the body of your argument?
7. How can you urge your audience to action?
8. What are three types of logical fallacies?
9. Why are brochures an effective marketing tool?
10. Why are fliers effective for marketing?

CHAPTER 14

Designing Web Sites

▸REAL PEOPLE *in the workplace*

Shannon Conner is a freelance technical communicator and Web site designer. He helps corporations create online and hard-copy content, including Help systems, user guides, and Web site workflow and layout. His primary goal is to enhance a company's communication strategy with its clients and associates.

To achieve his goal, Shannon must accomplish clear communication on two levels: (1) with the client to ensure he understands the client's needs and how he can best meet them, and (2) with the client's customers to ensure the client is presenting information in a clear, professional manner.

Shannon's communication strategy has a three-phase approach:

1. **Statement of Purpose**—The statement of purpose is a questionnaire-style document that Shannon walks through with the client. He uses this to gather pertinent information for the project.
2. **Mockup**—After completing the statement of purpose, Shannon creates a mockup of the product, whether it's a Web site, Web page, Help system, user guide, or brochure.
3. **Final Draft**—Shannon shares the mockup with the client. The client then has the opportunity to provide additional feedback to enable Shannon to fine-tune it for the final draft.

Employing clarity, conciseness, and consistency are essential in Web site design. Shannon says, "Consider the three C's the keystone of your communication goals. Always communicate with clarity, conciseness, and consistency. Without these three traits—especially when your medium is a Web site—the thoughts and ideas you worked hard to produce are worthless. What is the value of your content if no one reads it? Remember, users of Web sites typically scan the text; they don't read it. Provide them clean content so they can find exactly what they're looking for with as little effort as possible."

Shannon has taken on a new challenge. He and his wife Jeanne decided to open a photography studio. As entrepreneurs venturing into the photography field, they rely on Jeanne's expertise as a photographer and Shannon's skills as a Web site designer to create an attention-grabbing, creative, and effective Web site to help market their new small business startup. Thus, Shannon is challenged to meet his wife/business partner's needs and to design a successful Web site.◂

CHAPTER GOALS

When you complete this chapter, you will be able to do the following:

1. Understand the importance of the Internet.
2. Assess the worldwide growth of the Web.
3. Recognize the need for access to the Internet.
4. Distinguish among Web-accessibility problems, such as cognitive, hearing, and visual impairments.
5. Maintain ethical standards to ensure your Web site's credibility.
6. Choose criteria to design a successful Web site.

WWW

To learn more about designing Web sites, visit our Web Companion at http://www.prenhall.com/gerson.

Shannon's Communication Challenge

According to Shannon, "In my experience as a technical communicator, I have found that my client often has a preconceived notion as to the final product, and it may or may not be ideal for the client's customers. I need to drill-down to the most important elements that the client wants while still honoring clear and concise communication with the client's customers. In this situation with my wife's photography studio, I want to appeal to her artistic sensibilities and to be sure the Web site will actually help sell our new business.

To achieve these goals, I have devised a system called the Statement of Purpose to gather the most important information from the client about what it is they are trying to communicate to their audience. I then apply the information I've gathered as I design the layout and content for the Web site for the new photography studio.

My wife and I want a page included on her Web site to help current and potential clients learn about policy and procedures. In this way, clients can be prepared for the photo shoot and know what to expect. Jeanne already has her policy and procedures defined, so the content already exists. However, it is not ideal for Web publishing because it is linear, bulky, and non-hierarchical.

Refining Jeanne's existing content for use online and to ensure the success of our new photography studio is my communication challenge."

See pages 412–415 to learn how Shannon met his communication challenge.

WHY THE WEB IS IMPORTANT

Due to its speed, affordability (for end users as well as companies), and international access, the Internet is a key component for corporate communication. Like no other communication medium, the Web has changed the way companies do business. The Internet provides companies an international channel for

- Selling their products and services
- Updating employees and customers about corporate changes
- Providing employment opportunities
- Offering online forms and instructions for internal and external communication
- Creating a first point of contact for customers, through online call centers, order entry, and customer-service systems

To keep up with this electronic revolution, you and your company should learn how to communicate online also. With very low investments, practically any business can access a huge market, quickly and economically, regardless of the type of product being sold, the size of the company, or the location of the business ("Internet Growth").

The International Growth of the Internet

The extent to which the Internet has flooded the world market is startling. In 1995, only 16 million people had access to the Internet, representing just 0.4 percent of the world's population. By 2008, this number had grown to over one billion people or 21.9 percent of the world's population ("Internet World Stats"). Figure 14.1 compares the Internet usage breakdown by world region in 2007.

Asia now has an estimated 578 million Internet users. In North America, the Internet population currently is 248 million (See Table 14.1). These numbers represent an enormous and growing market for companies. No other communication channel (e-mail,

FIGURE 14.1 Internet Users in the World by Geographic Regions in 2007

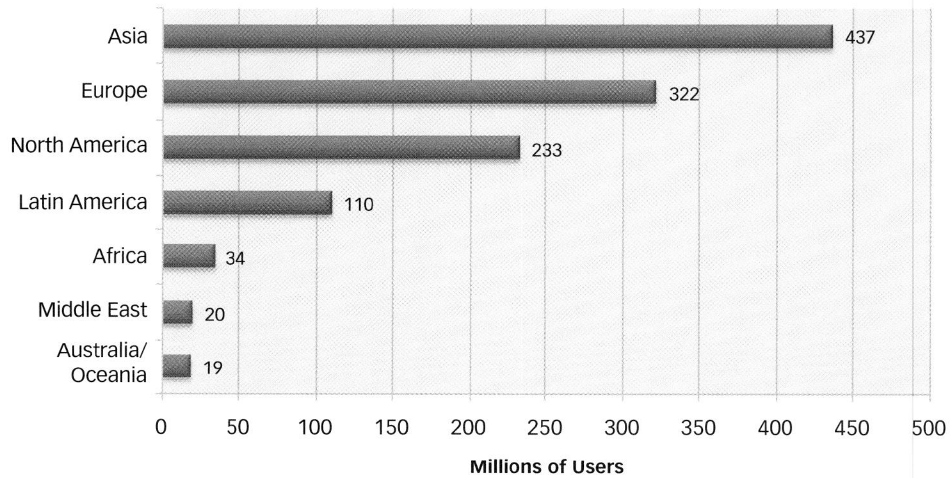

("Internet World Stats" http://www.Internetworldstats.com/stats.htm)

Table 14.1 World Internet Users and Population Stats in 2008
("Internet World Stats" http://www.internetworldstats.com/stats.htm)

World Regions	Population (2008 Est.)	Internet Users Dec. 31, 2000	Internet Usage, Latest Data	% Population (Penetration)	Usage % of World	Usage Growth 2000-2008
Africa	955,206,348	4,514,400	**51,065,630**	5.3%	3.5%	1,031.2%
Asia	3,776,181,949	114,304,000	**578,538,257**	15.3%	39.5%	406.1%
Europe	800,401,065	105,096,093	**384,633,765**	48.1%	26.3%	266.0%
Middle East	197,090,443	3,284,800	**41,939,200**	21.3%	2.9%	1,176.8%
North America	337,167,248	108,096,800	**248,241,969**	73.6%	17.0%	129.6%
Latin America/ Caribbean	576,091,673	18,068,919	**139,009,209**	24.1%	9.5%	669.3%
Oceania/Australia	33,981,562	7,620,480	**20,204,331**	59.5%	1.4%	165.1%
WORLD TOTAL	6,676,120,288	360,985,492	**1,463,632,361**	21.9%	100.0%	305.5%

memos, letters, reports, newsletters, or brochures) can reach so many people, in so many distant locations, 24 hours a day, seven days a week, as can the Internet.

Corporate Buy-In

Smart companies realize the impact of the Internet. Corporations are finding ways to integrate the Internet into how they are doing business, specifically in terms of e-commerce. In a *Business Communications Review*, 91 percent of employees stated that they worked for companies with publicly accessible Web sites (Knight). In addition, 74 percent of the respondents stated that their companies used a corporate Web site to promote products and services, as well as to convey a corporate image through product information and company access. An additional 72 percent of the survey respondents said their companies had integrated their corporate Web sites with their call centers.

Corporate Branding

Domain names offer another way to highlight how important the Internet has become, internationally, for companies. Business experts suggest that any company which plans to market a new product or begin a business should first register their domain name. According to intellectual property lawyers, registering an Internet domain name is as important as registering trademarks, patents, and company names. "As the internet embeds itself into daily life, any communication that doesn't include a website or company email address is a missed opportunity to build your brand" (Shearer).

As the Internet has grown in international importance, so too has the need for an international domain name. Recognizing this fact, the European Commission designated EURID—the European Registry for Internet Domains—as the Registry for the dot EU (.eu). The European Commission's goal is that ".eu" will become the unique, pan-European identification of Web sites and e-mail addresses, comparable to .org or .com. in the United States ("' .eu': A New Internet Top Level Domain").

WEB ACCESSIBILITY

An international goal is Internet accessibility for persons with disabilities. The Web Accessibility Initiative (WAI), in coordination with organizations around the world, pursues accessibility of the Web. Consider the following Web accessibility problems ("Internet Accessibility").

Cognitive

These include learning disabilities, reading disorders, and attention-deficit/hyperactivity disorder (ADHD). Web access can be enhanced with illustrations, graphics, and headings, which provide visual cues for easier Web understanding.

Hearing

Hearing impaired Web users may need assistive technology to read Web audio or captioned text for multimedia content.

Visual

Color blindness causes problems on the Internet. To combat this challenge, Web writers need to choose colors correctly, perhaps avoiding green and red fonts.

The U.S. government recognizes the importance of the Internet and the need for inclusiveness for both employees and external Web readers. In 1998, Congress amended the Rehabilitation Act, requiring federal agencies to make electronic and information technology accessible to people with disabilities. Section 508 for federal employees and the public eliminates barriers in information technology ("508 Law"). For information about Web accessibility, see Table 14.2.

Table 14.2 Sites for Web Accessibility

Web Site Name	URL
Web Accessibility Initiative (WAI)	http://www.w3.org/WAI/
U.S. Government Section 508 Website Accessibility Guidelines	http://www.access-board.gov/sec508/508standards.htm
The Society for Technical Communication Accessibility Special Interest Group	http://www.stcsig.org/sn/internet.shtml
National Center for Accessible Media	http://ncam.wgbh.org/
Great Britain Disability Rights Commission	http://www.drc-gb.org/publicationsandreports/report.asp
International Web Accessibility Concerns	http://www.stcsig.org/sn/international.shtml

THE CHARACTERISTICS OF ONLINE COMMUNICATION

Because e-readers are unique, you must alter the way you write online communication. This means changing your mindset as a writer. When you write a Web site, you must reconsider

- **Screen Versus Page**—Forget 8½″ × 11″. Online screens, accessed through hand-held computers and cell phones, are smaller. Well-designed Web screens rarely require scrolling.
- **Skimming Versus Linear Reading**—We read books "linearly," line by line. In contrast, Web screens are skimmed and scanned.
- **Hypertext Links Versus Chronological Reading**—We read books from beginning to end, sequentially. Web sites, however, allow us, even encourage us, to leap randomly from screen to screen.

Screen Layout

E-readers do not read online text the same way they read hard-copy text. For traditional hard-copy text, we can read from margin to margin on a traditional 8½″ × 11″ document. In contrast, reading from margin to margin online is not as easy, due to glare, varying screen sizes, and a desire to skim and scan. To help the e-reader access your Web content, you need to pay attention to the Web site's page length, margins, and font selection.

Margins

Hard-copy text, when read from margin to margin, is approximately 80 characters long (a character is every letter, punctuation mark, or space). The size and type of your font, of course, affects the number of characters you will type per line. Online, horizontal margins vary from monitor to monitor (21″, 19″, 17″, 15″, etc.). On a 17″ monitor, for example, text runs approximately 12″, left margin to right margin. On a 15″ monitor, text runs approximately 10″.

On a Web page, your audience is forced to read more words per line if you use the entire screen. That's difficult. Readers have trouble maintaining their focus (visually and mentally) when expected to read long lines of horizontal text. Vertically, the problem increases. Regardless of the monitor size, cybertext can run forever. The reader can scroll, and scroll, and scroll. Whereas paper sizes are controlled by convention and printing companies, you must control the size of a Web page.

A successful Web page should limit lines of text to perhaps one-third or two-thirds of the screen, with a graphic, white space, or hypertext links placed in the remaining third. Try for a third/third/third screen orientation: text, graphic, and white space. This breaks up the monotony of long lines into manageable chunks. Thus, the reader can maintain focus more easily (Eddings et al.).

Font

You can enhance your hard-copy text with designer fonts (stylized and cursive), boldface, or underlining for visual effect. In contrast, sans serif fonts, like Arial or Verdana, seem to work best for online reading. "Stylized, cursive, italicized, and decorative typefaces are difficult to read online" (Hemmi 11). Underlining is especially troublesome online, because online, hypertext links often are shown as underlined text. To avoid confusing your readers online, you must avoid using underlining as a highlighting technique.

Noise

Though hard-copy text is rigidly linear, it can be easier to read than online text. Paper is dull and absorbs light. Most hard-copy text is colorless and motionless; a book is composed of black ink on white paper. For online text, however, the screen is composed of glass, which reflects light, creating visual glare. To help e-readers, you need to consider a key challenge of reading online: computer "noise"—sound and visual distractions. Extended viewing of a computer screen is more demanding than continued reading of paper text. Web sites often contain lots of color, blinking text, animated graphics, frames of layered text, and sound and video. Noise—multiple distractions—inundates the screen. In such a busy communication vehicle as a Web site, the less we give the reader on one screen, the better.

ESTABLISHING CREDIBILITY IN A WEB SITE

As the creator of a Web site, you must be concerned about your site's credibility. When the audience accesses your site, he or she must be assured that the content within the site is credible, safe, and secure. For example, if a consumer enters credit card information when purchasing a product or service from your site, that consumer demands a sense of security. When people buy products from your site, they need to know that their purchase is warranted and that your customer service is accountable. To determine how the public views a Web site's credibility, Stanford University performed a study. In this study, 2,440 people evaluated the credibility of Web sites. Table 14.3 shows the results of their study, ranked one through ten.

Almost half of all respondents to the Stanford study (46.1%) ranked visual design most important for establishing credibility, focusing on layout, font type, font size, and color. However, Stanford also concluded that a Web site's visual appeal had limited value for long-term credibility. Though a Web site's design might initially attract an audience, the consumer will go to other Web sites if they are not satisfied with other aspects of the site, including the following key credibility concerns:

- **Identity of the site or its operator**—Prove your credentials by including your company's years in business, your staff's education and certifications, lists of satisfied customers, and customer testimonials.
- **The site's customer service**—Provide easy access for consumers when contacting people within the company for help. This includes 800 numbers,

Table 14.3 Evaluation of Web Site Credibility

	Percent (of 2,440 Comments)	Topics (Addressing Specific Credibility Issue)
1	46.1%	Design Look
2	28.5%	Information Design/Structure
3	25.1%	Information Focus
4	15.5%	Company Motive
5	14.8%	Information Usefulness
6	14.3%	Information Accuracy
7	14.1%	Name Recognition and Reputation
8	13.8%	Advertising
9	11.6%	Information Bias
10	9.0%	Writing Tone

hotlines, 24/7 assistance, e-mail addresses, rapid response, and the chance to speak to an actual person within the company.
- **Privacy policies**—Clearly and briefly state how your company ensures privacy related to credit cards and personal information.

(Fogg et al.)

ETHICAL CONSIDERATIONS IN A WEB SITE

As a business professional, you have an ethical responsibility to the audience viewing your Web site. Customers will input personal information, such as Social Security numbers, credit card numbers, expiration dates, and personal identification numbers, when purchasing products and services. This material must be safeguarded to avoid fraud and identity theft. The Federal Trade Commission, under the FTC Act, protects consumers against unfairness and deception by enforcing a company's promises about collections and use of personal information ("Privacy Initiatives").

Privacy Considerations

To ensure your audience's privacy and demonstrate your company's ethical practices, focus on the following:

- **Maintain records**—You must document personal information you have in your computer files and back up these files in archives.
- **Delete inactive files**—Don't stockpile files indefinitely. Only maintain files that your company needs to conduct business.
- **Secure the information**—Frequently update firewalls, anti-virus, and anti-spyware software.
- **Protect passwords**—Avoid sharing customer passwords online, over the phone, or in e-mail messages. Ask customers to change passwords frequently and to avoid using the same password for multiple accounts.

("Protecting Personal Information" *OnGuard Online*).

Copyright

If you use copyrighted information on your Web site, you have an ethical responsibility to credit the source. "Every element of a Web page—text, graphics, and HTML code—is protected by U.S. and international copyright laws, whether or not a formal copyright application has been submitted" (LeVie 20–21). If you and your company "borrow" from an existing Internet site, thus infringing upon that site's copyright, you can be assessed actual or statutory damages. The Copyright Act allows the owner of Web material to do the following: stop infringement and obtain damages and attorney fees. In addition to financial damages, your company, if violating intellectual property laws, could lose customers, damage its reputation, and lose future capital investments. To protect your property rights on your Web site, you should do the following:

Additional information:

See Chapter 1, "Communicating in the Workplace."

- Assume that any information on the Internet is covered under copyright protection laws unless proven otherwise.
- Obtain permission for use from the original creator of graphics or text.
- Cite the source of your information.
- Create your own graphics and text.
- Copyright any information you create.
- Place a copyright notice at the bottom of your Web site.

DOCUMENTATION IN A WEB SITE

When you create your Web site, be sure to credit sources of information, such as the following: content, statistics, photographs or other visual aids, music, videos, or any multimedia links. Giving source credit for this type of information ensures credibility of your Web site. In addition, you can verify and support content in the site by referencing acknowledged experts in a field and citing the source of the information. Your source citations could be in the form of MLA or APA documentation (see Chapter 16, "Research and Documentation"); in-text references ("according to Dr. Robert Cottrell, IBM Research Department"); or links to other Web sites. For copyrighted material (visuals and text), receive prior approval from the creator and then cite the source in your Web site.

Additional information:

See Chapter 16, "Research and Documentation."

SOFTWARE PROGRAMS FOR DESIGNING WEB SITES

You can create a Web site in several ways.

- Write text in any word processing program. Then, you can save your text as a Web page. From your word program's toolbar, click on "Save As." In the pop-up window, you'll see "Save as type." Click on the down arrow and then "Web page." Your text will be saved with an .html designation, allowing you to open it in any Web browser.
- Draft your text directly in an HTML editor, such as FrontPage, SharePoint, Dreamweaver, Ant HTML, HotDog Web Editor, Netscape Navigator 4.0 and above, Microsoft Internet Assistant for Word 6.0 for Windows and above, Corel WordPerfect 8.0, and others.
- Write your text using Hypertext Markup Language (HTML) coding or XHTML (Extensible Hypertext Markup Language).

CRITERIA FOR A SUCCESSFUL WEB SITE

Because Web sites are a different type of correspondence than traditional, paper-bound text, you need to employ different criteria when creating your site.

Home Page

The home page on a Web site is your welcome mat. Because it is the first thing your reader sees, the home page sets the tone for your site. A successful home page should consist of the following components.

Identification Information

Who are you? What is the name of your company, product, or service? How can viewers get in touch with you if they want to purchase your product? A good home page should provide your audience the following:

- The name of your company, service, or product.
- Contact information (or a link to corporate contacts), including phone numbers, e-mail addresses, fax numbers, street address, city/state/zip, and customer service contacts.

Amazingly, many Web sites omit this information. They provide the reader information about a product or service, but they prevent the reader from making a follow-up purchase or inquiry. That's not good business (Berst).

Graphics

Don't just tell the reader who or what you are. Show the reader and establish brand recognition (Redish 6). An informative, attractive, and appealing logo depicting your product or service could convey more about your company and be more memorable than words. Check out various Web sites to prove this point. Be careful, however, when you choose the size of your home page graphic. If the graphic is too large, it will take a long time to load. This delay might cause Web surfers to lose patience, stop the HTTP transfer, and visit another site. Then you've lost their business or interest.

Lead-In Introduction

In addition to a graphic, you might want to provide a one- or two-sentence lead-in, or a three- to eight-word tagline, explaining the purpose of the Web site. Few Web visitors will read a paragraph-long mission or vision statement (Redish 6).

For example, if your company is named "NovaTech," what do you market? The name alone does not explain the company's focus. A company named "NovaTech" could market telecommunications equipment, intergalactic armaments, or computer repairs. To clarify your company's purpose, provide a short lead-in (Olive 9). Quickly welcome readers to the site, and tell them what they can find in the following screens.

At this textbook's writing, Applebee's home page introduced itself with this sentence: "Applebee's Neighborhood Grill and Bar is the world's casual dining leader, with over 1,600 restaurants in forty-nine states and nine international countries. . . . Welcome to our neighborhood!" Covansys, a software development company, uses a three-word slogan to introduce itself: "Global Technology Services."

Navigation Links/Buttons

The "nav bar" is what distinguishes a Web site from all other types of written communication: the ability to click on a hot button and link to another screen. By providing the reader with either hypertext links or graphical user interfaces (GUIs), a home page acts as an interactive table of contents or index. The reader selects the topic he or she wishes to pursue, clicks on that link, and jumps to a new screen. Instead of being forced by the constraints of paper-bound text to read sequentially, line by line, page by page, the reader can follow a more intuitive approach. With hypertext links, you read what you want to read, when you want to read it.

Linked Pages

Once your reader clicks on the hypertext links from the home page, he or she will jump to the designated linked pages. These linked pages should contain the following.

Headings/Subheadings

The reader clicks on the home page link and arrives at a new screen. Will the reader know where he or she is now, or will the reader be lost in cyberspace? On paper-bound text, when readers turn the page, they know exactly where they are: on the next page. However, there is no "next" page in cyberspace. "Next" connotes sequence, which is clearly evident when pages are stitched together in a book or bound in a manual. In cyberspace, however, "next" could be any screen in any order. That's why the tool bar on your Internet browser reads "Back" and "Forward." To ensure that readers know where they are in the context of the Web site, you need to use headings. These give the readers visual reminders of their location.

Effective linked-page headings should be consistent. They should be the same font size, font type, and in the same location on the screen. For example, if one heading is centered and typed in 16-point Arial font, all headings should follow this format.

Navigation

Help the reader navigate through cyberspace. You can do this in two ways:

- **Home buttons**—The reader needs to be able to return to the home page easily from any page of a Web site. The home page acts as a table of contents or index for all the pages within the site. By returning to the home page, the reader can access any of the other pages. To ensure this easy navigation, provide a hypertext-linked "Home" button on each page.
- **Links between Web pages**—Why make the reader return to the home page each time he or she wants to access other pages within a site? If each page has hypertext links to all pages within the site, then the reader can access any page he or she chooses, in any order of discovery.

These hypertext links between pages can be provided in many ways. Many Web sites provide a vertical navigation bar or hypertext links presented horizontally at the bottom or top of each page. You select the approach that works best for your Web site.

Document Design

On a Web site, you can add distinctive backgrounds, colored fonts, different font faces and sizes, animated graphics, frames, and highlighting techniques (such as lines, icons, and bullets). However, just because you can doesn't mean you should! Document design should enhance your text and promote your product or service, not distract from your message.

Background

You can add backgrounds running the gamut from plain white, to various colors, to an array of patterns: fabrics, marbled textures, simulated paper, wood grains, grass and stone, psychedelic patterns, water and cloud images, and waves. They all look exciting, but some are not as effective as others.

When choosing a background, consider your corporate image. If you are an accounting firm, do you want a pink background? Perhaps a subtle stone or paper background would be more effective in portraying your corporate identity. If you are a childcare center, do you want a black background? This could convey a negative image. A white background with toys as a watermark might more successfully convey your center's mission statement.

In addition, remember that someone will attempt to read your Web site's text. Very few font colors or styles are legible against a psychedelic background—or against most patterned backgrounds, for that matter. Instead, to achieve readability, you want the best contrast between text and background. Despite the vast selection of backgrounds at your disposal, the best contrast is still black text on a white background.

Color

When you create a Web site, you can use any font color in the spectrum, but do you want to? What corporate image do you have in mind? The font color should be suitable for your Web site, as well as readable. A yellow font on a light blue background is hard to read. A red font on a black background is nightmarish.

The issue, however, isn't just esthetics. As noted, a primary concern is contrast. Red, blue, and black font colors on a white background are legible because their contrast is optimum. Other combinations of color do not offer this contrast, making readability difficult. In the following example, notice how yellow font on a blue background "strobes." The colors bleed into each other and become unreadable. Though black font on a white background might seem uninteresting, this color combination is the most readable choice.

Ineffective Contrast	Effective Contrast
Yellow font on a light blue background is unreadable.	Black font on a white background provides optimum contrast.
Red font on a black background is nightmarish.	White font on a blue background is readable.

Table 14.4 Highlighting Considerations

Effective Online Highlighting	Ineffective Online Highlighting
1. *Lines* (horizontal rules) can separate headings and subheadings from the text. 2. *Bullets and icons* enliven your text and break up the monotony of wall-to-wall words. 3. *First- and second-level headings,* achieved by changing your font size and style, separate key ideas. 4. *Boldface* also emphasizes important points. 5. *White space,* created by indenting, makes text more readable (Spyridakis 367; Yeo 12–14).	1. *Frames* are considered to be one of the worst highlighting techniques. Jesse Berst, editorial director of ZDNet, considers frames to be one of the "Seven Deadly Web Site Sins." He says, "Too many frames . . . produce a miserable patchwork effect." If you want to achieve the frame look, but without the hassle, use tables instead. They are easier to create and revise, they load faster, and they are less distracting to your readers. 2. *Italics and underlining.* Both are hard to read online. In addition, underlining looks like a hypertext link. 3. *Java applets* take a long time to load. 4. *Video* requires an add-on users must download before they can enjoy your creation. 5. *Animation* can be very sophomoric and distracting.

Graphics

Another problem is image size. Your Web site will be affected by the size of your graphics and the number of graphics you have included. For example, a "small" graphic (approximately 90 × 80 pixels, 1½″ × 1½″) consumes about 2.5 kilobytes. A "medium" graphic (approximately 300 × 200 pixels, 5″ × 4″) consumes about 25 kilobytes. A "large" graphic (approximately 640 × 480 pixels, 10″ × 8″) consumes about 900 kilobytes. When you use graphics, with varying amounts of color depth, your file size increases. The larger your file, the longer it will take for the images to load. The longer it takes for your file to load, the less interested your reader might be in your Web site. Limit the size and number of your graphics.

Highlighting Techniques

Font sizes and styles, lines, icons, bullets, frames, java applets, animation, video, audio! You can do it all on your Web page, but should you? Nothing is more distracting than a Web site full of highlighting techniques. See Table 14.4 for highlighting considerations.

EFFECTIVE PLANNING FOR WEB DESIGN

Storyboarding for Web Site Design

Prior to creating your Web site, determine its architecture. What will your Web pages look like? What will your Web site's linked pages consist of? Consider using storyboarding for a preliminary sketch of your Web site's layout.

- **Sketch the home page**—How do you want the home page to look? What color scheme do you have in mind? How many graphics do you envision? Where will you place your graphics and text—margin left, centered, margin right?
- **Sketch the linked page(s)**—How many linked pages do you envision? Will they have graphics, online forms, e-mail links, audio or video, headings and subheadings, or links to other Internet, intranet, or extranet sites? Will you use frames or tables?

See Figure 14.2 for a sample Web site storyboard.

FIGURE 14.2 Web Site Storyboard

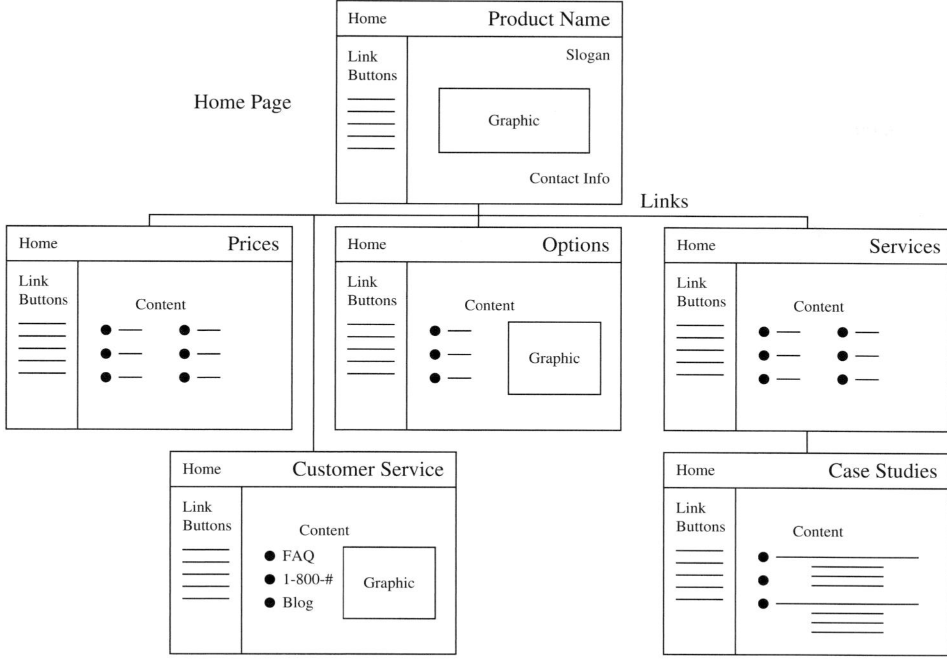

EFFECTIVE PACKAGING FOR WEB DESIGN

Conciseness

Conciseness is important in all workplace communication. Conciseness is even more important in a Web site. A successful Web page should be limited to one viewable screen, and a line of text should rarely exceed two-thirds of the screen.

Personalized Tone

The Internet is a relatively informal medium of communication. On a Web site, you can use pronouns, contractions, and positive words. Create a personalized tone that engages your readers.

Additional information:

See Chapter 3, "Meeting the Needs of the Audience."

Reader Involvement

In addition to achieving audience involvement through pronoun usage, a question/answer format is effective online. You can accomplish this in a Frequently Asked Question link (FAQ). You can also break up your text with headings that ask questions and text that provides answers. Not only is this format reader friendly, but also you will be actively engaging your reader (Redish 8). Figure 14.3 shows how a question/answer format engages the audience through personalized tone.

PERFECTING YOUR WEB SITE

Grammar

Perhaps because of the friendly, casual nature of the Internet, many Web sites are poorly proofread. That's a problem. Casualness is one thing; lack of professionalism is another. If your Web site represents your company, you don't want prospective clients to perceive your company as lacking in quality control. In fact, that's what poor grammar online denotes—poor attention to detail.

It's easy to see why many sites are grammatically flawed. First, anyone can go online almost instantaneously. All you do is code in your HTML and transfer the file to your Web

FIGURE 14.3 **Question/Answer to Achieve Audience Involvement**

This page is reader friendly due to the heading (Adopt-A-Highway Questions) and the boldfaced and italicized subheadings. Note how the text beneath each subheading is indented for easier access. Also, the questions and pronouns achieve audience involvement by speaking directly to the reader.

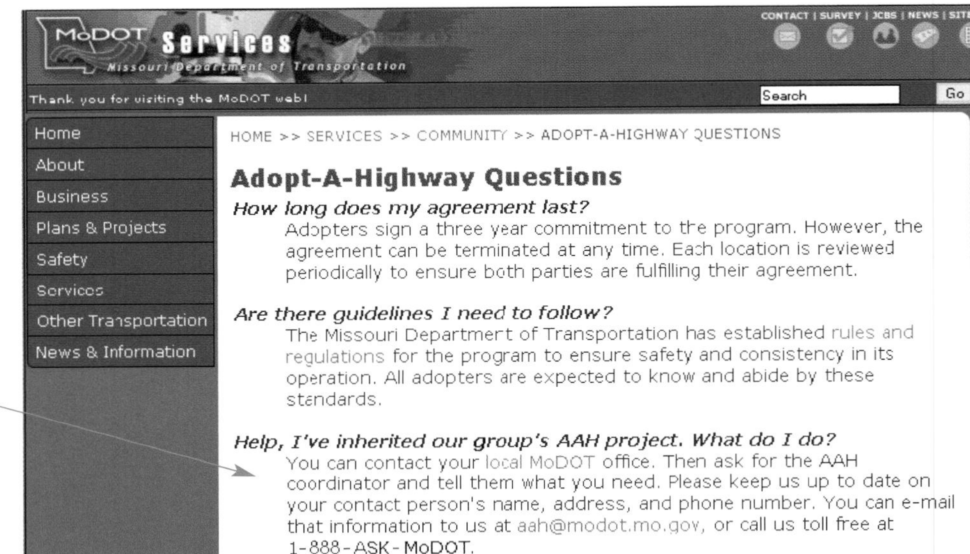

Courtesy of MoDot

server, and you are online. That's great for business, but it's bad for proofreading. Good proofreading takes time, a concept almost antithetical to the Internet. Next, though most word processing packages now have spell checks and grammar checks that tell you immediately when you have made an error, many Web conversion programs do not contain these tools. Thus, when constructing your Web site, you must be vigilant. The integrity of your Web site demands correct grammar.

Usability

Once you have created your Web site, is it what you had in mind? Do the links work? If not, edit and revise the site to ensure its usability. Usability is a way to determine whether your reader can "use" the Web site effectively, whether your site meets your user's needs and expectations. The reader wants to find information that helps him or her better understand the topic. In addition, the audience wants the Web site to be readable, accurate, up-to-date, and easy to access. Finally, the reader wants internal and external links to work. Thus, usability focuses on three factors (Dorazio):

- **Retrievability**—The user wants to find specific information quickly and be able to navigate easily between the screens and to exterior links that you might have included within your site. Your job is to make sure that all of your hypertext links are active.
- **Readability**—The user wants to be able to read and comprehend information quickly and easily.
- **Accuracy**—The user wants complete, correct, and up-to-date information.

The goal of usability testing is to solve problems that might make a Web site hard for people to use. By successfully testing a site's usability, a company can reduce customer and colleague complaints; increase employee productivity (reducing the time it takes to get work done); increase sales volume because customers can purchase products or services online; and decrease help desk calls and their costs. To achieve user satisfaction, revise your Web site to meet the goals listed in our Web Site Checklist. See Figure 14.4 for an effectively designed Web page.

FIGURE 14.4 An Effectively Designed Web Page

Courtesy of George Butler Associates

WEB SITE CHECKLIST

____ 1. Home Page
- Does the home page provide identification information (name of service or product, company name, e-mail, fax, telephone, address, etc.)?
- Does the home page provide an informative and appealing graphic?
- Does the home page provide a welcoming, informative, and concise introduction?
- Does the home page provide hypertext links connecting the reader to subsequent screens?

____ 2. Linked Pages
- Do the linked pages provide headings clearly indicating to the reader which screen he or she is viewing?
- Are the headings consistent in location and font size/style?
- Do the linked pages develop ideas thoroughly (providing the appropriate amount of detail, specificity, and value for the reader)?

____ 3. Navigation
- Does the Web site allow for easy return from linked pages to the home page?
- Does the Web site allow for easy movement between linked pages?

____ 4. Document Design
- Does the Web site provide an effective background, suitable to the content and creating effective contrast for reading the text?
- Does the Web site use color effectively, in a way suitable to the content and creating effective contrast for reading the text?
- Does the Web site use graphics effectively, so that they are suitable to the content, not distracting, and not causing delays for the site to load?
- Does the Web site use highlighting techniques effectively (lines, bullets, icons, audio, video, frames, font size and style, etc.) in a way that is suitable to the content and not distracting?
- Does the Web site's design allow for accessibility, meeting Web Accessibility Initiative and Section 508 mandates?

____ 5. Style
- Is the Web site concise?
  - Short words (one to two syllables)
  - Short sentences (10 to 12 words per sentence)
  - Short paragraphs (four typed lines maximum)
  - Line length not exceeding two-thirds of the screen
  - Text per Web page limited to one viewable screen—minimizing the need to scroll
- Is the Web site personalized, using positive words and pronouns?

____ 6. Grammar
- Does the Web site avoid grammatical errors?

Technology Tip

USING MICROSOFT WORD 2007 TO CREATE A WEB SITE

Formerly, Microsoft's FrontPage was a popular tool for creating Web sites. However, Microsoft's "Next Generation" of Web authoring tools replaces FrontPage (discontinued in 2006) with two new products, SharePoint and Expression Web.

SharePoint—The Next Generation Web Designing Tool

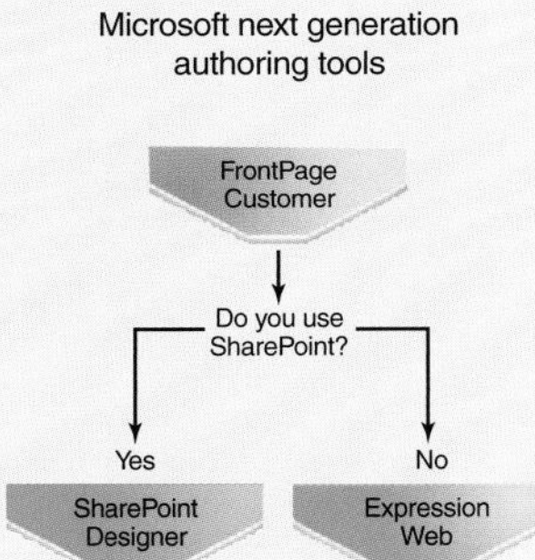

As of this textbook's writing, many companies have yet to decide on whether to adopt SharePoint. Instead of using either FrontPage, SharePoint, or Expression Web (which might not be loaded on your computers), you can use Microsoft Word 2007 to create your Web site. Follow these steps to do so:

1. From the **Insert** tab, insert a 3 × 2 table.

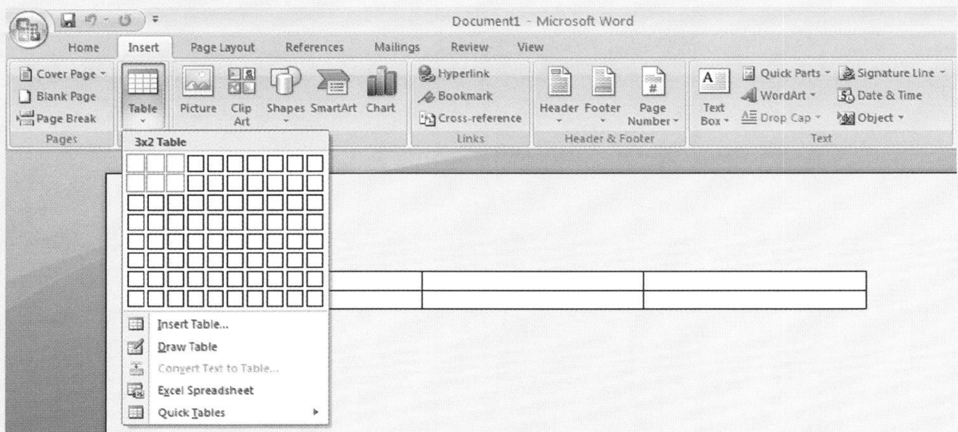

2. Merge the top right two columns.

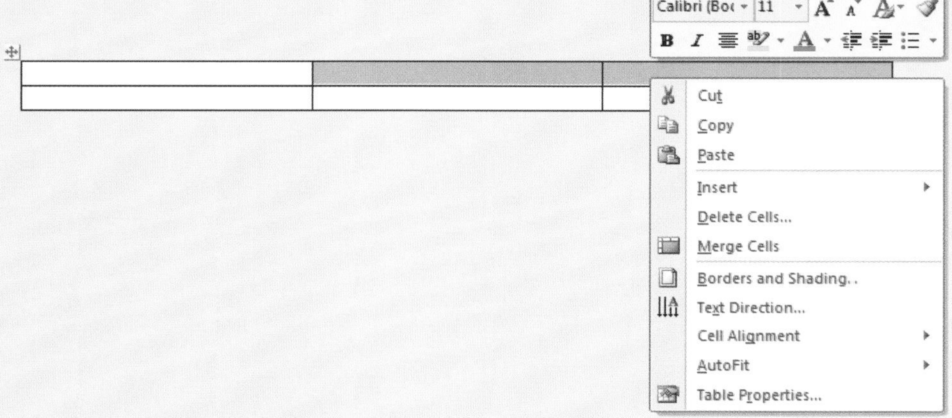

3. Place your cursor in the right rows and hit the "Enter" key to add space. At this point, you have the beginning design for a Web page.

4. Add graphics, text, color, and font types just as you would with any Word document.

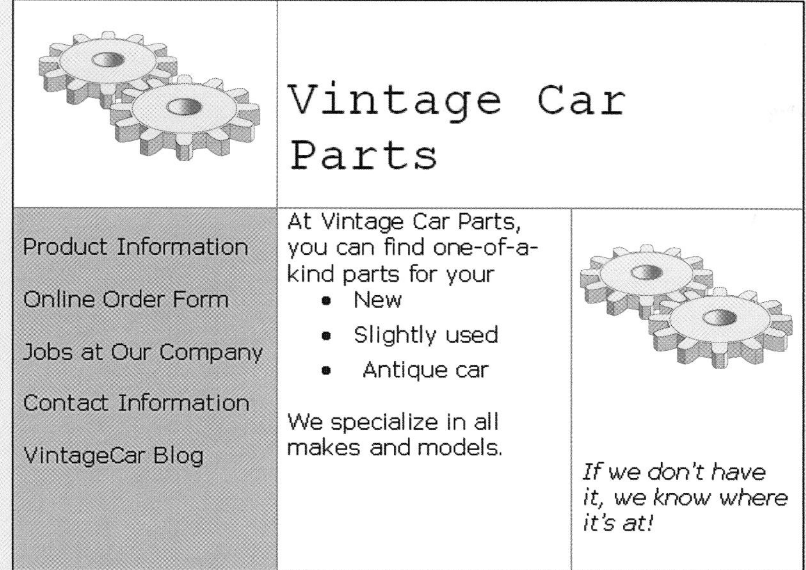

5. To create hypertext links, highlight selected text, right click, and scroll to ***Hyperlink***.

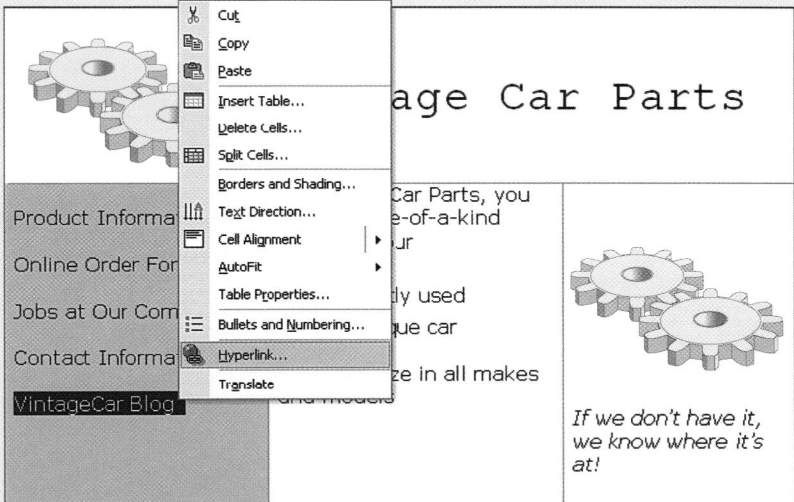

6. To create your Web site, click on the Word 2007 **Office Button**, scroll to **Save *As*,** and then to **Other Formats**.

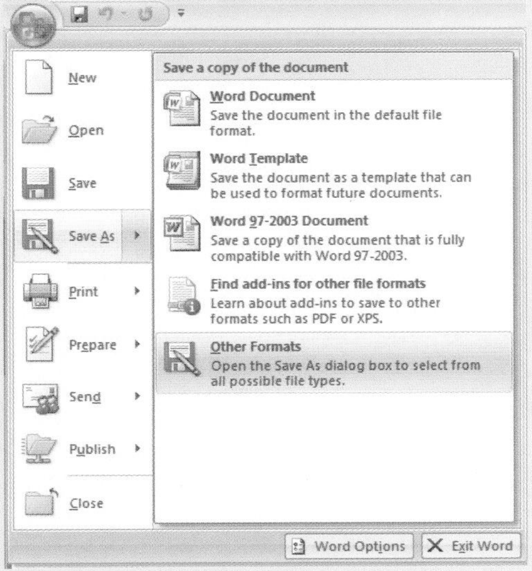

(Continued)

7. In the pop-up **Save As** window, click on the **Save as type** down arrow and click on **Web Page.** This saves your work as an HTML (Hypertext Markup Language) file which you can open in a Web browser.

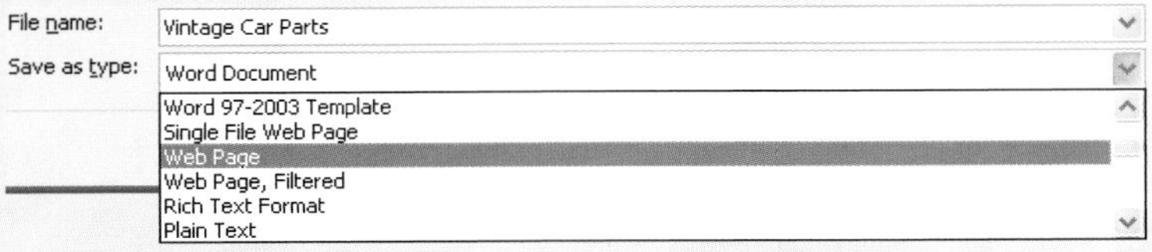

File name: Vintage Car Parts
Save as type: Word Document
Word 97-2003 Template
Single File Web Page
Web Page
Web Page, Filtered
Rich Text Format
Plain Text

Shannon Met His Challenge

To meet his communication challenge, Shannon used the P^3 process.

Planning

To plan his Web site, Shannon considered the following:

- Goal—provide policy and procedures for the photography studio's Web site
- Audience—existing and potential photography studio clients
- Channels—Statement of Purpose, Web page on Web site
- Data—policy and procedures including fees, ordering, pricing structure, and so on

To capture the most important elements of a project, Shannon says, "I worked with Jeanne to complete a Statement of Purpose. I use this tool to define the most important elements that she wants to communicate to her customers. This tool provides me with enough information to get started on a mockup of a Web page. The Statement of Purpose defines the audience, communication channel, and shelf life of the content. I plan to orchestrate Jeanne's needs with the needs of customers. In addition, I have a personal stake in the startup photography studio because its success will be good for our entire family." Figure 14.5 shows how Shannon used a Statement of Purpose to plan his Web site.

Packaging

After reviewing the Statement of Purpose and working with Jeanne, Shannon asked her, "What are the top three ideas you want your visitors to take with them when they leave a page on the Web site?" Jeanne's response was, "It's all important." After putting some serious thought to it, Jeanne was able to pick the top three most important ideas. However, she continued to want Shannon to include as many details as possible. Shannon says, "I then created a rough draft in the form of a mockup for my wife, who provided additional feedback to enable me to perfect the final version." Figure 14.6 shows how Shannon packaged his rough draft.

Perfecting

After studying the mockup, Jeanne said, "I see why you wanted me to pick the most important ideas for the potential customer to take away from the Web site. Let's make the Web site simpler, with far fewer words, and use lots more linked pages so that the customer can focus. I want the Web site to be as elegant as possible and illustrate some of my most treasured photographs so the customers can see what they will get when they hire me."

Shannon met his communication challenge by perfecting the Web page mockup. He and Jeanne are pleased with both the content and design of the Web site. Their business is up and running. As Shannon says, "Designing a Web site for myself and my wife had its challenges, but the almost instant profitability of the studio is a testament to the success of my design. We began our business in May 2007 and had a few clients. After I designed the Web site, business increased by 35 percent, and Jeanne is almost too busy. She now has an outlet for her many talents, and our family income is increasing. Business is great!" Figure 14.7 is Shannon's perfected Web page.

FIGURE 14.5 Statement of Purpose Completed by Client

Thank you for choosing J Conner Photography!

Session Fee

- Session fees are $125. This covers Jeanne's time, talent, and equipment and does not include any prints. Session fees are due at time of booking. Due to her busy schedule, session dates are only held once the session fee has been received.

Your Session

- Your session is individualized to fit your needs. I do not rush you through the session. We will take our time to get the best image possible. For small children and newborns, this may mean taking short breaks for feeding and snacks.

Ordering

- A minimum of 15 proofs will be provided in a private, password-protected online gallery within one week of your session date. J Conner will e-mail you a link to your gallery. You're welcome to forward your gallery to family and friends.

Print Prices

Wall Portraits (mounted and coated)	Gift Prints (coated)
30 × 40 $400	8 × 10 $40
20 × 24 $275	5 × 7 $25
16 × 20 $170	8 Wallets $20
Add $150 for gallery-wrapped canvas	

Boutique

- Boutique items are available after a minimum purchase of $300, unless otherwise noted.
- Session Album - $100
  A small 5×7, spiral bound album that includes watermarked proofs from your session. It's perfect for carrying around to show to friends and family.

Final prints

- Prints will be delivered in 2 to 3 weeks. Larger images, canvas prints, and specialty items may take 4 to 6 weeks. J Conner Photography will contact you for the best time to deliver the prints. All sales are final.

Reprints

- Reprints may be ordered up to one year from the session date. They are subject to the $45 re-hosting fee and a minimum order of $250 applies.

Copyright Information

- All images copyrighted by photographer, Jeanne Conner.

Referrals

- Thank you for your business. As a small business owner, I rely heavily on referrals and word-of-mouth advertising. As a thank you for spreading the word, I'd like to offer you a $20 print credit for each referred client who completes a session.

FIGURE 14.6 **Web Page Mockup**

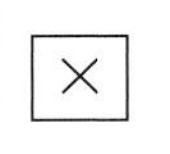

Thank you for choosing J Conner Photography! Please take a moment to look through our policies and procedures. If you have any questions, please contact Jeanne at jconner@jconnerphotography.com.

Session Fee
Session fees are $125. This covers Jeanne's time, talent, and equipment and does not include any prints. Session fees are due at time of booking. Due to her busy schedule, session dates are only held once the session fee has been received.

Your Session
Your session is individualized to fit your needs. I do not rush you through the session. We will take our time to get the best image possible. For small children and newborns, this may mean taking short breaks for feeding and snacks.

Ordering
A minimum of 15 proofs will be provided in a private, password-protected online gallery within one week of your session date. J Conner will e-mail you a link to your gallery. You're welcome to forward your gallery to family and friends. The ordering process is secure. Cash and credit/debit payments are made through the PayPal system.

Print Prices

Wall Portraits (mounted and coated)	Gift Prints (coated)
30 × 40 $400	8 × 10 $40
20 × 24 $275	5 × 7 $25
16 × 20 $170	8 Wallets $20
Add $150 for gallery-wrapped canvas	

Boutique
Boutique items are available after a minimum purchase of $300, unless otherwise noted. J Conner Photography also offers custom sterling silver photo jewelry and photo purses and handbags. Due to the handmade nature of these items, ask about current prices, styles, and availability.

Final prints
Prints will be delivered in 2 to 3 weeks. Larger images, canvas prints, and specialty items may take 4 to 6 weeks. J Conner Photography will contact you for the best time to deliver the prints. All sales are final. Please take care in handling your images. J Conner Photography is not responsible for damage incurred as a result of improper handling or framing.

Copyright Information
All images copyrighted by photographer, Jeanne Conner. It is illegal and unlawful to scan, copy, or reproduce Jeanne Conner's work in any manner or medium and punishable by law with fines starting at $150,000. As an artist, Jeanne Conner wishes to have complete control over the final look of her client's images, and scanned images damage her reputation as a photographer by distorting and devaluing the image quality. Thank you for respecting Jeanne Conner's work and livelihood by choosing not to reproduce her images.

Referrals
Thank you for your business. As a small business owner, I rely heavily on referrals and word-of-mouth advertising. As a thank you for spreading the word, I'd like to offer you a $20 print credit for each referred client who completes a session.

FIGURE 14.7 Perfected Web Page for Client

Home Page

Linked Page

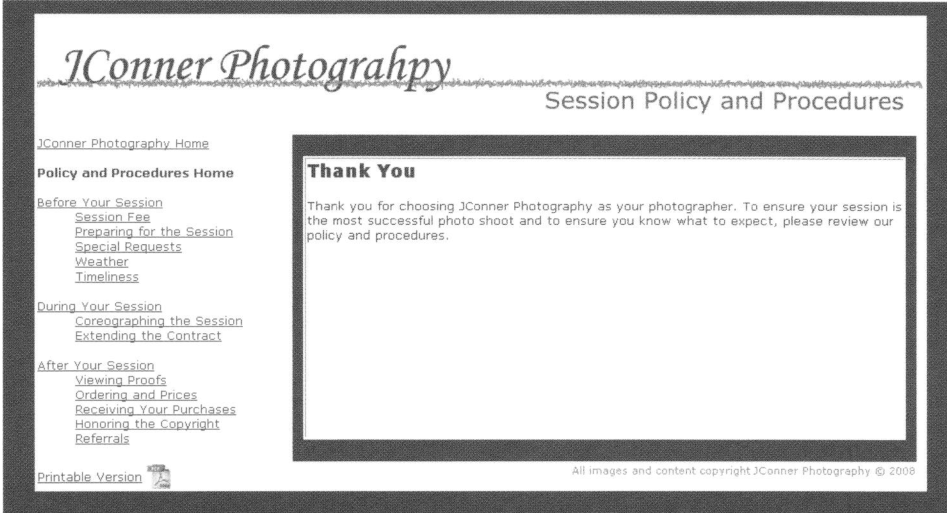

Shannon says, "It's been fun and interesting working with my wife, Jeanne. She has always been so creative as a photographer that I valued her input in the planning, packaging, and perfecting of the Web site. This experience has allowed us to enter each other's creative worlds. She has gotten interested in writing and wants to help me create a flier or a brochure for the photography business, and I am pleased that her business is rapidly growing. Good communication is essential for any business."

CHAPTER HIGHLIGHTS

1. The Internet, due to its speed, affordability, and international access, is a key component for corporate communication.
2. The Web Accessibility Initiative pursues accessibility to the Web for persons with disabilities such as cognitive, hearing, and vision impairment.
3. When creating a Web site, consider differences between paper text and online communication such as screen size, rapid reading (scanning), and hypertext links.
4. A successful Web site contains a home page and linked pages with easy navigation within the site, readable text, and effective document design.
5. An important consideration in Web site design is maintaining ethical standards and credibility.

WWW *Find additional Web site designing exercises, samples, and interactive activities at http://www. prenhall.com/gerson.*

CASE STUDIES

1. Future Promise is a not-for-profit organization geared toward helping at-risk high school students. This agency realizes that to reach its target audience (teens age 15–18), it needs an Internet presence. To do so, it has formed a team, consisting of the agency's accountant, sports and recreation director, public relations manager, counselor, graphic artist, computer and information systems director, two local high school principals, local high school students, and a representative from the mayor's office. Jeannie Kort, the PR manager, is acting as team leader.

   The team needs to determine the Web site's content, design, and levels of interactivity. Jeannie's boss, Brent Searing, has given the team a deadline and a few components that must be included in the site:

   - College scholarship opportunities
   - After-school intramural sports programs
   - Job-training skills (resume building and interviewing)
   - Service learning programs to encourage civic responsibility
   - Future Promise's 800-Hotline (for suicide prevention, STD information, depression, substance abuse, and peer counseling)
   - Additional links (for donors, sponsors, educational options, job opportunities, etc.)

Assignment

Form a team (or become a member of a team as assigned by your instructor) and design Future Promise's Web site. To do so, follow the criteria for Web design provided in this chapter. Then, plan, package, and perfect as follows:

- Plan
  - Research the topics listed previously (either in a library, online, or through interviews) to gather details for your Web site.
  - Consider your various audiences and their respective needs and interests.
  - Focus on your Web site's purpose—are you writing to inform, persuade, instruct, or build rapport?
  - Draw a storyboard of how you would like the Web site to look.
  - Divide your labors among the team members (who will research, write, create graphics, etc.?).
- Package
  - Draft your text either in a word processing program (you can save as an HTML file) or in FrontPage (or any other Web design program of your choice).
- Perfect
  - Review the Web criteria in this chapter.
  - Add, delete, simplify, move, enhance, and correct your Web site.
  - Test the site to make sure all links work and that it meets your audience's needs.

2. The short proposal on page 417 recommends that upper management approve the construction of a corporate Web site. The company's management has agreed with the proposal. You and your team have received a memo directing you to build the Web site.

Assignment

Using the proposal, build your company's Web site.

Date: November 11, 2009
To: Distribution
From: Paul Peterson
Subject: Recommendation Report for New Corporate Web Site

Introduction
In response to your request, I have researched the impact of the Internet on corporate earnings. Companies are still striving to position themselves on the World Wide Web and maximize their earnings potential. I concluded that from 2007 to 2009, 16 percent of companies with Internet sites were profiting from their online access. The time is right for our company, Java Lava, to go online.

Discussion
By going online, we can maximize our profits by going global. We can accomplish this goal as follows:

1. **International bean sales.** Currently, coffee bean sales account for only 27 percent of our company's overall profit. These coffee bean sales depend solely on walk-in trade. The remaining 73 percent stems from over-the-counter beverage sales. If we go global via the Internet, we can expand our coffee bean sales dramatically. Potential clients from every continent will be able to order our coffee beans online.
2. **International promotional product sales.** Mugs, T-shirts, boxer shorts, jean jackets, leather jackets, key chains, paperweights, and calendars could be marketed. Currently we give away items imprinted with our company's logo. By selling these items online, we could make money while marketing our company.
3. **International franchises.** We now have three coffeehouses, located at 1200 San Jacinto, 3897 Pecan Street, and 1801 West Paloma Avenue. Let's franchise. On the Internet, we could offer franchise options internationally.
4. **Online employment opportunities.** Once we begin to franchise, we'll want to control hiring practices to ensure that Java Lava's standards are met. Through a Web site, we could post job openings internationally and list the job requirements. Then potential employees could apply online.

In addition to the above information, used to increase our income, we could provide the following:

- A map showing our three current sites.
- Our company's history—founded in 1952 by Hiram and Miriam Coenenburg, with money earned from their import/export bean business. "Hi" and "Mam," as they were affectionately called, handed their first coffeehouse over to their sons (Robert, John, and William) who expanded to our three stores. Now a third generation, consisting of Rob, John, and Bill's six children (Henry, Susan, Andrew, Marty, Blake, and Stefani) could take us into the next millennium.
- Sources of our coffee beans—Guatemala, Costa Rica, Columbia, Brazil, Sumatra, and France.
- Freshness guarantees—posted ship dates and ground-on dates; 100 percent money-back guarantee.
- Corporate contacts (addresses, phone numbers, e-mail, fax numbers, etc.).

Conclusion
Coffee is a "hot" commodity now. The international market is ours for the taking. We can maximize our profits and open new venues for expansion. I'm awaiting your approval.

3. Tot's Toys, a start-up company owned and operated by Sandy Warner and Marissa Kline, plans to create a Web site to sell their handmade, hypoallergenic, and baby-safe toys. Their Web site needs to have at least five Web pages:
   - **Home Page,** complete with their company logo, an interesting background that suits the company's image, and easy navigation from screen to screen.
   - **Corporate Contact Page,** including phone and fax numbers, and their Web address (totstoys.com).
   - **Product Descriptions.** They want both thumbnail and full-image graphics of each toy. Their product description page will list materials of construction to show how the toys are safe and healthy.
   - **Online Shopping Cart.** Sandy and Marissa plan to ensure that their site is secure to allow for easy, credit-card purchasing.
   - **FAQs.** A frequently asked question page will engage the Web audience.

Assignment

Using the previous framework and following this chapter's criteria for successful Web sites, create the Tot's Toys Web site.

4. Review the following flawed home page. Determine why it does not succeed, based on the criteria for successful Web sites provided in this chapter.

Assignment

List the reasons for its lack of success. Then, suggest ways in which it could be improved. Finally, recreate it as an improved Web page.

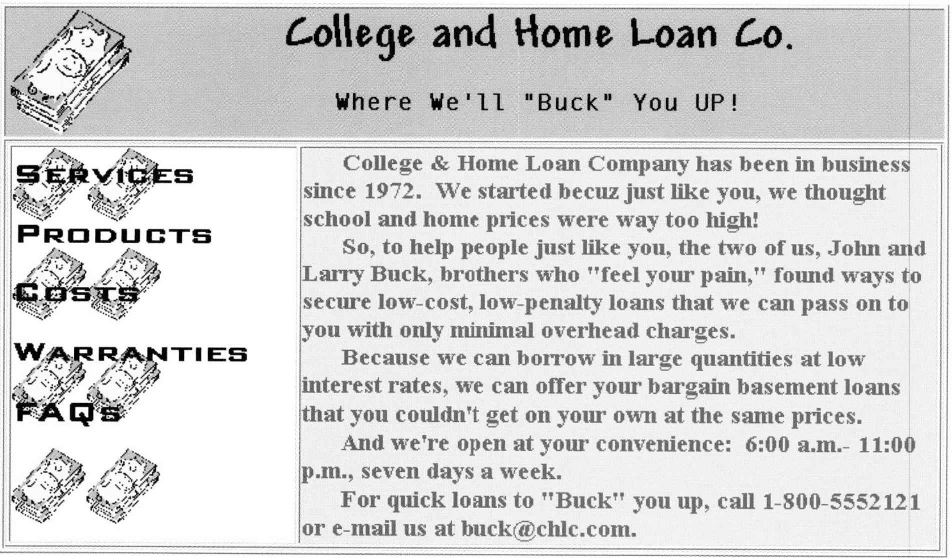

ETHICS IN THE WORKPLACE CASE STUDY

Sam Henry, a salesperson for a children's toy store, loves the Internet. He wrote a proposal to his company's CEO suggesting that the company's Web site could be improved by adding a corporate Web blog. This blog would allow for better communication between the company and its client base. The CEO agreed and allowed Sam to create the company blog site.

As part of the corporate blog, Sam included many links to related sites that he thought would appeal to customers. One of the links was to Sam's personal FaceBook account. Sam believed that this link would allow him to achieve a more personal contact with clients.

Sam's FaceBook provides personal information about his hobbies, his children, and his weekend activities, which include tailgating at sporting events. One photograph on the FaceBook site shows Sam sitting on his motorcycle and drinking beer.

Question: Is Sam's FaceBook profile an appropriate link for a corporate blog? Justify your answer based on information provided in Chapter 1.

INDIVIDUAL AND TEAM PROJECTS

1. Create a corporate Web site. To do so, make up your own company and its product or service. Your company's service could focus on dog training, computer repair, basement refinishing, vent cleaning, Web site construction, childcare, auto repair, personalized aerobic training, or online haute cuisine. Your company's product could be paint removers, diet pills, interactive computer games, graphics software packages, custom-built engines, flooring tiles, or duck decoys. The choice is yours. To create this Web site, plan, package, and perfect. Sketch your site through storyboarding. Draft your Web site, using HTML coding or FrontPage. Perfect by adding, deleting, simplifying, moving, reformatting, enhancing, correcting, and making sure your links work.
2. Create a Web site for your campus club; fraternity or sorority; your church, mosque, or synagogue; or your professional organization.
3. Create a personal Web site for yourself or for your family. Consider including biographies of your immediate family members, a family tree, family history, favorite vacation spots, favorite family outings or activities, and so forth.
4. Reconstruct an existing Web site you consider to be flawed. This might entail adding new information, deleting unnecessary details, simplifying text and graphics, moving information on the site for emphasis, reformatting the text for access, enhancing the tone, correcting errors, and ensuring that all links work.

DEGREE-SPECIFIC ASSIGNMENTS

Today, almost every corporation has a Web site. You can find corporate and professional organization Web sites about marketing and public relations, business information systems, accounting, management, and finance.

Assignment

1. Go online to find Web sites from your degree program. Review three to five Web sites for companies in your major field to determine types of content in a typical Web site. In an informal oral presentation to your class or e-mail message to your instructor, report on current issues you have discovered in your field.
2. Research several Web sites in your field of expertise. Find sites in your area of interest (academic discipline), or look at companies where you might eventually want to work. Use this chapter's Web Site Checklist to determine which sites excel and which sites need improvement. Then write a memo or report or give an oral presentation justifying your assessment. In this report or presentation, clarify what makes the sites successful. Next, explain why the unsuccessful sites fail. Finally, suggest ways in which the unsuccessful sites could be improved.

PROBLEM-SOLVING THINK PIECES

Access any company's Web site and study the site's content, layout (color, graphics, headings, use of varying font sizes/types, etc.), links (internal and external), ease of navigation, tone, and any other considerations you think are important. Then, determine how the Web site could be improved if you were the site's Webmaster. Once you have made this determination, write a memo recommending the changes that you believe will improve the site. In this memo

- Analyze the Web site's current content and design, focusing on what is successful and what could be improved
- Provide feasible alternatives to improve the site
- Recommend changes

WEB WORKSHOP

1. The Internet offers you many resources about how to design a successful Web site. Go online and look at the following sites:
   - Web Page Design for Designers (http://www.wpdfd.com/) provides a Web design FAQ.
   - Top Ten Mistakes in Web Design (http://www.useit.com/alertbox/9605.html) offers you exactly what the name implies.
   - The Web Design Group (http://www.htmlhelp.com/) provides an FAQ archive, information on Web design elements, and feature articles.
   - Web Design Guide (http://dreamink.com/) provides effective Web design principles, tutorials, a step-by-step beginner's guide, and Web resources.

   Review any of these sites for more information on Web design principles. What new and interesting information have you discovered? For example, what questions or concerns do most Web designers seem to have? What mistakes are common to Web design? Write a memo or e-mail to your instructor or give an oral presentation to share your findings.
2. Compare and contrast corporate home pages. Assess them according to this chapter's criteria for successful Web design. Which of the home pages is most effective, least effective, most interesting, or least interesting? Explain your decisions either in a memo, e-mail message, or an oral presentation.
3. Internet accessibility is the law. The ability for any Web user, regardless of disability, to access the Web is essential. Study any of the following Web sites dedicated to Web access issues. Summarize the major concerns and methods suggested to ensure Web access.

Web Site Name	URL
Web Accessibility Initiative (WAI)	http://www.w3.org/WAI/
U.S. Government Section 508 Website Accessibility Guidelines	http://www.access-board.gov/sec508/508standards.htm
The Society for Technical Communication Accessibility Special Interest Group	http://www.stcsig.org/sn/internet.shtml
National Center for Accessible Media	http://ncam.wgbh.org/
Great Britain Disability Rights Commission	http://www.drc-gb.org/publicationsandreports/report.asp
International Web Accessibility Concerns	http://www.stcsig.org/sn/international.shtml

QUIZ QUESTIONS

1. What are the characteristics of online communication?
2. According to the Web Accessibility Initiative, what types of problems impede Internet access?
3. What are characteristics of a successful Web site?
4. Why should you avoid "noise" on your Web site?
5. How can you achieve effective document design on a Web page?

CHAPTER 15

Descriptions, Process Analyses, and Instructions

▸REAL PEOPLE *in the workplace*

Stacy Gerson is a technical writer for GBA Master Series. As a documentation specialist, Stacy creates online user manuals for gbaMS (www.gbams.com), a company that develops computerized maintenance management software for public works and utilities infrastructure systems. The company's software helps cities, counties, and states track their inventory and assets relating to sewer and water; streets and roadway work orders; GIS; and fleet and equipment.

To write her user manuals, Stacy

- Interacts with her programmers, asking questions to clarify challenging content.
- Watches WebEx movies created by the software developers. These movies walk through programs and provide voice-over explanations of the content. Stacy uses the movie scripts for the basis of her text.
- Digs through the complex coding, provided by the subject matter experts, to find what will be most useful for her lay readers.
- "Learns how to use the software by using it."

Stacy focuses on the big picture when she creates her online manuals. Her goals are to create easy-to-read text, easy-to-follow steps, and text that "is not just well written but content that will help the end user."

Stacy has found that she needs to write well, have grammatical abilities, empathize with the end users' needs, and have computer/word processing skills. Stacy's number one task is to make things easily understandable for gbaMS's customers. The software can be complex at times, so if an instruction is unclear, "The customer can't do the job." To make the instructions understandable, "I try to write text that is logical, clear, simple, and visually appealing. If I look at something and say, 'What's that?' then I have to assume that it won't be clear to the client either."◂

CHAPTER GOALS

When you complete this chapter, you will be able to do the following:

1. Understand the purpose of description, process analysis, and instruction.
2. Apply the criteria for writing descriptions, process analyses, and instructions.
3. Use graphics effectively in descriptions, process analyses, and instructions.
4. Use the P^3 process—planning, packaging, and perfecting—to write effective descriptions, process analyses, and instructions.

WWW

To learn more about descriptions, process analyses, and instructions, visit our Web Companion at http://www.prenhall.com/gerson.

Stacy's Communication Challenge

Annually, gbaMS hosts a training conference for its end users, customers who have purchased software. These customers include city, county, and state employees, supervisors, and information technology specialists. Stacy says, "My role in this conference is to create 48 training manuals. As I write these manuals, I encounter three challenges:

1. Creating the agenda for the conference. The conference will last three days. Each day is divided into six or seven 50-minute time slots. Since the time allotted for training is limited, I must prioritize which topics to devote training time to.
2. gbaMS's manuals range in length from 15 to 40 pages. Entire manuals cannot be reviewed in 50 minutes, so I need to create instructional workbooks appropriate for the time allotted. I have to determine what's important and what can be realistically covered.
3. My audience creates the final challenge. The presentations are attended by beginning, intermediate, and advanced users of the software. With this in mind, I must adjust the content and visuals accordingly. Creating these manuals is a communication challenge."

See pages 440–442 to learn how Stacy met her communication challenge.

PURPOSE OF DESCRIPTIONS, PROCESS ANALYSES, AND INSTRUCTIONS

What products have you purchased that came packaged with instructions? A safe answer would be, "everything." Your iPod came with instructions. So did your cell phone, DVD, washing machine, printer, computer, and automobile. Manufacturers include instructions in the packaging of a mechanism, tool, or piece of equipment. The instructions help the end user construct, install, operate, and service the equipment.

In addition, instructions often include descriptions. Descriptions provide the end user with information about the mechanism's features, capabilities, and process analyses. A description helps the reader visualize the mechanism and may tell the user which components are enclosed in the shipping package, clarify the quality of these components, specify which function these components serve in the mechanism, allow the user to reorder any missing or flawed components, or tell the end user how something works.

DESCRIPTIONS

A *description* is a part-by-part depiction of the components of a mechanism, tool, or piece of equipment. Descriptions are important features in several types of workplace communication.

Types of Descriptions

Operations Manuals

Manufacturers often include an *operations manual* in the packaging of a mechanism, tool, or piece of equipment. This manual helps the end user construct, install, operate, and service the equipment. Operations manuals often include descriptions. Descriptions provide the end user with information about the mechanism's features or capabilities. Here is a brief description found in an operations manual:

example

> The Modern Electronics Tone Test Tracer, Model 77A, is housed in a yellow, high-impact plastic case that measures 1¼″ × 2″ × 2¼″, weighs 4 ounces, and is powered by a 1604 battery. Red and black test leads are provided. The 77A has a standard four-conductor cord, a three-position toggle switch, and an LED for line polarity testing. A tone selector switch located inside the test set provides either solid tone or dual alternating tone. The Tracer is compatible with the EXX, SetUp, and Crossbow models.

Product Demand Specifications

Sometimes a company needs a piece of equipment that does not currently exist. To acquire this equipment, the company writes a *product demand* specifying its exact needs:

example

This product demand specification is written for specialists. It assumes knowledge on the part of the reader, such as definitions for "high-speed" and "deep-hole," which a lay reader would not understand. In addition, the technical abbreviation "AISI" is not defined.

> Subject: Pricing for EDM Microdrills
>
> Please provide us with pricing information for the construction of 50 EDM Microdrills capable of meeting the following specifications:
>
> - Designed for high-speed, deep-hole drilling
> - Capable of drilling to depths of 100 times the diameter using 0.012″ to 0.030″ diameter electrodes
> - Able to produce a hole through a 1.000″ thick piece of AISI D2 or A6 tool steel in 1.5 minutes, using a 0.020″ diameter electrode
>
> We need your response by January 13, 2009.

Study Reports Provided by Consulting Firms

Companies hire a consulting engineering firm to study a problem and provide a descriptive analysis. The resulting *study report* is used as the basis for a product demand specification requesting a solution to the problem. One firm, when asked to study crumbling cement walkways, provided the following description in its study report:

example

> The slab construction consists of a wearing slab over a ½″ thick waterproofing membrane. The wearing slab ranges in thickness from 3½″ to 8½″ and several sections have been patched and replaced repeatedly in the past. The structural slab varies in thickness from 5½″ to 9″ with as little as 2″ over the top of the steel beams. The removable slab section, which has been replaced since original construction, is badly deteriorated and should be replaced. Refer to Appendix A, Photo 9, and Appendix C for shoring installed to support the framing prior to replacement.

FIGURE 15.1 Description of a Color Printer

HP Officejet Pro K550 series color printer

The title identifies the equipment to be described.

Impressionistic words such as "quick access," "easy troubleshooting," and "user-friendly control panel" are used for sales purposes to the lay audience.

Numbered callouts refer to the text and help the readers locate parts of the equipment.

1 Ink cartridge cover provides quick access to four snap-in ink cartridges.

2 Top cover flips open for easy troubleshooting or maintenance.

3 User-friendly control panel includes graphical icons on the buttons and lights. It provides printer and wireless networking status at-a-glance and simplifies wireless configuration with SecureEasySetup™.

4 150-sheet output tray with extender.

5 250-sheet and 350-sheet input trays for high-volume printing.

6 Automatic two-sided printing accessory for professional, two-sided documents.

7 Built-in (Full Speed) USB host connector enables wireless configuration through Windows® Connect Now.

8 Built-in wired Ethernet networking.

9 Built-in Hi-Speed USB 2.0 port enables fast and easy direct connections.

HP Officejet Pro K550dtwn color printer shown

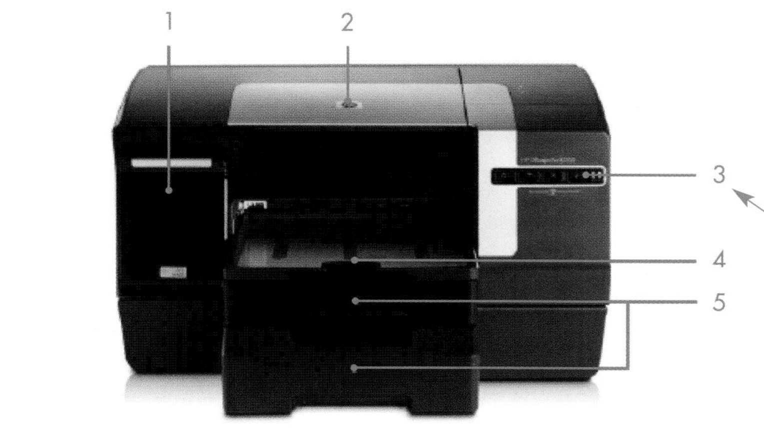

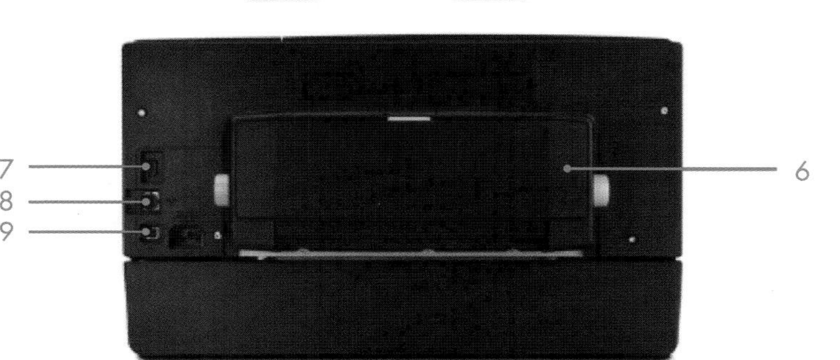

Courtesy of Hewlett-Packard

Sales Literature

Companies want to make money. One way to market equipment or services is to describe the product. Such descriptions are common in sales letters, proposals, and on Web sites. Figure 15.1 is a description from Hewlett-Packard.

Additional information:

See Chapter 13, "Persuasive Workplace Communication," and Chapter 19, "Internal and External Proposals."

DEFINING PROCESS ANALYSIS

What is a process analysis? Let us define a process analysis by comparing it to an instruction. Instructions provide a step-by-step explanation of how to do something. For example, in an instruction, you explain how to change the oil in a car, how to connect a DVD player to a computer, how to make polenta, or how to put together a child's toy. In an instruction, the audience wants to know how to do a job. A process analysis, in contrast, focuses not on how to do something but, instead, on how something works. For instance, a process analysis might explain how viruses attack our bodies, how airbags save lives, how metal detectors work, or how e-mail messages are transmitted.

CRITERIA FOR WRITING DESCRIPTIONS AND PROCESS ANALYSES

As with any type of workplace communication, there are certain criteria for writing descriptions and process analyses.

Title

Preface your text with a title precisely stating the topic of your description. This could be the name of the mechanism, tool, or piece of equipment you're writing about.

Overall Organization

In the **introduction** you specify and define what topic you are describing, explain the mechanism's functions or capabilities, and list its major components.

example

> The Apex Latch (#12004), a mechanism used to secure core sample containers, is composed of three parts: the hinge, the swing arm, and the fastener.

Technology Tip

How to Make Callouts Using Microsoft Word 2007

Callouts are a great way to label mechanism components in your technical description or process analyses. Follow these steps to create callouts.

1. Click on "**Insert**" in your Menu bar.

2. Click on "**Shapes.**" You will see a drop-down box.

3. Select the type of callout you want to use, such as the "**Rounded Rectangular Callout**" shown.

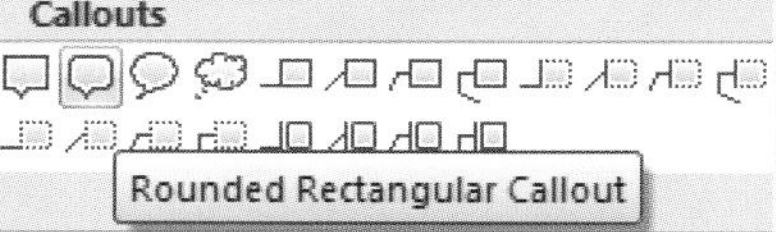

4. When you choose the callout, a crosshair will appear. Use it to draw your callout, as shown.

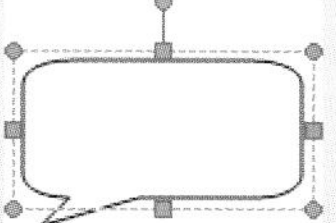

Note: When you draw the callout, Word 2007 opens up the "**Format**" tab, as shown.

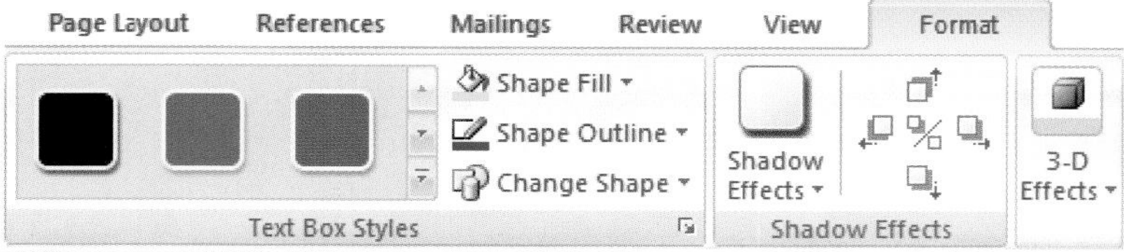

From the "**Format**" tab, you can change the color of the text box (from "**Text Box Styles**"), change its line type and density (from "**Shape Outline**"), create shadow effects, create 3-D effects, and add content.

The DX 56 DME (Distance Measuring Equipment) is a vital piece of aeronautical equipment. Designed for use at altitudes up to 30,000 feet, the DX 56 electronically converts elapsed time to distance by measuring the length of time between your transmission and the reply signal. The DX 56 DME contains the transmitter, receiver, power supply, and range and speed circuitry.

example

In the **discussion,** you use highlighting techniques (itemization, headings, underlining, white space) to describe each of the mechanism's components and how the mechanism works—its process.

Your **conclusion** depends on your purpose in describing the topic. Some options are as follows:

Sales. "Implementation of this product will provide you and your company . . ."	**Guarantees.** "The XYZ carries a 15-year warranty against parts and labor."	**Comparison/contrast.** "Compared to our largest competitor, the XYZ has sold three times more . . ."
Uses. "After implementation, you will be able to use this XYZ to . . ."	**Testimony.** "Parishioners swear by the XYZ. Our satisfied customers include . . ."	**Reiteration of introductory comments.** "Thus, the XYZ is composed of the aforementioned interchangeable parts."

Use graphics in your descriptions. Today, many companies use large, easy-to-follow graphics that complement the text. You can use line drawings, photographs, clip art, exploded views, or sectional cutaway views of your topic, each accompanied by call-outs (labels identifying key components of the mechanism).

Internal Organization

When describing your topic in the discussion portion of the description, itemize the topic's components in some logical sequence. Components of a piece of equipment, tool, or product can be organized by importance.

However, *spatial* organization is better for descriptions. When a topic is spatially organized, you literally lay out the components as they are seen, as they exist in space. You describe the components as they are seen either from left to right, from right to left, from top to bottom, from bottom to top, from inside to outside, or from outside to inside. In contrast, when writing your process analysis, you will use *chronological* organization to show how the tool or mechanism works.

Development

To describe your topic clearly and accurately, detail the following:

Weight	Materials (composition)
Size (dimensions)	Identifying numbers
Color	Make/model
Shape	Texture
Density	Capacity

Word Usage

Your word usage, either photographic or impressionistic, depends on your purpose. For factual, objective technical descriptions, use photographic words. For subjective, sales-oriented descriptions, use impressionistic words. Photographic words are denotative, quantifiable, and specific. Impressionistic words are vague and connotative. Table 15.1 shows the difference.

Figure 15.2 is a process analysis of why, when, and how automobile air bags inflate.

Table 15.1 Photographic versus Impressionistic Word Usage

Photographic	Impressionistic
6′9″	tall
350 lb	heavy
gold	precious metal
6,000 shares of United Can	major holdings
700 lumens	bright
0.030 mm	thin
1966 XKE Jaguar	impressive car

FIGURE 15.2 Description and Process Analysis of an Air Bag

The introduction explains what the topic is, why it is important, and where the mechanism is located.

Why, When, and How Does an Air Bag Inflate?

Air bags save lives. Driver and passenger air bags are designed to inflate in frontal or side crashes. Steering wheel, right front instrument panel, or side-panel air bags will not inflate on all occasions. If a car drives over a bump or if a crash is "minor," such as hitting an object while driving at a slow speed, an air bag will not deflate. However, when cars hit walls (or trucks, cars, or trees), air bags inflate to minimize injury and to save people's lives.

Before Inflation

How does the air bag detect whether the car has hit a bump versus being involved in a collision?

The air bag system's crash sensor can differentiate between head-on collisions and simple bumps in the road as follows:

1. A steel ball slides inside a smooth bored cylinder.
2. The ball is secured by a magnet or by a stiff spring. This inhibits the ball's motion when the car encounters minor motion changes, such as bumps or potholes.
3. If the car comes to a dramatic and rapid stop, a force equal to running into a brick wall at about 10 to 15 miles per hour, such as in a head-on crash, the ball quickly moves forward. This closes a contact and completes an electrical circuit, which initiates the process of inflating the air bag.

This description discusses the parts of the mechanism, the materials used, the location of these components, and the chemical compounds required for activation.

Parts of an Air Bag

Air bag systems consist of the following:

- **Air bag.** The air bag is made of thin nylon. A nontoxic powder (cornstarch or talcum powder) inside the air bag keeps it flexible and ensures the parts of the air bag do not stick together.
- **Air bag holding compartment.** The holding compartment is in the steering wheel, dashboard, seat, or door.
- **Sensor.** This device tells the bag to inflate.
- **Inflator canister.** This consists of sodium azide (NaN_3), potassium nitrate (KNO_3), and silicon dioxide (Sio_2), which produce nitrogen gas (N_2).

FIGURE 15.2 Continued

During Inflation

In a severe impact, the air bag sensing system will deploy in milliseconds. The following occurs:

1. The air bag's crash sensor triggers a switch that energizes a wire, sending electricity into a heating element in the propellant that releases gas from the inflator canister.
2. A pellet of sodium azide (NaN_3) is ignited, generating nitrogren gas (N_2).
3. A nylon air bag, folded into the dashboard, steering wheel, and/or side panels of the door, inflates at a speed of 150 to 250 mph, taking only about 40 milliseconds (about 1/20 of a second) for the inflation to be complete.
4. The sodium azide inside the air bag produces sodium metal, which is highly reactive, potentially explosive, and harmful when in contact with eyes, nose, or mouth. To render this harmless, the sodium azide reacts with potassium nitrate (KNO_3) and silicon dioxide (SiO_2)—also inside the air bag—to produce silicate glass, a harmless and stable compound.

Here, providing the mechanism's process, the text explains how an air bag works. Note the specificity of detail: 150 to 250 mph and 1/20 of a second.

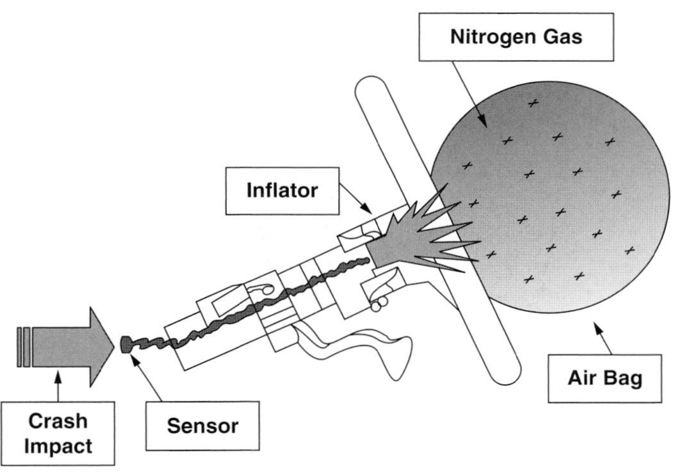

This illustration graphically depicts the process in action. When the crash sensor is triggered, it activates the inflator. Nitrogen gas explodes, inflating the air bag.

After Inflation

Air bag vents, minute holes in the bag, allow the deployed air bags to deflate immediately after impact. This ensures that the car's inhabitants do not smother.

Though air bags were first used in 1973, they have only been mandatory in all cars since 1998. Have air bags made a difference? Absolutely! The National Highway Traffic Safety Administration says air bags saved an estimated 1,043 lives in 1998 alone.

The conclusion sums up the process description by quantifying the significance of air bags as a means of saving lives.

REASONS FOR WRITING INSTRUCTIONS

Every day, at home and at work, you accomplish tasks by following instructions. At home, you use step-by-step instructions to fertilize your garden, assemble a Barbie Dream House, program your TiVo, create a personal information number for online shopping, or make your famous chili-cheese dip for Super Bowl Sunday. At work, you follow instructions to

DESCRIPTION AND PROCESS ANALYSIS CHECKLIST

____ 1. Does the description or process analysis have a title noting the topic's name (and any identifying numbers)?

____ 2. Does the description or process analysis's introduction (a) state the topic, (b) mention its functions or the purpose, and (c) list the components?

____ 3. Does the description or process analysis's discussion use headings to itemize the components for reader-friendly ease of access?

____ 4. Do you need to define the mechanism and its main parts?

____ 5. Is the detail within the description or process analysis's discussion precise? That is, does the discussion portion of the description or process analysis specify the following?

Colors	Capacities
Sizes	Textures
Materials	Identifying numbers
Shapes	Weight
Density	Make/model

____ 6. Are all of the calculations, measurements, or process timelines correct?

____ 7. Do you sum up your discussion using any of the optional conclusions discussed in this chapter?

____ 8. Does your description or process analysis provide graphics that are correctly labeled, appropriately placed, neatly drawn or reproduced, and appropriately sized?

____ 9. Do you write using an effective style and a personalized tone?

____ 10. Have you avoided biased language and grammatical and mechanical errors?

change your printer's toner, adjust the margins on a Word document, fill out online report forms, or help a new employee access his or her corporate e-mail account. You'll write instructions to help your audience complete the following tasks:

Operate a mechanism	Restore a product	Test components
Install equipment	Correct a problem	Use software
Manufacture a product	Service equipment	Set up a product
Package a product	Troubleshoot a system	Maintain equipment
Unpack equipment	Clean a product	Monitor a system
Repair a system	Assemble a product	Construct anything

Figure 15.3 is a sample workplace e-mail message giving a customer the steps to follow for computer technical support.

CRITERIA FOR WRITING INSTRUCTIONS

Follow these criteria for writing effective instructions.

Title Your Instructions

Preface your text with a title that explains two things: *what* you are writing about (the name of your product or service) and *why* you are writing (the purpose of the instructions). For example, to title your instructions "Digital Camera" would be uninformative. This title names the product, but it does not explain why the instructions are being written. Will the text discuss maintenance, setup, packing, or operating instructions? A better title would be "Operating Instructions for the XYZ Digital Camera."

Organize Your Instructions

Well-organized instructions help readers follow your directions. To organize your instructions effectively, include an *introduction*, a *discussion*, and a *conclusion*.

FIGURE 15.3 **E-Mail Instruction**

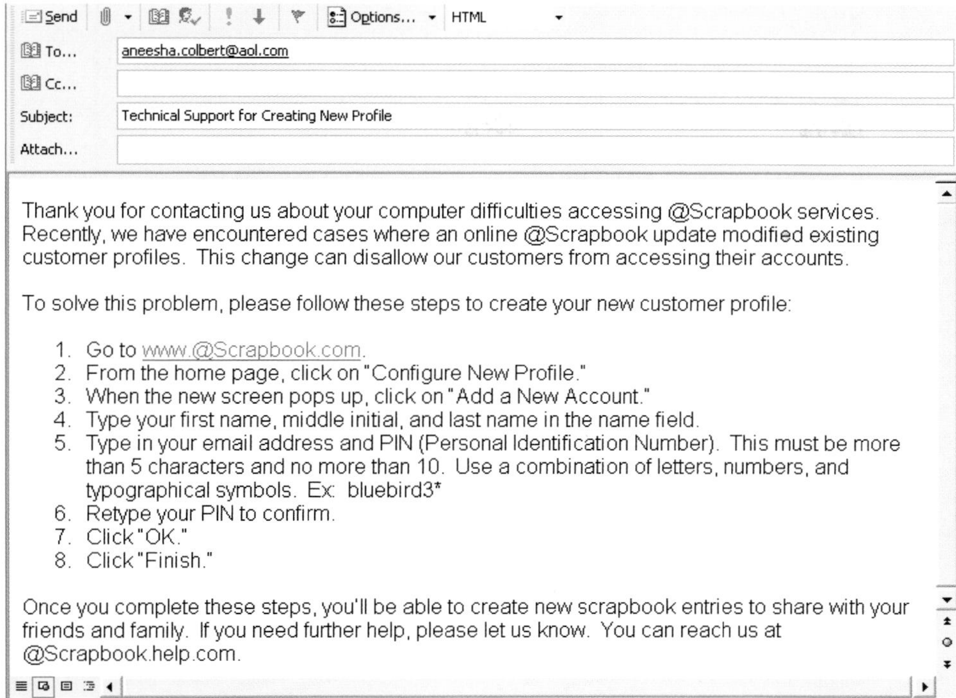

Send | Options... | HTML

To... aneesha.colbert@aol.com

Cc...

Subject: Technical Support for Creating New Profile

Attach...

Thank you for contacting us about your computer difficulties accessing @Scrapbook services. Recently, we have encountered cases where an online @Scrapbook update modified existing customer profiles. This change can disallow our customers from accessing their accounts.

To solve this problem, please follow these steps to create your new customer profile:

1. Go to www.@Scrapbook.com.
2. From the home page, click on "Configure New Profile."
3. When the new screen pops up, click on "Add a New Account."
4. Type your first name, middle initial, and last name in the name field.
5. Type in your email address and PIN (Personal Identification Number). This must be more than 5 characters and no more than 10. Use a combination of letters, numbers, and typographical symbols. Ex: bluebird3*
6. Retype your PIN to confirm.
7. Click "OK."
8. Click "Finish."

Once you complete these steps, you'll be able to create new scrapbook entries to share with your friends and family. If you need further help, please let us know. You can reach us at @Scrapbook.help.com.

Introduction

Begin your instructions by telling your readers two things: what *topic* you will be discussing and your *reasons* for writing. The topic names the product or service. Your reason for writing either explains the purpose of the instructions ("maintaining the machine will increase its longevity") or comments about the product's capabilities or ease of use.

Required Tools or Equipment

In addition to an introductory overview focusing on the topic and reasons for writing, you might want to tell your readers what tools or equipment they will need to perform the procedures. You can provide this information through a list or graphically.

Hazard Notations

Decide whether you should preface your instructions with dangers, warnings, cautions, or notes. This is an essential consideration to avoid costly lawsuits and to avoid potentially harming an individual or damaging equipment.

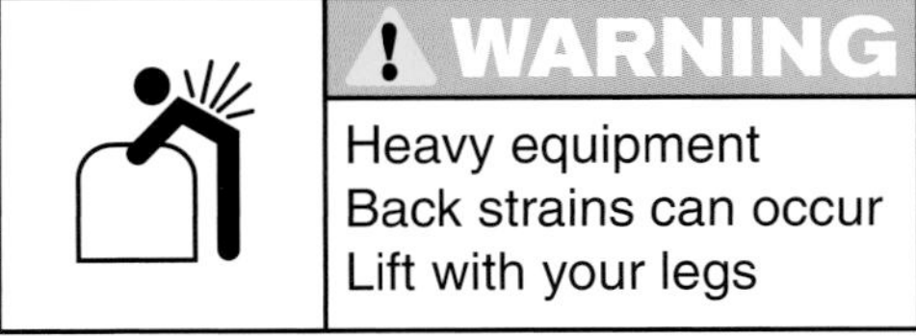

When including hazard alerts, consider the following:

- **Placement**—You can place hazard alerts before the text, in the text (next to appropriate steps), or after the text.
- **Access**—Make the caution, warning, or danger notice obvious. To do so, vary your typeface and type size, use white space to separate the warning or caution from surrounding text, or box the warning or caution.
- **Definitions**—What does *caution* mean? How does it differ from *warning*, *danger*, or *note*?

Follow the hierarchy of definitions, which clarifies the degree of hazard:

1. Danger—The potential for death.
2. Warning—The potential for serious personal injury.
3. Caution—The potential for damage or destruction of equipment.
4. Note—Important information, necessary to perform a task effectively or to avoid loss of data or inconvenience.

- **Colors**—Another way to emphasize your hazard message is through a colored window or text box around the word. *Note* is printed in blue or black, *Caution* in yellow, *Warning* in orange, and *Danger* in red.
- **Text**—To further clarify your terminology, provide the readers text to accompany your hazard alert. Your text should have the following three parts:
  1. A one- or two-word identification alerting the reader—Words such as "High Voltage," "Hot Equipment," "Sharp Objects," or "Magnetic Parts," for example, will warn your reader of potential dangers, warnings, or cautions.
  2. The consequences of the hazards, in three to five words—Phrases like "Electrocution can kill," "Can cause burns," "Cuts can occur," or "Can lead to data loss," for example, will tell your readers the results stemming from the dangers, warnings, or cautions.
  3. Avoidance steps—In three to five words, tell the readers how to avoid the consequences noted: "Wear rubber shoes," "Don't touch until cool," "Wear protective gloves," or "Keep disks away."
- **Icons**—Equipment is manufactured and sold globally; people speak different languages. Your hazard alert should contain an icon—a picture of the potential consequence—to help everyone understand the caution, warning, or danger.

Dangerous Equipment

Electrical Danger

Eye Protection

Discussion

Itemize and thoroughly discuss the steps in your instructions. Organize them *chronologically*—as a step-by-step sequence. To operate machinery, monitor a system, or construct equipment, your readers must follow a chronological sequence.

Conclusion

As with a description, you can conclude your instructions in various ways. You can end your instructions with a (1) comment about warranties; (2) sales pitch highlighting the product's ease of use; (3) reiteration of the product's applications; (4) summary of the company's credentials; (5) troubleshooting guide; (6) frequently asked questions (FAQs); and (7) corporate contact information.

Use Graphics to Highlight Steps

Clarify your points graphically. What the reader has difficulty understanding, your graphic can help explain pictorially.

Technology Tip

USING SCREEN CAPTURES TO DEPICT STEPS VISUALLY

Here's a good, simple way to visually depict steps in your instruction—screen captures. A screen capture lets you copy any image present on your computer monitor. To make screen captures, follow these steps:

Capturing the Image

1. Find the graphic you want to include in your instructions.
2. Press the "Print Screen" key on your computer keyboard.

Cropping the Image

Pressing the "print screen" key captures the *entire* image seen on the monitor. This might include images that you do not want to include in your instruction. To crop your image, follow these steps:

1. Paste the "print screen" image into a graphics program like MS Paint.
2. Click on the "Select" icon.

3. Click and drag the "Select" tool to choose the part of the captured graphic you want to crop.

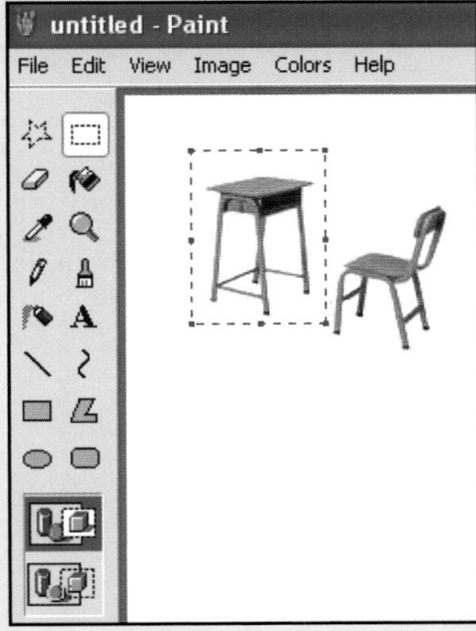

4. Copy and paste this cropped image into your user manual.

"Selected" portion of the graphic chosen for cropping.

Note: When capturing screenshots, be careful to avoid infringing upon a company's copyright to the image. Make sure that your use of the image meets the principle of "fair use" (which permits the use of images for criticism, comment, news reporting, teaching, scholarship, or research—Copyright Act of 1976, 17 U.S.C. § 107).

TECHNIQUES FOR WRITING EFFECTIVE INSTRUCTIONS

To write easy-to-understand instructions, follow these criteria for effective style:

1. **Number your steps**—Do not use bullets or the alphabet. Numbers, which you can never run out of, help your readers refer to a specific step. In contrast, if you use bullets, your readers have to count to locate steps—seven bullets for step 7, and so on. If you use the alphabet, you will be in trouble when you reach step 27.
2. **Use highlighting techniques**—Boldface, different font sizes and styles, emphatic warning words, color, and italics call attention to special concerns. A danger, caution, warning, or specially required technique must be evident to your reader. If this special concern is buried in a block of unappealing text, it will not be read. This could be dangerous to your reader or costly to you and your company. To avoid lawsuits or to help your readers see what is important, call it out through formatting.
3. **Limit the information within each step**—Don't overload your reader by writing lengthy steps.

BEFORE	AFTER
Overloaded Steps Start the engine and run it to idling speed while opening the radiator cap and inserting the measuring gauge until the red ball within the glass tube floats either to the acceptable green range or to the dangerous red line.	**Separated Steps** 1. Start the engine. 2. Run the engine to idling speed. 3. Open the radiator cap. 4. Insert the measuring gauge. 5. Note whether the red ball within the glass tube floats to the acceptable green range or up to the dangerous red line.

4. **Develop your points thoroughly**—Clarify your content by providing precise details.

BEFORE	AFTER
After rotating the discs correctly, grease each with an approved lubricant.	1. Rotate the disks clockwise so that the tabs on the outside edges align. 2. Lubricate the discs with 2 oz of XYZ grease.

5. **Use short words and phrases**—Conciseness in workplace communication helps to create easy-to-understand documents.
6. **Begin your steps with verbs—the imperative mood**—Note that each of the numbered steps in the following example begins with a verb.

> **Verbs Begin Steps**
>
> 1. *Number* your steps.
> 2. *Use* highlighting techniques.
> 3. *Limit* the information within each step.
> 4. *Develop* your points thoroughly.
> 5. *Use* short words and phrases.
> 6. *Begin* your steps with a verb.

7. **Personalize your text**—Involve your readers in instructions by using pronouns ("you," "your," "our," etc.).

8. Do not omit articles (a, an, the).

BEFORE	AFTER
1. Press right arrow button to scroll through list of programs. 2. Select program to scan. 3. Place item to scan face down on scanner glass in upper left corner.	1. Press **the** right arrow button to scroll through **a** list of programs. 2. Select **the** program you want to scan. 3. Place **an** item to scan face down on **the** scanner glass in **the** upper left corner.

See Figures 15.4 and 15.5 for sample instructions from Hewlett-Packard and Crate and Barrel.

SAMPLE INSTRUCTIONS

FIGURE 15.4 HP Officejet Pro K550 Series Color Printer

Selecting print media

The printer is designed to work well with most types of office paper. It is best to test a variety of print media types before buying large quantities. Use HP media for optimum print quality. Visit HP website at www.hp.com for details on HP Media.

- Load only one type of media at a time into a tray.
- For tray 1 and tray 2, load media print-side down and aligned against the right and back edges of the tray.

Loading media

This section provides instructions for loading media into the printer.

To load tray 1 or tray 2

1. Pull the tray out of the printer by grasping under the front of the tray.

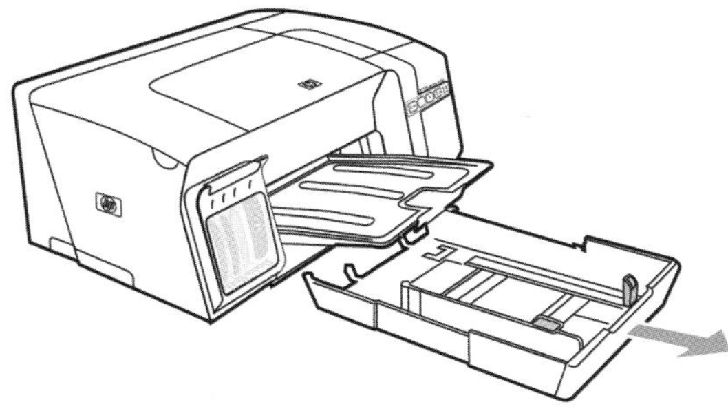

2. For paper longer than 11 inches (279 mm), lift the front cover of the tray (see shaded tray part) and lower the front of the tray.

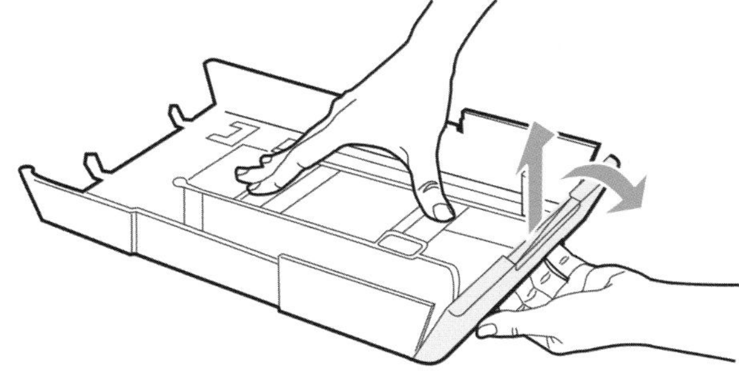

(Continued)

FIGURE 15.4 HP Officejet Pro K550 Series Color Printer (Continued)

3. Insert the paper print-side down along the right of the tray. Make sure the stack of paper aligns with the right and back edges of the tray, and does not exceed the line marking in the tray.

NOTE Tray 2 can be loaded only with plain paper.

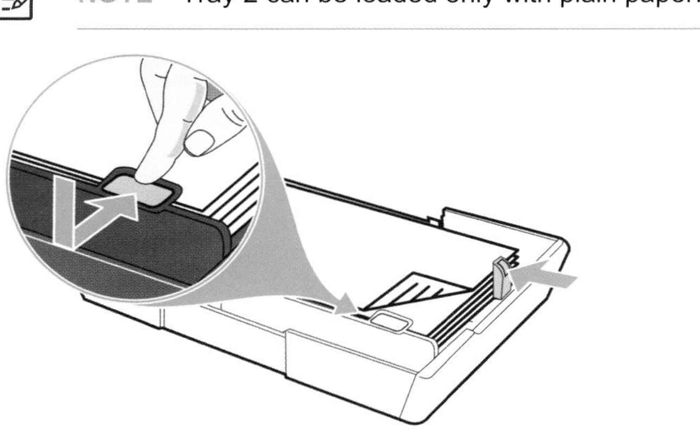

4. Slide the paper guides in the tray to adjust them for the size that you have loaded.
5. Gently reinsert the tray into the printer.

CAUTION If you have loaded legal-size or longer media, keep the front of the tray lowered. Damage to the media or printer might result if you raise the front of the tray with this longer media loaded.

6. Pull out the extension on the output tray.

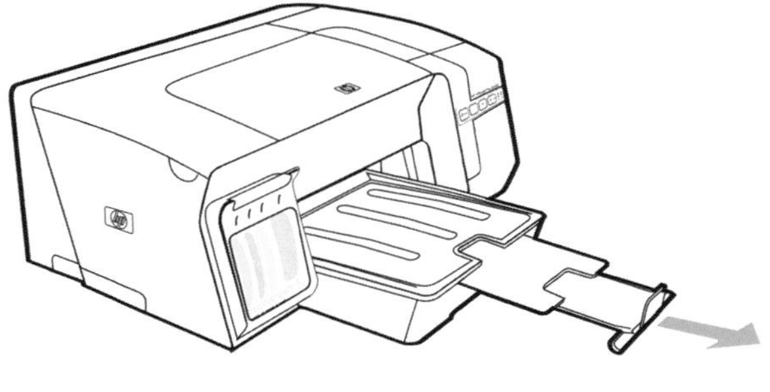

Courtesy of Hewlett-Packard.

FIGURE 15.5 Elements Bookcase Instructions Crate and Barrel

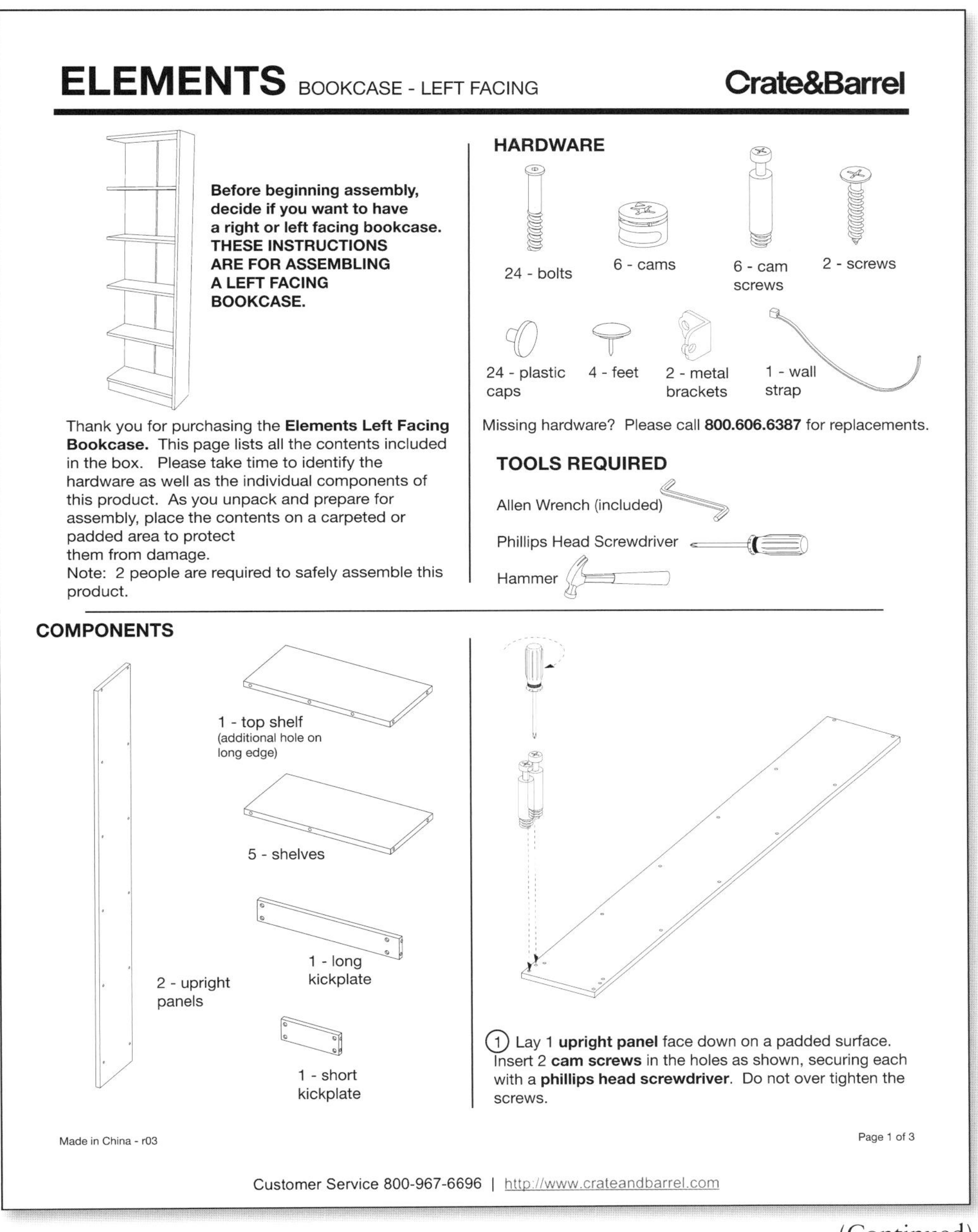

(Continued)

FIGURE 15.5 **Elements Bookcase Instructions Crate and Barrel (Continued)**

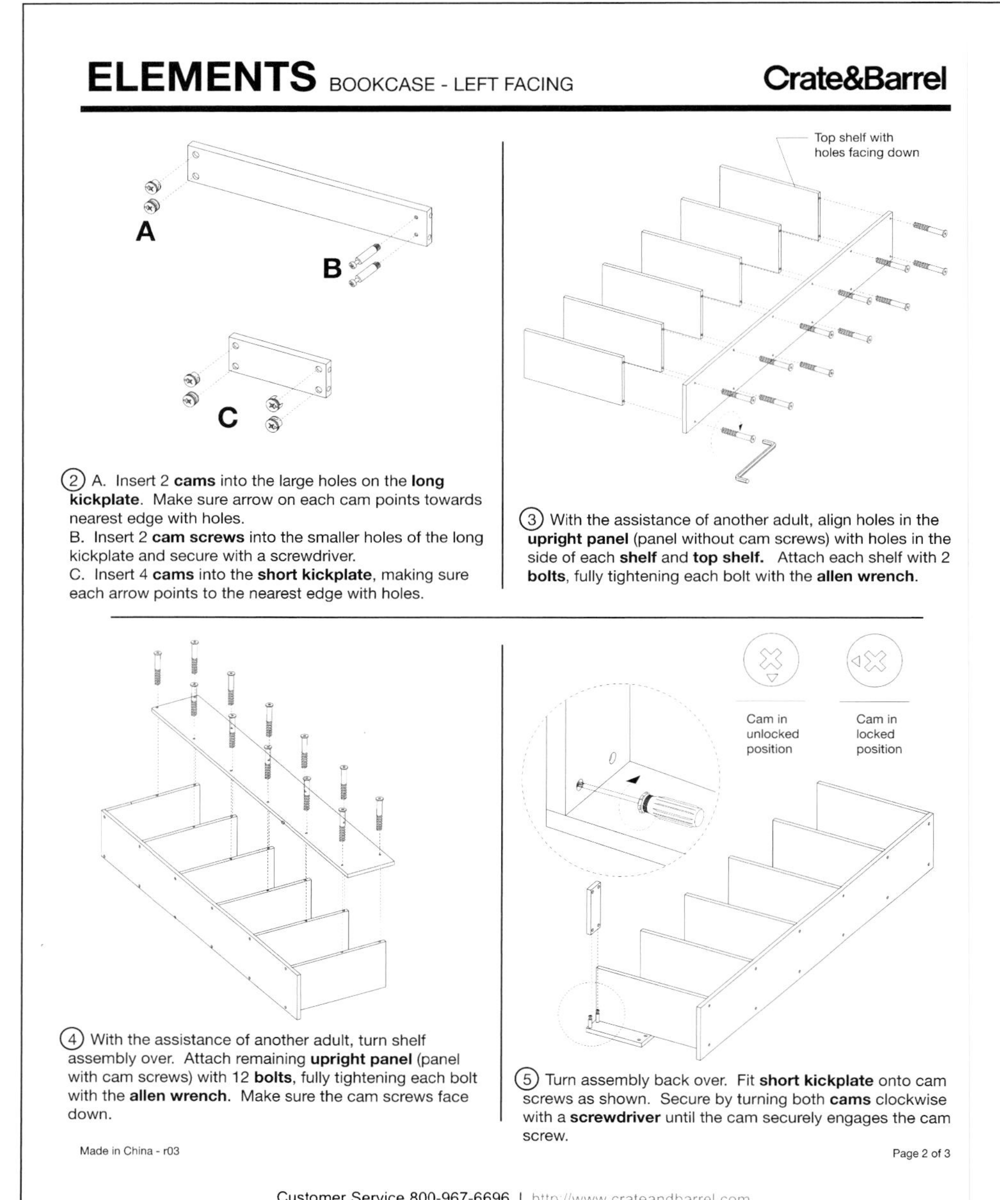

FIGURE 15.5 **Continued**

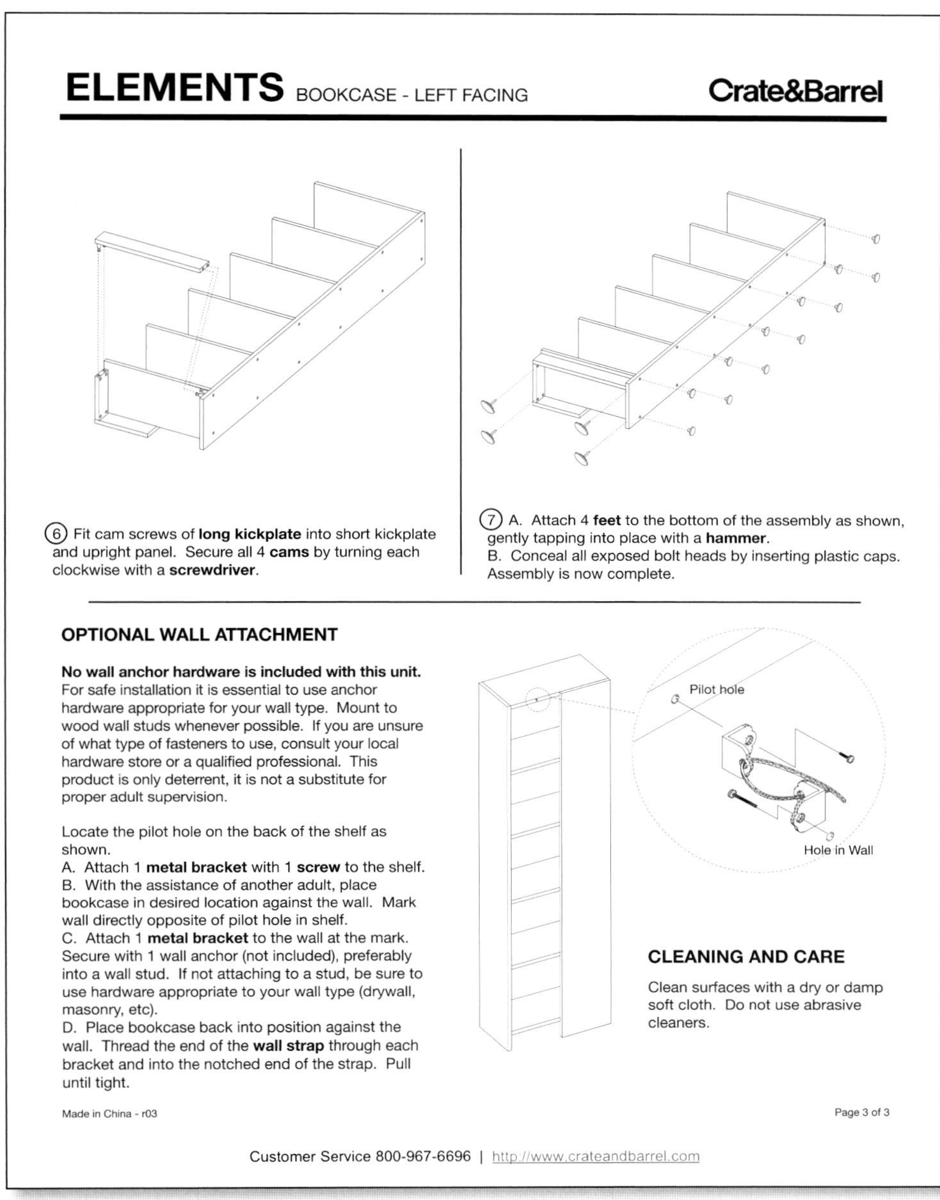

ELEMENTS BOOKCASE - LEFT FACING **Crate&Barrel**

(6) Fit cam screws of **long kickplate** into short kickplate and upright panel. Secure all 4 **cams** by turning each clockwise with a **screwdriver**.

(7) A. Attach 4 **feet** to the bottom of the assembly as shown, gently tapping into place with a **hammer**.
B. Conceal all exposed bolt heads by inserting plastic caps. Assembly is now complete.

OPTIONAL WALL ATTACHMENT

No wall anchor hardware is included with this unit. For safe installation it is essential to use anchor hardware appropriate for your wall type. Mount to wood wall studs whenever possible. If you are unsure of what type of fasteners to use, consult your local hardware store or a qualified professional. This product is only deterrent, it is not a substitute for proper adult supervision.

Locate the pilot hole on the back of the shelf as shown.
A. Attach 1 **metal bracket** with 1 **screw** to the shelf.
B. With the assistance of another adult, place bookcase in desired location against the wall. Mark wall directly opposite of pilot hole in shelf.
C. Attach 1 **metal bracket** to the wall at the mark. Secure with 1 wall anchor (not included), preferably into a wall stud. If not attaching to a stud, be sure to use hardware appropriate to your wall type (drywall, masonry, etc).
D. Place bookcase back into position against the wall. Thread the end of the **wall strap** through each bracket and into the notched end of the strap. Pull until tight.

CLEANING AND CARE

Clean surfaces with a dry or damp soft cloth. Do not use abrasive cleaners.

Made in China - r03

Page 3 of 3

Customer Service 800-967-6696 | http://www.crateandbarrel.com

Courtesy of Crate and Barrel.

INSTRUCTIONS CHECKLIST

_____ 1. Do the instructions have an effective title, clarifying the topic, and an introduction mentioning the topic and the reasons for performing the instructions?

_____ 2. Do the instructions include a list of required tools or equipment?

_____ 3. Are hazard alert messages used effectively?
- Are the hazard alert messages placed correctly?
- Is the correct term used ("Danger," "Warning," "Caution," or "Note")?

_____ 4. Are the instructions organized chronologically?

_____ 5. Does each step begin with a verb, avoid overloading by presenting one clearly defined action, and include articles?

_____ 6. Do the instructions clearly explain each point?

_____ 7. Do the instructions recognize the audience effectively?
- Are the instructions written for a specialist, semi-specialist, or lay audience?
- Is the text personalized using pronouns?

_____ 8. Does the instructions' document design use highlighting techniques effectively, such as graphics and headings?

_____ 9. Do the instructions use short words, short sentences, and short paragraphs?

_____ 10. Are the instructions free of grammatical errors?

Stacy Met Her Challenge

To meet her communication challenge, Stacy used the P^3 process.

Planning

To plan her manual, Stacy considered the following:

- Goals—create manuals for a software training conference
- Audience—approximately 200 beginning, intermediate, and advanced software users
- Channels—hard-copy text and computer-mediated training
- Data—drawn from existing online manuals and discussion with coworkers

Stacy plans each manual by outlining the content. "I had 48 manuals to create, and I didn't want to duplicate my efforts. I created an outline for each session for approval by my boss." Though Stacy draws from existing online manuals, she tailors her new training manuals to meet "specific, commonly used tasks." When Stacy is not sure which tasks her end users commonly use, she talks to her colleagues who work with the clients. The coworkers know which parts of the software cause problems for the users.

Stacy says, "One of my manuals focuses on how to maintain records of a city's street signs. To plan for this training manual, I'm creating a brief outline." Figure 15.6 shows Stacy's outline.

Packaging

To package the training manual, Stacy writes the rough draft in Figure 15.7. She organizes the text chronologically, step-by-step, from beginning to end. She also adds "notes" to explain difficult concepts or material.

Perfecting

When Stacy perfects her manuals, she performs usability testing to make sure the software works. Stacy says, "As I am writing the manuals, I am simultaneously performing the task in the software." Usability testing allows her to accomplish three goals:

- She adds screen captures so "the user can see what actually needs to be done."
- Usability testing ensures that "the steps are correct and in the right order."
- Stacy makes sure that "everything works and that no steps are omitted."

Stacy says, "We recently held our yearly conference with our clients. Overwhelmingly, they commented on how helpful the training manuals were proving to be. The conference follow-up evaluations especially commended the conciseness of the manuals; the visual appeal of graphics, space, and itemized tasks; and the feeling that I wrote with audience in mind." Figure 15.8 shows Stacy's perfected manual.

FIGURE 15.6 Street Signs Outline

Goals: Learn how to

1. Add records to the sign library
2. Add details to the record
3. Link to uniform manual
4. Add pictures
5. Link to streets
6. See how this works together with the other transportation programs to maintain your street assets

FIGURE 15.7 Rough Draft of Sign Library

The *Sign Library* module stores data and graphics for each street sign that exists in inventory. The information in the *Sign Library* module is then carried over to your *Sign Inventory* records to help you track and maintain your assets. This helps standardize your system and speed data entry. Follow the steps below to create a *Sign Library* record.

1. Select **Transportation>>Sign Master>>Sign Library** from the gbaMS main menu.
   - Click *GO* to bypass the filter. Then, click the Add button to create a record.
2. In the module header, you can enter the unique MUTCD code used to identify the record. This is important (and is a required field).
   - For our example, we've created a record for a stop sign. We've used the MUTCD code, R1-1, to identify this record.
3. Fill out the remaining fields with the class, subclass, shape, legend color, background color, and dimensions of the sign.
   - For our stop sign, we've indicated its standard features. Specifically, we've noted that it's a regulatory sign, octagonal in shape, with a white legend on a red background.
4. You can also include an image of the sign from a picture file (jpg, gif, etc.).
5. Save the record. You can now select this sign from a pick list in *Sign Inventory*.

FIGURE 15.8 Perfected Sign Library Instruction

The *Sign Library* module stores data and graphics for each street sign that exists in your city's inventory. The information in the *Sign Library* module is then carried over to your *Sign Inventory* records to help you track and maintain your assets. This helps standardize your system and speed data entry. Follow the steps below to create a *Sign Library* record.

1. Select **Transportation>>Sign Master>>Sign Library** from the gbaMS main menu.
2. Click *GO* to bypass the filter.
3. Click the Add button  to create a record.
4. In the module header, enter the unique MUTCD (Manual on Uniform Traffic Control Devices) code used to identify the record.
5. Fill out the remaining fields with the class, subclass, shape, legend color, background color, and dimensions of the sign.
6. Include an image of the sign from a picture file (jpg, gif, etc.).
7. Save the record. You can now select this sign from a pick list in *Sign Inventory*.

(Continued)

Stacy says, "I was on the elevator the other day during a break in our annual product information and training conference. My boss was also on the elevator and told me how impressed the customers were with the user manuals I had created for all the new software. I knew I had succeeded when my boss said, 'Thanks so much for your hard work, Stacy. We've gotten great reviews of your manuals.' He also said they had been hosting these annual conferences since the company began five years ago, and he had never heard such positive comments about any of the manuals. Because I have only been with the company for a few months, his comments were particularly important to me. They let me know that I was doing my job and that planning, packaging, and perfecting all the documents was really paying off."

FIGURE 15.8 Perfected Sign Library Instruction (Continued)

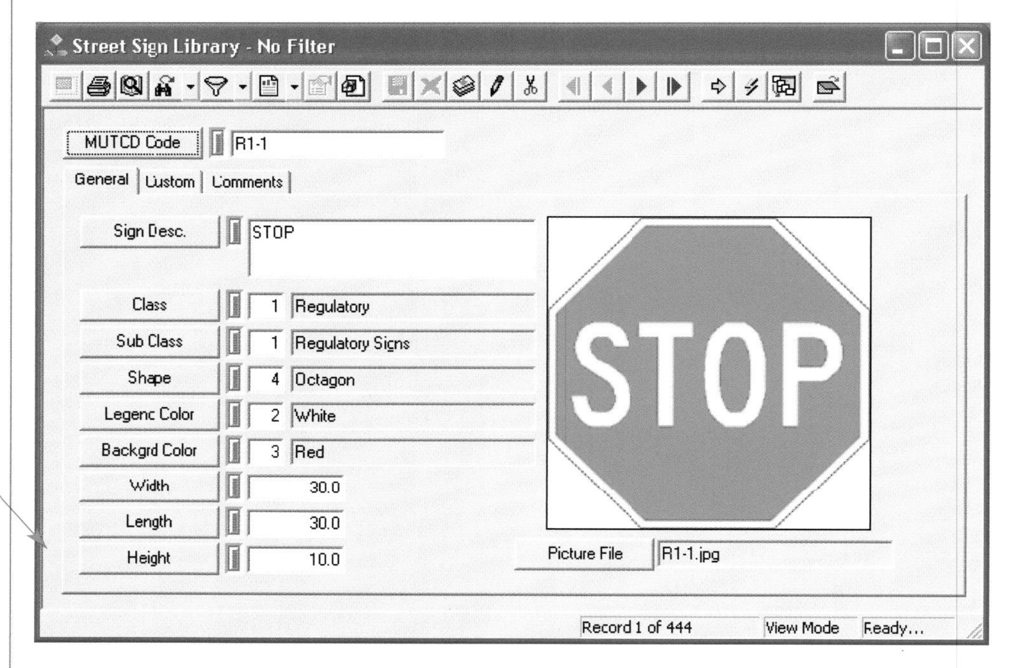

Courtesy of gbams.

CHAPTER HIGHLIGHTS

1. Technical descriptions are often used in user manuals, product specifications, reports, and sales literature.
2. Process analyses tell how a tool, mechanism, or piece of equipment works.
3. Use photographic or impressionistic words when writing a description.
4. Follow a spatial pattern of organization in a technical description and a chronological order for process analyses and instructions.
5. Number your steps and start each step with a verb when writing instructions.
6. Hazard alerts are important to protect your reader and your company. Place these alerts early in your instruction or before the appropriate step.
7. Graphics will help your reader see how to perform each step and understand a process analysis.
8. Highlighting techniques emphasize important points, thus minimizing damage to equipment or injury to its users.
9. Don't compress several steps into one excessively long and demanding step.
10. The P^3 process, complete with usability testing, will help you write an effective instruction, description, and process analysis.

WWW *Find additional description, instruction, and process analysis exercises, samples, and interactive activities at http://www.prenhall.com/gerson.*

CASE STUDIES

1. PhlebotomyDR needs to provide instructions for their staff (nurses, doctors, and technicians) regarding the correct procedures to ensure cleanliness and employee protection. These include hand washing procedures and the correct use of sterile equipment, such as gloves, masks, aprons, and shoe coverings.
   Following is a rough draft of one set of instructions for personnel safety.

Assignment

Revise the rough draft according to the criteria provided in this chapter. Correct the order of information, the grammar, and the instruction's content. In addition, improve the instruction by including appropriate graphics.

Hand washing

- Lather hands to cover all surfaces of hands and wrists.
- Wet hands with water.
- Rub hands together to cover all surfaces of hands and fingers. Pay special attention to areas around nails and fingers. Lather for at least 15 seconds.
- Dry thoroughly.
- Rinse well with running warm water.
- Avoid using hot water. Repeated exposure to hot water can lead to dermatitis.
- Use paper towels to turn off faucet.

Gloves

- Replace damaged gloves as soon as patient safety permits.
- Don gloves immediately prior to task.
- Remove and discard gloves after each use.

Masks

Wear masks and eye protection devices (goggles or eye shields) to avoid droplets, spray, or splashes and to prevent exposure to mucous substances. Masks are also worn to protect nurses, doctors, and technicians from infectious elements during close contact with patients.

Aprons and Other Protective Clothing

- Wear aprons or gowns to avoid contact with body substances during patient care procedures.
- Remove and dispatch aprons and other protective clothing before leaving the work area.
- Some work areas might require additional protective clothing such as surgical caps and shoe covers or boots.

2. Cathode Ray Tube Monitors versus Liquid Crystal Display Flat Panels. Nitrous Systems Biotechnology, Inc. plans to expand its company. Part of this expansion

includes hiring 50 new biotechnicians. Each, of course, will need his or her own desktop computer, complete with monitor. What kind of monitor should the company buy—CRTs or LCDs? LCD Flat Panels are sleek and take up less room, but CRTs are generally cheaper. To meet their computer technology needs, Nitrous has hired Alpha/Beta Consulting (A/B/C). Nitrous wants a brief, comparison/contrast report written, complete with a product description and a process analysis explaining the differences between these two monitor types.

Several questions need to be answered: Which monitor type provides the best resolution and viewing quality, refresh rates, and screen-viewable size options? Are CRTs cheaper, or has that price break changed? Do CRTs or LCDs provide greater energy savings? Finally, how do CRTs work versus LCDs?

Assignment

You are a CIS/IT tech who works for A/B/C. Research this topic and write the description, complete with process analysis, to meet the client's request.

ETHICS IN THE WORKPLACE CASE STUDY

Gary Porter is the human resources manager at a biomedical technology company. The company has an anti-nepotism policy for full-time employees. Gary's wife, Deb, is a self-employed consultant who makes presentations to corporations about changes in insurance benefits. She has an MBA from the state university, over 20 years experience in insurance, and a long list of well-known clients. Each participant in her training seminars receives an instructional manual on how to compare options and rates for benefits packages. As a consultant, Deb receives no benefits or regular pay from her clients. She is hired on an as-needed, contract basis.

Gary's company needs to review its insurance premiums and benefits in the hopes of lowering costs. He wants to bring in an expert in the field. Deb meets Gary's company's requirements for experience, knowledge, and ability to create user manuals.

Question: Is it ethical for Gary to hire his wife as a consultant? Justify your answer based on information provided in Chapter 1.

INDIVIDUAL AND TEAM PROJECTS

1. Write a set of instructions. To do so, first select a topic. You can write instructions telling how to monitor, repair, test, package, plant, clean, operate, manage, open, shut, set up, maintain, troubleshoot, install, use software, and so on. Choose a topic from your field of expertise or one that interests you. Follow the *P*³ process techniques to complete your instructions.

2. Find examples of instructions for consumer products. These can include instructions for assembling children's toys, refinishing furniture, insulating attics or windows, setting up stereo systems, flushing out a radiator, installing ceiling fans, and so on. Once you find some examples, bring them to class. Then, applying the criteria for good instructions presented in this chapter, determine the success of the examples. If they are successful, explain why and how. If they fail, show where the problems are and rewrite the instructions to improve them.

3. Microsoft Word's Help program provides instructional steps for hundreds of word processing operations. Access the Word Help Index to find instructions. Open 3 to 5 of these instructions and accomplish the following:

   - Discuss how they are similar to or different from the instructions discussed in this chapter.
   - Select one of the instructions and rewrite it, adding graphics (screen captures), cautions or notes, and a glossary of terms.

4. Write a description, either individually or as a team. To do so, first select a topic. You can describe any tool, mechanism, or piece of equipment. However, don't choose a topic too large to describe accurately. To provide a thorough and precise description, you will need to be exact and minutely detailed. A large topic, such as a computer, an oscilloscope, a respirator, or a Boeing airliner, would be too demanding for a two- to four-page description. However, do not choose a topic that is too small, such as a paper clip, a nail, or a shoestring. Choose a topic that provides you with a challenge but that is manageable. You might write about any of the following topics:

USB flash drive	Computer disk	DVD remote control
Wrench	Computer mouse	Mechanical pencil
Screwdriver	Light bulb	Ballpoint pen
Pliers	Calculator	Computer monitor
Wall outlet	Automobile tire	Cell phone

Once you or your team have chosen a topic, plan (listing the topic's components), write a draft to package the content (abiding by the criteria provided in this chapter), and perfect (revising your draft).

5. Write a process analysis, either individually or as a team, on the following topics:
   - How does blood coagulate?
   - How does an x-ray machine work?
   - How are viruses spread?
   - How do WiFi connections work?
   - How do rotary engines work?
   - How do fuel gauges work?
   - How is metal welded?
   - How does a metal detector work?
   - How do browsers work?
   - How do computer viruses work?

DEGREE-SPECIFIC ASSIGNMENT

In degree-specific teams, choose a topic from your area of interest and expertise to describe (including a process analysis) or to write an instruction for. For example, students majoring in BIS could describe how a software program, flash drive, or wireless mouse works. Biotech and nursing students could describe and provide a process analysis for a nebulizer, blood pressure monitor, or glucometer. Marketing students could explain the steps for marketing a new product or service. Accounting students could provide the steps for responding to an IRS challenge. Management students could provide the steps for hiring, training, or terminating employees. Finance students could focus on financial planning and create the steps necessary to achieve a client's financial goals.

Once the teams have decided on a topic, write the instruction, description, or process analysis required.

PROBLEM-SOLVING THINK PIECES

1. Good writing demands revision. Following are flawed instructions. Rewrite the text, following the criteria for instructions and the perfecting techniques included in this chapter.

Date: November 1, 2009
To: Maintenance Technicians
From: Second Shift Supervisor
Subject: Oven Cleaning

The convection ovens in kiln room 33 needs extensive cleaning. This would consist of vacuuming and wiping all walls, doors, roofs, and floors. All vents and dampers need to be removed and a tack cloth used to remove loose dust and dirt. Also, all filters need replacing. I am requesting this because when wet parts are placed in the ovens to cure the paint, loose particles of dust and dirt are blown onto the parts, which causes extensive rework. I would like this done twice a week to ensure cleanliness of product.

2. Read the following instructional steps. Are they in the correct chronological order? How would you reorder these steps to make the instructions more effective?

Changing Oil in Your Car

Run the car's engine for approximately ten minutes and then drain the old oil.

Park the car on a level surface, set the parking brake, and turn off the car's engine.

Gather all of the necessary tools and materials you might need.

Open the hood.

Jack up and support the car securely.

Place the funnel in the opening and pour in the new oil.

Replace the cap when you have finished pouring in new oil.

Locate the oil filler cap on top of the engine and remove the cap.

Tighten the plug or oil filter if you find leakage.

Run the engine for a minute, then check the dipstick. Add more oil if necessary.

Pour the used oil into a plastic container and dispose of it safely and legally.

3. Read the following process analysis and reorganize the numbered sentences to achieve a clear, chronological order. If necessary, research this process to learn more about how blood clots.

How does blood clot?

1. In our bodies, blood can clot due to platelets and the thrombin system.
2. When bleeding occurs, chemical reactions make the surface of the platelet "sticky." These sticky platelets adhere to the wall of blood vessels where bleeding has occurred.
3. Platelets, tiny cells created in our bone marrow, travel in the bloodstream and wait for a bleeding problem.
4. Soon, a "white clot" is formed, so called because the clotted platelets look white.
5. Blood clot consists of both platelets and fibrin. The fibrin strands bind the platelets and make the clot stable.
6. In the thrombin system, several blood proteins become active when bleeding occurs. Clotting reactions produce fibrin, long, sticky strings. These sticky strands catch red blood cells and form a "red clot."
7. Primarily, arteries clot due to platelets, while veins clot due to our thrombin system.

WEB WORKSHOP

1. Review any of the following Web site's online instructions. Based on the criteria provided in this chapter, are the instructions successful or not?
   - If the answer is yes, explain why and how the instructions succeed.
   - If the answer is no, explain why the instructions fail.
   - Rewrite any of the flawed instructions to improve them.

Web Sites	Topics
http://www.hometips.com/diy.html	Electrical systems, plumbing, kitchen appliances, walls, windows, roofing, and siding
http://www.hammerzone.com	Kitchen projects, tubs, sinks, toilets, showers, and water heaters
http://dmoz.org/Home/Home_Improvement/	Links to step-by-step procedures for painting, welding, soldering, plumbing, walls, windows, and door repair and installation
http://directory.google.com/Top/Home/Home_Improvement/	Links to sites for instructions on decorating, electrical, flooring, furniture, lighting, painting, plumbing, welding, windows, and doors
http://www.quakerstate.com/pages/carcare/oilchange.asp	Instructions for changing oil
http://www.csaa.com/yourcar/takingcareofyourcar	AAA instructions for car care, including general maintenance and checking fluids, hoses, drive belts, electrical systems, and tires
http://www.gateway.com/index.shtml	Access the FAQs for upgrading systems or correcting problems with printers, drivers, monitors, memory, and more

2. The U.S. Department of Labor article "Hazard Communication: A Review of the Science Underpinning the Art of Communication for Health and Safety" can be found at http://www.osha.gov/SLTC/hazardcommunications/hc2inf2.html#*.1.4. This article focuses on many important aspects of writing instructions, including the use of icons, readability, and audience variables. Access the article and report your findings either orally or in a memo, letter, or e-mail message.

QUIZ QUESTIONS

1. What are five reasons you would include instructions for your audience?
2. What should you include in the introduction of an instruction, description, or process analysis?
3. How can you create an accessible hazard alert?
4. What do graphics achieve in instructions, descriptions, and process analyses?
5. What is a description?
6. What is a process analysis?
7. What is the best organizational pattern for an instruction?
8. What is the best organizational pattern for a process analysis?
9. What is the best way to organize a description?
10. What is the difference between photographic and impressionistic words?

CHAPTER 16

Research and Documentation

▸REAL PEOPLE *in the workplace*

Tom Woltkamp is senior manager of information solutions for Teva Neuroscience, a global leader in the "development, production and marketing of generic and proprietary branded pharmaceuticals" (http://www.tevapharm.com/). Teva Pharmaceutical Industries is home based in Israel; Teva Neuroscience, a subsidiary of Teva Pharmaceuticals, is based in Kansas City, Missouri, with branches in Florida, Pennsylvania, Texas, North Carolina, California, Minnesota, and Canada.

Approximately 20 employees and consultants report to Tom regarding computer infrastructure (computer networks, help desks, call centers, hardware/software purchases) and application development for programming, software integration, installation, configuration, Web programming, and customized documentation.

On an ongoing basis, Teva employees in finance, human resources, sales, and the company's call centers need new software applications. They call on Tom's group to find solutions to their needs. The research and documentation process for Teva's software development is as follows:

- **Project charter—**These brief reports "drive the research." In a page or two, the Teva departments with whom Tom works "write down what they want." They clarify their needs, as Tom says, by stating "Here's my problem," "Here's what I think I want to do about it," and "Here's what I think the benefits will be."

That's when Tom begins his research into finding solutions for their software needs. Tom and his team conduct both primary and secondary research:

- **Questionnaires—**Tom and his group create FAQ checklists where they "play 20 questions to find what our end users want." They meet with their clients and discuss the project parameters.
- **Internet searches—**After the initial questionnaire, which helps zero in on the client's needs, Tom and his team search the Internet for "knowledge and solutions." They want to find out more about the topic of research and see if any products already exist that can meet their needs or if software is available they can "tweak" for customization.

CHAPTER GOALS

When you complete this chapter, you will be able to do the following:

1. Understand why to conduct research in your workplace communication.
2. Use both *primary* and *secondary* research in your workplace communication.
3. Locate information in the library and online.
4. Document your sources of information to avoid plagiarism.
5. Evaluate your researched material using a checklist.

WWW

To learn more about research, visit our Web Companion at http://www.prenhall.com/gerson.

- **Consultation—**Another avenue of research involves meeting with consultants—experts in the field—for their take on the topic.
- **Interviews—**Tom works with "key partners," including banks, vendors, professional organizations, and patients using the company's pharmaceuticals. By interviewing these constituents, Tom can learn more about how other entities might have solved similar problems or how the customers using a drug can be better served.
- **Online and hard-copy journals—**An excellent secondary source of research for Tom is professional journals, such as *PC Magazine*, *Call Center Magazine*, and *CRM Magazine* (which focuses on customer relations management).

Tom and his team spend approximately 20 to 30 percent of their work time on research. Why? Research allows Tom to make sure that the software solutions he and his team design and develop are correct. That is, research helps Tom ensure that his products are on target to meet the client's needs, will be time and cost efficient, and meet national, international, and industry standards. Research helps Tom make certain that he can achieve customer satisfaction and provide value to his organization.◀

Tom's Communication Challenge

Because Tom's projects usually impact many groups and departments in the company, his constant challenge is to keep an open line of communication with every person who might have a stake in the decisions he makes. Tom says, "I've learned from experience not to get too far into a project without the blessing of everyone who might bring the project to a grinding halt. My preferred method of communication is e-mail because it is fast, efficient, and it leaves a paper trail.

For one recent project, I was asked by sales management to provide wireless Internet connections for the company's 250 sales representatives and sales managers who live all over the country. Unfortunately, I couldn't just go to the nearest Verizon store and buy 250 wireless cards to distribute and call it a day. I was challenged to research and find the best possible, most cost-effective wireless options for the sales reps, and I had to perform the research in a timely manner."

See pages 463–465 to learn how Tom met his communication challenge.

WHY CONDUCT RESEARCH?

Research skills are important in your school or work environment. You may want to perform research to better understand a technical term or concept; locate a magazine, journal, or newspaper article for your supervisor; or find data on a subject to prepare an oral or written report. Technology is changing so rapidly that you must know how to do research to stay up-to-date.

You can research information using online catalogs; online indexes and databases; CD-ROM indexes and databases; reference books in print, online, and CD-ROM format; and by using Internet search engines and directories. Reference sources vary and are numerous, so this chapter discusses only research techniques.

Research, a major component of long, formal business reports, helps you develop your report's content. Often, your own comments, drawn from personal experience, will lack sufficient detail, development, and authority to be sufficiently persuasive. You need research for the following reasons:

- To create content
- To support commentary and content with details
- To prove points
- To emphasize the importance of an idea
- To enhance the reliability of an opinion
- To show the importance of a subject to the larger business community
- To address the audience's need for documentation and substantiation

USING RESEARCH IN REPORTS

Additional information:

See Chapter 18, "Long, Formal Reports."

You can use researched material to support and develop content in your formal reports.

Use quotes and paraphrases to develop your content. Workplace communicators often ask how much of a formal report should be *their* writing, as opposed to researched information. A general rule is to lead into and out of every quotation or paraphrase with your own writing. In other words

- Make a statement (your sentence).
- Support this generalization with a quotation or paraphrase (referenced material from another source).
- Provide a follow-up explanation of the referenced material's significance (your sentences).

If you follow this approach, you can successfully blend researched material with other content in the report. Doing so allows you to minimize excessive reliance on research.

Research Including Primary and Secondary Sources

Research material generally breaks down into two categories: primary and secondary research. *Primary research* is research performed or generated by you. You do not rely on books or periodicals for this type of research. Instead you create original research by preparing a survey or a questionnaire targeting a group of respondents, by networking to discover information from other individuals, by visiting job sites, or by performing lab experiments. You may perform this research to determine for your company the direction a new marketing campaign should take, the importance of diversity in the workplace, the economic impact of relocating the company to a new office site, the usefulness of a new product, or the status of a project. You may also need to interview people for their input about a particular topic. For example, your company might be considering a new approach to take for increased security on employee computers. You could ask employees for a record of their logs which would highlight the problems they have encountered with their computer security. With primary research, you will be generating the information based on data or information from a variety of sources that might include observations, tests of equipment, interviews, networking, surveys, and questionnaires.

When you conduct *secondary research*, you rely on already printed and published information taken from sources including books, periodicals, newspapers, encyclopedias, reports, proposals, or other business documents. You might also rely on information taken from a Web site or a blog. All of this secondary research requires parenthetical source citations (discussed later in this chapter).

CRITERIA FOR WRITING RESEARCH REPORTS

As with all types of workplace communication, writing using researched material requires that you

- Recognize your audience
- Use an effective style, appropriate to your reader and purpose
- Use effective formatting techniques for reader-friendly ease of access

Audience

When writing a research report, you first must recognize the level of your audience. Are your readers specialists, semi-specialists, or lay? If you are writing to your boss, consider his or her level of technical expertise. Your boss probably is at least a semi-specialist reader, but you must determine this based on your own situation. Doing so will help you determine the amount of definition, explanation, and examples necessary for effective communication.

You must decide next whether your reader will understand the purpose of your research report. This will help you determine the amount of detail needed and the tone to take (persuasive or informative). For instance, if your reader has requested the information, you will not need to provide massive amounts of background data explaining the purpose of your research. Your reader has probably helped you determine the scope and purpose. However, if your research report is unsolicited, your first several paragraphs must clarify your rationale. You will need to explain why you are writing and what you hope to achieve.

If your report is solicited, you probably know what your reader plans to do with your research. He or she will use it for a briefing, an article to be written for publication, a technical update, and so forth. Thus, your presentation will be informative. If your research report is unsolicited, however, your goal is to persuade your reader to accept your hypotheses substantiated through your research.

Additional information:

See Chapter 3, "Meeting the Needs of the Audience."

Effective Style

Research reports should be more formal than many other types of workplace communication. In a research report, you are compiling information, organizing it, and presenting your findings to your audience using documentation. Because the rules of documentation are structured rigidly (to avoid plagiarism and create uniformity), a research report is also rigidly structured.

The tone need not be stuffy, but you should maintain an objective distance and let the results of your research support your contentions. Again, considering your audience and purpose will help you decide what style is appropriate. For example, if your boss has requested your opinion, then you are correct in providing it subjectively. However, if your audience has asked for the facts—and nothing but the facts—then your writing style should be more objective.

Formatting

Additional information:

See Chapter 5, "Packaging Workplace Communication Through Effective Document Design."

Reading a research report is not always an easy task. As the writer, you must ensure that your readers encounter no difficulties. One way to achieve this is by using effective formatting. Reader-friendly ease of access is accomplished when you use highlighting techniques such as bullets, numbers, headings, subheadings, and graphics (tables and figures).

In addition, effective formatting includes the following:

- Overall organization (including an introduction, discussion, and conclusion)
- Internal organization (various organizational patterns, such as problem/solution, comparison/contrast, analysis, and cause/effect)
- Parenthetical source citations
- Works cited—documentation of sources

HOW TO PLAN YOUR RESEARCH

To begin a research project, follow these techniques:

1. **Select a general topic (or the topic you have been asked to study)**—Your topic may be a term, phrase, innovation, or work-related problem. If you are in public relations, for example, you might want to focus on the current problems a company is having with its image. If you're in accounting, you might want to research new tax laws for property valuation. If you work in banking, your boss might need researched information about challenges facing the Federal Reserve and prime lending rates. If your field is computer science, you could focus on wireless technologies.
2. **Locate sources of information**—Check a library or online sources to find material that relates to your subject. A quick review of your library's online periodical databases, such as ProQuest, Infotrac, or other equivalent sources, will help you locate periodical articles. Your library may also have a print or online edition of the *Reader's Guide to Periodical Literature* and other similar general or specialized periodical indexes or databases. Most libraries now have online catalogs to help you easily search for books and other materials, such as VHS, DVD, ebooks, and more on your topic. A keyword search of Internet metasearch engines will give you an idea of how much information might be readily available through the Internet.
3. **Establish a focus**—After you have chosen a topic for which you can find available source material, decide what you want to learn about your topic. A focus statement can guide you. In other words, if you are interested in marketing, you might write a focus statement such as the following:

example

I want to research current information about analyzing markets, including market size, growth rate, profitability, cost structure, and distribution channels.

For business information systems, you could write the following:

example

I want to research the challenges inherent in implementing wireless technologies, including standard or proprietary call control protocols, architectures designed for a constant Ethernet network connection, and certain protocol features relying on Ethernet implicit characteristics.

With focus statements such as these, you can begin researching your topic, concentrating on articles pertinent to your topic.

HOW TO RESEARCH YOUR TOPIC

You can find information about your research topic in the following sources.

Books

All books owned by a library are listed in catalogs, usually online. Books can be searched in online catalogs in a variety of ways: by title, by author, by subject, by keyword, or by using some combination of these. No matter how you search for a book, the resulting record will look something like the following example.

example

Artificial intelligence : robotics and machine evolution / David Jefferis.

Database:	DeVry Institute of Technology
Main Author:	Jefferis, David.
Title:	Artificial intelligence : ***robotics*** and machine evolution / David Jefferis.
Primary Material:	Book
Subject(s):	***Robotics***—Juvenile literature. Artificial intelligence—Juvenile literature. ***Robotics.*** Robots. Artificial intelligence.
Publisher:	New York : Crabtree Pub. Co., 1999.
Description:	32 p. : col. ill. ; 29 cm.
Series:	Megatech
Notes:	Includes index. An introduction to the past, present, and future of artificial intelligence and ***robotics,*** discussing early science fiction predictions, the dawn of AI, and today's use of robots in factories and space exploration.
Database:	DeVry Institute of Technology
Location:	Dallas Main Stacks
Call Number:	TJ211.2.J44 1999
Number of Items:	1
Status:	Not Changed

Periodicals

Use online databases, CD-ROM resources, or print or online periodical indexes to find articles on your topic. Online indexes can be searched in a variety of ways: by title, by author, by subject, by keyword, or by using a combination of these. No matter how you search, the resulting record will look something like the following example.

example

This example from an online periodical database tells us that an article on the subjects Athletes, Bicycling, Organizations, and Bicycle racing can be found in the March 2001 issue of Bicycling *magazine, volume 42, issue 2, beginning on page 14. The article is titled "DH Pro Eric Carter" and was written by Andrew Juskaitis. The article contains information about a person named Eric Carter. An abstract, or summary of the article, is given, followed by the full text of the article itself.*

Help | ?

DH Pro Eric Carter
Bicycling; Emmaus; Mar 2001; Andrew Juskaitis;

Volume:	42
Issue:	2
Start Page:	14
ISSN:	00062073
Subject Terms:	Athletes Bicycling Organizations **Bicycle racing**
Personal Names:	Carter, Eric

Abstract:
Juskaitis discusses bicycling with downhill national champion Eric Carter. Only after riding with Carter did he realize he's also at the forefront of changing the face of gravity-fed mountain ***bike racing****. Carter has joined with promoter Rick Sutton to push cycling's race organizations to implement four-to-six-rider racing on highly specialized courses.*

Full Text: . . . (omitted here for copyright issues)

Indexes to General, Popular Periodicals

Most libraries provide access to at least one of a number of indexes covering popular, nontechnical literature and newspaper articles from a variety of subject fields. There are online, CD-ROM, and print counterparts for most of these. The online and CD-ROM indexes provide the full text of many of the articles. Following are indexes to popular periodicals:

- Reader's Guide to Periodical Literature
- ProQuest Research Library
- Ebscohost
- SIRS Researcher (emphasizes social issues)
- Newsbank (emphasizes newspaper articles)

Databases of Specialized, Scholarly, or Technical Periodicals

Many libraries provide access to one or more specialized databases covering literature from a variety of disciplines. There are online, CD-ROM, and print counterparts for most of these. The online and CD-ROM databases provide the full text of many of the articles. Following are examples of databases to specialized periodicals:

- **Applied Science & Technology Index**—Covers engineering, aeronautics and space sciences, atmospheric sciences, chemistry, computer technology and applications, construction industry, energy resources and research, fire prevention, food and the food industry, geology, machinery, mathematics, metallurgy, mineralogy, oceanography, petroleum and gas, physics, plastics, the textile industry and fabrics, transportation, and other industrial and mechanical arts.
- **Business Periodicals Index**—Covers major U.S. publications in marketing, banking and finance, personnel, communications, computer technology, and so on.
- **ABI/Inform**—Covers business and management.
- **General Science Index**—Covers the pure sciences, such as biology and chemistry.
- **Social Sciences Index**—Covers psychology, sociology, political science, economics, and other social sciences topics.

Table 16.1 Internet Research Sources

Search Engines	Metasearch Engines	Subject Directories
Google	Clusty	Librarians' Index
Yahoo	Dogpile	Infomine
Ask.Com	SurfWax	Academic Info
	Copernic Agent	About.Com
		Google Directory
		Yahoo

- ERIC (*Education Resources Information Center*)—Provides bibliography and abstracts about educational research and resources. Available for free through the Internet.
- MEDLINE—Covers medical journals and allied health publications.
- PsycINFO—Covers psychology and behavioral sciences.
- LexisNexis—Includes the full text of newspaper articles, reports, transcripts, law journals, and legal reporters and other reference sources in addition to general periodical articles.
- SIRS (*Social Issues Resources*)—Provides bibliography and full-text articles about social issues.

The Internet

Perhaps one of the largest sources of research available today is the Internet. Millions of documents from countless sources are found on the Internet. You can find material on the Internet published by government agencies, organizations, schools, businesses, or individuals, as shown in Table 16.1. The list of options grows daily. For example, nearly all newspapers and news organizations have online Web sites.

Directories

Directories, like Yahoo, Librarians' Index, Infomine, Academic Info, About.com, and Google Directory, let you search for information from a long list of predetermined categories, including the following:

Arts	Government	Politics and Law
Business	Health and Medicine	Recreation
Computers	Hobbies	Science
Education	Money and Investing	Society and Culture
Entertainment	News	Sports

To access any of these areas, click on the appropriate category and then "drill down," clicking on each subcategory until you get to a useful site.

Search Engines

Search engines, like Google, Yahoo, and Ask.com, let you search millions of Web pages by keywords. Type a word, phrase, or name in the appropriate blank space and press the Enter key. The search engine will search through documents on the Internet for "hits," documents that match your criteria. One of two things will happen: Either the search engine will report "no findings," or it will report that it has found thousands of sites that might contain information on your topic.

In the first instance, "no findings," you'll need to rethink your search strategy. You may need to check your spelling of the keywords or find synonyms. For example, if you want to research information about online writing, you could try typing "writing online," "online writing," "electronic writing," "writing electronically," and other similar terms. In the second instance, finding too many hits, you'll need to narrow your search.

Metasearch Engines

A metasearch engine, like Dogpile, Clusty, SurfWax, and Copernic Agent, lets you search for a keyword or phrase in a group of search engines at once, saving you the time of searching separately through each search engine.

Considerations When Searching the Internet

Researching the Internet presents at least two problems other than finding information. First, is the information you have found trustworthy? Paperbound newspapers, journals, magazines, and books go through a lengthy publication process involving editing and review by authorities. Not all that's published on the Internet is so professional. Be wary. What you read online needs to be filtered through common sense. Second, remember that although a book, magazine, newspaper, or journal can exist unchanged in print form for years, Web sites change constantly. A site you find today online will not necessarily be the same tomorrow. That's the nature of electronic communication.

HOW TO DOCUMENT YOUR RESEARCH

Your readers need to know where you found your information and from which sources you are quoting or paraphrasing. A quotation must be exactly the same as the original word, sentence, or paragraph. You cannot haphazardly change a word, a punctuation mark, or the ideas conveyed. In contrast, when you paraphrase, you rewrite a quote using your own words. Whether you use a quote or a paraphrase, however, you must correctly cite the source. Therefore, you must document this information. Correct documentation is essential for the following reasons:

- You must direct your readers to the books, periodical articles, and online reference sources that you have used in your research report. If your readers want to find these same sources, they depend on your documentation. If your documentation is incorrect, readers will be confused. Instead, you want your readers to be able to rely on the correctness and validity of your research.
- The precise use of language and visuals also involves intellectual property laws and copyright laws as they relate to the Internet. If you "borrow" information from an existing Internet site, thus infringing upon that site's copyright, you or your employer can be assessed actual or statutory damages.
- *Plagiarism* is the appropriation of some other person's words and ideas without giving proper credit. Plagiarism is also an example of unethical workplace behavior. Writers are often guilty of unintentional plagiarism. This occurs when you incorrectly alter part of a quotation but still give credit to the writer. Your quotation must be *exactly the same* as the original word, sentence, or paragraph. You cannot change a word, a punctuation mark, or the ideas conveyed. Even if you have cited your source, an incorrectly altered quotation constitutes plagiarism.

  However, as discussed in Chapter l, avoiding plagiarism in workplace communication can sometimes be a gray area. According to Jessica Reyman, workplace communicators often use boilerplate information and templates—text that has been used before in other reports or proposals (61). Using boilerplate material can enhance productivity by saving time and is a common practice in corporations. The question is do you need to cite content that has already been used? The answer is yes . . . and no.

  For example, if you are writing a proposal or a report to a new client or a boss, and you draw from your company's library of existing proposals for material, that is an acceptable use of boilerplate content and templates. You do not need to cite the source of your information. However, if you find information on the Internet and use the material in your report or proposal without documentation, that is plagiarism and unethical behavior. You must cite the source of this original information.

To document your research correctly, you must (1) provide parenthetical source citations and (2) supply a works cited page (Modern Language Association) or a references page (American Psychological Association).

Sample MLA Works Cited Page

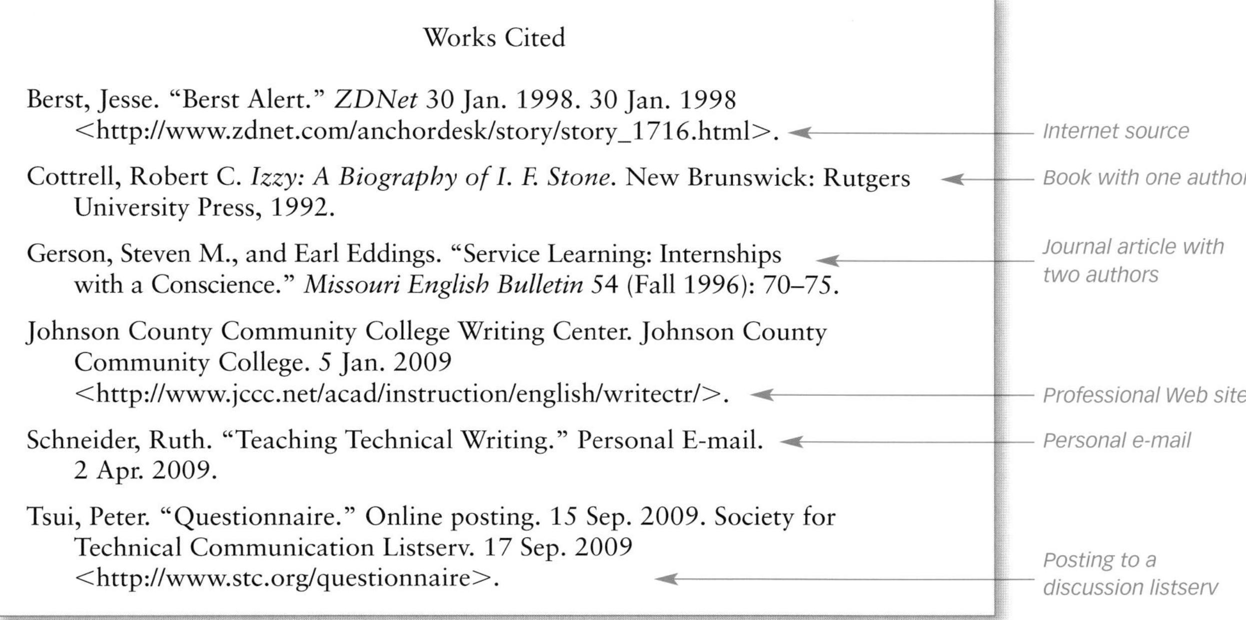

Works Cited

Berst, Jesse. "Berst Alert." *ZDNet* 30 Jan. 1998. 30 Jan. 1998 <http://www.zdnet.com/anchordesk/story/story_1716.html>.

Cottrell, Robert C. *Izzy: A Biography of I. F. Stone.* New Brunswick: Rutgers University Press, 1992.

Gerson, Steven M., and Earl Eddings. "Service Learning: Internships with a Conscience." *Missouri English Bulletin* 54 (Fall 1996): 70–75.

Johnson County Community College Writing Center. Johnson County Community College. 5 Jan. 2009 <http://www.jccc.net/acad/instruction/english/writectr/>.

Schneider, Ruth. "Teaching Technical Writing." Personal E-mail. 2 Apr. 2009.

Tsui, Peter. "Questionnaire." Online posting. 15 Sep. 2009. Society for Technical Communication Listserv. 17 Sep. 2009 <http://www.stc.org/questionnaire>.

Sample APA References Page

References

Berst, J. (1998, Jan. 30). Berst alert. *ZDNet.* Retrieved February 1, 1998, from http://www.zdnet.com/anchordesk/story/story_1716.html

Cottrell, R. C. (1992). *Izzy: A biography of I. F. Stone.* New Brunswick, NJ: Rutgers University Press.

Gerson, S. M., & Eddings, E. (1996, Fall). "Service learning: Internships with a conscience." *Missouri English Bulletin, 54,* 70–75.

Johnson County Community College Writing Center. Johnson County Community College. Retrieved January 5, 2009, from http://www.jccc.net/acad/instruction/english/writectr/

Schneider, R. (2009, Apr. 2). "Teaching technical writing."

Tsui, P. "Questionnaire." (2009, September 17). Questionnaire. Message posted to Society for Technical Communication Listserv, archived at http://www.stc.org/questionnaire

Internet source
Book with one author
Journal article with two authors
Professional Web site
Personal e-mail
Posting to a discussion listserv

Parenthetical Source Citations

The Modern Language Association (MLA) and the American Psychological Association (APA) use a simplified form for source citations. Source citations require only that you cite the source of your information parenthetically after the quotation or paraphrase.

MLA Format

One Author After the quotation or paraphrase, parenthetically cite the author's last name and the page number of the information.

> "Viewing the molecular activity required state-of-the-art electron microscopes" (Heinlein 193).

Note that the period follows the parenthesis, not the quotation. Also, note that no comma separates the name from the page number and that no lowercase *p* precedes the number.

Two Authors After the quotation or paraphrase, parenthetically cite the authors' last names and the page number of the information.

> "Though *Gulliver's Travels* preceded *Moll Flanders*, few scholars consider Swift's work to be the first novel" (Crider and Berry 292).

Three or More Authors Writing a series of names can be cumbersome. To avoid this, if you have a source of information written by three or more authors, parenthetically cite one author's name, followed by *et al.* (Latin for "and others") and the page number.

> "Baseball isn't just a sport; it represents man's ability to meld action with objective—the fusion of physicality and spirituality" (Norwood et al. 93).

Anonymous Works If your source has no author, parenthetically cite the shortened title and page number.

> "Robots are more accurate and less prone to errors caused by long hours of operation than humans" ("Useful Robots" 81).

APA Format

One Author If you do not state the author's name or the year of the publication in the lead-in to the quotation, include the author's name, year of publication, and page number in parenthesis, after the quotation.

> "Izzy's stay in Palestine was hardly uneventful" (Cottrell, 2003, p. 118).

(Page numbers are included for quoted material. The writer determines whether page numbers are included for source citations of summaries and paraphrases.)

Two Authors When you cite a source with two authors, always use both last names with an ampersand (&).

> "Line charts reveal relationships between sets of figures" (Gerson & Gerson, 2005, p. 158).

Three or More Authors When your citation has more than two authors but fewer than six, use all the last names in the first parenthetical source citation. For subsequent citations, list the first author's name last followed by *et al.*, the year of publication, and for a quotation, the page number.

> "Two-party politics might no longer be the country's norm next century" (Conners et al., 2002, p. 2).

Anonymous Works When no author's name is listed, include in the source citation the title or part of a long title and the year. Book titles are underlined or italicized, and periodical titles are placed in quotation marks.

> Two-party politics might be a thing of the past (*Winning Future Elections,* 2006).

> Many memos and letters can be organized in three paragraphs ("Using Templates," 2007).

Works Cited

Parenthetical source citations are an abbreviated form of documentation. In parentheses, you tell your readers only the names of your authors and the page numbers on which the information can be found. Such documentation alone would be insufficient. Your readers would not know the names of the books, the names of the periodicals, or the dates, volumes, or publishing companies. This more thorough information is found on the works cited page or references page, a listing of research sources alphabetized either by author's name or title (if anonymous). This is the last page of your research report.

Your entries should follow MLA or APA standards.

MLA Works Cited

A book with one author

Cottrell, Robert C. *Smoke Jumpers of the Civilian Public Service in World War II.* London: McFarland and Co., Inc., 2006.

A book with two or three authors

Heath, Chip, and Dan Heath. *Made to Stick: Why Some Ideas Survive and Others Die.* New York: Random House, 2007.

A book with four or more authors

Nadell, Judith, et al. *The Macmillan Writer.* Boston: Allyn and Bacon, 1997.

A book with a corporate authorship

Corporate Credit Union Network. *A Review of the Credit Union Financial System: History, Structure, and Status and Financial Trends.* Kansas City: U.S. Central, 2007.

A translated book

Phelps, Robert, ed. *The Collected Stories of Colette.* Trans. Matthew Ward. New York: Farrar, Straus Giroux, 1983.

An entry in a collection or anthology

Hamilton, Kendra. "What's in a Name?" *America Now: Short Readings from Recent Periodicals.* Ed. Robert Atwan. New York: Bedford/St. Martin's, 2005: 12–20.

A signed article in a journal

Davis, Rachel. "Getting—and Keeping Good Clients." *Intercom* (April 2007): 8–12.

A signed article in a magazine

Rawe, Julie. "A Question of Honor." *Time* 28 May 2007: 59–60.

A signed article in a newspaper

Gertzen, Jason. "University to Go Wireless." *The Kansas City Star* 29 Mar. 2007: C3.

An unsigned article

"Diogenes Index." *Time* 23 Sep. 1996: 22.

Encyclopedias and almanacs

"Internet." *The World Book Encyclopedia.* 2000 ed. Chicago: World Book.

Computer software

Drivers and Utilities. Computer software. Dell, Inc., 2002–2004.

Internet source

"Top Ten Qualities/Skills Employers Want." *Job Outlook 2006 Student Version.* National Association of Colleges and Employers, 2005: 5. http://career.clemson.edu/pdf_docs/NACE_JO6.pdf.

E-mail

Schneider, Ruth. "Teaching Workplace Communication." Personal E-mail. 2 Apr. 2008.

CD-ROM

McWard, James. "Graphics On-line." TW/Inform. CD-ROM. New York: EduQuest, 2008.

Personal Web site

Jones, Ellen. Home page. 29 Dec. 2009 <http://www.jccc.net/home/depts/1504>.

Professional Web site

Johnson County Community College Writing Center. Johnson County Community College. 5 Jan. 2009 <http://www.jccc.net/acad/instruction/english/writectr/>.

Posting to a discussion listserv

Tsui, Peter. "Questionnaire." Online posting. 15 Sep. 2009. Society for Technical Communication Listserv. 17 Sep. 2009 <http://www.stc.org/questionnaire>.

Library Database

Leo, John. "Culture and Ideas: On Society." *U.S. News and World Report* 26 May 2003: 47. LexisNexis Academic. LexisNexis. Billington Lib., Overland Park, KS. 27 May 2003 <http://web.lexis-nexis.com>.

APA References

A book with one author

Cottrell, R. C. (2006). *Smoke jumpers of the civilian public service in World War II.* London: McFarland and Co., Inc.

A book with two authors

Heath, C., & Heath, D. (2007). *Made to stick: Why some ideas survive and others die.* New York: Random House.

A book with three or more authors

Nadell, J., McNeniman, L., & Langan, J. (1997). *The Macmillan writer.* Boston: Allyn & Bacon.

A book with a corporate authorship

Corporate Credit Union Network. (2007). *A review of the credit union financial system: History, structure, and status and financial trends.* Kansas City, MO: U.S. Central.

A translated book

Phelps, R. (Ed.). (1983). *The collected stories of Colette* (M. Ward, Trans.). New York: Farrar, Straus & Giroux.

An entry in a collection or anthology

Hamilton, K. (2005). What's in a name? In R. Atwan (Ed.), *America now: Short readings from recent periodicals* (pp. 12–20). NewYork: Bedford/St. Martin's.

A signed article in a journal

Davis, R. (2007, April). Getting—and keeping good clients. *Intercom*, 8–12.

A signed article in a magazine

Rawe, J. (2007, May 28). A question of honor. *Time*, 59–60.

A signed article in a newspaper

Gertzen, J. (2007, March 29). University to go wireless. *The Kansas City Star*, p. C3.

An unsigned article

Diogenes index. (1996, September 23). *Time*, 22.

Encyclopedias and almanacs

Internet. (2000). *The world book encyclopedia.* Chicago: World Book.

Computer software

Drivers and Utilities [Computer software]. (2002–2004). Dell, Inc.

Internet source

Top ten qualities/skills employers want. (2006). *Job Outlook 2006 Student Version.* Retrieved May 31, 2007, from National Association of Colleges and Employers. http://career.clemson.edu/pdf_docs/NACE_JO6.pdf.

CD-ROM

McWard, J. (2008). Graphics on-line [CD-ROM]. TW/Inform. New York: EduQuest.

Personal Web site

Jones, E. Home page. Retrieved December 29, 2009, from http://www.jccc.net/home/depts/1504

Professional Web site

Johnson County Community College Writing Center. Johnson County Community College. Retrieved January 5, 2009, from http://www.jccc.net/acad/instruction/english/writectr/

Posting to a discussion listserv

Tsui, P. "Questionnaire." (2009, September 17). Questionnaire. Message posted to Society for Technical Communication Listserv, archived at http://www.stc.org/questionnaire

Library Database

Leo, J. (2003). Culture and ideas: On society. *U.S. News and World Report,* 47. Retrieved May 26, 2003, from LexisNexis database.

Alternative Style Manuals

Although MLA and APA are popular style manuals, others are favored in certain disciplines. Refer to these if you are interested or required to do so.

- *U.S. Government Printing Office Style Manual,* 29th edition. Washington, DC: Government Printing Office, 2000.
- *The Chicago Manual of Style*, 15th edition. Chicago: University of Chicago Press, 2003.
- Turabian, Kate L. *A Manual for Writers of Term Papers, Theses, and Dissertations,* 7th edition. Chicago: University of Chicago Press, 2007.

Technology Tip

Using Microsoft Word 2007 for Documentation

Microsoft Word 2007 provides students and business employees many new tools related to research and documentation. When you click on the "**References**" tab, you will find ways to create tables of contents, insert footnotes, insert citations, and create either bibliographies or works cited notations.

For example, to create parenthetical and bibliographic citations, using MLA, follow these steps:

1. Click on the "**Insert Citation**" down arrow and "**Add New Source.**"

The following "**Create Source**" screen will pop up (the screen fields will be blank; we have added the appropriate information—author's name, title, publishing company, city, and date).

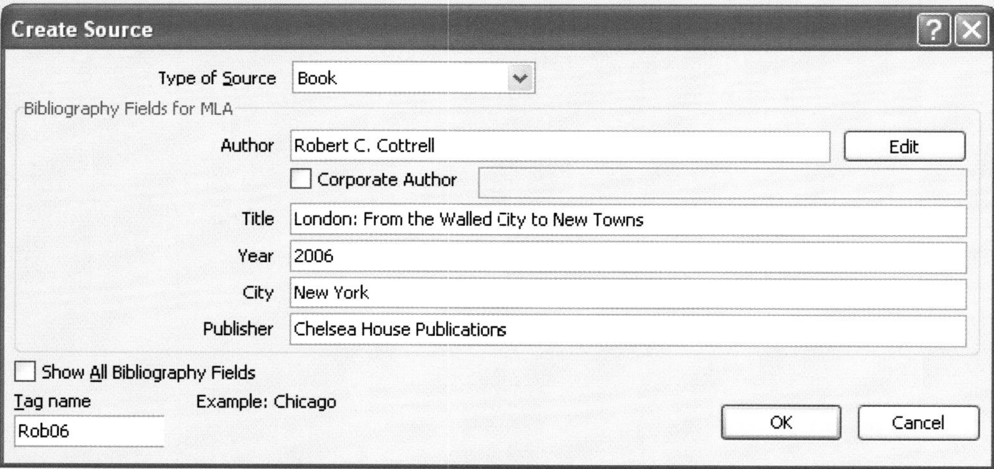

2. Click "**OK**" and the parenthetical source citation will be inserted.

3. To automatically insert the bibliographical information, click on the "**Bibliography**" down arrow and then on "**Bibliography**" (for MLA or "**Works Cited**" for APA, for example).

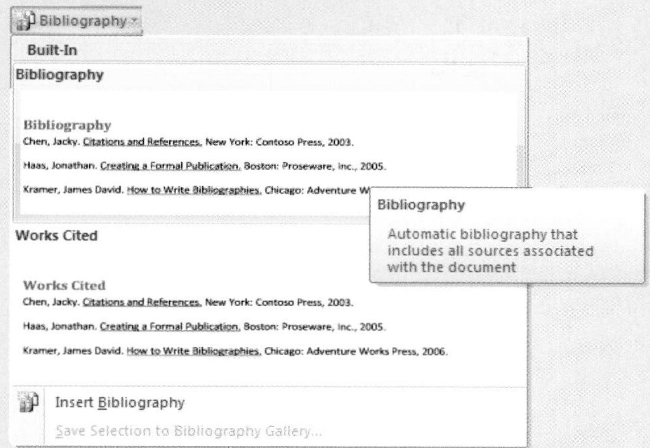

The following will appear:

> **Bibliography**
>
> Cottrell, Robert C. London: From the Walled City to New Towns. New York: Chelsea House Publications, 2006.

RESEARCH CHECKLIST

_____ 1. Have you considered your purpose and audience in choosing a research topic?

_____ 2. Have you limited the focus of your research topic?

_____ 3. Have you used primary and secondary research sources?

_____ 4. Have you researched using books, periodicals, and the Internet?

_____ 5. Have you documented your sources according to either the MLA or APA formats to avoid plagiarism?

Tom Met His Challenge

To meet his communication challenge, Tom used the *P*^{3} process.

Planning

Tom plans his project by considering the following:

- Goals—research wireless Internet options for sales representatives
- Audience—corporate employees and supervisors
- Channels—e-mail messages, face-to-face interviews, and a questionnaire
- Data—drawn from primary research

Tom says, "My first step was to make a list of all the questions I needed answered and another list of all the people I needed to involve in order to get the project off the ground. My next step was to send and answer a barrage of e-mails to those people to let them know that the request had been made and to ask for specific input. Here's a partial list of who I e-mailed and the questions I asked each person. With this primary research, I was then able to move forward with the project." Figure 16.1 shows how Tom used questions to plan his project.

(Continued)

FIGURE 16.1 Questions for Planning Research

SALES MANAGEMENT

- What is your timeline for this project?
- Each wireless card will cost approximately $XX per month. Does that fit into your budget?
- Under what circumstances will you expect the reps to use the wireless technology?
- Will there be consequences for reps who misuse the technology? If so, what will the consequences be and who will enforce them?
- What will the process be for replacing lost or damaged wireless cards?

IS PROJECT MANAGER RESPONSIBLE FOR CELL PHONES?

- Sales management has asked us to provide wireless Internet connections for the sales force. What vendors will we have to involve to get coverage for everyone?
- What are the plusses and minuses of each vendor's technology?
- What feedback have we gotten from the individuals we now have using the technology?
- What are the costs involved?
- What problems have we had with the vendors?
- What questions and/or recommendations do you have for this project?

IS PROJECT MANAGER RESPONSIBLE FOR STANDARD COMPUTER CONFIGURATION?

- Sales management has asked us to provide wireless Internet connections for the sales force. What impact might we expect this technology to have on our standard configuration?
- What questions and/or recommendations do you have for this project?

LEGAL

- Sales management has asked us to provide wireless Internet connections for the sales force. We will need you to review the contracts.

Packaging

While he waited for feedback from his colleagues, Tom says, "I assigned the project to Joe, one of my project managers, to collect all the responses and organize the information. Once Joe received the input, we drafted a list of recommendations and sent them out via e-mail for review by sales management and the legal department. Whenever I am working on a large, time-consuming, expensive project, I involve all concerned parties and find their input invaluable during this packaging phase of the *P*³ Process." Figure 16.2 shows Tom's e-mail draft.

Perfecting

After a few more meetings and a flurry of e-mails, the decision was made to grant sales management's request for wireless Internet cards. According to Tom, "The recommendations from all the respondents were modified and distributed as part of an e-mail to the entire sales force announcing the decision. The most successful business decisions are based on input from all concerned parties. After I receive this input, only then am I comfortable perfecting the e-mail and sharing it with management and the sales force." Figure 16.3 shows the perfected e-mail.

FIGURE 16.2 E-Mail for Review and Input

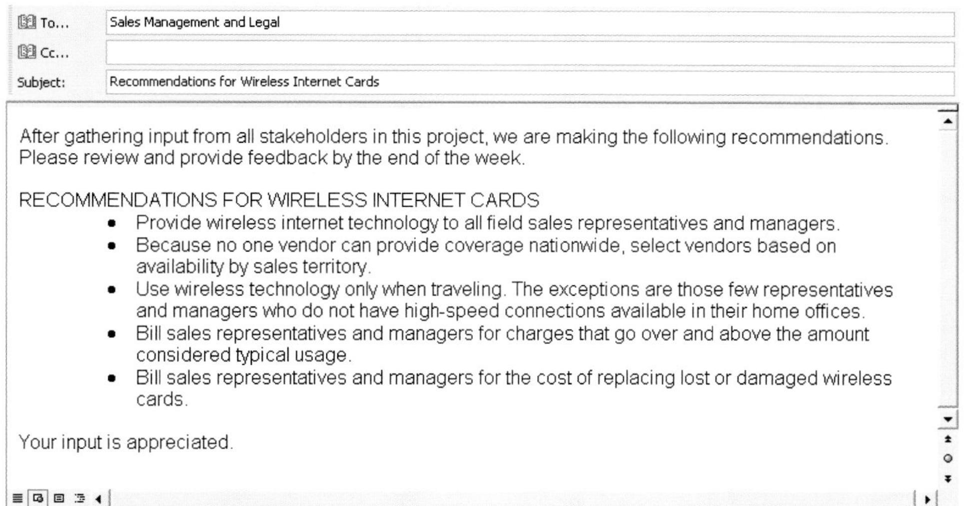
To... Sales Management and Legal

Cc...

Subject: Recommendations for Wireless Internet Cards

After gathering input from all stakeholders in this project, we are making the following recommendations. Please review and provide feedback by the end of the week.

RECOMMENDATIONS FOR WIRELESS INTERNET CARDS

- Provide wireless internet technology to all field sales representatives and managers.
- Because no one vendor can provide coverage nationwide, select vendors based on availability by sales territory.
- Use wireless technology only when traveling. The exceptions are those few representatives and managers who do not have high-speed connections available in their home offices.
- Bill sales representatives and managers for charges that go over and above the amount considered typical usage.
- Bill sales representatives and managers for the cost of replacing lost or damaged wireless cards.

Your input is appreciated.

FIGURE 16.3 Perfected E-Mail to Sales Associates

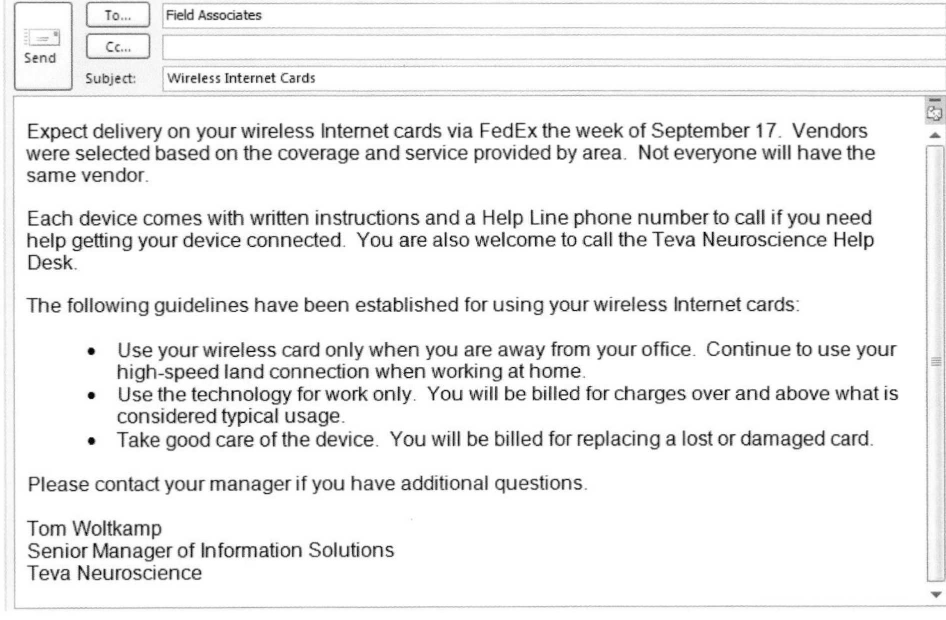
Send

To... Field Associates

Cc...

Subject: Wireless Internet Cards

Expect delivery on your wireless Internet cards via FedEx the week of September 17. Vendors were selected based on the coverage and service provided by area. Not everyone will have the same vendor.

Each device comes with written instructions and a Help Line phone number to call if you need help getting your device connected. You are also welcome to call the Teva Neuroscience Help Desk.

The following guidelines have been established for using your wireless Internet cards:

- Use your wireless card only when you are away from your office. Continue to use your high-speed land connection when working at home.
- Use the technology for work only. You will be billed for charges over and above what is considered typical usage.
- Take good care of the device. You will be billed for replacing a lost or damaged card.

Please contact your manager if you have additional questions.

Tom Woltkamp
Senior Manager of Information Solutions
Teva Neuroscience

Tom says, "It's so important to encourage all team members to participate in important decisions in the workplace. Since I became a manager 20 years ago, I have moved away from an authoritarian approach to decision making to an inclusive approach that allows input from even the newest members of the team. You can't have a great 'team' if anyone feels ignored. The P^3 process is the best approach to communication because it allows for input from team members. I never rush my communications. I want to give my team time to input comments and suggestions. Only then do I feel confident that we can meet the needs of our customers."

CHAPTER HIGHLIGHTS

1. You can research a topic either in a library or online at your computer.
2. Primary and secondary research can be conducted to assist you in developing content for reports.
3. A focus statement lets you determine the direction of your document.
4. Correct source citations help you avoid plagiarism and give credit to the writer.
5. The Modern Language Association (MLA) and the American Psychological Association (APA) are two widely used style manuals for citing sources.

WWW *Find additional information about research exercises, samples, and interactive activities at http://www.prenhall.com/gerson.*

CASE STUDY

The City of Oak Springs, Iowa, needs to improve a ten-mile stretch of road that runs east and west through the town. The winding, two-lane road was built in 1976. Since then, the city has grown with several businesses and new housing along this road, and the road no longer is sufficient for the increased traffic load. The road must be expanded to four lanes and straightened. To do so, new sewer lines, sidewalks, easements, esplanades, and lighting must be added. However, this construction will impact current homeowners (whose land, through eminent domain, can be expropriated without the owner's consent). In addition, part of the envisioned road construction will impact a wildlife refuge for waterfowl. Before construction plans can be made, city engineers must conduct research in order to determine needs and considerations of the public. Only after conducting primary and secondary research will the engineers be able to produce a construction plan to Oak Springs' city management for their consideration and approval.

Assignments

1. Perform primary research. Create a questionnaire for interviewing city residents, as well as employees from the city's parks and recreation, police, fire, and transportation departments.
2. Conduct secondary research. Read statutes online for your city and state regarding eminent domain, real estate, environmental considerations, state wildlife refuges, zoning, planning, and land use. Summarize your findings for your professor either in an e-mail message or oral presentation.

ETHICS IN THE WORKPLACE CASE STUDY

Hector Martinez, an accountant, is compiling information for a quarterly report on a new client's profit and loss statement. Prior to using Hector's accounting firm, the client had been with another accountant for many years. To facilitate Hector's compilation of the report, the client gives Hector access to prior records. When Hector writes his report, he cuts and pastes many paragraphs from the original accountant's report. Hector doesn't alter any words or cite that the content came from another writer.

Question: Is Hector committing intellectual theft or plagiarism by using another accountant's content? Justify your answer based on information provided in Chapter 1.

INDIVIDUAL AND TEAM PROJECTS

1. Correctly format and alphabetize a works cited page that contains the following entries:
   - An anonymously written magazine article
   - A magazine article written by two authors
   - A journal article written by one author
   - A book with three or more authors
   - A book with an editor
   - A signed newspaper editorial
   - A blog
   - An online publication
   - An e-mail message
   - A Web site

2. Select a topic from your major field or your job and write a research report. You might want to consider a controversy in your area of interest (such as the greenhouse effect, hazardous waste management, or computer viruses) or the impact of a technical innovation (such as micromachines or the Internet).

DEGREE-SPECIFIC ASSIGNMENTS

1. **Marketing and Public Relations**—Research how new marketing tools related to eCommerce are affecting both companies and customers. You could focus on how eCommerce affects a company's budget as it moves from traditional to online media marketing. You could focus on how eCommerce channels, such as RSS-feeds, review sites, blogs, and travel sites, affect customers' decision making and buying habits.
2. **Business Information Systems**—Over half of all United States employees work for small businesses. However, small businesses face challenges related to business information systems. These include electronic commerce, Internet adoption, network security, and IT planning methodologies. Imagine that a small business has hired you to build its information system. Research challenges that small businesses face in terms of their information technologies.
3. **Accounting**—Technology is constantly impacting accounting. Emerging technologies affect payroll, Web services, digital storage and retrieval, fraud detection, electronic monitoring for continuous assurance, and artificial intelligence in accounting. Choose one of these topics and report on your researched findings.
4. **Management**—Business managers face challenges daily. These include managing change, developing advertising strategies, managing new construction projects, adding information systems, managing investments, collaborating with business partners, and mergers and acquisitions. Choose any of these topics to find unique challenges presented and suggested ways to solve the problems.
5. **Finance**—Your company is considering an IPO (Initial Public Offering). Prior to going public, what financial considerations should you focus on? As the company's financial officer, research IPOs as they relate to a company's finances in terms of risk management, corporate restructuring, or corporate earnings potential.

PROBLEM-SOLVING THINK PIECES

1. Many communities have recycling projects that allow residents to recycle paper products, cans, and plastic. Not all businesses recycle, however. Research the benefits of recycling, determine how a business or businesses could implement a corporate recycling plan, and write a report recommending action based on your research.
2. In today's global economy, understanding and accommodating multiculturalism and cross-culturalism in business are important. Research the following:
   - The unique challenges that cultural diversity presents to businesses.
   - How companies have responded to these challenges.

   Write a report recommending why and how a business or businesses can help employees develop cultural awareness.
3. Many companies track the time their employees spend either surfing the Web or sending and receiving personal e-mail while at work. Research the following:
   - Software that companies use to track employee electronic communication usage.
   - Corporate guidelines for employee use of company-owned electronic communication hardware and software.
   - The legal and ethical ramifications of an employee's private use of corporate-owned e-mail and Internet access.
   - The legal and ethical ramifications of an employer eavesdropping on an employee's Web usage.

Write a research report on your findings and provide a corporate guideline for both employee and employer electronic communication responsibilities.

4. Corporate training is big business. Many companies hire outside consultants or staff company training departments to teach employees new skills. These could include training workshops on diet and exercise to improve work efficiency, techniques for avoiding e-mail viruses or screening e-mail spam, resume writing for transitional employees, time management, leadership skills, improved oral presentations, improved customer service skills, basic word processing, or business accounting for non-accountants.

   What training class does your company need? Research possible topics and training approaches. How have other companies offered this training? What benefits do employees derive from this training? How does the company benefit? What are the costs for this training (personnel, time, equipment, etc.)? Then, write a proposal or an instructional training module based on your research.

5. Entrepreneurialism is one of the fastest growing sectors in business. Many people are opening their own businesses. What does it take to open your own business? Before you can write an effective business plan and seek financing from a bank, you must research the project.

   Choose a new business venture, selling a product or service of your choice. What would it cost to open this business? What would be your best location, or should your business be online? What certifications or licensing is needed? How many personnel would you need? What equipment is necessary? Who would be your clientele?

   Based on research, write a proposal, appropriate for presentation to a bank. In this proposal, present your business plan for a new entrepreneurial opportunity.

WEB WORKSHOP

1. FirstGov.com allows you to research a wide variety of topics, such as education and jobs, benefits and grants, consumer protection, environment and energy, science and technology, and public safety and health. This Web site also provides information on breaking news. Access FirstGov.com and research a topic relevant to your career goals. Write a memo or report to your instructor summarizing your findings.
2. Go to an online news magazine, such as *Slate*, *Time*, or *U.S. News Online*. Type a topic of interest in the magazine's search engine. Research this topic and make a brief oral presentation to your class about the information you have gathered.
3. Access a search engine such as Google, Ask Jeeves, or Dogpile. Type in a topic relating to your major field. For example, for computer information technology, medical records and health information, or accounting, look for job openings in your field to learn about salary ranges, benefits, application requirements, and so on. Report your findings either in an oral presentation to the class or in a memo to your instructor.
4. Using an online newspaper, such as *The New York Times*, *CNN*, or *USA Today*, research business and technology news. Find out the major news stories of the day which relate to your career path. Report your findings in an oral presentation to the class or in a memo to your instructor.

QUIZ QUESTIONS

1. Explain why you need to consider audience when you write a researched report.
2. What is the difference between primary and secondary research?
3. What is the difference between a quote and a paraphrase?
4. How does a focus statement direct you when you write a research report?
5. Explain plagiarism. How can you avoid plagiarizing?

CHAPTER 17

Short, Informal Reports

▸REAL PEOPLE *in the workplace*

As **Linda M. Freeman, CPA**, says, she and her colleagues throughout the office of Marks, Nelson, Vohland & Campbell spend approximately "20 to 25 percent of our day on report writing, including background research time, drafting, and proofreading."

On a daily basis, Ms. Freeman and her colleagues write compilation reports, review reports, and audit reports. Because these reports are "subject to guidelines of the American Institute of Certified Public Accountants," the reports, in large part, are boilerplate with prescribed wording.

Linda also writes valuation reports on a daily basis. Some valuation reports are sent to the IRS for estate planning and gift tax reporting; others "go to the courts for settling disputes between business owners or divorcing couples." Because valuation reports are judgment calls on the part of the CPAs, these reports are not boilerplate and entail less prescribed language.

Private letter ruling requests are less-frequently written reports. They are "high-level tax-planning" documents that require intensive research. In these formal requests sent to the IRS, a client proposes a new business or activity structured in a unique way. The accountant projects that the venture will be taxed according to one standard and asks the IRS if they agree with this assumption.

Times are changing. Communication is "going more electronic," Linda says. Reports to federal agencies, for example, are "template-driven." Accountants are encouraged to use online forms, software from federal agencies, to help the government cut back on paperwork.

Writing takes up much of Linda's time. Her reports must address many precise accounting standards. Linda's ultimate goal in report writing is quality assurance: adherence to legalities.◂

CHAPTER GOALS

When you complete this chapter, you will be able to do the following:

1. Understand the purposes of a report.
2. Distinguish between short, informal reports and long, formal, researched reports.
3. Follow guidelines for writing short reports.
4. Use headings and talking headings to organize your short reports.
5. Write different types of short reports including incident reports, investigative reports, trip reports, progress reports, feasibility/ recommendation reports, and meeting minutes.
6. Use the P^3 process to write effective short, informal reports.
7. Evaluate your report using the report checklist.

WWW

To learn more about short, informal reports, visit our Web Companion at http://www.prenhall.com/gerson.

Linda's Communication Challenge

Approximately ten times a month, Linda must respond to IRS compliance issues. Linda says, "One of my clients received a notice from the IRS that tax information and documentation are missing from their files. Because of missing information, taxes have not been paid; therefore, the client can owe anywhere from $500.00 to $250,000.00 in penalties. PhotoFinish Inc. received numerous notices from the IRS regarding their profit-sharing plan. They ignored these messages for five years, due to bad management and high internal turnover. At the point when I was hired as the company's accountant, PhotoFinish owed $50,000 in penalties.

My challenge was to discover why the notices had been ignored for five years. I also had to write a convincing letter report to the IRS, persuading them to waive this penalty. This challenge was compounded by the fact that PhotoFinish provided me with no factual documentation. Helping clients to deal with the IRS is always a communication challenge."

See pages 496–499 to learn how Linda met her communication challenge.

WHAT IS A REPORT?

Report—this simple word is hard to define. Reports come in different lengths and levels of formality, serve different and often overlapping purposes, and can be conveyed to an audience using different communication channels. In fact, report types and characteristics can overlap. You could write a short, informal progress report that conveys information about job-related projects. For a short report, an e-mail format might be sufficient, directed to either a specialist, semi-specialist, or lay audience. However, you might write a longer, more formal progress report, in a letter format for an external lay audience, that provides facts, analyzes these findings, and recommends follow-up action.

Your reports will satisfy one or all of the following needs:

- Supply a record of work accomplished
- Record and clarify complex information for future reference
- Present information to a large number of people
- Record problems encountered
- Document schedules, timetables, and milestones

Additional information:

See Chapter 18, "Long, Formal Reports."

Table 17.1 Unique Aspects of Reports

Report Features	Distinctions	Definition of Unique Characteristics
Length and Scope		
	Short	A typical short report is limited to one to five pages. Short reports focus on topics with limited scope. This could include a limited timeframe covered in the report, limited financial impact, limited personnel, and limited impact on the company.
	Long	Long reports are more than five pages long. If a topic's scope is large, including a long timeframe, significant amounts of money, research, many employees, and a momentous impact on the company, a long report might be needed.
Formality (tone)		
	Informal	Most short reports are informal, routine messages, written as letters, memos, or e-mail.
	Formal	Formal reports are usually long and contain standardized components, such as a title page, table of contents, list of illustrations, abstract, appendices, and works cited/references.
Audience		
	Internal (specialist or semi-specialist)	Colleagues, supervisors, or subordinates within your company are an internal audience. Usually you would write an e-mail or memo report.
	External (multiple audience levels)	An external audience is composed of vendors, clients, customers, or companies with whom you are working. Usually you would write a report in letter format.
	Internal and external	If a report is being sent to both an internal and external audience, you would write either an e-mail, memo, or letter report.
Purpose		
	Informational	Informational reports focus on factual data. They are often limited in scope to findings. "Here's what happened."
	Analytical	Analytical reports provide information but analyze the causes behind occurrences. Then, these analytical reports draw conclusions, based on an interpretation of the data. "Here's what happened and why this occurred."
	Persuasive	Persuasive reports convey information and draw conclusions. Then, these reports use persuasion to justify recommended follow-up action. "Here's what happened, why this occurred, and what we should do next."
Communication Channels		
	E-mail	E-mail reports, written to internal and external audiences, are short and informal.
	Memo	Memo reports are written to internal audiences and are usually short and informal.
	Letter	Letter reports are sent to external audiences. These reports can be either long or short, formal or informal, depending on the topic, scope, purpose, and audience.
	Hard-copy PowerPoint printouts	PowerPoint slides can be printed and then bound for presentation. These can be formal or informal, depending on the topic, purpose, scope, and audience.
	Electronic (online)	Many reports can be accessed via a company's Web site. These reports can be downloaded and printed out and often are "boilerplate"—text that can be used repetitively. Electronic reports also provide interactivity, allowing end users to fill out the report online and submit it to the intended audience.

- Recommend future action
- Document current status
- Record procedures

Long? Short? Formal? Informal? To clarify unique aspects of different kinds of reports, look at Table 17.1.

TYPES OF REPORTS

Many reports fall into the following categories:

Incident reports	Investigative reports	Trip reports	Progress reports
Meeting minutes	Feasibility/recommendation reports	Research reports	Proposals

This chapter will focus on short, informal reports (incident reports, investigative reports, trip reports, progress reports, feasibility/recommendation reports, and meeting minutes). Chapter 18 will concentrate on long, formal reports requiring research. Chapter 19 will discuss long, formal proposals.

CRITERIA FOR WRITING REPORTS

Although there are many different types of reports and individual companies have unique demands and requirements, certain traits, including organization, development, audience, and style, are basic to all report writing.

Organization

Every short report should contain five basic units: identification lines, headings and talking headings, introduction, discussion, and conclusion/recommendations.

Identification Lines

You must identify the date on which your report is written, the names of the people to whom the report is written, the names of the people from whom the report is sent, and the subject of the report (as discussed in Chapter 10: Traditional Correspondence: Memos and Letters, the subject line should contain a *topic* and a *focus*). In a short, internal report written using a memo format, the identification lines will look like this:

Date: March 15, 2009

To: Rob Harken

From: Stacy Helgoe

Subject: Report on Usenet Conference

Headings and Talking Headings

To improve your page layout and make content accessible, use headings and talking headings. Headings—words or phrases such as "Introduction," "Discussion," "Conclusion," "Problems with Employees," or "Background Information"—highlight the content in a particular section of a document. Talking headings, in contrast, are more informative than headings. Talking headings, such as "Human Resources Committee Reviews 2009 Benefits Packages," informatively clarify the content that follows.

Introduction

The introduction supplies an overview of the report. It can include three or more optional subdivisions:

- **Purpose**—A topic sentence(s) explaining why you are submitting the report (rationale, justification, objectives) and exactly what the report's subject matter is.
- **Personnel**—Names of others involved in the reporting activity.
- **Dates**—What period of time the report covers.

In this introductory section, use headings or talking headings to summarize the content. These can include headings for organization, such as "Overview" or "Purpose," or more informative headings, such as "Purpose of Report," "Committee Members," "Conference Dates," "Needs Assessment," and "Problems with Machinery." The headings will depend on the type of report and the topic of discussion.

example

Introduction

Report Objectives: I attended the Southwest Regional Conference on Workplace Communication in Fort Worth, Texas, to learn more about how our company can communicate effectively. This report addresses the workshops I attended, consultants I met with, and pricing for training seminars.

Conference Dates: August 5–8, 2009

Committee Members: La'Tisha Lisk and Larry Rochelle

Some businesspeople omit the introductory comments in reports and begin with the discussion. They believe that introductions are unnecessary because the readers know why the reports are written and who is involved.

These assumptions are false for several reasons. First, it is false to assume that readers will know why you're writing the report, when the activities occurred, and who was involved. Perhaps if you are writing only to your immediate supervisor, there's no reason for introductory overviews. However, even in this situation you might have an unanticipated reader because

- Immediate supervisors change—they are promoted, fired, retire, or go to work for another company.
- Immediate supervisors aren't always available—they're sick for the day, on vacation, or off-site for some reason.

Second, avoiding introductory overviews assumes that your readers will remember the report's subject matter. This is false because reports are written not just for the present, when the topic is current, but for the future, when the topic is past history. Reports go on file—and are referenced at a later date. At that later date

- You won't necessarily remember the particulars of the reported subject matter.
- Your colleagues, many of whom weren't present when the report was originally written, won't be familiar with the subject.
- You might have outside, lay readers who need additional detail to understand the report.

An introduction—which seemingly states the obvious—is needed to satisfy multiple readers, readers other than those initially familiar with the subject matter, and future readers who are unaware of the original report.

Discussion

The discussion section of the report can summarize many topics, including your activities, the problems you encountered, costs of equipment, warranty information, and more. This is the largest section of the report requiring detailed development, discussed below. Use headings or talking headings to organize your content.

Conclusion/Recommendations

The conclusion section of the report allows you to sum up, to relate what you have learned, or to state what decisions you have made regarding the activities reported. The recommendation section allows you to suggest future action, such as what the company should do next. Not all reports require recommendations. You can use headings or talking headings to organize the content.

Conclusion/Recommendations

Benefits of the Conference:

The conference was beneficial. Not only did it teach me how to improve my workplace communication but also it provided me contacts for workplace communication training consultants.

Proposed Next Course of Action:

To ensure that all employees benefit from the knowledge I acquired, I recommend hiring a consultant to provide workplace communication training.

Talking headings, such as "Benefits of the Conference" and "Proposed Next Course of Action," provide more focus than simple headings, such as "Conclusion" and "Recommendation."

The conclusion shows how the writer benefited.

The recommendation explains what the company should do next and why.

Development

Now that you know what subdivisions are traditional in short reports, the next questions might be, "What do I say in each section? How do I develop my ideas?"

First, answer the reporter's questions.

1. *Whom* did you meet or contact, who was your liaison, who was involved in the accident, who was on your technical team, and so on?
2. *When* did the documented activities occur (dates of travel, milestones, incidents, etc.)?
3. *Why* are you writing the report and why were you involved in the activity (rationale, justification, objectives)?
4. *Where* did the activity take place?
5. *What* were the steps in the procedure, what conclusions have you reached, or what are your recommendations?

Second, when providing the foregoing information, *quantify.* Do not be vague or imprecise. Specify to the best of your abilities with photographic detail. The following justification is an example of vague, imprecise writing.

BEFORE

Installation of the machinery is needed to replace a piece of equipment deemed unsatisfactory by an equipment engineering review.

Which machine are we purchasing? Which piece of equipment will it replace? Why is the equipment unsatisfactory (too old, too expensive, too slow)? When does it need to be replaced? Where does it need to be installed? Why is the installation important? A department supervisor will not be happy with the preceding report. Instead, supervisors need information *quantified,* as follows:

AFTER

The *exposure table* needs to be installed by *9/09* so that we can *manufacture printed wiring products with fine line paths and spacing (down to 0.0005 inch).* The table will replace the *outdated printer* in *Dept. 76*. Failure to install the table *will slow the production schedule by 27 percent*.

Note that the italicized words and phrases provide detail by quantifying.

Audience

Since reports can be sent both internally and externally, your audience can be specialists, semi-specialists, lay readers, or multiple. Before you write your report, determine who will read your text. This will help you decide if terminology needs to be defined and what tone you should use. In a memo report to an in-house audience, you might be writing simultaneously to your immediate supervisor (specialist), to his or her boss (semi-specialist), to your colleagues (specialists), and to a CEO (semi-specialist). In a letter report to an external audience, your readers could be specialists, semi-specialists, or lay readers. To accommodate multiple audiences, use parenthetical definitions, such as Cash in Advance (CIA) or Continuing Property Records (CPR).

Additional information:

See Chapter 3, "Meeting the Needs of the Audience."

In reports, audience determines tone. For example, you cannot write directive reports to supervisors mandating action on their part. It might seem obvious that you can write directives to subordinates or a lay audience, but you should not use a dictatorial tone. You will determine the tone of your report by deciding if you are writing vertically (up to management or down to subordinates), laterally (to coworkers), or to multiple readers.

Style

Style includes conciseness, simplicity, and highlighting techniques. You achieve conciseness by eliminating wordy phrases. Say "consider" rather than "take into consideration;" say "now" rather than "at this present time." You achieve simplicity by avoiding old-fashioned words: "utilize" becomes "use," "initiate" becomes "begin," "supersedes" becomes "replaces."

Additional information:

See Chapter 6, "Perfecting Workplace Communication."

The value of highlighting has already been shown in this chapter. The parts of reports reviewed earlier use headings. Graphics can also be used to help communicate content, as evident in the following example. A recent demographic study of Kansas City predicted growth patterns for Johnson County (a large county south of Kansas City):

BEFORE

Johnson County is expected to add 157,605 persons to its 1980 population of 270,269 by the year 2010. That population jump would be accompanied by a near doubling of the 96,925 households the county had in 1980. The addition of 131,026 jobs also is forecast for Johnson County by 2010, more than doubling its employment opportunity.

The information is difficult to access readily. We are overloaded with too much data. Luckily, the report provided a table for easier access to the data. Through highlighting techniques (tables, white space, headings), the demographic forecast is made accessible at a glance, as shown in Table 17.2.

AFTER

Table 17.2 Johnson County Predicted Growth by 2010

	Population	Households	Employment
1980	270,269	96,925	127,836
2010	427,874	192,123	258,862
% Change	+58.3%	+98.2%	+102%

TYPES OF SHORT, INFORMAL REPORTS

All reports include identification information, an introduction, a discussion, and a conclusion/recommendations. However, different types of short reports customize these generic components to meet specific needs. Let's look at the criteria for six common types of reports: incident reports, investigative reports, trip reports, progress reports, feasibility/recommendation reports, and meeting minutes.

Incident Reports

Purpose and Examples

An *incident report* documents an unexpected problem that has occurred. This could be an automobile accident, equipment malfunction, fire, robbery, injury, or even problems with employee behavior. In this report, you will document what happened. If a problem occurs within your work environment that requires analysis (fact finding, review, study, etc.) and suggested solutions, you might be asked to prepare an incident report (also called a trouble report or accident report).

Criteria

Companies requiring reports do not always provide employees with easy-to-fill-in forms. To write an incident report when you have not been given a printed form, include the following components:

1. **Introduction**

    Purpose. In this section, document when and where the incident occurred. What motivated your visit to the scene of the problem?

    Personnel. *Who* was involved, and *what* role do you play in the report? That could entail listing all of the people involved in the accident or event. These might be people injured, as well as police or medical personnel answering an emergency call.

    In addition, *why* are you involved in the activity? Are you a supervisor in charge of the department or employee? Are you a police officer or medical personnel writing the report? Are you a maintenance employee responsible for repairing the malfunctioning equipment?

2. **Discussion** (**body, findings, agenda, work accomplished**)

    Using subheadings or itemization, quantify what you saw (the problems motivating the report). You should list your findings in chronological order. Be specific. Include the

    - Make or model of the equipment involved
    - Police departments or hospitals contacted
    - Names of witnesses
    - Witness testimonies (if applicable)
    - Extent of damage—financial and physical
    - Graphics (sketches, schematics, diagrams, layouts, etc.) depicting the incident visually
    - Follow-up action taken to solve the problem

3. **Conclusion/recommendations**

    Conclusion. Explain what caused the problem.

    Recommendations. Relate what could be done in the future to avoid similar problems.

Often, on the job you will write reports to colleagues and supervisors with whom you work on a day-to-day basis. Using e-mail allows the audience to quickly and conveniently access this short, informal report and respond to any issues.

Figure 17.1 presents an example of an incident report, written as an e-mail.

> **Dot.Com Updates**
>
> For more information about online incident reports, check out the following links:
>
> - US Consumer Product Safety Commission's online "Consumer Product Incident Report" form http://www.cpsc.gov/incident.html
> - National Fraud Information Center's "Online Incident Report Form" http://www.fraud.org/info/repoform.htm
> - Health & Safety Executive's "Incident Reporting" forms https://www.hse.gov.uk/forms/incident/index.htm

Additional information:

See Chapter 9, "Electronic Communication: E-mail Messages, Instant Messages, Text Messages, and Blogging."

FIGURE 17.1 E-mail Incident Report

The writer relies on a "problem/solution" analytical pattern of organization for this short, informal, internal e-mail report.

The writer includes a "signature file" that acts like an online business card, clearly identifying the writer and providing contact information.

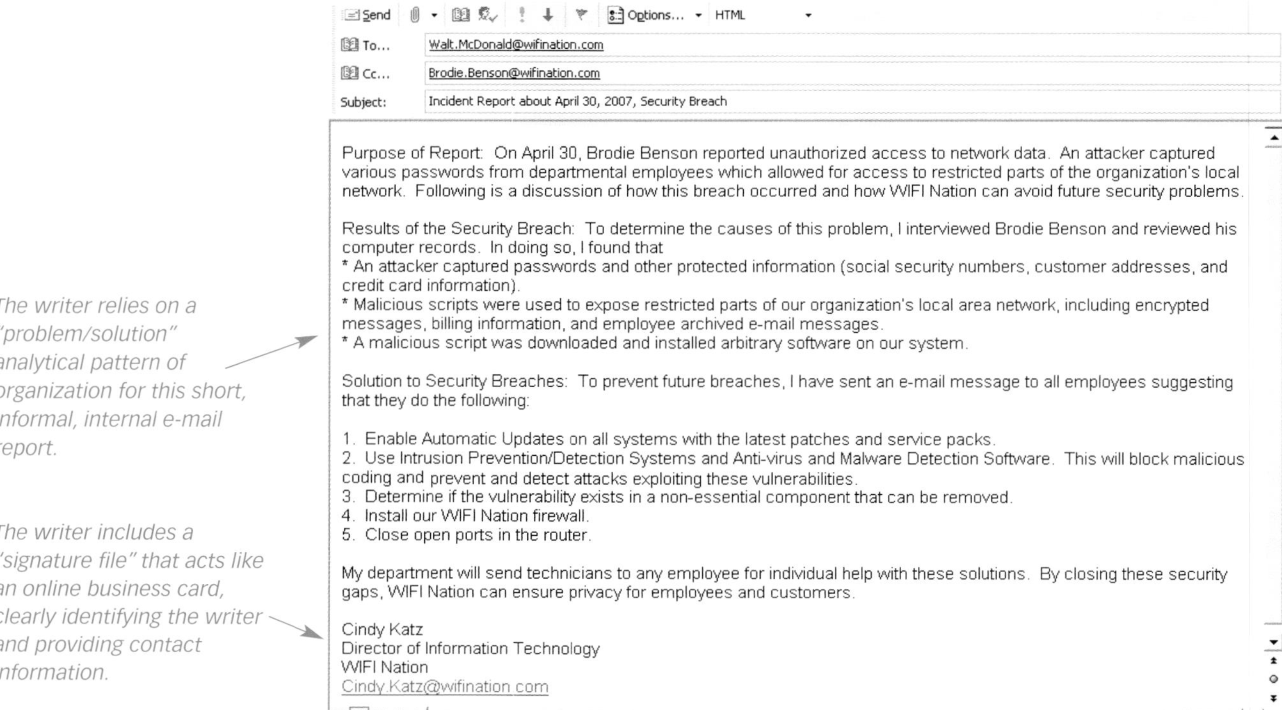

Send | Options... | HTML

To... Walt.McDonald@wifination.com

Cc... Brodie.Benson@wifination.com

Subject: Incident Report about April 30, 2007, Security Breach

Purpose of Report: On April 30, Brodie Benson reported unauthorized access to network data. An attacker captured various passwords from departmental employees which allowed for access to restricted parts of the organization's local network. Following is a discussion of how this breach occurred and how WIFI Nation can avoid future security problems.

Results of the Security Breach: To determine the causes of this problem, I interviewed Brodie Benson and reviewed his computer records. In doing so, I found that

* An attacker captured passwords and other protected information (social security numbers, customer addresses, and credit card information).
* Malicious scripts were used to expose restricted parts of our organization's local area network, including encrypted messages, billing information, and employee archived e-mail messages.
* A malicious script was downloaded and installed arbitrary software on our system.

Solution to Security Breaches: To prevent future breaches, I have sent an e-mail message to all employees suggesting that they do the following:

1. Enable Automatic Updates on all systems with the latest patches and service packs.
2. Use Intrusion Prevention/Detection Systems and Anti-virus and Malware Detection Software. This will block malicious coding and prevent and detect attacks exploiting these vulnerabilities.
3. Determine if the vulnerability exists in a non-essential component that can be removed.
4. Install our WIFI Nation firewall.
5. Close open ports in the router.

My department will send technicians to any employee for individual help with these solutions. By closing these security gaps, WIFI Nation can ensure privacy for employees and customers.

Cindy Katz
Director of Information Technology
WIFI Nation
Cindy.Katz@wifination.com

Investigative Reports

Purpose and Examples

As the word "investigate" implies, an *investigative report* asks you to examine the causes behind an incident. Something has happened. The report does not just document the incident. It focuses more on why the event occurred.

Criteria

Following is an overview of what you include in an effective investigative report.

1. **Introduction (overview, background)**

   Purpose. In the purpose section, document the date(s) of the incident. Then comment on your objectives or rationale. What incident are you reporting on and what do you hope to achieve in this investigation?

   You also might want to include these following optional subheadings:

   Location. Where did the incident occur?

   Personnel. Who was involved in the incident? This could include those with whom you worked on the project or those involved in the situation.

   Authorization. Who recommended or suggested that you investigate the problem?

2. **Discussion (body, findings, agenda)**

   This is the major part of the investigation. Using subheadings, document your findings. This can include the following:

   - A review of your observations. This includes physical evidence, descriptions, lab reports, testimony, and interview responses. Answer the reporter's questions: "Who," "What," "When," "Where," "Why," and "How."
   - Contacts—people interviewed
   - Difficulties encountered
   - Techniques, equipment, or tools used in the course of the investigation
   - Test procedures followed, organized chronologically

3. **Conclusion/recommendations**

   Conclusion. What did you accomplish? What did you learn? What discoveries have you made regarding the causes behind the incident? Who or what is at fault?

   Recommendations. What do you suggest next? Should changes be made in personnel or in the approach to a particular situation? What training is required for use with the current technology, or should technology be changed? What is the preferred follow-up for the patient or client? How can the problem be fixed?

Use letter formats to write short, informal reports to an external audience. Figure 17.2 illustrates an investigative report, written in letter format. This example is written at a specialist level because the city council needs to understand the scientific information about

FIGURE 17.2 Investigative Report in Letter Format (Specialist Level)

Frog Creek Wastewater Treatment Plant

9276 Waveland Blvd.
Bowstring City, UT 86554

Follow letter format when you write a short, informal report for an external audience.

September 15, 2009

Bowstring City Council
Arrowhead School District 234
Bowstring City, UT 86721

The subject line provides a topic (Frog Creek Wastewater Pollution) and a focus (Investigative Report).

Subject: Investigative Report on Frog Creek Wastewater Pollution

In the letter's introduction, reporter's questions clarify when and where the event occurred, who was involved, and why the report is being written.

On September 7, 2009, teachers at Arrowhead Elementary School reported that over a five-day period (September 2–6), approximately 20 students complained of nausea, lightheadedness, and skin rashes. On the fifth day, the Arrowhead administration called 911 and the Arrowhead School District (ASD 234) in response to this incident.

Bowstring City paramedics treated the children's illnesses, suggesting that the problems might be due to airborne pollutants. The Bowstring City Council contacted the Frog Creek Wastewater Treatment Plant (FCWTP) to investigate the causes of this problem. This report is submitted by Mike Moore (Director of Public Relations), Sue Cottrell (Wastewater Engineer), and Fred Starr (Wastewater Engineer), in response to Bowstring City's request.

The findings use chronological organization to document the incidents.

Committee Findings

Impact on Schoolchildren: Arrowhead Elementary School administrators reported the following:

- Monday, September 2—two children reported experiencing nausea.
- Tuesday, September 3—two children reported experiencing nausea, and one child experienced lightheadedness.
- Wednesday, September 4—three children experienced skin rashes.
- Thursday, September 5—two children complained of nausea, one child was lightheaded, and two children showed evidence of skin rashes.
- Friday, September 6—two children reported nausea, three reported skin rashes, and two lightheadedness.

After Friday's occurrences, Arrowhead Elementary School administrators called 911. Bowstring paramedics reported that (with parental approval) the children were treated with antacids for nausea, antihistamines for skin rashes, and oxygen for their lightheadedness. No other incidents were reported in the neighborhood surrounding the school.

Report on Frog Creek Pollutants: Wastewater Engineers Sue Cottrell and Fred Starr took samples of Frog Creek on September 7–10 and found the following:

Typical Frog Creek Readings	Readings from September 7–10
Low alkalinity (generally <30 mg/l) (milligrams per liter)	High alkalinity readings: <45 mg/l
Low inorganic fertilizer nutrients (phosphorous and nitrogen) <20 mg/l)	High phosphorous and nitrogen readings: <25 mg/l
Limited algae growth: 1 picometer	High algae readings: 4 picometers

(Continued)

FIGURE 17.2 Investigative Report in Letter Format (Specialist Level) Continued

The findings not only investigate the causes of the incident, but also document with specific details. To achieve a readable format, the text is made accessible through highlighting techniques—underlined and italicized subheadings, bulleted details, and a comparison/contrast table.

Mike Moore
Page 2
September 15, 2009

Explanation of Findings:
Despite normally low readings, in late summer, with heat and rain, these readings can escalate. Higher algae-related odors above the 3–6 picometer thresholds, along with increased alkalinity (<50 mg/l), can create health problems for youth, elderly, or anyone with respiratory illnesses. These studies showed that algae, alkalinity, and fertilizer nutrients were higher than usual.

The above elevated readings were caused by three factors (heat, rain, and northeasterly winds).

- Heat—On the days of the Arrowhead Elementary School incident, the temperature ranges were 92°F–95°F, unusually high for early September. Algae and chemical growth increases in temperatures above 84°F.
- Rain—In addition, on September 4–6, Bowstring City received 2" of rain, swelling Frog Creek to 3' above its normal levels. Studies show that rain-swollen creeks and rivers lead to increased pollutants, as creek bottom silt rises.
- Wind—On September 4–5, a prevailing northeasterly wind blew from Frog Creek toward Arrowhead Elementary School's playground.

Follow-up Studies: On September 11–14, our engineers rechecked Frog Creek, finding that the chemical levels had returned to a normal, acceptable range.

Conclusion about Incident

Frog Creek normally has acceptable levels of algae, alkalinity, and fertilizer nutrient levels. The heat and higher water levels temporarily led to elevated pollutant readings. These levels subsequently returned to normal. Wind directions during the school incidents also had an impact on the children's illnesses. On follow-up questionnaires, FCWTP employees found that the school children's ailments had subsided.

The Arrowhead Elementary School situation appears to have been an isolated incident due to atmospheric changes.

Recommendations for Future Course of Action

Though we constantly monitor Frog Creek for safety, FCWTP HAZMAT employees would be happy to work with parents and teachers to provide additional health information. In a one-hour workshop, presented during the school day or at a Parent-Teacher Organization meeting, FCWTP could offer the following information:

- Scientific data about stream and creek pollutants
- The effects of rain, wind, and heat on creek chemicals
- Useful preventive medical emergency techniques

This information would explain real-world applications for science classes, as well as provide valuable health tips for parents and teachers. We have enclosed for your review information about our proposed training sessions.

The letter's conclusion provides options for the readers and a positive tone appropriate for the intended audience.

Please let us know if you would like to benefit from this free-to-the-public workshop. We would be happy to schedule one at your convenience.

Sincerely,

Mike Moore
Frog Creek Wastewater Treatment Plant Director of Public Relations

Enclosure

the incident. Moreover, this report will be kept on file not only for documentation but also for potential litigious situations.

In contrast to Figure 17.2, the specialist investigative report, Figure 17.3 is written at a lay level for an external audience consisting of the elementary school administration, the teachers, and the parents of the school children affected by the incident in Frog Creek. The highly technical and scientific information is omitted, the material is condensed, and the writer includes sufficient details to alleviate the concerns of the audience. In addition, the writer emphasizes actions taken and assurances of how and why this incident will not recur.

FIGURE 17.3 Investigative Report in Letter Format (Lay Audience)

Frog Creek Wastewater Treatment Plant
"Safety Is Our Number 1 Concern."
9276 Waveland Blvd.
Bowstring City, UT 86554

September 15, 2009

Attention: Parents and Arrowhead Elementary School Teachers and Administrators

Subject: Report on September 2–6 Frog Creek Incident

Last week, your children at Arrowhead Elementary School experienced nausea, lightheadedness, and skin rashes. This was due to an unusual environmental situation at Frog Creek, a rare case of airborne pollutants caused by high winds and rain.

How Did This Happen?

Algae is a good thing. The crayfish, snails, and minnows that your children love seeing in Frog Creek thrive on algae and lichen (the small, green plants that are the food base for most marine life). However, when temperatures rise above 84°F, algae can grow to an unhealthy level.

The same thing applies to the acid level in water. When acid is regulated by alkalinity, algae growth is controlled. When acid levels rise, however, algae can bloom or marine creatures can die. Both of these problems lead to unusual odors.

That's what happened last week. Our studies showed that algae and alkalinity were higher than usual. The causes were increased heat, rain, and wind:

- Heat—On the days of the Arrowhead Elementary School incident, the temperature ranges were 92°F to 95°F, unusually high for early September.
- Rain—In addition, on September 4–6, Bowstring City received 2" of rain, swelling Frog Creek to 3' above its normal levels. The rain forced the creek bottom silt to rise, which led to increased pollutants.
- Wind—On September 4–5, a northeasterly wind blew from Frog Creek toward Arrowhead Elementary School's playground.

What Can We Do to Help?

Could heat, rain, and wind lead to similar situations in the future? Yes. But . . . the incidents from last week were very rare. Please do not expect a repeat occurrence any time soon. Tell your children to enjoy Frog Creek for its beauty and natural resources.

We would be happy to meet with you and your children to explain the science of this environmental event. Plus, we'd like to provide techniques for managing simple ailments like nausea, skin rashes, and lightheadedness.

In a one-hour workshop, presented during the school day or at a Parent-Teacher Organization meeting, FCWTP could offer the following information:

- Scientific data about stream and creek pollutants—with hands-on tutorials for your students.
- The effects of rain, wind, and heat on creek chemicals—complete with graphics and an age-appropriate PowerPoint presentation.
- Useful preventive medical emergency techniques—which every parent and child should know.

Because this letter report is sent to multiple audiences, no reader address is given. Instead, an "attention line" is used.

To communicate effectively with the public, the writer used pronouns and a pleasant tone. Words like "unusual" and "rare" are used to lessen the parents' concerns.

For headings, the writer used questions that an audience of concerned parents and teachers might have.

Because the audience is lay, the writer uses simple sentence structure ("Algae is a good thing"; "The same thing applies to acid levels in water.") and avoids complex discussions on science, focusing on temperature, rain levels, and wind direction.

The writer emphasizes the positive by using words such as "please," "enjoy Frog Creek for its beauty," and "We would be happy to meet with you."

The writer focuses on the audience's future concerns and suggests age-appropriate material to explain the incident to the elementary school children. This attempt at community outreach connects the wastewater treatment plant to concerned constituents.

(Continued)

FIGURE 17.3 Investigative Report in Letter Format (Lay Audience) Continued

> Mike Moore
> Page 2
> September 15, 2009
>
> This information would explain real-world applications for science classes, as well as provide valuable health tips for parents and teachers. We have enclosed for your review information about our proposed training sessions.
>
> Please let us know if you would like to benefit from this free-to-the-public workshop. We would be happy to schedule one at your convenience. Frog Creek Wastewater Treatment Plant (FCWTP) wants to assure you that your child's "safety is our number 1 concern."
>
> Sincerely,
>
> Mike Moore
> Frog Creek Wastewater Treatment Plant Director of Public Relations
>
> Enclosure

Trip Reports

Purpose and Examples

A *trip report* allows you to report on job-related travel. When you leave your work site and travel for job-related purposes, your supervisors not only require that you document your expenses and time while off-site, but they also want to be kept up to date on your work activities.

Criteria

Following is an overview of what you will include in an effective trip report.

1. **Introduction (overview, background)**

   Purpose. In the purpose section, document the date(s) and destination of your travel. Then comment on your objectives or rationale. What motivated the trip, what did you plan to achieve, what were your goals, why were you involved in job-related travel?

   You might also want to include these following optional subheadings:

   Personnel. With whom did you travel?

   Authorization. Who recommended or suggested that you leave your work site for job-related travel?

2. **Discussion (body, findings, agenda)**

   Using subheadings, document your activities. This can include a review of your observations, contacts, seminars attended, or difficulties encountered.

3. **Conclusion/recommendations**

   Conclusion. What did you accomplish—what did you learn, whom did you meet, what sales did you make, and what of benefit to yourself, colleagues, or your company occurred?

   Recommendations. What do you suggest next? Should the company continue on the present course (status quo) or should changes be made in personnel or in the approach to a particular situation? Would you suggest that other colleagues attend this conference in the future, or was the job-related travel not effective? In your opinion, what action should the company take?

Figure 17.4 presents an example of an informal trip report written in memo format. The memo format, providing identification lines (date, to, from, subject), is written to an internal audience and creates a hard-copy document.

Technology Tip

Creating Headers and Footers in Microsoft Word 2007

To create headers and footers (useful for new-page notations in reports), follow these steps:

1. Click on the "**Insert**" tab on your toolbar. You will see the following ribbon.

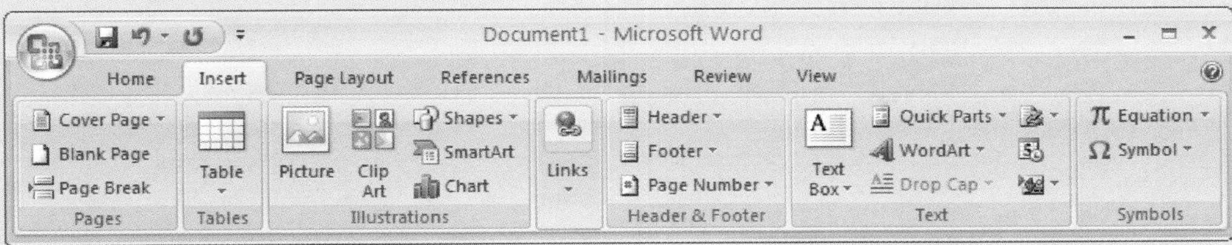

2. Click on either "**Header**" or "**Footer**." When you click on your choice of either "**Header**" or "**Footer**," you will see a drop-down menu, such as the one that follows.

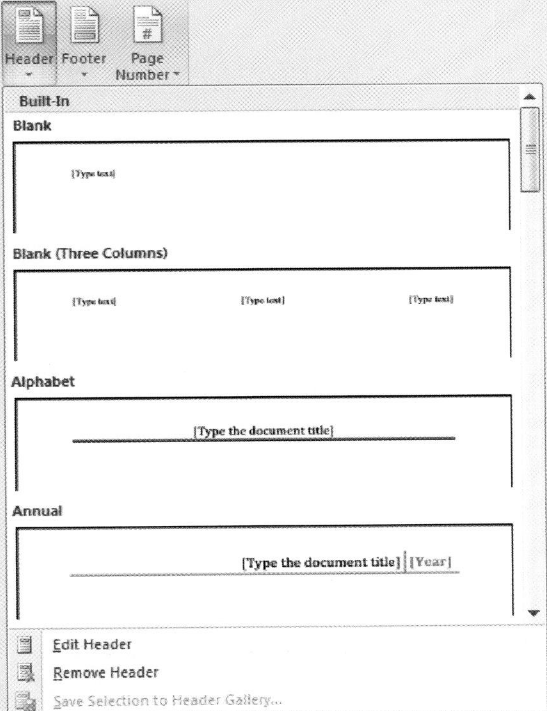

3. Choose the type of "**Header**" or "**Footer**" you want to use in your document and type your content (date, name, page number, etc.), as shown in the following example.

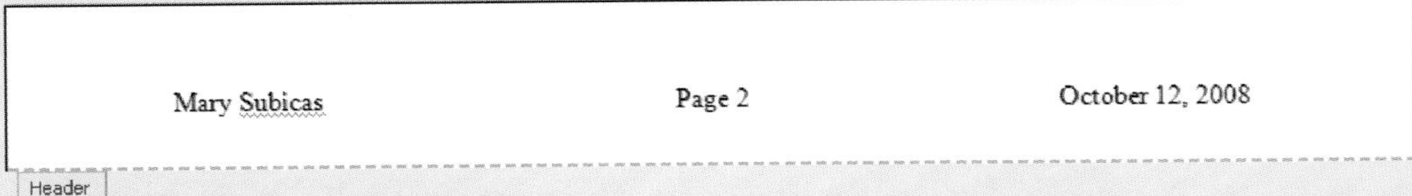

FIGURE 17.4 Trip Report in Memo Format

The "focus"

The "topic"

The introduction section answers the "reporter's questions"—Who, What, When, Where, and Why.

Different "heading levels" and highlighting techniques are used to make the information more accessible, to help the readers navigate your text.

DATE: February 26, 2009
TO: Debbie Rulo
FROM: Oscar Holloway
SUBJECT: Trip Report—Unicon West Conference on Electronic Training

Introduction

Purpose of the Meeting: On Tuesday, February 23, 2009, I attended the Unicon West Conference on Electronic Training, held in Ruidoso, New Mexico. My goal was to acquire hands-on instruction and learn new techniques for electronic training, including the following:

- Online discussion groups
- E-based tutorials
- Intranet instruction
- Videoconference lecture formats

Conference Participants: My coworkers, Bill Cole and Gena Sebree, also attended the conference.

Discussion

Presentations at the Conference:
Gena, Bill, and I attended the following sessions:

- ***Online Discussion Groups***
  This two-hour workshop was presented by Dr. Peter Tsui, a noted instructional expert from Texas State University, San Marcos, Texas. During Dr. Tsui's presentation, we reviewed how to develop online questions for discussion, post responses, interact with colleagues from distant locales, and add to streaming chats. Dr. Tsui worked individually with each seminar participant.

- ***E-Based Tutorials***
  This hour-long presentation was facilitated by Debbie Gorse, an employee of Xenadon E-Learning Inc. (Colorado Springs, Colorado). Ms. Gorse used video and overhead screen captures to give examples of successful E-based, computer-assisted instructional options.

- ***Intranet Instruction***
  Dr. Randy Towner and Dr. Karen Pecis led this hour-long presentation. Both are professors at the University of Nevada, Las Vegas. Their workshop focused on course development, online instructional methodologies, customizable company-based examples, and firewall-protected assessment. The professors provided workbooks and hands-on learning opportunities.

- ***Videoconferencing***
  Denise Pakula, Canyon E-Learning, Tempe, Arizona, spoke about her company's media tools for teleconferenced instruction. These included lapel microphones, multi-directional pan/tilt video cameras, plasma display touch screens, wideband conference phones, recessed lighting ports with dimming preferences, and multimedia terminals.

FIGURE 17.4 Continued

Oscar Holloway
Page 2
February 26, 2009

A "new-page notation" helps your readers avoid losing or misplacing pages.

Conclusion

The conclusion focuses on the primary findings to give the audience direction.

Presentation Benefits:
Every presentation we attended was beneficial. However, the following information will clarify which workshop(s) would benefit our company the most:

1. Dr. Tsui's program was the most useful and informative. His interactive presentation skills were outstanding and included hands-on activities, small-group discussions, and individual instruction. In addition, his online discussion techniques offer the greatest employee involvement at the most cost-effective pricing. We met with Dr. Tsui after the session, asking about his fees for on-site instruction. He would charge only $90 per person (other people we researched charged at least $150 per person). Dr. Tsui's fees should fit our training budget.

2. E-based tutorials will not be a valid option for us for two reasons. First, Xenadon's products are prepackaged and allow for no company-specific examples. Second, Ms. Gorse's training techniques are outdated. Videos and overhead projections will not create the interactivity our employees have requested in their annual training evaluations.

3. The Towner/Pecis workshop was excellent. Intranet instruction would be ideal for our needs. We will be able to customize the instruction, provide participants individualized feedback, and ensure confidentiality through firewall protection. Furthermore, Drs. Towner and Pecis used informative workbooks and hands-on learning opportunities in their presentation.

4. Ms. Pakula's presentation on videoconferencing focused more on state-of-the-art equipment rather than on instruction. We believe that the price of the equipment both exceeds our budget and our needs. Our current videoconference equipment is satisfactory. If the purchasing department is looking for new vendors, they might want to contact Canyon.

Recommendations

The recommendation suggests the next course of action.

Gena, Bill, and I suggest that you invite Dr. Tsui to our site for further consultation. We also think you might want to contact Drs. Towner and Pecis for more information on their training.

Dot.Com Updates

For more information about trip reports, check out the following links:

- This Disney Trip Report Archive lets you read reports for all trips taken during an indicated period, plus you can add your own trip report. http://www.mouseplanet.com/dtp/trip.rpt/
- Vanguard, an independent consulting company specializing in customer contact and convergence, provides their employees a sample trip report template. http://www.vanguard.net/documents/CCT%20Trip%20Report%206-02.doc

Progress Reports

Purpose and Examples

A *progress report* lets you document the status of an activity, explaining what work has been accomplished and what work is remaining. Supervisors and customers want to know what progress you are making on a project, whether you are on schedule, what difficulties you might have encountered, and what your plans are for the next reporting period. Because of this, your audience might ask you to write progress (or activity or status) reports—daily, weekly, monthly, quarterly, or annually.

Criteria

Following is an overview of what you will include in an effective progress report.

1. **Introduction** (**overview, background**)

   Objectives. These can include the following:
   - Why are you working on this project (what's the rationale)?
   - What problems motivated the project?
   - What do you hope to achieve?
   - Who initiated the activity?

   Personnel. With whom are you working on this project (i.e., work team, liaison, contacts)?

   Previous activity. If this is the second, third, or fourth report in a series, remind your readers what work has already been accomplished. Bring them up-to-date with background data or a reference to previous reports.

2. **Discussion** (**findings, body, agenda**)

   Work accomplished. Using subheadings, itemize your work accomplished either through a chronological list or a discussion organized by importance.

   Problems encountered. Inform your reader(s) of any difficulties encountered (late shipments, delays, poor weather, labor shortages) not only to justify your possibly being behind schedule but also to show the readers where you'll need help to complete the project.

   Work remaining. Tell your reader what work you plan to accomplish next. List these activities, if possible, for easy access. A visual aid, such as a Gantt chart or a pie chart, fits well after these two sections. The chart will graphically depict both work accomplished and work remaining.

3. **Conclusion/Recommendations**

   Conclusion. Sum up what you've achieved during this reporting period and provide your target completion date.

   Recommendations. If problems were presented in the discussion, you can recommend changes in scheduling, personnel, budget, or materials that will help you meet your deadlines.

Figure 17.5 presents an example of a progress report.

Feasibility/Recommendation Reports

Purpose and Examples

A *feasibility/recommendation report* accomplishes two goals. First, it studies the practicality of a proposed plan. Then, it recommends action. Occasionally, your company plans a project but is uncertain whether the project is feasible. Will the plan work, does the company have the correct technology, will the idea solve the problem, or is there enough money? One way a company determines the viability of a project is to perform a feasibility study, to document the findings, and then to recommend the next course of action.

FIGURE 17.5 Progress Report in Memo Format

TO: Juan Cisneros
FROM: Pat Smith
DATE: April 2, 2009
SUBJECT: First Quarterly Report—Project 80 Construction

Purpose of Report

In response to your December 20, 2008, request, following is our first quarterly report on Project 80 Construction (Downtown Airport). Department 93 is in the start-up phase of our company's building plans for the downtown airport and surrounding site enhancements. These construction plans include the following:

1. *Airport construction*—terminals, runways, feeder roads, observation tower, parking lots, maintenance facilities.
2. *Site enhancements*—northwest and southeast collecting ponds, landscaping, berms, and signage.

Work Accomplished

In this first quarter, we have completed the following:

1. *Subcontractors.* Toby Summers (Project Management) and Karen Kuykendahl (Finance) worked with our primary subcontractors (Apex Engineering and Knoblauch and Sons Architects). Toby and Karen arranged site visitations and confirmed construction schedules. This work was completed January 12, 2009.

2. *Permits.* Once site visitations were held and work schedules agreed upon, Toby Summers and Wilkes Berry (Public Relations) acquired building permits from the city. They accomplished this task on January 20, 2009.

3. *Core Samples.* Core sample screening has been completed by Department 86 with a pass/fail ratio of 76.4 percent pass to 23.6 percent fail. This meets our goal of 75 percent. Sample screening was completed January 30, 2009.

4. *Shipments.* Timely concrete, asphalt, and steel beam shipments this quarter have provided us a 30-day lead on scheduled parts provisions. Materials arrived February 8, 2009.

5. *EPA Approval.* Environmental Protection Agency (EPA) agents have approved our construction plans. We are within guidelines for emission controls, pollution, and habitat endangerment concerns. Sand cranes and pelicans nest near our building site. We have agreed to leave the north plat (40 acres) untouched as a wildlife sanctuary. This will cut into our parking plans. However, since the community will profit, we are pleased to make this concession. Our legal department also informs us that we will receive a tax break for creating this sanctuary. EPA approval occurred on February 15, 2009.

Problems Encountered

Core samples are acceptable throughout most of our construction site. However, the area set aside for the northwest pond had a heavy rock concentration. We believed this would cause no problem. Unfortunately, when Anderson Brothers began dredging, they hit rock, which had to be removed with explosives.

The introduction explains why the report has been written and what topic will be discussed.

The discussion provides quantified data and dates for clarity, such as "76.4 percent pass" and "January 30, 2009." The discussion also clarifies who worked on the project and lists other primary contacts.

A "Problems Encountered" section helps justify delays and explain why more time, personnel, or funding might be needed to complete a project.

(Continued)

FIGURE 17.5 Progress Report in Memo Format Continued

Numbered points with italicized sub-headings help readers access the information more readily. This is especially valuable when an audience needs to refer to documents at a later date.

The conclusion sums up the overall status of the project: "15 percent" complete.

The recommendation explains how the problems discussed in the "Problems Encountered" section can be solved and what is needed to complete the job: "additional personnel" and "increased funds." It also states who to contact for help, "Darlene Laughlin."

Pat Smith
Page 2
April 2, 2009

Since this northwest pond is near the sand crane and pelican nesting sites, EPA told us to wait until the birds were resettled. The extensive rock removal and wait for wildlife resettlement have slowed our progress. We are behind schedule on this phase.

This schedule delay and increased rock removal will affect our budget.

Work Remaining

To complete our project, we need to accomplish the following:

1. *Advertising*. Our advertising department is working on brochures, radio and television spots, and highway signs. Advertising's goal is to make the construction of a downtown airport a community point of pride and civic celebration.

2. *Signage*. With new roads being constructed for entrance and exit, our transportation department is working on street signage to help the public navigate our new roads. In addition, transportation is working with advertising on signage designs for the downtown airport's two entrances. These signs will juxtapose the city's symbol (a flying pelican) with an airplane taking off. The goal is to create a logo that simultaneously promotes the preservation of wildlife and suggests progress and community growth.

3. *Landscaping*. We are working with Anderson Brothers Turf and Surf to landscape the airport, roads, and two ponds. Our architectural design team, led by Fredelle Schneider, is selecting and ordering plants, as well as directing a planting schedule. Anderson Brothers also is in charge of the berms and pond dredging. Fredelle will be our contact person for this project.

4. *Construction*. The entire airport must be built. Thus, construction comprises the largest remaining task.

Project Completion

Though we have just begun this project, we have completed approximately 15 percent of the work. We anticipate a successful completion, especially since deliveries have been timely.

Only the delays at the northwest pond site present a problem. We are two weeks behind schedule and $3,575 over cost. With approximately ten additional personnel to speed the rock removal and with an additional $2,500, we can meet our target dates. Darlene Laughlin, our city council liaison, is the person to see about corporate investors, city funds, and big-ticket endowments. With your help and Darlene's cooperation, we should meet our schedules.

(Continued)

FIGURE 17.5 **Continued**

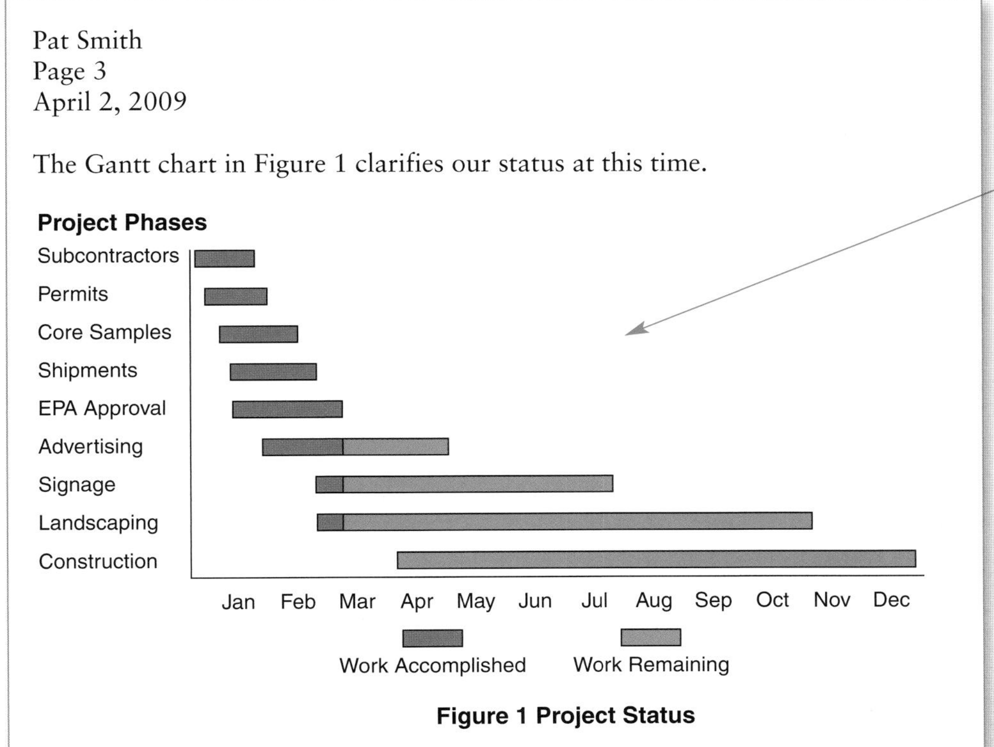

A graphic not only adds interest to the report, but also helps quantify information and clarify details visually.

FAQs

Q: Do companies actually ask employees to write short reports? If so, when, why, to whom, and how often do you write reports?

A: The answer is "yes"! You'll be asked to write reports more often than you'd imagine. Look at what one employee says:

> "I work for a small engineering and IT consulting company that contracts to various federal government agencies. My engagement is with the USDA Farm Service Agency. That is the agency that handles most of the agricultural subsidies derived from the Farm Bill. I work for a department there called the Architecture and Management Center, which is part of the CIO's organization.
>
> The reason that I do two status reports each week is that they are for different audiences. The individual report is for my consulting firm. They manage the allocation of resources to our task elements, so they want to keep an eye on how each person is utilized. The team report is for our USDA client. They are more interested in how the task elements are proceeding. Each task element has a number of people assigned, and one of those people updates the client weekly. Beyond that, the task elements are broken down into activities that can take anywhere from a few weeks to a whole year to complete. For each of those activities, according to its project plan, we will have various reports that are either about progress, issues, findings and recommendations, research, or major decisions. We also frequently produce 'position papers' that advise the client on points of strategy, usually of a technical nature.
>
> Although these products are developed for various audiences, my consulting company is very interested in all of the reports and other documents we produce. They glean material from any source available to use in future proposals."

Criteria

The following are components of an effective feasibility report.

1. **Introduction (overview, background)**

   Objectives. Under this subheading, you can answer any of the following questions:

   - What is the purpose of this feasibility report? Until you answer this question, your reader doesn't know. As mentioned earlier in this chapter, it's false to assume prior knowledge on the part of your audience. One of your responsibilities is to provide background data. To answer the question regarding the report's purpose, you should provide a clear and concise statement of intent.
   - What problems motivated this study? To clarify for your readers the purposes behind the study, *briefly* explain either
     - What problems cause doubt about the feasibility of the project (i.e., is there a market; is there a piece of equipment available that would meet the company's needs; is land available for expansion?).
     - What problems led to the proposed project (i.e., current equipment is too costly or time consuming; current facilities are too limited for expansion; current net income is limited by an insufficient market).
   - Who initiated the feasibility study? List the name(s) of the manager(s) or supervisor(s) who requested this report.

   Personnel. Document the names of your project team members, your liaison between your company and other companies involved, and your contacts at these other companies.

2. **Discussion (body, findings)**

   Under this subheading, provide accessible and objective documentation.

   Criteria. State the criteria upon which your recommendation will be based. Criteria are established so you have a logical foundation for comparison of personnel, products, vendors, costs, options, schedules, and so on.

   Analysis. In this section, compare your findings against the criteria. In objectively written paragraphs, develop the points being considered. You might want to use a visual such as a table to organize the criteria and to provide easy access.

3. **Conclusion/recommendations**

   Conclusion. In this section, you go beyond the mere facts as evident in the discussion section: You state the significance of your findings. Draw a conclusion from what you have found in your study. For example, state that "Tim is the best candidate for director of personnel" or "Site 3 is the superior choice for our new location."

   Recommendations. Once you have drawn your conclusions, the next step is to recommend a course of action. What do you suggest that your company do next? Which piece of equipment should be purchased, where should the company locate its expansion, or is there a sufficient market for the product?

Figure 17.6 presents an example of a feasibility/recommendation report.

Meeting Minutes

Purpose and Examples

Meeting minutes document the results of a meeting—what was discussed, proposed, voted on, and planned for future meetings. As the recording secretary for your project team, a community organization, your city council, or a departmental meeting, your job is to record the meeting's minutes. In meeting minutes, you record who attended the meeting, when it began, when it ended, and where it took place. You report on the topics discussed, decisions arrived at, and plans for the future.

FIGURE 17.6 Feasibility Report in Memo Format

FROM: Cindy Katz, Director of Information Technology
TO: Shamir Rammalah, Accounts Payable
DATE: August 13, 2009
SUBJECT: Feasibility Study for Technology Purchases

Purpose of the Report

The purpose of this report is to study which technology will best meet your communication needs and budget. After analyzing the feasibility of various technologies, we will recommend the most cost-effective technology options.

Technology Problems

According to your memo dated August 1, 2009, your department needs new communication technologies for the following reasons:

- Your department has hired three new employees, increasing your headcount to ten.
- Currently, your department has only five laptops. This slows down the registration process because we cannot have all ten employees registering students simultaneously.
- We also must update training because WIFINation has purchased new software.
- Accounts Payable's current printer is also insufficient due to your increased headcount.

Vendor Contacts

Our vendor contacts for the laptops, printers and software are as follows:

Electek	**Tech On the Go**	**Mobile Communications**
Steve Ross	Jay Rochlin	Karen Allen
stever1@electek.com	jrochlin@tog.com	karen.allen@mobcom.net

Criteria for Vendors

The following criteria were considered to determine which communication technology would best meet your department's needs:

1. *Trainers*—Because of the unique aspects of the software, we need trainers who are familiar with password creation, privacy laws, malware, spyware, and corporate e-mail policy.
2. *Maintenance*—We need to purchase equipment and software complete with either quarterly or biannual service agreements (at no extra charge).
3. *Service Personnel*—The service technicians should be certified to repair and maintain whatever hardware we purchase. In addition, the vendors must also be able to train our personnel in hardware usage.
4. *Warranties*—The warranties should be for at least one year with options for renewal.
5. *Cost*—The total budgeted for your department is $15,000.

Needs Assessment

Purchasing agrees that the Accounts Payable hardware and software needs exceed their current technology. Not only are the department's laptops and printer insufficient in number, but also they do not allow the personnel to access corporate e-mail, the Internet, or word processing packages. Updated equipment is necessary.

The "Purpose" reminds the reader why this report has been written and what the report's goal is.

The "Problem" details what issues have led to this report.

The "Vendor" section provides contact information (names and e-mail addresses).

The "Criteria" states the topics used to research the report and includes precise details. In this way, the audience can understand the rationale for later decisions.

(Continued)

FIGURE 17.6 Feasibility Report in Memo Format Continued

Cindy Katz
Page 2
August 13, 2009

Vendor Evaluation

- **Electek**—Having been in business for ten years, this company is staffed by highly trained technicians and sales staff. All Electek employees are certified for software training. The company promises a biannual maintenance package and subcontracted personnel if employees cannot repair hardware problems. They offer manufacturers' guarantees with extended service warranties costing only $100 a year for up to five years. Electek offers 20 percent customer incentives for purchases of over $2,000.
- **Tech On the Go** (**TOG**)—This company has been in business for two years. TOG provides only subcontracted service technicians for hardware repair. TOG's employees are certified in software training. The owners do not offer extended warranty options beyond manufacturers' guarantees. No special customer pricing incentives are offered though TOG sells retail at a wholesale price.
- **Mobile Communications**—Having been in business for five years, Mobile has certified technicians and sales representatives. All repairs are provided in house. The company offers quarterly maintenance at a fee of $50 ($200 per year). Mobile offers a customer incentive of 10 percent discounts on purchases over $5,000.

The organizational mode comparison/contrast is used to analyze the strengths and weaknesses of each vendor.

Cost Analysis

- Laptops—Accounts Payable requests one laptop per departmental employee. Our analysis has determined that the most affordable laptops we can purchase (with the requested software and wireless Internet connections) would cost $1,500 per unit. Thus, ten laptops would cost $15,000. Even with discounts, this exceeds your department's budget.
- Printers—Accounts Payable requests three additional printers, with the capability to print double-side pages, staple, collate, print in color, and print three different sizes of envelopes and three different sizes of paper. Our analysis has found that the most affordable printers meeting your specifications would cost $2,500 each. Again, when combining this cost with that of laptops, you exceed your budget.
- Training—The manufacturer can provide training on our new software. Training can be offered Monday through Friday, starting in September. The manufacturer says that effective training must entail at least 20 hours of hands-on practice. The cost for this would be $5,000.

The discussion provides specific details to prove the feasibility of the plan or project.

The following table compares the three vendors we researched on a scale of 1–3, 3 representing the highest score.

Graphics depict the findings more clearly and more concisely than a paragraph of text.

Table 1: Criteria Comparison

Criteria	Electek	TOG	Mobile
Maintenance	3	2	3
Personnel	3	3	3
Warranties	3	2	2
Cost	3	2	2
Total	12	9	10

FIGURE 17.6 Continued

Cindy Katz
Page 3
August 13, 2009

Summary of Findings

We cannot purchase the number of laptops and printers you have requested. Doing so exceeds the budget. Training is essential. Your department must adjust its budget accordingly to accommodate this need.

All three vendors have the technology you require. However, TOG and Mobile do not meet the criteria. In particular, these companies do not provide either the maintenance packages, warranties, or pricing required.

Recommended Action

Given the combination of cost, maintenance packages, warranties, and service personnel, Electek is our best choice.

We suggest the following options for printers and laptops: purchasing five laptops instead of ten; purchasing one additional printer instead of three; and/or sharing printers with nearby departments.

The conclusion sums up the findings, explaining the feasibility of a course of action—why a plan should or should not be pursued.

The recommendation explains what should happen next and provides the rationale for this decision.

Criteria

Meeting minutes should include the following key components:

1. **Introduction**
   Include the following:
   - **Date, time, and place**—At the beginning of the minutes, list the date on which the meeting is held, what time the meeting began, and where the meeting was located.
     September 7, 2009
     7:00 p.m.
     Conference Room C
   - **Attendees**—List the names of those who attended the meeting.
   - **Approval of last meeting's minutes**—After asking participants to read the last meeting's minutes, vote to accept them.
2. **Discussion (findings, agenda)**
   This is the most important part of your minutes. In this long section, you report on the agenda items. Recording secretaries have shared with us a major concern. They believe that their job requires them to report *every* word that has been spoken. This is not true. In fact, *Robert's Rules of Order* says exactly the opposite:
   - A reporting secretary should keep a record of the proceedings, stating what was done and *not* what was said. Your job is not to report every committee members' comments. Instead, focus on decisions made, conclusions arrived at, issues confronted, opposing points of view, and votes taken.
   - If resolutions are agreed upon, the reporting secretary must document the wording of this resolution exactly.
   - Finally, all content within this discussion section must be reported objectively, without "criticism, favorable or otherwise" (*Robert's Rules of Order Revised*).

3. Conclusion
   - **Old/new business**—The last topic of a meeting's agenda should be a review of any "old" topics still unresolved and needing further discussion. Similarly, toward the end of a meeting, you will need to report on new topics, perhaps ones that will need to be covered in future meetings.
   - **Next meeting**—Report when the committee will meet next, providing the date, time, and location.
   - **Time of adjournment**—Report when the meeting ended.
   - **Signature**—A typical lead-in statement reads "Respectfully submitted by ______." You can sign your name beneath the typed signature (unless the minutes will be submitted electronically).

Organization

Typically, meeting minutes are organized chronologically. When you write your minutes, you can organize the content according to the sequential order of the topics discussed. This sequential organization allows readers to follow a meeting's agenda. However, minutes also can be organized according to importance. Though chronology is an easy approach to documenting a meeting, your audience might better understand the meeting's focus if you document which topics received the greatest discussion or which topics will have the largest impact.

Layout

Reporting secretaries often contend that they must write essay-like reports, following a paragraph format. This is neither true nor advisable. In fact, an essay-like report, including wall-to-wall words, will turn off your readers. Instead, use headings, subheadings, and bulleted lists to make the minutes more accessible. Figure 17.7 shows an example of meeting minutes.

Use the Short, Informal Report Checklist to help you in writing your short report.

SHORT, INFORMAL REPORT CHECKLIST

____ 1. Have you chosen the correct communication channel (e-mail, letter, or memo) for your short, informal report?

____ 2. Does your subject line contain a topic and a focus? If you write only "Subject: Trip Report" or "Subject: Feasibility/Recommendation Report," you have not communicated thoroughly to your reader. Such a subject line merely presents the focus of your correspondence. But what's the topic? To provide both topic and focus, you need to write "Subject: Trip Report on Solvent Training Course, ARCO Corporation—3/15/09" or "Subject: Feasibility Report on Company Expansion to Bolker Blvd."

____ 3. Does the introduction explain the purpose of the report, document the personnel involved, or state when and where the activities occurred?

____ 4. When you write the discussion section of the report, do you quantify what occurred? In this section, you must clarify precisely. Supply accurate dates, times, calculations, and problems encountered.

____ 5. Is the discussion accessible? To create reader-friendly ease of access, use highlighting techniques, such as headings, boldface, underlining, and itemization. You also might want to use graphics, such as pie charts, bar charts, or tables.

____ 6. Have you selected an appropriate method of organization in your discussion? You can use chronology, importance, comparison/contrast, or problem/solution to document your findings.

____ 7. Does your conclusion present a value judgment regarding the findings presented in the discussion? The discussion states the facts; the conclusion decides what these facts mean.

____ 8. In your recommendations, do you tell your reader what to do next or what you consider to be the appropriate course of action?

____ 9. Have you effectively recognized your audience's level of understanding (specialist, semi-specialist, lay, management, subordinate, colleague), multiple, internal, or external and written accordingly?

____ 10. Is your report accurate? Correct grammar and calculations make a difference. If you've made errors in spelling, punctuation, grammar, or mathematics, you will look unprofessional.

FIGURE 17.7 **Meeting Minutes**

Employee Benefits Project Team
Meeting Minutes

Date: April 12, 2009

Time: 7:00 p.m.–9:30 p.m.

Objectives: In this second project team meeting, the goal was to discuss employee benefits options that could be added to the company's cafeteria plan.

Attendees: Stacy Helgoe, Christie Pieburn, Darren Rus, Andrew McWard, Bill Lamb, Jessica Studin

Team Leader: Phyllis Goldberg, Director of Employee Benefits

Agenda:

1. Flex-Time—Stacy and Christie reported on flex-time options. They stated that in a survey dated March 15, 76 percent of employees favored flex-time benefits. These included beginning our workdays at 7:00 a.m. and extending work hours until 6:30 p.m. Doing so would allow employees to either arrive for work earlier than standard and/or work later than usual. This work-schedule flexibility would allow employees to manage childcare needs, appointments, and other concerns that might fall within the traditional workday.

   The downside of flex-time, according to management, includes increased security coverage, increased utilities (heating, lighting, etc.), and inconsistent coverage of office hours. The benefits of flex-time include increased employee morale, less congestion in the parking lots at traditionally prime times, and employee empowerment.

   Our project team voted 6–1 to promote flex-time to management at next month's meeting.

2. Stock Options—At last month's meeting, Darren was asked to study the possibility of adding stock options to our cafeteria plan. He met with accounting, our external auditor, and three proprietary stock brokerage companies (Bull and Bear, Market Trend, and BuyItNow). Please see the attached brokerage firm proposals.

   However, both accounting and our auditor convinced Darren that this year would not be the best time to offer our employees stock options. Our profit margin ratio is down, we cannot meet our quarterly forecasts, and we cannot distribute dividends to stockholders.

   Future European expansion plans should increase our revenues. Therefore, our project team voted 5–2 to table this issue until the end of next quarter.

3. Personal Leave Days—Currently, our employees are allotted five sick leave days a year. Our proposal asks for an additional two personal leave days. These days would have to be used within the calendar year. The personal leave days would not roll over, nor could they be banked. Andrew McWard surveyed the employees regarding this topic. Ninety-eight percent of those surveyed favored the personal leave proposal.

(Continued)

FIGURE 17.7 **Meeting Minutes Continued**

Page 2
Meeting Minutes
April 12, 2009

Our project team voted 6–1 to promote personal leave days as a cafeteria plan option.

4. Sick Bank—Last year, a survey suggested that employees might be willing to share their unused sick leave days with other employees facing catastrophic illnesses or family emergencies. This is a common practice at other corporations of our size. This not only gives employees a feeling of ownership, but also it is a method of team building.

   Bill revisited this issue with employees. He attended departmental meetings across the company, asking for input. Bill reported that employees do not seem to favor this proposal at the moment. His informal survey shows only a 30 percent interest.

   Unless these numbers change, our committee sees no reason to pursue a sick bank. We voted 4–3 against proposing a sick bank option.

5. Short- and Long-Term Disability—Jessica visited three insurance carriers (In-Need Insurance, Evergreen Insurance Inc., and Employee Associates General). Her goal was to get quotes on short- and long-term disability insurance. Our employee survey ranked this as a number one priority.

   Our committee agreed with the survey and highly recommended this as an add-on to our employee cafeteria plan (7–0 vote). In fact, without this benefit, employees with illnesses longer than five days or any banked hours have no salary or job insurance. This fact will negatively impact both employee retention and recruitment. Most major firms offer this benefit. To stay competitive, we must do so also.

Old Business—Andrew reminded us that the company holiday party and family picnic needed to be planned. We will contact the hospitality committee to determine their status.

New Business—In next month's meeting, we will focus on ways to implement our recommendations.

Respectfully submitted by Jessica Studin, Recording Secretary

Attachment:
Brokerage Firm Proposals
Insurance Proposals

Linda Met Her Challenge

To meet her communication challenge, Linda used the *P*³ process.

Planning

To plan her short report, Linda considers the following:

- Goal—explain to the IRS her client's tax issues and request an abatement
- Audience—specialists (IRS representatives)
- Channels—short report
- Data—reasons for her client's failure to meet the IRS deadlines, including the previous accountant's actions

Linda's first challenge was to find out what led to the IRS penalty. First, she read and analyzed the IRS letter. Next, she met with the client for the purpose of fact finding. Linda says, "I asked questions such as the following: 'How did this happen?' 'Who was in charge when this started?' 'What outside parties had access to your documents?' 'Why are we missing the stream of paperwork that documents this occurrence?' Then, I called the other companies involved, including previous accounting firms and the company which managed my client's profit-sharing plan. Once I discovered all the facts, I started outlining the letter." Linda's outline appears in Figure 17.8.

Packaging

In Linda's problem-to-solution rough draft, she began with a "Sob Story" to diminish her client's culpability. For example, she noted that her client had relied on the credibility of those whom they had hired to file the forms, pay the taxes, and manage the profit-sharing account. The second part of the draft conveyed to the IRS what her client planned to do to avoid future problems. This included changing accounting firms, implementing rules and regulations, and changing the profit-sharing management company. Finally, Linda closed the draft by requesting that the IRS waive the penalty. Figure 17.9 shows her rough draft.

Perfecting

Linda follows a precise procedure for perfecting her letter report. First, Linda focuses on tone and style. Here's her communication philosophy: "I try to communicate with a client the way I communicate with my parents. I write respectfully but simply enough for them to understand."

Next, she has a colleague read her draft. As Linda says, "You're so deep in the story that you might have left out facts that make the story flow." The colleague can also tell Linda if anything is unclear or if information is missing. Once Linda is happy with the revisions suggested by her colleague, she sends the letter report to her company's production staff. Their job is to assemble, type, and save the report in the company's files. The production staff makes sure that the report meets the company's standards for font, margin, and design. See Figure 17.10 for Linda's perfected letter report.

FIGURE 17.8 Outline of Linda's Letter Report

I. Administrative changes in the company
  A. Third-party administrator hired to handle profit-sharing plan
  B. Change in third-party administrator of profit-sharing plan
  C. Company administrative assistant did not notify owner of IRS notice
  D. Company administrative assistant fired
  E. New administrative assistant did not know rules

II. IRS sent second notice
  A. Company contacted former third-party administrator
  B. Original accounting firm had also undergone changes
  C. New accounting firm fired

III. New accounting firm hired
  A. Owner made changes to handling of correspondence and filing of all tax reports and returns

IV. Request for waiving of penalties due to
  A. Unfortunate failure of office assistant
  B. Third-party administrator was supposed to file an extension request
  C. Third-party administrator has been terminated

FIGURE 17.9 Linda's Rough Draft Letter Report with Colleague Comments

Replace "Dear Sirs" with a more informative subject line.

Linda, let's break this text up with headings.

I notice some repetition such as "several" (shown in red). Also, phrases like "made a decision in the year 2006" can be revised for conciseness. Let's check for these style concerns throughout the letter.

Let's avoid such an excessively long paragraph and make the information more accessible. How about trying bullets?

Linda, I think that this should read "staff . . . was" for agreement.

September 29, 2009

Internal Revenue Service
Ogden, UT 84201

RE: PhotoFinish, Inc. Profit-Sharing Plan
3900021
Year ended: April 30, 2007

Dear Sirs:

PhotoFinish has received notices indicating that there is an outstanding penalty for late filing of the profit-sharing plan return. They have asked for our assistance in resolving this issue.

PhotoFinish has been through several administrative changes over the past several years. Several years ago, a profit-sharing plan was established by PhotoFinish. A third-party administrator was hired to handle the necessary calculations, recordkeeping, preparation of returns, and other administrative functions. This third-party administrator handled these functions for the plan for several years, without incident. PhotoFinish made a decision in the year 2006, to terminate this profit-sharing plan and utilize another vehicle for employee savings. The plan was terminated as of December 31, 2006, and the third-party administrator handled the filing of the final returns, and the transfer of funds to another qualified account.

During this same time period, PhotoFinish underwent several personnel changes. This is a small company. The office administrative functions have been handled by the owner along with an assistant. The owner was primarily responsible for sales calls, purchasing, and marketing responsibilities. The assistant handled the daily bookkeeping, correspondence, and phone calls. During 2007, the IRS issued a notice indicating that a late filing penalty had been assessed regarding the profit-sharing plan for the April 30, 2006 year. The assistant received this correspondence, but did not call this to the owner's attention, or call on the third-party administrator for assistance. Due to job performance problems, this assistant was terminated, and another office worker was hired to take her place. At that time, numerous items of correspondence, unfiled paperwork, and other issues were uncovered. The new assistant was unaware of the full extent of any late filings, or other open issues, and was only able to locate small portions of related correspondence regarding this late filing over a period of several months.

Subsequently, in 2007, another notice was issued by the IRS, indicating that the penalty was still outstanding, along with corresponding interest charges. The company contacted their former third-party administrator for assistance at that time. The accounting firm that had handled this filing had also undergone significant personnel changes. The staff who had serviced this account were no longer there. While the accounting office's records indicated that the Form 5500's had been filed timely each year, the actual mailing receipts and original extension applications could not be located. By this time, the accounting firm's involvement with administering this plan had been completed for slightly over four years, so it was difficult to provide any further investigation into the problem. However, the accounting firm had always provided timely service, and did not have any reason to suspect that the returns had not been submitted timely or had not been received by the IRS.

FIGURE 17.9 Continued

In order to provide better office procedures and office controls for the company, the owner consulted a new accounting firm for assistance with their accounting review, tax return preparation, and other consulting work. Changes have been made to the handling of the office correspondence, and the filing of all tax reports and returns for the company.

The company asks for these circumstances to be considered in requesting an abatement of this late filing penalty. They did not have a history of late filings for their profit-sharing plan Form 5500, and they had been utilizing the services of a reputable third-party administrator to handle the return filings, extension applications, and maintain the necessary documentation. Due to the unfortunate failure of the office assistant to take action on the notices, or alert anyone of the notices, the company was unaware of this issue until a long period of time had passed. As soon as they were aware of the outstanding penalty, they took appropriate action. They have every reason to believe that an extension was properly filed, due to the normal course of action taken by their former third-party administrator, but due to changes in that office, are unable to provide further details. PhotoFinish has hired a new accounting firm to provide oversight to prevent future omissions in dealing with compliance filings and other paperwork. The profit-sharing plan has already been terminated and no other filings are outstanding at this time.

PhotoFinish does feel that reasonable efforts were made by them, and that reasonable cause should apply in this situation. We ask for abatement of the late filing penalty as well as any related interest charges. Your consideration in this matter is greatly appreciated.

Sincerely,

Linda

This would be a good place for a heading, something like "Corrective Action Taken."

This sentence has 41 words; I think it's too long and difficult to understand.

Linda, my English teachers always told me to avoid using passive voice.

FIGURE 17.10 Perfected Letter Report

MARKS, NELSON, VOHLAND & CAMPBELL, LLC
7701 College Blvd., Suite 150
Overland Park, KS 66210

September 29, 2009

Internal Revenue Service
Ogden, UT 84201

Subject: **PhotoFinish, Inc. Profit-Sharing Plan**
Case Number: 3900021
Year ended: April 30, 2007

(Continued)

FIGURE 17.10 Perfected Letter Report Continued

Page 2 September 29, 2009

PhotoFinish received notices indicating that there is an outstanding penalty for late filing of the profit-sharing plan return. They have asked for our assistance in resolving this issue.

History of PhotoFinish's Profit-Sharing Plan

PhotoFinish has been through several administrative changes over the past five years. Ten years ago, a profit-sharing plan was established by PhotoFinish. A third-party administrator was hired to handle the necessary calculations, recordkeeping, preparation of returns, and other administrative functions. This third-party administrator handled these functions for the plan for five years, without incident. PhotoFinish decided in December 31, 2006, to terminate this profit-sharing plan and utilize another vehicle for employee savings. The third-party administrator handled the filing of the final returns, and the transfer of funds to another qualified account.

During this same time period, PhotoFinish underwent two personnel changes. The sequence leading to PhotoFinish's problems is as follows:

- The office administrative functions have been handled by the owner along with an assistant. The owner was primarily responsible for sales calls, purchasing, and marketing responsibilities. The assistant handled the daily bookkeeping, correspondence, and phone calls.
- During 2007, the IRS issued a notice indicating that a late filing penalty had been assessed regarding the profit-sharing plan for the April 30, 2006, year. The assistant received this correspondence, but did not call this to the owner's attention, or call on the third-party administrator for assistance.
- Due to job performance problems, this assistant was terminated, and another office worker was hired to take her place. At that time, numerous items of correspondence, unfiled paperwork, and other issues were uncovered. The new assistant, unaware of the full extent of any late filings or other open issues, was only able to locate small portions of related correspondence regarding this late filing over a period of several months.

Subsequently, in 2007, another notice was issued by the IRS, indicating that the penalty was still outstanding, along with corresponding interest charges. PhotoFinish contacted their former third-party administrator for assistance at that time. The accounting firm that had handled this filing had also undergone significant personnel changes. The staff who had serviced this account was no longer there. While the accounting office's records indicated that the Form 5500's had been filed timely each year, the actual mailing receipts and original extension applications could not be located. By this time, the accounting firm's involvement with administering this plan had been completed for slightly over four years, so it was difficult to provide any further investigation into the problem. However, the accounting firm had always provided timely service, and did not have any reason to suspect that the returns had not been submitted timely or had not been received by the IRS.

Corrective Action Taken by PhotoFinish

To provide better office procedures and office controls for PhotoFinish, the owner consulted a new accounting firm for assistance with their accounting review, tax return preparation, and other consulting work. Changes have been made to the handling of the office correspondence, and the filing of all tax reports and returns for the company.

FIGURE 17.10 Continued

Page 3 September 29, 2009

PhotoFinish's Request for Penalty Waiver

PhotoFinish asks you to waive the filing penalty for the following reasons:

1. They did not have a history of late filings for their profit-sharing plan Form 5500.
2. They had been using the services of a reputable third-party administrator to handle the return filings, process extension applications, and maintain the necessary documentation.
3. Due to the failure of the office assistant to take action on the notices, or alert anyone of the notices, PhotoFinish was unaware of this issue until five years had passed. As soon as they were aware of the outstanding penalty, they took appropriate action.
4. PhotoFinish believed that an extension was properly filed, due to the normal course of action taken by their former third-party administrator.
5. Due to changes in the third-party administrator, PhotoFinish cannot provide further details.
6. PhotoFinish has hired a new accounting firm to prevent future omissions in dealing with compliance filings and other paperwork.
7. The profit-sharing plan administrator has already been terminated, and no other filings are outstanding.

PhotoFinish has made reasonable efforts to solve this filing issue, and reasonable cause should apply in this situation. We ask for abatement of the late filing penalty as well as any related interest charges. Thank you for helping us with this problem.

Sincerely,

Linda M. Freeman, C.P.A.

Linda says, "Following the P^3 process leads to my success as a written communicator. Planning, packaging, and perfecting allow me to create well-written, carefully edited documents. I have a 95 percent 'win ratio' when communicating with the IRS. What a great testament to my success as a writer."

CHAPTER HIGHLIGHTS

1. Reports are used to document many different occurrences on the job and are written to both internal and external audiences.
2. Use headings and talking headings, such as "Introduction," "Discussion," "Conclusion/Recommendations," "Projected Cost of the Project," "Information About Laptops," and "Needs Assessment for Travel-Related Expenses" when designing your report.
3. E-mail, letters, and memos are effective communication channels for short, informal reports.
4. Progress reports recount work accomplished and work remaining on a project.
5. Feasibility/recommendation reports are used to determine the viability of a proposed project.
6. Outlining is a planning technique that will help you write effective short, informal reports.
7. An incident report documents an unexpected problem that has occurred.
8. An investigative report asks you to examine the causes behind an incident.
9. Meeting minutes document the results of a meeting.
10. Using the P^3 process helps you write more effective reports.

WWW *Find additional short, informal report exercises, samples, and interactive activities at http://www.prenhall.com/gerson.*

CASE STUDIES

Read the following case studies and write the appropriate report.

1. Edith Isaacson is a salesperson at HFA Insurance. She and her coworker, Joan O'Brien, made a presentation to potential clients at Commerce Bank and Trust (2981 Riverwalk Dr., San Antonio, TX 77229). When Edith and Joan returned to their 1999 Ford Taurus company car, parked in the bank visitor's parking lot, they saw that their car had been sideswiped. Underneath the car's driver-side window wiper was a note signed by Judy Towsky, stating "I'm sorry, but I accidentally hit your car while backing up. I had to pick up my child from daycare, so I couldn't wait for you. I've left my insurance company name and number (SafetyNet Insurance, 1-888-555-2121) and my cell phone number is 1-888-555-1212. Sorry for the inconvenience. Please call my insurance company."

The car's left rear bumper was dented to such an extent that the car was not drivable. Edith called HFA's roadside assistance. The tow truck arrived within 30 minutes to take the car in for repairs. Two days later, Edith learned that ABC Auto Repair estimated that the car needed $2,459 in repair (rear panel replacement, paint, new bumper, new tire, and realignment). Repairs would take 14 days minimum. Edith needed $39 a day for a rental car.

Assignment

Write an incident report for Edith. Begin by stating the purpose of the report and introducing the individuals involved. Develop the report's content by explaining the causes and results of the incident. Finally, conclude by suggesting a follow-up action.

2. Cindy Kaye, Director of Administrative Technology at EFA Incorporated, needs to write a trip report. She just returned from her weekly meeting in Philadelphia and must report on her activities. While in Philadelphia, Cindy accomplished many of her regular duties. These included attending the weekly update meeting (where all involved parties report on their last week's activities); the ongoing technology training sessions, teaching employees how to use the company's intranet system for online reporting and to access online benefits information; breakout sessions on problems encountered; and meetings with vendors to assess options for new hardware and software purchases. Cindy also was involved in two new initiatives: discussions of how to train employees on the new technology improvements and ways to increase revenue to pay for technology changes.

Cindy encountered one new problem in her last trip to Philadelphia. The airline carrier she typically traveled on announced its bankruptcy. She had been traveling for free on this airline due to her banked air miles. She will lose this advantage if she can no longer travel on her preferred carrier. Cindy must find affordable options.

Assignment

Based on the information provided, write a trip report for Cindy, addressed to her supervisor, Dr. Susan Hart, Vice President of Administrative Services. In this report, document Cindy's activities. In addition, through research, provide technology training options for her employees and suggest ways to increase revenue. You can go online to a search engine and type in "technology training" and "how to increase revenue" (or similar synonyms) to find options for Cindy. In addition, brainstorm with your classmates reasonable ways that Cindy can travel affordably from Miami to Philadelphia.

ETHICS IN THE WORKPLACE CASE STUDY

Jason White is the purchasing agent for a computer software company. All of the over 100 computer consultants who sell the company's software travel throughout the country to meet with clients. The consultants need PDAs, handheld computers, and cell phones to stay in touch with management, coworkers, and customers.

Jason knows that he can purchase these PDAs from many sources. Usually, Jason's company requires a minimum of three bids for major purchases. However, a good friend of his has promised to sell Jason the PDAs at "the lowest price possible." Jason wants to save time, save money, and do his friend a favor. To explain his decision, Jason will write a recommendation report to his boss favoring his friend.

Question: Is it ethical for Jason to go against company policy, even if it saves money and time? Justify your answer based on information provided in Chapter 1.

INDIVIDUAL AND TEAM PROJECTS

1. Write a progress report. The subject of this report can involve a project or activity at work. Or, if you haven't been involved in job-related projects, write about the progress you're making in this class or another course you're taking. Write about the progress you're making on a home improvement project (refinishing a basement, constructing a deck, painting and papering a room). Write about the progress you're making on a hobby (rebuilding an antique car, constructing a computer, making model trains, etc.). Whatever your topic, first plan, then package a draft, and, finally, perfect, revising the text. Follow the criteria presented in this chapter regarding progress reports.
2. Write a feasibility/recommendation report. You can draw your topic either from your work environment or home. For example, if you and your colleagues were considering the purchase of new equipment, the implementation of a new procedure, expansion to a new location, or the marketing of a new product, you could study this idea and then write a report on your findings. If nothing at work lends itself to this topic, then consider plans at home. For example, are you and your family planning a vacation, the purchase of a new home or car, the renovation of your basement, or a new business venture? If so, study this situation. Research car and home options, study the market for a new business, and get bids for the renovation. Then write a feasibility/recommendation report to your family documenting your findings. Whether your topic comes from business or home, gather your data in planning, draft your text in packaging, and then revise in perfecting. Follow the criteria for feasibility/recommendation reports provided in this chapter to help you write the report.
3. Write an incident report. You can select a topic either from work or home. If you have encountered a problem at work, write an incident report documenting the problem and providing your solutions to the incident. If nothing has happened at work lending itself to this topic, then look at home. Has your car broken down, did the water heater break, did you or any members of your family have an accident of any sort, did your dog or cat knock over the vase your mother-in-law gave you for Christmas? Consider such possibilities, and then write an incident report documenting the incident. Follow the criteria for incident reports provided in this chapter, and use the P^3 process (planning, packaging, and perfecting).
4. Revision is the key to good writing. Figure 17.11 shows an example of a seriously flawed progress report that needs revising. To improve this report, form small groups and first decide what's missing according to this chapter's criteria for good progress reports. Then use the perfecting techniques presented in this chapter to rewrite and revise the report.

DEGREE-SPECIFIC ASSIGNMENTS

1. **Marketing and Public Relations.** Your company, PR&U, has been hired to provide advertising options for a client's line of office supply products. Your public relations team has brainstormed options, including building an Internet site, producing

FIGURE 17.11 Flawed Progress Report that Needs Revision

Site Visit—Alamo Manufacturing
November 1

Sam, I visited our Alamo site and checked on the following:

1. Our plant facilities suffered some severe problems due to the recent wind and hail storms. The west roof lost dozens of shingles, leading to water damage in the manufacturing room below. The HVAC unit was submerged by several feet of water, shorting out systems elsewhere in the plant. In addition, our north entryway awning was torn off its foundation due to heavy winds. This not only caused broken glass in our front entrance door and a few windows bordering the entrance, but also the entryway driveway now is blocked for customer access.

2. Maintenance and security failed to handle the problems effectively. Security did not contact local police to secure the facilities. Maintenance responded to the problems far too many hours late. This led to additional water and wind damage.

3. Luckily, the storm hit early in the day, before many of our employees had arrived at work. Still, a few cars were in the parking lot, and they suffered hail damage. Are we responsible?

4. The storm, though hurting manufacturing, will not affect sales. Our 800-lines and e-mail system were unaffected. But, we might have problems with delivery if we can't fix the entryway impediment. I think I have some solutions. We could reroute delivery to one of our new plants, or maybe we should consider direct delivery to our sales staff (short term at least). Any thoughts?

5. Meanwhile, I have gotten on Maintenance's case, and it is working on repairs. Brownfield HVAC service has given us a quote on a new sump pump system that has failsafe programs (not too bad an extra cost, given what we can save in the long run). Plainview Windows & Doors is already on the front entrance problem. It promises replacement soon.

6. As for future plans, I think we need to reevaluate both Maintenance and Security. Training is an option. We could also add new personnel, reconsider our current management in those departments, or maybe just a few, good, hard, strongly stated comments from you would do the trick.

Anyway, we're up and running again. Upfront repair costs are covered by insurance, and long-term costs are minimal because sales weren't hurt. Our only remaining challenges are preventive, and that depends on what we do with our Maintenance and Security staff. Let me know what you think.

television and radio spots, and writing advertising brochures, newsletters, and fliers.

Your client has given you the following criteria: cost effectiveness (their budget is $35,000), ability to update, customer accessibility, quick turnaround time, and visual appeal (the company wants to present an attractive and youthful look).

PR&U has on-site Internet developers and technical writers who can provide Web design and create hard-copy text for brochures, newsletters, and fliers. Consultants would have to be hired to produce the television and radio spots. Thus, the Web and hard-copy marketing collateral would be most cost effective. PR&U believes that an Internet site would provide the best options for updating content and allowing customer access. In contrast, brochures, newsletters, and fliers, though visually attractive, would take more time to develop and cost more to revise.

Assignment

Write a recommendation report for the client. Follow the instructions provided in this chapter. Include criteria, an analysis of options, and your suggestions for the client's next step. Do additional research as needed to support your conclusions.

2. **Business Information Systems.** Your BIS department faces the challenge of developing and implementing a company-wide information system. Various software applications, accessed via the company's intranet, will provide the following services:
   - The accounting department will use software for risk management assessments, business valuations, comparative analyses, and evaluation of debt structure and receivables. (Original implementation due date: September 2007; revised plan for implementation: November 2009)
   - The marketing department will use software to create highly targeted prospect lists, sales plans, and marketing evaluations. (Original implementation due date: September 2007; revised plan for implementation: October 2009)
   - Manufacturing will use software to optimize internal productivity and regulate shipping. (Original implementation due date: November 2007; revised plan for implementation: January 2009)
   - Human resources will use optical character recognition (OCR) software for assessing job candidates and benefits software for compensation and insurance calculations. (Original implementation due date: November 2007; revised plan for implementation: December 2009)
   - All departments will use in-house developed communication software (e-mail, instant messaging, listservs, and chat rooms). (Original implementation due date: December 2007; revised plan for implementation: February 2009)

   The system is behind schedule due to a variety of causes, including loss of financial support, personnel problems, revised priorities, and system malfunctions.

Assignment

Write a progress report detailing the implementation schedule for the new, system-wide software platform. Follow the criteria provided in this chapter. Invent new information as needed or research additional content to substantiate your comments.

3. **Accounting.** As an accountant for Liss, Levin, and Myers Accounting, you have been asked to audit a client's businesses. Reggie King is CEO of King Associates, a parent company for a dozen wholly owned subsidiaries. Mr. King's subsidiaries include three construction firms, six building supplies companies (manufacturing and marketing wall board, lumber, hardware, stucco, paint, and carpeting), two plumbing companies, and one land development firm.

   You must travel to these companies, located in three different cities, to perform your audits. Primarily, you will focus on the following risk assessments:
   - An assessment of the company's financial records and account balance levels
   - An evaluation of the company's internal control structure
   - An assessment of the company's procedures for controlling and detecting risk

   Your audit found the following problems. At three of the visited sites, management was not communicating effectively with Mr. King about the perceived risks (offshore production, limited workforce, slow delivery of goods, and increased costs of services). The problem was due to what you have defined as an overly complex organizational structure, involving unusual managerial lines of authority. At four sites, you found material misstatements in the financial statements. These were due to high turnover rates in the companies' accounting and information technology staffs. At two sites, you found that accounting practices were not uniform. Most troubling, at three sites, you found inadequate or missing controls related to the approval of payroll transactions. This was due to inadequate record keeping of assets, an inadequate system for authorizing transactions, and an inadequate system for approving transactions.

Assignment

Write a trip report detailing your auditing findings at the King subsidiaries. Follow the criteria provided in this chapter. Invent new information as needed or research additional content to substantiate your comments.

4. **Management.** You are the events planning manager for the Lewis Clark Convention Center in Galveston, Texas. The convention center is attached to the Surf Winds Hotel located on Seawall Boulevard. Your supervisor has asked you to write a feasibility report to upper-level management, detailing your sales forecast for the quarter and explaining how you plan to meet your sales goals.

   Your goal for the quarter is $300,000 in convention sales. You plan to achieve this goal as follows:

   - Make ten cold calls a day, generating $70,000 in business
   - Visit six to ten companies each month and make formal sales presentations, generating $50,000 in business
   - Attend three to five networking events (at the chamber of commerce, bar association meetings, trade shows, and assorted "meet and greets"), generating $60,000 in business
   - Touch base with 10 to 12 former customers, generating $80,000 in business
   - Mass mail fliers and place brochures statewide to generate $40,000 in business

Assignment

Write a feasibility report, documenting your forecasted sales plans for the convention center. Report on who you have already called and who you plan to contact next. Follow the criteria provided in this chapter. Invent new information as needed or research additional content to substantiate your comments.

5. **Finance.** You work for a mortgage company as a loan officer. A client asks you to recommend the best loan option for his planned home purchase. The client, Randall Bridge, hopes to purchase a $250,000 home in Naples, Florida. Randall, a first-time buyer, earns $60,000 a year as a biomedical engineer. He does not have any money for a down payment, so he hopes to finance 100 percent of the loan.

   Your company, FirstFinancingOrg, specializes in first-time buyers just like Randall. You can suggest at least four options for this client: a zero-down 80/20 mortgage; an interest-only loan for a fixed period of time (up to 10 years); a 30-year fixed loan; or a negative amortization loan.

Assignment

Write a feasibility/recommendation report for Randall, explaining to him the four options and recommending the best choice. Follow the criteria provided in this chapter. Invent new information as needed or research additional content to substantiate your comments.

PROBLEM-SOLVING THINK PIECES

1. Angel Guerrero, a computer information technologist at HeartHome Insurance, has traveled from his home office to a branch location out of town. While on his job-related travel, he encountered a problem with his company's remanufactured laptop computer. He realized that the problem had been ongoing not only for this laptop but also for six other remanufactured laptops that the company had recently purchased.

   Angel thinks he knows why the laptops are malfunctioning and plans to research the issue. When he returns to the home office from his travel, he needs to write a report. What type of report should he write? Explain your answer, based on the information provided in this chapter.
2. Minh Tran is a special events planner in the marketing department at Thrill-a-Minute Entertainment Theme Park. Minh and her project team are in the middle of a long-term project. For the last eight months, they have been planning the grand opening of the theme park's newest sensation ride—*The Horror*—a wooden roller coaster that boasts a 10 g drop.

During one of the team's weekly project meetings, the team hit a roadblock. The rap group "Bite R/B Bit," originally slated to play at the midnight unveiling, has cancelled at the last minute. The team needs to get a replacement band. One of Minh's teammates has researched the problem and presented six alternative bands (at varying prices and levels of talent) for consideration. Minh needs to write a report to her supervisor.

What type of report should she write? Explain your decision, based on the criteria for reports provided in this chapter.

3. Toby Hebert is human resources manager at Crab Bayou Industries (Crab Bayou, Louisiana), the world's largest wholesaler of frozen Cajun food. She and her staff have traveled to New Orleans to attend meetings held by five insurance companies that planned to explain and promote their employee benefits packages.

   During the trip, Toby's company van is sideswiped by an uninsured driver. When Toby returns to Crab Bayou, she has to write a report. What type of report should she write? Base your decision on the criteria for reports provided in this chapter.

4. Bill Baker, claims adjuster for CasualtyU Insurance Company (CUIC), traveled for a site visit. Six houses insured by his company were in a neighborhood struck by a hailstorm and 70 mile per hour winds. The houses suffered roof and siding damages. Trees were uprooted, with one falling on a neighbor's car. Though no injuries were sustained, over $50,000 worth of property damage occurred.

   When he returns to CUIC, what type of report should he submit to his boss? Defend your decision based on criteria for reports provided in this chapter.

WEB WORKSHOP

1. More and more, companies and organizations are putting report forms online. The reason for doing this is simple—ease of use. Go online and access online report forms. All you need to do is use any Internet search mechanism, and type in "online ______ report form." (In the space provided, type in "trip," "progress," "incident," "investigative," or "feasibility/recommendation.") Once you find examples, evaluate how they are similar to and different from the written reports shown in this chapter. Share your findings with others in your class, either through oral presentations or written reports.

2. Every company writes reports—and you can find examples of them online. Use the Internet to access Report Gallery, found at http://www.reportgallery.com/. Report Gallery bills itself as "the largest Internet publisher of annual reports." From this Web site, you can find the annual reports from 2,200 companies, including many Fortune 500 companies.

   Study the annual reports from five to ten different companies. What do these reports have in common? How do they differ from each other? Discuss the reports' page layout, readability, audience involvement, content, tone, and development. How are the reports similar to and different from those discussed in this chapter?

   Share your findings with others in your class, either through oral presentations or written reports.

QUIZ QUESTIONS

1. What are some purposes of a report?
2. What are four common types of short, informal reports?
3. List the four basic units of a report.
4. What types of material are included in the discussion section?
5. Explain how the reporter's questions can help you prepare your report.
6. Why would you use e-mail, memo, or letter formats for short, informal reports?
7. What organizational modes are useful for writing short, informal reports?
8. How do visual aids assist the reader in a report?
9. What do you include in the identification lines?
10. What do you include in the conclusion and recommendation?

CHAPTER 18

Long, Formal Reports

▸REAL PEOPLE *in the workplace*

Cities, states, and the federal government constitute the largest employers in the United States. The government hires engineers, financial analysts, accountants, scientists, administrative assistants, fire fighters, police officers, employees in public works and utilities, and parks and recreation workers.

Do these employees have to write long, formal reports? The answer is "yes." **Charles Worth**, the assistant city manager/city clerk of Round Rock, Alabama, says that his city's engineers, police officers, fire chiefs, financial advisors, and parks and recreation employees often write formal reports to city council members and the city's mayor. These long reports help the city council members "make informed decisions" that impact the city's residents and business owners.

Sometimes topics for these reports are generated within city hall. Other times, topics for the reports come from concerns expressed by citizens. A citizen contacts one of the city council members, for example, to share a concern about streets, sidewalks, animal control, traffic, or taxes. Maybe a citizen wants to expand a business, improve safety guard training for school crosswalks, or build a skateboard park for resident recreation. City council members ask the city's staff to research the issues, write a report that provides information on relevant statutes and/or other cities' best practices for handling issues, analyze options, and then recommend follow-up actions.

In one recent instance, the city needed to expand an existing road. The two-lane, winding road, initially built in the early 1940s, no longer met the city's growing population needs. To accommodate city growth, the road had to be widened to four lanes and straightened to meet increased traffic and speed limits. This change would necessitate significant research and resources. City engineers needed to study the extent to which a road change would impact

- An adjacent animal refuge
- Current homeowners' land
- Sewage lines
- Easements
- Electrical and water lines
- New signage
- Traffic signals

CHAPTER GOALS

When you complete this chapter, you will be able to do the following:

1. Understand the purposes of writing long, formal reports.
2. Distinguish among informational, analytical, and recommendation reports.
3. Follow guidelines for writing long, formal reports.
4. Include the major components of long, formal reports: front matter, text, and back matter.
5. Distinguish between primary and secondary research to develop your content.
6. Evaluate your long, formal reports using the report checklist.
7. Follow the P^3 process—planning, packaging, and perfecting—to create effective long, formal reports.

WWW

To learn more about long, formal, research reports, visit our Web Companion at http://www.prenhall.com/gerson.

The roadway project would be time consuming and costly. Only through a long, formal report could city engineers thoroughly provide the city council information about this project, analyze the options, and recommend appropriate construction steps.

Charles says that poor communication damages the city employees' credibility and diminishes their professionalism. Flawed reports—ones that aren't well written or clearly developed—suggest that the writers "lack interest in the city's residents and business owners," are unresponsive to their constituents, or "just don't want to help" the city's taxpayers. That's bad PR, potentially leading to citizen complaints and inefficient government. It's not surprising that Charles says, "One of the most important assets you can have in government is writing skills."◄

Charles's Communication Challenge

As assistant city manager/city clerk of Round Rock, Alabama, Charles Worth is responsible for city purchases. Charles says, "My accounting manager has informed me that the city's financial software is outdated. To get approval for the acquisition of new financial software, I need to write a long, formal report, which will be submitted to the city manager and members of the city council. The report must provide sufficient documentation to the city council to justify large, capital expenditures.

My communication challenge is as follows:

1. I must inform the city manager and the city council of the current software shortcomings. My audience does not consist of accountants or computer specialists. City council members are elected citizens from the community with varied backgrounds. Therefore, the report must be written at a lay person's level.
2. I must use details, language, and illustrations that persuade the audience to approve the new software.
3. I must analyze options for solving the city's problem with outdated software. This will require that I research optional vendors, taking into consideration the allotted budget for software replacement and our requirement to allow for open bidding.
4. Based on my research, I must persuasively recommend the best vendor."

See page 536–538 to learn how Charles met his communication challenge.

WHY WRITE A LONG, FORMAL REPORT?

In some instances, your subject matter might be so complex that a short report will not thoroughly cover the topic. For example, your company asks you to write a report about the possibility of an impending merger. This merger will require significant commitments regarding employees, schedules, equipment, training, facilities, and finances. Only a long report, complete with research, will convey your content sufficiently and successfully.

You could have to write a long, formal report requiring research for a variety of reasons, such as the following:

1. When a subject is important
2. When documentation will reveal the importance of the topic
3. When large sums of money are involved
4. When large numbers of people are affected
5. When time and resources are devoted to development of the report
6. When research will explain and support the topic

Additional information:

See Chapter 17, "Short, Informal Reports."

TOPICS FOR LONG, FORMAL REPORTS

When you devote considerable time and energy to long, formal reports, you will be dealing with topics that often have serious and complex issues to be considered. The following are titles of long, formal, research reports:

- The Effect of Light Rail on Wyandotte County: An Economic Impact Study on Infrastructure
- The Increasing Importance of Mobile Communications: Security, Procurement, Deployment, and Support
- Managing Change in a Technologically Advancing Marketplace at EBA Corp.
- Information Stewardship at The Colony, Inc.: Legal Requirements for Protecting Information
- Remote Access Implementation and Management for New Hampshire State University

These examples share several common denominators including the significant costs required for implementation of large scale changes, the impact on large numbers of people, the time and resources required to research the topics, and the challenges of writing the long, formal reports expected by the intended audiences. Short reports would not suffice.

TYPES OF LONG, FORMAL REPORTS: INFORMATIVE, ANALYTICAL, AND RECOMMENDATION

Reports can provide information, be analytical, and/or recommend a course of action. Occasionally, you might write a report that only informs. You might write a report that just analyzes a situation. You could write a report to recommend action persuasively. Usually, however, these three goals will overlap in your long, formal reports.

For example, let's say that your company is expanding globally and will need a WAN (a wide area network that spans a large geographical area). In a long, formal report to management, you might first write to inform the audience that this WAN will meet the strategic goals of storing and transmitting information to your global coworkers. The report also will analyze the ways in which this network will be safe, reliable, fast, and efficient. Finally, the report will recommend the best designs for providing secure communications to clients, partners, vendors, and coworkers. Ultimately, you must persuade the audience, through research, that your envisioned network's design, hardware specifications, software, and estimated budget will satisfy both internal and external needs. Such a report would have to be informative, analytical, and persuasive to convince the audience to act on the recommended suggestion.

Information

When you provide information to your audience, focus on the facts. These facts will help your readers better understand the situation, the context, or the status of the topic. For example, Nitrous Solutions asked Alpha/Beta Consulting, (A/B/C) to report on backup data storage options for software archives. To so do, A/B/C needed to present factual information about the importance of backup storage. In a long, formal report, A/B/C informed their client about the following: the threats to data, the cost of lost data, how often data should be backed up, and what data should be archived. See Figure 18.1 for information about backup data storage.

Analysis

When you analyze for your audience, you begin with factual information. However, you expand on this information by interpreting it and then drawing conclusions. Once Alpha/Beta Consulting presented the informational findings about backup data storage facilities, they followed this information with a more in-depth analysis of backup options and drew a conclusion for their client. See Figure 18.2 for analysis of backup data storage options.

Recommendation

After providing information and analysis, you can recommend action as a follow-up to your findings. The recommendation allows you to tell the audience why they should purchase a product, use a service, choose a vendor, select a software package, or follow a course of action. Alpha/Beta Consulting presented findings and analyzed information for their client Nitrous Solutions. Based on their analysis, the consulting company then made a recommendation. See Figure 18.3 for their analysis and recommendation.

FIGURE 18.1 Information About Backup Data Storage

Information is presented in a variety of ways. The consulting company details data in bulleted lists, asks questions and provides answers, and explains facts and findings in paragraphs.

Description of Backup Data Storage:

The following information provides facts about permanent storage to help Nitrous Solutions maintain a high degree of security in an off-site backup data storage location.

Facts:

Due to events such as 9/11 and Hurricane Katrina, the importance of maintaining mission-critical electronic data and backups has become evident for companies that wish to avoid a catastrophic outcome. Moving data to a separate off-site location dilutes risk and protects against data loss when, for example, the company's IT infrastructure or its critical electronic information is damaged by the following:

Threats to Your Data	Dangers and Costs of Data Loss
➤ Fire ➤ Flood ➤ Hurricane or tornado ➤ Lightning strike ➤ Earthquake ➤ Heat, sunlight ➤ Humidity, moisture, spilled liquids ➤ Smoke, dust and dirt ➤ Electrical surge or power failure ➤ Media failure ➤ Hard drive failure	➤ Loss of mission-critical files ➤ Corrupted database ➤ Corrupted operating system ➤ Loss of all files on hard drive ➤ Laptop damage, theft, or loss ➤ Damaged computer hardware ➤ Backup media loss or theft ➤ Total loss of all data at your site ➤ Compromised data security ➤ Competitor access to your data ➤ Lost business records

(Continued)

FIGURE 18.1 Information About Backup Data Storage Continued

When and What Data Should Be Backed Up?
Some basic questions that are commonly asked when determining when and what data should be backed up are as follows:

How often should desktop PCs or Macs be backed-up?

- Data on desktop PCs and Macs should be backed-up at a minimum of at least once a week. This will prevent anyone losing more than one week's worth of information.

Should everything on my PC or Mac be backed-up once a week?

- This is not necessary. Only data needs to be backed-up once a week. Software, operating systems, and even static data do not need to be backed up that often. Backing-up everything on your PC or Mac adds time to the process and requires substantially more backup tapes, backup disks, or other types of media.

How long should I keep computer backups?

- Some companies keep original backups up to seven years and have a set schedule that includes five different increments of backups: daily, weekly, monthly, quarterly, and yearly.

This is a topic for discussion with your key management, legal staff, information systems technicians, records management, and key personnel. Balancing the needs of a company's computer system protection with good records management takes informed decisions on the parts of all these departments.

FIGURE 18.2 Analysis of Backup Data Storage Options

Analysis of Backup Data Storage Options

Industry analysts claim that two out of five businesses that experience a disaster will go out of business within five years of the event due to information and service loss as a consequence of disasters.

This is an alarming statistic considering the high cost of not "expecting the unexpected." In the event of a disaster, a corporation could lose hundreds, thousands, or even millions of dollars through lost productivity. At worst the corporation could go out of business without the possibility of a second chance.

The two main types of off-site data storage are the following:

Off-site Data Storage Facility	Internet Data Storage Facility
Tapes, CDs, DVDs, or hard drives are sent to a predetermined off-site location for security and disaster recovery purposes, thus allowing a business to be safe in the knowledge that they will be able to recover quickly and seamlessly from any disaster. With simply a walk across the street or a drive downtown to the storage facility, the business can quickly begin to rebuild from the data they have stored.	Off-site Internet data storage, although relatively new, has quickly become one of the fastest growing arenas in the records and information management industry. A business can simply select the files and the time that they wish to have their files backed up with the off-site backup software. Their files are compressed, encrypted, and sent to a backup server over their existing Internet connection. The business can upload or download your files as often as needed, with usually no additional charges.

FIGURE 18.2 Continued

Off-Site Location for Data Storage Options	
Off-Site Residential Location	**Off-Site Facility Location**
Costs	
• Completely free	• $5–$650 per month depending on space and features
Reliability	
• Data are accessible whenever you need it • Save and archive as much as you need with no restrictions or extra fees	• Insurance on data is automatic with contract • Data are fully accessible during normal business hours, or after hours by emergency only
Security	
• More secure than leaving your data in your office building	• Vault and/or building monitored by 24-hour surveillance and alarm systems • Security patrolled
Scalability	
• Can store as little or as much data as needed, with no restrictions, guidelines, rules, or regulations, free of charge	• Just like the Internet storage, as your company grows, so can data storage; of course, an increase of storage means an increase of monthly payment
Ease of Use	
• Your house, your rules • Maintenance of data management can be completed in your spare time	• Easy to set up and manage • Step-by-step instructions available • Some companies offer free technical help

Conclusion

When analyzing information, A/B/C presented facts and figures, pros and cons, and specific details. Then, the consulting company interpreted this information in a conclusion.

Regardless of how much or how little a business uses a computer, your company will create important and unique data. The unique data can include financial and project budget records, digital images, client profiles, and marketing sales. The data are priceless and constantly at risk.

Data loss is likely to occur for many reasons:

- Hardware failure = 42 percent
- Human error = 35 percent
- Software corruption = 13 percent
- PC viruses = 7 percent
- Hardware destruction = 3 percent

For these reasons, Nitrous Solutions should choose backup storage for its valuable data.

FIGURE 18.3 Analysis and Recommendation

Off-Site Location for Permanent Storage:
Analysis and Recommendation

WEIGHTED ANALYSIS:

Type		Off-Site Residential		Off-Site Facility	
Criteria	Weight	*Rating	Score	*Rating	Score
Cost	15%	5	0.75	3	0.45
Performance	25%	-	-	-	-
Reliability	30%	4	1.20	4	1.20
Security	15%	3	0.45	4	0.60
Scalability	10%	5	0.50	5	0.50
Ease of use	5%	4	0.20	4	0.20
Total	100%		3.1		2.95

*Rating: The weighted analysis is based on a 1 to 5 scale; 1 is poor and 5 is excellent.

Recommendation:

Because Nitrous Solutions does not currently have a network infrastructure in place, the data storage decision should be considered a high priority since all new data that the company creates will be significant to the company's overall success. Alpha/Beta Consulting recommends that Nitrous Solutions use the residential off-site data storage option due to the slight advantage it has over a professional storage facility. The one and only real advantage between the two is cost. Whereas residential storage is free with no contracts or monthly charges, a storage facility will charge anywhere from $5 to $650 per month, depending on the size of data and equipment that needs to be accompanied with such data.

If Nitrous Solutions determines the need to have multiple backup solutions, rather than a single reinforcement to the backup data, another highly recommended means of storage would be that of an Internet storage facility. Due to the popular demand and increasing technology, many different companies are now offering free trials of their Internet storage facilities and software. Due to the unknown intentions of Nitrous Solutions, only the features for Internet storage will be researched because almost all options for Internet companies are either compatible or identical.

MAJOR COMPONENTS OF LONG, FORMAL REPORTS

Since short reports run only a few pages, you can assume that your readers will be able to follow your train of thought easily. Thus, short reports merely require that you use headings such as "Introduction," "Discussion," and "Conclusion/Recommendation" to guide your readers through the document or talking headings to summarize the content more thoroughly. Long reports, however, place a greater demand on readers. Your readers could be overwhelmed with many pages of information and research. A few headings won't be enough to help your readers wade through the data.

Additional information:

See Chapter 17 "Short, Informal Reports."

FIGURE 18.4 Components of a Long, Formal Report

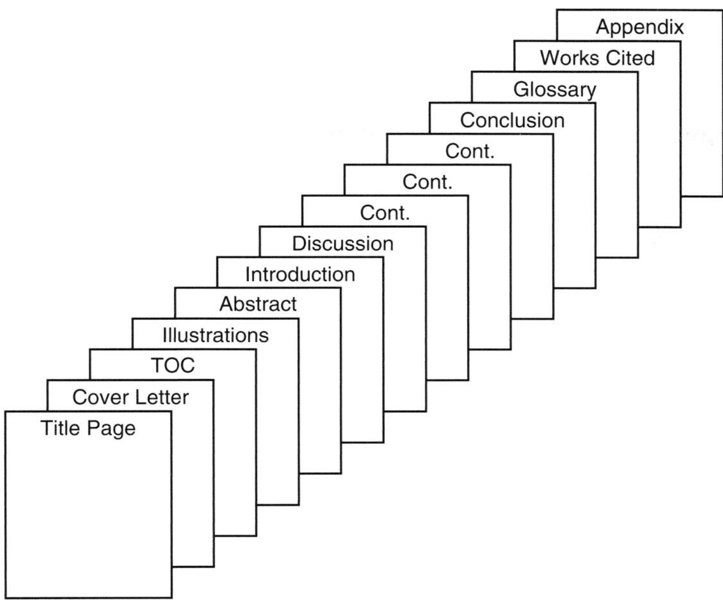

In addition to the basic components of a short report, a long, formal report includes the following:

- **Front matter** (title page, cover letter, a table of contents, list of illustrations, and an abstract or executive summary)
- **Text** (introduction including purpose, issues, background, and problems; discussion; and conclusion/recommendation)
- **Back matter** (glossary, works cited or references page, and an optional appendix) See Figure 18.4 for the components of a long, formal report.

Title Page

The title page serves several purposes. On the simplest level, a title page acts as a dust cover or jacket keeping the report clean and neat. More important, the title page tells your reader the

- Title of the long report (thereby providing clarity of intent)
- Name of the company, writer, or writers submitting the long report
- Date on which the long report was completed

If the long report is being mailed outside your company to a client, you also might include on the title page the audience to whom the report is addressed. If the long report is being submitted within your company to peers, subordinates, supervisors, or owners, you might want to include a routing list of individuals who must sign off or approve the report. Following are two sample title pages. Figure 18.5 is for a long report with routing information; Figure 18.6 is for a long report without routing information.

Cover Letter

Your cover letter of memo prefaces the long report and provides the reader an overview of what is to follow. It tells the reader the following:

Additional information:

See Chapter 10, "Traditional Correspondence: Memos and Letters."

- Why you are writing
- What you are writing about (the subject of this long report)
- What exactly of importance is within the report
- What you plan to do next as a follow-up
- When the action should occur
- Why that date is important

FIGURE 18.5 Title Page for Long Report (with Routing Information)

Provide routing information for the long report.

Report on Multicultural Workforce at StartCo Insurance

Prepared by: ______________________________ Date: __________

Pete Niosi
Assistant Director, Human Resources

Reviewed by: ______________________________ Date: __________

Leah Workman
Manager, Accounting

Recommended by: ___________________________ Date: __________

Greg Foss
Department Supervisor, Customer Service

Recommended by: ___________________________ Date: __________

Shirley Chandley
Director, Human Resources

Approved by: ______________________________ Date: __________

Ralph Houston
Vice President

FIGURE 18.6 Title Page for a Long Report (without Routing Information)

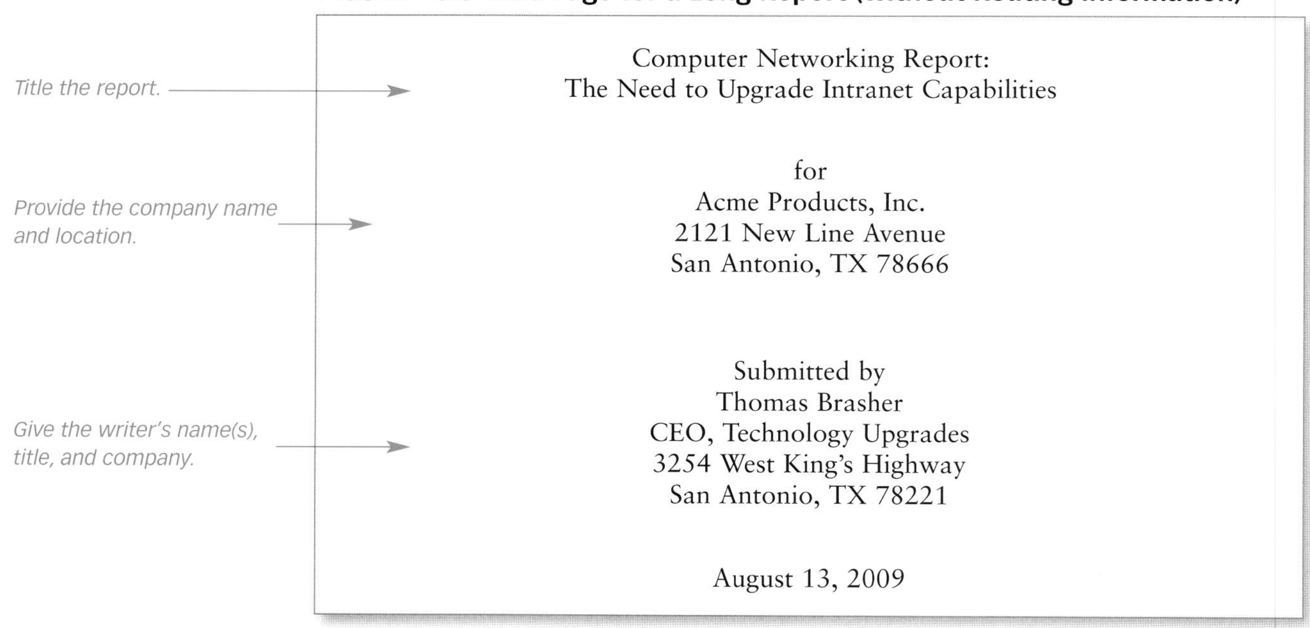
Title the report.
Provide the company name and location.
Give the writer's name(s), title, and company.

Computer Networking Report:
The Need to Upgrade Intranet Capabilities

for
Acme Products, Inc.
2121 New Line Avenue
San Antonio, TX 78666

Submitted by
Thomas Brasher
CEO, Technology Upgrades
3254 West King's Highway
San Antonio, TX 78221

August 13, 2009

Table of Contents

Long reports are read by many different readers, each of whom will have a special area of interest. For example, the managers who read your reports will be interested in cost concerns, timeframes, and personnel requirements. Technicians, in contrast, will be interested in technical descriptions and instructions. Not every reader will read each section of your long report.

Your responsibility is to help these different readers find the sections of the report that interest them. One way to accomplish this is through a table of contents. The table of contents should be a complete and accurate listing of the main *and* minor topics covered in the report. In other words, you don't want just a brief and sketchy outline of major headings. This could lead to page gaps; your readers would be unable to find key ideas of interest. In the table of contents in Figure 18.7, we can see that the discussion section contains approximately 16 pages of data. What is covered in those 16 pages? Is anything of value discussed? We don't know.

BEFORE

FIGURE 18.7 Flawed Table of Contents

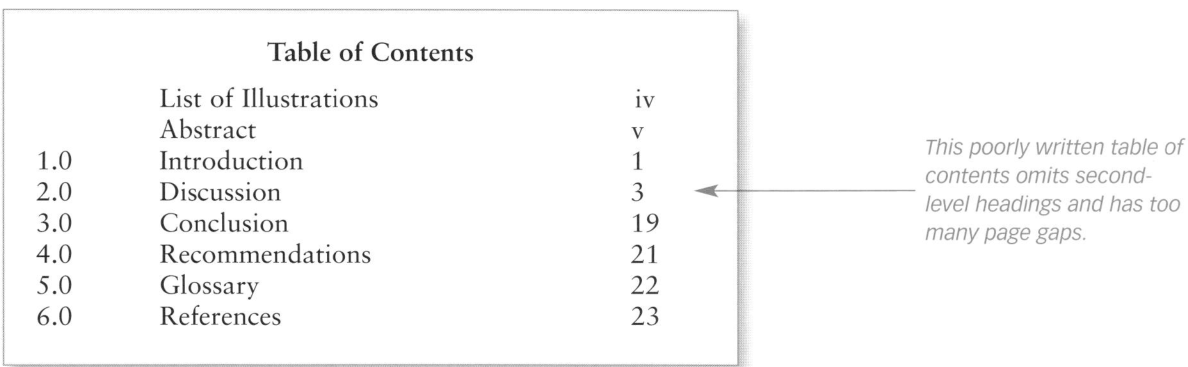

Table of Contents

This poorly written table of contents omits second-level headings and has too many page gaps.

In contrast, an effective table of contents fleshes out this detail so your readers know exactly what is covered in each section. By providing a thorough table of contents, you will save your readers time and help them find the information they want and need. Figure 18.8 is an example of a successful table of contents.

In the example, note that the actual pagination (page 1) begins with the introductory section. Page 1 begins with your main text, not the front matter. Instead, information prior to the introduction is numbered with lowercase Roman numerals (i, ii, iii, etc.). Thus, the title page is page i, and the cover letter is page ii. However, you don't need to print the numbers on these two pages. Therefore, the first page with a printed number is the table of contents. This is page iii, with the lowercase Roman numeral printed at the foot of the page.

AFTER

FIGURE 18.8 Effective Table of Contents

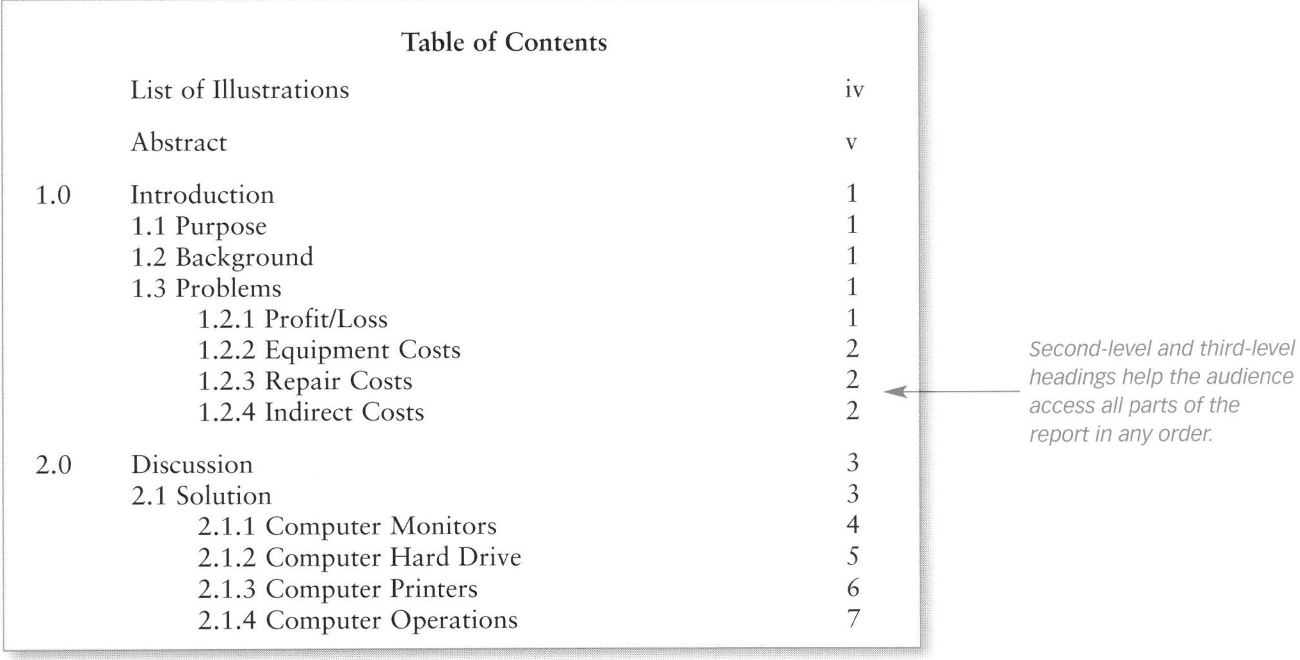

Table of Contents

Second-level and third-level headings help the audience access all parts of the report in any order.

(Continued)

FIGURE 18.8 Effective Table of Contents Continued

iii

Technology Tip

Creating a Hierarchy of Headings Using Microsoft Word 2007

When writing your long, formal reports, you will want to include headings and subheadings. To help your readers navigate the text, create a clear hierarchy of headings (also called "cascading headings") that distinguish among "first-level headings," "second-level headings," "third-level headings," and "fourth-level headings."

Microsoft Word 2007 will help you create this hierarchy of headings. You can apply a "style" to your entire document in Word's "formatting text by using styles" tool.

For example, let's say that you want your headings and subheadings to look as follows:

HIERARCHY OF HEADINGS	DESCRIPTION
HEADING ONE	First-level headings: Times New Roman, all cap, boldface, 16 pt. font.
Heading two	Second-level headings: Times New Roman, boldface, Italics, 14 pt. font.
Heading three	Third-level headings: Arial, boldface, 12 pt. font.
Heading four	Fourth-level headings: Times New Roman, italics, 12 pt. font.

To create a hierarchy of headings, follow these steps:

1. Click on the **Home** tab and then click on the down arrow to the right of **Heading 2.**

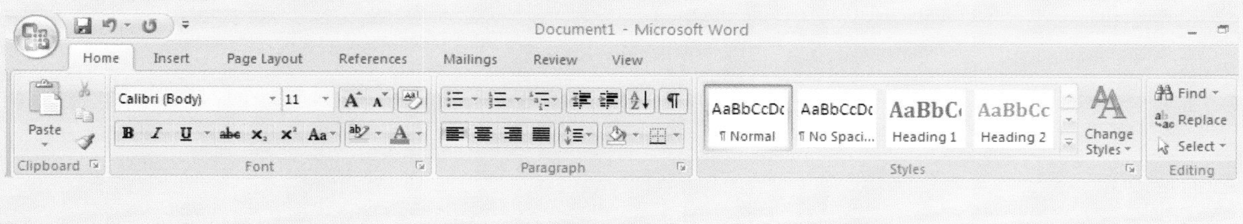

The following styles and formatting box will pop up.

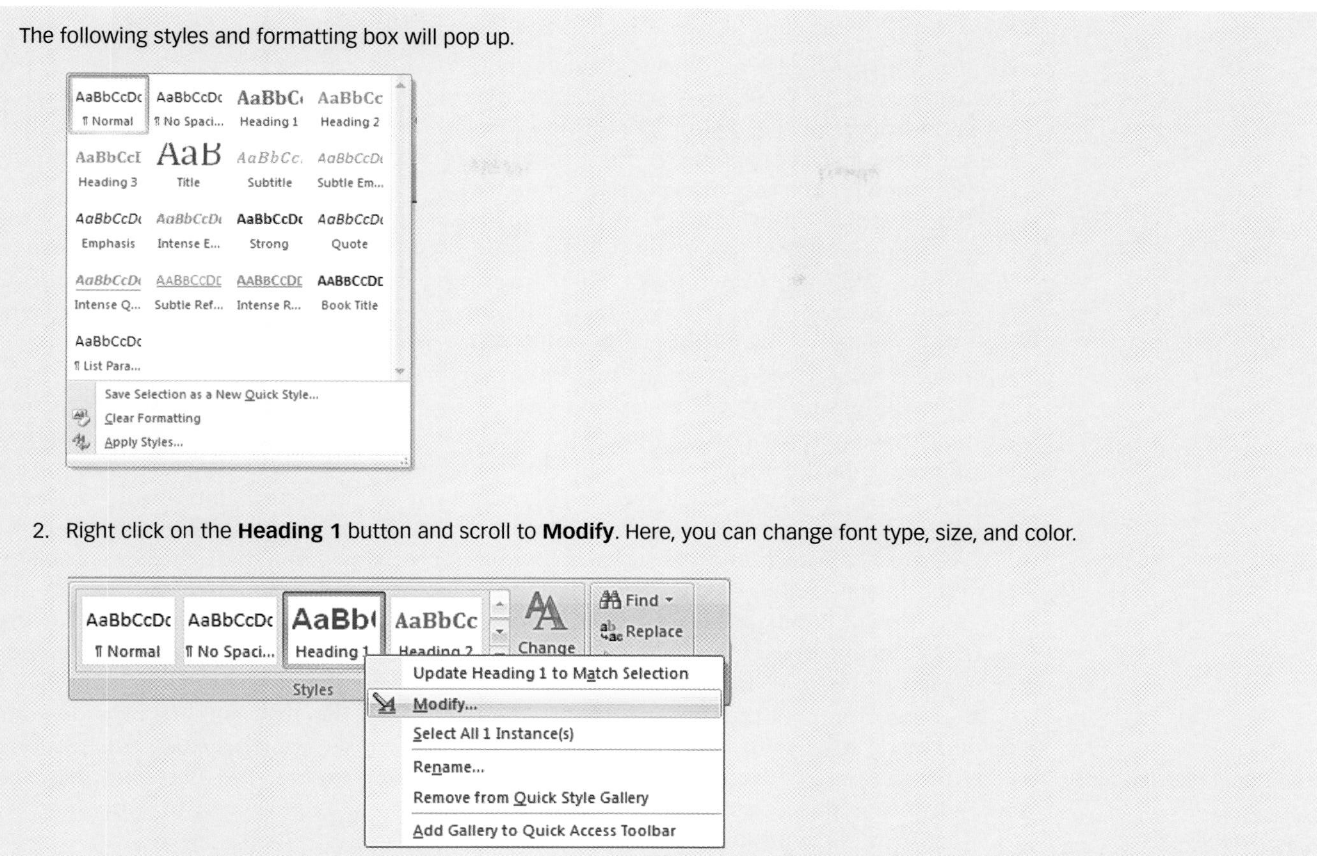

2. Right click on the **Heading 1** button and scroll to **Modify**. Here, you can change font type, size, and color.

List of Illustrations

If your long report contains several tables or figures, you will need to provide a list of illustrations. This list can be included below your table of contents, if there is room on the page, or on a separate page. As with the table of contents, your list of illustrations must be clear and informative. Don't waste your time and your reader's time by providing a poor list of illustrations like Figure 18.9. Instead, complete an effective list of illustrations like Figure 18.10.

BEFORE

FIGURE 18.9 Flawed List of Illustrations

List of Illustrations

Fig. 1	2
Fig. 2	4
Fig. 3	5
Fig. 4	5
Fig. 5	9
Table 1	3
Table 2	6

AFTER

FIGURE 18.10 Effective List of Illustrations

List of Illustrations

Figure 1.	Revenues Compared to Expenses	2
Figure 2.	Average Diesel Fuel Prices Since 2009	4
Figure 3.	Mainshaft Gear Outside Face	5
Figure 4.	Mainshaft Gear Inside Face	5
Figure 5.	Acme Personnel Organization Chart	9
Table 1.	Mechanism Specifications	3
Table 2.	Costs: Expenditures, Savings, Profits	6

Abstract

As mentioned earlier, a number of different readers will be interested in your long report. Because your readers are busy with many different concerns and might have little technical knowledge, they need your help in two ways: they need information quickly, and they need it presented in terminology for a semi-specialist. You can achieve both these objectives through an abstract or executive summary.

The abstract is a brief overview of the proposal or long report's key points geared toward a semi-specialist. To accomplish the required brevity, you should limit your abstract to approximately one to two pages. Each long report you write will focus on unique ideas. Therefore, the content of your abstracts will differ. Nonetheless, abstracts might focus on the following: (a) the *problems* necessitating your report, (b) your suggested *solutions*, and (c) the *benefits* derived when your suggestions are implemented.

For example, let's say you are asked to write a formal report suggesting a course of action (limiting excessive personnel, increasing your company's workforce, improving your corporation's physical facilities, etc.). First, your abstract should specify the problem requiring your planned action. Next, you should mention the action you are planning to implement. This leads to a brief overview of how your plan would solve the problem, thus benefiting your company.

Another approach to an abstract for a long report is to summarize the content being analyzed in the report, setting the stage for the major issues to be discussed. In this abstract, you will highlight all the main points in the long report, omitting any supporting facts and documentation (to be developed in the discussion section of the report). You not only want to be brief, focusing on the most important issues, but also you should avoid terminology and concepts for specialists. The purpose of the abstract is to provide your readers with an easy-to-understand summary of the entire report's focus. Your readers want the bottom line, and they want it quickly. Therefore, either avoid all terminology for specialists, completely define your terms parenthetically, or refer readers to a glossary.

Additional information:

See Chapter 3 "Meeting the Needs of the Audience."

Figure 18.11 is a sample abstract for a long report.

Executive Summary

An executive summary, similar to an abstract but generally longer, is found at the beginning of either a formal report or a proposal and summarizes the major topics covered in the document. By reading the executive summary, your audience (the decision makers) gets an overview of the much longer report or proposal. Executive summaries can even take the

FIGURE 18.11 A Sample Abstract for a Long Report

Summarizing the report in the abstract allows the reader to get a "snapshot" of the content. Since most people think that Microsoft Windows is the only option for desktop computing, this report discusses an alternative—Linux.

Abstract

The Microsoft Windows desktop has established a dominant presence in the enterprise desktop marketplace. Currently, the Windows desktop has nearly total control over the enterprise desktop market primarily due to

- An aggressive marketing strategy
- Its intuitive ease of use
- Its apparent ease of installation and administration
- Readily available support

Windows, however, is not the only viable desktop solution in the enterprise marketplace. Because of potential risks and benefits associated with any technology product, as IT manager I must be alert for alternatives that are secure, reliable, and cost effective.

The IT manager who wrote this abstract lets the reader see that a recommendation will be made based on the facts in the report.

This report provides technical information regarding the issue of migration to a Linux enterprise desktop solution, analyzes the pros and cons of migrating to a Linux solution, and recommends appropriate action for our company, Design International, Inc.

place of the much longer formal report or proposal. In this way, you save your reader's time and communicate only the most relevant parts of the report or proposal. To create an effective executive summary, follow these key points:

1. Begin with a purpose statement for the report or proposal.
2. Give an overview of the key ideas discussed in the report or proposal.
3. Inform the reader if any problems will affect the outcome of the report or proposal.
4. Suggest solutions to the problems, or encourage the reader to suggest solutions.
5. Conclude the executive summary and recommend a course of action.

Key Parts of an Executive Summary

To write an effective and concise executive summary, use only the most important details and supporting statistics or information. Omit details or technical content which your reader either does not need to know or which would unnecessarily confuse the reader. Details are used in the discussion section of the report or the proposal. To limit the length of your executive summary to one or two pages, choose the most important elements, and omit those that are secondary. An effective summary can include the following:

- Purpose and scope of the report or proposal, stating the problem, or need and your ability to improve the situation
- Research or methods used to develop your content
- Conclusions about your analyses of the topic
- Your qualifications showing that you can resolve the issue
- A project management plan and timetable
- The total project budget
- Recommendations based on your findings

Introduction

Your introduction should include the following: (1) purpose or overview and (2) background or problem.

Purpose or Overview

In one to three sentences or a short paragraph, tell your readers the purpose of your long report. This purpose statement informs your readers *why* you are writing or *what* you hope to achieve. This statement repeats your abstract to a certain extent. However, it's not redundant; it's a reiteration. Although numerous people read your report, not all of them read each line or section of it. They skip and skim.

The purpose statement, in addition to the abstract, is another way to ensure that your readers understand your intent. It either reminds them of what they have just read in the abstract or informs them for the first time if they skipped over the abstract. Your purpose statement is synonymous with a paragraph's topic sentence, an essay's thesis, the first sentence in a letter, or the introductory paragraph in a shorter report.

Background or Problem

Whereas the purpose statement should be limited to one to three sentences or a short paragraph for clarity and conciseness, your discussion of the problem must be much more detailed. For example, if you are writing a report about the need for a new facility, your company's current work and storage space must be too limited. Your company has a problem that must be solved.

Your introduction's focus on the problem and background, which could average one to two pages, is important for two reasons. First, it highlights the importance of your report and establishes a context for the reader. The introduction emphasizes for your readers the report's priority. In this section, you persuade your readers that a problem truly exists and needs immediate attention.

Second, by clearly stating the problem and background information, you also reveal your knowledge of the situation. This section reveals your expertise. Thus, after reading this section of the introduction, your audience should recognize the importance of the issue and trust you to solve it or understand the complexity of the topic being discussed.

Discussion

The discussion section of your long report constitutes its body. In this section, you develop the detailed content of the long report. As such, the discussion section represents the major portion of the long report, perhaps 85 percent of the text. What will you focus on in this section? Because every report will differ, we can't tell you exactly what to include. However, your discussion can contain any or all of the following:

- Analyses
  - Existing situation
  - Solutions
  - Benefits
- Product specifications of mechanisms, facilities, or products
- Comparison/contrast of options
- Assessment of needs
- Features of the systems or products
- Optional approaches or methodologies for solving the problems
- Managerial chains of command (organizational charts)
- Biographical sketches of personnel
- Corporate and employee credentials
  - Years in business
  - Satisfied clients
  - Certifications
  - Previous accomplishments
- Schedules
  - Implementation schedules
  - Reporting intervals
  - Maintenance schedules
  - Delivery schedules
  - Completion dates
  - Payment schedules
  - Projected milestones (forecasts)
- Cost analyses
- Profit and loss potential
- Documentation and researched material
- Survey results
- Lab report results
- Warranties
- Maintenance agreements
- Online help
- Training options
- Impact on the organization (time, personnel, finances, customers)
- Definitions

You will have to decide which of these sections will be geared toward specialists, semi-specialists, or a lay audience. Once this decision is made, you will write accordingly, defining terms as needed.

Defining Acronyms, Abbreviations, and Technical Terminology

We have all become familiar with common acronyms such as *scuba* (*s*elf-*c*ontained *u*nderwater *b*reathing *a*pparatus), *radar* (*r*adio *d*etecting *a*nd *r*anging), *NASA* (*N*ational *A*eronautics and *S*pace *A*dministration), *FICA* (*F*ederal *I*nsurance *C*ontributions *A*ct), and *MADD* (*M*others *A*gainst *D*runk *D*riving)—single words created from the first letters of multiple words. We are comfortable with abbreviations like *CPR* (*C*ardio*p*ulmonary *R*esuscitation), *FBI* (*F*ederal *B*ureau of *I*nvestigation), *ATM* (*A*utomated *T*eller *M*achine), *NFL* (*N*ational *F*ootball *L*eague), *IBM* (*I*nternational *B*usiness *M*achines), and *LA* (*L*os *A*ngeles).

Sometimes, however, abbreviations and acronyms can have two or more definitions. CPR can also mean "Continuing Property Records." ATM can mean "Asynchronous Transfer Mode." More often than not, acronyms, abbreviations, and technical terminology cause problems, not because they are too common but because people misunderstand them. You have to decide when to use acronyms, abbreviations, and technical terminology and how to use them effectively. One simple rule is to define your terms. You can do so as follows:

- parenthetically
- in a sentence
- in an extended paragraph
- in a glossary (discussed below)

Defining terms parenthetically Rather than just writing *CIA*, write *CIA* (*Cash in Advance*). Such parenthetical definitions, which are only used once per correspondence, don't take a lot of time and won't offend your readers. Instead, the result will be clarity. If you use many potentially confusing acronyms or abbreviations, or if you need to use a great deal of technical jargon, then parenthetical definitions might be too cumbersome. In this case, supply a separate glossary.

Defining your terms in a sentence If you provide a sentence definition, include the following: ***Term* + *Type* + *Distinguishing characteristics*.** For example, using a sentence to define *HTTP*, you would write the following:

Term *Type*

example

Hypertext Transfer Protocol is a computer access code that provides secure communications on the Internet, an intranet, or an extranet.

Distinguishing characteristics

Using extended definitions of one or more paragraphs When you need to provide an extended definition of a paragraph or more, in addition to providing the term, type, and distinguishing characteristics, also consider including examples, procedures, and descriptions. Look at the following definition of a voltmeter.

example

The voltmeter is an instrument used to measure voltage. The voltmeter usually consists of a magnet, a moving coil, a resistor, and control springs. Types of voltmeters include the microvoltmeter, millivoltmeter, and kilovoltmeter, which measure voltages with a span of 1 billion to 1. By connecting between the points of a circuit, voltmeters measure potential difference.

Conclusion/Recommendation

You must sum up your long report in a page or so, providing your readers with a sense of closure. The conclusion can restate the problem, the important implications of your analysis, your solutions, and the benefits to be derived. In doing so, remember to quantify. Be specific—state percentages and amounts.

Your recommendation will suggest the next course of action. Specify when this action will or should occur and why that date is important. The conclusion/recommendation section can be made accessible through highlighting techniques, including headings, subheadings, underlining, boldface, itemization, and white space.

Glossary

If you have not defined your terms parenthetically, in a sentence, or in a paragraph, you should use a glossary. A glossary is an alphabetized list of terms placed after your conclusion/recommendation. See Figure 18.12.

Because you will have numerous readers with multiple levels of expertise, you must be concerned about your use of specialist's language (abbreviations, acronyms, and terms). Although some of your readers will understand your terminology, others won't. However, if you define your terms each time you use them, two problems will occur: you might insult specialists, or you will delay your audience as they read your text. To avoid these pitfalls, use a glossary. A glossary is an alphabetized list of highly technical, specialized terminology placed after your conclusion/recommendation.

A glossary is invaluable. Readers who are unfamiliar with your terminology can turn to the glossary and read your definitions. Those readers who understand your word usage can continue to read without stopping for unneeded information.

Works Cited (or References)

If you use research to write your long report, you will need to include a works cited or references page. This page(s) documents the sources (books, periodicals, interviews, computer software, Internet sites, etc.) you have researched and from which you have quoted or paraphrased. Correct documentation and source citations are essential in your long report to enhance your credibility and demonstrate your ethical behavior. Remember that boilerplate content and templates, already created material in your company's document library, do not necessarily require documentation. However, if you research the material on the Internet or from any other published source, you must document this material or be guilty of unethical behavior.

Additional information:

See Chapter 1, "Communicating in the Workplace," and Chapter 16, "Research and Documentation."

Appendix

A final, optional component is an appendix. Appendices allow you to include any additional information (survey results, tables, figures, previous report findings, relevant letters or memos, etc.) that you have not built into your long report's main text. The contents of your appendix should not be of primary importance. Any truly important information should be incorporated within the report's main text. Valuable data (proof, substantiation, or information that clarifies a point) should appear in the text where it is easily accessible. Information provided within an appendix is buried, simply because of its placement at the end of the report. You don't want to bury key ideas. An appendix is a perfect place to file

FIGURE 18.12 Glossary

Glossary

CPA	Certified Public Accountant
FICA	Federal Insurance Contributions Act (Social Security taxes)
Franchise	Official establishment of a corporation's existence
Gross pay	Pay before deductions
Line of credit	Amount of money that can be borrowed
Net pay	Pay after all deductions
Profit and loss statement	Report showing all incomes and expenses for a specified time

nonessential data that provides either documentation for future reference or further explanation or support of content in the report.

USING RESEARCH IN LONG, FORMAL REPORTS

You can use researched material to support and develop content in your long, formal report. Use quotes and paraphrases to develop your content. Workplace communicators often ask how much of a long, formal report should be *their* writing, as opposed to researched information. A general rule is to lead into and out of every quotation or paraphrase with your own writing. In other words, do the following:

- Make a statement (your sentence).
- Support this generalization with a quotation or paraphrase (referenced material from another source).
- Provide a follow-up explanation of the referenced material's significance (your sentences).

Research Includes Primary and Secondary Sources

Primary research is research performed or generated by you. You do not rely on books or periodicals. Instead you create original research by preparing a survey or a questionnaire targeting a group of respondents, by networking to discover information from other individuals, by visiting job sites, or by performing lab experiments. You may perform this research to determine for your company the direction a new marketing campaign should take, the importance of diversity in the workplace, the economic impact of relocating the company to a new office site, the usefulness of a new product, or the status of a project. You also may need to interview people for their input about a particular topic. For example, your company might be considering a new approach to take for increased security on employee computers. You could ask employees for a record of their logs which would highlight the problems they have encountered with their computer security. With primary research, you will be generating the information based on data or information from a variety of sources that might include observations, tests of equipment, interviews, networking, surveys, and questionnaires.

When you conduct *secondary research*, you rely on already printed and published information taken from sources including books, periodicals, newspapers, encyclopedias, reports, proposals, or other business documents. You might also rely on information taken from a Web site or a blog. All of this secondary research requires parenthetical source citations. In some instances, however, you will use already existing (boilerplate) material from your company's document files. This in-house material does not necessarily require documentation. Figure 18.13 is an example of a long, formal report.

Additional information:

See Chapter 16, "Research and Documentation."

FAQs

Q: Why do long, formal reports include research? Aren't research papers just assignments we have in freshman composition classes?

A: Research, a major component of long, formal reports, helps you develop your report's content. Often, your own comments, drawn from personal experience, will lack sufficient detail, development, and authority to be sufficiently persuasive. You need research for the following reasons:

- To create content
- To support commentary and content with details
- To prove points
- To emphasize the importance of an idea
- To enhance the reliability of an opinion
- To show the importance of a subject to the larger business community
- To address the audience's need for documentation and substantiation

LONG, FORMAL REPORT CHECKLIST

_____ 1. Have you included the major components for your long, formal report (front matter, text, and back matter)?

_____ 2. In your long, formal report, have you written to inform, analyze, and/or recommend?

_____ 3. Have you used research effectively, incorporating quotes and paraphrases successfully?

_____ 4. Have you used primary and/or secondary research?

_____ 5. Did you correctly give credit for the source of your research?

_____ 6. Have you correctly used a hierarchy of headings to help your audience navigate the text?

_____ 7. Did you use tables and/or figures to develop and enhance content?

_____ 8. Have you met your audience's need for definitions of acronyms, abbreviations, and specialist's terms by providing a glossary?

_____ 9. Have you achieved clarity and conciseness?

_____ 10. Is your text grammatically correct?

FIGURE 18.13 Long, Formal Report Recommending the Use of VoIP

VoIP at Crenshaw Retail, Inc.

Submitted by

John Staples

Manager of Information Technology

March 2, 2009

FIGURE 18.13 Continued

Date: March 2, 2009
To: Tiffany Steward, CEO
From: John Staples, Manager Information Technology
Subject: Recommendation for Migration to VoIP

Ms. Steward, thank you for allowing me to report on a possible solution to Crenshaw Retail's Internet Protocol challenges. This report is in response to a survey my team conducted and the information technology department's focus on improving internal and external communication.

Currently, because we use analog phone lines, we are having numerous communication problems both internal and external. Internally we are unable to handle the need for rapid response to customers. Externally we cannot respond quickly to either customer requests for information or placement of orders. Our revenues are down approximately 27 percent for the fiscal year. Based on substantial research, my team and I would like to present an option—VoIP. This report will provide you the following information:

- An explanation of VoIP
- How VoIP originated
- Justification for implementing VoIP
- Planning VoIP implementation
- Startup costs
- Return on investment (ROI)
- Current events that are happening around the VoIP community

I look forward to meeting with you to discuss the findings in this report on March 10 in your office at 10:00 a.m. Your assistance and support of this report's recommendations will prove invaluable in increasing our revenues for next quarter.

This effective cover memo uses bullets to highlight the key components of the report. It also achieves a personalized tone through positive words ("thank you," "look forward," and "prove invaluable") and pronouns. Finally, the cover memo is persuasive by showing how implementation of a new system will reverse the downward trend of revenues.

Table of Contents

iii

(Continued)

List of Illustrations

iv

Abstract

This report about Voice over Internet Protocol (VoIP) will help management better justify why or why not to implement VoIP at Crenshaw Retail, Inc. The topics in the report that will be discussed are as follows:

- An explanation of VoIP
- How VoIP originated
- Justification for implementing VoIP
- Planning VoIP implementation
- Startup costs
- Return on investment (ROI)
- Current events that are happening around the VoIP community

v

In the introduction, the writer establishes interest by emphasizing the company's lost revenue and its impact on the workforce. This "hooks" the reader by clarifying the report's importance. In addition, the writer highlights challenges created by the current telephone system.

Introduction

Purpose

Crenshaw Retail, Inc. needs to address decline in revenue. This report will explain how Crenshaw's analog phone lines caused the revenue decline. The company's antiquated phones should be replaced by Voice over Internet Protocol lines.

Background

Since 2008, Crenshaw Retail, Inc. has experienced a significant drop in revenues of approximately 27 percent from $44,000,000 to $33,660,000. Because of this severe decline in revenue, we have had to cut operating expenses primarily by terminating employees. Some of these employees have been with Crenshaw Retail since it began doing business in 1978. To terminate over 120 of these long-term employees, we have had to offer generous buyout packages including termination amounts of one-month's pay for every year of service, benefits to include healthcare and term-life insurance for as long as 12 months, and assistance with job-search preparation for these employees.

Challenges Created by Reliance on Analog-Based Telephony

Our research has discovered that revenues have declined to such an extent because of our dependence on analog phone lines. These lines have been causing the following problems:

1. Burned out Key, PBX, and Data equipment
2. Cut-offs and squealing on lines
3. Crosstalk, echo, or hollow sounds on lines
4. Off-site problems with equipment
5. Garbled data
6. Modems that can't connect

1

7. Customer "can't hear" complaints
8. Dropped calls

We need to upgrade from these analog phone lines that are antiquated, inefficient, costly to maintain, and creating decreased productivity in our workforce.

Discussion

What Is VoIP?

Voice over Internet Protocol (VoIP) entails a phone system, either computer or traditional handset based, that resides on a network and uses data packages, instead of having to be bound by the traditional restraints of analog phone lines.

Figure 1 shows the method of connectivity for VoIP. The gateway is the first part of the network that is recognized. Although firewalls and other devices are not shown in the picture, it is primarily how the data stream gets to the user's desk.

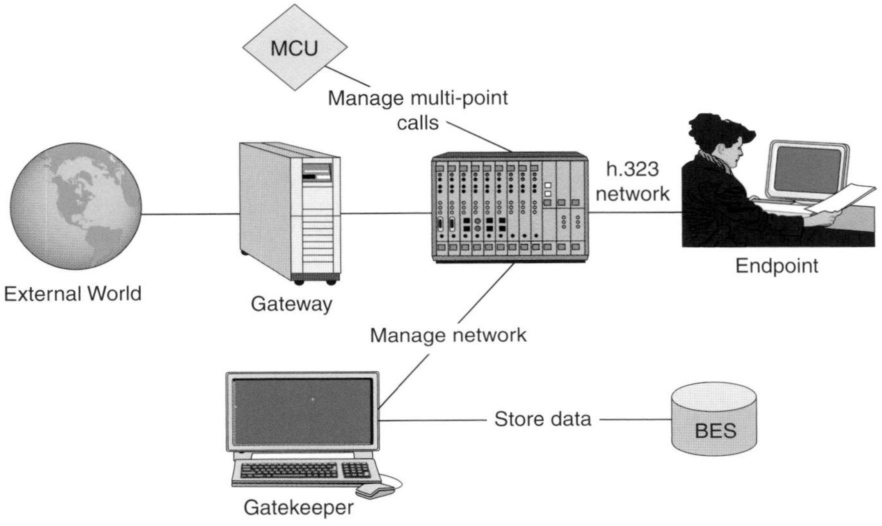

Figure 1 VoIP Connectivity (Source: Dept. of Commerce)

In this long, formal report, research contributes to the development of the topic. Both the Department of Commerce graphic and the wireless howto *steps help the writer explain a complex concept.*

Compression methods make packets travel faster and are easier to manage. Sending a signal to a remote destination could be done digitally. It must first be digitalized with an ADC (analog to digital converter), transmitted, and at the end, converted again in analog format with DAC (digital to analog converter) to be able to use the Voice signal (Valdes). Roberto Arcomano, author of *Wireless Howto*, explains the steps of VoIP compression.

1. ADC converts analog voice to digital signals.
2. The bits are compressed and formatted for transmission.
3. The voice packets are inserted in data packets using a real-time protocol.
4. The packets use the signaling protocol ITU-T H323 to call to users.
5. At RX packets are disassembled. The data is extracted, then converted to analog voice signals and sent to sound card, or phones.
6. This is all done in a real-time fashion because people don't want to be waiting to hear a vocal answer.

2

(Continued)

How VoIP Originated

The journey of Voice over Internet Protocol is a long one. VoIP is almost as old as the Internet, but it really started being utilized around 1995 when only PC-to-PC communication was being handled. Later on during 1995, Vocaltec, Inc. released Internet Phone Software. This particular software was intended to run on a home PC (486/33 MHz). This worked as long as both the caller and the receiver had the same tools and software. However, the sound quality was inferior to that of the standard equipment in use at that time.

Once switches were made that were capable of switching these types of packets, around 1998, voice traffic picked up significantly and is still growing ("Cheaper"). The IDC estimates VoIP has grown at a rapid pace. About 1.5 million Americans use the phones now, but 27 million could be making calls over the Internet by 2009 (Suzukamo). According to Infonetics research, North American VoIP service revenue was up to $1.3 billion in 2004. Sales are expected to continue ringing up to $19.9 billion in 2009 (Francis).

Justification of Implementation

Crenshaw Retail should look at whether we want to switch from traditional PBX-based telephone systems totally over to VoIP systems, or if we want to just include VoIP in certain parts of our infrastructure. The main motivator for switching to VoIP is cost savings. VoIP is not nearly as heavily regulated as traditional telephony and so avoids the enormous burden of federal, state, and local taxes that is placed on traditional telephony. There is an 84 percent cost reduction in the already 26 percent of 2,000 global companies that have implemented VoIP (Senia). Even the NASDAQ estimates that it has cut $40 million of its $100 million annual network costs by using VoIP to consolidate its 15 networks into one.

There are many places in our network where we can save money. Jeffrey T. Hicks and John Q. Walker break down the overall basic premise of the cost savings, or return on investment (ROI). VoIP customers are able to make any call that they need, whether it be local or long distance, and are charged one flat rate. With an analog phone line, Crenshaw pays one flat rate also for local calls. A company will see a big difference in expenses with long distance rates. *Network World* discusses making a budget and budget case for an IT project. "A proposal for a project requires a five-step process, opportunity analysis, infrastructure analysis, process/organization analysis, tool analysis and project analysis" (Denoia and Randall). In the article the writers discuss how each one of these points is important, but opportunity analysis is the main factor when determining the cost, savings, and resulting ROI for the VoIP.

This bulleted list highlights the benefits the company could derive from implementation of VoIP. Providing the detailed list allows the audience to consider how implementation will apply to current and even future budgets.

Denoia and Randall list a guideline of cost saving elements to consider:

- Conference calling
- Long distance
- PBX avoidance
- Video/multimedia
- Adds and changes
- Lower cabling costs
- Toll bypass

3

FIGURE 18.13 **Continued**

Paul Desmond from *Network World* explains that extending phone services to new locations is one of the biggest returns that users report on their IP telephony investments, even when they don't drop a PBX in the process. "Savings from reduced moves, adds and changes is probably the benefit touted most loudly—with good reason," according to Johna Till Johnson, president of Nemertes Research and a *Network World* columnist. "A Nemertes survey of 100 companies with average IT budgets of $10 million or more shows that employees move an average of 0.87 times per year—or almost once every year—at a cost of $100 per move. For a company with 1,200 employees, that amounts to $104,400 in annual costs that literally go away with IP telephony because users can move their own phones" (Desmond).

Planning for Implementation of VoIP

Crenshaw Retail needs to evaluate where there could be a possibility of added expenses before implementing a VoIP network; these are called the hidden costs. Ray Kriss, from NetGain Communications, points out a few hidden costs that some companies' networks overlook:

> Unlike traditional telephone systems that operate within their own networks, VoIP relies on the LAN and WAN, which means a VoIP system shares bandwidth with existing business communications assets. While there is usually enough bandwidth in the LAN for internal use, the WAN is often the most constricted point for any network, which leads to increased potential for congestion and delays in information. For a business to execute a full conversion to VoIP, it must consider an increase in network bandwidth. (Kriss)

Because of security reasons, most VoIP systems will be put on virtual local area networks, so the system can be better monitored. This is effective, but a security assessment costs money (Kriss).

To persuade effectively, this writer anticipates questions the audience might have by factoring in hidden costs of implementing a new system. For more information about the importance of refutation and persuasive writing, see Chapter 13.

Startup Costs for VoIP

Using headings breaks the long, formal report into manageable "chunks" for easy reading. In addition, these "talking headings" are informative and aid navigation for the audience.

The initial cost of implementing a new system can be expected to be expensive. The startup cost can be high, if you begin with a big project. "There's new network equipment, servers, IP phones, management software, and diagnostic tools to buy" (Hicks and Walker). A network infrastructure upgrade may be necessary also, because Crenshaw's equipment may not be able to handle the new VoIP system. In addition, there must be training done for the equipment and service (Hicks and Walker).

An Explanation of ROI

Barbara DePompa Reimers, from *Computerworld Magazine*, explains how achieving a speedy return on investment from VoIP technology has been hard, in part because upfront costs can be high and traditional long-distance phone charges have dropped in recent years. Most analysts have pushed out the time it takes to gain a return on VoIP to an average of 19 months, compared with earlier estimates of less than a year (Reimers). To name a few companies that have seen a good ROI from VoIP, Kevin Lopez, the national manager of telecommunications at Grant Thornton, explains how his network has been in place for 9 to 10 months. Grant Thornton has cut communication costs by routing voice traffic over its wide area network and eliminating toll charges. It has also reduced network management requirements and consolidated 28 stand-alone systems into four hubs all without replacing its existing private branch exchange (PBX) and digital phones. The payoff is Grant Thornton has saved $800,000 in its first year on intra-company long-distance toll charges and $160,000 on equipment lease

4

(Continued)

payments, Lopez says (Reimers). There are many occasions where companies have saved money from VoIP, from local school districts, like the Saugus Union School District, to multi-billion dollar companies (Hochmuth).

ROI does not have to be just financial; it can also deal with the morale of a company. These are the hard and soft benefits of implementation. Crenshaw must try to predict what type of impact the implementation of VoIP would have on its employees. This could improve productivity or slow it down; it could excite workers or turn them off to the idea. The soft benefits can be harder to predict than the hard benefits, because there are no numbers to check with to know that you are benefiting. "According to analysts and corporate customers, the key to achieving an ROI on VoIP is to plan carefully. Businesses should conduct an IT audit of their network infrastructures, including current data and voice traffic patterns. If users skip this step and don't know their current costs, they won't know whether the VoIP migration has saved them money" (Reimers).

In my opinion, an audit of Crenshaw's entire network should be done before implementing a project as large as this. As Reimers stated, you won't know how much you have saved, if any, and a company trying to save money will be doing so in vain. Planning is the key when it comes to all projects, and it is the key to getting a good ROI. Walker and Hicks suggest that part of the planning process starts with calculating a good ROI, by taking the expected returns from a project, subtracting the cost of implementing the project, and dividing by the amount of time required. The divisor is usually given in years, so that the resulting units are measured in annual ROI.

Table 1 is an example of an ROI for a fictitious business, which is provided by Quintums VoIP calculator.

The table makes complex information easier to understand. The audience then can make a financial comparison to their own company.

Table 1 Return on Investment Calculator

Table 1 Return on Investment Calculator for VoIP Installation		
Group	Amount	Cost
Average number of calls per hour	100	
Average call duration in minutes	5	
Number of working hours per day	8	
Number of working days per week	5	
Number of working weeks per month	4	
Total number of minutes per month	80,000	
Percentage of national calls	70	0.28 per minute
Percentage of international calls	30	0.80 per minute
Total number of national minutes	56,000	$15,680.00
Total number of international minutes	24,000	$19,200.00
Percentage of completion over IP	90	0
Percentage of completion over PSTN (Public Switched Telephone Network)	10	$1,673
Savings in call charges with VoIP		$36,553

5

FIGURE 18.13 **Continued**

A Fair and Balanced Look at ROI

So far, the return on investments that have been discussed can be considered to be an optimistic ROI analysis. Robin Harris Foster, from the *Business Communications Review,* explains how companies often ignore considerations of human nature, accounting principles, and the true use and benefit of the investment in a given enterprise. Foster lists a few things to include in an ROI that should make the end result come out more fair and balanced.

- Credible ROI analysis depends on reasonable expectations of people, not on what's possible in an idealized model. A promising source of return from payroll savings may require compliance of workers. When human behavior can diminish or negate the benefits, you should assume less than 100 percent compliance and therefore less than 100 percent of the possible savings.
- Costs recur, but they also evolve, and the pace of evolution makes a big difference over the course of an investment. Purchasing the hardware and software for a proposed communications solution is only a beginning.

  To implement and live with the solution, Crenshaw will also incur maintenance charges after warranty; upgrade purchases or subscription costs, and administrative or technical staff costs (wages, benefits, bonuses, and costs such as training and real estate). ROI analysis should take account of the fact that labor and maintenance costs often increase year over year and must be factored into an analysis.
- Never give a benchmark the benefit of the doubt. They raise the red flag in any ROI analysis, as they can misstate the operational and financial impact of a solution, because they look at other company's increases and compare them to their own.

Current Issues in VoIP

VoIP in the workplace is still a fairly new concept, so there are still many legal issues to be resolved. One type of legality is the federal ruling to require Internet phone service to provide 911 emergency capabilities with their services. "After a 17-year-old girl couldn't call 911 when home intruders shot her parents, the state of Texas sued VoIP provider Vonage for deceptive trade practices for failing to offer built-in 911 services" (Suzukamo). VoIP services lack the essential 911 service it needs for critical emergency features like location detection. Vonage says they would like to have the 911 services but that the local exchange carriers (ILEC) won't offer the right network access. The ILEC's rebuttal was, "The real issue is cost. Vonage doesn't want to pay for integrating into the 911 infrastructure," they say. "There are existing procedures to connect to the 911 network today, which several [VoIP] providers utilize" (Suzukamo).

As achieved earlier in the report, the writer presents fair and unbiased information to anticipate questions. This allows the audience to make a decision based on pros and cons.

The key to the FCC's ruling is that Emergency-911 automatically provides emergency dispatchers with the street address of the caller. Some Internet phone services allow customers to move their phones to different cities or even countries and still use the same phone number. Combining both features would pose

6

(Continued)

The writer uses the conclusion to reiterate the key points. This helps the audience remember what has been discussed. In addition, the writer recommends the next course of action.

a special challenge to VoIP providers and probably contribute to raising their costs. For companies like Time Warner Cable, this ruling will not affect them since they already offer the services, but the smaller VoIP will most likely be the ones that will feel the effect of the ruling.

Conclusion/Recommendation

Crenshaw needs to fully investigate the options when deciding if we should or shouldn't implement VoIP into our business network. VoIP can be compared to the use of the Internet in companies. It's here to stay, and it is going to get nothing but larger. The benefits of companies that have implemented VoIP outweigh the negatives.

The topics that have been presented are the following: what VoIP is, where VoIP came from and where it is now, justification of implementation, planning, summary, startup costs, return on investment, a fair and balanced look at ROI, and current events that are happening around the VoIP community. My recommendation is that we perform further research on the possibility of migrating to VoIP beginning with a review of our entire communication network.

7

A glossary is an excellent place to define terms and abbreviations so that readers can quickly access the terms. Using a glossary also allows the writer to avoid disrupting content in the long report with definitions. Finally, a glossary allows the writer to prepare a report for multiple audience levels.

Glossary

ADC	analog to digital converter
Analog	a variable signal continuous in both time and amplitude
DAC	digital to analog converter
LAN	local area network
PSTN	Public Switched Telephone Network—analog phone systems
ROI	return on investment
Voice signal	allows users to access phone options with a single voice command
VoIP	Voice over Internet Protocol
WAN	wide area network

8

Works Cited

Arcomano, Roberto. *Wireless Howto*. 31 Jul. 2002. 14 Aug. 2007. <http://tldp.org/HOWTO/Wireless-HOWTO.html>.

Cheaper Changes. 18 Feb. 2005. Vonage. 24 May 2005 <http:// http://www.vonage-forum.com/printout553.html>.

Denoia, Lynn, and Tom Randall. "The Case for VoIP." *Network World*. 2 Aug. 2004. 28 June 2005 <http://proquest.umi.com/pqdweb?did=677270081&sid=3&Fmt=3&clientId=1571&RQT=309&VName=PQD>.

Desmond, Paul. "THE ROI OF CONVERGENCE." *Network World*. 14 June 2004. 28 May 2005 <http://proquest.umi.com/pqdweb?did=652227571&sid=3&Fmt=4&clientId=1571&RQT=309&VName=PQD>.

Duffy, Jim. "More VoIP Issues Bubbling Up." *Network World*. 9 Feb. 2004. 26 May 2005 <http://proquest.umi.com/pqdweb?did=544375901&sid=3&Fmt=4&clientId=1571&RQT=309&VName=PQD>.

9

FIGURE 18.13 **Continued**

Foster, Robin H. "Building a Credible and Conservative ROI for VoIP." *Business Communications Review* May 2004. 19 May 2005 <http://proquest.umi.com/pqdweb?did=639952941&sid=3&Fmt=4&clientId=1571&RQT=309&VName=PQD>.

Francis, Bob. *IT Managers Are Ringing Up VoIP*. 13 June 2005. *InfoWorld*. 27 June 2005 <http://www.networkworld.com/news/2005/051305-voip.html>.

Hicks, Jeffrey T., and John Q. Walker. *Taking Charge of Your VoIP Project: Strategies and Solutions for Successful VoIP Deployments*. Boston, Mass: Pearson Education, 2004. 1–294.

Himmelsbach, Vawn. "VoIP: Diving into IP telephony." *Computer Dealer News*. 18 Mar. 2005. 24 May 2005 <http://proquest.umi.com/pqdweb?did=816930851&sid=4&Fmt=4&clientId=1571&RQT=309&VName=PQD>.

Hochmuth, Phil. "Costs, Security Vex VoIP Users." *Network World*. 24 Feb. 2003. 25 May 2005 <http://proquest.umi.com/pqdweb?did=293953291&sid=3&Fmt=4&clientId=1571&RQT=309&VName=PQD>.

Hochmuth, Phil. *School District Saves with VoIP, Open Source*. 16 May 2005. Network World. 19 May 2005 <http://www.networkworld.com/news/2005/051605-renie-saugus.html>.

Johnson, Johna T. *911 Isn't a Negotiable Service*. 18 Apr. 2005. Network World. 25 May 2005 <http://www.networkworld.com/columnists/2005/041805eyejohnson.html>.

Kriss, Ray. *Hidden Costs, and Savings, of VoIP*. 17 Jan. 2005. NetGain Communications. 26 Apr. 2005 <http://www.computerworld.com/networkingtopics/networking/story/0,10801,98935,00.html?from=story_package>.

Reimers, Barbara D. "Wringing Savings from VOIP." *Computerworld Magazine*. 20 Jan. 2004. 29 May 2005 <http://proquest.umi.com/pqdweb?did=279401651&sid=3&Fmt=4&clientId=1571&RQT=309&VName=PQD>.

Senia, Al. "VoIP Strengthens Foothold in Enterprise Accounts." *America's Network*. 15 Nov. 2004. 25 June 2005 <http://proquest.umi.com/pqdweb?did=739552731&sid=2&Fmt=4&clientId=1571&RQT=309&VName=PQD>.

Suzukamo, Leslie B. "Federal Ruling to Require Internet Phone Service to Provide 911 Emergency Data." *Knight Rider Tribune Business News*. 21 May 2005. 27 May 2005 <http://proquest.umi.com/pqdweb?did=842707361&sid=1&Fmt=3&clientId=1571&RQT=309&VName=PQD>.

Valdes, Robert. "How VoIP Works." *How Stuff Works*. 14 August 2007. <http://communication.howstuffworks.com/ip-telephony.htm>.

VOICE.SKY Selects NexTone's iVMS Management System for Advanced VoIP Interconnects; Partnership Enables VOICE.SKY to Expand Its Service Reach and Support Growing VoIP Deployments in the Middle East and Africa. 23 May 2005. *Business Wire*. 30 May 2005 <http://proquest.umi.com/pqdweb?did=843637961&sid=1&Fmt=3&clientId=1571&RQT=309&VName=PQD>.

10

Charles Met His Challenge

To meet his communication challenge, Charles used the P^3 process.

Planning

To plan his long report, Charles considers the following:

- Goal—inform the audience of the need for new financial software, analyze findings from research, and recommend a preferred vendor
- Audience—lay readers (city manager and city council members)
- Channels—long, formal report
- Data—vendor options; software applications; cost of software, training, and implementation; payment schedules

To gather content for the long, formal report, Charles says he must do the following primary research:

1. "Meet with my accounting manager to determine the required software applications for the finance department.
2. Seek proposals from software vendors.
3. Determine how to allocate finances to pay for the software purchases."

To accomplish these steps, Charles plans by brainstorming a list of project requirements, as shown in Figure 18.14.

Packaging

After using primary research to find what the city employees needed in terms of financial software and receiving proposals from vendors, Charles drafted an executive summary. Because of the large amount of information to be covered in the report, Charles merely wanted to get the content down on paper. He says, "I knew that my accounting manager would review the draft and make suggestions for improvement before I perfected the long, formal report. We always try to edit each others' written communication." Figure 18.15 shows how Charles packaged his executive summary rough draft.

Perfecting

Charles met his communication challenge by perfecting the executive summary draft. He added talking headings, corrected grammar errors, achieved conciseness by avoiding semicolons, and added another table to clarify MEGS's fee of an additional $70,000. See Figure 18.16 for Charles' perfected executive summary.

FIGURE 18.14 Charles's Brainstorming List of End User Requirements

Prioritized List of Needs

1. End user ease of use, complete with extensive online help for inexperienced end users
2. Ability to accomplish specific finance tasks
   a. Automated interest allocations
   b. Pooled cash
   c. Seamless purchasing card interaction
   d. Superior budget module
   e. Project accounting
3. Lotus Notes compatibility
4. Seamless interaction with Excel and Word

FIGURE 18.15 Rough Draft with Errors of Executive Summary and Comments from the Accounting Manager

Re: Procurement of new financial software

The financial software (ACS) that the Finance Department is currently utilizing is 16 years old. The current software is not Windows based and is extremelly cumbersome to accomplish the simplest tasks. Many staff hours are wasted retyping information into Excel spreadsheets for reporting purposes; dual and even triple data entry is common due to the shortcomings of the current software. Additionally, the current software currently resides on an AS/400. Many technology experts believe the AS400 is becoming obsolete. In their management letter for year ended December 31, 2009, the City's auditors, Cottrell and Kreisler, LLC made several references to the inadequacy of the current software. Attached is an excerpt from the management letter referencing these weaknesses. They site multiple internal control issues, inability to track vial information and redundant data entry. During the 2010 budget discussions a Decision Package was approved for $90,000 for hardware or software that might be required for a new financial software package. It was determined at that time to wait to make the decision on how to proceed.

Requests for proposals for new financial software were sent to seven software vendors. The following table illustrates the five responses received on March 14, 2009.

New Land	Utica, NY	$254,295
MEGS	St. Louis, MO	$159,538
MoonRay	Jupiter Beach, FL	$129,884
ACS	Springfield, IA	$123,800
Roedell	Taos, NM	$109,318

After a review of the responses, three vendors were selected to provide on-site demonstrations. Staff from finance and from other departments spent many hours comparing these vendor's products. The highest priorities were identified as end-user ease of use, ability to accomplish specific finance tasks and Lotus Notes compatibility. MEGS software met all of the department's needs including the following: automated interest allocation, pooled cash, seamless purchasing card interaction, superior budget module, project accounting, etc. Unfortunately MEGS is $70,000 over our allotted budget of $90,000. MEGS has offered to divide the payment over two years so the remaining $70,000 would be included in the General Fund in the 2010 budget. With the implementation of the software the Finance Department will have some cost reductions related to the constant need for temporary help to accomplish current requirement for redundant data entry.

After reading the draft, Charles's financial manager suggested the following changes:"Charles, great content! Thanks for your hard work on this project. My department really needs this update. Before we send the report to the city council, please consider these ideas for revision: (1) Add headings to break the text into more manageable units. (2) Avoid lengthy sentences. (3) You use the word 'current' a lot. (4) Let our admin. asst. check the text for grammar. (5) You need to explain MEGS's cost analysis to show why we need to go $70,000 over budget. I'd be very happy to review this again if you'd like."

FIGURE 18.16 Perfected Executive Summary for City Manager and City Council Members

Executive Summary

Background of Finance Department's Software Requirements

The finance department's current ACS software is 16 years old. Because the software is outdated, it has the following problems:

- The software is not Windows based.
- The software is not time efficient. Many staff hours are wasted retyping information into Excel spreadsheets for reporting purposes. Dual and even triple data entry is common due to the shortcomings of the current software.

(Continued)

FIGURE 18.16 Perfected Executive Summary for City Manager and City Council Members Continued

- The ACS financial software resides on the AS400 hardware system. Many technology experts believe the AS400 is becoming obsolete.
- In their management letter for year ended December 31, 2009, the city's auditors, Cottrell and Kreisler, LLC, made several references to the inadequacy of the current software, including the following: multiple internal control issues, inability to track vital information, and redundant data entry.

During the 2010 budget discussions, a decision package was approved for $90,000 for hardware or software that might be required for a new financial software package.

Comparison of Vendor Proposals

Requests for proposals for new financial software were sent to seven software vendors. Table 1 illustrates the five responses received on March 14, 2009.

Table 1 Vendor Comparison		
Vendors	**Locations**	**Proposed Fees**
New Land	Utica, NY	$254,295
MEGS	St. Louis, MO	$159,538
MoonRay	Jupiter Beach, FL	$129,884
ACS	Springfield, IA	$123,800
Roedell	Taos, NM	$109,318

After a review of the responses, three vendors were selected to provide on-site demonstrations. Staff from finance and other departments compared these vendors' products. The highest priorities were identified as end user ease of use, ability to accomplish specific finance tasks, and Lotus Notes compatibility. MEGS's software met all of the department's needs including the following:

- Automated interest allocation
- Pooled cash
- Seamless purchasing card interaction
- Superior budget module
- Project accounting
- Extensive online help for inexperienced end users
- Seamless interaction with Excel and Word

Recommendation for Purchasing MEGS's Software

Although MEGS is $70,000 over our allotted budget of $90,000, MEGS has offered to divide the payment over two years. The remaining $70,000 would be included in the General Fund in the 2010 budget. With the implementation of the software, the finance department will have some cost reductions related to the constant need for temporary help to accomplish current requirement for redundant data entry. Table 2 illustrates MEGS's cost analysis of our software needs. MEGS was the only vendor to explain their bid through a cost analysis.

Table 2 MEGS Cost Analysis	
Software Needs	**Costs**
Software	$94,058
Training/Data Conversion	$47,040
Implementation	$18,440
Total	**$159,538**

Charles says, "Whenever I write to the city council, I know that they are extremely busy and not necessarily proficient in all areas discussed. I try to write text that is easy to access, clearly stated, and persuasive. Without good communication, I can't help the City to stay on budget and to meet our citizen's needs. The P³ process lets me be a successful communicator."

We recommend purchasing MEGS's financial software, including training, data conversion, and implementation. Doing so will benefit our finance department, allowing it to provide cost- and time-efficient services to our city.

CHAPTER HIGHLIGHTS

1. Long, formal reports require time, people, and money, and have far-reaching effects.
2. Long, formal reports often deal with topics that consider serious and complex issues.
3. Long reports can provide information, be analytical, and/or recommend a course of action.
4. Long, formal reports include front matter (title page, cover letter, table of contents, list of illustrations, and abstract); text (introduction, discussion, conclusion/recommendation); and back matter (glossary, works cited, and optional appendix).
5. To help your readers navigate the report, create a clear hierarchy of headings.
6. When you write your long report, consider your audience and enhance your discussion with figures and tables for clarity, conciseness, and cosmetic appeal.
7. Primary research is research performed and generated by you.
8. Surveys or questionnaires are primary sources of research.
9. When you conduct secondary research, you rely on already printed and published information.
10. Secondary research is from books, periodicals, databases, and online searches.

WWW *Find additional long, formal research report exercises, samples, and interactive activities at http://www.prenhall.com/gerson.*

MEETING WORKPLACE COMMUNICATION CHALLENGES

CASE STUDY

Alpha/Beta Consulting (A/B/C) solves customer problems related to hardware, software, and Web-based applications. The consulting company helps clients assess their computing needs, provides training, installs required peripherals to expand computer capabilities, and builds e-commerce applications, such as Web sites and corporate blogs. Their client base includes international companies and academic institutions. A/B/C works extensively with FIRE industries (Finance, Insurance, and Real Estate).

A new client, Home and Hearth Security Insurance Company, has asked A/B/C to build its e-commerce opportunities. Home and Hearth's current Web site is outdated. Since the Web site was created, Home and Hearth has added new services, including insurance coverage for electronic commerce, employment-related practices liability, financial institutions, management protection, and medical professional liability. In addition, Home and Hearth has expanded globally. It now has a presence in the Far East (Japan, Taiwan, and Singapore), the Mid East (Dubai, Jordan, and Israel), and South Africa. Home and Hearth's new services and new service locations are not evident on the current Web site.

An additional challenge for Home and Hearth is client contact. Currently, Home and Hearth depends on a hard-copy newsletter. However, corporate communications at Home and Hearth believes that a corporate blog would be a more effective means of connecting with its modern client base.

A/B/C can meet Home and Hearth's needs by building a new Web site and by creating a corporate blog.

Assignments

1. Write the abstract that will preface A/B/C's long, formal report to Home and Hearth Security Insurance Company.
2. Write the introduction that will preface A/B/C's long, formal report to Home and Hearth Security Insurance Company.
3. Create a survey questionnaire asking Home and Hearth employees what they would like to see on the company's Web site and corporate blog.

4. Outline what you believe Home and Hearth's new Web site should include to cover its new international focus. This constitutes the discussion section of the report.
5. Conduct any research you need to find more information about either Web design or the development of a corporate blog.

ETHICS IN THE WORKPLACE CASE STUDY

Jordan Benjamin is a marketing manager for an online distributor of healthcare-related equipment and supplies. As manager of 300 employees worldwide, she has to write a report documenting employee use of company-owned PCs. This report will be delivered to both management and company stockholders at the annual meeting. Jordan decides that the only way to obtain information about PC use is by e-mailing a questionnaire to her employees. Some of the questions she asked in the questionnaire are as follows:

1. Do you use your PC for personal activities?
2. Do you maintain a personal blog?
3. Do you send and receive personal e-mails over your PC?
4. How much time each day do you spend on your PC for personal use?
5. Do you order from any online stores using your company-owned PC?

When Jordan writes her report, she bases her analyses on the 77 percent response rate she received from employees. She also used many of their verbatim comments in her written report.

Question:

Is it ethical for Jordan to require employees to respond to personal questions in a questionnaire she disseminated through the company? By e-mailing the questionnaire to the employees, does Jordan suggest that the company "expects" employees to respond? Is this ethical? Is it ethical for Jordan to use comments solicited from the employees on the questionnaire? Justify your answer based on information provided in Chapter 1.

INDIVIDUAL AND TEAM PROJECTS

1. Read a long, formal report. It might be one your instructor provides you or one you obtain from a business or online (see the links provided in this chapter's Web Workshop). If the report contains an abstract or executive summary, is it successful, based on the criteria provided in this chapter? Explain your decision. If the abstract or executive summary could be improved upon, revise it to make it more successful. If the report does not have an abstract or executive summary, write one.
2. Read the following abstract from Alpha/Beta Consulting, written for their client Nitrous Systems. It needs revision. How would you improve its layout and content?

> Nitrous Systems, a multi-discipline architectural design firm, is a start-up business with plans to begin operation in October 2009. The company currently has no information technology network for internal and external communication. Alpha/Beta Consulting will recommend a server to meet Nitrous Systems' unique needs, a backup solution for archiving architectural materials and critical communications, software for digital asset management, and a proposed network design. After implementing these suggestions, Nitrous Systems will have an IT network that will be safe, secure, dependable, and prompt. The system that Alpha/Beta Consulting is recommending also can be restructured with simplicity as the company's needs grow. By installing the IT network recommended in this report, Nitrous Systems will be able to fulfill their company's mission statement.

Table 18.1 Time Sheet

Staff	Hours Worked Each Week														
	wk 1	wk 2	wk 3	wk 4	wk 5	wk 6	wk 7	wk 8	wk 9	wk 10	wk 11	wk 12	wk 13	wk 14	Total Hours for Project
Kenyon Patel	6.0	10.5	7.0	7.5	7.0	8.5	6.0	11.0	8.0	8.0	7.0	6.5	12.0	8.0	**113.00**
Randy Butler	6.5	5.5	5.5	3.0	5.5	8.0	4.0	5.5	6.0	8.0	7.0	9.5	15.0	8.0	**97.00**
Maria Villatega	0.0	0.0	3.5	4.0	4.5	5.0	3.0	1.5	10.0	17.0	8.0	9.0	8.0	4.0	**77.50**
James Soto	4.5	3.5	4.0	5.5	5.0	5.0	3.0	2.0	14.0	12.0	8.5	10.5	7.5	3.5	**88.50**
Total	**17.00**	**19.50**	**20.00**	**20.00**	**22.00**	**26.50**	**16.00**	**20.00**	**38.00**	**45.00**	**30.50**	**35.50**	**42.50**	**23.50**	**376.00**

3. Alpha/Beta Consulting, as agreed upon in a meeting with their client Nitrous Systems, determined that their research for a contracted long, formal recommendation report would take no less than 350 hours and no more than 400 hours. As detailed in Table 18.1, A/B/C's four employees assigned to the project worked 376 hours. Using the information provided in the time sheet, create a line graph that tracks the employees' hours and a pie chart showing the percentage of time each employee worked.

4. At the conclusion of their long, formal report, the employees at A/B/C wrote a recommendation. Their recommendation, though excellent in terms of its content, needed to be reformatted for easier access and better emphasis of key points. Read the employees' recommendation and improve its layout.

Recommendations to Meet Nitrous Systems' Network Needs

To ensure a secure and dependable network that can be restructured to meet Nitrous Systems' developing needs, Alpha/Beta Consulting recommends the following:

You can improve your computing system's "Application Layer." The applications software programs that Alpha/Beta Consulting recommends consist of the Adobe Creative Suite 2, for imaging, editing, illustrations, file sharing, Web designs, and digital processing; Microsoft Office, for spreadsheet and Excel documents, word processing files, project timelines, and maintenance; and Internet communication utilities, such as Web browsers and FTP clients, to guarantee rapid deployment of company communications and contract sales. All these applications will be run on both the current Macintosh and Microsoft operating systems.

You also need to add a "Data Layer" to your new computing system. The data transferred on the new network will consist of large graphic files and fonts being accessed from a centralized server, postscript printing, voice, and messaging. TCP/IP will be the main communication protocol, but Appletalk may also be used.

To meet Nitrous Systems' network infrastructure requirements, Alpha/Beta Consulting recommends Category 6 (Cat 6) cables connected to 1 Gbps switches. These constitute your "Network Layer." The logical star topology network will enable a higher degree of performance for all users associated with the production of architectural designs, meet all organizational duties and communications requirements, and provide a centralized server as the main point of access for all the client machines.

Finally, once installed, the physical network must have a "Technology Layer." This will enable the company to perform at a competitive level. This communication system will be a vital piece of the overall success of the company. The server will provide quick and reliable data storage, meet Nitrous Systems' data backup needs, and allow for streamlined work flow.

DEGREE-SPECIFIC ASSIGNMENTS

Restful Inn Hoteliers is a growing international company with corporate headquarters in Seattle, Washington; Boston, Massachusetts; Barcelona, Spain; and Singapore. As it builds more hotels around the world, Restful Inn is confronting numerous issues related to growth in emerging markets. Restful Inn Hoteliers' employees in management, business information systems, accounting, international sales and marketing, and finance must consider these new challenges, such as the need for

- Managing dispersed teams
- Expanding computer systems and networks
- Handling international accounting regulations and standards
- Marketing goods and services in the new locations
- Meeting international finance and banking standards

Assignments

Drawing from these challenges and any others that you find through research, choose a topic unique to your major field as an employee of Restful Inn Hoteliers. Based on your research, write an abstract or executive summary to your boss at Restful Inn. Refer to the chapter for samples and criteria.

PROBLEM-SOLVING THINK PIECES

Read the following text from a long, formal report written by Alpha/Beta Consulting. Based on the explanation in this chapter, decide whether the text informs, analyzes, and/or recommends. Explain your decisions.

Purpose Statement

This report will recommend the design parameters, hardware specifications, and estimated expenses needed to build and install external and internal computer network technology to meet Nitrous Systems' communication needs.

This report's recommendation will focus on the following key areas:

- A solution for Nitrous Systems' server needs
- Recommendations for a backup solution
- Possible software for digital asset management
- A proposed network design

Needs Analysis

1. <u>Company Background</u>

Nitrous Systems, a multi-discipline architectural design firm located in Raleigh, North Carolina, is a start-up company. Nitrous Systems will occupy a 2,000 square foot office space. They plan to begin operations in October 2009. The company will provide industrial, interior, landscape, and green architecture services for an international client base. Communication is a major component of their business plan. Thus, a reliable, efficient, creative communications/IT technology system is mandatory.

2. <u>Basic IT Requirements</u>

Currently, Nitrous Systems has no network. The company needs an infrastructure plan that allows for the following:

- An IT platform that will be compatible with Macintosh and Windows.
- Eight Windows workstations and four Macintosh platforms.
- A network based on WiMAX impact.
- An internal and external network that is safe, secure, reliable, and fast.
- An internal and external network that can store, execute, and transmit architectural materials.
- A network that will allow for successful communications to clients, partners, and vendors.

- A data backup solution.
- Digital asset management software.

3. **Detailed IT Requirements**

In addition to the basic network requirements for Nitrous Systems, the company also has asked for recommendations to meet the following micro-network needs:

- Information about the most efficient server operating system for reliable communication in a cross-platform environment.
- Requirements for different file and storage server hardware and software.
- An analysis of which FTP server software would transfer secure data reliably to external clients.
- The need for scripts to scan recently modified files and back them up to a server.
- Reliable backup solutions and backup media types.
- A comparison of different software applications that would prevent data duplication in the permanent backup archives.
- A comparison of software that could catalog backup storage.
- A software solution allowing Microsoft Word documents to be converted and catalogued into one PDF file.

WEB WORKSHOP

You can find many examples of long, formal reports online. Compare the format of the online reports with the criteria for long, formal, researched reports discussed in this chapter. Explain where the online reports are similar to or different from this chapter's criteria. Do the online reports successfully communicate information and analyze issues and problems? Check out these sites, for example:

- Formal Reports: Carnegie Commission on Science, Technology, and Government (http://www.carnegie.org/sub/pubs/ccstfrep.htm)
- Reports from the American Society of Civil Engineers Following Hurricane Katrina (http://www.asce.org/static/hurricane/whitehouse.cfm)
- International Narcotics Control Strategy Report, Released by the Bureau for International Narcotics and Law Enforcement Affairs (http://www.state.gov/p/inl/rls/nrcrpt/2003/vol2/html/29910.htm)
- The Use of Radio Frequency Identification by the Department of Homeland Security to Identify and Track Suspicious Individuals (http://www.dhs.gov/xlibrary/assets/privacy/privacy_advcom_rpt_rfid_draft.pdf)
- Report on "GreenDC Week: Earth Day Celebration" (http://dceo.dc.gov/dceo/lib/dceo/Green_DC_Week.pdf)
- The United States' Government Accountability Offices' Report on "Women's Participation in the Sciences" (http://www.gao.gov/new.items/d04639.pdf)

QUIZ QUESTIONS

1. What are the reasons for writing long, formal reports?
2. Explain informative formal reports.
3. Explain analytical formal reports.
4. Explain recommendation formal reports.
5. What are the components of a long, formal report?
6. What do you include in an abstract?
7. What are the reasons for using research in a long, formal report?
8. Explain primary research.
9. Explain secondary research.
10. What is the purpose for using a hierarchy of headings in a long, formal report?

CHAPTER 19

Internal and External Proposals

▸REAL PEOPLE *in the workplace*

Mary Woltkamp jokingly calls herself the "Queen" of Effective Business Communications, Inc. "Why not?" she laughs. "Technically, I'm the owner, but I'm also the receptionist, the accountant, the marketing department, and the janitor. I'm a one-person company, so I do it all. The title 'Owner' or 'President' sounds pretentious under the circumstances. Anyway, the most important thing I do is help clients accomplish the goals they have for their print and electronic materials. I'm not sure what the title is for that."

Whatever her title, Mary creates a variety of materials for her clients including requirements documentation, user manuals, training materials, job aids, newsletter articles, marketing copy, proposals, presentations, and more. Her clients are just as varied as the materials.

On any project, the client's goal is Mary's primary focus. Before getting started, she encourages clients to be very clear about who their audience is and what the objectives of the finished materials are. Then she uses her training and experience to create written pieces that address the audience and meet the goals.

During her work day, if Mary is not in her home office writing or doing research, she is on the phone or in a client's office attending project meetings or interviewing subject matter experts. She also spends a good deal of time reading existing documentation in order to fully understand the subject matter of a project. Although tackling new topics can be a challenge at times, "It's what I love best," she says. "I'm always learning something new."

According to Mary, no matter what the project is, clear two-way communication is critical at every step. "Obviously, clients hire me to share my expertise with them, but if I don't listen to my clients carefully in return, I run the risk of creating materials that fail to get the job done. In that case, nobody is happy."

And since she is a small company with a small budget for advertising, she counts on satisfied customers and word of mouth for new projects. So far, it has worked well. Even so, when given an opportunity, she makes sure to effectively communicate what she does professionally and to pass out her business card. Her title on the card? Well, it's not "Queen," but someday it might be.◂

CHAPTER GOALS

When you complete this chapter, you will be able to do the following:

1. Format proposals effectively using components such as a title page, cover letter, table of contents, list of illustrations, abstract, introduction, discussion, conclusion/ recommendation, and glossary.
2. Write effective internal proposals to persuade corporate decision makers to address issues and provide resources.
3. Write effective external proposals to sell a new service or product to a potential customer.
4. Distinguish among common proposal terms including RFP, T&C, SOW, boilerplate, solicited proposals, and unsolicited proposals.
5. Apply research techniques to gather information for proposals.
6. Evaluate your proposals using the proposal checklist.

WWW

To learn more about proposals, visit our Web Companion at http://www.prenhall.com/gerson.

Mary's Communication Challenge

Mary says, "When clients approach me about creating materials for them, they have a problem they are trying to solve. Although the client's concerns vary, my challenge is always to make sure the actual problem is clearly identified, so that I can propose the appropriate materials to resolve it.

Recently a client asked me to redesign the training manual they use to train new finance department employees on the company's expense reporting application. Costly mistakes were being made, and the client assumed that the manual wasn't clearly communicating what employees needed to do.

I suggested that the first step be a front-end analysis including a review of the existing manual and interviews with recent trainees and a manager, a step the client had not yet taken, to pinpoint the exact causes and to avoid wasting time and money on the wrong solution. The client agreed, and I was able to conduct a thorough analysis before submitting a project proposal."

See pages 561–563 to learn how Mary met her communication challenge.

WHY WRITE A PROPOSAL?

When you write a proposal, your goal is to sell an idea persuasively. Consider this scenario: Your company is growing rapidly. As business increases, you believe that several changes must occur to accommodate this growth. For example, you think that the company needs a larger facility. This new building could be located in your city's vibrant new downtown expansion corridor, be placed in a suburban setting, or entail the expansion of your current site. A new building or expansion should include amenities to improve recruitment of new employees, such as workout facilities, daycare, restaurant options, and even a gaming room. Finally, as part of new employee recruitment, you believe that the company must increase its diversity hiring practices.

Internal Proposals

How will you convey these ideas to upper level management? The topic is large and will require extensive financial obligations, time for planning, and a commitment to new staffing. A short, informal report will not suffice. In contrast, you will have to write a type of longer, formal report—an *internal proposal*—for your company's management.

Additional examples of internal proposals include the following:

- Your company needs to improve its mobile communication abilities for employees who work at diverse locations. To accomplish this goal, you write an internal proposal requesting the purchase of WiFi-compatible laptops, handheld computers, and cell phones with Internet access.
- Your company's insurance coverage is skyrocketing. As a member of the human resources staff, you have researched insurance carriers and now will propose insurance options to upper-level management.
- Your company is migrating to a new software platform. Employees will need training to use the software. In a long, formal report, you propose consulting companies who can offer the training, optional schedules, funding sources, and post-training certification.

External Proposals

While *internal proposals* are written to management within your company, *external proposals* are written to sell a new service or product to an audience outside your company. Your biotechnology company, for example, has developed new software for running virtual cell cultures. The software simulates cell runs and displays synchronous strip charts for sterile monitoring. Data from the runs is graphed for comparison purposes. Not only will your company sell the software but also the company provides consulting services to train clients in the software use. Your responsibility is to write an *external proposal* selling the benefits of this new corporate offering to a prospective client.

Requests for Proposals

Many external proposals are written in response to RFPs (*requests for proposals*). Often, companies, city councils, and state or federal agencies need to procure services from other corporations. A city, for example, might need extensive road repairs. A government agency needs Internet security systems for its offices. A hospital asks engineering companies to submit proposals about facility improvements. An insurance company needs to buy a fleet of cars for its adjusters. To receive bids and analyses of services, the city will write an RFP, specifying the scope of its needs. Competing companies will respond to this RFP with an external proposal.

In each of these instances, you ask your readers to make significant commitments regarding employees, schedules, equipment, training, facilities, and finances. Only a long, formal proposal, possibly complete with research, will convey your content sufficiently and successfully.

FAQs

Q: Whenever I read about proposals, I hear terms like RFP, T&C, SOW, boilerplate, solicited, and unsolicited. What do these words mean?

A: Here's a table defining these common proposal terms:

Proposal Terms	Definitions
RFP	**Request for proposals**—means by which external companies and agencies ask for proposals
T&C	**Terms and conditions**—the exact parameters of the request and expected responses
SOW	**Scope of work *or* statement of work**—costs, dates, deliverables, personnel certifications, and/or company history
Boilerplate	Any content (text or graphics) that can be used in many proposals
Solicited Proposal	A proposal written in response to a request
Unsolicited Proposal	A proposal written on your own initiative

CRITERIA FOR PROPOSALS

To write an effective proposal, you will need to provide the following:

- Title page
- Cover letter
- Table of contents
- List of illustrations
- Abstract or executive summary
- Introduction
- Discussion (the body of the proposal)
- Conclusion/recommendation
- Glossary
- Works cited (or references) page
- Appendix

For detailed information about title pages, cover letters or cover memos, tables of contents, lists of illustrations (pie charts, tables, photographs, line drawings, etc.), discussion, glossaries (to define terminology), works cited (to credit source material), and appendices (for additional material or technical specifications), look at Chapter 18 in this textbook. Each of these components typical of long, formal reports is thoroughly covered in the chapter and illustrations are provided. Following is additional information specifically related to a proposal's abstract, introduction, discussion, and conclusion/recommendation.

Additional information:

See Chapter 10, "Traditional Correspondence: Memos and Letters," and Chapter 18, "Long, Formal Reports."

Abstract

Your audience for the proposal will be diverse. Accountants might read your information about costs and pricing, technicians might read your technical descriptions and process analyses, human resources personnel might read your employee biographies, and shipping/delivery might read your text devoted to deadlines. One group of readers will be management—supervisors, managers, and highly placed executives. How do these readers' needs differ from others? Because these readers are busy with management concerns and might have little technical knowledge, they need your help in two ways: they need information quickly, and they need it presented in easy-to-understand terminology. You can achieve both these objectives through an abstract or executive summary.

The abstract is a brief overview of the proposal's key points geared toward a semi-specialist. If the intended audience is composed of upper-level management, this unit briefly contains specific details. To accomplish the required brevity, you should limit your abstract to approximately three to ten sentences. These sentences can be presented as one paragraph or as smaller units of information separated by headings. Each proposal you write will focus on unique ideas. Therefore, the content of your abstracts will differ. Nonetheless, abstracts should focus on the following: (1) the *problem* necessitating your proposal, (2) your suggested *solution*, and (3) the *benefits* derived when your proposed suggestions are implemented. These three points work for external as well as internal proposals.

For example, let's say you are asked to write an internal proposal suggesting a course of action (limiting excessive personnel, increasing your company's workforce, improving your corporation's physical facilities, etc.). First, your abstract should specify the problem requiring your planned action. Next, you should mention the action you are planning to implement. This leads to a brief overview of how your plan would solve the problem, thus benefiting your company. If you were writing an external proposal to sell a client a new product or service, you would still focus on problem, solution, and benefit. The abstract would remind the readers of their company's problem, state that your company's new product or service could alleviate this problem, and then emphasize the benefits derived.

In each case, you not only want to be brief, focusing on the most important issues, but also you should avoid highly technical terminology and concepts. The purpose of the abstract is to provide your readers with an easy-to-understand summary of the entire proposal's focus. Your readers want the bottom line, and they want it quickly. Therefore, either avoid all technical terminology completely or define your terms parenthetically.

The following is an example of an abstract for an internal proposal.

example

An effective abstract highlights the problem, possible solutions, and benefits in the proposal.

ABSTRACT

Due to deregulation and the recent economic recession, we must reduce our workforce by 12 percent.

Our plan for doing so involves

- Freezing new hires
- Promoting early retirement
- Reassigning second-shift supervisors to our Desoto plant
- Temporarily laying off third-shift line technicians

Achieving the above will allow us to maintain production during the current economic difficulties.

Executive Summary

The terms "abstract" and "executive summary" often are used interchangeably. However, the distinction between the two terms is usually dependent upon length. Whereas an effective abstract should be very brief, an executive summary can be one to five pages long. Not every proposal requires an executive summary. In some instances, the more concise abstract is sufficient. If you choose to include a longer and more detailed executive summary, in addition to the abstract, this summary will accomplish the following goals:

- Highlight the proposal's objectives.
- Detail the proposal's significance.
- Outline the proposal's main points.
- Help the readers understand the recommendations.
- Let the reader know what to do next.

Key Parts of an Executive Summary

To accomplish these goals, an executive summary can include the following topics. To limit your summary to no more than five pages, choose the most important elements and omit those that are secondary:

- Purpose and scope of the proposal, stating the problem or need and your ability to improve the situation
- Research or methods used to develop your content
- Conclusions about your analyses of the topic
- Your qualifications showing that you can resolve the issue
- A project management plan and timetable
- The total project budget
- Recommendations based on your findings

Introduction

Your introduction in the proposal should include two primary sections: (1) purpose (overview, background, or goal) and (2) problem.

Purpose

In one to three sentences, tell your readers the purpose of your proposal. This purpose statement informs your readers *why* you are writing or *what* you hope to achieve. This statement repeats your abstract to a certain extent. However, it's not redundant; it's a reiteration. Although numerous people read your report, not all of them read each line or section of it. They skip and skim.

The purpose statement, in addition to the abstract, is another way to ensure that your readers understand your intent. It either reminds them of what they have just read in the abstract or informs them for the first time if they skipped over the abstract. Your purpose statement is synonymous with a paragraph's topic sentence, an essay's thesis, the first sentence in a letter, or the introductory paragraph in a shorter report.

The following is an effective purpose statement.

example

> **Purpose Statement:** The purpose of this report is to propose the immediate installation of the 102473 Numerical Control Optical Scanner. This installation will ensure continued quality checks and allow us to meet agency specifications.

A purpose statement expresses the goal of the proposal.

Problem (Needs Analysis)

Whereas the purpose statement should be limited to one to three sentences for clarity and conciseness, your discussion of the problem must be much more detailed. For example, if you are writing an internal proposal to add a new facility, your company's current work space must be too limited. You have got a problem that must be solved. If you are writing an external proposal to sell a new piece of equipment, your prospective client must need better equipment. Your proposal will solve the client's problem.

Your introduction's focus on the problem, which could average one to two pages, is important for two reasons. First, it highlights the importance of your proposal. It emphasizes for your readers the proposal's priority. In this problem section, you persuade your readers that a problem truly exists and needs immediate attention. Second, by clearly stating the problem, you also reveal your knowledge of the situation. The problem section reveals your expertise. Thus, after reading this section of the introduction, your audience should recognize the severity of the problem and trust you to solve it. One way to help your readers understand the problem is through the use of highlighting techniques, especially headings and subheadings. Figure 19.1 shows an introduction with purpose statement and needs analysis.

Discussion

When writing the text for your proposal, you need to sell your ideas persuasively, develop your ideas thoroughly through research, observe ethical workplace communication standards, organize your content so the audience can follow your thoughts easily, and use graphics.

Communicating Persuasively

A successful proposal will make your audience act. Writing persuasively is especially important in an *unsolicited* proposal since your audience has not asked for your report. A *solicited* proposal, perhaps written in response to an RFP, is written to meet an audience's specific request. Your audience wants you to help them meet a need or solve a problem. In contrast, when you write an *unsolicited* proposal, your audience has not asked for your assistance. Therefore, in this type of proposal, you must convincingly persuade the audience that a need exists and that your proposed recommendations will benefit the reader.

To write persuasively, you should accomplish the following:

- Arouse reader interest—focus on your audience's needs that generated this proposal.
- Refute opposing points of view in the body of your proposal.

FIGURE 19.1 Introduction with Purpose Statement and Needs Analysis

> 1.0 INTRODUCTION
>
> **1.1 Purpose Statement**
>
> This is a proposal for a storm sewer survey for Yakima, Washington. First, the survey will identify storm sewers needing repair and renovation. Then it will recommend public works projects that would control residential basement flooding in Yakima.
>
> **1.2 Needs Analysis**
>
> 1.2.1. Increased Flooding. Residential basement flooding in Yakima has been increasing. Fourteen basements were reported flooded in 2008, whereas 83 residents reported flooded basements in 2009.
>
> 1.2.2. Property Damage. Basement flooding in Yakima results in thousands of dollars in property damage. The following are commonly reported as damaged property:
>
> - Washers
> - Dryers
> - Freezers
> - Furniture
> - Furnaces
>
> Major appliances cannot be repaired after water damage. Flooding also can result in expensive foundation repairs.
>
> 1.2.3. Indirect Costs. Flooding in Yakima is receiving increased publicity. Flood areas, including Yakima, have been identified in newspapers and on local newscasts. Until flooding problems have been corrected, potential residents and businesses may be reluctant to locate in Yakima.
>
> 1.2.4. Special-Interest Groups. Citizens over 55 years old represent 40 percent of the Yakima, Washington, population. In city council meetings, senior citizens with limited incomes expressed their distress over property damage. Residents are unable to obtain federal flood insurance and must bear the financial burden of replacing flood-damaged personal and real property. Senior citizens (and other Yakima residents) look to city officials to resolve this financial dilemma.

Provide specific details to explain the problem. Doing so shows that you understand the reader's need and highlights the proposal's importance.

Additional information:

See Chapter 13, "Persuasive Workplace Communication."

- Give proof to develop your content, through research and proper documentation.
- Urge action—motivate your audience to act upon your proposal by either buying the product or service or adopting your suggestions or solutions.

Researching Content for Proposals

As in any long, formal report, consider developing your content through research. This can include primary and secondary sources such as the following:

- Interviewing customers, clients, vendors, and staff members
- Creating a survey and distributing it electronically or as hard-copy text
- Visiting job sites to determine your audience's needs

- Using the Internet (online access to libraries, search engines, corporate Web sites, and so on)
- Reading journals, books, newspapers, and other hard-copy text

Additional information:

See Chapter 16 "Research and Documentation."

Communicating Ethically

When you write a proposal, your audience will make decisions based on your content. They will decide what amounts of money to budget, how to allocate time, what personnel will be needed to complete a task, and if additional equipment or facilities will be required. Therefore, your proposal must be accurate and honest. You cannot provide information in the proposal that dishonestly affects your decision makers. To write an ethical proposal, you must provide accurate information about credentials, pricing, competitors, needs assessment, and sources of information and research. When using research, for example, you must cite sources accurately to avoid plagiarism, as discussed in Chapter 16. In addition, as discussed in Chapters 1 and 16, when using in-house, company boilerplate material (content and templates), you may not need to provide documentation. However, any material taken from external sources (the Internet, books, journals, etc.) must be documented. Documentation ensures that you are communicating ethically.

Organizing Your Content

Your proposal will be long and complex. To help your audience understand the content, use modes of organization. These can include the following:

- **Comparison/contrast**—Rely on this mode when offering options for vendors, software, equipment, facilities, and more.
- **Cause/effect**—Use this method to show what created a problem or caused the need for your proposed solution.
- **Chronology**—Show the timeline for implementation of your proposal, reporting deadlines to meet, steps to follow, and payment schedules.
- **Analysis**—Subdivide the topic into smaller parts to aid understanding.

The discussion section of your proposal lends itself to many different modes of organization. Consider applying various modes of organization to the key components of your proposal's body. See Table 19.1.

Using Graphics

Graphics, including tables and figures, can help you emphasize and clarify key points. For example, note how the following graphics can be used in your proposal's discussion section:

- **Tables**—Your analysis of costs lends itself to tables.
- **Figures**—The proposal's main text sections could profit from the following figures:
    - **Line charts** (excellent for showing upward and downward movement over a period of time. A line chart could be used to show how a company's profits have decreased, for example.)

Table 19.1 Key Components of the Proposal's Discussion Section

Analyses of the existing situation, your suggested solutions, and the benefits your audience will derive.	Spatial descriptions of mechanisms, tools, facilities, or products	Process analyses explaining how the product or service works	Chronological instructions explaining how to complete a task
Comparative approaches to solving a problem	Comparing and contrasting purchase options	Analyzing managerial chains of command	Chronological schedules for implementation, reporting, maintenance, delivery, payment, or completion
Corporate and employee credentials	Years in business	Satisfied clients	Certifications
Previous accomplishments	Biographical sketches of personnel	Projected milestones (forecasts)	Comparative cost charts

- **Bar charts** (effective for comparisons. Through a bar or grouped bar chart, you could reveal visually how one product, service, or approach is superior to another.)
- **Pie charts** (excellent for showing percentages. A pie chart could help you show either the amount of time spent or amount of money allocated for an activity.)
- **Line drawings** (effective for technical descriptions and process analyses)
- **Photographs** (effective for technical descriptions and process analyses)
- **Flowcharts** (a successful way to help readers understand procedures)
- **Organizational charts** (excellent for giving an overview of managerial chains of command)

Additional information:

See Chapter 8, "Visual Aids in Workplace Communication."

Conclusion/Recommendations

Sum up your proposal, providing your readers closure. The conclusion can restate the problem, your solutions, and the benefits to be derived. Your recommendation will suggest the next course of action. Specify when this action will or should occur and why that date is important. Figure 19.2 is a conclusion/recommendation from an internal proposal.

FIGURE 19.2 Conclusion/Recommendation for Internal Proposal

Summarize the key elements of the proposal.

Recommend follow-up action and show the benefits derived.

3.0 SOLUTIONS FOR PROBLEM

Our line capability between San Marcos and LaGrange is insufficient. Presently, we are 23 percent under our desired goal. Using the vacated fiber cables will not solve this problem because the current configuration does not meet our standards. Upgrading the current configuration will improve our capacity by only 9 percent and still present us the risk of service outages.

4.0 RECOMMENDED ACTIONS

We suggest laying new fiber cables for the following reasons. They will

- Provide 63 percent more capacity than the current system
- Reduce the risk of service outages
- Allow for forecasted demands when current capacity is exceeded
- Meet standard configurations

If these new cables are laid by September 1, 2009, we will predate state tariff plans to be implemented by the new fiscal year.

PROPOSAL CHECKLIST

Have you included the following in your proposal?

____ 1. Title page (listing title, audience, author or authors, and date)

____ 2. Cover letter (stating why you're writing and what you're writing about; what exactly you're providing the readers; what's next—follow-up action)

____ 3. Table of contents (listing all major headings, subheadings, and page numbers)

____ 4. List of illustrations (listing all figures and tables, including their numbers and titles, and page numbers)

____ 5. Abstract (stating in semi-specialist terms the problem, solution, and benefits)

____ 6. Introduction (providing a statement of purpose and a lengthy analysis of the problem)

____ 7. Discussion (solving the readers' problem by discussing topics such as procedures, specifications, timetables, materials/equipment, personnel, credentials, facilities, options, and costs)

____ 8. Conclusion (restating the benefits and recommendation for action)

____ 9. Glossary (defining terminology)

____ 10. Appendix (optional additional information)

SAMPLE INTERNAL PROPOSAL

Figure 19.3, an internal proposal emphasizing a company's need for technology upgrades and new computers, was created in PowerPoint. This software package can be used for both oral and hard-copy proposal presentations. Many companies rely on PowerPoint for their proposals. Commerce Bank's Trust Department, for example, prepares its proposals using PowerPoint. They say that PowerPoint makes proposals easier to read as well as to write. Whereas traditional 8½″ × 11″ hard-copy text can be wordy and demanding to read, PowerPoint printouts, using six to ten lines of text, are more reader friendly. In addition, once the PowerPoint slides have been made and bound for a hard-copy handout, the same slides can be used for the bank's oral presentation.

For a sample external proposal, go to our WebCompanion at http://www.prenhall.com/gerson.

FIGURE 19.3 Proposal Written in PowerPoint

Bio
Staffing
Your One-Stop Shop for Biomedical Needs
Submitted to
Leann Towner
Chief Financial Officer
By
Jonathan Bacon
Manager, Information Technology Department
June 12, 2009

Many companies create proposal boilerplates using PowerPoint. Companies have found that PowerPoint is a valuable tool for writing and submitting proposals for the following reasons:

- *Easy to create*
- *Easy to transport as a disk vs. multiple-paged documents*
- *Easy to send either electronically or as hard-copy mail*
- *Easy for an audience to read and access*
- *Allows for color and animation*
- *Allows for easy-to-create handouts*

Bio
Staffing
Date: June 12, 2009
To: Leann Towner, Chief Financial Officer
From: Jonathan Bacon, Information Technology Manager
Subject: Proposal for New Corporate Technology Support
Leann, in response to your request, the IT Department is happy to propose a new technology support system.
Among detailed analyses of our current system, this proposal presents the following:
✓ **BioStaffing's** current technology challenges
✓ Hardware purchasing suggestions
✓ Technology vendor recommendations
We are confident that our suggestions will maintain **BioStaffing's** competitive edge and increase employee satisfaction. If I can answer any questions, either call me (ext. 3625) or e-mail me at **jbacon@biostaff.com.**

(Continued)

FIGURE 19.3 Proposal Written in PowerPoint Continued

Table of Contents

List of Illustrations

Both Figure/Table numbers and titles are given for easy reference and clarity.

1.0 Abstract

An abstract often presents the problem necessitating the proposal, provides the proposed solutions, and shows the potential benefits.

1.1 Problem

BioStaffing's current technology policies are not responsive to our evolving needs, our hardware is outdated, and our current vendor is changing ownership.

1.2 Solution

To solve these problems, the Information Technology (IT) Department suggests the following:

- Upgrading our technology needs biannually rather than every five years
- Purchasing new computers, printers, scanners, and digital cameras
- Hiring a new vendor to supply and repair our hardware

1.3 Benefits

A biannual replacement schedule will allow us to stay current with hardware advancements. Purchasing new technologies will help our staff more effectively meet client needs. Hiring a new vendor is key to these goals. Our current vendor agreement ends June 30, 2009. A new vendor will provide better turnaround, pricing, maintenance, and merchandise.

FIGURE 19.3 **Continued**

2.0 Introduction

2.1 Purpose

The purpose of this proposal is to improve **BioStaffing's** current technology. By revising our policies, purchasing new hardware, and hiring a new technology vendor, we can increase both customer and worker satisfaction.

Pronouns such as "our" and "we" are appropriate for an internal proposal written to coworkers.

2.2 Problem

BioStaffing faces three key challenges:

2.2.1 *Technology Policies*
2.2.2 *Outdated Hardware*
2.2.3 *Vendor Changes*

2.0 Introduction cont.
Problems

2.2.1 *Technology Policies*
Since 1987, we have replaced hardware and software on a five-year, rotating basis. This was an effective policy initially since new-computer costs were expensive.

However, the current policy is no longer responsive to our needs, for the following reasons:

- <u>Prices have gone down</u>. Today's hardware costs are more affordable. In addition, vendors will offer **BioStaffing** buyer-incentives. We can buy hardware on an as-needed basis by taking advantage of special sales pricing.

  In contrast, our current five-year hardware replacement policy is not responsive. It disregards today's changes in technology pricing, and it disallows us from taking advantage of dealer incentives.

A variety of highlighting techniques, such as boldface headings, underlined subheadings, and bullets, make the text easier to read.

2.0 Introduction cont.
Problems

- <u>Repair costs have gone up</u>. By replacing technology only every five years, **BioStaffing** has resorted to repairing and retrofitting outdated equipment. Costs for these repairs have increased more than five-fold over the last ten years, as noted in the following line graph (Figure 1).

Figure 1: Increase in Costs Due to Repairs and Retrofitting per Workstation

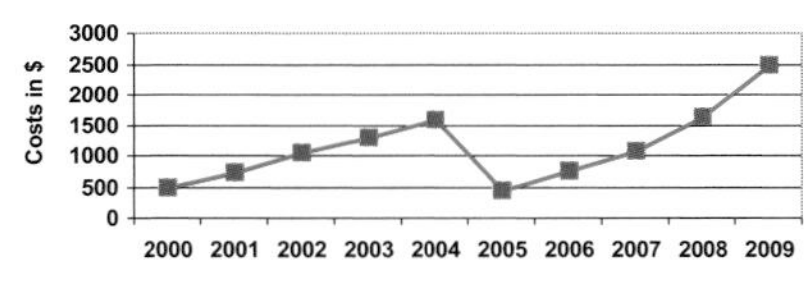

The line graph adds clarity and provides visual appeal.

(Continued)

FIGURE 19.3 **Proposal Written in PowerPoint Continued**

2.0 Introduction cont.
Problems

2.2.2 Outdated Hardware

We need to purchase new technology items because our current hardware and software are outdated.

- *Our computers have insufficient memory*. Most of our computers have only 2.0 GHz processors with approximately 256 MB of RAM versus the current minimal standard 3.0 GHZ and 512 MB. This negatively impacts speed of document retrieval.*

The proposal uses specific details to persuade the audience.

2.0 Introduction cont.
Problems

- *Our ink jet printers are slow and produce poor quality documents*. Our current black and white printers produce only 7 ppm with a dpi resolution of 1440 x 720.* In contrast, new black and white laser printers will produce up to 20 ppm at a dpi resolution of 5760 x 1440.

NOTE: At 7 ppm (times 60 minutes per hour), **BioStaffing** personnel can print 420 pages per hour. Current printing requires **9.5 hours**.

In contrast, at 20 **ppm** (times 60 minutes per hour), improved printers can produce 1200 pages per hour, requiring only **3.3 hours.**

Laser printers can save our company over 6 hours of lost time each day.

2.0 Introduction cont.
Problems

To clarify the importance of **ppm**, look at Figure 2.

Figure 2: Time Spent Printing Based on PPM

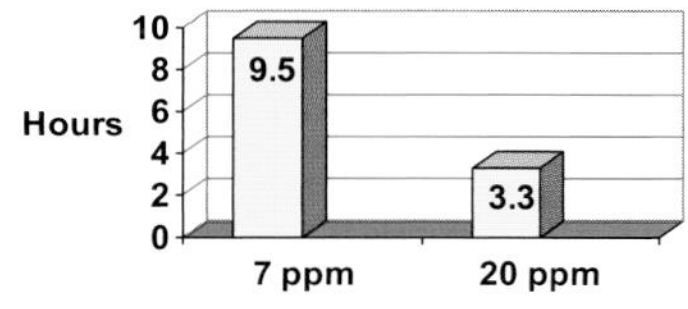

FIGURE 19.3 **Continued**

2.0 Introduction cont. Problems

2.2.3 Vendor Changes

Our current technology vendor is changing ownership. After having been in business for 20 years, *Business Sourcing* is being sold to an offshore company, *International Technologies*. Our technology agreement will end, as of June 30, 2009. We must renegotiate a technology contract.

This offers is a window of opportunity. Through a new vendor, we can improve quality of service as well as pricing.

3.0 Discussion

3.1 Revised Technology Replacement Policy

We must revise **BioStaffing's** technology replacement policy. It is not responsive to our growing technology needs. We suggest this approach.

3.1.1 Biannual Rotating Replacements

We propose that BioStaffing divide its 12 departments into two categories. The "Red" team would receive technology upgrades one year (even numbered years), while the "Blue" team would receive technology upgrades the next year (odd numbered years).

IT has divided the departments into teams alphabetically for fairness, as shown in Table 1:

3.0 Discussion cont.

Table 1: Biannual Rotating Technology Replacements

Red Team (*even* numbered years)	Blue Team (*odd* numbered years)
Accounting	Manufacturing
Administrative Services	Personnel
Corporate Communication	Sales
Information Technology	Shipping and Receiving

The table, color, and white space show how formatting aids readability.

(Continued)

FIGURE 19.3 **Proposal Written in PowerPoint Continued**

3.0 Discussion cont.

3.2. Hardware Purchases

After researching department requirements, IT determined that **BioStaffing** needs to upgrade business application hardware as follows:

3.2.1 Hardware Allocation Analysis

Based on survey results from each department, IT suggests that the hardware listed in Table 2 be allocated as follows:

- **Accounting—**2 desktops, 1 laptop, 5 handhelds, and 1 laser printer
- **Administrative Services—**3 desktops, 1 laptop, 1 handheld, and 1 laser printer
- **Corporate Communication**—1 laptop, 7 handhelds, 1 scanner, 1 laser printer, and 2 digital cameras
- **Information Technology**—1 laptop, 7 handhelds, 1 scanner, 1 laser printer, and 2 digital cameras
- **Manufacturing**—2 desktops, 1 laptop, 2 handheld, and 1 laser printer
- **Personnel**—2 desktops, 1 laptop, 4 handhelds
- **Sales**—2 desktops, 1 laptop, 10 handhelds, and 1 digital camera
- **Shipping and Receiving—**1 desktop, 1 laptop, 2 handhelds

3.0 Discussion cont.

3.2.2 Hardware Costs

Table 2: Business Application Hardware Costs

Hardware	Individual Costs	Total Costs per Item
12 Desktop Computers	$1,000 each	$12,000
8 Laptop Computers	$700 each	$5,600
38 Handheld computers	$150 each	$5,700
5 Laser printers	$400 each	$2,000
3 Scanners	$170 each	$510
5 Digital cameras	$200 each	$1,000
Total Costs for all items		$26,810

3.0 Discussion cont.

3.3 Technology Vendor Options

Because **BioStaffing's** vendor agreement with *Business Sourcing* ends June 30, 2009. Information Technology has researched new vendor options. Our primary criteria included

- *pricing*
- *quality*
- *delivery time*
- *maintenance*
- *warranties*
- *customer service*

Based on these criteria, following is an overview of our findings:

FIGURE 19.3 **Continued**

3.0 Discussion cont.

Table 3: Technology Vendor Options						
Companies	**Pricing**	**Quality**	**Delivery Time**	**Maintenance**	**Warranties**	**Customer Service**
BizTech Warehouse	20% bulk discount Hardware: $21,448	Re-manufactured name-brand hardware	Same-day delivery at no extra cost.	24-hours maintenance Options at additional hourly cost—negotiable 8:00 a.m.-5:00 p.m. Maintenance for products under warranty	90 days Renewable at negotiable price	24/7 hotline, Fleet of maintenance trucks.
Technologies Today	No discount Hardware $26,810	New name-brand hardware	Same-day delivery for home office, Overnight FedEx for other sites.	Manufacturer's maintenance (1-800 number and/or nearest dealership)	1-year Mnf's warranty	8:00 a.m–5:00 p.m. M-F office help.
SOS (Super-Saver Office Supplies)	10% bulk discount Hardware $24,129	In-house, store-brand hardware	Overnight FedEx.	1-800 number and/or nearest dealership	3-year extended warranties, parts and labor	8:00 a.m.–5:00 p.m. M-F Office help. Answering service on weekends.

3.0 Discussion cont.

3.3.1 Assessment of Vendor Options

- **Pricing**—BizTech Warehouse provides the best pricing.
- **Quality**—Technologies Today provides new, name-brand hardware. This is appealing. However, remanufactured hardware can be as good as new.
- **Delivery Time**—With 25 locations nationwide, including Topeka, our home office's city. BizTech provides same-day delivery.
- **Maintenance**—BizTech provides the most prompt maintenance service. They charge a maintenance fee, but we will receive a substantial discount as well as product rebates if we choose their company.
- **Warranties**—SOS's three-year warranty is outstanding, as it Technologies Today's 1-year guarantee. However, with plans to upgrade our hardware biannually, a three-year warranty loses its value.
- **Customer Service**—BizTech's 24/7 accessibility is perfect for our workplace.

3.0 Discussion cont.

Information Technology evaluated the vendor options as follows:

5 = Excellent 4 = Good 3 = Average 2 = Poor 1 = Unacceptable

Based on this scale, Figure 3 shows our evaluation findings:

Figure 3: Vendor Evaluation Matrix

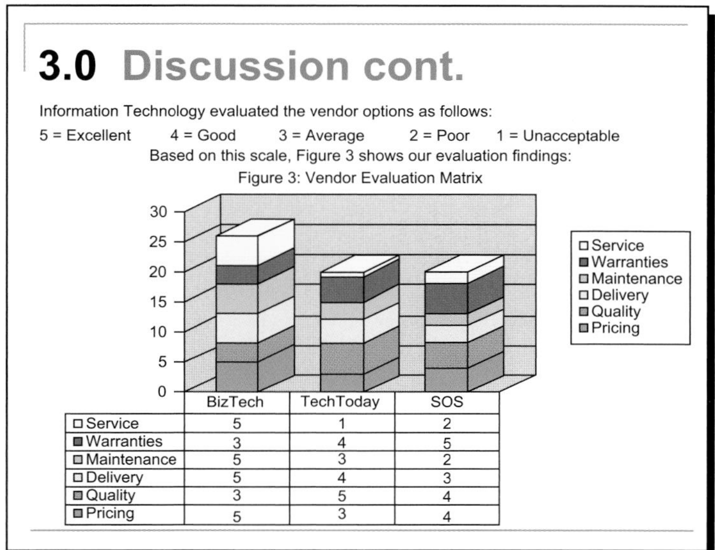

	BizTech	TechToday	SOS
Service	5	1	2
Warranties	3	4	5
Maintenance	5	3	2
Delivery	5	4	3
Quality	3	5	4
Pricing	5	3	4

(Continued)

FIGURE 19.3 Proposal Written in PowerPoint Continued

4.0 Conclusion

4.1 Conclusion

BioStaffing's technology needs are critical.

- *Repair costs* due to old hardware have risen five-fold per workstation, from as low as $500 to as high as $2,500.
- *Outdated printers*, producing only 7 pages per minute, require up to 9.5 hours a day to print BioStaffing's 4,000 pages of documentation.
- *Expired technology vendor contracts* need to be renegotiated.

Key to the above challenges is our current 5-year technology replacement policy. Technology advances demand a more responsive policy.

4.0 Conclusion cont.

4.2 Recommendation

The Information Technology Department, based on research and surveys, proposes the following:

- *A biannual technology replacement policy*—this would allow BioStaffing to meet employee technology needs more responsively as well as benefit from vendor pricing incentives.
- *New communication hardware*—this would include desktop, laptop, and handheld computers; laser printers; scanners; and digital cameras.
- *A new technology vendor*—based on our research, we suggest that BizTech Warehouse will best meet our technology needs.

4.0 Conclusion cont.

4.3 Benefits

These changes must occur before June 30, 2009, when our current vendor contract ends. By acting now, BioStaffing will benefit in several ways.

Positive words and phrases such as "benefit," "save," "competitive edge," "provide employee satisfaction," and "increase productivity" help sway the audience toward the writer's stance.

- We can save up to 20% on hardware and maintenance costs.
- We will maintain our competitive edge in the marketplace.
- Most importantly, our employee satisfaction and productivity will increase as they work with the latest technology and software upgrades.

FIGURE 19.3 **Continued**

5.0 Glossary

Acronym/Abbreviation	Definition
dpi	Dots per inch
GHz	Gigahertz
GB	Gigabyte
HD	Hard drive
MB	Megabyte
MHz	Megahertz
ppm	Pages per minute

Though all readers might not need terms defined, a glossary provides clarity to a less specialized audience.

Mary Met Her Challenge

To meet her communication challenge, Mary used the P^3 process.

Planning

To plan her proposal, Mary considered the following:

- Goals—rewrite a manual to improve its ease of use
- Audience—customer
- Channels—face-to-face interviews, questionnaires, and a proposal
- Data—drawn from interviews and questionnaires

Conducting the front-end analysis is the planning portion of Mary's P^3 approach to writing her proposal. As Mary says, "I can't even begin to write without some information. Getting the appropriate information from the client is an essential part of my communication process. I find that I usually need to meet with the client face-to-face and then use primary research in the form of discussion or questionnaires as the basis for the proposal."

In order to get the information she needed, Mary says, "I did a complete review of the existing manual and spent time talking with and observing several employees and their manager. I created the questionnaire, limited to no more than ten questions. I keep my questionnaires short so that I don't overwhelm my client with additional work." Figure 19.4 shows Mary's questionnaire.

Packaging

After reviewing the notes from her front-end analysis, Mary drafted the "Findings" section of her proposal, as shown in Figure 19.5. Mary says, "I submitted this rough draft to my client for input and commentary. I wanted to make sure we were in agreement about my approach to the project. Whenever I work with clients, I always find their suggestions invaluable to the success of our effort. This interaction allows us to arrive at a mutually agreeable 'perfected' proposal that meets the needs of the client. A happy client helps to ensure the continued success of my small business."

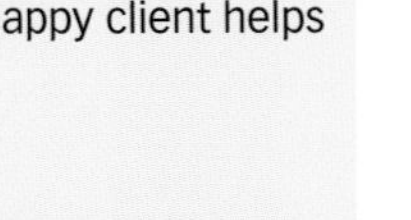

Perfecting

Before incorporating the "Findings" section into her proposal, Mary sent the draft of the section to the individuals she interviewed. She wanted to be sure that she had included everything they felt was important.

Mary says, "With their approval of the draft, I completed a proposal for the client that included, among the legal paragraphs that I include in all proposals, the following sections:

- A summary of the client's request
- The goal of the project
- The findings

(Continued)

FIGURE 19.4 Questionnaire for Planning a Proposal

ABOUT THE MANUAL

- Are instructions for completing tasks set off clearly from the other text and written in a numbered, step-by-step format?
- Are graphics and screen shots used when appropriate to enhance the instructions?
- Does the material have a table of contents and/or index to help learners find the information they need?
- Are terms and acronyms clearly defined?

FROM INTERVIEWS WITH EMPLOYEES

- What tasks do you perform without making mistakes? How often do you perform those tasks?
- What tasks do you struggle with? How often do you perform those tasks?
- Do you refer to the training manual when you are having trouble?
- If you do, can you find the information you need easily? Or at all?

FROM INTERVIEW WITH MANAGER

- What mistakes are causing the most problems?
- How often are these mistakes made?

FIGURE 19.5 Partial Draft of "Findings" Section

FINDINGS:

- Employees perform 80 percent of the tasks in Application X with 100 percent accuracy. Each of these tasks is performed at least once a week if not daily.
- Employees consistently identified five tasks that they have trouble completing. Three of these tasks are complicated, multi-step tasks with costly consequences when errors are made. Two of the tasks are performed only once a quarter.
- The instructions in the existing training manual are well-written. Graphics and screen shots are used often to enhance the clarity of the instructions.
- Employees found the manual to be very thorough and helpful in the classroom training sessions.
- The manual does not have an index, and the table of contents is skeletal.
- Employees do not use the manual as a reference when they are having trouble with a task because the manual is cumbersome to handle and instructions are hard to find.

- The recommended solution
- Description of the recommended solution (including a prototype)
- A timeline for completing the project
- An estimate of the time and cost for each step of the project

Before any document is ready to be sent out the door, I then have a colleague proofread every word for clarity, spelling, and punctuation. I value this part of the *P*[3] process, for no matter how experienced I am as a writer, I can always benefit from another person's opinion." Figure 19.6 is Mary's perfected proposal excerpt.

FIGURE 19.6 Perfected Proposal Excerpt—Four Sections of the Total Proposal

CLIENT REQUEST

Company ABC (Client) has asked Effective Business Communications, Inc. (EBC) to submit a proposal for the redesign of an existing training manual. The manual is used to train new finance department employees on Application X, the company's expense reporting application.

Employees are currently making costly mistakes, and the client believes that the manual is failing to communicate what employees need to know or do. The client's request for proposal (RFP) does not indicate what information the client relied on when deciding to have the manual revised.

PROJECT GOAL

The goal of the project is to eliminate the mistakes employees make when using Application X.

FINDINGS

In response to the RFP, EBC asked for and received permission to conduct a front-end analysis to confirm the cause of the mistakes being made in the application.

The findings include

- Employees perform 80 percent of the tasks in Application X with 100 percent accuracy. Each of these tasks is performed at least once a week if not daily.
- Employees consistently identified five tasks that they have trouble completing. Three of these tasks are complicated, multi-step tasks with costly consequences when errors are made. Two of the tasks are performed only once a quarter.
- The instructions in the existing training manual are well-written. Graphics and screen shots are used often to enhance the clarity of the instructions.
- Employees found the manual to be very thorough and helpful in the classroom training sessions.
- The manual does not have an index, and the table of contents is skeletal.
- Employees do not use the manual as a reference when they are having trouble with a task because the manual is cumbersome to handle and instructions are hard to find.

RECOMMENDATION

Although the manual would benefit from the addition of an index and a more complete table of contents, a complete revision of the manual is unnecessary. Instead, EBC recommends the creation of job aids for those tasks that are complicated, multi-step processes or for those tasks that are performed infrequently.

Mary says, "I always try to break text into small, readable amounts of content so that the lengthy proposal is easy to read. I know how busy my clients are, and I want to make the documents I submit to them pleasant to read—the less time they have to spend reading the proposal, the more likely it is that we'll be doing business together."

CHAPTER HIGHLIGHTS

1. You might have multiple readers for a proposal, including internal and external audiences. Consider your audiences' needs. To communicate with different levels of readers, include abstracts, glossaries, and parenthetical definitions.
2. A proposal could include the following:
   - Title page
   - Cover letter
   - List of illustrations
   - Abstract
   - Introduction
   - Discussion
   - Conclusion
   - Recommendation
   - Glossary
   - Works cited (or references)
   - Appendix
3. Subheadings and visual aids will make your proposal more accessible.
4. Use primary and secondary research to develop your content.
5. Write persuasively to convince your audience to act. To accomplish this goal, arouse reader interest, refute opposing points of view, gather details to support your argument, and urge action.
6. Write ethically by documenting sources and making sure your content (prices, timelines, credentials, etc.) are accurate.

WWW *Find additional proposal exercises, samples, and interactive activities at http://www.prenhall.com/gerson.*

MEETING WORKPLACE COMMUNICATION CHALLENGES

CASE STUDIES

1. BEST (Bellaire Educational Supplies/Technologies) manufactures and markets school supplies for all levels of education (K–12, private/parochial, and college). BEST has been in business for 15 years, has a Triple A rating from the Educational Marketing Group, and is owned and operated by Sam and Scott Wilcox, two former educators. Some of BEST's clients include the New York, Los Angeles, Chicago, and Miami independent school districts. Supporting their logo—*Your BEST buy in the market!*—BEST promises to meet or beat any competitor's pricing.

   Turtle Creek Independent School District (TCISD), 3476 Schoolway, Houston, TX 77902, has written an RFP, asking for proposals for educational supplies for their faculty, paraprofessionals, and staff. TCISD needs the following before July of their upcoming school year:

   - 1,200 desks and chairs
   - 180 modular computer workstations
   - 54 blackboards and 26 whiteboards
   - 16 cases of chalk
   - 100 boxes of dry erase markers
   - 30 overhead projectors
   - 35 pull-down screens
   - 500 gradebooks

Assignment

Write an external proposal for BEST. To do so, research pricing for the previously listed items, include warranty information, and provide delivery methods and dates.

2. Maxwell Publishing Co. has 1,500 employees who have offices in two, four-story companion buildings. The company currently has two cafeterias, although one cafeteria could support both buildings. It would be a cost savings to use the second cafeteria space for other purposes. The personnel committee has a wonderful idea for this space. Based on employee surveys, the committee plans to write an unsolicited internal proposal asking management to renovate this unused space, turning it into a daycare center for employee usage.

   Maxwell Publishing Co. prides itself on being employee friendly, family friendly, and one of the city's best places to work. It already has an on-site fitness center and corporate recreational program. However, due to childcare needs, many of Maxwell's employees cannot take advantage of these benefits. Over 45 percent of the employees have children under the age of 4, 63 percent of the employees have children still in elementary school, and 12 percent of the employees are single mothers.

Assignment

Write the internal proposal to Maxwell's CEO, Wes Maxwell. In this proposal, include the following:

- Licensing requirements for a daycare facility
- Required credentials for a licensed daycare worker
- Insurance regulations and costs
- The number of children who can be cared for in a space measuring 2,400 square feet
- Preferred hours of operation

Research to find this information.

3. RavenWood Homes, a neighborhood housing development, wants to upgrade its property. The homes association has decided to enhance its entrance with a fountain, improve the area's landscaping, and add more street lighting for security as well as esthetics. RavenWood wants the entrance fountain to be made out of black granite, to stand 12 feet high, and to have the water cascade over the housing development's name, which will be etched into the fountain's base. The name must be at least three feet high for visibility. Lighting should illuminate the fountain for night viewing.

   The homes association believes it needs at least 125 ten-foot tall red maple trees to line both sides of RavenWood's main boulevard, Woodline Dr. This street also must have 50 standard street lights. RavenWood also will need a maintenance contract for grass cutting and tree pruning, as well as for the fountain's upkeep.

Assignment

As a landscaping architectural/engineering company, write the solicited external proposal for RavenWood. Include pricing, a timetable, your credentials, and a list of satisfied customers.

4. Your company, Foods & More, provides individual and corporate party-planning services. This includes food, beverages, plates, utensils, decorations (flowers, balloons, banners, and party favors), and entertainment (music and games). The company has been in business for ten years, and is owned by Rochelle Kraft, Sally Gorton, and Lisa Clark, all of whom have design degrees from the Powhattan State College of Design.

   Foods & More has provided services for as few as ten guests to as many as 1,000. The company has worked with Microsoft, Boeing, Sears, The Gap, H&R Block, AMC Theaters, and more. Pricing starts at $15 per person, but individual services and menus alter this base fee. For example, the company's "Cowboys and

Cowgirls" BBQ (with hamburgers, hotdogs, potato salad, and beans) costs $15 per person. In contrast, their more gourmet meal, "A Taste of Italy," includes chicken Alfredo, baby vegetables, Caesar salad, marble cheesecake, and wine. It lists for $30 per person. A DJ costs $25 per hour, while a country-western band costs $100 per hour. Other fees are negotiable.

Assignment

Write an unsolicited external proposal for Foods & More. Create menu items and other services. As an audience, write this proposal to your current employer, college, fraternity or sorority, or your business communication class.

ETHICS IN THE WORKPLACE CASE STUDY

Candice Miller, manager of human resources, is preparing an internal proposal to promote new training classes for her company's employees. To discover where employees can improve their job skills, Candice conducts primary research. She interviews staff and tells them how their comments will be used. After completing her research but before she writes the proposal, Candice gets the employees' approval and has them sign a release statement: "I give the company permission to use my statements in this proposal."

Months later, Candice is invited to attend the national convention of human resources managers and is asked to make a PowerPoint presentation about her company's new training options. In this presentation, she uses the interviews from her earlier proposal. She did not get permission from her fellow employees since they had already signed release forms.

Question: Is it ethical for Candice to use her colleagues' interview comments in her speech at the convention? Justify your answer based on information provided in Chapter 1.

INDIVIDUAL AND TEAM PROJECTS

1. Write an external proposal. To do so, create a product or a service and sell it through a long report. Your product can be an improved radon detection unit, a new fiber optic cable, safety glasses for construction work, bar codes for pricing or inventory control, an improved four-wheel steering system, computer graphics for an advertising agency, and so on. Your service may involve dog grooming, automobile servicing, computer maintenance, home construction (refinishing basements, building decks, room additions, and so on), freelance technical writing, at-home occupational therapy, or telemarketing. The topic is your choice. Draw from your job experience, college coursework, or hobbies. To write this proposal, follow the P^3 process provided in this chapter—plan, package your draft, and then revise to perfect.
2. Write an internal proposal. You can select a topic from either work or school. For example, your company or department is considering a new venture. Research the prospect by reading relevant information. Interview involved participants or survey a large group of people. Once you have gathered your data, document your findings and propose to management the next course of action. If you choose a topic from school, you could propose a daycare center, on-campus bus service, improved computer facilities, tutoring services, coed dormitories, pass/fail options, and so on. Write an internal proposal to improve your company's Web site; expand or improve the security of your company's parking lot; create or improve your company's physical site security in light of post 9-11 issues; improve policies for overtime work; improve policies for hiring diversity; improve your company's policies for promotion. Research your topic by reading relevant information or by interviewing or surveying students, faculty, staff, and administration. Once you have gathered your data, document your findings and recommend a course of action. In each instance, be sure to plan, package your draft, and then revise to perfect.

DEGREE-SPECIFIC ASSIGNMENTS

1. **Marketing and Public Relations.** You are the marketing manager at the Overland Park Convention Center. You have to grow your business by $25,000 this quarter. To expand your customer base, you will write external proposals to the local bar association, chamber of commerce, junior league, banking association, insurance organization, and so on.

Assignment

Through brainstorming, list what needs or problems these organizations might have, how your convention center can meet these needs, how your proposal can be persuasive, and what benefits your proposal might offer. Outline your findings and write a brief abstract, which would preface the proposal.

2. **Business Information Systems.** You work at Design Today, an engineering and architectural firm. Employees at Design Today depend on e-mail for their daily correspondence and on the Internet for their research. However, the employees are being bombarded by constant spam and pop-up advertisements, disrupting their work. As the business information systems director, you need to solve this problem by writing an internal proposal.

Assignment

Through brainstorming, list what problems spam and pop-up advertisements create, how your department could resolve these problems, how your proposal can be persuasive, and what benefits your proposal might offer. Outline your findings and write a brief abstract, which would preface the proposal.

3. **Accounting.** Your chain of grocery stores, FoodTime, is experiencing financial problems. Sales are down and expenses are up. Currently, the chain of ten grocery stores has 500 employees. FoodTime operates 24 hours a day, and has an in-store gourmet French bakery, sushi bar, and home food delivery service. Management has asked you and your staff of accountants to write an internal proposal for cost-cutting measures.

Assignment

Through brainstorming, list FoodTime's problems, how you propose to solve these problems, how your proposal can be persuasive, and what benefits your proposal might offer. Outline your findings and write a brief abstract, which would preface the proposal.

4. **Management.** Technology is ruling business and industry. However, your company has fallen behind the technology curve. Your employees are not as computer savvy as they need to be. For example, none of your colleagues knows how to use the company's new software platform, consisting of e-mail functions, spreadsheet capabilities, online reporting, or archiving of e-mail messages. They don't even know how to access the company's online database and data library. You have been asked to write a solicited, internal proposal up to management. The proposal will suggest ways to solve this lack of technology know-how on the part of company employees.

Assignment

Through brainstorming, list what problems employee lack of technology skills can create, how your company could resolve these problems, how your proposal can be persuasive, and what benefits your proposal might offer. Outline your findings and write a brief abstract, which would preface the proposal.

5. **Finance.** A local builder is developing a subdivision with 100 home sites. The builder already has a loan from a mortgage company, but mortgage rates have

been increasing for months. Your mortgage company offers refinancing options. You know you can offer the builder better rates. To convince the builder to use your services, you must write an unsolicited external proposal.

Assignment

Through brainstorming, list what problems the builder's current mortgage loan presents, how your company could solve these problems, how your proposal can be persuasive, and what benefits your proposal might offer. Outline your findings and write a brief abstract, which would preface the proposal.

PROBLEM-SOLVING THINK PIECES

1. Stinson, Heinlein, and Brown Accounting, LLC, employs over 2,000 workers, including accountants, computer information specialists, a legal staff, paralegals, and office managers. The company requires a great deal of written and oral communication with customers, vendors, government agencies, and coworkers. For example, a sample of their workplace communication includes the following:
   - Written reports to judges and lawyers.
   - Letters and reports to customers.
   - E-mail and memos to coworkers.
   - Oral communication in face-to-face meetings, videoconferences, and sales presentations.

   Unfortunately, not all employees communicate effectively. The writing companywide is uneven. Discrepancies in style, grammar, content, and format hurt the company's professionalism. The same problems occur with oral communication. George Hunt, a mid-level manager, plans to write an internal, unsolicited proposal to the company's principal owners, highlighting the problems and suggesting solutions.

   What must Mr. Hunt include in his proposal—beyond the obvious proposal components (a title page, table of contents, abstract, introduction, and so forth)—to persuade the owners to accept his suggestions? Suggest ways in which the problem can be solved.

2. Toby Hebert is sales manager at Crab Bayou Industries (CBI), located in Crab Bayou, Louisiana. In his position, Toby manages a sales staff of 12 employees who travel throughout Louisiana, Texas, Arkansas, and Mississippi. Currently, the sales staff members use their own cars to make sales calls, and CBI pays them 40 cents per mile for travel expenses. Each staff member currently travels approximately 2,000 miles a month, with cars getting 20 miles per gallon.

   Gasoline prices, at the moment, are over $3.75 a gallon. With gasoline and car maintenance costs higher than ever, the current rate of 40 cents per mile means that CBI's sales employees are losing money. Something must be done to solve this problem. Toby has met with his staff, and they have decided to write a proposal to Andre Boussaint, CBI's CEO.

   What must Toby and his staff include in the proposal—beyond the obvious proposal components (a title page, table of contents, abstract, introduction, and so forth)—to persuade the CEO to accept the suggestions? Suggest ways in which the problem can be solved.

WEB WORKSHOP

By typing "RFP," "proposal," "online proposal," or "online RFP" in an Internet search engine, you can find tips for writing proposals and requests for proposals (RFPs), software products offered to automatically generate e-proposals and winning RFPs, articles on how to write proposals, samples of RFPs and proposals, and online RFP and proposal forms.

To perform a more limited search, type in phrases like "automotive service RFP," "computer maintenance RFP," "desktop publishing RFP," "Web design RFP," and many more topics. You will find examples of both proposals and RFPs from businesses, school systems, city governments, and various industries.

To enhance your understanding of business and industry's focus on proposal writing, search the Web for information on RFPs and proposals. Using the criteria in this chapter and your knowledge of effective workplace communication techniques, analyze your findings.

- How do the online proposals or RFPs compare to those discussed in this textbook, in terms of content, tone, layout, and so on?
- What information provided in the textbook is missing in the online discussions?
- What are some of the industries that are requesting proposals, and what types of products or services are they interested in?
  a. Report your findings, either in an oral presentation or in writing (e-mail message, memo, letter, or report).
  b. Imagine that you are requesting a proposal for a product or service. Create your own online form to meet this need.
  c. Rewrite any of the proposals that you think can be improved, using the criteria in this chapter as your guide.
  d. Respond to an online RFP by writing a proposal. To complete this assignment, go online to research any information you need for your content.

QUIZ QUESTIONS

1. Explain the purpose of an internal proposal.
2. Explain the purpose of an external proposal.
3. What content do you include in the introduction of a proposal?
4. What is an abstract in a proposal?
5. What do you include in a purpose statement?
6. What are the typical components of the discussion section?
7. What do you include in the recommendation?
8. What is an RFP?
9. What can you include in an appendix?
10. What type of visuals can you include in a proposal?

APPENDIX

Grammar, Punctuation, Mechanics, and Spelling

Correct organization and development of your memos, letters, or reports are important for the success of your workplace communication. However, no one will be impressed with the quality of your work, or with you, if your writing is riddled with errors in sentence construction or punctuation. Your written correspondence is often your first contact with business associates. Many people mistakenly believe that only English teachers notice grammatical errors and wield red pens, but businesspeople also take note of such errors and may see the writer as less competent because of them.

We were working recently with a young executive who is employed by a branch of the federal government. This executive told us that whenever his supervisor found a spelling error in a subordinate's report, this report was paraded around the office. Everyone was shown the mistake and had a good laugh over it, and the report was then returned to the writer for correction. Our acquaintance assured us that all of this was in good-natured fun. However, he also said that employees quickly learned to edit and proofread their written communication to avoid such public displays of their errors. He went on to say that his dictionary was well thumbed and always on his desk.

Your writing at work may not be exposed to such scrutiny by coworkers. Instead, your writing may go directly to another firm, and those readers will see your mistakes. To avoid this problem, you must evaluate your writing for grammar, punctuation, and spelling errors. If you don't, your customers, bosses, and colleagues will.

GRAMMAR RULES

To understand the fundamentals of grammar, you must first understand the basic components of a sentence.

A correctly constructed sentence consists of a subject and a predicate (some sentences also include a phrase or phrases).

example

The *meeting*	began	at 4:00 p.m.
subject	predicate	phrase

Subject: The *doer* of the action; the subject usually precedes the predicate.
Predicate: The *action* in the sentence

example

He	ran	to the office to avoid being late.
doer	action	phrases

If the subject and the predicate (1) express a complete thought and (2) can stand alone, you have an *independent clause*.

example

The meeting began	at 4:00 p.m.
independent clause	phrase

A *phrase* is a group of related words that does not contain a subject and a predicate and cannot stand alone or be punctuated as a sentence. The following are examples of phrases:

at the house
in the box
on the job
during the interview

If a clause is dependent, it cannot stand alone.

example

Although he tried to hurry,	he was late for the meeting.
dependent clause	independent clause

example

He was late for the meeting	although he tried to hurry.
independent clause	dependent clause

NOTE: When a dependent clause begins a sentence, use a comma before the independent clause. However, when an independent clause begins a sentence, do not place a comma before the dependent clause.

Agreement Between Pronoun and Antecedent (Referent)

A pronoun has to agree in gender and number with its antecedent.

Susan went on *her* vacation yesterday.
The *people* who quit said that *they* deserved raises.

Problems often arise when a singular indefinite pronoun is the antecedent. The following antecedents require singular pronouns: *anybody*, *each*, *everybody*, *everyone*, *somebody*, and *someone*.

incorrect

Anyone can pick up *their* applications at the job placement center.

correct

Anyone can pick up *his or her* applications at the job placement center.

Problems also arise when the antecedent is separated from the pronoun by numerous words.

incorrect

Even when the best *employee* is considered for a raise, *they* often do not receive it.

correct

Even when the best *employee* is considered for a raise, *he or she* often does not receive it.

Agreement Between Subject and Verb

Writers sometimes create disagreement between subjects and verbs, especially if other words separate the subject from the verb. To ensure agreement, ignore the words that come between the subject and verb.

incorrect

Her *boss* undoubtedly *think* that all the employees want promotions.

correct

Her *boss* undoubtedly *thinks* that all the employees want promotions.

incorrect

The *employees* who sell the most equipment *is* going to Hawaii for a week.

correct

The *employees* who sell the most equipment *are* going to Hawaii for a week.

If a sentence contains two subjects (a compound subject) connected by *and*, use a plural verb.

incorrect

Joe and Tiffany *was* both selected employee of the year.

correct

Joe and Tiffany *were* both selected employee of the year.

incorrect

The bench workers and their supervisor *is* going to work closely to complete this project.

correct

The bench workers and their supervisor *are* going to work closely to complete this project.

Add a final *s* or *es* to create most plural subjects or singular verbs, as follows:

Plural Subjects	**Singular Verbs**
boss*es* hire	a boss hir*es*
employe*es* demand	an employee demand*s*
experiment*s* work	an experiment work*s*
attitud*es* change	the attitude chang*es*

If a sentence has two subjects connected by *either . . . or, neither . . . nor,* or *not only . . . but also*, the verb should agree with the closest subject. This also makes the sentence less awkward.

example

> Either the salespeople or the warehouse worker deserv*es* raises.
> Not only the warehouse worker but also the salespeople deserv*e* raises.
> Neither the salespeople nor the warehouse worker deserv*es* raises.

Singular verbs are used after most indefinite pronouns such as the following:

another	everything
anybody	neither
anyone	nobody
anything	no one
each	nothing
either	somebody
everybody	someone
everyone	something

Anyone who works here *is* guaranteed maternity leave.
Everybody want*s* the company to declare a profit this quarter.

Singular verbs often follow collective nouns such as the following:

class	organization
corporation	platoon
department	staff
group	team

The *staff is* sending the boss a bouquet of roses.

Comma Splice

A *comma splice* occurs when two independent clauses are joined by a comma rather than separated by a period or semicolon.

incorrect

Sue was an excellent employee, she got a promotion.

Several remedies will correct this error.

1. Separate the two independent clauses with a semicolon.

correct

Sue was an excellent employee; she got a promotion.

2. Separate the two independent clauses with a period.

correct

Sue was an excellent employee. She got a promotion.

3. Separate the two independent clauses with a comma and a *coordinating conjunction (and, but, or, for, so, yet).*

correct

Sue was an excellent employee, *so* she got a promotion.

4. Separate the two independent clauses with a semicolon (or a period), a conjunctive adverb, and a comma. *Conjunctive adverbs* include *also, additionally, consequently, furthermore, however, instead, moreover, nevertheless, therefore*, and *thus*.

correct

Sue was an excellent employee; therefore, she got a promotion.

or

Sue was an excellent employee. *Therefore*, she got a promotion.

5. Use a *subordinating conjunction* to make one of the independent clauses into a dependent clause. Subordinating conjunctions include *after, although, as, because, before, even though, if, once, since, so that, though, unless, until, when, where*, and *whether*.

correct

Because Sue was an excellent employee, she got a promotion.

Faulty or Vague Pronoun Reference

A pronoun must refer to a specific noun (its antecedent). Problems arise when (1) there is an excessive number of pronouns (causing vague pronoun reference) and (2) there is no specific noun as an antecedent. Notice that there seems to be an excessive number of pronouns in the following passage, and the antecedents are unclear.

example

Although Bob had been hired over two years ago, *he* found that *his* boss did not approve *his* raise. In fact, *he* was also passed over for *his* promotion. The boss appears to have concluded that *he* had not exhibited zeal in *his* endeavors for their business. Instead of being a highly valued employee, *he* was not viewed with pleasure by those in authority. Perhaps it would be best if *he* considered *his* options and moved to some other company where *he* might be considered in a new light.

The excessive and vague use of *he* and *his* causes problems for readers. Do these words refer to Bob or to his boss? You are never completely sure. To avoid this problem, limit pronoun usage, as in the following revision.

example

Although Bob had been hired over two years ago, he found that his boss, Joe, did not approve his raise. In fact, Bob was also passed over for promotion. Joe appears to have concluded that Bob had not exhibited zeal in his endeavors for their business. Instead of being a highly valued employee, Bob was not viewed with pleasure by those in authority. Perhaps it would be best if Bob considered his options and moved to some other company where he might be considered in a new light.

To make the preceding paragraph more precise, we have replaced vague pronouns (*he* and *his*) with exact names (*Bob* and *Joe*).

Fragments

A *fragment* occurs when a group of words is incorrectly used as an independent clause. Often the group of words begins with a capital letter and has end punctuation but is missing either a subject or a predicate.

incorrect

Working with computers.

(lacks a predicate and does not express a complete thought)

The group of words may have a subject and a predicate but be a dependent clause.

incorrect

Although he enjoyed working with computers.

(has a subject, *he*, and a predicate, *enjoyed*, but is a dependent clause because it is introduced by the subordinate conjunction *although*)

It is easy to remedy a fragment by doing one of the following:

- Add a subject.
- Add a predicate.
- Add both a subject and a predicate.
- Add an independent clause to a dependent clause.

correct

Joe found that working with computers used his training.

(subject, *Joe,* and predicate, *found*, have been added)

correct

Although he enjoyed working with computers, he could not find a job in a computer-related field.

(independent clause, *he could not find a job*, added to the dependent clause, *Although he enjoyed working with computers*)

Fused Sentence

A *fused sentence* occurs when two independent clauses are connected with no punctuation.

incorrect

The company performed well last quarter its stock rose several points.

There are several ways to correct this error.

1. Write two sentences separated by a period.

correct

The company performed well last quarter. Its stock rose several points.

2. Use a comma and a coordinating conjunction to separate the two independent clauses.

correct

The company performed well last quarter, *so* its stock rose several points.

3. Use a subordinating conjunction to create a dependent clause.

correct

Because the company performed well last quarter, its stock rose several points.

4. Use a semicolon to separate the two independent clauses.

correct

The company performed well last quarter; its stock rose several points.

5. Separate the two independent clauses with a semicolon, a conjunctive adverb or a transitional word or phrase, and a comma.

correct

The company performed well last quarter; *therefore*, its stock rose several points.

correct

The company performed well last quarter; *for example*, its stock rose several points.

The following are transitional words and phrases, listed according to their use.

To Add

again	in addition
also	moreover
besides	next
first	second
furthermore	still

To Compare/Contrast

also	nevertheless
but	on the contrary
conversely	still
in contrast	

To Provide Examples

for example	of course
for instance	put another way
in fact	to illustrate

To Show Place

above	here
adjacent to	nearby
below	on the other side
elsewhere	there
further on	

To Reveal Time

afterward	second
first	shortly
meanwhile	subsequently
presently	thereafter

To Summarize

all in all	last
finally	on the whole
in conclusion	therefore
in summary	thus

Modification

A *modifier* is a word, phrase, or clause that explains or adds details about other words, phrases, or clauses.

Misplaced Modifiers

A *misplaced modifier* is one that is not placed next to the word it modifies.

incorrect

He had a heart attack *almost* every time he was reviewed by his supervisor.

correct

He *almost* had a heart attack every time he was reviewed by his supervisor.

incorrect

The worker had to *frequently* miss work.

correct

The worker *frequently* had to miss work.

Dangling Modifiers

A *dangling modifier* is a modifier that is not placed next to the word or phrase it modifies. To avoid confusing your readers, place modifiers next to the word(s) they refer to. Don't expect your readers to guess at your meaning.

incorrect

While working, tiredness overcame them.
(Who was working? Who was overcome by tiredness?)

correct

While working, the staff became tired.

incorrect

After soldering for two hours, the equipment was ready for shipping. (Who had been soldering for two hours? Not the equipment!)

correct

After soldering for two hours, the technicians prepared the equipment for shipping.

Parallelism

All items in a list should be parallel in grammatical form. Avoid mixing phrases and sentences (independent clauses).

incorrect

We will discuss the following at the department meeting:

1. Entering mileage in logs (phrase)
2. All employees have to enroll in a training seminar. (sentence)
3. Purpose of quarterly reviews (phrase)
4. Some data processors will travel to job sites. (sentence)

correct

We will discuss the following at the department meeting:

1. Entering mileage in logs
2. Enrolling in training seminars
3. Reviewing employee performance quarterly
4. Traveling to job sites

(phrases)

correct

At the department meeting, you will learn how to

1. Enter mileage in logs
2. Enroll in training seminars
3. Review employee performance quarterly
4. Travel to job sites

(phrases)

PUNCTUATION

Apostrophe (')

Place an *apostrophe* before the final *s* in a singular word to indicate possession.

example

Jim's tool chest is next to the furnace.

Place the apostrophe after the final *s* if the word is plural.

example

The employees' reception will be held next week.

Don't use an apostrophe to make singular abbreviations plural.

incorrect

The EXT's will be shipped today.

correct

The EXTs will be shipped today.

Colon (:)

Use a *colon* after a salutation.

example

> Dear Mr. Harken:

In addition, use a colon after an emphatic or cautionary word if explanations follow.

example

> Note: Hand-tighten the nuts.
> Caution: Wash thoroughly if any mixture touches your skin.

Finally, use a colon after an independent clause to precede a quotation, list, or example.

example

> She said the following: "No comment."
> These supplies for the experiment are on order: plastic hose, two batteries, and several chemicals.
> The problem has two possible solutions: hire four more workers, or simply give everyone a raise.

NOTE: In the preceding examples, the colon follows an independent clause.

A common mistake is to place a colon after an incomplete sentence. Except for salutations and cautionary notes, whatever precedes a colon *must* be an independent clause.

incorrect

The two keys to success are: earning money and spending wisely.

correct

The two keys to success are earning money and spending wisely.

or

The two keys to success are as follows: earning money and spending wisely.

or

The two keys to success are as follows:

1. Earning money
2. Spending wisely

Comma (,)

Writers often get in trouble with *commas* when they employ one of two common "words of wisdom."

- When in doubt, leave it out.
- Use a comma when there is a pause.

Both rules are inexact. Writers use the first rule to justify the complete avoidance of commas; they use the second rule to sprinkle commas randomly throughout their writing. On the contrary, commas have several specific conventions that determine usage.

1. Place a comma before a coordinating conjunction *(and, but, or, for, so, yet)* linking two independent clauses.

example

> You are the best person for the job, *so* I will hire you.
> We spent several hours discussing solutions to the problem, *but* we failed to decide on a course of action.

2. Use commas to set off introductory comments.

example

First, she soldered the components.
In business, people often have to work long hours.
To work well, you need to get along with your coworkers.
If you want to test equipment, do so by 5:25 p.m.

3. Use commas to set off sentence interrupters.

example

The company, started by my father, did not survive the last recession.
Mrs. Patel, the proprietor of the store, purchased a wide array of merchandise.

4. Set off parenthetical expressions with commas.

example

A worker, it seems, should be willing to try new techniques.
The highway, by the way, needs repairs.

5. Use commas after each item in a series of three or more.

example

Prakash, Mirren, and Justin were chosen as employees of the year.
We found the following problems: corrosion, excessive machinery breakdowns, and power failures.

6. Use commas to set off long numbers.

example

She earns $100,000 before taxes.

NOTE: Very large numbers are often written as words.

example

Our business netted over $2 million in 2009.

7. Use commas to set off the day and year when they are part of a sentence.

example

The company hired her on September 7, 2009, to be its bookkeeper.

NOTE: If the year is used as an adjective, do not follow it with a comma.

example

The 2009 corporate report came out today.

8. Use commas to set off the city from the state and the state from the rest of the sentence.

example

The new warehouse in Austin, Texas, will promote increased revenues.

NOTE: If you omit either the city or the state, you do not need commas.

example

The new warehouse in Austin will promote increased revenues.

Dash (—)

A *dash,* typed as two consecutive hyphens with no spaces before or after, is a versatile punctuation mark used in the following ways.

1. After a heading and before an explanation.

> **example**
> Forecasting—Joe and Joan will be in charge of researching fourth-quarter production quotas.

2. To indicate an emphatic pause.

> **example**
> You will be fired—unless you obey company rules.

3. To highlight a new idea.

> **example**
> Here's what we can do to improve production quality—provide on-the-job training, salary incentives, and quality controls.

4. Before and after an explanatory or appositive series.

> **example**
> Three people—Sue, Luci, and Tom—are essential to the smooth functioning of our office.

Ellipses (...)

Ellipses (three spaced periods) indicate omission of words within quoted materials.

> **example**
> "Six years ago, prior to incorporating, the company had to pay extremely high federal taxes."
> "Six years ago, . . . the company had to pay extremely high federal taxes."

Exclamation Point (!)

Use an *exclamation point* after strong statements, commands, or interjections.

> **example**
> You must work harder!
> Do not use the machine!
> Danger!

Hyphen (-)

A *hyphen* is used in the following ways.

1. To indicate the division of a word at the end of a typed line. Remember, this division must occur between syllables.
2. To create a compound adjective.

> **example**
> He is a well-known engineer.
> Until her death in 2009, she was a world-renowned chemist.
> Tom is a 24-hour-a-day student.

3. To join the numerator and denominator of fractions.

example Four-fifths of the company want to initiate profit sharing.

4. To write out two-word numbers.

example Twenty-six people attended the conference.

Parentheses ( )

Parentheses enclose abbreviations, numbers, words, or sentences for the following reasons.

1. To define a term or provide an abbreviation for later use.

example We belong to the Society for Technical Communication (STC).

2. To clarify preceding information in a sentence.

example The people in attendance (all regional sales managers) were proud of their accomplishments.

3. To number items in a series.

example The company should initiate (1) new personnel practices, (2) a probationary review board, and (3) biannual raises.

Period (.)

A *period* must end a declarative sentence (independent clause).

example I found the business trip rewarding.

Periods are often used with abbreviations.

D.C.	Mrs.	A.M. or a.m.
e.g.	Ms.	P.M. or p.m.
Mr.		

It is incorrect to use periods with abbreviations for organizations and associations.

incorrect

S.T.C. (Society for Technical Communication)

correct

STC (Society for Technical Communication)

State abbreviations do not require periods when you use two capital letters.

incorrect

KS. (Kansas)
MO. (Missouri)
TX. (Texas)

correct

KS
MO
TX

Question Mark (?)

Use a *question mark* after direct questions.

Do the lab results support your theory? Will you work at the main office or at the branch?	**example**

Quotation Marks (" ")

Quotation marks are used in the following ways.

1. When citing direct quotations.

He said, "Your division sold the most compressors last year."	**example**

NOTE: When you are citing a quotation within a quotation, use double quotation marks (" ") and single quotation marks (' ').

Kim's supervisor, quoting the CEO, said the following to explain the new policy regarding raises: "'Only employees who deserve them will receive merit raises.'"	**example**

2. To note the title of an article or a subdivision of a report.

The article "Robotics in Industry Today" was an excellent choice as the basis of your speech. Section III, "Waste Water in District 9," is pertinent to our discussion.	**example**

When using quotation marks, abide by the following punctuation conventions:

- Commas and periods always go *inside* quotation marks.

She said, "Our percentages are fixed."	**example**

- Colons and semicolons always go *outside* quotation marks.

He said, "The supervisor hasn't decided yet"; however, he added that the decision would be made soon.	**example**

- Exclamation points and question marks go inside the quotation marks if the quoted material is either exclamatory or a question. However, if the quoted material is not exclamatory or a question, then these punctuation marks go outside the quotation marks.

John said, "Don't touch that liquid. It's boiling!"	**example**

(Although the sentence isn't exclamatory, the quotation is. Thus, the exclamation point goes inside the quotation marks.)

example

How could she say, "We haven't purchased the equipment yet"?

(Although the quotation isn't a question, the sentence is. Thus, the question mark goes outside the quotation marks.)

Semicolon (;)

Semicolons are used in the following instances.

1. Between two independent clauses *not* joined by a coordinating conjunction.

example

The light source was unusual; it emanated from a crack in the plastic surrounding the cathode.

2. To separate items in a series containing internal commas.

example

When the meeting was called to order, all members were present, including Susan Johnson, the president; Ruth Schneider, the vice president; Harold Holbert, the treasurer; and Linda Hamilton, the secretary.

MECHANICS

Abbreviations

Never use an abbreviation that your reader will not understand. A key to clear workplace writing is to write on a level appropriate to your reader. You may use the following familiar abbreviations without explanation: *Mrs.*, *Dr.*, *Mr.*, *Ms.*, and *Jr.*

A common mistake is to abbreviate inappropriately. For example, some writers abbreviate *and* as follows:

example

I quit my job & planned to retire young.

This is too colloquial for professional workplace writing. Spell out *and* when you write.

The majority of abbreviation errors occur when writers incorrectly abbreviate states and technical terms.

States

Writers often abbreviate the names of states incorrectly. Use the U.S. Postal Service abbreviations in addresses.

Abbreviations for States

AL	Alabama	IN	Indiana
AK	Alaska	IA	Iowa
AZ	Arizona	KS	Kansas
AR	Arkansas	KY	Kentucky
CA	California	LA	Louisiana
CO	Colorado	ME	Maine
CT	Connecticut	MT	Montana
DE	Delaware	NE	Nebraska
FL	Florida	NV	Nevada
GA	Georgia	NH	New Hampshire
HI	Hawaii	NJ	New Jersey
ID	Idaho	NM	New Mexico
IL	Illinois	NY	New York

Abbreviations for States (Continued)

NC	North Carolina	MD	Maryland
ND	North Dakota	MA	Massachusetts
OH	Ohio	MI	Michigan
OK	Oklahoma	MN	Minnesota
OR	Oregon	MS	Mississippi
PA	Pennsylvania	MO	Missouri
RI	Rhode Island	VT	Vermont
SC	South Carolina	VA	Virginia
SD	South Dakota	WA	Washington
TN	Tennessee	WV	West Virginia
TX	Texas	WI	Wisconsin
UT	Utah	WY	Wyoming

Technical Terms

Units of measurement and scientific terms must be abbreviated accurately to ensure that they will be understood. Writers often use such abbreviations incorrectly. For example, "The unit measured 7.9 cent." is inaccurate. The correct abbreviation for centimeter is *cm*, not *cent*. Use the following abbreviation conventions.

Technical Abbreviations for Units of Measurement and Scientific Terms

absolute	abs	cubic centimeter	cc
alternating current	AC	cubic feet per second	cfs
American wire gauge	AWG	cubic foot	cu ft (or ft^2)
ampere	amp	cubic inch	cu in. (or $in.^3$)
ampere-hour	amp-hr	cubic meter	cu m (or m^3)
amplitude modulation	AM	cubic yard	cu yd (or yd^3)
angstrom unit	Å	current (electric)	I
atmosphere	atm	cycles per second	CPS
atomic weight	at wt	decibel	dB
audio frequency	AF	decigram	dg
azimuth	az	deciliter	dl
barometer	bar.	decimeter	dm
barrel, barrels	bbl	degree	deg
billion electron volts	BeV	dekagram	dkg
biochemical oxygen demand	BOD	dekaliter	dkl
		dekameter	dkm
board foot	bd ft	dewpoint	DP
Brinell hardness number	BHN	diameter	dia
British thermal unit	Btu	direct current	DC
bushel	bu	dozen	doz (or dz)
calorie	cal	dram	dr
candela	cd	electromagnetic force	emf
Celsius	C	electron volt	eV
center of gravity	cg	elevation	el (or elev)
centimeter	cm	equivalent	equiv
circumference	cir	Fahrenheit	F
cologarithm	colog	farad	F
continuous wave	CW	faraday	f
cosine	cos	feet, foot	ft
cotangent	cot	feet per second	ft/sec

(Continued)

Technical Abbreviations for Units of Measurement and Scientific Terms (Continued)

Term	Abbreviation	Term	Abbreviation
fluid ounce	fl oz	megacycle	mc
foot board measure	fbm	megahertz	MHz
foot-candle	ft-c	megawatt	MW
foot-pound	ft lb	meter	m
frequency modulation	FM	microampere	μamp
gallon	gal	microinch	μin.
gallons per day	GPD	microsecond	μsec
gallons per minute	GPM	microwatt	μw
grain	gr	miles per gallon	mpg
grams	g (or gm)	milliampere	mA
gravitational acceleration	g	millibar	mb
hectare	ha	millifarad	mF
hectoliter	hl	milligram	mg
hectometer	hm	milliliter	ml
henry	H	millimeter	mm
hertz	Hz	millivolt	mV
high frequency	HF	milliwatt	mW
horsepower	hp	minute	min
horsepower-hours	hp-hr	nautical mile	NM
hour	hr	negative	neg or –
hundredweight	cwt	number	no.
inch	in.	octane	oct
inch-pounds	in.-lb	ounce	oz
infrared	IR	outside diameter	OD
inner diameter or inside dimensions	ID	parts per billion	ppb
intermediate frequency	IF	parts per million	ppm
international unit	IU	pascal	pas
joule	J	positive	pos or +
Kelvin	K	pound	lb
kilocalorie	kcal	pounds per square inch	psi
kilocycle	kc	pounds per square inch absolute	psia
kilocycles per second	kc/sec	pounds per square inch gauge	psig
kilogram	kg	quart	qt
kilohertz	kHz	radio frequency	RF
kilojoule	kJ	radian	rad
kiloliter	kl	radius	r
kilometer	km	resistance	r
kilovolt	kV	revolution	rev
kilovolt-amperes	kVa	revolutions per minute	rpm
kilowatt-hours	kWH	secant	sec
lambert	L	second	s (or sec)
latitude	lat	specific gravity	sp gr (or SG)
length	l	square foot	ft^2
linear	lin	square inch	$in.^2$
linear foot	lin ft	square meter	m^2
liter	l	square mile	mi^2
logarithm	log.	tablespoon	tbs (or tbsp)
longitude	long.	tangent	tan
low frequency	LF	teaspoon	tsp
lumen	lm	temperature	t
lumen-hour	lm-hr	tensile strength	ts
maximum	max		

Technical Abbreviations for Units of Measurement and Scientific Terms (Continued)

thousand	m	volume	vol
ton	t	watt-hour	whr
ultra high frequency	UHF	watt	W
vacuum	vac	wavelength	WL
very high frequency	VHF	weight	wt
volt	V	yards	y (or yd)
volt-ampere	VA	years	y (or yr)
volts per meter	V/m		

Capital Letters

Capitalize the following:

1. Proper nouns.

example

people	cities	countries	companies	schools	buildings
Susan	Houston	Italy	Bendix	Harvard	Oak Park Mall

2. People's titles (only when they precede the name).

example

Governor Sally Renfro
or
Sally Renfro, governor
Technical Supervisor Todd Blackman
or
Wes Schneider, the technical supervisor

3. Titles of books, magazines, plays, movies, television programs, and CDs (excluding the prepositions and all articles after the first article in the title).

example

Grey's Anatomy
The Colbert Report
The Catcher in the Rye
Walk the Line
American Idol

4. Names of organizations.

example

Girl Scouts
Kansas City Regional Home Care Association
Kansas City Regional Council for Higher Education
Programs for Technical and Scientific Communication
American Civil Liberties Union
Students for a Democratic Society

5. Days of the week, months, and holidays.

example

Monday
December
Thanksgiving

6. Races, religions, and nationalities.

example

American Indian
Jewish
Polish

7. Events or eras in history.

example

the Gulf War
the Vietnam War
World War II

8. North, South, East, and West (when used to indicate geographic locations).

example

They moved from the North.
People are moving to the Southwest.

NOTE: Don't capitalize these words when giving directions.

example

We were told to drive south three blocks and then to turn west.

9. The first word of a sentence.
10. Don't capitalize any of the following:

example

Seasons—spring, fall, summer, winter
Names of classes—sophomore, senior
General groups—middle management, infielders, surgeons

Numbers

Write out numbers one through ten. Use numerals for numbers 11 and above.

11	12
104	2,093
536	5,550,286

Although the preceding rules cover most situations, there are exceptions.

1. Use numerals for all percentages.

example

2 percent   18 percent   25 percent

2. Use numerals for addresses.

example

12 Elm   935 W. Harding

3. Use numerals for miles per hour.

example

5 mph   225 mph

4. Use numerals for time.

3:15 a.m.	example

5. Use numerals for dates.

May 31, 2009	example

6. Use numerals for monetary values.

$45 $0.95 $2 million	example

7. Use numerals for units of measurement.

14′ 6¾″ 16 mm 10 V	example

8. Do not use numerals to begin sentences.

incorrect

568 people were fired last August.

correct

Five hundred sixty-eight people were fired last August.

9. Do not mix numerals and words when writing numbers. When two or more numbers appear in a sentence and one of them is 11 or more, numerals are used.

We attended 4 meetings over a 16-day period.	example

10. Use numerals and words in a compound number adjective to avoid confusion.

The worker needed six 2-inch nails.	example

SPELLING

The following is a list of commonly misspelled or misused words. You can avoid many common spelling errors if you familiarize yourself with these words. Remember to run spell check; also remember that spell check will not understand context. The incorrect word contextually could be spelled correctly.

accept, except
access, excess
addition, edition
advise, advice
affect, effect
all ready, already
assistants, assistance
bare, bear
brake, break
cite, site, sight
coarse, course
council, counsel
desert, dessert
disburse, disperse
fiscal, physical
forth, fourth
incite, insight
its, it's
loose, lose
miner, minor
passed, past

patients, patience
personal, personnel
principal, principle
quiet, quite
rite, right, write
stationery, stationary
their, there, they're
to, too, two
whose, who's
your, you're

PROOFREADER'S MARKS

Figure Appendix 1.1 illustrates proofreader's marks and how to use them.

FIGURE APPENDIX 1.1 Proofreader's Marks

PROOFREADER'S MARKS			
Symbol	Meaning	Mark on Copy	Revision
ℓ	delete	hire better peoplesℓ	hire better people
(tr) or ∩∪	transpose	computer systme	computer system
⁀‿	close space	we ne ed 40	we need 40
#	insert space	threemistakes	three mistakes
¶	begin paragraph	¶The first year	The first year
(RUN IN)	no paragraph	financial. We earned twelve dollars.	financial. We earned twelve dollars.
[	move left	[Next year	Next year
]	move right	]Next year	Next year
^.	insert period	happy employee ^.	happy employee.
^,	insert comma	For two years^,	For two years,
^;	insert semicolon	We need you^; come to work.	We need you; come to work.
^:	insert colon	dogs^: brown, black, and white.	dogs: brown, black, and white.
^-	insert hyphen	first rate	first-rate
∨'	insert apostrophe	Marys car	Mary's car
∨" / ∨"	insert quotation marks	Like a Rolling Stone	"Like a Rolling Stone"
(SP)	spell out	(4) chips	four chips
(cap) or ≡	capitalize	the meeting	The meeting
(lc)	lower case	the Boss	the boss
^	insert	the ^left margin	the left margin

MEETING WORKPLACE COMMUNICATION CHALLENGES

Spelling

In the following sentences, circle the correctly spelled words within the parentheses.

1. Each of the employees attended the meeting (accept except) the line supervisor, who was out of town for job-related travel.
2. The (advise advice) he gave will help us all do a better job.
3. Management must (affect effect) a change in employees' attitudes toward absenteeism.
4. Let me (site cite sight) this most recent case as an example.
5. (Its It's) too early to tell if our personnel changes will help create a better office environment.
6. If we (lose loose) another good employee to our competitor, our production capabilities will suffer.
7. I'm not (quite quiet) sure what she meant by that comment.
8. (Their There They're) budget has gotten too large to ensure a successful profit margin.
9. We had wanted to attend the conference (to too two), but our tight schedule prevented us from doing so.
10. (You're Your) best chance for landing this contract is to manufacture a better product.

In the letter below, correct the misspelled words.

March 5, 2009

Joanna Freeman
Personel Director
United Teletype
1111 E. Street
Kansas City, MO 68114

Dear Ms. Freeman:

Your advertizmemt in the Febuary 18, 2009, Kansas City Star is just the opening I have been looking for. I would like to submit my quallifications.

As you will note in the inclosed resum, I recieved an Enginneering degree from the Missouri Institute of Technology in 2005 and have worked in the electronic enginneering department of General Accounts for three years. I have worked a great deal in design electronics for microprocesors, controll systems, ect.

Because your company has invented many extrordinary design projects, working at your company would give me more chances to use my knowlege aquired in school and through my expirences. If you are interested in my quallifications, I would be happy to discuss them futher with you. I look foreward to hearing from you.

Sincerly,

Bob Cottrell

Bob Cottrell

Fragments and Comma Splices

In the following sentences, correct the fragments and comma splices by inserting the appropriate punctuation or adding any necessary words.

1. She kept her appointment with the salesperson, however, the rest of her staff came late.
2. When the CEO presented his fiscal year projections, he tried to motivate his employees, many were not excited about the proposed cuts.
3. Even though the company's sales were up 25 percent.
4. The supervisor wanted the staff members to make suggestions for improving their work environment, the employees, however, felt that any grievances should be taken directly to their union representatives.
5. Which he decided was an excellent idea.
6. Because their machinery was prone to malfunctions and often caused hazards to the workers.
7. They needed the equipment to complete their job responsibilities, further delays would cause production slowdowns.
8. Their client who was a major distributor of high-tech machinery.
9. Robotics should help us maintain schedules, we'll need to avoid equipment malfunctions, though.
10. The company, careful not to make false promises, advertising their product in media releases.

In the letter below, correct the fragments and comma splices.

May 12, 2009

Maurene Pierce
Dean of Residence Life
Mann College
Mannsville, NY 10012

Subject: Report on Dormitory Damage Systems

Here is the report you authorized on April 5 for an analysis of the current dormitory damage system used in this college.

The purpose of the report was to determine the effectiveness of the system. And to offer any concrete recommendations for improvement. To do this, I analyzed in detail the damage cost figures for the past three years, I also did an extensive study of dormitory conditions. Although I had limited manpower. I gathered information on all seven dormitories, focusing specifically on the men's athletic dorm, located at 1201 Chester. In this dorm, bathroom facilities, carpeting, and air conditioning are most susceptible to damage. Along with closet doors.

Nonetheless, my immediate findings indicate that the system is functioning well, however, improvements in the physical characteristics of the dormitories, such as new carpeting and paint, would make the system even more efficient.

I have enjoyed conducting this study, I hope my findings help you make your final decision. Please contact me. If I can be of further assistance.

Rob Harken

Rob Harken

Punctuation

In the following sentences, circle the correct punctuation marks. If no punctuation is needed, draw a slash mark through both options.

1. John took an hour for lunch (, ;) but Joan stayed at her desk to eat so she could complete the project.
2. Sally wrote the specifications (, ;) Randy was responsible for adding any needed graphics.
3. Manufacturing maintained a 93.5 percent production rating in July (, ;) therefore, the department earned the Golden Circle Award at the quarterly meeting.
4. In their year-end requests to management (, ;) supervisors asked for new office equipment (, ;) and a 10 percent budget increase for staffing.
5. The following employees attended the training session on stress management (, :) Steve Janasz, purchasing agent (, ;) Jeremy Kreisler, personnel director (, ;) and Prakash Patel, staff supervisor.
6. Promotions were given to all sales personnel (, ;) secretaries, however, received only cost-of-living raises.
7. The technicians voted for better work benefits (, ;) as an incentive to improve morale.
8. Although the salespeople were happy with their salary increases (, ;) the technicians felt slighted.
9. First (, ;) let's remember that meeting schedules should be a priority (, ;) and not an afterthought.
10. The employee (, ;) who achieves the highest rating this month (, ;) will earn 10 bonus points (, ;) therefore (, ;) competition should be intense.

In the following letter, no punctuation has been added. Instead, there are blanks where punctuation might be inserted. First, decide whether any punctuation is needed (not every blank requires punctuation). Then, insert the correct punctuation—a comma, colon, period, semicolon, or question mark.

January 8_ 2009

Mr_ Ron Schaefer
1324 Homes
Carbondale_ IL_ 34198

Dear Mr_ Schaefer_

Yesterday_ my partners_ and I read about your invention in the Herald Tribune_ and we want to congratulate you on this new idea_ and ask you to work with us on a similar project_

We cannot wait to begin our project_ however_ before we can do so_ I would like you to answer the following questions_

- Has your invention been tested in salt water_
- What is the cost of replacement parts_
- What is your fee for consulting_

Once_ I receive your answers to these questions_ my partners and I will contact you regarding a schedule for operations_ We appreciate your design concept_ and know it will help our business tremendously_ We look forward to hearing from you_

Sincerely_

Elias Agamenyon

Elias Agamenyon

Agreement (Subject/Verb and Pronoun/Antecedent)

In the following sentences, circle the correct choice to achieve agreement between subject and verb or pronoun and antecedent.

1. The employees, though encouraged by the possibility of increased overtime, (was were) still dissatisfied with their current salaries.
2. The supervisor wants to manufacture better products, but (they he) doesn't know how to motivate the technicians to improve their work habits.
3. The staff (was were) happy when the new manager canceled the proposed meeting.
4. Anyone who wants (his or her their) vote recorded must attend the annual board meeting.
5. According to the printed work schedule, Susan and Tom (work works) today on the manufacturing line.
6. According to the printed work schedule, either Susan or Tom (work works) today on the manufacturing line.
7. Although Tamara is responsible for distributing all monthly activity reports, (she they) failed to mail them.
8. Every one of the engineers asked if (he or she they) could be assigned to the project.
9. Either the supervisor or the technicians (is are) at fault.
10. The CEO, known for her generosity to employees and their families, (has have) been nominated for the humanitarian award.

In the following memo, find and correct the errors in agreement.

MEMO

DATE:	October 30, 2009
TO:	Tammy West
FROM:	Susan Lisk
SUBJECT:	REPORT ON AIR HANDLING UNIT

There has been several incidents involving the unit which has resulted in water damage to the computer systems located below the air handler.

The occurrences yesterday was caused when a valve was closed creating condensation to be forced through a humidifier element into the supply air duct. Water then leaked from the duct into the room below causing substantial damage to four disc-drive units.

To prevent recurrence of this type of damage, the following actions has been initiated by maintenance supervision:

- Each supervisor must ensure that their subordinates remove condensation valves to avoid unauthorized operation.
- Everyone must be made aware that they are responsible for closing condensation valves.
- The supply air duct, modified to carry away harmful sediments, are to be drained monthly.

Maintenance supervision recommend that air handlers not be installed above critical equipment. This will avoid the possibility of coil failure and water damage.

Capitalization

In the following memo, nothing has been capitalized. Capitalize those words requiring capitalization.

date: december 5, 2009
to: jordan cottrell
from: richard davis
subject: self-contained breathing apparatus (scba) and negative pressure respirator evaluation and fit-testing report

the evaluation and fit-testing have been accomplished. the attached list identifies the following:

- supervisors and electronic technicians who have used the scba successfully.
- the negative pressure respirators used for testing in an isoamyl acetate atmosphere.

fit-testing of waste management personnel will be accomplished annually, according to president chuck carlson. new supervisors and technicians will be fit-tested when hired.

all apex corporation personnel located in the new york district (12304 parkview lane) must submit a request form when requesting a respirator or scba for use. any waste management personnel in the north and south facilities not identified on the attached list will be fit-tested when use of scba is required. if you have any questions, contact chuck carlson or me (richard davis, district manager) at ext. 4036.

Grammar Quiz

The following sentences contain errors in spelling, punctuation, verb and pronoun agreement, sentence structure (fragments and run-ons), and modification. Circle the letter corresponding to the section of the sentence containing the error.

example

When you recieve the salary increase, your family will celebrate the occasion.
(A) B C

1. Each department manager should tell his subordinates to advise of any negative occurrences regarding in-house training.
A B C D

2. The lawyer's new offices were similar to their former ones ________ the offices were on a quiet street.
A B C D

3. New York City is divided into five boroughs ; Manhattan, the Bronx, Queens, Brooklyn, and Staten Island.
A B C D

4. Fashion consultants explain that clothes create a strong impression ________ so they advise executives ________ to choose wardrobes carefully.
A B C D

5. Everyone should make sure that they are well represented in union meetings ; otherwise , management could become too powerful.
A B C D

6. The employment agency , too busy to return the telephone calls from prospective clients , are harming business opportunities.
A B C D

7. The supervisory staff are making decisions based on scheduling , but all employees want to ensure quality control.
(A: are; B: ,; C: employees; D: ensure)

8. Because the price of cars has risen dramatically ___________ most people keep their cars longer ___________ to save capital expenditures.
(A: ___; B: their; C: ___; D: capital)

9. The department's manager heard that a merger was possible ___________ but he decided to keep the news quiet.
(A: department's; B: ___; C: to; D: quiet)

10. The manager of the department believed that her employee's were excellent , but she decided that no raises could be given ___________ because the price of stocks was falling.
(A: employee's; B: ,; C: ___; D: was)

11. The detailed report from the audit ___________ of the department gave good advise about how to restructure , so we are all ready to do so.
(A: ___; B: advise; C: ,; D: all ready)

12. When my boss and me looked at the books , we found these problems : lost invoices, unpaid bills , and late payments.
(A: me; B: ,; C: :; D: ,)

13. The most dedicated staff members are accepted for the on-site training sessions because : they are responsive to criticism, represent the company's future, and strive for improvements.
(A: are; B: accepted; C: :; D: company's)

14. Despite her assurances to the contrary , there is still three unanswered questions : who will make the payments, when will these payments occur, and why is there a delay?
(A: ,; B: is; C: :; D: there)

15. The reputation of many companies often depend on one employee who represents that company.
(A: companies; B: depend; C: who; D: represents)

16. Either my monthly activity report or my year-end report are due today , but my computer is broken ; therefore, I need to use yours.
(A: year-end; B: are; C: ,; D: ;)

17. Today's American manpower , according to many foreign governments, suffers from lack of discipline.
(A: manpower; B: ,; C: foreign; D: suffers)

18. Rates are increasing next year because : fuel, maintenance, and insurance are all
A B C
higher than last year.
D

19. Many colleges have long-standing football rivalries , one of the most famous ones
A B
is between KU and KSU (two universities in Kansas).
C D

20. Everyone who wants to enroll in the business school should do so before June if
A B
they can to ensure getting the best classes.
C D

21. Because John wanted high visibility for his two businesses, he paid top dollar
A B
___________ and spent long hours looking for appropriate cites.
C D

22. Many people apply for jobs at Apex , however , only a few are accepted.
A B C D

23. Because most cars break down occasionally , all drivers should know how to
A B
change a flat tire , and how to signal for assistance.
C D

24. Arriving on time, working diligently, and closing the office securely—everyone
A B
needs to know that they are responsible for these job duties.
C D

25. Mark McGwire not only hit the ball further than other players, but also he hit
A B
more dingers than other players who hit fewer homers.
C D

REFERENCES

Chapter 1

Adams, Rae, et al. "Ethics and the Internet." *Proceedings: 42nd Annual Technical Communication Conference,* April 23-26, 1995: 328.

Adrian, Merv. "IT First Look." *Forrester's Research,* June 6, 2005.

"America's Young Entrepreneurs." *National Association for the Self Employed,* 2004. 7 June 2005. http://www.nase.org/fey/youngentrepreneurs_stats.htm.

Barker, Thomas, et al. "Coming Into the Workplace: What Every Technical Communicator Should Know—Besides Writing." *Proceedings: 42nd Annual Technical Communication Conference,* April 23-26, 1995: 38-39.

Beem, Kate. "Educate employees on company e-mail policies." *The Kansas City Star,* Feb. 2, 2002: D26.

Bottorff, Dean L. "Ethics and Culture Management Services." *Ethics Quality,* 18 Oct. 1997. 14 Jun. 2004. http://www.ethicsquality.com/about.htm#ethics.

Boyett, J. H., and D.P. Snyder. "Twenty-first Century Workplace Trends." *On the Horizon,* 2 (1998): 1, 4-9.

Bremer, Otto A, et al. "Ethics and Values in Management Thought." *Business Environment and Business Ethics: The Social, Moral, and Political Dimensions of Management,* Ed. Karen Paul. Cambridge, MA: Ballinger, 1987: 61-86.

"Business Identity." *Crane's,* 2007. 28 July 2007. http://www.crane.com/business/businessidentity/.

"Code for Communicators." *Rocky Mountain Chapter, Society for Technical Communication,* 2007. 1 Aug. 2007. http://www.stcrmc.org/resources/resource_code.htm

Cerner, 26 Dec. 2007. http://www.cerner.com/public.

"Code of Ethics." *IABC,* 2003. 8 Oct. 2003. http://www.iabc.com/members/joining/code.htm.

"Generation Y Communication Trends: Harnessing the Growth of Instant Messaging in Delivering Business Communication Solutions." Pika Technologies, Inc. 2006, 24 Dec. 2007. www.ctipro.cz/download/pikageneration_communication_trends.pdf.

Girill, T. R. "Technical Communication and Ethics." *Technical Communication,* 34 (Aug. 1987): 178-79.

Guy, Mary E. *Ethical Decision Making in Everyday Work Situations.* New York: Quorum Books, 1990.

Hartman, Diane B., and Karen S. Nantz. "Send the Right Messages About E-mail." *Training and Development,* (May 1995): 60-65.

Johnson, Dana R. "Copyright Issues on the Internet." *Intercom,* (Jun. 1999): 17.

LeVie, Donald S. "Internet Technology and Intellectual Property." *Intercom,* (January 2000): 20-23.

"Microsoft Standards of Business Conduct." Microsoft, 2007. 28 Jun. 2007. http://www.microsoft.com/about/legal/buscond/default.mspx#values

"People Principles." McDonalds, 2007. 1 Aug. 2007. http://www.mcdonalds.com/corp/values/people/people_principles.html.

Reyman, Jessica. "Rethinking Plagiarism for Technical Communication." *Technical Communication,* (Feb. 2008): 61-67.

Safer, Morley. "The 'Millennials' Are Coming." *CBSNews,* 11 Nov. 2007. http://www.cbsnews.com/stories/2007/11/08/60minutes/main3475200.shtml.).

Sherman, Ruth. "Understanding Your Communication Style." *Online Women's Business Center,* 10 Aug. 2001. 12 Jun. 2004. http://www.au.af.mil/au/awc/awcgate/sba/comm_style.htm.

Turner, John R. "Ethics Online: Looking Toward the Future." *Proceedings: 42nd Annual Technical Communication Conference,* Apr. 23-26, 1995: 59-62.

Tyson, Laura D'Andrea. "Outsourcing: Who's Safe Anymore?" *BusinessWeek.com,* 23 Feb. 2004. 7 June 2005. http://www.businessweek.com/@@fqQGZ4YQxSYVgBUA/magazine/content/04_08/b3871032_mz007.htm.

"Vision, Mission & Values." *Black and Veatch,* 2007. 28 June 2007. http://www.bv.com/about_us/vision_mission_3_values.aspx.

Wal-Mart. 26 December 2007, http://walmartstores.com/GlobalWMStoresWeb/.

Wilson, Catherine Mason. "Product Liability and User Manuals." *Proceedings: 34th International Technical Communication Conference,* May 10, 1987: WE—68-71.

"Work trends of the future." *PageWise,* 2002. 11 Jun. 2004 http://azaz.essortment.com/worktrendsfutu_rrgq.htm.

"Writing: A Ticket to Work . . . Or A Ticket Out, A Survey of Business Leaders." *National Commission on Writing,* 2004. www.writingcommission.org.

"Writing Skills Necessary for Employment, Says Big Businesses." *National Commission on Writing,* 2007. 1 Aug. 2007. www.writingcommission.org/pr/writing_for_employ.html.

Chapter 2

"Achieving the Promise of the Mobile Enterprise." *Motorola: Position Paper,* Aug. 2005. 31 Aug. 2007. http://www.motorola.com/Enterprise/contentdir/en_US/Enterprise/Files/Enterprise_Mobility_Solutions_Position_Paper.pdf.

Finney, Paul. "Telepresence TV." *The New York Times,* 29 May 2007. 27 Aug. 2007. http://www.nytimes.com/2007/05/29/technology/29video.html?ex=1338091200&en=34d6f327495b1e25&ei=5088&partner=rssnyt&emc=rss.

Fisher, Lori, and Lindsay Bennion. "Organizational Implications of the Future Development of Technical Communication: Fostering Communities of Practice in the Workplace." *Technical Communication,* 52 (August 2005): 277-288.

Hughes, Michael A. "Managers: Move from Silos to Channels." *Intercom,* (March 2003): 9–11.

"Individual's and Teams' Roles and Responsibilities." *GOAL/QPC*, February 24, 2003. http://www.goalqpc.com/index.htm.

Mader, Stewart. "Using Wiki in Education." *The Science of Spectroscopy*, www.scienceofspectroscopy.info/edit/index.php?title=Using_wiki_in_education. 14 Dec. 2005.

Nesbitt, Pamela, and Elizabeth Bagley-Woodward. "Practical Tips for Working with Global Teams." *Intercom*, (June 2006): 25-30.

"Top Ten Qualities/Skills Employers Want." *Job Outlook 2006 Student Version*. National Association of Collegesand Employers, 2005: 5. http://career.clemson.edu/pdf_docs/NACE_JO6.pdf.

"TWiki—Enterprise Wiki & Collaboration Platform." *TWiki*, 4 Mar. 2007. http://twiki.org.

"We Are Smarter Than Me." 18 Dec. 2006. www.wearesmarter.org.

"What is Wikipedia." *Wikipedia*, 9 Jan. 2007. http://en.wikipedia.org/wiki/Wikipedia:Introduction.

"Wiki." *Wikipedia*, 9 Jan. 2007. www.wikipedia.org.

Chapter 3

"Around the World." Coca Cola, 31 Aug. 2007. http://www.thecoca-colacompany.com/ourcompany/aroundworld.html.

Cardarella, Toni. "Business." *The Kansas City Star*, September 30, 2003. http://www.kansascity.com/mld/kansascity/business/6877765.htm?lc.

"Civilian Labor Force by Age, Sex, Race, and Hispanic Origin, 1994, 2004, and Projected 2014." The United States Department of Labor, 7 Dec. 2005. 6 Mar. 2007. http://www.bls.gov/news.release/ecopro.t07.htm.

Courtis, John K., and Salleh Hassan. "Reading Ease of Bilingual Annual Reports." *Journal of Business Communication*, 39 (October 2002): 394–413.

Flint, Patricia, et al. "Helping Technical Communicators Help Translators." *Technical Communication*, 46 (May 1999): 238–248.

Gardner, Amanda. "Language a Widening Barrier to Health Care." *Health On the Net Foundation*, 9 Jan. 2007. 6 Mar. 2007. www.hon.ch/News/HSN/533896.html.

"Generation Y Communication Trends: Harnessing the Growth of Instant Messaging in Delivering Business Communication Solutions." Pika Technologies, Inc., 2006. 24 Dec. 2007. www.ctipro.cz/download/pika generation_communication_trends.pdf.

Grimes, Diane Susan, and Orlando C. Richard. "Could Communication Form Impact Organizations' Experience with Diversity?" *Journal of Business Communication*, 40 (January 2003): 7–27.

Horton, William. "The Almost Universal Language: Graphics for International Documents." *Technical Communication*, 40 (November 1993): 682–693.

Hussey, Tim, and Mark Homnack. "Foreign Language Software Localization." *Proceedings: 37th International Technical Communication Conference*, May 20–23, 1990: RT-44–47.

Melgoza, Cesar M. "Geoscape Findings Confirm Hispanic Market Growth." *Hispanic MPR.com*, 6 Mar. 2007. www.hispanicmpr.com/2006/02/14/.2006.

"Multilingual Features in Windows XP Professional." August 24, 2001. http://www. microsoft.com/windowsxp/pro/techinfo/planning/multilingual/default.asp.

Nethery, Kent. "Let's Talk Business." February 16, 2003. http://www.cuspomona.edu/~cljones/powerpoints/chap 02/sld001.htm.

"Occupational Employment in Private Industry by Race/Ethnic Group/Sex and by Industry, United States 2005." *The U.S. Equal Employment Opportunity Commission*, 9 May 2005. 6 Mar. 2007. http://www.eeoc.gov/stats/jobpat/2003/national.html.

Rains, Nancy E. "Prepare Your Documents for Better Translation." *Intercom*, 41.5 (December 1994): 12.

Sanchez, Mary. "KC hospitals seek to overcome language barriers." *The Kansas City Star*, January 7, 2003: A1, A4.

Scott, Julie S. "When English Isn't English." *Intercom*, (May 2000): 20–21.

St. Amant, Kirk R. "Communication in International Virtual Offices." *Intercom*, (April 2003): 27–28.

Swenson, Lynne V. "How to Make (American) English Documents Easy to Translate." *Proceedings: 34th International Technical Communication Conference*, May 10, 1987: WE-193–195.

"Swiss Fight Encroachment of English." *The Kansas City Star*, December 7, 2002: A16.

Walmer, Daphne. "One Company's Efforts to Improve Translation and Localization." *Technical Communication*, 46 (May 1999): 230–237.

Weiss, Edmund H. "Twenty-five Tactics to 'Internationalize' Your English." *Intercom*, (May 1998): 11–15.

Chapter 4

Campbell, Kim S., et al. "Leader-Member Relations as a Function of Rapport Management." *Journal of Business Communication*, 40 (2003): 170–194.

Chapter 6

Labbe, J.R. "A Post-Literate World Will Leave Much to be Desired." *The Kansas City Star*, Aug. 29 2007: B9.

U.S. Census Bureau. 2005. Aug. 31 2007. http://www.census.gov/PressRelease/www/releases/archives/education/004214.html.

Chapter 8

Horton, William. "The Almost Universal Language: Graphics for International Documents." *Technical Communication*, (November 1993): 682–693.

Reynolds, Michael, and Liz Marchetta. "Color for Technical Documents." *Intercom*,(April 1998): 5–7.

"WiFi Gaining Traction." *Business Week*, 3 Oct. 2005. 15 Aug. 2007http://www.businessweek.com/technology/tech_stats/wifi051003.htm.

Chapter 9

Bradley, Tony. "Policing Instant Messaging: Create A Policy To Govern IM Applications." *Processor*, 5 Aug. 2005 12 Apr. 2008. http://www.processor.com/editorial/article.asp?article=articles%2Fp2731%2F32p31%2F32p31.asp.

Clark, Robert. Interview. 15 Dec. 2007.

Foremski, Tom. "IBM is preparing to launch a massive corporate wide blogging initiative." *Silicon Valley Watcher,* 13 May 2005. 23 May 2005. http://www.siliconvalleywatcher.com/mt/archives/2005/05/can_blogging_bo.php.

Gard, Lauren. "The Business of Blogging." *BusinessWeek Online,* 13 Dec. 2004. 6 Jun. 2005. http://www.businesswekk.com/magazine/content/04_50/b3912115_mz016.htm.

"Gates backs blogs for businesses." *BBC News,* 21 May 2004. 23 May 2005. http://news.bbc.co.uk/2/hi/technology/3734981.stm.

"Generation Y Communication Trends: Whitepaper." Pika Technologies, Inc. 2006. 10 Apr. 2008. http://www.ctipro.cz/download/pika-generation_communication_trends.pdf.

Green, Heather. "Blogspotting." 9 Jul. 2007. 10 Jul. 2007. http://www.businessweek.com/the_thread/blogspotting/archives/2007/04/blogging_growth.html.

Hoffman, Jeff. "Instant Messaging in the Workplace." *Intercom,* (February 2004): 16–17.

"How To: Instant Messaging Security." *PC World,* 14 Jul.2006. 12 Apr. 2008. http://www.pcworld.com/businesscenter/article/135205/how_to_instant_messaging_security_.html.

Kharif, Olga. "Blogging for Business." *BusinessWeek Online,* 9 Aug. 2004. 23 May 2005. www.businessweek.com/technology/content/aug2004/tc2004089_3601_tco24.htm.

Lenhart, Amanda, and Susannah Fox. "Bloggers." *Pew Internet and American Life Project,* 19 Jul. 2006. 10 Apr. 2008. http://www.pewinternet.org/pdfs/PIP%20Bloggers%20Report%20July%2019%202006.pdf.

Li, Charlene. "Blogging: Bubble or Big Deal?" *Forrester*, 5 Nov. 2004. 23 May 2005. http://www.forrester.com/Research/Print/Document/0,7211,35000,00.html.

Maddock, Jeremy. "Verizon Wireless Reports Record Text Messaging Volume in June." 25 Jul. 2007. 19 Apr. 2008. http://www.teleclick.ca/2007/07/verizon-wireless-reports-record-text-messaging-volume-in-june/.

McAlpine, Rachel. "Passing the Ten-Second Test." *Wise-Women.com.* April 22, 2002. http://www.wise-women.org/features/tenseconds/index.shtml.

Moore, Linda E. "Serving the Electronic Reader." *Intercom,* (April 2003): 16–17.

Munter, Mary, et. al. "Business E-mail: Guidelines for Users." *Business Communication Quarterly,* 66 (March 2003): 29.

Noguchi, Yuki. "Life and Romance in 160 Characters or Less: Brevity Gains New Meaning as Popularity of Cell Phone Text Messaging Soars." *Washington Post,* http://www.washingtonpost.com/wpdyn/content/article/2005/12/28/AR2005122801430.html. 29 Dec. 2005. 2 Nov. 2007.

Ollman, Gunter. "Instant Message Security." *Technical Info,* March 2004. 12 Apr. 2008. http://www.technicalinfo.net/papers/IMSecurity.html.

Ray, Ramon. "Blogging for Business." *Inc.com,* Sep. 2004. 23 May 2005. http://www.inc.com/partners/sbc/articles/20040929-blogging.html.

Shinder, Deb. "Instant Messaging: Does it Have a Place in Business Networks?" *WindowSecurity.com,* 2005. 8 Jun. 2005. www.windowsecurity.com/articles/Instant-Messaging-Business-Networks.html.

"Text messaging." *Wikipedia,* http://en.wikipedia.org/wiki/Text_messaging. 2 Nov. 2007.

"Welcome to Technorati." *Technorati,* http://technorati.com/about/. 10 Apr. 2008.

Wuorio, Jeff. "Blogging for business: 7 tips for getting started." Microsoft, 23 May 2005. http://www.microsoft.com/smallbusiness/issues/marketing/online_marketing/blogging_for_business.

—."5 Ways Blogging Can Help Your Business." Microsoft, 23 May 2005. http://www.microsoft.com/smallbusiness/issues/marketing/online_marketing/5¬_ways_blog.htm.

"Writing: A Powerful Message from State Government." *Report of The National Commission on Writing for American's Families, Schools, and Colleges,* College Entrance Examination Board, 2005.

"Writing: A Ticket to Work . . . Or A Ticket Out, A Survey of Business Leaders." *National Commission on Writing,* 2007. www.writingcommission.org.

Chapter 10

"Writing: A Ticket to Work . . . Or A Ticket Out, A Survey of Business Leaders." *National Commission on Writing,* 2007. www.writingcommission.org.

Chapter 11

Bloch, Janel M. "Online Job Searching: Clicking Your Way to Employment." *Intercom,* (September/October 2003): 11–14.

Dikel, Margaret F. "The Online Job Application: Preparing Your Resume for the Internet." September 1999. http://www.dbm.com/jobguide/eresume.html.

Dixson, Kirsten. "Crafting an E-mail Resume." *BusinessWeek Online,* November 13, 2001. http://www.businessweek.com.

Drakeley, Caroline A. "Viral Networking: Tactics in Today's Job Market." *Intercom,* (September/October 2003): 5–7.

Hartman, Peter J. "You Got the Interview. Now Get the Job!" *Intercom,* (September/October 2003): 23–25.

"Human Resources Department." KCMO. 12 Nov. 2007. http://www.kcmo.org/hr.nsf/web/home

Isaacs, Kim. "Resume Critique Checklist." *MonsterTRAK Career Advice Archives,* 5 Sep. 2006. www.monster.com.

Kallick, Rob. "Research Pays Off During Interview." The *Kansas City Star,* March 23, 2003: D1.

Kendall, Pat. "Electronic Resumes," December 9, 2003. http://www.reslady.com.

Ralston, Steven M., et al. "Helping Interviewees Tell Their Stories." *Business Communication Quarterly,* 66 (September 2003): 8–22.

"Resume Guidelines." Career Services Center. Overland Park, KS: Johnson County Community College, 2003.

Stafford, Diane. "Show Up Armed with Answers." *The Kansas City Star*, August 3, 2003: L1.

Stern, Linda. "New Rules of the Hunt." *Newsweek*, (February 17, 2003): 67.

"To Lie or Not to Lie." *EasyJob*, 2003. 6 Apr. 2004. http://www.easyjob.net/resume/to-lie-in-you-resume.html.

"Top Ten Qualities/Skills Employers Want." *Job Outlook 2005 Student Version*. National Association of Colleges and Employers, 2005: 1-13. www.jobweb.com/joboutlook/2005outlook/JO5Student.pdf.

Vogt, Peter. "Your Resume's Look Is as Important as Its Content." *MonsterTRAK Career Advice Archives*, 6 Mar. 2007. www.monster.com.

"Writing: A Powerful Message from State Government." *Report of the National Commission on Writing*, College Board, 2005.

Chapter 12

Anthony, Joseph. "Communicating Bad News: 10 Tips." 17 Feb. 2004. 12 Sep. 2007. http://www.bcentral.com/articles/anthony/209.asp.

Braud, Gerard. "Crafting a Crisis Communication Plan." *International Association of Business Communicators Online Newsletter*, 10 Apr. 2008. http://www.iabc.com/cwb/archive/2007/0707/Braud.htm.

"Delivering Bad News by E-mail Is More Accurate, Less Painful, Study Suggests." *Informs*, 22 Jun. 1999. 19 Feb. 2004. http://www.informs.org/Press/BadNews.html.

Droste, Therese. "Using Email Efficiently for Clear Communication." *Monster Forums*, 2003. 20 Feb. 2004. http://forums.monster.com/forum.asp?forum=2716.

Griffin, Chip. "Using New Media to Tame a Crisis." *International Association of Business Communicators Online Newsletter*, 10 Apr. 2008. http://www.iabc.com/cwb/archive/2007/0707/Griffin.htm.

Patel, Amita, and Lamar Reinsch. "Companies Can Apologize: Corporate Apologies and Legal Liability." *Business Communication Quarterly*, 66 (March 2003): 9-25.

Chapter 13

"CDC and CPSC Warn of Winter Home Heating Hazards." *Centers for Disease Control and Prevention*, 6 Mar. 2007. http://www.cdc.gov/nceh/pressroom/2006/COwarning.htm.

Chapter 14

Berst, Jesse. "Seven Deadly Web Site Sins." ZDNet, January 30, 1998. 15 Dec. 2005. http://www.zdnet.com/anchordesk/story/story_1716.html.

Dorazio, Pat. Interview. June 2000.

Eddings, Earl, Kim Buckley, Sharon Coleman Bock, and Nathaniel Williams. Software Documentation Specialists at PDA. Interview. January 12, 1998.

".eu": A New Internet Top Level Domain." *The Official Journal of the European Union*, 24 May 2003. 4 Jun. 2004. http://europa.eu.int/information_society/topics/ecomm/all_about/todays_framework/public_resources/names_addresses/eu_creation/index_en.htm.

"508 Law." Section 508. 15 Aug. 2002. 10 Jun. 2004. http://www.section508.gov/index.cfm?FuseAction=Content&ID=3.

Fallows, Deborah. "China's Online Population Explosion." *Pew/Internet*, 12 July 2007. 29 July 2007. http://www.pewinternet.org/PPF/r/218/report_display.asp

Hemmi, Jane A. "Differentiating Online Help from Printed Documentation." *Intercom*, (July/August 2002): 10–12.

"Internet Accessibility." *STC AccessAbility SIG*, 9 Jun. 2004. 15 Jun. 2004. http://www.stcsig.org/sn/internet.shtml.

"Internet Growth: Today's Road to Business and Trade." *Internet World Stats*, 2004. May 30, 2004. http://www.internetworldstats.com/emarketing.htm.

"Internet World Stats." *Internet World Stats*, 12 Sep. 2008. http://www.internetworldstats.com/stats.htm.

Knight, Fred. S. "The Internet and Web Go Mainstream." *Business Communications Review*, (October 1998): 42-46.

LeVie, Donald S. "Internet Technology and Intellectual Property." *Intercom*, (January 2000): 20-23.

Olive, Eric G. "Usability: Making the Web Work." *Intercom*, (November 2002): 8-10.

OnGuard Online. 1 Aug. 2007. http://onguardonline.gov/stopthinkclick.html.

"Privacy Initiatives." Federal Trade Commission, 1 Aug. 2007. http://www.ftc.gov/privacy/index.html.

"Protecting Personal Information." Federal Trade Commission, 1 Aug. 2007. http://www.ftc.gov/bcp/conline/edcams/infosecurity/index.html.

Redish, Janice C. "Writing for the Web: Letting Go of the Words." *Intercom*, (June 2004): 4-10.

Shearer, Richard. "Internet Growth Reinforces Reasons for Online Branding." *The National Business Review*, 30 May 2004. 3 June 2004. http://www.nbr.co.nz/home/column_article.asp?id=9045&cid=3&cname=Technology.

"World Internet Usage." 2007. 1 Aug. 2007. http://www.internetworldstats.com/stats.htm

Chapter 16

The American Psychological Association, 6 Mar. 2007. www.apa.org.

The Modern Language Association: MLA, 6 Mar. 2007. www.mla.org.

Reyman, Jessica. "Rethinking Plagiarism for Technical Communication." *Technical Communication*, (Feb. 2008): 61-67.

Chapter 17

"Robert's Rules of Order Revised." *Constitution Society*, September 15, 2004. http://www.constitution.org/rror/rror—00.htm.

INDEX